Introduction to Social Systems I

Huijiong Wang · Shantong Li

Introduction to Social Systems Engineering

Huijiong Wang
Development Research Center
of the State Council
Beijing
China

Shantong Li
Department of Development Strategy
and Regional Economy
Development Research Center
of the State Council
Beijing
China

ISBN 978-981-13-3908-0 ISBN 978-981-10-7040-2 (eBook)
https://doi.org/10.1007/978-981-10-7040-2

Softcover re-print of the Hardcover 1st edition 2018

Printed on acid-free paper

This Springer imprint is published by Springer Nature
The registered company is Springer Nature Singapore Pte Ltd.
The registered company address is: 152 Beach Road, #21-01/04 Gateway East, Singapore 189721, Singapore

Foreword I

With the rise of industrial revolution and scientific management movement, the emphasis on specialized labor division has become one of the significant features in economic and social development. However, problems facing science today are far more complicated than any other time in human history. To deal with the problems requires not only the application of in-depth specialized knowledge, but also the synthesizing of comprehensive knowledge, including social sciences, natural sciences, and systems engineering. *Introduction to Social Systems Engineering* by Mr. Wang Huijiong is a monograph on the theory and methodology of social systems engineering that is based on his long-term theoretical and practical studies in the field and contains significant academic value in China's systems engineering researches.

As a famous engineering expert and economist, Mr. Wang Huijiong has witnessed China's independence and growth, participated in the national construction, and made contributions to China's economic and social prosperity since the reform and opening-up drive. After striving in the field of electrical engineering for over 33 years, Wang has built up thoughts and theories on social systems engineering from his practice in project construction and management. In the 1980s, Wang began to work in the Chinese Academy of Social Sciences (CASS) as a research fellow, and later in the Technological and Economic Research Center of the State Council (one of the predecessors of Development and Research Center of the State Council or DRC) as a research fellow and executive director. He then successfully redirected his research field from project construction and management to social sciences and policy analysis, such as economic growth, industrial planning. Besides, he participated in a number of significant national strategic research projects, from which he has accumulated a wealth of theoretical and practical experience on socio-economic policy studies. Over the last three decades, Wang has been in charge of a series of major research programs. For example, *China in 2000* (led by Ma Hong), in which he worked as a member of the leaders team, won the first prize of National Scientific and Technological Progress Award in 1987. *Integrated Economic Development Policies and Planning*, another grand policy research project, was accomplished through the cooperation with UN Development

Program, in which internationalization, openness, and practicality were highlighted. It focused systematically and profoundly on several major issues in China, including China's industrialization and policy-making, economic system reform, and regional economic development. Its broad range, covering research organization, coordination, and contents, was rarely seen in domestic studies.

In his broad research fields, Mr. Wang Huijiong has conducted long-term and deep practical and theoretical studies on social systems engineering. In 1980, he published *Introduction to Systems Engineering*, systematically concluding his studies on engineering design and management for more than 30 years. After joining CASS and DRC, he applied the research theories and methodologies on systems engineering to social and economic systems studies, continuously enriching and improving the theories and methodologies on systems engineering. In 2015, his *Methodology of Social Systems Engineering* was published, which further facilitates the research on and application of social systems engineering in China. This book written in English is his latest monograph in the field and is an expansion and improvement of his work in 2015. It elaborates the inevitability of social systems engineering progress from the perspective of development history of natural sciences and social sciences, which contains grand theoretical vision and profound historical depth. In the methodology part of social systems engineering, he reviewed and analyzed several crucial kinds of quantitative and qualitative research methods and conducted monographic studies on several key subsystems, which reflected Wang's solid theoretical foundation, rich practical experience, and systematic thinking.

One of the major contribution of the book is to provide a more expansive and comprehensive view for the research on social sciences and public policies by applying theories and methodologies from systems engineering studies. Yet, state and society form a huge and complicated system with a series of economic, social, cultural, scientific and technological issues intertwined with each other. Especially in the following period from now on, human society will be faced with more complicated problems due to the dramatic development and extensive application of information technology. Conducting researches on social systems engineering will help not only to analyze problems and cope with challenges for the community of shared future for mankind in a more comprehensive manner, but also to find more reasonable solutions and policy schemes. Admittedly, professionals are needed in real practice, but the problems facing mankind in production activities and daily livelihoods cannot be solved solely by limited knowledge in one aspect alone. To be specific, the design, construction, and management in engineering area rely not only on natural sciences and expertise in engineering fields, but also on the economic and management knowledge from social sciences. Similarly, reforms and development in economic sector can also be integrated systematically and managed intensively by utilizing concepts and methods from engineering field like planning, designing, testing, executing.

China has made remarkable achievements in economic and social development since the founding of the People's Republic of China, especially since the initiation of the Reform and Opening-up Policy in the late 1970s. As a consequence, more and more countries expect to seek knowledge and wisdom suitable for their own development from China's development history and experience. As Chinese President Xi Jinping once said, 'The world is so big, and the problems are so many. The international community expects to hear China's voice and see China's plans. China cannot be absent.' To be in line with the requirements of the time, to respond to international concerns and to perform the responsibility as a major power, President Xi Jinping declared at the UN Sustainable Development Summit in September 2015 that China would establish the Center for International Knowledge on Development (CIKD) to study and communicate with other countries on development theories and practices suitable to their respective national conditions. The DRC is a policy research and consultation agency serving the central committee of CPC and the State Council and is also one of the 25 high-end national think tanks in China. The DRC has clearly blueprinted its strategy to build a high-level international think tank and to take conducting international communication and cooperation as one of its key functions. The Chinese government thus granted the DRC to undertake the founding of CIKD based on its functional position and operating characteristics. In March 2017, the Chinese government officially approved the DRC to launch CIKD. In August 2017, President Xi Jinping and UN Secretary-General Antonio Guterres sent congratulatory letters to the official launch of CIKD. In his letter, President Xi Jinping asked CIKD to make positive contributions to researching and exchanging development theories, facilitating international development and cooperation and promoting the implementation of the 2030 Agenda for Sustainable Development. One month later, President Xi Jinping stressed at the Dialogue of Emerging Market and Developing Countries that China would strengthen the cooperation in experience sharing and capability building with the international community through CIKD platform. Therefore, the CIKD has set case studies on China's development as its most important task at the beginning of its official operation. *Introduction to Social Systems Engineering*, written by Mr. Wang Huijiong, is not only a professional output on systems engineering, but also a summary of his participation in China's development and research on national policies, containing rich Chinese elements, experience and wisdom, and will be a good vehicle to tell Chinese stories. Hence, we are pleased to offer the book as one of CIKD's products to international society. Mr. Wang Huijiong is a veteran expert on China's policy researches who worked over 30 years in DRC, so we would like to, taking the opportunity of the book's publication, to extend our sincere gratitude to all senior scholars represented by Mr. Wang for their contributions to China's construction, reform and development undertakings. We also expect more experts and scholars to summarize the theories and practices on China's development and to make positive contributions to international development cooperation and global sustainable development with their counterparts from other countries.

Beijing, China
October 2017

Li Wei
President of Development Research Center of the State Council

Foreword II

It is an honor, as well as a pleasure, to write a foreword to this book. An honor, because I regard Prof. Wang Huijiong as one of my teachers about China's economic development, to which both of the book's authors have made important contributions through their research. A pleasure, because my work on China since 1980 has given me the opportunity to know the authors personally and to discover dimensions of their insights and experiences that transcend the technicalities of social science. Without them, what I have done in China would have been both less useful and less enjoyable.

We first met in the early 1980s through the coincidence of two separate but basically similar projects on China's development prospects to the year 2000, which are described and discussed in Sect. 4.3.2 of the book, and both of which benefited from the support of governmental leadership at that time. The authors were working in the Technical-Economic Research Center (TERC) of the State Council with many other Chinese colleagues in many other institutions on a comprehensive study that included both an overall general framework (in which they played key roles) and detailed analyses of individual industries and provinces. I was the deputy chief of a smaller World Bank team led by Edwin Lim that was covering the same range of issues but with more emphasis on international comparisons and lessons from foreign experience.

The Bank team relied heavily on information, advice, and guidance from Chinese colleagues, without which it would have been impossible to connect our knowledge of the rest of the world with Chinese reality in a fruitful way. In this regard, the TERC rendered us a great service by organizing no less than four days of seminars (March 11–14, 1985) to discuss various aspects of our draft report. As I write these words, I have on my desk the notes I made at those seminars and at an 8 March meeting with Mr. Ma Hong, director general of the TERC, accompanied by Prof. Wang Huijiong and another brilliant and articulate old friend, Prof. Li Boxi. One of my responsibilities in the Bank report was for the modeling, and looking back through my notes, I can see how much we learned both from the TERC's comments on our model and from a presentation on the TERC's own model and projections made by a 'young lady,' whom I guess was Prof. Li Shantong.

The TERC report and the World Bank report, though different in many ways, arrived at broadly similar conclusions, and both made useful contributions to China's development to 2000 and beyond by charting the unknown terrain ahead and suggesting ways of navigating through it. The TERC–World Bank seminars also exemplified the way in which close interaction between foreign and Chinese experts added hugely to the productivity, interest, and enjoyment of policy-oriented research on China. Much has changed since those far-off days, with the TERC now merged into the Development Research Center, and China a vastly more prosperous country, but the legacy of the work we did and the friendships we made have endured.

The present book spans almost all areas of scientific knowledge, almost all periods of human history, and almost all parts of the world. Among other things, it relates historical experience in China to experience in the West (Europe and its offshoots), with the purpose of reducing the current asymmetry between China's good knowledge of the West and the West's more limited knowledge of China. The other purpose of the book, as the authors describe it in their preface, is 'to try to fill the gap among natural science, social science and engineering.' Social Systems Engineering (SSE) is a discipline designed for this purpose, and more precisely for combining the insights of different scientific disciplines to improve human well-being, whether at the level of the firm, the locality, the nation, or the world. The book reviews the main theories of SSE and the techniques and models used in its application, especially to planning. It concludes with two chapters that view the progress of China since 1950 through the lens of the long series of five-year plans, whose use has continued, though their methods and content have evolved.

Oxford, UK
October 2017

Adrian Wood
Professor Emeritus of International Development
University of Oxford

Preface

We are glad to present to you this book *Introduction to Social Systems Engineering* (SSE).

The purpose of this book is two.

The first purpose of this book is to try to fill the gap among natural science, social science, and engineering. It may seem to be a little bit ambitious.

Since the scientific revolution, industrial revolution post the middle part of last millennium, there is rapid progress of science and technology (S&T). Various branches of S&T are emerged to be studied with in-depth. The development of science is characterized by its ever-increasing specialization necessitated by huge amount of data, the complexity of technique within every field. This is also true for various types of engineering; for example, there was no clear distinction between civil engineering and architecture in traditional civil engineering before modern times, but currently, there are branches of civil engineering such as structural engineering, architecture, highway engineering, harboring engineering, coastal engineering. The results of this increasing specialization have caused difficulty of coordination of large industrial projects generally including many different disciplines. It is also detrimental for further development of science and engineering, because the academic study of in-depth and in-breadth are in mutual promotion with each other. Interdisciplinary studies emerged in the early twentieth century. General system theory and systems engineering were established and became popular since mid of the twentieth century.

Both authors of this book are with background of science and engineering and worked in the Development Research Center for policy and planning studies more than thirty years, but Wang had published *An Introduction to Systems Engineering* (in Chinese version) in 1980. We felt deeply that the issues of social science in dealing with planning and public policy are even much more complicated than that of natural science and systems engineering. There are needs to have such new discipline *Social System Engineering* (SSE) to meet the new demand. We had searched the programs of universities globally; there is only University of Tsukuba with a master's program in social systems engineering, and doctoral program in policy and planning sciences. The University of Pennsylvania, which has an undergraduate program of Networked and SSE, its curriculum is consisted of computer science, economics, systems engineering and sociology. It can be seen that the contents of study of this emerging discipline are only defined basically. We wish to contribute our exploration of this emerging discipline with our perception, knowledge, and working experience.

The second purpose of this book is also derived from our impression in dealing with Western world. We have the impression that the Chinese intellectuals and a part of people know the Western situation relatively better than the knowledge with China of their Western counterpart. This is a natural phenomenon due to the backward state of society, economy, science and technology (S&T) of China before establishment of PRC. Therefore, most Chinese learned Western experience urgently in order to catch up. But China is a large unique country with five thousand years of civilization. There were many philosophers and scholars in the Chinese history, especially in the Spring and Autumn and the Warring State period, who had contributed a lot of valuable knowledge to the mankind, for example, Confucius and Confucianism, Taoism, Legalism.

Currently, China is emerged to be one of the largest nations in economy and trade within a short period of no more than seven decades. As policy researchers, we think it is necessary to share some of our knowledge to global interested readers related to China's cultural system, its socio-economic and S&T development since the First Five-Year Plan up to the present Thirteenth Five-Year Plan as well as some experience of growth of our Center, one of China's national think tanks. We believe that exchange of information to promote a mutual understanding among countries is a pre-requisite toward a harmonious global society. That is our second purpose to present this book.

Then we shall introduce briefly the structure and contents of this book. It is divided into three parts:

Part I is *Emergence of Social Systems Engineering* (*SSE*) with four chapters and deals with basic theories of SSE and its current growth in some countries.

Chapter 1 is an overview of the whole book.

Chapter 2 presents growth of science around 2500 years and resulted in three trends of science in the early twentieth century, overspecialization, impact of mechanistic view, and neglect of nonphysical science. General systems theory and systems engineering were emerged to respond to these trends. Key concepts of general systems theory and some aspects of systems engineering are summarized to be the theoretical foundation of social systems engineering.

Growth of social science both in China and Western countries is presented in Chap. 3. Based upon purpose of this book, several schools of social studies of ancient China are briefed. Authors of this book believe that their influence to contemporary East Asian Continent is the same as the impact of Greek philosophy to current Western civilization. Development of social science in Western countries is also presented since the Hellenistic period up to present. Three major thoughts effect the development of social science in the twentieth century are analyzed; they are impact of Marxism and others, influences of Freud and growth in parallel of further specialization, and cross-disciplinary approach. Emergence of social systems theory and method is discussed. Psychology and various aspects related to human behavior and its interaction with environment are discussed. Parsons' theory of social system and his AGIL framework of systems of action are identified to be the other theoretical foundation of SSE.

Chapter 4 presents the development of SSE since the twentieth century. There is a trend of development of SSE in USA, Japan, China, and Europe, etc. Clarification of concepts of design and planning of SSE is presented in relative detail. Three case studies in the nature of social systems engineering are given for illustration. They are: *Nation as large-scale system* by Chestnut, H., two-large-scale research projects done by Development Research Center, *China toward the year 2000* and *Integrated Economic Development Policies and Planning.*

Part II is *Outline of Social Systems Engineering* (*SSE*) with six chapters.

Chapter 5 is methodology and principle of planning of SSE. Due to the broadness of various disciplines covered by the study of SSE, methodology of qualitative analysis of two important studies is quoted. The first is Hall's Morphology Box for systems engineering; it includes phases, steps, and disciplines which can be applied universally to all types of engineering programs and plans. The second is Warfield's *societal systems*, which focused specifically on planning and public policy, the essential areas of study of SSE. His methodology of Π-Σ process and concept of TOTOs[1] the organized conduct of inquiry are introduced. And a concrete policy study project of OECD. *Expo 2000 Forum for the Future* with four successive conferences related to S&T, economy, society, governance from 1997–2000 based on TOTOs is described in detail to illustrate the art of integration of different disciplines, which is also one difficult action 'integration' of Parsons' AGIL framework of systems of actions. Systems engineering logic, a detailed treatment of 'steps' is presented to complement Hall's Morphology Box.

Principle of planning based upon concepts of SSE is also given methodologically. Three frameworks of planning are introduced; they are based on AGIL framework of systems of actions, balanced approach of three subsystems (social, economic, and S&T) of a national system, vertical hierarchy of subsystem (macro, meso, and micro), respectively. Three planning models are also given; they are the system environmental model; the function-structure model, and the process behavioral model. Major concepts of structure of T-21 model are quoted to provide concrete components of function-structure model.

Chapter 6 is methodology of quantitative approach of SSE. Measurement and quantification are two major difficulties in the development of SSE compared to the development of systems engineering in later part of twentieth century. There are improvements of conditions of these two aspects in recent years. Progress of development of social indicators is described in detail. Case studies of preparation of *systems of indicators for sustainable development* by UN and its attached organizations and *framework of preparation of indicators of national competitiveness* by IMD are described. Models are discussed in a broader perspective to

[1] *Note* TOTOs is the abbreviation of Task-Oriented Transient Organization.

avoid mathematical determinism. Three decades of experience of policy modeling of DRC is summarized. A case study of DRC-CGE model is presented with the history of its formation, the structure of the model with its various modules, and its application to the analysis of China's Thirteenth Five-Year Plan.

Chapter 7 gives a comprehensive discussion of various types of planning system which is one major aspect of application of SSE. There are debates on planning both in China and other countries due to the failure of central planning system in later period of former USSR and the ideology of economic liberalism. But the rapid growth of Chinese economy is partly due to its continuity and improvement of its Five-Year development planning. EU, Japan, USA, a part of the world, and large enterprises have kept the practice of planning in different types. This chapter starts with a definition of planning and its role and gives the story of a military plan of Qin dynasty to establish a unified empire in ancient China to illustrate the point that planning does exist throughout the human history. Various types of planning are discussed. Evolution of development of modern national planning system in global society is described. Development of planning theories is briefed. The current trend, emergence of Future Studies and Scenario is presented with case study of *Europe 2020* and case study of *Mapping the Global Future Project* of National Intelligence Council of USA. By the meantime, two detail case studies of Medium-Term National Plans of France and India are given. This gives a relatively complete information of long-range scenario planning and medium-term planning.

Chapter 8 deals with Boundary and Environment and Social Change of a Social System. All systems are located in space and time of certain environments and interact with them. The boundary between a physical system and its environment is relatively easier to be identified, but the determination of a boundary line between social system is generally complicated, because it is identified through norms, institutions, and regulations. This chapter has given a general discussion of boundary and environment. A case study of business planning, which illustrates the interaction between an enterprise and its environment-Growth of *Yantian International Container Terminals* (*YICT*) in Shenzhen, SEZ of China is given. This case study serves for dual purpose; the first is to provide an example of business planning which should also be an aspect of application of SSE; the second is to illustrate the basic view of authors to modify A of Parsons AGIL framework of systems of actions. We have also modified A (adapt) into AA (active adapt) so that there are both changes for the business enterprise and its external environment. Based upon purpose of this book, the formation and change of culture are discussed in detail; especially the dominant influence of Confucianism to China's social system, the basic concepts of Analects are introduced to foreign readers. Finally, one pervasive characteristic of social system,-the social change, is discussed, and the megatrend of contemporary social change, globalization and regionalization, from the perspective of the authors are presented.

Chapter 9 deals with regulation of social system. It seems to be a common sense that the social system needs to be regulated in order to keep order, security, and development of the social system. Public policy, law, regulation, etc., all perform the functions of regulation, and all of them belong to the discipline of public administration. Therefore, a brief discussion of public administration is given to provide a broad context. Then, a Western legal textbook with unique chapter of policy analysis, the growth of legal system of China since the Spring and Autumn and Warring State period up to Qing dynasty, and also the establishment of modern legal system of China is briefed. A relatively detail discussion of public policy formed the core part of this chapter. Clarification of terminology, history and schools of public policy and policy science both in China and Western countries are summarized. A social system approach of public policy-making process with elements and their relationships to be routinized is described. Two case studies are presented to conclude this chapter. One is applying Analytic Hierarchy Process to project 'Differentiated Policy to construct the Functional Zones,' and the other is 'Industrial Policy of China.'

Chapter 10 deals with global think tanks. There is rapid growth of think tanks since the establishment of first think tank the British Royal United Service Institute (RUSI) in 1831. This growth process was strengthened post-WWII and reached its peak around 1992. This chapter gives a brief discussion of role of think tanks, types of think tanks, its process of development, etc. Relatively detailed case studies are given to several selected think tanks: Brookings Institution, Rand Corporation, International Development Center of Japan. And also case studies of international think tanks such as OECD, the World Bank and its activity in China are described. Think tanks with unique features such as International Center for Economic Growth (ICEG) with network of member communication organization and African Economic Research Consortium (AERC) which is also consisted of networked member organization are also introduced. Finally, development of China's think tanks, especially case of Development Research Center of the State Council, which both authors of this book are attached is presented to conclude this chapter.

Part III is *Application of Social Systems Engineering* with 2 chapters.

Chapter 11 deals with development of China since its First Five-Year Plan up to Fifth Five-Year Plan. China had no experience of national planning in the period of 1950s and followed the former Soviet model of centralized planning before 1980. *The First Five-Year Plan*, the construction of 156 core projects and the *Outline of the Long-Term Plan of Development of Science and Technology 1955–1967* is described in very detail; these have laid the foundation of China's planning, industry and development of Science and Technology. Analysis of China's planning performance (1953–1980) from the perspective of social systems engineering is given in last part of this chapter.

Chapter 12 deals with development of China since the Sixth Five-Year Plan up to present. All planning documents of Five-Year Planning are briefed. It can be seen in the picture how China is in transition from a former centralized planning to an indicative planning system and establishment of the socialist market economy with Chinese characteristics through gradualist approach.

Finally, several features and issues of this book should be clarified.

1. The discipline of social systems engineering is a very new subject with very broad coverage of various concepts. In order to prove our view and certain concept, we generally quote available similar concept of other scholars or studies directly without paraphrase. We think this approach is better to strengthen our concept. Due to the nature of this subject, there is also a part of concepts of others with paraphrase.
2. Based upon the original definition of social system given by Parsons or given by us in chapter one, it should include the political system. And study of social systems engineering should include the study of political system, its function, structure and behavior. But both authors had working experiences only in the policy consultative institution; we have no experience at all in political-administrative position. Therefore, political system is not discussed; it is classified roughly in the social system. And we separate economic, S&T system from the social system to facilitate the discussion of planning and public policy. Because economic system is the foundation of national socio-economic development, while science and technology are the driving force of socio-economic development of all countries.
   Based upon above point, we focus more on planning function rather than design in the study of SSE, because planning is generally based on existed political system and power structure., while design will be more detailed and focused more on ideal objective. But there is a general discussion of design and planning in this book.
3. Chapter 8 with the title of 'Boundary and Environment of the Social System' is not well focused, because the initial intention of writing this chapter is to study the role of environment and its interaction with the system. But discussion of cultural system which is also a part of social system and very important part of social system is also included in this chapter. It is better to be separated to become an independent chapter, but it is not done due to time constraints.

Both authors of this book acknowledged frankly that our initial background is science and engineering. We get our knowledge of strategic planning, public policy, economics, and sociology through learning and doing in Development Research Center more than 30 years; we have also a lot of chances to learn from our foreign friends in many international conferences and through cooperation of many projects. Our knowledge is very limited; mistakes and errors in this book are unavoidable. All comments and criticism on this book are welcome sincerely. Progress is made through learning from and correction of mistakes! We do believe!

Beijing, China

Huijiong Wang
Shantong Li

Acknowledgements

Robert W. Gutkin
Michael X.YE
Ping Wang
Jeff Dodd
James Lillard
WenChong Weng
Faye Lazo
Boxi Li
Xiaofeng Hua
Jian Qin
Qi Wang
Jingning Guo (Secretary)
Jiachuan Chen (Graduated Student)
Jing Wen (Graduated Student)

Thanks should be given to above people listed. This book can hardly be completed without the contribution of their efforts.

Huijiong Wang
Shantong Li

Contents

Part I
Emergence of Social Systems Engineering (SSE)

Chapter 1
Overview

1.1 Introduction

Ever since their existence on earth, humans have struggled to recognize and study their natural and social environment, and themselves within it. It took a long historical period for philosophy, natural science, and social science to emerge as categories of studies. However, their respective pace of development accelerated in the West during the Renaissance Period (the 14th–17th centuries). The pattern of such inquiry and development was the emergence of multiple sub-disciplines within each category of study. For example, astronomy, physics, chemistry, biology, and geology, etc. became robust disciplines in natural science while economics, politics, sociology, and anthropology, etc. were disciplines under social science. In the 20th Century, there was a new trend of development in natural and social sciences which further divided each discipline vertically into specific fields of studies or integrated multiple disciplines horizontally into broader studies. The emergence of cultural anthropology, social psychology, social statistics, etc., illustrates the former while the emergence and development of interdisciplinary studies in a broader scope represents the latter. Systems science and systems engineering are perfect examples of interdisciplinary study.

As a comprehensive discipline, the systems science was developed around the 3rd decade of the 20th century. The development of systems engineering thereafter has furthered the breadth of interdisciplinary study and application between natural science and social science. The complexity of issues affecting mutual existence and development faced by the mankind is increasing rapidly alongside the development of human society. The construction and operation of large engineering projects require not only the application of specific knowledge, but also the development and application of broad and comprehensive social knowledge. Systems engineering is emerged to close the gap of integration and coordination among the evolved, specific fields of nature and society.

H. Wang and S. Li, *Introduction to Social Systems Engineering*,
https://doi.org/10.1007/978-981-10-7040-2_1

Natural science and engineering fields are more fact-based and objective in nature, and therefore, are less affected by ideologies and political prejudice. Also, the feasibility and proofs of such studies can be tested through experiments and practices. Comprehensive and very complex systems engineering principles have been developed and applied broadly and successfully during the past century. Examples include the lunar landing project, electric system grid, communication system, transportation system, and internet system, etc. However, can such a comprehensive, linear, and integrated study also incorporate the abstract but critical application of the social system? That is, events and actions of human behaviors that are subjective and interpretive in nature and manifested in government and corporate system? This book will give a general literature review of relevant theories, provide the authors' personal perspective on social systems engineering (SSE), present empirical evidence based on China's experience in planning and policies development, and provide critique and comparison of current practices with general planning and SSE-based planning.

1.2 Social System and Social Systems Engineering

The social system has existed ever since the existence of human beings on earth. All events and activities, power and politics, war and peace, exploration and empire, trade and industry, science and technology, culture and religion, belief and thought (Black, 2005), recorded as the history of the mankind serve as inputs and outputs of exploration of social system and SSE. There were many studies relevant to social system and SSE by scholars in various periods. Some of them may not be applicable in current societies as they may seem preposterous from contemporary perspective. But many may still as valid study today, because the development of human history is a continuous spectrum of inheritance and innovation. Learning from history is an important source for SSE. Many valuable explorations of social system and SSE occurred at various stages of the human development history.

1.2.1 Exploration of Social System and SSE Before the 20th Century

In the Axial Age[1] of world history, the Greek philosopher Plato (427–347 BC) established the concept of an ideal government and the structure of society. He thought that societies had a tripartite class structure corresponding to the appetite/spirit/reason structure of the individual soul. The three classes are Productive

[1]Note: This is a term raised by German philosopher Karl Jaspers, which characterized the period of ancient history during about the 8th to 3rd centuries BC.

(Workers), Protective (Warriors or Guardians) and Governing (Rulers or Philosopher Kings). During China's Spring and Autumn and the Warring States Periods (771–476 BC and 475–221 BC), several famous philosophers and scholars, such as Confucius (551–479 BC)—the founder of Confucianism, Lao-tse, and Mo-tse, have had enduring influence on China's political and cultural systems. Confucius' successor Mencius defined the human relationship of society in his essay as "affection between father and son, righteousness between ruler and subject, distinction between husband and wife, orderly sequence between old and young, and fidelity between friends." (Zhao, Zhang, & Zhou, 1999, p. 119). Above cited thoughts are only a small example to show exploration of social systems throughout human history.

1.2.2 Exploration of Social System in the 20th Century

The concept of an integral social "system" was established in the middle of the 20th century and there has been systematic exploration of social system by scholars and practitioner ever since.

1.2.2.1 Exploration by Sociologists

Talcott Parsons (1902–1979), an American sociologist, published *The Structure of Social Action* in 1937 and then *The Social System* in 1951, which made a preliminary systematic exploration of social systems. Robert Merton (1910–2003), a student of Parsons and a leading sociologist in America, tried to bridge the divide between the abstract theory of Parsons' and the empirical survey work that typified much of modern American sociology. His most influential general work is the collection of essays *Social Theory and Social Structure* (Merton, 1968). Niklas Luhmann (1927–1998), a German sociologist, who went to Harvard and studied under Parsons, published a German version of *Social System* in 1984 which was translated into English and published by Stanford University Press in 1995. His work made a highly academic and philosophical exploration of social system.

In *The Social System,* Parsons (1951) defined social system as follows:

> A social system consists in a plurality of individual actors interacting with each other in a situation which has at least a physical or environmental aspect, actors who are motivated in terms of a tendency to the "optimization of gratification" and whose, relations to the situations, including each other, is defined and mediated in terms of a system of culturally structured and shared symbols. (pp. 5–6)

The above definition of social system has embedded three other systems: the personality system (plurality of individual actors); a cultural system (wider values giving coherence to the norms attached to status roles), and the environmental system to which the society must adjust.

In *Economy and Society*, Parsons and Smelser (1956), tried to synthesize theory in economics and sociology and created a framework to analyze action systems. In their model, four fundamental system problems are identified as latency or latent pattern maintenance (including tension management) (L), goal attainment (G), adaptation (A), and integration (I). This AGIL Paradigm will be discussed in detail in Chap. 3 of this book.

Parsons is considered to represent the school of functionalism in the sociological field. There are several other schools in sociology, such as conflict theory, social exchange theory etc. Different schools have different views on social systems and different opinions on Parsons' theory. Some scholars criticized his theory as conservative or flawed. After reviewing several works on functionalist theory, Von Bertalanffy (1968), the founder of General Systems Theory (GST), found that Parsons' work "overemphasizes maintenance, equilibrium, adjustment, homeostasis, stable institutional structures, and so on, with the result that history, process, sociocultural change, inner-directed development, etc., are underplayed and at most appears as 'deviants with a negative value connotation'. The theory therefore appears to be one of conservatism and conformism defending the 'system' (or the mega machine of the present society, to use Mumford's term as is, conceptually neglecting and hence obstructing social change." (p. 196).

The authors of this book concur with above comments on Parsons' theory and its imperfections, but we keep the view that every theory is a product of the process of development of society and history, its truth is relative to its time-frame, its imperfections can be improved or supplemented with other theories. Parsons might be influenced too much by Darwin's theory of evolution by natural selection; therefore, 'adaptation' was very much emphasized in his AGIL functional framework. Of course, the existence and development of any social system need to be adaptive to the natural and social environment where it exists. Active rather than passive adaptation is necessary for survival. Both Bertalanffy's GST and Parsons' AGIL schema can be understood from various perspective and applied integrally and complementally to study the social issues facing the mankind.

There are further studies of social systems in the late 20th century. Many systems scientists, computer engineers, and electrical engineers engaged in studying social systems; J. Forrester, K. Boulding, P. Checkland, R. L. Ackoff, J. N. Warfield, just to name a few. Their involvement has not only improved the breadth and depth of study of social systems, but also established the basis of SSE. Below section discusses some of their contributions.

1.2.2.2 Exploration by System Scientists and Electrical Engineers

In late 20th century, there were increasing numbers of system scientists and electrical engineers involved in the study of social systems. Due to their engineering and scientific background, they approached the social systems with an engineering perspective, thus contributing to the emergence of SSE. Below briefly describe three of such studies.

(1) *Study of societal systems by John Nelson Warfield (1925–2005)*

Warfield, an American systems scientist, was a professor and director for advanced study in the Integrative Sciences (ASIS) at George Mason Society and president of the Systems, Men and Cybernetics Society of IEEE. He also served as Editor of IEEE Transactions on Systems, Men and Cybernetics. In *Society System: Planning, Policy and Complexity,* Warfield (1976) explored the major application of SSE, i.e. plan preparation and policy making, and introduced several concrete methods of SSE. One method of qualitative approach to social systems is task oriented transient organization (TOTO) which is a form of organization of collective inquiry. Interpretative structural modeling based upon Boolean Algebra to study the structure of social systems was also explored. He pointed out the complexity and deficiencies in dealing with societal problems and stated that much of the knowledge available in many social and behavioral sciences has not been applied to complex societal problems. Warfield concluded:

> The solutions to most of current societal problems depend on a host of disciplines, such as sociology, politics, economics, and institutional arrangements, in addition to technology. Superposition; that is, separately and sequentially looking at, say, the economic considerations, then the political ones, then the social aspects, etc., does not work. They must be integrated. (p. 13)

His point of view illustrates the importance of Integration—the "I" in Parsons' AGIL paradigm. Warfield recognized the main complexity in dealing with societal problems because "Political society wants things simple. Political scientists know them to be complex. This is no small matter". (p. 15)

(2) *Study of social system by Jay Wright Forrester (1918–)*

Forrester is an American system scientist and the founder of system dynamics. In his paper "Counterintuitive Behavior of Social Systems", Forrester (1971) explained that the features of the system dynamics model were not derived statistically from time series data, but rather, they were statements about system structure and the policies that guide decisions. It's evident that Forrester, a computer engineer and Parsons, a well-known sociologist of structural functionalism, worked in the same direction with social structure and function in mind. Forrester also discovered some unique counterintuitive natures of complex social systems. According to him, three counterintuitive behaviors of social systems are especially problematic and are primary causes for failures in government programs and policies if not recognized and dealt with adequately. The first such behavior is that social systems are inherently insensitive to most policy changes chosen by policy maker to alter system behavior. The second behavior is that social systems seem to have a few sensitive influences through which behavior can be changed, but generally these influences are not what most people expect. The third behavior is that social systems exhibit a conflict between short-term and long-term consequences of a policy change. A policy producing improvement in the short run is usually the one that degrades a system in the long run. This point of view is manifested by the current global concern over the environment issue and sustainability of development due to inappropriate high economic growth policy pursued by many countries. Forrester was disappointed that the social sciences in his time retreated into small corners of research when they should be dealing with the great challenges of society (p. 2)

(3) *Designing social system in a changing world by Bela H. Banathy (1919–2003)*

Banathy, a professor at San Jose University, was an educator and system scientist. He founded the International Systems Institute (ISI) in Carmel, California in 1981 and was president of the International Federation for Systems Research in 1994–1998. In his *Designing Social Systems in a Changing World*, Banathy (1996) pointed out: "we are entering the twenty-first century with organizations designed during the nineteenth. Improvement or restructuring of existing systems, based on the design of the industrial machine age, does not work anymore. Only a radical and fundamental change of perspectives and purposes, and the redesign of our organizations and social systems, will satisfy the new realities of our era". (p. 1)

His book focuses on the role of design in a world of new reality. This new reality is a major societal transformation from the industrial machine age to the postindustrial information/knowledge age. He classified people in the society into four groups based upon their response to social change. The first group—Reactive Orientation: "Back to the Future" refer to people who are dissatisfied with the present, long for the past, and want to return to what it was. The second group—The Inactive Style: "Don't Rock the Boat" categorize people who are satisfied with things as they are. The third group—The Proactive Style: "Riding the Tide" describe people who anticipate change, prepare for it when it arrives, and exploit its opportunities. The fourth group—The Interactive Style: "Shooting the Rapids" denote people who believe that it is within their power to attain the future they envision and desire to bring about, provided they learn how to do it and have the willingness to do the steering (pp. 38–40). Different approaches of planning should be designed for different groups of people. This concept to classify people into groups of different behaviors is important in SSE.

Banathy cited twenty-four definitions of design to show the thinking of design scholars of the last three decades. Similarly, he discussed six different perspectives on social systems. Among them, the perspectives on characteristics of social systems by Peter Checkland were highlighted below. According to Banathy, Checkland maintains:

> Natural and engineered systems cannot be other than what they are. Human activity systems, on the other hand, are manifested through the perceptions of human beings who are free to attribute a variety of meanings to what they perceive. There will never be a single (testable) account of human activity systems, only a set of possible accounts, all valid according to particular Weltanschauungen.
>
> Given man and his abilities, we have the huge range of human activity systems, from the one-man-with-a-hammer at one extreme to the international political systems at the other. (as cited in Banathy, 1996, p. 14)

These three studies serve to reinforce our premise, that social system has existed ever since humans came into existence on earth and become increasingly complex as human society progresses.

1.2.3 Exploration of SSE in the 20th Century

The process of exploration and establishment of SSE seems lagging behind systems engineering, which deals with large and complex but defined and focused engineering projects. Systems engineering emerged around the 1930s, largely initiated through exploration of "Methodology of Systems Engineering" by Bell Labs with the aim to identify and manipulate the properties of a system as a whole that may differ greatly from the sum of properties of its parts. Bell Labs was generally acknowledged as the first organization to use the terminology 'Systems Engineering'. Scholars of operation research also contributed philosophical concepts and technologies to this emerging discipline during the period of Second World War. Rand Corporation formed the concept and technique of "system analysis" in 1946. The first *Handbook of Systems Engineering* was published in USA in 1965 to meet the demand of complex projects in defense, aerospace, logistics, and transportation. While the theoretical foundation for systems engineering has been well established through the joint efforts of corporations, academics, and researchers, such a foundation for SSE is lacking as it is still at the stage of being explored. Below briefly describe a few examples of SSE exploration.

1.2.3.1 Exploration by Universities in Japan

Based on available information, the University of Tsukubu might be the first university in Japan to offer master courses on SSE in the 20th century. Its major curriculum covers three fields: social economics, management science, and urban planning. Professors with civil engineering and architecture background predominantly teach the courses.

1.2.3.2 Exploration by Institute of Electrical and Electronics Engineers (IEEE)

Among the theoretical system of different engineering disciplines, electrical and electronic engineering is relatively close to SSE, which is interdisciplinary. Many papers related to SSE were published in annual conference proceedings of IEEE in the 1970s. These papers generally analyzed certain social issues from the perspective of control and automation. However, there were several excellent explorations on social systems from the SSE perspective as well. Some are discussed in the previous section. And here is another example:

Exploration by Harold Chestnut (1917–2001). Chestnut, an American electrical and control engineer, had his lifelong career with the G.E. Company since 1940. He worked in the Aeronautics and Ordnance Department and the Systems Engineering and Analysis Branch of the Advanced Technology Laboratory, where he served as manager from 1956–1972. He was the first president of the International Federation of Automatic Control (IFAC) from 1965–1967. He published two books related to systems engineering, *Systems Engineering Tools* and *System Engineering Methods* in 1965 and 1967, respectively. In *Nation as Large Scale System*, Chestnut (1982) provided an overall analysis of a nation, which can be considered as a critical exploration of SSE. Part of his paper will be discussed in Chap. 4 of this book.

1.2.4 Exploration of Social System in the 21st Century

In the process of development of human society into the 21st century, mankind has a deepened recognition of the relationship between social system and the ecological environment. Ecology is the scientific analysis and study of interactions among organisms (including human beings) and their environment. These interactions determine the territorial distribution and abundance. This term was first coined in 1869 by German Biologist Haeckel in a study of plant ecology. Later, it evolved into an interdisciplinary field including biology, geography, and earth science. The main influence of ecological concept on sociological theory occurred between 1920s and 1940s in the U.S. Initially, through the urban ecology by Chicago sociologists. Concern about the impact of human activities on the environment in the late twentieth century has promoted further study of social system and the ecological environment, accompanied by a further broadening of interdisciplinary study within the field of sociology. Difficulties in dealing with social changes have propelled continuous studies by systems scientists. Three cited books below highlight this general trend of study of social systems in the 21st century.

1.2.4.1 Human Ecology-Basic Concepts for Sustainable Development by Marten (2001)

Marten has raised the concept of co-evolution and co-adaptation of the human social system within the ecological environment where they co-exist. Based on Marten, the social system is all about people and their population; values and knowledge that form their worldview as individuals and as a society and which guide them to process and interpret information and translate it into action; technology that facilitates their range of possible actions; social organization and institutions that specify socially acceptable behavior and influence what they do. "Like ecosystems, social systems can be on any scale—from a family to the entire human population of the planet." (p. 18). In Fig. 1.1, Marten illustrates the relationship between the human social system and the ecosystem.

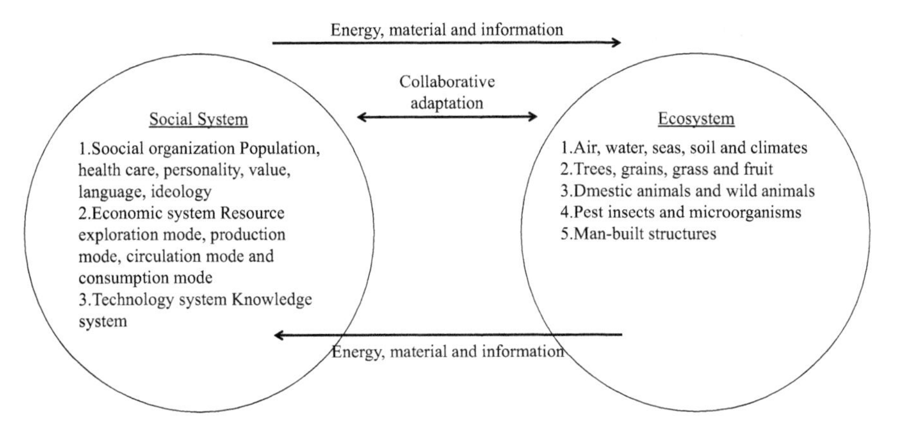

Fig. 1.1 Interaction of the human social system with the ecosystem (Marten, 2001, p. 18)

1.2.4.2 The Handbook of Economic Sociology (2nd Ed.) by Smelser and Swedberg (2005)

This handbook includes many topics that are also useful components of SSE, for example, the topic of 'the State and the Economy' is an important component in social system. The topics of 'Behavior Economics' and 'Emotions and the Economy' provide a better understanding of interaction of individual actors and human societies. And the topic 'Networks and Economic Life' is an updated exploration of economic life of the social system in the Internet age.

1.2.4.3 Systems Thinking for Social Change by Stroh (2015)

Throughout human history, changes in ways of human living and production are continuous which range from hunting and collecting, agricultural society, industrial society, to the current transition into a post-industrial or information/knowledge society. There is social inertia to various degrees, resisting the social change process. At a relatively lower micro level, in dealing with educational issues, reducing poverty or developing environmental sustainability, we may find that the organizations and systems we want to change have a life of their own. Despite our repeated attempts to improve them, they essentially continue to operate the same way as if our effort makes no difference. Scholars and practitioners concerned have been working relentlessly in search of ways and means to achieve better results and more lasting impacts with fewer resources in the process of social change and development. This book is aimed at contributing to these same efforts by applying systems thinking, principles, and tools and sharing experiences in solving real world problems.

1.2.5 Exploration of SSE in the 21st Century

In recent year, a variety of explorations in SSE have emerged as scholars, researchers, and practitioners seek different ways to conceptualize and incorporate SSE into the broader social arena to study large-scale human behavioral dynamics and complex social systems, develop digital solutions to predict behavior, and assist policy makers to solve societal problems. Below describe a few examples.

- SSE related research initiatives and academic programs in US. Engineering Social System (ESS) is a joint research initiative (https://www.hsph.harvard.edu/ess/about.html) between Harvard University, MIT, Northeastern University, Santa Fe Institute, and other corporate and non-corporate research partners such as IBM, Microsoft, Nokia, Google, Intel, UNICEF, Rockefeller Foundation, etc. Its mission is to "seek solutions to important global problems and to better inform leaders and policy makers who will turn great ideas into real changes that will make the world a better place." By analyzing large behavioral datasets, generated from mobile phones and credit cards from around the world, ESS researchers attempt to (1) generate new knowledge about large-scale human behavioral dynamics and complex social systems via machine learning and network analysis algorithms; (2) transform that knowledge into actionable recommendations; (3) ensure resources for implementation; and (4) measure the results. At its core, ESS integrates computational social science and physics of society in big data mining to generate insights into the "behavior of people ranging from individual to nations" and to "leverage these insights to radically re-think how traditional social systems are engineered."

Networked and Social Systems (NETS) Engineering Program at University of Pennsylvania, School of Engineering and Applied Science. (http://www.nets.upenn.edu/). This undergraduate degree program is named after Rajendra and Neera Singh, Owners of Telecom Venture LLC. NETS, began in 2011 and with its first class graduated in 2015 with a Bachelor of Engineering degree, is a multidisciplinary program that connects the study of networks with the study of human behavior. By mixing courses in engineering, mathematics and science with courses in sociology, game theory, economics and policy, NETS not only provide fundamental knowledge about "various strategies and decisions that make up human interaction" but also train students to "build the digital solutions that anticipate the way people—and systems—will act."

1.2.5.1 Book Publications About SSE

Since SSE is still at its preliminary explorative stage, there are not many books specifically devoted to the subject. Below cite three books that we consider most relevant.

(1) *Social System Engineering by (Oosawa)* (2007)

This book may be the first one ever published with the name SSE. In Japan, the higher education and research institutions have explored and studied unremittingly the subjects of "social engineering" and "social systems engineering". University of Tsukuba, for example, has offered graduate studies on SSE. However, such a field has not gained a convincing recognition and dominant influence in the international academic arena. *Social System Engineering* consists of 15 chapters including Orientation; Problem Definition; Data Collection; Data—Information; Modeling; Discrete Model; System Dynamics; Predication; Game Theory; Probability and Bayesian Theory; Uncertainty and Human Judgment; Reflection on Network; Information Technology and Social Systems; Methods for Generation of Ideas; and Conclusion. Based on the chapter topics, this book has mainly collected several existing quantitative tools for analyzing social systems with more emphasis on dealing with uncertainty. It also covers topics of contemporary society. For example, the influence of the development of information technology and network on social systems. Since the book is intended for academic use (a textbook for higher educational institutions), its coverage on the operation and management of social system, such as planning and policies in the real world is limited.

(2) *Introduction to Social Systems Engineering edited by Yang and Li (*2008*)*

This may be the first book published under the name of SSE in China. It is a textbook for higher education institutions. *Introduction to Social Systems Engineering* consists of nine chapters, namely, Introduction, Principles of Systems Science, Self-organization System Theories, Complex System Theories, System Cybernetics, System Operations Research, Systems Engineering Approach, Classic Social System Theories, and Modern Social System Theories. Compared with the *Social System Engineering* by Oosawa Hikaru, this book is more about various theories related to systems engineering and less on applied mathematics. Contents related to social systems are discussed in Sect. 4 of this chapter. Chapter 8 discusses social systems theories of Marx, Spencer, Parsons, Luhmann etc. Chapter 9 presents the components of social system and their mutual relations, and the organizational modes of social system with the self-organizational mode in particularly. In the end, it states the nature of complexity, chaos, fractal, etc. of the social system. The characteristics of the book are encapsulated in the summary: "This book is a combination of the specifics and realities of universities/colleges of science and technology. It aims to strengthen the creation on the systems thought, systems thinking, and systems engineering working skills of students majoring in social work and public management." (Yang & Li, 2008, p. 253)

(3) *Methodology of Social System Engineering (Chinese edition) by Wang* (2014)

This book has almost the same structure as this current English book, but the focus and content are different. The Chinese edition *Methodology of Social Systems Engineering* is written for the Chinese professionals and readers, especially for the planners at government or corporate level. The purpose is to provide a new concept and balanced approach to planning. This book consists of two parts with ten chapters. Part 1, focusing on theoretical foundations, has four chapters: (i) Overview; (ii) General System Theory and Systems Engineering; (iii) Social Systems; and (iv) Outlines of SSE. Part 2, concentrating on methodology of SSE, has six chapters: (v) Overall Discussion of Methodology which discussed in detail A. D. Hall's (1969) Three-Dimensional Morphological box for systems engineering along with Bardach's (2005) eightfold path to show the common steps in planning and policy analysis; (vi) Planning System explained and compared planning experiences of France, Japan, India and China, and briefly discussed scenario analysis in planning in *Global Trend 2030: Alternative Worlds* (National Intelligence Council, 2012) and spatial development planning of Japan and European Commission; (vii) Cultural Systems and Its Impact on Development discussed progress of cultural theory and studies on culture and development by UNESCO, selected developed countries, and China; (viii) System of Qualitative and Quantitative Analysis explored some ways and means for regulation and control of social system, described the performance measurement[2] in detail, and discussed selected models appropriated to study the social system and model application techniques; (ix) Analysis and Evaluation of Public Policy System discussed public policy system as an important means to keep a social system in order and in progress; (x) Growth of Think Tank discussed selected think tanks from both developed and developing countries as case studies, some international organizations, and China's Development Research Center of the State Council where the two authors had worked more than 30 years. Think tanks have been around for 180 years. The authors maintain that a well-organized think tank with groups of qualified experts from various disciplines can perform a proper role in planning, designing and improvement of a social system.

1.2.5.2 Description of Social-Systems Approach

To date, there is no terminology for "Social Systems Engineering" in any dictionary of social science or encyclopedia, which is an indicator that it is still a discipline to be acknowledged by all. However, Nishet (2008) described *social-systems approach* in his "Social Science" contribution to *Encyclopedia Britannica*; it is quoted below:

> Still another major tendency in all of the social sciences after World War II was the interest in "social systems." The behavior of individuals and groups is seen as falling into multiple interdependencies, and these interdependencies are considered sufficiently unified to warrant use of the word "system." Although there are clear uses of biological models and

[2]How to measure performance of the social system was once a major obstacle in the establishment of social systems engineering, but much progress has been made in recent years to establish various types of indicators.

> concepts in social-systems work, it may be fair to say that the greatest single impetus to development of this area was widening interest after World War II in cybernetics—the study of human control functions and of the electrical and mechanical systems that could be devised to replace or reinforce them. Concepts drawn from mechanical and electrical engineering have been rather widespread in the study of social systems.
>
> In social-systems studies, the actions and reactions of individuals, or even of groups as large as nations, are seen as falling within certain definable, more or less universal patterns of equilibrium and disequilibrium. The interdependence of roles, norms, and functions is regarded as fundamental in all types of group behavior, large and small. Each social system, as encountered in social-science studies, is a kind of "ideal type," not identical to any specific "real" condition but sufficiently universal in terms of its central elements to permit useful generalization.

He also stated that "Structuralism in the social science is closely related to the theory of social system." We concur with Nishet's viewpoint and consider that it has encapsulated some fundamental concepts of SSE.

1.2.5.3 Description of Social Engineering

(1) *Terminology of social engineering*

Although there is no terminology of social systems engineering found in any dictionary, there is description of social engineering in *Oxford Dictionary of Sociology* (Scott, 2014). It defines social engineering as:

> Planned social change and social development; the idea that governments can shape and manage key features of society, in much the same way as the economy is managed, assuming that adequate information on spontaneous trends is available through social indicator and social trends reports. For example, the extent of women's employment is clearly determined in part by government policy to promote or impede women's paid work. (p. 695)

(2) *Origin and evolution of social engineering*

In 1894, a German industrialist J. C. van Marken introduced the term sociale engeniers (social engineers) in an essay. His functional parallel idea was that at today's workplace, employers needed both social engineers (specialists) to deal with human problems and technical engineers to deal with mechanical problems. This term and concept spread to America in 1899 and the notion "social engineering" was introduced as the task name for social engineers. Later, US sociologists adopted the usage of social engineer as one building block on a metaphor of social relation as "machineries", to be dealt with in the similar fashion by technical engineers. The concept of social engineering was further broadened by Germany philosopher Theodor W. Adorno (1903–1963) for observation of mass culture in his study of cultural industry. Both Japan and US have continued the study and development of social engineering ever since.

1.3 Social Systems Engineering

1.3.1 Necessity of Developing SSE. *Two Aspects Necessitate the Development of SSE*

1.3.1.1 Necessity to Promote and Improve the Development of Human Society

The *socio-economic* development of mankind has progressed over a long historical period. For over 18 centuries and well before the industrial revolution, the average annual compound growth rate of global population and GDP was insignificant. These growth rates reached 0.98 and 2.21%, respectively, in the period of the post-industrial revolution (Maddison, 2001). However, divergence in growth rates varied significantly in different countries in their historical development. In China's example, its population and economy was very much affected by the political condition in her long history of feudalism. Two famous dynasties, Han and Tang, in China's history, are briefly discussed below.

In the beginning period of Han Dynasty (202 BC–2nd Century AD), the population and economy had declined considerably compared to the period of Warring State (inhabitants around 20 million) due to civil wars since the second emperor of Qin Dynasty. Several emperors of Han Dynasty adopted series of policies to encourage population growth and promote agricultural production, by the end of Han Dynasty, it had around 60 million inhabitants. Similar situation happened in Tang Dynasty (618 AD–907 AD), the most powerful dynasty in China's history. The Tang Dynasty had no more than three million households in the beginning due to wars and unstable political and social environment that lasted for four centuries between Han and Tang Dynasties. The number of households and inhabitants reached the highest in the year 754 AD of Tang Dynasty—around nine million of households and 53 million of inhabitants, respectively.

Since the Industrial Revolution, although economic development has greatly improved people's living conditions and material life; for example, people's medical and education conditions and average life span have improved significantly. Humans are facing ever more complicated social issues such as wars, crimes, poverty, urban problems, regional problems, international problems, malnutrition, hunger and diseases. Warfield (1976) regretfully pointed out that, "Experience shows how imperfectly we deal with these problems" (p. 1). Japanese Oosawa (2007) listed over 160 social issues faced by society; such as unbridled mammonism, lack of medical personnel, increase of medical cost, issues about immigrant workers, crimes of police officers and civil servants, aging population, illegal discharge of industrial wastes, hereditary systems in enterprises and organizations, pollution of lakes and rivers, increase of natural disasters, social injustice and unbalance, local cultural recession and increased crime rate among juveniles (p. 2). These social issues also exist widely in China. Although China has achieved great socio-economic development since the establishment of PRC in 1949 and the economic system has achieved an extraordinarily fast development since the reform and opening-up in late 1970s, conflicts and new issues in the process of economic and social development are becoming more prominent. In Premier Wen Jiabao's Report on the Work of the Government presented to National People's Congress on March 5, 2013, he acknowledged that China still faces many difficulties and problems in economic and social development, particularly:

> Unbalanced, uncoordinated and unsustainable development remains a prominent problem.
>
> There is a growing conflict between downward pressure on economic growth and excess production capacity.
>
> Enterprises' operating costs are increasing and their capacity for innovation is weak.
>
> The growth of government revenue is slowing down while fixed government expenditures are increasing.
>
> There are potential risks in the financial sector.
>
> The industrial structure is unbalanced.
>
> The agricultural foundation is still weak.
>
> Economic development is increasingly in conflict with resource conservation and environmental protection.
>
> The development gap between urban and rural areas and between regions is large, and so are income disparities between individuals.
>
> Social problems have increased markedly, and many problems in the areas of education, employment, social security, medical care, housing, the environment, food and drug safety, workplace safety, and public order affect people's vital interests.
>
> Some people still lead hard lives.
>
> There are many systemic and institutional obstacles to developing in a scientific way.
>
> The transformation of government functions has not been fully carried out, and some areas are prone to corruption. (Wen, 2013, pp. 14–15)

Chinese leaders were aware that many complicated problems fall into the category of SSE. From the structural perspective, families, one of the component parts of the social system, are also undergoing changes. They evolved from the big families where several generations lived under one roof in the agricultural society to today's small families made of couples and their children, single parent families, or families of LGBT[3]. Different types of families have different kinds of problems. There are many studies and publications titled *Family Problems*, *Family in Transition*, *Modern Family* etc. done by sociologists. They can illustrate fully the complexity and urgency of study of SSE. However, the development of SSE and its theories falls behind the demands. Therefore, it is necessary to promote its development with more participation from academics and experienced professionals working in socio economic field.

---

[3]Note: LGBT is an initialism that stands for lesbian, gay, bisexual, and transgender.

1.3.1.2 Necessity for Advancement in Education System to Train Qualified Personnel

Disciplines in academic and education fields can be classified into basic, special, and comprehensive. Although people trained in special disciplines are in demand, broad issues facing mankind whether at work or in personal life cannot be solved just by specific knowledges applied to today's highly complex society. This weakness of traditional education system in providing qualified generalist to deal complex engineering project had been shown in last century. Planning, design, construction and operation of any large engineering project demands not only knowledge from natural science and related engineering disciplines, but also knowledge of economy, sociology and management from the field of social science in order to achieve better result of the project. Systems engineering was emerged in last century to meet the demand. This situation is even more complex in social systems engineering. It can be seen from the development and reform of a nation, the phases and steps to deal with social issues had been generalized by Hall, a pioneer of systems engineering in his morphology Box (refer to Fig. 5.2a of Chap. 5 of this box). Currently, the terminology and concept of 'top level design' is suggested by the Chinese economic field to be applied to economic reform, this includes implicitly applying the planning design concept and approach established in engineering fields to the realm of economic reform. But the current education system and its curriculum are biased towards the training of specialized professionals. It seems that the training and development of professionals with a broad-based knowledge and comprehension are inadequate. Leaders from large multinational corporations, academic fields, and government agencies reported intense shortage of qualified engineers with capacity of system thinking and integration. International Council on System Engineering (INCOSE) was established in 1995 in US to address the needs. But INCOSE focuses only on training qualified systems engineers when there is an even greater demand for training qualified social systems engineers. Therefore, promotion of the study of SSE is necessary to mend the deficiency in the current educational system. This is a relatively complicated subject to be discussed. Because the demand of generalist may be few in number compared to the demand of specialist, but they are indispensable in the division of labor of modern society, especially, the society of the mankind is in transition towards a knowledge society. In Japan, such programs are in existence for more than 30 years while only a handful of institutions of higher learning in the West have recently established the program of SSE. And in China, such a study seems only at its infancy, even although China has some practical experience in the field for a long time. It is necessary to develop this new discipline through collaboration with learning communities and related organizations both domestic and abroad.

Education is an important subsystem of any social system. The contemporary global education system especially the universities have existed for one thousand years.[4] Changes must be made to adapt to the current and future needs. There is already emergence and development of distance education, online learning, Open University, etc., to facilitate needed change. More can be said on this subject, but it is out of the scope of this book.

1.3.2 Time Is Ripe for Development of SSE

1.3.2.1 There Is an Urgent Demand for the Development of SSE

As described in Sect. 1.3.1, although there was rapid growth in the global economy and great improvement of people's living condition post the Industrial Revolution, in which developing counties (a fair part of them were colonized countries before the WW II) even enjoy a higher average annual economic growth rate than that of developed countries in general, many social issues remain unsolved. One of them is severe income disparity. Foroohar (2016) pointed out that "Financialization has funneled wealth up, not down" but "Average hourly pay has increased $1.25 in 50 years after adjusting for inflation" (p. 24). Nearly the same conclusion can be found in Capital in the Twenty-First Century by Piketty (2015) as he states that the top decile share in US national income rose from less than 35% in the 1970s to 45–50% in the 2000–2010 (p. 24). He notes that "...a market economy based on private property, if left to itself, contains powerful forces of convergence, ...but it also contains powerful forces of divergence, which are potentially threatening to democratic societies and to the value of social justice on which they are based." (p. 571).

1.3.2.2 The Scope of Study of SSE is Basically Defined and Practiced

It can be seen from previous descriptions that the scope of study of SSE is planning and policy making by a large part of researchers and available publications. Although there is no shortage of planning at national or corporate level, it was reported in the World Development Report 1983 (World Bank, 1983) that:

> After 1945, however, the establishment of planning agencies represented a major institutional departure in developing countries. These agencies were intended to provide medium and long-term perspectives on development, supplementing the short-term preoccupations of finance ministries. There is now common place in developing countries. A recent World Bank survey of some eighty countries indicated that four out of five have multiyear development plans, over the past ten years, approximately 200 plan documents have been prepared.

[4]Note: The first universities in Europe were the University of Bologna (1088), the University of Paris (1150) and the University of Oxford (1167).

> In most countries, however, planning agencies have not lived up to expectations. By the late 1960s there was widespread talk of a crisis in planning. (p. 66)

It should also be pointed out that there is no shortage of studies of development planning by economists, including Nobel laureates. Most of a nation's planning and corporate planning are conducted by economists in general. Piketty (2015) frankly discerns that he sees economics as a sub-discipline of the social sciences, alongside history, sociology, anthropology, and political science (p. 573). He also dislikes the expression "economic science" because it suggests that economics has attained a higher scientific status than the other social sciences (pp. 573–575). He points out correctly that "For far too long, economists have sought to define themselves in terms of their supposedly scientific methods…" (p. 574). He also suggests that "social scientists in other discipline should not leave the study of economics…" (p. 575)

The authors maintain what is cited above can partially explain the failure of planning to meet the demand expectation and the cumulative social issues in development. Exploration of SSE by experts from various disciplines may be helpful in better planning and policy making. This can be shown by Cast Study: **Expo 2000 OECD Forum for The Future** in Chap. 5 and Case Study of **China toward the year 2000** done by Development Research Center of China presented in Chap. 4.

1.3.2.3 Some Difficulties in SSE are Resolved and Some are Being Explored

It's generally understood that many difficulties exist in exploring social system, such as, measurement and quantitative approach as are prevalent in any hard science. But great progress has been made by United Nations and other international institutions in establishing social indicators which will be discussed in Chap. 6 of this book. With respect to quantitative approaches, the difficulty is to quantify human behaviors. But *Behavioral Model and Simulation—From Individual to Societies* (National Research Council, 2008) has provided updated information about available models from individual to societies even though no concrete description (equations, variables and parameters) of various models is given. This book is based on a study supported by a joint grant between the National Academy of Science and the US Department of Air Force. It is a common practice in US science and technology (S&T) policy to let the military take the lead in research and disseminate results appropriately for civilian uses. The World Development Report 2015 (World Bank, 2015) centers its theme on "Mind, Society, and Behavior." In the Forward section, President Jim Yong Kim maintains that "Many development economists and practitioners believe that the 'irrational' elements of human decision making are inscrutable or they cancel each other out when large numbers of peoples interact, as in markets, yet, we now know this is not the case. Recent research has advanced our understanding of the psychological, social and cultural influences on decision and human behavior and has demonstrated that they have a significant impact on development outcomes." (p. xi). The World Bank is the largest international organizations with large number of highly experienced researchers and experts in policy studies for many countries. The findings in this report are significant. With existing literature paving the way and more relevant research in the pipeline, we suggest that it is the right time to establish and develop the discipline of SSE.

1.4 Exploration of Definition of Social Systems Engineering

1.4.1 Current State

From previous discussions, it is evident that the theoretical foundation for SSE are still being explored. The curriculum of University of Tsukuba covers social economics, management and urban planning. On the other hand, the University of Pennsylvania stresses the integration of social networks, technologies and economics into social systems. There is also the terminology social engineering which overlaps SSE in scope of study. The famous Chinese scientist, Mr. Qian Xuesen defined social engineering as the science for construction and management of a country. It is not unusual to have these different perceptions because different scholars may uphold their own perspectives and understandings at the exploratory stage of SSE and it is difficult to come up with a perfect definition for SSE from a single perspective, due to its complexity.

A metaphor of the blind men and an elephant parable can be used in the emergence of SSE. The moral of the parable is that humans tend to project their partial experiences as the whole truth, ignore other people's partial experiences, and one should consider that one may be partially correct and may have only partial information. Compared with systems engineering, SSE has a broader coverage of various disciplines. Thus, to understand and define SSE as completely as possible, we need to examine closely a variety of definitions by some other scholars.

There are two available definitions in published books which will be described below for reference and comparative purpose.

1.4.1.1 Definition of Social System Engineering by Oosawa (2007)

"Society" is the assembly of the human being. Society depends upon the social system constructed and operated by systems, institutions and mechanisms. Therefore, "social system engineering" is the knowledge to understand and explore plans for social system and the society from engineering or practical perspective.

But the current university curriculum and textbooks frequently associate the above vocabularies in the areas of regions, urban, civil engineering, construction, land, environment, welfare, economy, industry and society; and the knowledge composed of perceptions, conceptions and methodologies is related to key terminologies such as 'system', 'model', 'simulation', 'analysis', 'design', 'management' 'evaluation', 'forecasting' and 'optimization'. It seems that all these cannot clearly define the knowledge of 'social system engineering', and neither can they identify the boundary and scope of 'social system engineering'. (translation, p. iii)

1.4.1.2 Definition of SSE by Yang and Li (2008)

Yang and Li (2008) briefly defined that "SSE is a discipline of engineering technology to study social system, its task is to integrate various knowledge and technology related to society, and combines with the practice of social engineering to proceed toward innovation, to construct new control system of society, to satisfy the demand of the human being to reform and construct the society consciously. Study of SSE is helpful to strengthen the close integration of theory and practice of social science and to promote the development of social science." Yang and Li classified modern science and technology and natural sciences into three hierarchical levels, i.e. basic science—technological science—engineering science, among them, the application of engineering science and technology has the most direct and close relationship with the practice of production and engineering. They stated, "Engineering is a form of expression to reform the nature of the mankind, it is a compound of interaction between the mankind and the nature. Engineering science in its primary sense is a concept of natural science, it is a technology to reform the objective world directly, it is an important link and path for science and technology going toward production and practice. Therefore, SSE is a discipline belonging to the engineering technology category (translation, p. 22).

In Yang and Li's conceptual framework, they cited Chinese scientist Qian Xuesen who stated, "social science should proceed from social science to social technology in similar order as natural science to engineering technology. Applied social science should follow the same pattern as engineers designing a new architecture in order to design and reform our objective world scientifically." Qian also stated that "social engineering is a science to reform, construct and manage society, one of its objectives is to combine the social science with other sciences; it is a type of practical technology." Hence, Yang and Li gave a definition based upon Qian Xuesen's view that social engineering is the technology of organization, construction, and management of society. Such a technology falls into the scope of systems engineering and its object is the whole social system (translation, pp. 22–23).

In general, we concur with Qian's statements about social engineering and its functionality. However, although Qian expressed his view on the direction of development of social science and defined what social engineering was, he did not use or define the term SSE.

1.4.2 *Definition of SSE Given in This Book*

1.4.2.1 Principal Underlining the Definition

Authors of this book follow the basic principal that the progress of mankind is through inheritance and creation. In the process of creating definitions adaptable to current and future needs, we aim to inherit and carry forward the best and avoid the pitfalls of the past. To achieve the aim requires broad surveys and readings of various literatures, and learning by doing at the same time.

1.4.2.2 Components of SSE

The definition of SSE can be defined through two of its components: the social system and engineering.

(1) *Definition of social system*

Parsons' definition of social system was relatively concise but abstract. To make the meaning more concrete and easily understood, we define the social system as follows:

The social system is composed of individuals and social units or groups (such as families, corporations, organizations, government agencies, etc.) connected through permanent relations which can be replicated to form structures with functions. The social system exists in certain physical and non-physical environment. With the aim towards "optimal gratification", individuals and groups within the external boundary of a social system interact with each other and with those outside the boundary via exchange of information and action. Their interactions within and beyond the boundary are influenced by cultural systems, shared value system, existing regulations and institutions. The social system is dynamic and subject to changes caused by internal and external forces. This definition can be expressed in a diagram shown in Fig. 1.2.

(2) *Definition of engineering*

Two definitions for engineering are cited below as references.

According to Smith (2012), engineering is: "the application of science to the optimum conversion of the resource of the nature to the use of humankind. The field has been defined by the Engineers Council for Professional Development in the U.S., as the creative application of 'scientific principles to design or development structures, machines, apparatus, or manufacturing process, or works utilizing them singly or in combination; or to construct or operate the same with full cognizance of the design; or to forecast behavior under specific operating condition; all as respects an intended functions, economics of operation and safety to life and property.

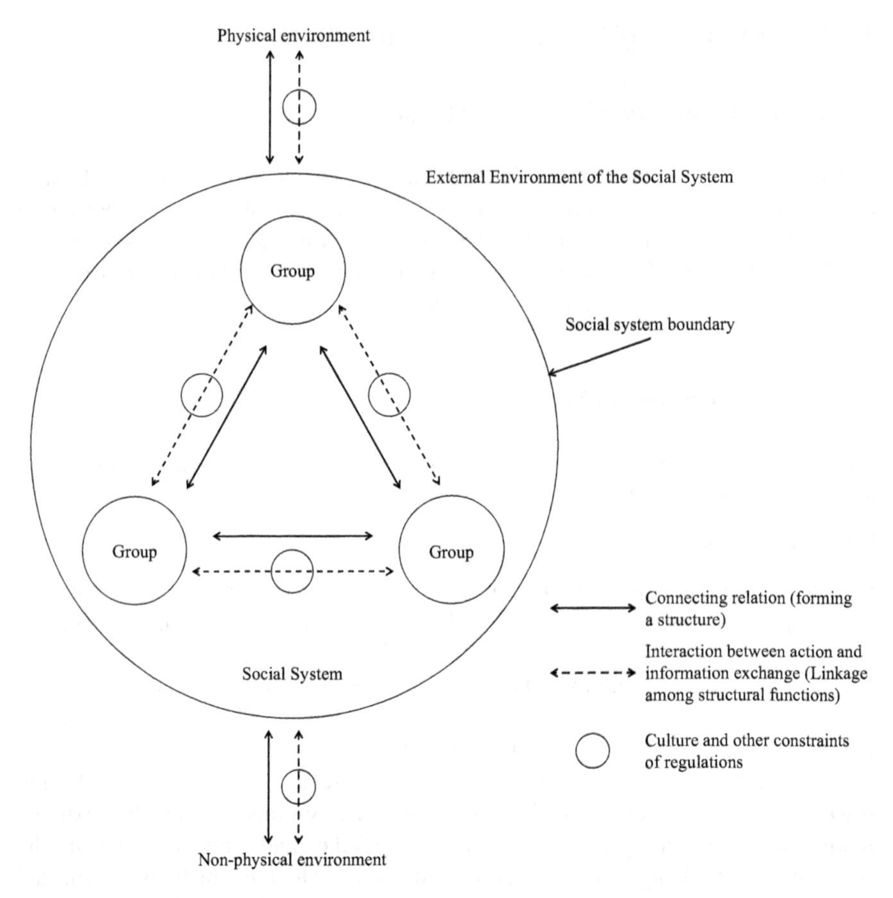

Fig. 1.2 Social system and its relationships

In *Introduction to Engineering*, Mayne and Margolis (1982) described engineering as "The art of the practical application of scientific and empirical knowledge to the design, production or accomplishments of various parts of construction projects, machines and material of use to man." (p. 4)

These definitions of engineering are written for conventional and hard systems. We can modify them to describe engineering for social system which is more of a soft system. Below is our first attempt:

- *Engineering is the creative application of scientific principles to Plan, design, develop and improve the structure and parts of the social system, or the social process; or to operate the same with full cognizance of the design; or to forecast behavior under specific operating condition. It should achieve the intended functions of economic prosperity and social justice.*

1.4.2.3 Definition of SSE Given in This Book

Due to the complexity of SSE, this book will provide two definitions of SSE. Definition "1" and "2" complement each other.

(1) *Definition 1*

SSE is interdisciplinary engineering based on the application of general systems theory, systems engineering, Parsons' theory of social system and social action, and behavior science. SSE extensively applies principles and concepts of various sciences and knowledge domains to plan, create, design, operate, manage, and improve complex social systems for achieving expected effects and goals that are mutually satisfactory. The expected effects and goals are measured by multi-dimensional indicators including social, economic, S&T and ecological environment. In a narrow sense, the SSE is applicable to the improvement and management of large complex organizations, i.e., countries with established political power structures and dominant values. It can also be applied to the construction and management of large corporations.

(2) *Definition 2*

SSE guides the plan, design, construction, operation and improvement of a social system and its process with certain value. The social system is composed of individuals and social units or groups (such as families, corporations, organizations, government agencies, etc.) connected through permanent relations which can be replicated to form structures with functions. The social system exists in certain physical and non-physical environment with boundaries. With the aim towards "optimal gratification", individuals and groups within the external boundary of a social system interact with each other and with those outside the boundary via exchange of information and action. Their interactions within and beyond the boundary are influenced by cultural systems, shared value system, existing regulations and institutions.

1.4.2.4 Explanation of the SSE Definition

Since SSE is an emerging discipline, there is no consensus on its definition but a great variance in how the terminology is used, for example, some scholars considered social engineering as equal to 'SSE'. Some may have high expectations in its result when applying SSE without considering its potential risks. Therefore, it's necessary to provide some explanation on the above SSE definitions and their wording.

(1) *Definition 1 consisting of three parts.* The first part emphasizes the basic theories of SSE. Based on the authors' work experience in the field of systems engineering, not many engineers pay due attention to the basic theories which includes behavior science that underpinning Parsons' core concept of 'action'

in social systems. The second part expresses a more realistic aim when applying SSE in real world and it is for achieving expected effects and goals that are satisfactory rather than optimal. The last part is SSE's applicability. The authors assume that SSE can be applied to countries with established political power structures and dominant values. Brikland (2005) described the political system as "The Black Box" (p. 202) as it is too complicated to be treated with engineering measures or means. So, rather than designing a new social or political system, we can use SSE to improve or reform existing systems.

(2) *Definition 2* combines definitions for engineering and social systems into an integrated whole. Thus, Definition 1 and 2 can be used to complement each other.

(3) *Application of SSE.* Based on the two definitions, SSE can be applied broadly to areas such as family, organization, corporation, national or regional planning and policies. This book will limit its study of application to national, corporate, and regional planning and related institutions.

1.4.2.5 Qualifications for a Social System Engineer

Machol (1965) described system engineer qualifications in *System Engineering Handbook.* We maintain that competent social system engineers should have the same qualities of a system engineer. Namely, he/she must (1) be a generalist, rather than a specialist but not an amateur; (2) be a "T-shaped man", broad yet deep in one field—scholarly experience, such as Ph.D. (or equivalent), or work experience in multiple engineering or management fields can provide depth and extensive interests and abilities can facilitate breadth; (3) have courage and ability to enter new field and become so called "6-month expert"; (4) have strong foundation in mathematics and logic; and (5) learn continuously and prudently, especially mindful learning from mistakes. This last attribute is crucial for a social system engineer now and in the future. Due to the explosive growth of knowledge in an information/knowledge age, a fair amount of knowledge that one has learned in one's college education may become obsolete quickly. Therefore, life-long learning and how-to-learn are critical.

1.5 Content Structure of This Book

This book is divided into three parts as illustrated in Fig. 1.3.

Part I-Emergence of SSE discusses the necessity and inevitability of emergence of SSE from the historical development perspective of both natural and social sciences. Two fundamental theories related to SSE are explained in detail, one is general system theory and the other is Parsons' "theory of action" and AGIL framework. The AGIL framework is modified partially by the authors to better

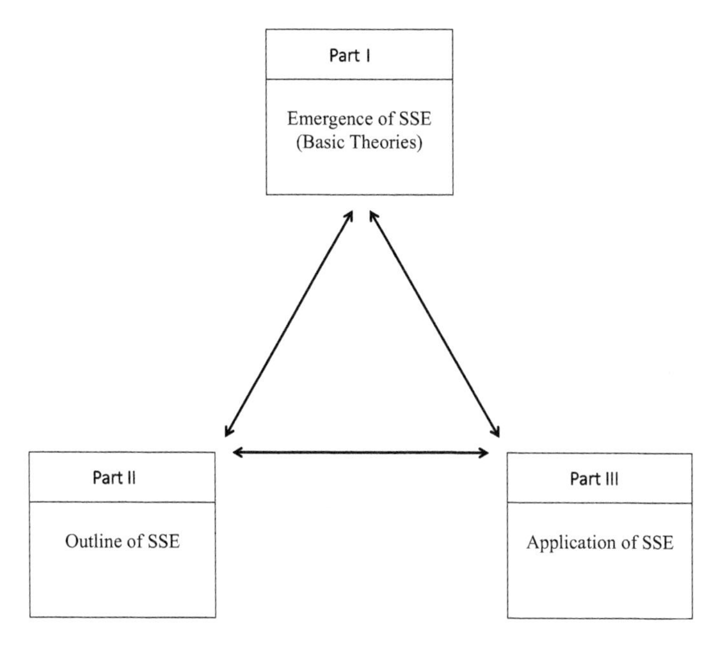

Fig. 1.3 Book content structure

match the basis of SSE. Trend of development of SSE in global society is presented. Difference between plan and design is discussed to clarify these two activities. Design is discussed in relatively detail both for physical system and social system. because in later part of this book, will focus more on discussion of planning three case studies of research projects and study in the nature of SSE are given to have a preliminary idea of what SSE is.

Part II-Framework or Outline of SSE. Due to the broad disciplines covered by SSE, this part has selected topics based upon two concepts: one is methodology in nature, which can be applied universally to approach all issues and disciplines; the other is cored topic relevant most to SSE. Therefore, there are two chapters, one is methodology qualitative in nature, the other is quantitative in nature. The other four cored topics are regulation; planning; boundary and environment, within which, cultural system is described certain detail; and think tanks, which can perform the role for studying and application of SSE.

Part III-Application of SSE presents the growth story of China as a case study. China has experienced a relatively fast economic and social development since the establishment of PRC in 1949. It implemented five 5-Year Plans from 1953 to 1980 with an average annual economic growth rate around 6%. The name of its national plan was changed into national economic and social development plan in 1981. Since then, seven 5-Year Plans have been implemented from 1981–2015. A brief introduction of planning experience in these two periods will be discussed. Major focus will be on the first 5-Year Plan and the first S&T Development Plan (Also known as "12-Year S&T Plan 1956–1967") during the first period, because these two plans laid the foundation for China's development. Seventh 5-Year Plan in the second period and the recent 13th 5-Year Plan will be discussed to show the progress and features of China's planning system.

References

Banathy, B. H. (1996). *Designing social systems in a changing world* (pp. 1, 14, 38–40). New York: Plenum Press.
Bardach, E. (2005). *A Practical guide for policy analysis: The Eightfold path to more effective problem solving*. Washington, DC: CQ Press.
Black, J. (Ed.). (2005). *World history*. Bath, UK: Parragon Inc.
Brikland, T. A. (2005). *An introduction to the policy process* (2nd ed., p. 202). New York: M.E. Sharpe Inc.
Chestnut, H. (1982). Nations as large scale systems. In Y. Y. Haimers (Ed.), *Large scale systems* (pp. 155–183). Amsterdam: North-Holland Publishing Company.
Foroohar, R. (2016). American capitalism's great crises. *Time, 187*(19), 24.
Forrester, J. W. (1971, January). Counterintuitive behavior of social system (p. 2). *Issue of the Technology Review*.
Hall, A. D. (1969). Three-dimensional morphology of systems engineering. *IEEE Transactions of Science and Cybernetics* (pp. 156–160), SSC-5.
Machol, R. E. (1965). *Systems engineering handbook*. New York: McGraw Hill Book Company.
Maddison, A. (2001). *The world economy: A millennial perspective* (X. Y. Wu, X. C. Xu, Y. F. Ye, & F. Q. Shi, Trans.). Paris: OECD. Beijing: Beijing University Press.
Marten, G. G. (2001). *Human ecology: Basic concepts for sustainable development* (p. 18). New York: Earthscan.
Mayne, R., & Margolis, S. (1982). *Introduction to engineering* (p. 4). New York: McGraw Hill Book Company.
Merton, R. K. (1968). *Social theory and social structure*. New York: Free Press.
National Intelligence Council. (2012). *Global trends 2030: Alternative worlds*. Retrieved from https://www.odni.gov/files/documents/GlobalTrends_2030.pdf.
National Research Council. (2008). *Behavioral modeling and simulation: From individuals to societies*. Washington, DC: The National Academies Press.
Nishet, R. A. (2008). Social science. In *Encyclopedia Britannica*. Retrieved from https://www.britannica.com/topic/social-science/The-20th-century.
Oosawa, H. (2007). *Social system engineering* (Japanese version) (pp. iii, 2). Tokyo: Ohmsha. Co. Jp.
Parsons, T. (1951). *The social system* (pp. 5–6). New York: Free Press.
Parsons, T., & Smelser, N. J. (1956). *Economy and society*. New York: Free Press.
Piketty, T. (2015). Capital in the twenty-first century. (Goldhammer, A., pp. 24, 51, 573–575, Trans.). USA: The Belknap Press of Harvard University Press.
Scott, J. (Ed.). (2014). *Oxford dictionary of sociology* (4th ed., p. 695). Oxford: Oxford University Press.
Smelser, N. J., & Swedberg, R. (Eds.). (2005). *The handbook of economic sociology* (2nd ed.). Princeton, NJ: Princeton University Press.
Smith, R. J. (2012). Engineering. In *Encyclopedia Britannica*. Retrieved from https://www.britannica.com/technology/engineering.
Stroh, D. P. (2015). *Systems thinking for social change: A practical guide to solving complex problems, avoiding unintended consequences, and achieving lasting results*. White River Junction, VT: Chelsea Green Publishing.

The World Bank. (1983). *World development report 1983* (p. 66). New York: Oxford University Press.

The World Bank. (2015). *World development report 2015* (p. xi). Washington, DC: The World Bank.

Von Bertalanffy, L. (1968). *General system theory* (p. 196). New York: George Braziller Inc.

Warfield, J. N. (1976). *Societal systems: Planning, policy, and complexity* (pp. 13–15). New York: Wiley Interscience.

Wang, H. J. (2014). *Methodology of social systems engineering* (Chinese version). Beijing: China Development Press.

Wen, J. B. (2013). *Report on the work of the government* (pp. 14–15). Retrieved from https://china.usc.edu/sites/default/files/legacy/AppImages/wen-jiabao-2013-work-report.pdf.

Yang, B. W., & Li, Z. G. (Eds.). (2008). *Introduction to social systems engineering* (p. 22) (Chinese version). Beijing: Petroleum Industry Publisher.

Zhao, Z. T., Zhang, W. T., & Zhou, D. Z. (1999). *Mencius* (Chinese-English translation, p. 119). Library of Chinese classics. China: Hunan People's Publishing House.

Chapter 2
General Systems Theory and Systems Engineering

2.1 Science, Technology (S&T) and Society

2.1.1 Economic History Shows That the Economic Growth Is Closely Related to S&T, Especially Post the First Industrial Revolution

2.1.1.1 The Higher Economic Growth Performance Since the First Industrial Revolution

(1) It is recognized by all historians worldwide that the First Industrial Revolution that erupted first in England had a great impact to the growth of global economy. Although there are different views of the duration of the First Industrial Revolution, generally it is considered to have begun in the late eighteenth century and ended by the beginning of twentieth century. Its impact on the global economy can be shown from the following two tables by historical economist Maddison (1995) (Table 2.1).
Madissen emphases that the growth performance from 1820 to 1992 is caused mainly by the technological progress of the Industrial Revolution.

H. Wang and S. Li, *Introduction to Social Systems Engineering*,
https://doi.org/10.1007/978-981-10-7040-2_2

Table 2.1 **a** Levels of world economic performance 1500–1992.[a] **b** Rate of world economic growth 1500–1992 annual average compound growth rate (%)

Item	1500	1820	1992
(a)			
World population (million)	425	1068	5441
GDP per capita (1990$)	565	651	5145
World GDP (billion 1990$)	240	695	27,995
World exports (billion 1990$)	n.a.	7	3786

Item	1500–1820	1820–1992
(b)		
World population	0.29	0.95
GDP per capita	0.04	1.21
World GDP	0.33	2.17
World exports	n.a.	3.73

[a]*Source* Maddison (1995)

(2) There are successive Industrial Revolutions driven by technology. New Industrial Revolution describes that there are three industrial revolutions following the first revolution. However, historian Stavrianos (1999) claims that those changes can be grouped together as a second Industrial Revolution caused by the post-WWII breakthroughs in nuclear energy, computer and robot, space science, genetic engineering, information revolution, green revolution, etc (pp. 760–762). We are in agreement with the view of the historian. We think that the term 'Revolution' needs a more precise definition.

Economist Sachs (2015) explains that the endogenous economic growth is due to the great waves of technological change. Sachs illustrates this view with a graph of Kondratiev[1] waves (Fig. 2.1, Sach, p. 83).

2.1.1.2 The Role of S&T in Economic Development

It is clear from the above discussion that technology played a major role to drive the economic growth; how about the role of science? It is necessary to understand several definitions of them and the different roles played by them in the history of mankind and the changing relationship between them. We had discussed the definitions of

[1]Note: Kondratiev cycle or long wave. A theoretical long term cycle ranging from boom to recession over a period of fifty years or so upon which short term business cycles are superimposed. This long wave was proposed by Kondratiev, Nikolai D (1892–1938) in his publication the *Long Wave in Economic Life* in 1925. He was a Russian economist who did work in agricultural economics and in the development of economic planning in former USSR. He identified long cycles from the end of the 1780s to 1844–1851, from 1844–1851 to 1890–1896, and an upswing from 1890–1896 to 1914–1920. Figure 2.1 may be done by Shiller, Robert J. to continue his work.

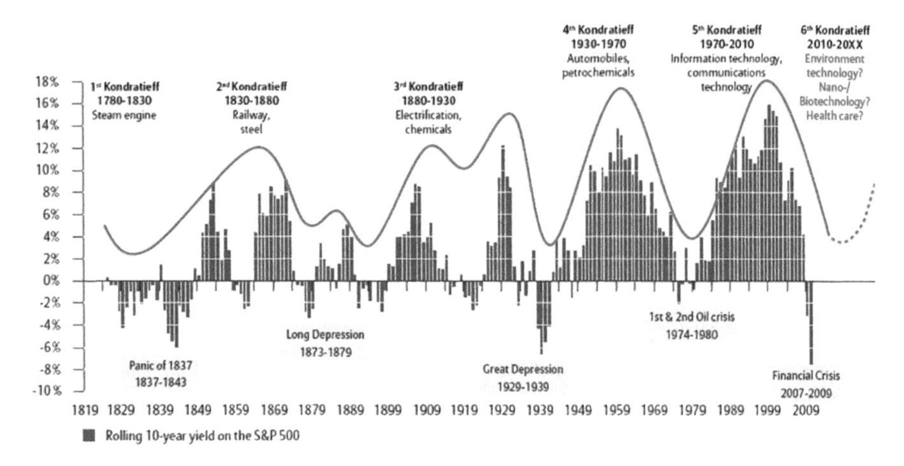

Fig. 2.1 Kondratiev waves

"Engineering" in Sect. 1.4.2.2(2) of Chap. 1. Hereunder, several definitions of S&T will be given and their role in different stages of the history will be discussed.

(1) *Definitions of S&T*

(i) [2]Science is, in the "(most general sense) the study of physical and social phenomena." (Jary & Jary, 1991, p. 350);

(ii) Technology is the "practical application of knowledge and use of techniques in the productive activities." Technology is a social product which involves both the "hardware" of human artefacts such as tools and machines, and also the knowledge and ideas involved in different productive activities. (Jary & Jary, 1991, p. 651).

(2) *Relationship between Science, Technology and Engineering*

There are different views between historians and historians of science of Western scholars with regard to the status of science before 600 BC. For example, historian Stavrianos describes that the "Sumerians developed sciences and mathematics, as well as writing in response to the concrete needs of their increasingly complex society." In his book, World to 1500-A Global History (p. 53), he also describes that a strictly Sumerian empire lasted from 2113–2016 BC. But in the Timetable of Science, the authors describes that science "as an organized body of thought is generally considered to have begun with the Ionian School of Greek philosophers about 600 BC. Discoveries or inventions prior to 600 BC are nearly always examples of technology, since they are specific device of technique." (p. 1)

[2]There are many definitions related to science. For example, the definition of science given by *Concise Encyclopedia (1996)* of Random House is "any systematic field of study or body of knowledge that aims, through experiment, observation, and deduction, to produce reliable explanation of phenomena, with reference to the material and physical world". This definition is a post Galileo (1564–1642), modern definition of science. Galileo introduced experimentation into science.

However, this book still classifies those activities before 600 BC as astronomy, mathematics, and life science wherever possible.

Although science does not begin until 600 BC, technology begins much earlier. For example, the stone tools were used around 2,400,000 BC by hominids in Africa. This is also the case for engineering projects. The Egyptians built a canal from Lake Timsaeh (the Nile) to the Red Sea in 1250–1201 BC.

Therefore, it can be said that technology and engineering predate science. The close linkage between science and technological and engineering activities began to form around the nineteenth century, the period of the Industrial Revolution. There was direct application of science to technology beginning around 19th century.

(3) *The Scientific Revolution that Preceded the Industrial Revolution Changed the Relationship between Science and Technology*

There was a period of Scientific Revolution in the world history: the emergence of modern science during the early modern period (European history, late 15th century–late 18th century) when development in astronomy, biology, chemistry, earth science, mathematics, medicine, and physics transformed the view of the society and the nature. The Scientific Revolution began in Europe toward the end of the Renaissance period and continued through the 18th century. Its impact to socioeconomic development of the mankind can be understood from the abstract of the conclusion of Chapter 10 in Stavianos' book (The World Since 1500-A Global History (1980)). Stavianos states that:

> As the nineteenth century passed, science became an increasingly important part of Western society. At the beginning of the century science still was on the periphery of economic and social life. But by the end it was making basic contributions to the old-established industries, it was creating entirely new industries, and it was affecting profoundly the way of thinking as well as the way of living of Western man. It made Europe's hegemony over the globe technological possible, and it determined to a large extent the nature and effects of this hegemony. European art or religion or philosophy did not greatly affect non-Western people because they had made comparable contributions in these fields. But there was no such parity in science and technology. Only the West had mastered the secrets of nature and had exploited them for the material advancement of mankind. (p. 203)

In 1951, Conant, J. B. defined science as "[3]an interconnected series of concepts and conceptual schemes that have developed as a result of experimentation and observation and are fruitful of further experimentation and observations." (p. 25) This new definition of science can distinguish itself from technology or engineering.

[3]Note: This quotation from Conant is for the illustration that there are many definitions of 'science' due to progress of history and improvements of recognition.

2.1.2 *Major Events in the History of Science and Technology (S&T)*

The achievements of S&T in the past two centuries have greatly promoted the well-being of the human society in general. But there are still many social issues faced by the mankind which have been outlined in Sect. 1.3 of Chap. 1. A brief study of the S&T history may provide a broad context for understanding the relationship between S&T and the society where the former existed. It can also provide the background for understanding the emergence of the general systems theory and systems engineering. There are several books which illustrate the time line of important S&T events. For example, the *Concise History of the World* (2013) published by the National Geographic Society; *Important Events of National Science in Annual Tables* (1975) published by Shanghai People's Press (in Chinese), and Philip's *Millennium Encyclopedia-History of the World* 12,000 years in timeline. However, none of these books can satisfy the quantification requirement in. Therefore, we have chosen Alexander Hellemans and Bryan Bunch's (1991) the *Timetables of Science* to be the main reference for our book. Hellemans and Bunch classify periods of the world history differently than a typical historian. Their clarification is from the perspective of scientific and technological growth, Hellemans and Bunch group the history of the world into ten periods. Because the purpose of this section is to illustrate the relationship between S&T, economy and society, and to provide the background of the emergence of the general systems theory and systems engineering, here we only describe eight of the ten periods. Our presentation of the history of S&T begins from the period of 2,400,000–599 BC, and ends at 1945 to accommodate the emergence of the general systems theory and systems engineering. Below are brief introductions of eight historical periods in the development of scientific and technological development.

2.1.2.1 Science Before There Were Scientists: 2,400,000–599 BC

The mankind evolved from H. habilis through H. erectus to H. sapiens in this period. They were hunter-gatherers in most of this periods. After the agricultural revolution about 10,000 years ago, civilizations began to emerge. The ancient Egyptians developed medicine, calendar, and engineering techniques and technology; for example the building of the Great Pyramid of Giza as the tomb of Egyptian pharaoh Cheops in 2900–2801 BC. There was Mesopotamian mathematics, and the Indus culture (centered around Harappa and Mohenjo-Daro in the present Punjab region in Pakistan. The early Chinese civilizations had also flourished by about 3000 BC. A primitive form of plow is in use in China around 4000–3501 BC, e.g. silkworms and mulberry trees are domesticated in China around 2800 BC., and the Duke of Chou built either an early-version of mechanic compass ("south-pointing carriage") or a magnetic compass around 1050–1001 BC.

During this period, the following were recorded: 76 events of technology, 16 events of mathematics, 10 events of astronomy, and 42 events of life science (mainly related to domestication of dog, pig, sheep and potatoes, wheat etc.). Medicine events were also included in the life science category. (Hellemans & Bunch, 1991, pp. 1–19)

2.1.2.2 Greek and Hellenistic Science 600 BC–529 AD

(1) *Features of this Period*

The Greeks were a seafaring people with a decentralized economy who lived in city-states controlled by upper-class citizens. This led to relative freedom of expression and thought. Greek science may be a continuation of ideas and practices developed by the Egyptians and the Babylonians, but the Greeks were the first to look for general principles beyond the observation. They were the first to introduce a scientific method based on reasoning and observation, although that method was not based on systematic experimentation which is the foundation of science as we know to-day. The Greek manuscripts were translated into Arabic by Arabs and eventually passed to Europe around 1453, which were the origin of science in Europe.

(2) *Representative Figures and their Achievements*

There are many eminent Greek scholars, but here we only introduce a few of them. **Thales of Miletus** is considered to be the founder of the Ionian school of natural philosophy. He and his two followers are regarded to be the First Scientists[4] in the history of Science and Technology. **Pythagoras** is another philosopher and mathematician, and his Pythagoras theorem is well known. **Democritus** is regarded to be the earliest atomist. He thought that substances differed from each other because of the shapes, positions, and grouping of the constituting atoms. **Empedocles** raised the concept of the elements. This philosopher believed that all matter was made up of four primary substances or elements: earth, air, fire and water. All appearances of matter could be explained by the commingling and separation of these elements under two forces in opposition. **Hippocrates** is referred as the 'Father of Western Medicine'. He explained health and disease by the balance of four humors (black bile, yellow bile, blood and phlegm), although this theory eventually hampered the development of Western medicine. Nevertheless, Hippocrates freed medicine from religion and superstition and put it on a more scientific footing.

Aristotle is considered to be the father of life sciences. He classified organisms into a hierarchy ranging from the most imperfect (plants) to the most perfect (man). His writings covered both natural and social sciences. **Archimedes** is often classed among the first rank of mathematicians in history. He is also known for his work in

[4]Note: In fact, the term 'scientist' is coined with physicist by Willan Whewell at a meeting of the British Association for the Advancement of science in 1933, he popularizes his new word in his *The philosophy of the inductive sciences* in 1840.

physical science, founding hydro statistics and statics, including an explanation of the principle of lever. He is credited to be an engineer because of his designing of innovative machines. **Ptolemy** is a Greco-Egyptian writer, and is known as a mathematician, astronomer and geographer. He is best known for his Almagest. The most important part of the Almagest is its description of the Ptolemaic system, a model of planetary motion in which the earth is the center of the universe and the sun and moon move around the earth in perfect circles.

2.1.2.3 Medieval Science 530–1452

(1) *Features of this Period*

This period is characterized by three aspects:

There was decline of science in Europe, it was also called Dark Ages in historiography of Europe, i.e. the period in Europe following the collapse of the Western Roman Empire. The European culture was still strongly influenced by the Romans, who notoriously had little interest in theoretical science except medicine.

The emergence of Arab science which flourished from about 700 to 1300, played a key role in bridging the gap between the Hellenistic period and the Renaissance era. Due to intense commercial activity, the Arabs came into contact with a large number of different cultures. These civilizations had new ideas to contribute to Arab thinking. And the Arabic language was a strong unifying factor in the Islamic empire. Many of the works of the ancients have been preserved because they (Greek and Hindu works) were translated into Arabic.

A number of centers of learning appeared throughout the Islamic empire. Baghdad was founded in 762 AD by the Muslim caliph (civil and religions leader). The seventh Assasid caliph Al-Ma'mun (786–833) founded a "House of Wisdom" with an astronomical observatory there. Many of the important Greek works were translated at that "House".

(2) *Major Achievements and Representative Figures*

Arabian mathematics blossomed in part because it combined mathematical knowledge of the Greeks with that of the Indians. Al-Khwarizmi (780–850) was a Persian mathematician, astronomer and geographer who worked as a scholar in the House of Wisdom. He is considered to be the father of algebra who presented the first systematic solution of linear and quadratic equations in Arabic. Chemistry was developed into an experimental science among the Arabs. In physics, Arab scholars excelled in the craft of instrument making. Alhazen (965–1040) was an Arab or Persian scientist, mathematician, astronomer and philosopher. He made significant contributions to the principle of optics (His book titled *Treasury of Optics*), astronomy, mathematics, meteorology and scientific method. He was an early proponent of the concept that hypothesis must be proved by experiments based on confirmable procedures or mathematical evidence. Medicine was highly developed. But the science of anatomy did not progress because dissection of corpses was not allowed by Islamic Law.

(3) *Science in China*

According to Hellemans & Bunch (1991):

> The Chinese society has been known throughout history for the stability of its traditions and its bureaucracy. Yet until about the fifteenth century, China was more successful than Western Europe in applying scientific knowledge to.....
>
> The Chinese discovered paper as early as 105 AD. The first printed text (in block printing) appeared in 865 AD. Around 100 BC the Chinese discovered that a magnet orients itself towards the North Pole, but they did not use magnets for navigation at seas until the tenth century. In 969 AD rockets were used in warfare. The Venetian traveler Marco Polo reported that the Chinese used firearms in 1237.
>
> The escapement, the most important part of the mechanical clock, was invented in 725AD. The Chinese were familiar with many other mechanical devices, such as the eccentric and the connecting rod, as well as the piston rod, long before they became known in Europe. (p. 59)

The knowledge gathered by the Arabs diffused to Christian Europe through translations made in the 12th century from Arabic into Latin.

2.1.2.4 The Renaissance and the Scientific Revolution 1453–1659

(1) *Features of this Period*

On May 29, 1453, the Turks captured the city of Constantinople and many Greek-speaking scholars escaped to the West brought with them classical manuscripts in Greeks along with the ability to translate the ancient writings into Latin, the common language of learning in Europe at that time. This was the beginning of the Renaissance. The date of beginning of the Scientific Revolution is credited to the publication of Copernicus's heliocentric theory and Vesalius's anatomy in 1543. Thereafter, scientists in the Renaissance era gradually began to perform experiments more frequently.

In the early Renaissance of the 15th century, it was primarily a time of absorption of classical learning and the adoption of Arabic mathematics. Alchemy and astrology still played a dominant role more than chemistry and astronomy during this time. Physics had to absorb Aristotle's work. It was much more a period of change in arts and letters than in science. In continuation of the trend that predates the Renaissance, artists, especially painters, made significant contributions to science due to the development of perspective drawing, and the dissection of human bodies became valuable to both arts and science.

The later Renaissance that began in Italy reached northern Europe in the 16th century. There the movement was joined by the new values of the Reformation. With the Renaissance came the economic transformations.

Major technological achievements in this period include printing from movable type, and adding machine was invented by Pascal in 1645.

(2) *Achievements of Science in the Period of the Scientific Revolution (1453–1659) (Table* 2.2*)*

Table 2.2 Major achievement of science in the period of scientific revolution (1453–1659)[a]

Serial No.	Year	Name of scientists	Events
1	1543	Nicolaus Copernicus	Publishes *De Revolutionibus Orbium Coelestium,* outlining a heliocentric Universe
2	1600	William Gilbert (Astronomer)	Publishes *De Magnete,* a treatise on magnetism, and suggests that earth is a magnet
3	1609	Johannes Kepler	Suggests that Mars has an elliptical orbit
4	1610	Galileo Galilei	Observes the moons of Jupiter and experiments with balls rolling down slopes
5	1620s	Francis Bacon	Publishes *Novum Organum Scientiarum and The New Atlantis* outlining the scientific method based on induction and experimentation
6	1639	Jeremiah Horrocks	Observes the transit of Venus
7	1643	Evangelista Torricelli	Invents the barometer

[a]*Source* Palffy (2014)

2.1.2.5 The Newtonian Epoch 1660–1734

(1) *Features of this Period*

The Royal Society (a learned society for science) was founded in England in 1660. The creation of this institution stimulated scientific inquiry, and improved scientific communication through meetings between scientists.

The achievements in the development of science in this period is extraordinary due to the breakthroughs in physics, such as Newton's law of universal gravitation which led to an understanding that different phenomena have the same underlying cause. Newton's *Principia* (in three books) became the foundation of physics for the next 200 years. It also formed the basis of the scientific method that slowly made its way into the study of natural phenomena. Newton formed the "method of analysis and synthesis", a procedure that includes both an inductive and a deductive stage. Newton's theories were formulated from observation. These theories were used to predict other phenomena. One of Newton's most important discoveries was probably the product of intuition, which he later backed with experiments, reasoning and mathematics. Newtonian epoch lasted up to 19th century-Formation of "Age of Determinism".

The success of Newton's theories, the validity of which became well established during the first decades of the 18th century, not only had a profound impact on science itself, but also induced a profound change of philosophical thinking, resulting in the emergence of a "mechanical philosophy" and the "Age of Determinism" around more than two centuries.

The first steam engine (Newcomen) was used in England in 1712 for pumping water out of coal mines. The new technology allowed more coal to be mined from deeper underground levels. The flying shuttle patented by John Kay in 1733. All these had cleared the way for Britain's Industrial Revolution.

(2) *Major Advances of Science (1660–1734) (Table* 2.3*)*

Table 2.3 Major events of advance of science (1660–1734)[a]

Serial No.	Year	Name of scientists	Events
1	1660s	Robert Boyle	Publishes New Experiments Physico-mechnical: Touching the Spring of the air, and its effects investigating air pressure
2	1665	Robert Hooke	Introduces the world to the anatomy of fleas, bees, and cork in *Micrographia*
3	1669	Nicolas Steno	Writes about solids (fossils and crystals) contained within solids
4	1669	Jan Swammerdam	Describes how insects develop in stages in *Historia Insectorum Generalis*
5	1670s	Antonie van Leeuwenhoek	Observes single-celled organisms, sperm, and even bacteria with simple microscopes
6	1676	Ole Rømer	Uses the moons of Jupiter to show that light has a finite speed
7	1678	Christiaan Huygens	Announces in first his wave theory of light
8	1686	John Ray	Publishes *Historia Plantarum*, an encyclopedia of the plant kingdom
9	1687	Isac Newton	Outlines his laws of motion in Philosophiae Naturalis Principia Mathematica
10	1727	Stephen Hales	Publishes *Vegetable Statick* demonstrating root pressure

[a]*Source* Palffy (2014)

2.1.2.6 The Enlightenment and the Industrial Revolution 1735–1819

(1) *Features of this Period*

The Enlightenment is an European intellectual movement that reached the highest point in the 18th century. Enlightenment thinkers were believers in social progress and in the liberating possibilities of rational and scientific knowledge. This period shows clearly the interaction among technology, science, economy and society. Influenced by Newton's law of gravitation, the philosophers believed in the existence of natural laws that regulated both the physical universe and also the human society. In economics, the key slogan was laissez faire, i.e. to let the people do what they will. It was the opposition to governmental regulation and was based on theory and practice of mercantilism which was dominant in modernized part of Europe during the 16th–18th century. The classic formulation of laissez faire in this period was the famous work of Adam Smith who published *An Inquiry into the Nature and Causes of the Wealth of Nations* published in 1776. In religion, the philosophers rejected the traditional belief that God controls the universe and determines the fate of man. In government, the philosophers had a key phrase-"the social contract" which viewed government as a social contract between the rulers and the ruled.

Industrial Revolution began in the Great Britain in the second half of the 18th century, Amida general turn toward the development of machines to make work faster or more efficient, especially in England. There were also some changes in the use of material, such as the turn to coal for both heating and making iron due to limitation of supply of wood. Abraham Darby (1678–1717) developed a method of producing pig iron in a blast furnace fueled by coke. From the 1760s onwards James Watt (1736–1819) improved steam engine invented by Newcomen in 1712. Richard Arkwright (1732–1792), invented the spinning frame, which could produce strong cotton thread. He was also the first person to use one of James Watt's steam engine to drive machines in a cotton mill. Manchester became the center of the cotton industry in Britain by that time. Industrial Revolutions started in England, and because of its success in England it was quickly taken up in other European nations and in the United States. During the 18th century, technology in the modern sense of the word appeared, there was the direct application of science to technology. Technology also had its impact on development of sciences. Concepts of work, power and energy began to be explored and formalized. Teaching of science and technology was recognized as an important part of the technological development of a country. Many scientific institutions and universities were founded in this period such as Gőttingen University (Germany), the Royal Danish Academy of Sciences, Princeton University, the University of Moscow, the Lunar Society (England), the American Philosophical Society, etc. The Ecole Polytechnique in Paris, founded in 1794, became an important source of development of science during the 19th century.

(2) *Major Advances of Science (1735–1819) (Table* 2.4*)*

2.1.2.7 Nineteenth-Century Science 1820–1894

(1) *Features*

Science in the 19th century had four features:

(i) Science and teaching of science was giving much of the form it has to-day. After the term "scientist" was established in England in 1833 through the suggestion of William Whewell, the term "natural philosopher" began to be replaced.
The occupation of scientist became a paid profession. This happened first in Germany, where in a few decades the universities developed into centers where science flourished. Justus von Liebig (1803–1873), a German chemist, as a professor at the University of Giessen, devised the modern laboratory-oriented teaching method. After that the teaching of science became linked to scientific research, a practice that was followed by most of the universities in the world later. The universities and scientific societies also started to publish scientific information, a development that was imitated in other countries.

Table 2.4 Major achievement of science (1735–1819)[a]

Serial No.	Year	Name of scientists	Events
1	1735	Carl Linnaeus (Swedish botanist)	Publishes *Systems Nature*, the beginning of his classification of natural world into animal kingdom, the plant kingdom and the mineral kingdom, some preliminary concepts of system
2	1735	George Hadley	Explains the behaviour of the trade winds in a short paper that remains unknown for decades
3	1738	Daniel Bernoulli	Publishes *Hydrodynamica*, which lays the foundation for the kinetic theory of gases
4	1749	Georges-Louis Lecherc, later the Comte de Buffon	Publishes the first volume of *Histoire Naturelle*
5	1754	Joseph Black	His doctoral thesis on carbonates is a pioneering work in quantitative chemistry
6	1766	Henry Cavendish	Makes hydrogen, or inflammable air, by reacting zinc with acid
7	1770	Benjamin Franklin	Publishes a chart of the Gulf Stream
8	1774	Joseph Priestley	Makes oxygen by heating mercuric oxide, using sunlight and a magnifying glass; he calls it dephlogisticated air
9	1774	Antoine Lavoisier	After learning the technique from Priestley, makes the same gas, and goes on to call it oxygène
10	1774	Nevil Maskelyne	Calculates the density of Earth by measuring the gravitational attraction of a mountain
11	1779	Jan Ingenhousz	Discovers that green plants in sunlight give off oxygen; this is photosynthesis
12	1788	James Hutton	Publishes his theory concerning the age of Earth
13	1793	Christian Sprengel	Describes plant sexuality in his book on pollination
14	1798	Thomas Malthus	Produces his first essay on human population, which later influences Charles Darwin and Alfred Russel Wallace
15	1799	Alessandro Volta	Invents the electric battery
16	1800	William Herschel	Discovers infrared radiation
17	1803	John Dalton	Introduces the idea of atomic weights
18	1811	Marry Anning	Finds the skeleton of the first known ichthyosaur

[a]*Source* Palffy (2014)

(ii) The philosophical basis of 19th century. There were several schools of philosophy in the 19th century. Hegel's philosophy of nature was not based on experimentation but on a priori concepts, which remained strong throughout the first half of the century in Germany. The idea that science would ultimately explain all phenomena of nature became stronger.
There was emergence of "positivism" of the philosophical school by French philosopher Auguste Comte (1798–1857). He had established law of three stages, which was one of the first theories of social evolutionism. His three stages were: the theological stage in which man's place in society and society's restrictions upon man were referenced to God; the metaphysical stage, or the stage of investigation, when people started reasoning and questioning authority and religion; and the scientific stage, in which science started to answer questions in full stretch. Comte published the *Course in Positive Philosophy,* a series of texts between 1830 and 1842. The first three volumes of the *Course* dealt chiefly with the physical sciences, i.e. existed mathematics, astronomy, physics, chemistry, biology, whereas the latter two volumes emphasized the inevitable coming of social science.

(iii) Linkage between science and technology became much closer. The linkage between science and technology became much closer in the 19th century. The discoveries of science of chemistry has progressed further that chemistry became the archetypal 19th century science. It was in Germany, which became the leading country in theoretical chemistry, that chemical research had the biggest impact on industry. By the end of the 19th century, Germany had developed into the largest manufacturer of several chemicals such as dyes, fertilizers, and acids used in industrial processes. Scientists in the 18th century had become increasingly fascinated with electricity, especially as better means of generating and storing it became available. There were a series of inventions by Alessandro Volta, Hans Christian, Faraday, E. M. Clarke, Steinmetz, Edison and Joseph Swan in this area. In 1880, Thomas Alva Edison's first electric generation station, designed mainly for lighting, was opened in London. The first electric street car was introduced in Berlin in 1881. And in 1882, Edison patented a three-wire system for transporting electric power. In the same year the Pearl Street power station went into operation. Electric energy was becoming a major source of energy in the later part of the period. The relationship between scientific education and technological progress became fully understood and popularized during the 19th century and later.

(iv) National differences in style of research in the 19th century. During the 19th century, difference in style of research between countries became obvious. In Germany, where science was strongly organized, the university system favored pure science; in England scientists traditionally dealt with practical problems. In the United States, relatively little attention was given to science in the 19th century because the interests of immigrants were largely oriented toward practical ventures rather than the pursuit of the theoretical knowledge. In the United States, the attention was on technology especially the labor saving devices because of the comparatively low population density and the resulting high salaries for manual workers. Therefore technological inventors and entrepreneurs played a big role in industrial development: many major U.S. industries were founded by inventors such as Alexander Graham Bell (1847–1922), George Westinghouse (1846–1914), Thomas Edison (1847–1931) and George Eastman (1854–1932), founder of Eastman Kodak. In 1848 the American Association for the Advancement of Science was founded, it played an important role in the development of American science. But theoretical science remained as a secondary field in American education well into the 20th century.

(2) *Major Advances of Science (1820–1894) (Table* 2.5*)*

(3) *Technology*

The impact of electromagnetism on the technology of communications and people's daily life was great. Telegraph and telephone system were developed in the 19th century as well as electrical power plants and their distribution system.

The transportation system was also revolutionized through steamboats, development of locomotive service and railroads in Europe and America. And Jean

Table 2.5 Major advances of science (1820–1894)[a]

Serial No.	Year	Name of scientists	Events
1	1820	Hans Christian Ørsted	Discovers that when a current is switched on, a nearly compass needle flickers
2	1821	Michael Faraday	Discovers the principle behind the electric motor
3	1837	Louis Agassiz	Describes an ice age
4	1842	Christian Doppler	Explains why binary stars are colored
5	1845	Alexander von Humboldt	Introduces the idea of ecology
6	1859	Charles Darwin	Outlines his theory of evolution in *On the Origin of Species* by means of natural selection
7	1859	Louis Pasteur	Disproves spontaneous generation of life
8	1865	August Kekule	Describes the chemical structure of the benzene molecule
9	1866	Gregor Mendel	Publishes his work on the genetics of peas
10	1869	Dmitri Mendeleev	Lays out the periodic table of the elements
11	1873	James Clerk Maxwell	Publishes his laws of electromagnetism

[a]*Source* Palffy (2014)

Lenoir (1822–1900) developed the first internal combustion engine in 1855. But Karl Benz' (1844–1929) patented Motorcar from 1885 is considered to be the first practical motorcar.

New construction techniques and materials were developed, for example, the Portland cement and the cheap steel from the Bessemer process. The suspension bridge was pioneered as early as 1825. Modular prefabricated design was also used to build the first skyscrapers in Chicago starting in 1885.

The chemical industry arouse during the 19th century, especially in Germany.

New agricultural machines contributed to the ability of few farmers to feed and clothe a growing population.

Certain technological developments had a direct impact on other sciences. Mechanical calculators became practical and available near the end of the century. Charles Babbage designed a machine to calculate various functions automatically which he called the Difference Machine in 1822 (but he abandoned it in 1832). In 1890, Herman Hollerith (1860–1929) found a way to simplify handling the information from the US census based on punched cards. Hollerith later founded the company IBM.

2.1.2.8 Science in the 20th Century Through Two World Wars 1895–1945

(1) *Features*

There are three features in this period. The first is the growth of number of scientists and change of nature of scientific research from individual into a communal effort; the second is the emergence of the philosophy of "New Realism"; and the third is two major achievements in physics, the quantum mechanics and the theory of relativity. These three progresses are elaborated below.

(i) During the 20th century, there was a high growth of the number of scientists: more scientists lived in the 20th century than in all previous eras together. The nature of scientific research also changed into a communal effort. The progress was not only determined by the great discoveries of a talented few, but also by the numerous small steps made by specialized researchers.
    Many of the discoveries and observations made during the 19th century were without explanations such as the periodic table and Mendel's laws etc. These were explained by new scientific theories emerged in the 20th century. During the 20th century several industries founded their own research laboratories engaged in both basic and applied research. Among the most important of these was Bell Labs in the United States, which is one of the birthplaces of systems engineering.
    During the 19th century, the western society was transformed through technology. But in the 20th century, science itself started having a direct effect on the society. And the timespan between a discovery and its technical application

became much shorter, e.g. the discovery of the electron and the growth of information communication technology.

(ii) New philosophies: Science in the first half of the 20th century became highly successful in explaining the nature of matter, mechanism of chemical reactions, fundamental process of life, and the general abstraction of universe. These successes of science started to exert profound influence on philosophical thinking. The American philosophers C. S. Peirce and William James, became founders of the philosophical school of pragmatism, they believed that reality could be understood by experience. But in 2012, Maurizo Ferraris proposed so called "New Realism". In fact, Darwin's theory of evolution also had a major impact on philosophers, especially Herbert Spencer. And changes in mathematics and the development of psychology also had important philosophical implications.

(iii) Two Major Achievements: There were two major achievements of science which changed the paradigm dominated by Newton's epoch for more than two centuries. The quantum mechanics (major contributor: Wolfgang Pauli, Werner Heisenberg, Max Planck and Erwin Schrodinger) is essential to understand the behavior of systems at atomic length scale and smaller where Newton's mechanics failed to apply. Uncertainty becomes the norm rather than cause and effect. The other is Relativity: The Special and the General Theory by Albert Einstein. The special theory of relativity put forward in 1905 is limited to events as they appear to observers of uniform relative motion. The unique import and consequences of the theory are: the velocity of light is absolute; the mass of a body increases with its velocity, although appreciably only at velocities approaching that of light; the mass and energy are equivalent; and the Lorentz-FitzGerald contraction, that is bodies contract as their velocity increases, again only appreciably near the velocity of light, etc. The general theory of relativity is completed in 1915, which generalizes special relativity and Newton's law of universal gravitation, providing a unified description of gravity as a geometric property of space and time.

(2) *Major advances of science (1895–1945) (Table* 2.6)

(3) *Technology*

In the 20th century, the extent of the dependence of technology became clear. The discovery of the electron gave rise to an entirely new technology, i.e. electronics. The technological device that transformed society mostly during this period was the electronic vacuum tube that became the heart of the development during the first half of the 20th century. It was applied to long-distance telephone connection, the radio broadcasting in the 1920s and 1930s and of television in the 1940s.

Some of the first working electronic computers were also the result of military needs. The mathematician Norbert Wiener developed an electronic gun-pointing device based on the feedback mechanism. He is also a major contributor to Cyber tics, which is very relevant to the study of systems. The first electronic digital

Table 2.6 Major advances of science (1895–1945)

Serial No.	Year	Name of scientists	Events
1	1895	Wilhelm Rontgen	Discovers X-rays
2	1898	Marie Curie	Isolates radioactive polonium
3	1900	Max Planck	Describes discrete packets, or quanta of energy
4	1905	Albert Einstein	Produces his paper on special relativity
5	1906	J. J. Thomson	He is awarded the Nobel prize in physics for his discovery of electron
6	1912	Alfred Wegener	Proposes a theory of continent drift
7	1915	Thomas Hunt Morgan	Introduces the chromosome theory of inheritance
8	1926	Erwin SchrÖdinger	Unleashes wave mechanics
9	1927	Werner Heisenberg	Sets out his uncertainty principle
10	1928	Paul Dirac	Introduces quantum electrodynamics
11	1929	Edwin Hubble	Finds that the Universe is expanding
12	1930	Subrahmanyan Chandrasekhar	Describes black holes
13	1931	Georges Lemaître	Suggests that the Universe began as a primeval atom
14	1934	Fritz Zwicky	Proposes the existence of dark matter
15	1935	Konrad Lorenz	Explain the basis of animal instinct
16	1936	Alan Turing	Describes the Universal Turing Machine, a programmable computer
17	1939	Linus Pauling	Writes *The Nature of the Chemical Bond*, which uses the idea of quantum physics to explain chemistry
18	1942	J. Robert Oppenheimer	Takes on the Manhattan Project to develop the atomic bomb

[a]*Source* Palffy (2014)

computer (ABC) was developed in the late 1930s and early 1940s by John V. Atanas off and Clifford E. Berry for the purpose of solving systems of equations although the war prevented completion of the development. Nevertheless, ideas based on the ABC were used in creating ENIAC, the first general purpose electronic digit computer, operational by 1945.

2.1.2.9 Summarized Tables of History of Growth of Science and Technology

Table 2.7a, b summarize the growth of a number of major events of various branches of science and technology in different historical period. These two tables

Table 2.7 **a** Number of important events of science in history 600 BC–1945 AD.[a] **b** Supplementary table of Table 2.6a

<table>
<tr><th rowspan="2">Period</th><th colspan="8">Name of branches of science</th><th rowspan="2">Total in period (A)</th></tr>
<tr><th colspan="2">Astronomy</th><th colspan="2">Life science</th><th>Mathematics</th><th colspan="2">Physical science</th><th>Technology</th></tr>
<tr><td colspan="10">(a)</td></tr>
<tr><td>1
2
3</td><td colspan="2">46
42
90</td><td colspan="2">33
26
102</td><td>55
49
111</td><td colspan="2">39
28
73</td><td>92
78
82</td><td>265
223
458</td></tr>
<tr><td rowspan="2">4</td><td colspan="2" rowspan="2">58</td><td>Biology</td><td>Medicine</td><td rowspan="2">100</td><td>Physics</td><td>Chemistry</td><td rowspan="2">46</td><td rowspan="2">410</td></tr>
<tr><td>62</td><td>51</td><td>69</td><td>24</td></tr>
<tr><td rowspan="2">5</td><td>Astronomy</td><td>Earth science</td><td rowspan="2">110</td><td rowspan="2">92</td><td rowspan="2">85</td><td rowspan="2">133</td><td rowspan="2">161</td><td rowspan="2">174</td><td rowspan="2">953</td></tr>
<tr><td>115</td><td>82</td></tr>
<tr><td>6</td><td>194</td><td>106</td><td>180</td><td>197</td><td>143</td><td>193</td><td>219</td><td>230</td><td>1462</td></tr>
<tr><td>7</td><td>190</td><td>72</td><td>228</td><td>212</td><td>93</td><td>336</td><td>230</td><td>270</td><td>1631</td></tr>
</table>

Period	Name of period	Year beginning-end	No. of years in this period (B)	No. of S&T events/year (A/B)
(b)				
1	Greek and Hellenic tic science	600 BC–AD 529	1529	0.17
2	Medieval science	530–1452	923	0.24
3	The renaissance and the scientific revolution	1453–1659	207	2.21
4	The Newtonian Epoch	1660–1734	74	5.54
5	The Enlightenment and the industrial revolution	1735–1819	84	11.34
6	Nineteenth century science	1820–1894	75	19.49
7	Science in the 20th century through World War II	1895–1945	51	31.98

[a]*Note* Table 2.7a, b are prepared based on the number of events of various branches of science by Alexander Hellemans & Bryan Bunch (1991) *The Time Tables of Science-A Chronology of the most Important People and Events in the History of Science*, New updated Edition (1991). New York: Simon Schuster. It should be noted that table. It takes "Greek and Hellenistic Science: 600 BC–529 AD" to be period 1

are prepared based on information of *the Timetables of Science*. It can be seen from Table 2.7b that the number of S&T events per year increased rapidly since the period of the Renaissance and Scientific Revolution.

2.2 Rise and Development of General Systems Theory

2.2.1 Rise of General Systems Theory

2.2.1.1 Trends of Science in the Early 20th Century

The development of science and technology around more than 2500 years since the beginning of Greek and Hellenistic science, there were several trends of science in the early 20th century.

(1) *Issue of High Specialization*

Science is characterized by its ever-increasing specialization necessitated by huge amount of data, the complexity of technique within every field. Even within one field, it had been divided into many branches that no longer could anyone be competent in all branches of the subject. And in consequence, specialists of one discipline are encapsulated in their private universes, and it is difficult for them to exchange views with specialists of other disciplines. While a part of modern science is big science in nature, it requires interdisciplinary teams to fulfill the requirement.

(2) *The Lasting Impact of Mechanistic View Against Emerging New Views*

According to Bertalanffy (1968), the mechanistic view, a secondary effect created in Newtonian epoch was not altered, even the introduction of relativity theory and the advent of quantum mechanics in early part of the 20th century have no effect on this view, simply the deterministic laws in physics were displaced by statistical laws. But new problems have been emerged in various branches of modern physics, such as wholeness, dynamic interaction and organization etc. these new problems cannot be dealt with by mechanistic view.

(3) *Nonphysical Science*

Modern science focused mainly to laws of nature, few attempts are given to state exact laws in nonphysical fields. There are necessities to make expansion of conceptual schemes to the impact and progress of these nonphysical fields such as biological, behavioral and social sciences.

2.2.1.2 Rise of General Systems Theory

To overcome the aspects of negative trends of science in the early 20th century, some scientists and philosophers had the awareness of the existence of certain common system properties in the appearance of structural similarities or isomorphism. They tried to find principles, laws and models existed to generalized systems, i.e. an effort for the unity for science. The prominent people among them were Ludwig von Bertalanffy, Kenneth Boulding, and Bogdanov. Alexander Bogdanov (1873–1928), a Russian Physician and philosopher, published in Russia between 1912–1917 his theory *Tektology: Universal Organization Science,* which tried to establish a discipline of unifying all social, biological and physical science by considering them as systems of relationships. Tektology foreshadowed general systems theory (GST) and used many of the same concepts as modern systems theorists. Ludwig von Bertalanffy (1901–1972), an Austrian biologist, known as one of the founder of GST. Although Bertalanffy formulated his ideas in 1930s, he was not recognized until one of his papers *the Theory of Open systems in Physics and Biology* was published in the American Journal Science in 1950. Kenneth Boulding (1910–1993) published his *General Systems Theory-The Skeleton of Science* in 1956. In 1954, the International Society for General System Theory ISGST was founded. This society later become the International Society for Systems Science, ISSS. Both Bertalanffy and Boulding were founders of this Society.

According to Bertalanffy (1968), the major aims of GST include:

(1) There is a general tendency towards integration in the various sciences, natural and social.
(2) Such integration seems to be centered in a general theory of systems.
(3) Such theory may be an important means for aiming at exact theory in the nonphysical fields of science.
(4) Developing unifying principles running vertically through the universe of the individual sciences, this theory brings as nearer to the goal of the unity of science.
(5) This can lead to a much-needed integration in scientific education (p. 38).

2.2.2 *Systems and Key Concepts of General Systems Theory*

2.2.2.1 Definition of Systems

(1) *Qualitative Definitions of Systems*

There are various definitions of systems from different professions. Several qualitative definitions of systems are quoted for comparative and complementary purposes at below:

(i) Bertalanffy (1968) defines system as “sets of elements standing in interrelation” (p. 55)
(ii) Ellis and Ludwig (1962) define that “A system is a device, procedure, or scheme which behaves according to some description, its function being to operate on information and/or energy and/or matter.” (p. 3)

(iii) Jary and Jary (1991) maintain that systems are consisted of any set or group of elements or parts (e.g. an organism or a machine) organized for a definite purpose and in relation to an external environment. Such systems may be natural or man-made, and may be taken to include social systems. Hence a society or a social organization may be deemed a system in this sense. This definition is important in social theory, which has often treated social relations, groups or societies as a set of interrelated parts which function so that to maintain their boundaries with their wider environment. (p. 648)

(iv) Paynter (1961) states that the description of a system necessary begins with the identification of a universe (u)-or "universe of discourse" as it is often called in logic-which is a domain or set of sufficient scope to include elements within the system, plus all exterior elements with which the system may be interacting. The system (S) is then a well-defined subset of U, the element of S possessing properties and interrelationships which happen to be of particular interest to be identified. The complementary set, U less S is labeled to be the environment E of Fig. 2.2. Hence, E is the set of elements which interact with the elements of S but are not in S. (pp. 11–12)

(v) A combination of interacting elements organized to achieve one more stated purpose (taken from ISO/IEC15288).

(2) *Mathematical Definitions of Systems*

System can be defined mathematically in various ways. Two examples are quoted below:

(i) A system can be expressed by a system of simultaneous differential equations (Bertalanffy, 1968).

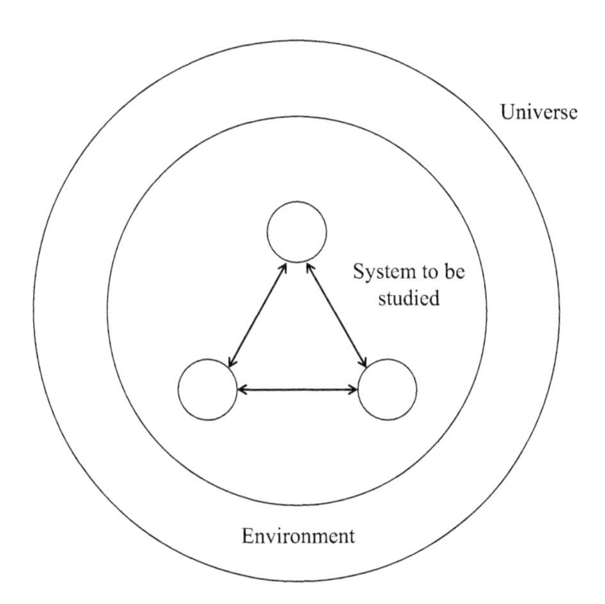

Fig. 2.2 Description of a system

Denoting some measures of elements, pi (i = 1, 2, …n), by Qi, these, for a finite number of elements and in the simplest case, will be of the form:

$$\left.\begin{aligned} \tfrac{dQ_1}{dt} &= f_1(Q_i, Q_2, \ldots Q_n) \\ \tfrac{dQ_2}{dt} &= f_2(Q_i, Q_2, \ldots Q_n) \\ \ldots \\ \tfrac{dQ_n}{dt} &= f_n(Q_1, Q_2, \ldots Q_n) \end{aligned}\right\} \quad (2.1)$$

Change of any measure Q_i, therefore is a function of all Q's, from Q_1 to Q_n; conversely, change of any Q_i entails change of all other measures and of the system as a whole. (p. 56)

(ii) Systems defined by extension of George Klir's (1991) formula (Figs. 2.3 and 2.4).
(iii) Brief discussion of Set and Relation. It can be seen from above definitions of systems, the word 'set' is appeared in three definitions of systems and the word 'relation' is appeared in Klir's formula. Both are mathematical terms used here in the analysis of systems.

Set (Syn: Class) A collection of things. The individual objects of a set are the elements, or members of the set. Sets are subject to operations and relations such as union, intersection, inclusion, etc.; the algebra of sets is that part of the general theory of sets, which studied these. There are several symbols used for the set. For example, A is the symbol of a set, then

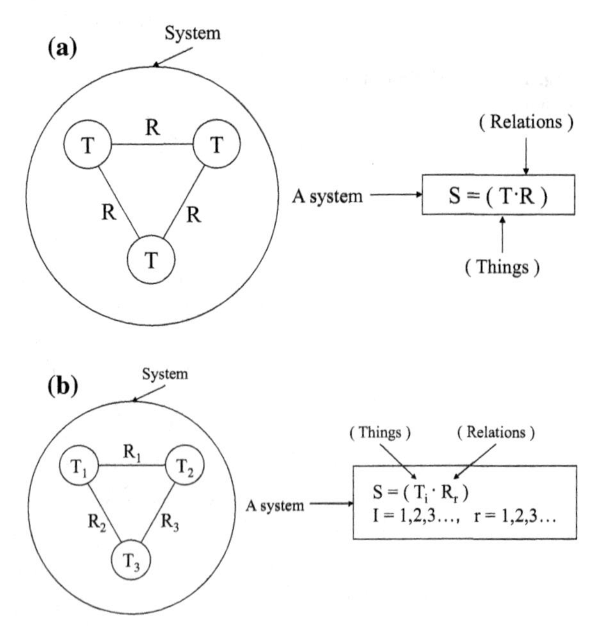

Fig. 2.3 **a** Definition of a system George Klir's formula (1991). **b** Definition of a system (adapted by the authors from Klir's formula)

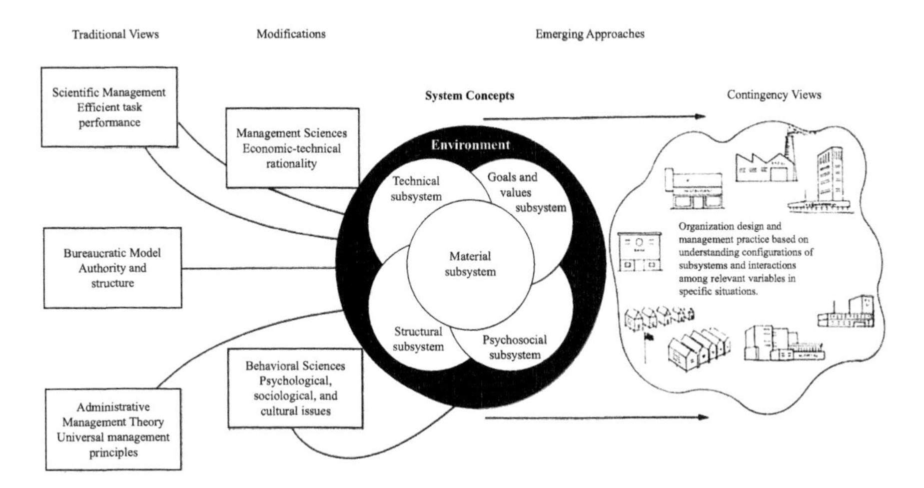

Fig. 2.4 Organization as an open system and evolution of management theory (*Source* Kast and Rosenweg (1979)

$a_1 \in A$ a_1 is a member or an element of set A
$b_1 \notin A$ $b_1$is not a member of A
$C \subset A$ C is a subset, it is a part of set A

Relations: There are several specific relationships, the membership ($a_1 \in A$), containment ($C \subset A$) and identity or equivalence. If the Roman capital is used to stand the symbol of relations, then

$R \equiv S$ it means the two relations R and S are precisely the same
xRy or $R(x, y)$ means X bears the relation R to Y.

The study of "relations" had been an important part of study in hard systems engineering. This had been part of class notes for M.I.T. course 2.751 in 1960s. This mathematical tool of set and relation may be developed with its full potential in the study of social systems engineering.

2.2.2.2 Nature of Structure of Systems and Its Classification

(1) *Changing Nature of Structure of Systems*

The structure of a "Hard" system is the spatial layout or connection of subsystems, and components in three-dimensional space at a certain amount of time. But the structure of a "soft" system, especially the traditional social system, it is the hierarchical[5] arrangement of subsystems, and components based upon their functions. It

[5]Note: In modern social system the spatial connection of subsystems of components may be in the networked or other forms of topology.

should be emphasized that the GST is an outgrowth of accumulated knowledge of science around 2500 years. It is also a product of industrialized society. But the mankind is in the stage of transition from an industrialized society to an information society or knowledge society, the society becomes increasingly networked. The traditional GST focused mainly on hierarchical systems without enough focus on networked systems, both physical and non-physical. The GST shall also be updated and supplemented, there are already emergence of social physics in a networked society which describes mathematical connection between information and idea flow on the one hand and people's behavior on the other. It is expected emergence of various theory may promote the growth of this discipline, social systems engineering.

(2) *Classification of Systems*

Systems can be classified in various ways based upon criteria chosen, selected types will be discussed.

(i) Systems can be classified into closed systems and open systems depending upon whether they have exchanges of energy, material and information with its external environment.
A closed system generally has no exchange of material and information with its external environment except energy. Closed systems, tend to increase their entropy, tend to run down and become "dying system."
The immediate environment is the next higher system minus the system itself. The entire environment includes the plus all systems at higher levels (part of U of Fig. 2.2) which contain it. Generally, the environment has certain influence to the system, i.e. it can exert a degree of control to the system, but it cannot be controlled by the system.
An open system (all living systems) has exchanges of energy, material and information with its external environment. Its organization is controlled by information and fueled by some forms of energy. A part of this energy will become negative entropy to preserve order of the system.
The nature of "open" or "closed" is relative in sense, nearly all social system and biological system are open system in sense. And the advantage of open system can be shown from China's rapid economic and social development since its reform and opening to outside world in late 1970s compared to its record of development in pre-reformed era, when Chine was in a state of semi-isolation with outside world.

(ii) Systems can also be classified into non-living or living. Generally, the non-living systems are man-made, for example the mechanical system (including its sub-systems and components), the electrical system, the communication system, the transportation system etc. A living or organic system is subject to the principles of natural selection and is characterized by the features of self-regulation, organization, metabolism and growth, reaction capacity, adaptability, reproduction capability and development capability.

Living systems theory pioneered by James Miller will be briefed in Sect. 2.2.3.1 later.

(iii) There are also other types of classification of systems, such as static and dynamic, simple and complex, decomposable and non-decomposable, organized complexity and unorganized complexity etc.

2.2.2.3 Key Concepts of General Systems Theory

The key concept of GST listed from (2) to (15) were formed by Bertalanffy, Boulding, Joseph Litterer, and other scholars. But it is necessary to raise the point that those distinguished scientists are also influenced by thoughts of Fredrich Hegel (1770–1831), which is described in (1) as follows:

(1) *Hegel's Thoughts*

Hegel had formulated the statements concerning the nature of the system: the whole is more than the sum of the part; the whole defines the nature of the parts; the parts cannot be understood by studying the whole; and the parts are dynamically interrelated or interdependent.

(2) *Property of Wholeness*

The system has a property of wholeness, i.e. the property of wholeness requires that it should be observed from the whole rather than partial. For example, to understand the nation as a system, it is necessary to observe it from various aspects to synthesize to get its whole property. The nation as a system should be observed and judged upon the synthesis of observations from its social performance, economic performance, S&T capability etc. The judgement should not be established by observing only its parts.

(3) *Subsystem or Component*

A complex system is generally composed of subsystems and components. These subsystems are dependent mutually, this is also true for subsystem and its components. Independent subsystems grouped together also cannot form a system. Independent components grouped together cannot form a system or subsystem. This feature is applicable to all types of systems, mechanical, biological or social. Every system should be composed of two mutually dependent components at least.

With a simple power system for example, it can be composed by three subsystems, the generation subsystem, the distribution subsystem and the utilization (consumers) subsystem. These three subsystems are mutually interdependent. This power system cannot be operated in lacking one of these three subsystems.

(4) *Open Systems View*

It had been described in 2.2.2.2-(2)-(i) that there are two types of system: closed and open. Biological and social systems are inherently open system. GST and open system view had also influenced the organization theory and practice around 1970s. It was stated by Kast and Rosenzweg (1979) that "Systems theory provides a new paradigm for the study of organizations and their management, a basis for thinking of the organization as an open system in interaction with its environment. It also helps us understand the interrelationships between the major components of the organization—its goals, technology, structure and psychosocial relationships." (p. 107). Figure 2.4 shows the systems concept of organization, the central circle is an organization within an environment. (The black part). The system of organization is composed of five sub-systems: the managerial subsystem, the goals and values subsystem, the structural subsystem, the technical subsystem and the psychosocial subsystem. Figure 2.4 also shows the evolution of theory of organization and management. The system concept of organization is evolved directly from scientific management, bureaucratic model, administrative theory and behavioral sciences. It also receives the impact from scientific management to management sciences. This diagram also shows the system concept of organization and management is in progress toward contingency management.

(5) *Goal Seeking*

Systematic interaction must result in some goal or final state to be reached or some equilibrium point be approached.

(6) *Input—Transformation-Output Model.*

All systems, if they are to attain their goal, must transform inputs into outputs. The open system can be viewed as a transformation model in dynamic relationship with its environment, it receives various inputs, transform these inputs in some way, and export outputs. Figure 2.5a is a general input-transformation-output model of an open system.

Figure 2.5b is a specific 'Input-Transformation-Output' model of an economic system. The input is factors of production, i.e. labor, capital and other factors of production. The production function of a specific firm is the transformation function, which transforms input into goods and service output of this firm.

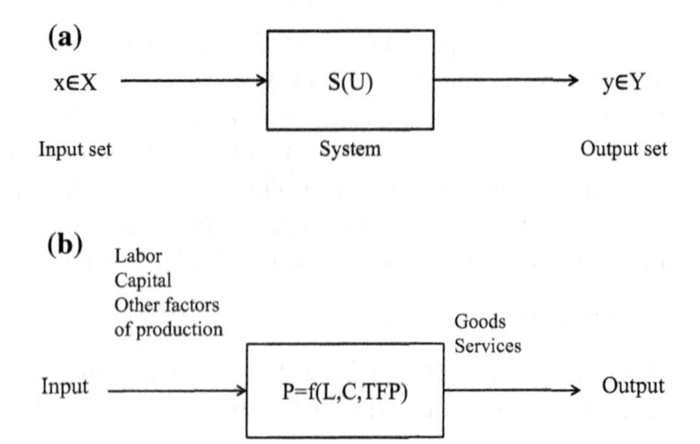

Fig. 2.5 **a** An input-transformation-output model. **b** Specific 'input-transformation-output' model of a firm

(7) *System Boundaries*

Systems have boundaries that separate them from their environments. It is crucial to identify the boundaries in understanding the relationships of systems and their environment. Because there are both an internal structure and performance of its function of any system, while the performance of functions of any system is closely related to its external environment. Analysis of any system should analyze also its external environment. To implement an economic analysis of a country, it is necessary to analyze the economy, trade, demand of investment and consumptions of its neighboring countries and those analyses of a broader global environment. In the analysis of system of a multinational corporation (MNC), it is necessary to analyze the natural environment (climate, water supply and disposal, geological condition etc.) and the social environment (current market and potential market, accessibility of labor force and financial service, legal condition and market entry of hosting country etc.).

The concept of boundaries will help to clarify the distinction between closed and open systems, the former has rigid, impenetrable boundaries while the later has permeable boundaries between itself and a broader supra system. It is easier to identify the boundaries of physical and biological system, but it is difficult to define exactly the boundaries of organizations of a social system. For example, the scope of sales and production of a multi-national corporation may cover many regions of global continents. But from the perspective of theoretical analysis, the abstract boundaries can be assumed based upon the scope of study.

(8) *Negative entropy*

Entropy is one of the basic concept of GST. Originally it is imported from thermodynamics, it is defined as the energy not available for work after its transformation from one form to another. In systems, it is defined as a measure of the relative degree of disorder existed within a closed system at a defined moment of time, the natural tendency of the force of entropy will increase until eventually the entire system fails, i.e. the change of entropy in a closed system is always a positive (+) value; however, in open biological or social systems, this increase of entropy can be arrested, and even it can be transformed into negative entropy, because the open system can import resources (energy, material, information) from the environment, and transform those resources to improve the system.

(9) *Steady State, Dynamic Equilibrium, and Homeostasis*

The concept of steady state and dynamic equilibrium is the state of change of the system. Both are closely related to entropy and negative entropy. The steady state reached by change of a closed system is the state of maximum entropy, i.e. a state of death or totally in disorder. While an open system may attain a state that the system remains in dynamic equilibrium through the continuous inflow of resources. Homeostasis stands for the sum of all control functions creating the state of dynamic equilibrium in a healthy organism. It is the ability of the body to maintain a narrow range of internal conditions despite environmental changes.

(10) *Feedback*

The concept of feedback is important to understand how a system maintains a steady state. Physical system with feedback can be made in the direction of a specific output by regulating the process with a controlling mechanism. This controlling mechanism is operated based on the principle of feeding back the information of the output. There are two types of feedback, the positive feedback and the negative feedback. The negative feedback is to feedback the information of output to the controlling mechanism to control the output, the multiplier of the new output will be less than one, in other words, this kind of feedback is to control the state of system in reverse direction, therefore the term negative feedback is derived. Increase the level of feedback will cause the decrease of level of output, thus to achieve the steady state of the system. If the multiplier is greater than one, it is called positive feedback. In the regulating process of positive feedback, each new output will be larger than the previous one due to the multiplier is greater than one.

The concept of feedback is very important in social system engineering. In the implementation of an economic policy to regulate the output of a national economy, it requires correct, in time feedback information of the real situation to policy makers.

(11) *Regulation*

The system and its elements (subsystem, components) must be regulated in its operation to achieve the predetermined goals. The process of regulation including detection of deviation compared to preset goals and taking action to correct in time. In the process of regulation, feedback is prerequisite to achieve effective control. Many troubles of the real world of the social system is due to insufficient feedback or incorrect and distorted feedback. Figure 2.6 shows the basic cycle of regulation,

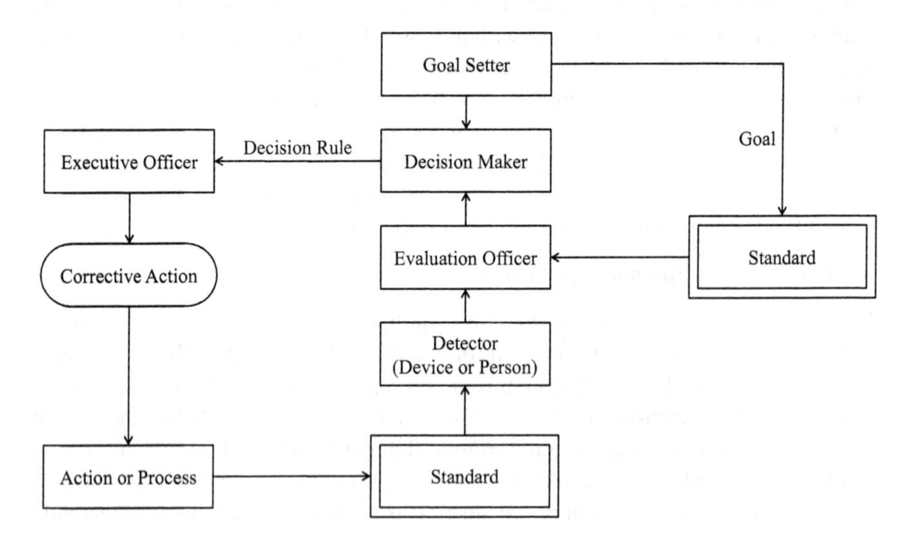

Fig. 2.6 Basic cycle of regulation

although it is applicable to automatic control of engineering systems, but it is more adaptable to management of organizations including government and corporations. For example, the planning department should collect various information of output from the operational department to compare them with goals or objectives preset, and make recommendation to modify those goals or take supplementary measures to decision makers. The decision maker will determine to adopt certain legal or policy measures based upon the decision rules, corrective actions will be taken by executer to achieve the pre-determined or revised output.

Regulation is an important element of social systems engineering, it is closely related to structure of organization, clarification of power and responsibility, pattern of centralization and decentralization of decision making mechanism and hierarchical structure of organizations. It includes laws, public policies and other regulations to direct the behavior and relationship of people and various social groups.

(12) *Hierarchical Structure*

The concept of hierarchical structure is one of the basic concept of systems theory. The wholeness of a system is generally consisted of nesting smaller subsystems. To analyze and clarify in hierarchical order is an important step to understand the nature of the system. It is also an important step to analyze the structure of system and its relations. The power structure of control and the proper channel of information flow is important aspect of analysis of social systems engineering.

One of the founding father of ISSS, Kenneth Boulding presented his paper *General Systems Theory-The Skeleton of Science* in 1956. He is concerned about the existing over specialization and lack of communications. He arranges theoretical systems and constructs in a hierarchy of complexity. His original paper has given nine systems arranged in the hierarchical order of complexity. The higher the level, the more the complexity. Different books describe his concepts with different expression (diagram, table, etc.). We chose the table from Khisty and Amekudzi (2012) with some modifications. (See Table 2.8).

In the concepts of hierarchical level of systems, generally the lower level of system is included in the upper level of systems. The concepts of hierarchy can be seen from *Systemanaturae* of Linnacus, he classified the system of organism into three levels, the animal kingdom, the plant kingdom and the mineral kingdom, this example can show the process of evolution of knowledge.

The concept of hierarchy has many applications, it can be applied to design of measuring indicators, for example, in U.N.'s the 2030 Agenda for sustainable development, it has 17 goals, such as no poverty, zero hunger, good health and well-being, quality education, gender equality, clean water and sanitation, affordable and clean energy, decent work and economic growth, industry, innovation and infrastructure, etc., it has also 169 targets to be the second level and the third level of 304 indicators for measurement.

Table 2.8 Boulding's hierarchy of systems based on their complexity

Level	Boulding's system	Characteristics	Example
1	Structure (Frameworks)	Static relation of function or position	Atom, crystal, bridge
2	Clockworks	Simple dynamic system with predetermined motion	Solar system, clocks, machines
3	Control (cybernetics)	Closed-loop control mechanism	Thermostats
4	Open systems	Self-maintaining structure	Cells
5	Genetic-system	Society of cells	Plants
6	Animal	Nervous systems, self-awareness	Birds and beasts
7	Humans	Self-consciousness, knowledge accumulation and transmission, language	Human being
8	Sociocultural system	Roles, values, communication	Family, community, society
9	Transcendental system	Beyond our knowledge	Religion

Note Khisty, Mohammadi, and Amekudzi (2012)

(13) *Internal Elaboration*

Change of closed systems is moving toward increase of entropy and disorganization. But the open systems appear to move in the direction of greater differentiation, elaboration, and higher level of organization.

This feature of internal operation is consistent with the evolutionary paradigm by R. Fivaz in 1989. It stated that spontaneous general evolution from the uncomplicated to the complex is universal, i.e. simple systems will become differentiated and integrated both within the system and with the environment outside of the system.

(14) *Multiple Goal Seeking*

Biological and social system generally have multiple goals or purposes, especially the social system. Because the social system is composed by individuals and various subsystems (various social groups, the workers, the farmers, the intellectuals, the businessman and businessman etc.) with different value system, different cultural and professional backgrounds, their goals and purposes differ greatly. Study of social systems requires a detail survey of social groups classified into proper categories. Their goals, demands and behaviors must be carefully studied and various appropriate response should be planned and properly coordinated.

(15) *Equifinality and Multifinality of Open Systems*

In the physical system, there existed a direct cause-and-effect relationship between the initial conditions and the final state. But biological and social systems, operated differently. The open systems have two different features. The first feature is equifinality, that is the systems have equally valid alternative ways of attaining the same objective from different initial conditions (convergence). With China as an example, she is one of the low income developing countries, and she had per capita national income less than 400 RMB in 1980. China also had very different initial conditions compared to developed G7[6] countries. But China has implemented the development strategy of reform and opening in late 1970s, and through implementation of seven Five Year Plans post 1980, Chinese Gross National Income at 2014 reached 10,097.0 billion $ and per capital income around 7400 $ in Atlas method, it is 17,966.9 billion $ and 13,170 $/capita in terms of PPP. Chinese economic output has surpassed six countries of G7 except U.S.A. The second feature is multifinality, which means open systems from a given initial state, can achieve different and mutually exclusive objectives. (divergence).

(16) *Simple System and Complex System and their Behaviors*

Systems can also be classified into simple and complex systems. Their features have been studied and compared by Flood and Jackson (1991):

(i) Simple systems are characterized by a small number of elements; few interactions between the elements; attributes of the elements are predetermined; interaction between elements is highly organized; well defined laws governing behavior; the system does not evolve over time; subsystems do not pursue their own goals; the system is unaffected by behavioral influence; the system is largely closed to the environment.
(ii) Complex systems are characterized by many elements; many interactions between the elements; attributes of the elements are not predetermined; interaction between elements is loosely organized; they are probabilistic in their behavior; the system evolves over time; subsystems are purposeful and generate their own goals; the system is subject to behavioral influences and the system is largely open to the environment.

The above features of complex system are generally existed in nearly all social systems, especially it is very much applicable to the planning and improvement of a national system. Many failures in public policies to achieve its goals are due to insufficient understanding and study the related aspects of the complexity of the social system.

2.2.3 *Evolution of General Systems Theory*

There are many systems theories evolved after the publication of concepts of Bertalanffy in 1950. In Skyttner's (2005) *General Systems Theory*, sixteenth

[6]Note: G7 is the name of seven developed countries, i.e. United Kingdom, United States of America, Japan, Germany, Italy, France and Canada.

systems theories are described. Four of them will be abstracted and commented below which are relevant to the theme of this book.

2.2.3.1 The General Living Systems Theory by Miller, J. G.

(1) The basic element of the social system is the person, who belongs to a living system which is characterized by purposive. It maintains within its boundary a process of metabolism by interaction with its environment, that is through the continuous exchange of material and energy across the system boundary. This process also gives the material and energy necessary for all essential activities, such as production, reproduction, and other activities, including the repair of parts of the living system. The metabolism or processing of information is also very important in making possible regulation and adjustment of both internal stress and external strain. Information processing and programmed decisions are major means to control the processing of material and energy. This living system can be represented by an Input-Transformation-Output model based upon key concepts of GST. [Section 2.2.2.3(6)] (Fig. 2.7).
It is not difficult to understand the performance of a living system by with some basic knowledge of systems engineering. It is described by Chestnut (1965) that "If the system has adequate energy available, has materials capable of meeting the system environment needs, and uses the necessary information adequately, the basic ingredients are present for a successful system." (p. 61) If the word 'environment' is deleted in above quotation, the whole quotation can also apply well to the living system.

(2) Miller has also classified all living systems into eight hierarchical level based upon key concepts of GST [Sect. 2.2.2.3(12)]. Each level has its typical individual structure and processes. These eight levels are cells, organs, organisms, groups, organizations, communities, societies and supranational systems. It is the later five levels that should be studied in social systems engineering.

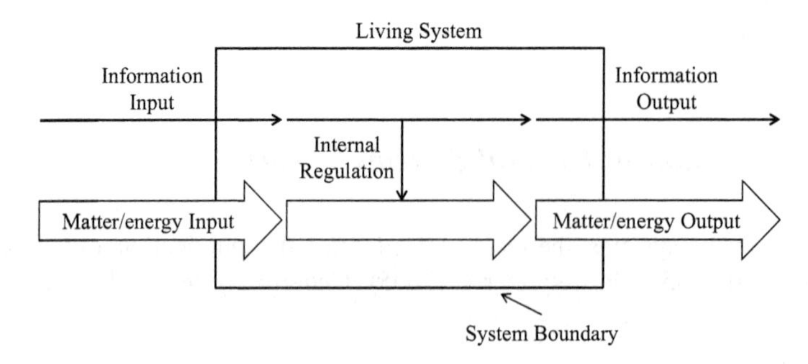

Fig. 2.7 Input-transformation-output model of the living system (*Note* Figure 2.7 is adapted from Fig. 3.4 of General Systems Theory by Lars Skyttner)

Table 2.9 Name of subsystems for processing input of living system

Types of subsystem	Name of subsystem
Subsystems processing material/energy/information	#1 Reproducer; #2 Boundary
Subsystems processing material/energy	#3 Ingester; #4 Distributor; #5 Converter; #6 Producer; #7 Storage; #8 Extruder; #9 Motor; #10 Supporter
Subsystems processing information	#11 Input transducer; #12 Internal transducer; #13 Channel and net; #14 Timer; #15 Decoder; #16 Associator; #17 Memory; #18 Decider; #19 Encoder; #20 Output transduces

(3) He has also suggested 20 subsystems related to the processing of material/energy/information, which is shown in Table 2.9. The function of these 20 sub-system is self-explanatory especially those sub-system related to processing of information, nearly all names of these subsystems are terms from ICT.

(4) Miller has also prepared a Level and Subsystem table of living systems with 160 cells (8 levels × 20 subsystems). In his table, the components of the subsystems are listed for the eight levels, for example, 'DNA and RNA molecules' is located in level "11" cell and subsystem 'Reproducer,' altogether 153 components are listed with 7 missing as not recognized. The arrangement has a remarkable resemblance to the periodic table of the elements and in a sense it has a similar function for living systems. His original work published in 1978 has more than 1000 page. (pp. 120–121, 123–125)

2.2.3.2 Bowen and Family Systems Theory

Family system is an important subsystem of the social system. Due to shortage of time and space, this book will focus mainly on the national system. Therefore, this part is briefed from Skyttner's (2005) book for expectation that there will be more studies on family system to have a better harmonious society.

Family system is a particular kind of social system and viewed as an emotional system which is a network of interlocking relationship valid also among biological, psychological and sociological process.

Bowen's theory is a theory of the family group and its internal relationships. Generally, the function of the individuals is often incomprehensive without the context of relation with a group, the group can be a family, a business unit, associations or a country. Relationships between members of a species are directed on common principles of universals.

A family is a coherent unit because it operate in ways consistent with it being a system. Following the rules of the emotional system, generally man responds sometimes with a basis of self-interest and sometimes on the interests of the group. The ability to remember the past and plan for the future allows man to engage in acts of reciprocal altruism. Here co-dependence, the other-centeredness that results in excessive abandonment is a key concept and this is crucial factor to make a stable and happy family.

Emotionally determined function of a family member will create an atmosphere or field which in turn influences the emotional function of the whole family.

A main part of Bowen's theory is the concept of triangling which says that the relationship in families and other groups consists of interlocking triangles. It is never possible to adequately explain an emotional process if links to other relationships are ignored.

In the original paper, Bowen has explained the role of triangle in the emotional system between three persons and between four persons (father, mother, daughter and son). But the authors think that triangle a third person outside the family may be important subject for discussion (pp. 192–197).

2.2.3.3 Powers and the Control Theory

William T. Powers is an American researcher of a psychologist cybernetician. He pondered the question: 'why does the same disturbance sometimes result in different responses.' He gave the answer in his book: *Behavior: The Control of Perception* (1973). He concluded: Human behavior is based on the concepts of control of reference perceptions and of feedback.

Because of the importance of behavior in social systems engineering, several of his related concepts are quoted below:

(1) Behavior is governed by internal reference signals and there exists a hierarchy of negative feedback control mechanism which are discernible in a person's behavior.
(2) Within this hierarchy, the higher level mechanism set the reference conditions for lower levels and receive information about deviations between controlled conditions and their reference values. (Power has established at least nine different levels).
(3) In all hierarchies of control, the lowest level system must have the fastest response. While the higher the level in the hierarchy, the slower the adjustment and the longer the endurance of a disturbance.
(4) The behavior of an organism is organized around the control of perceptions. Perceptions have no significance outside of the human brain. A presumed external reality is not the same as experienced. Reference signal for natural control systems are established inside the organism, and cannot be effected from outside.
(5) A conflict is an encounter between two control systems which try to control, but according to two different reference levels. A conflict is only likely to occur between systems belonging to the same orders. (pp. 185–189)

2.2.3.4 Taylor and the Geopolitical Systems Model

Alastair MacDonald Taylor (1915–2005), a Canadian historian, professor of geography and political science at Queen's University of Canada, was among the first to apply systems theory to the historical development of human societies, which he designated as Time-Space-Technics (TST). This TST framework recognizes human societies as an open natural system in equilibrium with their environments in a hierarchy of integrative levels.

His two-part model was cited in Laszlo's book The systems view of the world: A holistic vision for our time (1996).

The first part of the model is shown in Fig. 2.8, which includes the following information:

(1) The evolution of society is divided into five levels, from S1 to S5.
(2) The properties for technology, science, transportation, communication, and government at each level are defined.
(3) The five main levels of environmental control are presented as a geometrical sequence: point (Fig. 2.8, S2, Particulate Universal), line (Fig. 2.10 S3, one-dimensional), plane, volume as control capabilities grow. All these levels are built upon previous lower levels. As time passes, an increased complexity

SOCIETAL LEVEL	SYSTEM OF ENVIRONMENTAL CONTROL		EMERGENT QUALITIES					
	EXPLETED SPACE	IMPLETED SPACE	PROPERTIES	TECHNOLOGY	SCIENCE	TRANSPORTATION	COMMUNICATIONS	GOVERNMENT
S_5	Three-dimensional (extra-terestrial)	Megalopis ('Ecumenopolis')	BELOW +	Electrical-nuclear energy Automation Cybernetics	Einsteinian relativity Quantum mechanics Systems theory	Supra-surface Inner space systems Outer space explorations Surface systems Sub-surface vehicles	Electronic transmission (simultaneously) throughout expleted space	'Ecumenocracy' (Supra-national policies) Multi-level transaction Sovereignty invested in global mankind
S_4	Two-dimensional (oceans, continents)	City	BELOW +	Transformation of energy (steam) Machine technology Mass production	'Greek miracle' Scientific method Newtonian world-view	Maritime technology and navigation Thalassic and oceanic networks Highway networks Railroad technology	Mechanical transmission (printing) Alphabet	National state system Emergence of democracy Sovereignty of state (as primary actor)
S_3	One-dimensional (riverine societies)	Town	BELOW +	Non-biological prime movers (wind, water) Metal tools Continuous rotary motion (wheel) Irrigation technics	Mathematics Astronomy	Sailboats Riverine transport Wheeled vehicles Intra- and inter-urban roads	Writing	Ancient bureaucratic empires Theocratic politics Sovereignty of god-kings
S_2	Particulated Universal (sedentary)	Village	BELOW +	Animal energy Domestication of plants and animals Polished stone tools Spinning Pottery	Neolithic proto-science	Animal transport Paths, village routes Neolithic seafaring	Ideograms	Biological-territorial nexus Tribal level of organization and decision-making
S_1	Undifferentiated Universal (Nomadic)	Cave/tent (intraterrestrial)		Human energy Control of fire Stone and bone tools Partial rotary action		Human transport Sleds Dug-outs, canoes	Pictograms	Biological nexus (family, hunting band, clan)

Fig. 2.8 Levels of human societal organization (*Note* From Ervin Laszlo's *The World System*, Copyright@1972 by Ervin Laszlo)

and heterogeneity takes place. This figure can be interpreted as a matrix, the model provides a time/space grid, giving societal/environment quantization in vertical dimension and stabilization in horizontal dimension.

The quantization takes place at a very slow rate on low level (S1, S2) and gains speed as it passes through the different levels.

The stabilization shift from reactive-adaptive to active-manipulative as it progresses through the different levels.

The second part of the model is expressed in Fig. 2.9 which shows the importance of positive and negative feedback processes and interplay between them on all levels of sociocultural organizations. Figure 2.9 shows the following phenomena.

(1) A sociocultural system is consisted of many subsystems, political, economic, social, etc. The functioning of the existing sociocultural system is to be a converter with its numerous subsystems.
(2) There are biospheric and sociocultural inputs from the total environment.
(3) The regulation of the sociocultural system's material and societal output is through the negative and positive feedback.
(4) The interaction between the material and societal technique results in systematic self-stabilization (Cybernetics I) and transformation. (Cybernetics II).

Negative feedback ensuring overall stability is created by social institutions and predominant general morality. Example of this can be detected in mature sub-hominid societies where the struggle for life is fully operative and the prevailing technique exercise maximum environmental control.

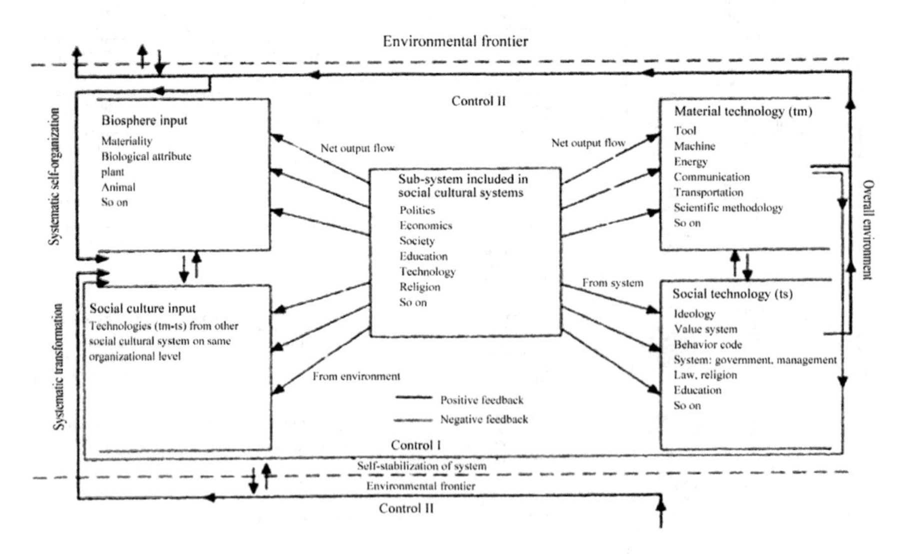

Fig. 2.9 Systemic self-organization, transformation and stabilization in human society (*Note* From Ervin Laszlo's *The World System*, Copyright@1972 by Ervin Laszlo)

Science and technology have an important role in creating the mechanism of positive feedback. Taylor has defined this positive feedback as follows: "Quantization occurs when deviation is amplified to the point where no deviation-correcting mechanism can prevent the rupturing of the basic systematic framework, that is when the latter can no longer contain and channel the energies and thrust which have been generated." This systematic transformation with a new technique can be seen from S4 of Fig. 2.8, the transformation of energy and mass production technology has formed the driving force in forming S4.

In ancient China and Egypt, whenever a new technique and favorable environment occurred, there were surplus food and better societal organization which opened the way for urban revolution at level S3 for these two earliest civilizations.

Both Figs. 2.8 and 2.9 provide an important framework of social systems engineering, these two figures show the relationship between the sociocultural system and its environment, the many subsystems of the sociocultural system and the role of S&T as a driving force of the process of transformation of a sociocultural system.

In addition to extrapolating Taylor's systems model, Laszlo, a Hungarian philosopher and a system scientist, also presented his concept of Natural Systems in his book *Introduction to Systems Philosophy* (1972) which contains a system concept of public policies.

2.2.3.5 Comment: Taylor's Study Is Quoted because of Two Reasons

(1) He is the first to apply system theory to study the historical development of human society. From the perspective of the authors of this book, lessons and experiences of the human history is very important. The basic human behavior and its relation to others recorded since 5. (societal level) will give important guidance to build up a better society.
(2) He has quoted two figures from Ervin Laszlo's The World System, these two figures are also reprinted in this book because these figures provide a broad and concise information for Social Systems Engineering. In addition, it is worthwhile to introduce Laszlo who is also a system scientist of public policy.

2.2.4 Application of General Systems Theory

2.2.4.1 General

General system theory(GST) as a unity of sciences has many applications. In Bertalanffy's (1968) original publication, he described many areas of applications, such as biological applications, system concepts in psychopathology, etc. He also stated: the practical application in systems analysis and engineering of systems theory to problems arising in business, government, international politics, demonstrates that the approach "Work" (p. 196). And in this book, we shall emphasize more of its application to engineering and governmental aspect.

2.2.4.2 Application of GST to Social Science and Planning

Nearly half a century has passed since the publication of *General System Theory* by Bertalanffy. Although application of GST has made some progress, there is still barrier due to ideological differences. Although there are progresses on several aspects of difficulties in its application to social science.

(1) *Progress on the Measurement of Social Science*

With respect to the measurement and social indicators, there are a lot of progress in this area in past several decades, for example, United Nations published *Indicators of Sustainable Development—Guidelines and Methodologies*in 2001, there are six themes and 19 indicators related to social side; the World Bank Group with the assistance of 40 partners had published 2016 *World Development Indicators,* there are seven sustainable development goals with 62 targets related to social side; and there is also exploration of indicators related to measuring wellbeing by OECD on 2011. In short, there are a lot of effort by global institutions for this work. Therefore, it seems that measurement by social indicator and quality of life is no more a serious problem.

(2) *Progress in Quantification*

Forty-seven years ago, Professor I. G. Kemeny had a paper titled *The Social Sciences Call on Mathematics* published in the book the *Mathematical Sciences-A Collection of Essays* (1969) edited by the National Research Council's Committee. Kemeny described that "There is a long tradition of cooperation between mathematics and the physical science." and "The connection between mathematics and the social science is a more recent development, and it is still in an embryonic stage." (p. 2). This situation has changed greatly 39 years later. The Nation Research Council's Committee on Organizational Modelling: From Individuals to Societies published a book *Behavioral Modelling and Simulation-From Individuals to Societies* (2008) which is the result of over 3 years of effort by a committee of 13 experts to respond to the request by U.S. Air Force. That book contains rich information related to IOS (Individual, organizational and societal) models. Therefore, the progress of quantification of social sciences is very fast that various social modeling is available currently.

(3) *Progress in Planning and Issues*

We are in agreement with Bertalanffy's view on the meaning of GST to the social science that "the concept of society as a sum of individuals as social atoms, e.g., the model of Economic Man, was replaced by the tendency to consider society, economy, nation as a whole super ordinated to its parts." But there is no shortage of planning at national or corporate level, which had been described in Chap. 1, Sect. 1.3.2, the failure of planning is due to the failure to consider society, economy, and nation as a whole. The process of implementation is also very complicated, generally the response to feedback information is not in time. There are also barriers existed in economic fields and political economy. It is worthwhile to quote a part of description from J. P-van Gigch's Applied General Systems Theory (1978):

> In general, people are afraid of planning because they associate it with socialistic or communistic political systems where economics and production are guided and controlled by governmental agencies. In the United States, at the time of writing, some form of National Economic Planning is a distinct possibility. However, the prevailing view holds that the competitive system, laissez-faire economics, and Adam Smith's "Invisible hand" can still guide the economy from crisis to crisis without obvious human intervention. Into-day's complex world, where international events have national repercussions and vice versa. It is sheer folly to believe in the myth and magic of the "invisible hand" that resolves all problems. Planning is indispensable and mandatory. (p. 533)

That truth can be seen from China's rapid development since 1950's. Although its national planning system needs further improvement from the perspective of GST.

2.3 Rise and Development of Systems Engineering (SE)

2.3.1 History of Formation of Systems Engineering and Its Definitions

2.3.1.1 History of Formation of Systems Engineering

(1) *The First Book of Systems Engineering*

The first book with title of systems engineering was contributed by professor Goode and Machol (1957), *System Engineering: An Introduction to the Design of Large-Scale Systems*. Their book had laid the foundation of academic aspects of systems engineering. Box 2.1 shows the major content of this book which may be helpful to understand better its evolution around 60 years in later period.

Box 2.1 Outline of *System Engineering: An Introduction to the Design of Large Scale System* Brief Description: This book contains six parts and 31 chapters. Because computer was a new tool in the age of 1950s, Part IV (Chaps. 14–20) gives a detail description of digital and analog computer, therefore, this part is omitted in the following contents of the book:

Part I Introduction

1. Complexity-The problem
2. Example of a large-scale system
3. An integrated approach to system design.

Part II Exterior System Design

4. Environment Origin and formulation of the problem
5. Mathematical model
6. Design of experiment, data gathering
7. Analysis of experiments-mathematical statistics
8. An example of exterior system design.

Part V Interior System Design

21. Solution of the problem-Steps and tools
22. Single Thread System logic
23. High traffic Queueing theory
24. Competitive Aspect-Game theory
25. Guide to system design-Linear programming, Group dynamics and Cybernetics
26. Simulation
27. Component parts of the system
28. Communications-Information theory
29. Reflexive Control-Information theory
30. Input-Output Human engineering.

Part VI Epilogue

31. Economics, Test and Evaluation, and Management.

(2) *A Methodology for Systems Engineering (1962) Written by Arthur D. Hall (1925–2006)*

Although this book is published five years later than the first book, but it is recognized by the academic and professional field that the term systems engineering can be traced back to Bell Telephone Laboratories in 1940s where Hall started his career as electrical engineer. His book is consisted of four major parts: part 1 is philosophical foundations, which includes explanation of systems engineering, case history of TD-2 radio relay system, fundamental concepts of systems engineering, and an operational analysis of the system engineering process; part 2 is titled problem definition, which approaches problem definition through the systematic study of needs and the environment; part 3 is decision making which concerns the design of value systems and decision making, with special emphasis on problem of setting good objectives; part 4 is system synthesis and analysis, which presents several different approaches to system synthesis and analysis. The content of this book is very comprehensive and original. Arthur D. Hall has further elaborated his methodology and published a paper *Three-Dimensional Morphology of System Engineering in IEEE Trans on April 1969.* This book has laid a very pragmatic foundation of systems engineering.

(3) *Later Development*

There are many studies, development and publications related to systems engineering. Machol, R. E. & Tanner, W. P. & Alexander, S. N. published the first *Systems Engineering Handbook* in 1965 in memory of Professor Goode due to the fact that Goode had agreed to prepare a System Engineering Handbook for McGraw-Hill Company in 1960, he prepared a brief outline, but died untimely in the same year. The National Aeronautics and Space Administration (NASA) also developed its application of systems engineering and published *NASA Systems Engineering Handbook* in 1992. The other major study and publications, such as *System Engineering Tools* (1965) and *Systems Engineering Methods* (1967) by Chestnut, H. (1917–2001), who was an electrical engineer and manager at General Electric, and Professor Sage, A. P (1993–2014), who wrote several papers and books on Systems Engineering, his relatively recent publication is *An Introduction to Systems Engineering.* (2000). There are many publications related to Systems Engineering and Systems Think Series by Wiley publishers, Plenum Press and others.

(4) *Establishment of International Council of Systems Engineering*

(i) *General*

The International Council on Systems Engineering (INCOSE) is a non-profit membership organization dedicated to the advancement of systems engineering and the professional status of systems engineering. It was founded in 1990, it has over ten thousand members which is the world's largest Professional Network of Systems Engineers.

(ii) *History*

The beginning of the INCOSE may be traced back to a meeting in 1989, hosted by General Dynamics at the University of California. The purpose of the meeting was to discuss the apparent shortage of qualified engineers who could think in terms of total system, rather than a specific discipline. Boeing hosted a follow-up meeting in 1990 in Seattle, Washington which was attended by about 30 people. The group adopted a charter, formed ad hoc committees to tackle the systems engineering issues, and formed the National Council on Systems Engineering. More meetings and workshops were held in 1990 and 1991 sponsored by the Aerospace Corporation, IBM, and TRW. The organization officially changed its name to the International Council on Systems Engineering (INCOSE) to reflect the growing participation of professionals from ten different countries.

(iii) *Organization and Activities*

INCOSE is governed by a Board of Directors. Its technical activities are led by a Technical operation board with oversight over 39 working groups under 7 Technical Committees. They focused on Education and Research, Modeling and Tools, Process and Improvement, SE Management, SE Initiatives, Standards and SE Applications. Its major activities include publication of periodicals, manuals and instructional material; sponsor of conferences; development of standards; and works to promote education and registration of systems engineers.

2.3.1.2 Definitions of Systems Engineering

There are more than several tens of definitions of S. E. from different authors with different backgrounds. We shall choose two definitions from two pioneers and three definitions from INCOSE for comparative purpose. Original sentences from the authors will be quoted in order to avoid misinterpretation.

(1) Definition from Chap. 1 of S. E. Handbook (1965) by Machol, R. E. "Thus, when I speak of system engineering. I refer to the design of systems, the word 'design' being significant. It implies that the output of the system engineer is a set of specifications suitable for use in constructing a real system out of hardware". (p. 1–4).
(2) Definition from A Methodology for Systems Engineering (1962) by Arthur D. Hall:

> "Systems engineering probably is not amenable to a clear, sharp, one sentence definition. Systems engineering has many facets; therefore, a complete definition of it must have many facets too. To get a reasonably comprehensive and useful view of all sides, we might define the systems engineering function in terms of:
>
> (i) Its evolution
> (ii) The systems engineering process (as generalized from case histories)
> (iii) Its objectives
> (iv) The kind of work done and called systems engineering
> (v) Organizational arrangement for carrying out the function
> (vi) The tools and techniques which it uses
> (vii) The kind of people who do it
> (viii) Its relation to other fields such as administration, application engineering, operation research etc." (p. 4)

(3) Definitions from INCOSE. INCOSE has published several versions of Systems Engineering Handbook. The following three definitions of S. E. are quoted from S. E. Handbook (2007) Version 3.1:

(i) "Systems engineering is a discipline that concentrates on the design and application of the whole (system) as distinct from the parts. It involves looking at a problem in its entirety, taking into account all the facets and all the variables and relating the social to the technical aspect (Ramo).
(ii) Systems engineering is an iterative process of top-down synthesis, development and operation of a real-world system that satisfies, in a near optimal manner, the full range of requirements for the system (Eisner).
(iii) Systems engineering is an inter-disciplinary approach and means to enable the realization of successful system (INCOSE) (2.1 of 10).

(4) Discussion of Definitions of SE

(i) Discussion of Definitions of SE

Five definitions of SE have been quoted above. There are even more definitions on SE in other SE publications. It is necessary to clarify the basic concepts of SE. It is a new concept raised early by Professor Goode, his concept on SE is clear, it is design of a new system. But he introduced several new disciplines to be included in the process of design, those are Operation Research and Cybernetics. And his view is confirmed to certain extent by other pioneers of SE. Hall (1962) discussed this point in his book and maintained that "engineering design, or simply design, is more closely related to systems engineering than to operation research. Professor Goode made no distinction between the two fields at all; in his book the terms "systems engineering" and "design" are used interchangeably. Other systems philosophers agree with this substitution and go on further to say that what is now called systems engineering will eventually be recognized as part of the general design process." (p. 20)

The three definitions from INCOSE only put emphasis on 'whole', 'iterative process' and 'interdisciplinary approach', all of them are complementary in nature, while 'design' is the key.

(ii) Authors of this book agree with above view. The following definition is English translation of definition of SE given in *Methodology of Social Systems Engineering* (Chinese version) written by Wang (2014):

> Systems engineering is a discipline of comprehensive science and technology, its object studied is design and operation of complex large scale (generally focus more on design) system. The purpose is to study and design of new engineering project, manage and improve the engineering object existed to achieve the overall most satisfactory result. By the meantime, it is necessary to analyze fully the relations between the system and its environment, explore the possibility of co-adaptation and co-evolution; in the time domain, it focuses on the inter-relationship and linkage following the order of planning, program, research, design, testing, manufacture, installation and operation. It is an overall engineering discipline with interdisciplinary approach, it is also a methodology of thinking and working. Its theoretical basis such as GST and other principles is also applicable to social system, administrative system and other systems. (p. 32).

2.3.2 *Development of Systems Engineering*

Since the first book of systems engineering published in 1957, there are many publications and definitions of systems engineering which had been described in Sect. 2.3.1. But a large part of studies is focused on "Hard" systems engineering. It seems that there will be more studies and development focused on "Soft" systems engineering. This section will discuss and explore this point. But we shall start the discussion by introduction of the major content of Systems Engineering Handbook Version 3.1 by INCOSE.

2.3.2.1 Introduction of Major Content of Systems Engineering Handbook Version 3.1 of INCOSE

This Handbook represents current state of art of SE. It is published in 2007 which is 50 years later than the first book of SE contributed by Professor Goode and Professor Machol. We had already given the outline of the first book in Box 2.1, and will present the outline of *SE Handbook* version 3.1 at below. We can have a general understanding of the trend of study of SE by comparing these two books.

(1) *Brief Description*

This *SE Handbook* differs with the first *SE Handbook* by Machol, R. E.& other published in 1965, it also differs from the first book of SE written by Goode and Machol in 1957. This handbook is prepared to be reference for examination of certified systems engineering professionals (CSEP). It is pragmatic and focused on key process activities. The content of this handbook can be divided into two parts. Three fifth of the book is the main part which contains ten chapters, the remaining two fifth are appendices. (Appendix A to Appendix O). The major content will be briefed as follows.

(2) *Major Content*

(i) Chapter 1 SE Handbook scope. This chapter defines the purpose and scope of this handbook.
(ii) Chapter 2 SE Overview. This chapter 2 provides an overview of the goals and value of using systems engineering throughout the systems life cycle.
(iii) Chapter 3 Generic Life Cycle Stages, which describe an informative life cycle model with six stages: Concept, Development, Production, Utilization, Support, and Retirement.
(iv) Chapters 4–6 will be represented by Fig. 2.10. These four process to support SE is identified by [7]ISO/IEC15288.

---

[7]Note: ISO/IEC 15288 is a system engineering standard covering process and life cycle stages. Initial planning for the ISO/IEC 15288: 2002 (E) standard started in 1994 when the need for a common Systems Engineering process was recognized. In 2004 this standard was adopted as IEEE 15288. ISO/IEC 15288 has been updated twice. It is managed by ISO/IECJTCI/SC7 which is the ISO committee responsible for developing ISO standard for Software and Systems Engineering

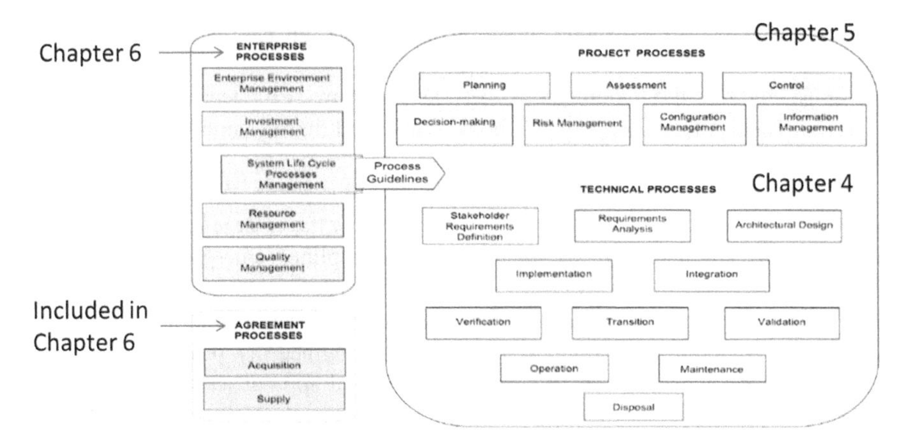

Fig. 2.10 System Life Cycle Process Overview per ISO/IEC15288 (*Source SE Handbook* (2007). p. 12 of 6)

(v) Chapters 7–9 are activities that support the above four process. Chapter 7 Enabling systems engineering process activities. It includes Decision management, Requirement Management and Risk and Opportunity Management. Chapter 8 Systems Engineering Support Activities. It includes 8.1–8.10. They are Acquisition and Supply, Architectural Design, Configuration Management, Information Management, Resource Management, Validation and Verification respectively. Chapter 9 Specialty Engineering Activities. It includes 9.1–9.9 they are Design for Acquisition, Logistics-Integrated Logistics Support, Electromagnetic Compatibility Analysis, Environmental Impact Analysis, Human Systems Integration, Mass properties Engineering Analysis, Modeling Simulation and Prototyping, Safety and Health Hazards Analysis, Sustainment Engineering Analysis and Training Needs Analysis.

(vi) Tailoring Overview. It includes Introduction, Tailoring Process and Traps in Training.

(3) *Selected Appendix with Abstraction*

This SE Handbook is a pragmatic handbook intended to be applied internationally, therefore, its appendices are also important. Appendix D is selected and abstracted for illustrative purpose.

Appendix D: The context for Appendices E to N.

These appendices are the first step in providing elaboration of systems engineering practice. The intent is to expand the practices defined in version 3.0 of the *SE Handbook* provide in depth information, "how to" guidance, and example for SE practitioners. It is envisioned that existing material and expand these appendices with additional SE practice. It is also envisioned that industrial specific practices in areas such as aerospace, transportation … and the U.K. Ministry of Defense will provide supplementary materials to the handbook, … the common practices identified will become world class practices …

Appendix E: The Hierarchy within a system …

2.3.2.2 Systems Engineering to Be a General Design Engineering

In consideration, the relationships between Systems Engineering and General Systems Theory, the authors of this book considered that Systems Engineering can be identify to be a General Design Engineering.

General System Theory is a theory applied to all systems. While the emergence of systems engineering is nearly in the same period of the emergence of GST. There are so many engineering projects of various disciplines emerged post the First Industrial Revolution, for example, electrical engineering, chemical engineering, mechanical engineering, transportation engineering, communication engineering etc. There are enormous amounts of design of engineering projects of various disciplines. But they have common logical steps and approaches to the process of design, we can give the term of the approach 'General Design Engineering' or Systems Engineering, because it is applicable to design of all types of engineering projects with different disciplines. Table 2.10 is a schematic diagram showing this common approach.

2.3.2.3 Explore the Concept of General Design Engineering from "Hard" Systems Theory to "Soft" Systems Theory

Establishment of INCOSE marked that systems engineering based on "Hard Systems Theory" has entered a matured stage.

It seems that both general systems theory and systems engineering are growing more or less independently under the same historical background. The rapid growth of modern science and its ever-increasing specialization in the 19th centuries has established intangible barriers among specialists in different disciplines, but these is existence of general system properties in the appearance of structural similarities or isomorphism's in different fields. Then we have emergence of concept of general

Table 2.10 Elements of general design engineering (or SE) for "Hard" systems engineering

<table>
<tr><td colspan="3">Goal of project (Analysis in frame of space and time)
Setting up performance measurement indicators</td></tr>
<tr><td rowspan="4">Analysis of various related environmental factors related to systems (Natural and social factors)</td><td>Design of main system and subsystem
Flow diagram of main system and subsystem</td><td rowspan="4">Feasibility analysis (technical, economic and social-including political)</td></tr>
<tr><td>Selection of various equipment for the main system and subsystem. Make connections</td></tr>
<tr><td>Spatial arrangements of all equipment of main system and subsystem</td></tr>
<tr><td>Selection of regulation and control of systems and equipment</td></tr>
</table>

system theory early around 1930s. In the meantime, the rapid growth of various engineering projects post the First Industrial Revolution have also created many different engineering disciplines, but there is existence of General Design Engineering (or Systems Engineering) in the appearance of similarities of structure, process, and logical steps in thinking. Both general systems theory and systems engineering require some philosophical thinking. And there are needs further to break up the barrier between natural science and social science, while an in-depth study of *General Systems Theory* will be helpful.

(1) *A Taxonomy of Sciences and Systems and Future prospects of systems engineering*

Figure 2.11 is partly derived from Gigch's (1978) concept, but modified by us. The taxonomy may be debatable due to its classification of social sciences and behavioral science. Therefore, it is titled to be tentative. It is used for the convenience to have a general concept related to various branches of science.

(2) *Progress and Prospects of Systems Engineering*

Based upon available information, there are several publications focused on social systems with interdisciplinary approach for example, Professor Evans at Concordia

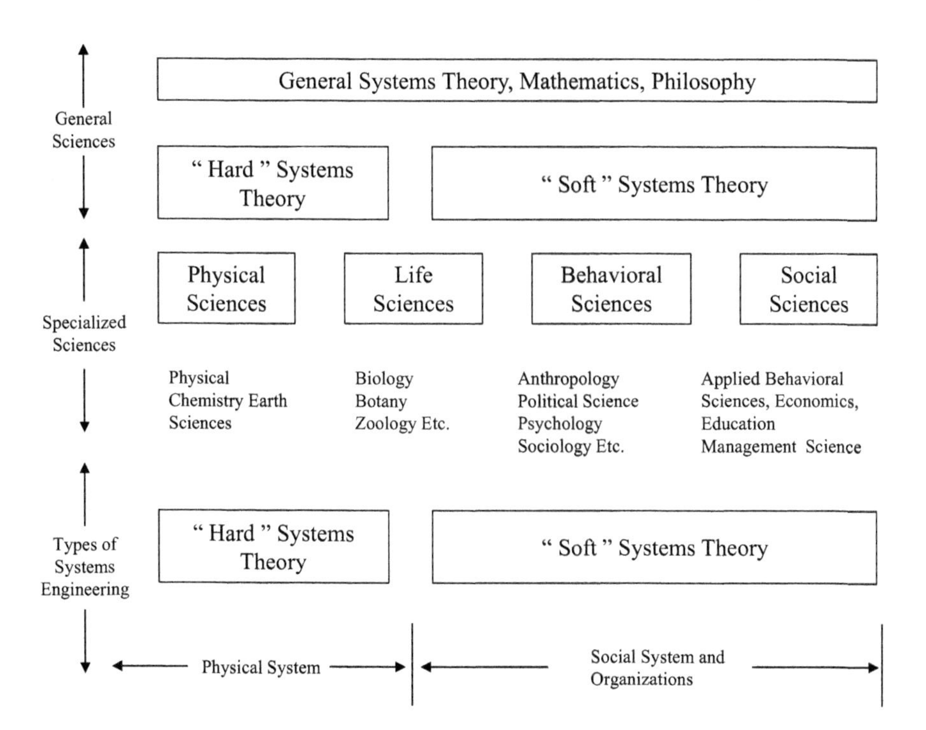

Fig. 2.11 Tentative Taxonomy of Sciences and Systems (*Source* Adapted from Gigch, J. P. (1978). Applied General Systems Theory (2nd Edition) (Fig. 2.1, p. 39). New York: Harpercollins Publishers)

University College of Alberta published the first Edition of the book *Using Statistics in the Behavioral and Social Science* in 1984. This book is now in its fifth edition. It is known in the politics and public policy that these fields are characterized by uncertainty, while statistical inference is one of the means to study the complex behavioral and social sciences.

There is also a publication *Systems Engineering with Economics, Probability and Statistics written* by professor Khisty, C. J., & Mohammati, J. & Amekudzi, A. A. Its first edition was published in 1982. The revised second edition is published in 2012. Three new chapters are added, these three new chapters are "Systems Thinking: Case Studies", "Sustainable Development, Sustainability, Engineering and Planning" and "Case Studies in Engineering and Planning for Sustainability". This book although with the title of Systems engineering-with Economics, Probability, but authors of this book already have the awareness that both hard systems engineering and soft systems engineering should be used in combination for problem solving, which is shown in a diagram on the cover of the book.

2.3.3 Systems Engineering and Concepts Related to General Systems Theory in Ancient China

2.3.3.1 Systems Engineering in China

(1) It is a historical fact that in the history of development of S&T, China had lacked a historical period of scientific revolution. China even missed to certain extent the dispersion of the impact of the First Industrial Revolution. In that sense, China became a latecomer of modern science and technology as well as the modern education system. Its catching up process to modern S&T and education system may be only around one hundred years.
China established a Systems Engineering Society in 1980 based upon suggestions of several famous scholars and professor. Its members reached around eleven ten thousand currently. Nearly every provinces and autonomous regions has a branch society. It has several publications: *Theory and Practice of Systems Engineering* (monthly magazine), *Journal of Systems Science and Systems Engineering* (Quarterly Journal), *Fuzzy System & Mathematics* (Quarterly), *Transportation Systems Engineering and Information* (Quarterly) etc.

(2) Education of Systems Engineering. An influential scholar Mr. Qian Xuesen and others published a paper *Systems Engineering-A Technology of Management* in 1978. Since then, nearly all universities established the discipline of Systems Engineering in their Management Colleges including post graduate studies.

2.3.3.2 Exploration of General Systems Theory in Ancient China

The Brilliant Age of Hellenistic science of Greeks is more or less similar to the period of Spring and Autumn and also the Warring State (770 BC–221 BC) of China. There are many eminent scholars contributed various valuable philosophical, and scientific (natural and social) thoughts. As Talcott Parsons says, in this brilliant period, the four sources of the world civilization, namely, Greece, Babylonia, India and China, all underwent a "philosophical breakthrough". Although we have described "Major Events in the History of Science and Technology" in 2.1.2, based upon Hellenman and Bunch's publication (1988), two facts related to the theme of GST must be added.

(1) *I-Ching or the Book of Change*

This is a very famous Chinese classics published in late 9th century BC. It is an ancient divination text and the oldest Chinese classics. It is an influential text read throughout the worlds of religion, psychoanalysis, business, literature, and art. It was transformed into a cosmological text with a series commentaries known as the "Ten Wings" through the Spring and Autumn and Warring State period. Confucius had studied I-Ching with full effort and wrote two books titled Copulative Biography to explain the phenomena of natural and social changes. It contains many concepts similar to GST. Hereunder, only very few descriptions are selected for illustrative purpose and for raising reader's interest to read the original.

(i) Yin and Yang are in opposition mutually, but they are also in dependence mutually, this is the basic principle or so called Dao[8] in Chinese. Who masters this basic principle can achieve a state of universal De, it may also achieve success to certain events through study and understanding of this rule. Dao is in existence objectively, but to understand the meaning of Dao depends upon the perspective of the observers. The kind people see its kindness, the clever people see its wisdom ...". (from I-Ching, Copulative Biography Upper book, Chap. 5)

(ii) "Confucius said: "Why one thinks so much about the changes of the world? People walk on different paths, but they will end up at the same final destination; people may have thousands of thoughts and considerations, but all of which will converge to the uniformity in the end regardless of one's many thoughts on the changes of the world! The sun sets, then moon rises; the moon sets, then the sun rises, the sky and the earth are filled with brightness due to alternative rise and set of the sun and the moon; cold winter passes, hot summer comes, hot summer passes cold winter comes, the year is formed through alternative coming and going of summer and winter .... The complementarity of submitting and expanding will produce benefits .... To study

[8]Note: Dao is the metaphysical core of Chinese. It derives from the perspicacious generalization from human life, society, politics to ontology; while De is the revelation and expansion of Dao, the application of Dao to give guidance to social, political and human life.

the secrets of things and understand the changes of Dao, it is the highest De". (from I-Ching, Copulative Biography Lower book, Chap. 5).

(iii) "... There are rules governing the steady and dynamic state of everything in the universe, the distinction between masculine (yang gang) and feminine (yin rou) can be judged clearly. People of the same kind will gather together, things of different kind will be divided into different groups, thereby good and bad luck will be resulted...." (from I-Ching, Copulative Biography Upper Book, Chap. 1).

(2) *The Chinese Medicine*

It had been described in 2.1.2.2(2) that Hippocrates in the period of Greek and Hellenistic Science is referred to be the "Father of Western Medicine." It is interesting to compare the theory and practice of Chinese Medicine which is shown fully in the ancient Chinese book *Huang Di Nei Jing* dated back in the beginning of Eastern Han Dynasty (around 32–90 AD). A very primitive and simplified copy was unearthed from a Han tomb. One unique feature of the Chinese medical theory is its systems perspective. It treats the human body as a whole system, the cause of disease is attributed to the change of external environment and unbalanced internal conditions. The major force of change is based on yin-yang theory. It is also interesting to note the similarities of ancient philosophical and medical thoughts, Hippocrates believed that all matter was made up of four primary substances: earth, air, fire and water. While it was thought that the universe consisted of five transformative phases in ancient times of China. It was said that Tzu-Szu (492–431 BC) had advanced this theory and later was redeveloped by Mencius (372–289 BC). The world was viewed as a constant interaction and combination of the five transformation phases (the principle of WuXing in Chinese), they are wood, fire, earth, metal and water. Yin-Yang and WuXing have become part of Chinese medical theory which may require further scientific exploration.

2.4 Summary Points

1. The purpose of this chapter is:

(1) Give essential points of General Systems Theory (GST) and Systems Engineering (SE), and trend of their growth and applications.
(2) Description of development of GST and SE is presented from the historical context of development of science and technology.
(3) A broad discussion of relationships between science, technology, economy and society is given. The mutual interdependence of them is emphasized, rather than a simple view of technology determinism, or economic determinism, etc.

2. The relationships between science, technology, economy and society are explored in two ways:

(1) Economic history shows that the growth rate of world population, GDP/capita and world population is quite low before 1920, economist concluded that technology played a major role to drive the economy. But in the global history, there is a period of scientific revolution before the First Industrial Revolution erupted in U.K. The close linkage between science and technological and engineering began around the nineteenth century, in the period of First Industrial Revolution.

(2) Relationships between S&T, economy and society is further explored through major events in the history of science and technology. Based on the history of S&T of a Western study, eight periods are identified and major events of S&T are described. These eight periods are Science before there were scientists: 2400,000–599 BC, Greek and Hellenistic Science (which is the origin of the Western science) 600 BC–529 AD, Medieval science 530–1452, the Renaissance and the Scientific Revolution 1453–1659, the Newtonian epoch 1660–1734, the Enlightenment and the Industrial Revolution 1735–1819, Nineteenth century science 1820–1894 and Science in the 20th century through two World Wars 1895–1945 respectively. Features and major achievements are described for each of these periods. Several major conclusions are derived from the study of above eight periods:

(i) Technology predates science due to the needs of basic survival of the mankind to adapt to the natural environment.

(ii) Before the period of scientific revolution, there were philosophers who observe and study the natural and social phenomena. There were no experiments in general for the natural science. Much of the knowledge was accumulated over several thousand years and some were proven correct and some were wrong. Several branches of natural science were established primarily.

(iii) Through the period of scientific revolution and Newtonian epoch, scientific method is established through observation, combination of induction and deduction and proof through experiments. A system of modern science is established. The secondary effect of Newtonian epoch is mechanistic determinism of a world view.

(iv) Since the period of the Enlightenment and Industrial Revolution, the rapid accumulation of knowledges requires further in-depth studies. Thus, various branches of science are further developed and explored. Along with the creation of many sciences of specific fields, scientists become a profession. Modern social science is also developed post the period of Enlightenment.

(v) The application of scientific knowledge to technological development becomes direct and the period between scientific exploration and its application to technological and engineering fields becomes shorter increasingly in the 20th century.

(vi) The discovery of theory of relativity and quantum mechanics has broken the world view of mechanistic determinism. Uncertainty and probabilistic view are emerged.

(vii) As modern society becomes more complex, modern engineering projects become larger and complex increasingly. Coordination and integration become a necessity in contemporary science and engineering. General Systems Theory and Systems Engineering were developed nearly simultaneously to meet the needs.

(viii) Study of history of S&T demonstrates that its development is also closely related to the economic and social conditions of its historical period.

(ix) A summary table is prepared by the authors of this book which shows the quantitative relationship of growth of S&T with the various historical periods.

3. The section of General Systems Theory is approached from five aspects:

(1) Emergence of GST in the early 20th century is analyzed from two issues of development of science around 2500 years since the period of Greek and Hellenistic science. The two issues are "negative effect due to ever-increasing specialization" and "lasting impact of mechanistic determinism". Bertalanffy, the founder's five aims in creation of GST are quoted.
(2) Definitions of System and its classification. Several definitions and several types of classifications are described to provide a relatively in depth understanding.
(3) key concepts of GST are described, they are: subsystem or components, holism, open systems view, input-transformation-output model, system boundaries, steady state dynamic equilibrium and homeostasis negative entropy, feedback, hierarchy, internal elaboration, multiple goal seeking, equifinality and multifinality of open systems, simple system and complex system and their features.
(4) Progress of application of GST is described, especially its application to planning.

4. Section 2.3 deals with the rise and development of Systems Engineering. It is interesting to note the fact that the emergence of General Systems Theory and Systems Engineering are nearly occurred in the same period. The former is for the unification of science while the latter is for a 'General Design Engineering' (Hall, 1962). Section 2.3 is further divided into 3 parts:

(1) History of formation and its development of systems engineering around fifty years are described, five definitions from international scholars and one definition given by one author of this book is also presented for comparative purpose.
(2) Development of system is explored by focusing on the transition from 'Hard' systems engineering to 'Soft' systems engineering, there is a need to break up further the barrier between natural science and social science. Some progresses in Western countries are described. This book represents an effort of the process of exploration.
(3) Brief introduction of development of systems engineering of China is provided. China has not established the concept of modern general systems theory, but some of ancient Chinese literatures had several concepts of GST is introduced briefly including the concept of Chinese medicine to supplement the information of part 1 of this chapter, i.e. science in ancient China.

References

Bertalanffy, L. V. (1968). *General System Theory* (pp. 38, 55–56). New York: George Braziller.
Chestnet, H. (1965). *Systems engineering tools* (p. 65). U.S.A.: John Wiley & Sons, Inc.
Connate, J. B. (1951). *Science and common sense* (p. 25). New Haven: Yale University.
Ellis, D. O., & Ludwg, F. J. (1962). *Systems philosophy* (p. 13). New Jersey: Prentice Hall.
Evans, A. N. (2014). *Using basic statistics in the behavioral and social sciences* (5th ed.). London: Sage Publication Inc.
Gigch, J. P. (1978). *Applied general systems theory* (2nd Ed.) (pp. 39, 533). New York: HarperCollins Publishers.
Goode, H. H., & Machol, R. E. (1957). *System engineering: An introduction to the design of large scale systems*. New York: McGraw Hill Book Co., Inc.
Hall, A. D. (1962). *A methodology for systems engineering* (p. 4). New Jersey: D. Van Nostrand Co., Inc.
Haskins, C. (Ed.) (2007). *Systems Engineering Handbook*. Version 3.1 Seattle: INCOSE.
Hellemans, A., & Bunch, B. (1991). *The timetables of science-A chronology of the most important people and events in the history of science* (pp. 1–19, 59). New York: Simon & Schuster Inc.
Incose. (2007). *Systems engineering handbook*. Version 3.1 Seattle: INCOSE.
Jary, D., & Jary, J. (1991). *Collins dictionary of sociology* (pp. 350, 648, 651). Great Britain: HarperCollins Publisher.
Kast, F. M., & Rosenzweg, J. E. (1979). *Organization and management* (3rd Ed.) (pp. 107, 117). New York: McGraw Hill Inc.
Khisty, C. J., Mohammadi, J., & Amekudzi, A. A. (2012). *Systems engineering-with economics, probability and statistics* (2nd ed., p. 7). Richmond: Ross Publishing.
Klir, G. (1991). *Facets of systems science*. New York: Plenum.
Luck, S., & Brown, M. (Eds.). (1999). *Philips millennium encyclopedia*. London: George Philip Ltd.
Maddison, A. (1995). *Monitoring the world economy 1820–1992*. Paris: OECD.
Neil Kagan (Ed.). (2013). *Concise history of the world-An illustrated time line* (Revised Ed.). Washington, D.C.: National Geographic Society.
Palffy, G. (Senior Ed). (2014). *The science book* (pp. 32–33, 72–73, 108–109, 200–201). Great Britain: Dorling Kindereley Ltd.
Paynter, H. M. (1961). *Analysis and design of engineering system* (pp. 11–12). Massachusetts: The MIT Press.
Sachs, J. D. (2015). *The age of sustainable development* (p. 83). New York: Columbia University Press.
Stavrianos, L. S. (1979). *The world to 1500-A global history* (p. 53). USA: Prentice Hall.
Stavrianos, L. S. (1980). *The world since 1500-A global history* (p. 203). USA: Prentice Hall.
Stavrianos, L. S. (1999). *A global history: From prehistory to the 21st century* (7th Ed.) (pp. 760–762) (Chinese version). Beijing: Peking University Press.
Skyttner, L. (2005). *General systems theory* (2nd ed.), (pp. 120–121, 123–125, 192–197, 185–189). USA: World Scientific Publishing Co.
UNDESA. (2001). *Indicators of sustainable development: Guidelines and methodologies* (2nd ed.). New York: United Nations.
Wang, H. J. (2014). *Methodology of social systems engineering* (p. 32) (Chinese Version). Beijing: Development Press.
World Bank Group. (2016). *World development indicators*. Washington, D.C.: The World Bank.
Xu, Zihong. (1990). *Complete translation of Zhou Yi (Chinese version)*. GuiZhou: GuiZhou People's Publisher.
Zi, Yuan (Ed.). (2012). *Pictorial Huang Di Nei Jing (Chinese version)*. Xian: Shaan Xi Normal University Publisher.

Chapter 3
Social Science and the Social System

3.1 Definition of Social Science and Its Development

3.1.1 Definition of Social Science and Its Scope of Study

3.1.1.1 Definition and Scope of Social Science

According to R. A. Nisbet, the well-known sociologist and former Albert Schweitzer Professor of the Humanities at Columbia University,[1] social science is a branch of science which studies mainly the human behavior in the social and cultural aspect. Its scope of study includes cultural (or social) anthropology, sociology, social psychology, political science and economics. Generally, it also includes related parts of social and economic geography and education. History has also been a part of social science, but some scholars consider history to be part of the discipline of humanities.[2] Comparative law may be also included in the study of social science.

The term "behavioral science" has also been applied to cover the study of social science since the beginning of the 1950s. According to Nisbet, the terms "social science" and "behavioral science" are synonymous.

---

[1]Note: R. A. Nisbet is the author of 'Social Science' section (2012) of Encyclopedia Britannica.
[2]Note: Humanities is a study of the human culture. It emphasizes the exchange of ideas, but for example applying the method of speculation and fantasy rather than applying the experimental method of natural science. Humanities is a large independent discipline that includes amongst of the studies ancient and modern linguistics, literature, philosophy, religious studies and performance arts (such as music and drama). Sometimes, humanities is also classified as social science, including history, anthropology, regional studies, communication studies, cultural studies, law and linguistics etc.

H. Wang and S. Li, *Introduction to Social Systems Engineering*,
https://doi.org/10.1007/978-981-10-7040-2_3

3.1.1.2 Debate on the Term of Social Science Between Sociologists

It is necessary to point out that sociologists have different views about what constitutes social science. The following quotation from *The Concise Oxford Dictionary of Sociology* (1994) by Marshall (Editor) shows the background of ambiguity with the classification and scope of the study of social science.

> Discipline boundaries are by no means always clear and the generic term social science usually covers most or all of the discipline mentioned. All, to various degrees, are engaged in debates about the nature of science and scientific status. Are the social sciences directly comparable to the natural sciences, or does the fact that their object of study is human make them different, in what sense (if any) are they scientific? Sociologists, in particular, have addressed these questions more or less continuously from the time of the classical theorists onwards. (p. 494)

3.1.2 Development of Social Studies and Social Science Before the 20th Century

3.1.2.1 Development of Social Studies in China

From the perspective of the authors of this book, with the existence of groups of human beings, human society was also in existence, as was the social and cultural behavior of human beings. Therefore, there is a vast amount of information and knowledge accumulated in the area of social studies. As recognized by Talcott Parsons, China is one of the four sources of world civilization that underwent a "philosophical breakthrough" in the axial[3] period of global history. A brief description of selected aspects of social studies of China in the axial period that relate to the theme of this book will be described.

(1) *Development of Social Studies in the Spring and Autumn Period (722–481 BC) and the Warring States Period (403–221 BC)*

Although the history of Chinese civilization begins much earlier than the above two periods, this period can be compared to a certain extent to the Greek Hellenistic period. There was the Contention of a Hundred Schools of Thought, the Confucian, Taoist school, Mohist school, Legalist, and Theory of Yin-Yang and Five Elements and many others. Many of these schools explored rationally the relationship between individual, the state and morality. Hereunder, several major schools and

---

[3]Note: Axial Age (from German: Achsenzeit) is a term coined by German philosopher Karl Jaspers in the sense of a "pivotal age" charactering the period of ancient history during about the 8th to the 3rd century BC. But in the recent *The World, a history*, by Fernandez-Armesto (2010), the Axial Age is designated to be 500–100 BC.

the representative figures will be discussed and some of their thoughts related to social systems will be described.

(i) *Confucius (551–479 BC)*

He is highly respected not only in China, but also in global history to be a leader of the birth of philosophy in the period of 700–300 BC together with Laozi (604–531 BC), Plato (428–348 BC) and Aristotle (384–322 BC).

Confucius taught social ethics in China. He had many students and followers, and formed the Confucianism, which had great impact not only in China, but also had influence the Northeast Asia and Southeast Asia. His teaching was recorded by his students, and formed the *Four Books*, *i.e.*, *Great Learnings*, *Doctrine of the Mean*, *Mencius*[4] *and Analects*. He had also been responsible for editing the Five Classics, *i.e.*, *Classic of Poetry*, *Book of Documents*, *Book of Rites*, *I Ching* (*Book of Changes*) and *Spring and Autumn Annals*. He also edited four books, *The Book of Songs, The Book of Changes*, *The Book of History, The Book of Rites*, and he compiled *The Spring and Autumn Annals*. The *Four Books* and the *Five Classics* became necessary readings to pass the imperial examinations to become governmental officials from the Sui Dynasty to Qing Dynasty, the system of examinations lasted in China around 1300 years.

There are abundant thoughts from Confucius. The following description of a harmonious society which is quoted from *The Book of Rites* shows the long term goal pursued by Confucius.

"When the high morality (Da Dao) is popularized, everybody will serve for the public interest. Good and capable people will be chosen to serve the public, everybody shall focus on creditability and be friendly to others. Therefore, the people will not only support their parent, tend their own children, so that the older shall be died in happiness, the adults can serve fully for the society, the infants can be grown up soundly. All special groups of people, such as the old man without wife, the old widow, the orphan, the single, the handicapped can be supported properly. Every man shall have a job and every girl shall be married in time. The people shall collect the goods or assets by others, but they are not for the purpose to hold to enjoy themselves, the people shall also hate those lazy behavior in working for the public, everybody should work with full effort not for private interest. Under the above circumstances, nobody will play conspiracy, nobody will steal goods and assets, and nobody will start war and insurrection. This is so called the Great Unity Society (or Da Dong society in Chinese)." (Translated from *Li Yun* of *Book of Rites*).

Nearly all social thoughts of Confucius were related to his ideal goal of society. Several sentences approached the thought of holism which is an essential element of GST (General System Theory) as quoted below:

[4]The word Mencius is the same with Mengzi. Different publications may use either of these two to refer the same person.

> The Master said, "To go too far is as bad as not to go far enough" (Chapter Xian Jin of The Analects, p. 115)
>
> The Master said, "Do I regard myself as a possessor of Wisdom? Far from it. But even if a simple peasant comes in all sincerity and asks me a question, I am ready to thrash the matter out, with all its pros and cons, to the very end." (Chapter Zi Han of The Analects, p. 91)

Confucius said: "If one love to learn, one shall approach the wisdom; to act with full effort, one shall approach the kind heart and tenderness; if one knows shame then one shall approach bravery. Knowing above three, then one knows how to cultivate one's action of morality. If one knows the cultivation of action of morality, then one knows to govern the people, then one shall know how to govern a country.' There are generally nine principles to govern a country, they are cultivation of one's own morality, respect knowledgeable people of eminent goodness, loves one's family, highly esteems the senior officials, considerate all officials, love the people like one's own son, draws all kinds of workers, gives preference treatment to remote different races and comforts the vassals on all sides respectively." (Translated from Chap. 19 of *The Doctrine of the Mean*) (p. 16).

The word "mean" in Chinese is "Zhong Yong", which means the middle, "no bias", balance, or not go to the extreme. This is one of the major philosophies of Confucius. The above three points nearly have shown the same concept of holism of GST. The last paragraph is translated from the "*The Doctrine of the Mean*" which also gives the thoughts of politics of Confucius.

(ii) *The Confucianism*

Confucius had many followers. Two famous follower were Mencius (372–289 BC) and Xunzi (298–238 BC). *The Work of Mencius* is one of the *Four Books*. The thoughts of Xunzi have attracted the attention of policy studies in Western countries in recent years. Some brief selected economic and legal thoughts of Xunzi are quoted below:

> The means to enrich the country is to economize the expenditure of the government to enrich the people, and save the surplus grain, foodstuff and assets well. Expenditure saving should be based on the provisions of rite and morality while policy measures are required to make people to become rich and abundant.[5]
>
> Moderate the use of goods, let the people make a generous living, and be good to storing up the harvest surplus. Moderate the use of goods by means of ritual principles, and let the people make a generous living through the exercise of government. (*Xunzi* Book 10, p. 267).
>
> It is necessary to determine the boundary of a country based upon the size of its territory, to calculate the amount of income to raise the people, to distribute the work according to the strength and weakness of the labor force. Utilization of the labor force should consider fully that they are qualified to do their work, and benefits must be received by them in doing their work, benefits received must be sufficient enough to support themselves and cover all

[5]Note: XunZi (1993, p. 99).

expenses of food, cloth, etc., by the meantime, there may be surplus and savings continuously, this is so called in accordance with the law.[6]

If one taxes lightly the cultivated fields and outlying districts, imposes excises uniformly at the border stations and in the marketplaces, keeps statistical records to reduce the number of merchants and traders, initiates only rarely projects requiring the labor of the people, and does not take the farmers from the fields except in the off-season, the state will be wealthy. This may indeed be described as "allowing the people a generous living through the policy of government. (*Xunzi* Book 10, p. 271)

Thus if one only reproves and does not instruct, then punishment will be numerous but evil will still not be overcome. If one instructs but does not reprove, then dissolute people will not be chastened. If one reproves but does not reward, then applying harsh discipline to the people will not exhort them to good; if the reproofs and rewards are not of the category proper to the occasion, subjects will be suspicious, vulgar, and venturesome, and the Hundred Clans will not be unified. (*Xunzi* Book 10, pp. 295–297)

(iii) *Taoist School (or Taoism)*

Taoist school was established mainly by Laozi (581–500 BC). His major thoughts were expressed in his book *Dao De Jing*, which contains 81 chapters, and is the first comprehensive philosophical system in the history of Chinese philosophy. The philosophy of Laozi is first about the universe, then human life and next, politics.

Fu Huisheng's introduction to *Laozi* (1999) expresses the following views:

Dao derives from the perspicacious generalization from human life, society, politics to ontology; while De is the revelation and expansion of Dao, the application of Dao to give guidance to social, political and human life. The relation between Dao and De is that between a body and function...The origin of the universe and myriad things comes from Dao... (p. 37)

The ambiguous description of Dao in the book derived different interpretations in the later generations and also brought into existence of philosophical schools. Dao as core of Laozi's philosophy denies the absolute and divine power of heaven, and theorizes the atheistic thoughts, this historic progress produces profound influence on later generations. (p. 38)

The political thought of Laozi can be shown by quotation of several sentences in Chapter 57 of *Laozi*.

Kingdoms can only be governed if rules are kept; Battles can only be won if rules are broken. But the adherence of all under heaven can only be won by letting alone. (p. 117)

The core thought of Laozi on politics is "To govern by doing nothing." This can also be shown in several statements in Chapter 57 of Laozi.

Therefore a sage has said:

So long as I 'do nothing' the people will of themselves be transformed.

So long as I love quietude, the people will of themselves go straight.

So long as I act only by inactivity the people will of themselves become prosperous. (p. 117)

---

[6]Note: XunZi (1993, p. 100).

The following paragraph is also selective quotation from Fu's *Introduction to Laozi* (1999), which can show that the philosophy of Laozi has many elements similar to dialectics or General Systems Theory of contemporary society.

> Pairs of opposites and universality of the concept is emphasized. "These ten thousand creatures cannot turn backs to the shade without having the sun on their bellies (the myriad things shoulder the yin and embrace the yang)." (Chap. 42)

He emphasized the balance and harmony of contradictions, "It is on this blending of breaths that their harmony depends (through coalescing of yin and yang, they attain a state of harmony)." (Chap. 42).

He also emphasized interdependence of the opposite. "It is because everyone under Heaven recognizes beauty as beauty, that the idea of ugliness exists. And equally is everyone recognized virture (goodness) as virture (goodness) this would merely create conceptions of wickedness." (Chap. 2) At the same time, the myriad things under Heaven are not only opposite and interdependent, but also mutually changeable… "It is upon bad fortune that good fortune leans, upon good fortune that bad fortune rests." (Chap. 58) And the change is usually gradual from quantity to quality. "The tree big as a man's embrace began as a tiny sprout, the tower nine stories high began with a heap of earth, the journey of a thousand leagues began with what was under the feet." (Chap. 64) (pp. 39–40).

There were translations of thoughts of Confucius and Taoism by Western scholars such as Waley, Yang. Those translations were published by China's publisher.

(iv) *Zou Yan*

He is classified to be Taoism. He lived in the later period of the Warring State. While his date of birth and death is not quite clear, it may be around 324–250 BC. According to Lan (1950), his major thought is the establishment of a comprehensive theory of Yin and Yang and Five Elements (Wuxing in Chinese), which became one school (p. 241), but his school of thought is also classified to be a branch of Taoism by later scholars. The theory of Yin and Yang has been explained previously, the Theory of Five Elements is shown in Fig. 3.1.

The Theory of Five Elements is not only derived from the natural phenomena, but the five natural elements also have the relationship of mutual reinforcement and mutual counteraction between each other. For example, metal is derived from earth, which is shown by the green line and the arrowhead, and the dotted red line shows the counteractive relations, for example, that fire can melt the metal, etc.

Theory of Yin Yang and Five Elements is also closely related to Chinese medicine. The Five Elements are in correspondence with respective organs of the human body, which is also shown in Fig. 3.1. Chinese medicine treats the symptom of sickness as a whole, for example, if there is disorder of the function of liver (wood), it will affect the stomach (earth), and cause disorder of function of digestion. Therefore, for the treatment of uncomfortable stomach, there is also need to treat the liver, etc. (Based upon *Huang Di Nei Jing* p. 55 or *The Inner Canon of the Yellow Emperor*).

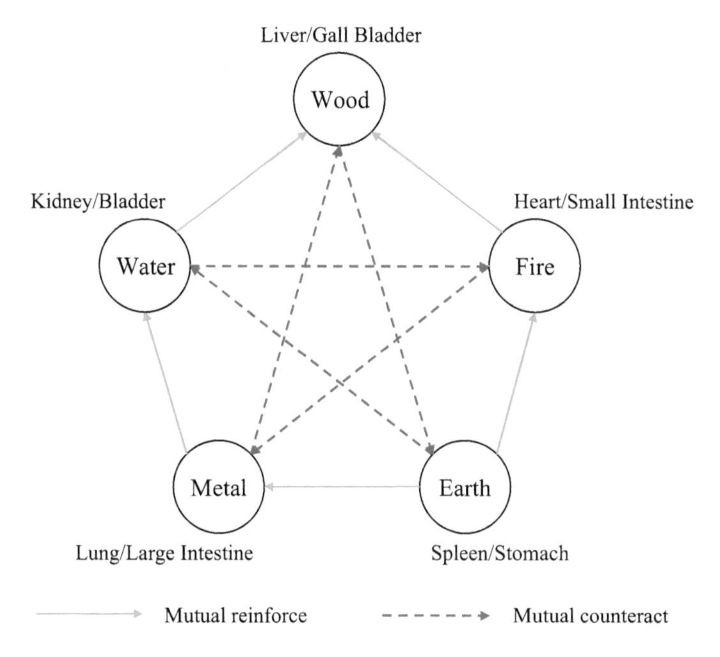

Fig. 3.1 Relations between five elements including concepts of Chinese medicine

(v) *Sunzi*

He was a military theorist by the end of the Spring and Autumn period. He contributed to the famous military treatise *The Art of War* written by his purported descendent Sun Bin around 400 BC, which contains 13 chapters. Although *The Art of War* is a military treatise, it also contains many concepts related to GST and dialectics. Hereunder, two paragraphs[7] are translated to show the concept of 'wholeness', 'synthesis' and 'analysis'.

Sunzi said: "War is an important event of a country, it is related to the live or die of the people, survive or perish of a country. Therefore, it is necessary to study in detail and consider prudently before the launch of war." *The Art of War* by Sunzi (p. 40).

> Therefore, comparison and estimation of five aspects of situation between the enemy and our side should be explored. The first is political condition, the second is the climate, the third is geographical environment, the fourth is quality of generals, and the fifth is the military institutional and legal system. With regard to the political condition, it means there is an enlightened government supported fully by the people, they can live and die together with the government and are not afraid to sacrifice themselves. With respect to the climate, it refers to the condition of climate, weather and season such as overcast and rainy, clear day, severe winter and hot summer, etc. The so called geographical environment refers to

[7]Note: These two paragraphs are translated by authors of this book based upon a Chinese version of *The Art of War by* Sunzi published by Zhonghua Publisher, Shanghai 1962.

> long distance, nearby, rugged terrain, plain, broad or narrow area, a place that is dead or alive, etc. Quality of generals means the character and quality of a general, clever and intelligent, prestige and authority, kind or cruel, brave or coward, serious or relaxing, etc. The military institutional system refers to the institutional system of gong and drum (the communication system of ancient war) flags, post of generals and officers, logistical supply, etc. (p. 40)

(vi) *Others*

(a) Legalists. The representative figure was Han Fei (280–233 BC), who was a student of Xunzi. He absorbed Confucianism from Xunzi, Taoism from Laozi, and Legalism since the Eastern Zhou Dynasty (770–256 BC). He summarized thoughts of three schools and deemed that the monarch should take hold of power and the liegeman should abide by the law and receive orders.
(b) Mohism. The representative figure was Mozi (around 470–399 BC). He had been a carpenter with working knowledge, therefore he was not only a philosopher, but advocated 'Love each other and benefit each other', 'Non-attack', 'Attach importance to clever people with morality', etc. He also laid the ground work for empirical method. The School of Mohism compiled Mozi, which originally contained 71 chapters, from which today only 53 chapters are left. Parts of these chapters recorded Chinese mechanics, and the physics of China in their early stages.

(2) *Development of Social Studies Post the Period of the Warring State*

(i) The application of theories of Legalists was one of the factors to help the First Emperor of Qin (250–210 BC) establish a unified empire in his time. In the beginning and middle period of the Western Han Dynasty (206 BC–8 AD), Taoism was appreciated by the rulers as the guiding thought. However a scholar statesman Dong Zhongshu (179–104 BC), made a proposal to Martial Emperor[8] of Han Dynasty (134 BC) that there should be a unified School of Thoughts. Hundred Schools should be ousted while Confucianism should be kept dominant. The Emperor adopted his proposal. Thereafter, Confucianism and all their social thoughts became a dominant doctrine in China up to the end of Qing Dynasty, lasting around 2000 years. Thoughts of Confucianism became an important part of Chinese culture.
(ii) Taoism still played an important part in Chinese history in spite of the dominance of Confucianism. There was a series of famous philosophers of Taoism in the Han Dynasty, such as Wang Chong (27–47 AD), and Wang Fu (78–163). Therefore, Taoism kept its influence and even became one of the religions of China, the Taoist religion. In the Wei, Jin and Six Dynasties (220–589 AD), scholar Wang Bi (226–249 AD), who was a philosopher and

[8]Note: Martial Emperor or Han Wudi is the seventh emperor of the Han Dynasty (156–87 BC). China had become strong and prosperous under his rule during this period, and the term Han Chinese was derived.

official and a representative figure and founder of metaphysics in Wei Jin Dynasty, and Guo Xiang (252–312 AD), a philosopher and senior official of the government (teacher of the emperor), both advocated that there was convergence of thoughts between Taoism and Confucianism. In the Sui (581–618 AD) and Tang (618–907 AD) Dynasties, with the preference of the rulers, Taoist thought and religion flourished. Xuan Zong, the emperor of Tang Dynasty (685–762 AD) had annotated and expounded Dao De Jing by himself, and even took a series of measures for raising the social position of Taoist thought and religion. This situation also lasted to the Song Dynasty (960–1279 AD) due to the Emperor's preference of Taoist religion, as well as the theory of Taoism.

(iii) To summarize the above, both Confucianism and Taoism became dominant social thought in China, lasting around 2000 years, but the former is even more dominant, since all officials in the past dynasties were required to pass the imperial examination of reading the Four Books and Five Classics.

(3) *Discussion of Religions of China*

Although there are various religions in China, there was no system of unification of the State and religion in Chinese history. Religion has not played any dominant role in Chinese politics in past history. There are several reasons for this, because only the Taoist religion was home grown. Although Buddhism is relatively popular among the Chinese people, according to Fu (1999) "Buddhism was introduced into China between the Han Dynasties. It attached itself to Yellow-Emperor-and-Laozi School, then to the Metaphysical School. Even after Kumarajiva[9] tried his best in his translation to give a true picture of Indian Buddhism, in fact, there still existed many similarities between Buddhism and Lao-Zhuang and the Metaphysical School" (p. 57).

From the perspective of authors of this book, due to the domination of Confucianism, the scholars were not fetish to Heaven or to Buddha, for example, Xunzi advocated "regulating what Heaven has mandated and using it" in his paper "Discourse on Nature". This means man can give full play to exploit nature in following the natural law. Han Yu (768–824 AD), an eminent litterateur, philosopher and official of the Tang Dynasty was demoted due to his proposal to the Emperor "stopping to greet the bone of Buddha."

The rural people even utilized the name of folk religion to rebel the government. There were no shortages of such events in Chinese history.

In short, the influence of religion to politics or to people's daily lives was relatively minor in China compared to Western countries, the Middle East and other parts of Asian countries.

[9]Note: Kumarajiva (334–413 AD) was a Buddhist monk, scholar, and translator from the Kingdom of Kucha. He had been at Chang'an and translated many Buddhist texts written in Sanskrit to Chinese.

3.1.2.2 Development of Social Studies in Western Countries

(1) *Social Studies in Hellenistic period*

It is necessary to recognize that modern social science developed more thoroughly in Western countries, even though the civilization and early thoughts date back to the origin of ancient Greece. Therefore, a brief discussion of social studies around the Hellenist period will be given.

(i) *Sophism*

According to Stevenson (1998), Athens was a city state, democracy was gaining momentum, and every free adult male was expected to participate in government. The democratic system involved more participation and responsibility on the part of its citizens. Debate and determining courses of action, were frequently part of assemblies, consisting of large groups of the populace of the city. Many Greek citizens were expected to give great attention to society and politics, and Sophism emerged.

Sophism is a specific method of teaching both in Ancient Greece and in the Roman Empire. Major figures included Protagoras, and Thrasy machus (pp. 6, 38–39).

(a) Protagoras (around 490–420 BC). He said: "man is the measure of all things", which was interpreted by Plato to mean that there is no absolute truth, but that which individuals deem to be the truth. Protagoras also said "we can't be sure about whether the gods exist". Therefore, he believed people should do what they think is best for them, without looking to a higher power. This is considered to be a type of relativism taught by the Sophists. (pp. 39–41)
(b) Thrasy machus (around 459–400 BC). He was also a Sophist, but believed that social order is imposed on everyone else by the most power. He said "Justice is nothing else but the advantage of the stronger". His view was disturbing. People who are in power to make the rules, tend to maintain their own advantages, making it difficult for those at a disadvantage to do anything about it. (pp. 40–41)

Sophism had a practice of charging money for education, and providing wisdom only to those who could pay. This led to condemnations by Socrates, who, with his followers, started an important school of philosophy in the Hellenistic period.

(ii) *Socrates, Plato and Aristotle*

(a) Socrates (around 470–399 BC). He is credited as one of the founders of Western philosophy. His most important contribution to Western thought is his dialectic method of inquiry, known as the Socratic Method. To solve a problem, it is dissected into a series of questions, from which the answers gradually distill the answer a person is seeking. This method was applied by Socrates to teach his students. He just raised a series of questions, not only to draw individual answers, but also to encourage insight into the issues at hand.

The students obtain final answers by themselves through answering questions. The influence of this approach is most apparent in the use of scientific method today. Socrates also provided an example of ethics, by sacrificing himself to keep truth (pp. 4, 20–23).

(b) Plato (428–347 BC). He was the founder of the Academy in Athens, the first institution of higher learning in the Western world. Along with his teacher, Socrates and his student, Aristotle, Plato laid the very foundations of Western philosophy and science. He was the innovator of the written dialogue and dialectic forms in philosophy. He seems to be the founder of Western political philosophy, with his Republic, and Laws, only his idea of an ideal state of government will be described below for illustrative purpose.

Plato asserts that societies have a tripartite class structure corresponding to the appetite/spirit/reason structure of individual soul. The components of this structure are: Productive (works), the laborers, carpenters, merchants, farmers etc., these correspond to the appetite of the soul; Protective (Warriors or Guardians), those who are strong and brave in the armed forces, these correspond to the "spirit" part of the soul; Governing (Rulers or Philosopher Kings), those who are intelligent, self- controlled, well suited to make decisions for the community.

Plato also described his ideal state where property is owned by all, and labor is specialized (pp. 4, 49–53).

(c) Aristotle (384–322 BC). He was a Greek philosopher and scientist. He joined Plato's Academy in Athens around for twenty years. His writings cover many subjects, including physics, biology, metaphysics, logic, ethics, poetry, music, linguistics, politics and government, These constitute the first comprehensive system of Western philosophy. He had been tutor of Alexander the Great, after the death of Plato. His view on physical science profoundly shaped medieval scholarship. He was respected by medieval Muslim intellectuals and revered as "The First Teacher."

According to Stevenson (1998), Aristotle's philosophy has two particular aspects. One aspect is his view that everything has its own purpose, which is part of a larger purpose. His philosophy consists largely of trying to figure out which these purposes are, and how they fit together. The second aspect is his logic, the way he looks at words and gets them to do what he wants to accomplish.

The economic perspective of Aristotle was in favor of private property, but against accumulating money for its own sake (pp. 59–62).

(iii) *Social studies in the Medieval Period*

(a) The progress of social studies was slow in the medieval world, due to dominance of scholasticism. The emergence of scholasticism and its continued existence had a complex historical background. The first scholastic philosopher was Boethius (480–524 AD), who translated and preserved those writings of Aristotle writings that dealt with logic. Scholastic philosophers became even more prominent in the 10th century from a number of Islamic philosophers, whore discovered important lost writings by Aristotle, and spread his ideas to Christian and Jewish philosophers. Scholasticis reached its peak around the 13th century, due to heavy influence from the Church. Although scholasticism kept the continuity of Greek thinkers, religion was considered the superior interest, and prevailed in the priority in a clash of interests.

(b) Emergence of primitive sociology based upon Atkinson (2015). Ibn Khaldun (1332–1406) was a North African Arab historiographer and historian. He is held to be a forerunner of the discipline of sociology for his book, the *Muqaddimah* ("Introduction"). He conceived a theory of social conflict, and raised the concept of 'asabiyyah', which has been translated into social cohesion (or solidarity). According to Ibn Khaldun, asabiyyah exists in societies as small as clans and as large as empires, but the sense of a shared purpose and destiny wanes as a society grows and ages, and the civilization weakens. Ultimately, such a civilization will be taken over by a smaller younger one with a stronger sense of solidarity. A nation may experience-but will never be brought down by-a physical defeat but when it "becomes the victim of a psychological defeat....that marks the end of a nation". (p. 18)

(iv) *Development of Social Studies in the Period of Scientific Revolution (around 17th–18th century)*

Although the development of social studies lagged behind natural science during the period of Scientific Revolution, the diffusion of thoughts of social studies was nearly the same as the natural science around the 18th century. Due to the rapid growth of international trade, Western people had more information and understanding of people of the non-Western world. The Western intellectuals generally no longer accepted the 'ethnocentrism'. Also people in social studies recognized the fact that the cultural behavior or social behavior of the human beings is affected by history, and not affected by the basis of biology.

There was an emergence of several schools of economics. Thomas Mun (1571–1641), an English writer, advocated a mercantilist policy around 1630, using foreign exports to increase a nation's wealth. Francǫis Quesnay, the physiocrat, argued that land and agriculture are the only sources of economic prosperity.

Economics was becoming more scientific in early 1682 through the work of William Petty (1623–1687). He was an English economist, scientist and philosopher. He was highly influenced by Francis Bacon and believed that mathematics and the senses must be the basis of all rational sciences. He showed how the economy can be measured in *Quantulumcunque Concerning Money,* and Quesnay produced the "Tableau economique" (Economic Table) in 1785 which provided the foundations of Physiocrats. And this was also the work in an analytical way to macroeocnomics.

The first modern work of economics, *An Inquiry into the Nature and Cause of the Wealth of Nations* was contributed by Adam Smith (1723–1790) in 1776.

Therefore, social studies was no longer an accessory part of biology, it was separated from biology and became an independent discipline.

Social studies in 18th century was also effected by two theoretical developments. One development was the structural concept from English philosopher and political theorist Thomas Hobbes (1588–1679), who considered society, and human beings who are its simplest elements, a machine. To understand how society works, one must take it apart in imagination, resolve it into its simplest elements, and then recompose it to a healthy functioning part according to the laws of motion of these components. Social studies related to national political structure emerged from French social philosopher Jean-Jacques Rousseau (1712–1778). The theoretical impact from the philosophy of developmentalism took shape, and became the social evolutionism in the 19th century. It can be found in writings of Rousseau and Adam Smith who both recognized: the present is formed through development of the past, it is the result of development in long term of the dimension of time, and the line of development has not been caused by God or fortuitous factors, it is resulted by inherent conditions and causes of the human society. This might be called social speciation-that is, the emergence of one institution from another in time, and a differentiation of function and structure that goes with this emergence.

(v) *Themes of Social Thoughts and Development of Social Science in the 19th Century*

The 19th century was a unique period, there is famous opening sentence in the Dickens classic, *A Tale of Two Cities,* that "It was the best of time, it was the worst of times, it was the age of wisdom, it was the age of foolishness, it was the epoch of belief, it was the epoch of incredulity…". The basic concepts, themes and issues of social science were that the European society was impacted greatly by the twin blows of French Revolution and Industrial Revolution. The old order breakup, an order that had been based upon kinship, land ownership, social class, religious authority and monarchy was discarded. The complex combination of status, authority and wealth that had long been consolidated, was also discarded. In fact, the history of politics, industry and trade in the 19th century was the history of recombination and consolidation of all these factors, but the social thoughts of the 19th century gave these a new social reality and new meaning.

The influence of these two revolutions on the thoughts and value systems of the human beings was so massive and widespread that it cannot be compared to any other period in human history. Changes in politics, society and culture which came from France and England near the end of the 18th century, spread almost immediately through Europe and America in the 19th century, and then on to Asia, Africa and Oceania in the 20th century. There were two great effects of these two revolutions, one was the overwhelmingly democratic wave, the other was the formation of the industrial capitalist. Institutions established for centuries, even millennia, together with their systems of authority, status, belief and community were toppled.

There were seven major social issues that emerged in the 19th century beginning with the French Revolution and the Industrial Revolution in England.

(a) Population

Social issues caused by the rapid growth of the population. The European population increased from 140,000,000 in 1750 to 266,000,000 in 1850, while the global population increased from 728,000,000 in 1750 to more than 1,000,000,000 in 1850. This explosive growth in population had serious impact to economy, society and the government, and this concern formed the basis of Malthus's theory of population.

(b) Living and working condition of the laborers

Although there were improvements in living and working conditions for urban laborers, beyond those for laborers in the rural areas in the beginning of 19th century, slums formed in the new centers of industry, and low wages paid to of the many laborers migrating from the rural areas to new industrial towns or cities were a frequent theme in the social thought of the 19th century. Scholars, from David Ricardo to Karl Marx, could see little possibility for the condition of labor improving under capitalism.

(c) Transformation of property

Throughout the prior history of mankind, the ownership of property encompassed 'hard' or tangible property, such as land and money, or other visible property such as precious stones, etc. However, intangible kind of property emerged, such as shares of stocks, bonds etc., which assumed greater influence in the economy, led to the dominance of financial interests, and to a widening disparity between the propertized and the masses. This transformation of the character of property made it easier for the concentration of property in the hands of a relative few, and might even have created the economic domination of politics and culture.

(d) Urbanization

The rapid process of urbanization in Western Europe changed the features of the cities and towns. In earlier centuries of Europe, the cities had been regarded as a setting of civilization, culture and freedom of mind. But scholars found the emergence of other side of cities in the 19th century: broken families, the atomization of human relationship, alienation and disrupted values.

(e) Technology

The process of the production of mechanization in the factories, and into the agricultural sector. Social thinkers began to awaken to the changes in the relationship between man and nature, between man and man, and even between man and God. Karl Marx and other social thinkers such as Thomas Carlyle, Alexis de Tocqueville were of the view that technology seemed to lead to the dehumanization of the worker and to the exercise of a new kind of tyranny over human life.

(f) The factory system

Industrialization and mechanization promoted the rapid process of urbanization. This system of work, whereby masses of workers left home and family to work long hours in the factories, also became a major theme of social concern in the 19th century.

(g) Development of political masses

The French Revolution provided both positive and negative lessons to be learned, and a democratic political system was established in America. This new system also offered much to be studied. Democracy in America (1835–40), was written by Alexis de Tocqueville (1805–59), a French political scientist, who was widely considered to be one of the first comparative political and historical sociologists. His analysis of the political experience of America in 1830s was undertaken in the belief that lessons could be learned that would be applicable to Europe. This is one of the major themes of social study in the 19th century.

(vi) *Establishment of Sociology in the 19th Century*

Modern society was the product of the Age of Reason, the application of rational thought and scientific discoveries. Table 3.1 shows some of the major contributors to the formation of the discipline of sociology.

Table 3.1 Formation of sociology

Contributor	Year	Events
Adam Ferguson	1767	*Essay on the History of Civil Society* Explains the importance of civic spirit to counteract the destructive influence of capitalism in society
Henride Saint-Simon	1813	*Essay on the Science of Man* Proposes a science of society
Auguste Comte	1830–1842	*Course in Positive Philosophy* (6 volumes) details the evolution of sociology as a science
Karl Marx and Friedrich Engels	1848	*The Communist Manifesto* Predicts social change to be a result of proletarian revolution
Karl Marx	1867	*Das Kapital* Comprehensive analysis of capitalism
Herbert Spencer	1874–1885	*System of Synthetic Philosophy* (multi volume) Argues that societies evolve like life forms, only the strongest survive
Emile Durkheim	1893	*The Division of Labor in Society* Describes the organic solidarity of interdependent individuals
Emile Durkheim	1895	*The Rules of Sociological Method* Founds the first European department of sociology at the University of Bordeaux

Note Table 3.1 is derived from pp. 18–19 of Atkinson's (2015) *The Sociology Book*

Auguste Comte (1798–1857) was a French philosopher, a founder of the discipline of sociology and of the doctrine of positivism. His contribution of *Course in Positive Philosophy* and the three stages of progress in human understanding of the world has been described. Hereunder, his framework for the new science of sociology, based upon existing "hard" science will also be described. This framework is a basic concept of social systems engineering. Figure 3.2 shows a framework of hierarchy of sciences proposed by Comte. It is arranged logically, so that each science contributes to those following it, but not to those preceding it (Fig. 3.2a). The beginning of this hierarchy is mathematics, while the apex is sociology. From Comte's perspective, it was necessary to have a thorough grasp of the other sciences and methods before attempting to apply these to society. Figure 3.2b shows the order of complexity of these sciences.

Augusts Comte developed a comprehensive approach to the study of society based upon scientific principles, which he initially called "social physics", but later described as sociology. He also divided sociology into two broad fields of study: "social statics", the forces that hold societies together; and "social dynamics", the forces that determine social change. It is clear that he applied the concept of "statics" and "dynamics" in mechanics of physics to study the social phenomena.

Herbert Spencer (1820–1903) He was a British social theorist. He was highly influential in the 19th century due to his contribution to the study of social change from evolutionary perspective. He interpreted society as a living, growing organism which as it became more complex, must self-consciously understand and control the mechanism of its own success. The most important of those mechanisms was the intense competition for resources, which Spencer labeled "the survival for the fittest" (this concept led Darwin's natural selection by several years). He believed that the unrestricted application of this principle would eventually lead to the best possible society. Therefore his work was labeled 'social Darwinism'. He made some valuable contributions to early sociology and influenced structural functionalism, he and Auguste Comte nearly have the same concepts in the development of sociology.

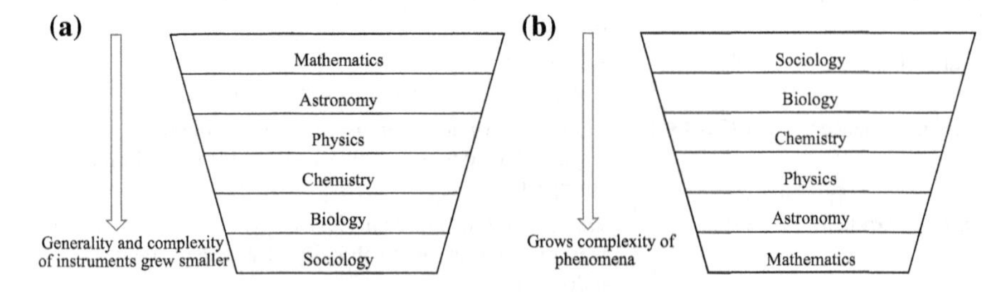

Fig. 3.2 Order of complexity of various disciplines of science

3.2 Development of Social Science in 20th Century and Emergence of the Concept of a Social System

3.2.1 Development of Social Science in 20th Century

Development of social science in the 20th century is affected mainly by the impact of the Democratic and Industrial Revolution of Western society, these two revolutions have sustainable influence both in depth and in extent. These two revolutions extended their influences to non-Western world in extent; and they impacted the traditional morality and culture in depth. Along with the emergence of Western nationalism, capitalism and industrialization, rapid development of science and technology, weakening of religious belief and growth of individualism, heavily impacted the ancient tribe, and also the small economy and religious concepts. The 20th century is also a period of the turbulence of mankind in history. It is a period of two World Wars with massive destruction and the loss of human life and assets. It is also a period of the emergence of Fascism and the Nazis, although both were ultimately destroyed by the people. It is also a period of the emergence of communism in the former Soviet Union, and which became wide spread to parts of Europe and East Asia. It is a period in which colonialism was destroyed, and a growth of large numbers of newly emerging countries. It is also a period of the eruption of global financial crisis on two occasions, caused basically by a depressed global economy. It is also a period of the emergence of socialism in China and a Socialist Market Economy in the EU. The rise of Islamic doctrine reflected the force and impact of religion, while the appearance of terrorism reflected the emergence of thought of extremism. In the later part of 20th century, there was a growth of both the process of globalization and regional integration. In short, various turbulences occurred in politics, economy and society that show that the development of social science is facing huge challenges, but it also shows clearly the fact that the theoretical development and the practice of the application of social science offers tremendous opportunities for further exploration.

A part of social science, once considered in the past that society can improve all issues related to moral, psychological and other issues post the solution of production and organization. However, the reality is far from that. When basic needs, demands on clothing, food, housing of human beings are satisfied, higher and more meaningful demands that can improve people's lives will be raised to a dominant position.

The other important event post WWII, is the establishment of the supra-national institution, the United Nations, which has an Economic and Social Council within its organization, concerned with global economic and social issues, and which provides guidance to their appropriate development. UN declared its "First Decade of Development" in 1961. It prepared a document launch a "Third Decade of Development" in 1979. The contents of that document covered aspects of agriculture, energy and raw material, industry, development and environment, technology, social goal and distribution and specific theme related to low income countries. Several World Summits were organized by the UN, with the theme of sustainable development in 1992, social development in 1995, millennium development goals in 2000, and Transforming our world: the 2030 Agenda for Sustainable Development. These activities have coordinated the consensus for social development of government, business groups and academic field of global society to a certain extent.

3.2.2 *Development of Social Science Since the 20th Century*

3.2.2.1 Three Major Thoughts Effect the Development of Social Science in the 20th Century

(1) *Impact of Marxism and Others*

Marxism had a worldwide influence in the 20th century. His ideas inspired numerous revolutionaries, in the former Soviet Union, in China, in part of Latin America, etc. According to Atkinson (2015): "Marx's analysis of how capitalism had created socioeconomic classes in the industrial world was based on more than mere theorizing, and as such was one of the first "scientific" studies of society, offering a comprehensive economic, political, and social explanation of modern society." (p. 31) Marx saw sociology as a way of understanding society in order to bring about social change. He said famously, "The philosophers have only interpreted the world, in various ways. The point, however, is to change it."

One of Marx's ideas which influenced the 20th century is his concept of directed planning of a society to replace the blind forces of competition and struggle among economic elements. This impressed the English economist John Maynard Keynes (1883–1946) to create Keynesian economics.

Charles Wright Mills (1916–62), an American sociologist, argued that sociologists should suggest means of society in his book *The Sociological Imagination* (1959), and Michel Foucault (1926–84) who was a French post-structuralism philosopher, professor of 'the history of systems of thought', began his study of the nature of power in society in his book *Discipline and Punish: The Birth of Prison* (1975). Both examined the nature of power in society and its effects on the individual-the way in which society shapes our lives, rather than the way we shape society.

It can be seen from the perspectives of the above social scientists that the study of sociology, or more broadly the social science is more than simply the quest to analyze and describe social structure and system, the major goal is to improve the social systems that exist.

This is also the goal of social systems engineering in exploration of this theme.

(2) *Influences of Freud*

Person is the basic unit of society. Social groups are formed from various persons. It is hardly imaginable that one can understand social science, and apply it well without general understanding of the areas of personality, mind and character of the individual, the writings of Sigmund Freud (1856–1939) had important influence on 20th century culture and thought. He was one of the most important figures in the development of psychology.

(3) *Further Specialization and Cross-Disciplinary Approaches were Developed in Parallel*

There was significant development of social science in the 20th century. The number of people engaged in teaching and studying of social science increased tremendously in the world. During the stage of transition from the 20th century into the new millennium, there has been an increase in further specialization. The new branches of social science in universities and post graduate studies number more than one hundred.

In the meantime, there has been a development of cross-fertilization and interdisciplinary cooperation in parallel with the development of specialization of social science. This trend happened post WWII. There was a new course developed cooperatively by historians and sociologists, who proceeded to research in cooperation. There was no such phenomenon happening before 1945. There has been an emergence of many new areas of work as political sociology, economic anthropology, psychology of voting, and industrial sociology, etc. Several specific terms, "structure", "function", "motivation", and "alienation" are used commonly in many branches of social science. Those working in social science have left the ivory tower of academics, and large numbers of economists, sociologists, political scientists and demographic experts now work in government and firms. Policy science and policy analysis have become an important specific area for national development, while the basis of policy science is systems science and behavior science.

3.2.2.2 Development of New Theory and Method

(1) *Mathematics and Quantitative Methods*

Mathematics and quantitative methods have been applied universally, especially in the later part of 20th century. In the area of economics, econometrics, Input Output technique and Computable General Equilibrium model have been commonly used. Economics has nearly become a branch of mathematics. Quantitative method has also been applied in sociology, political science, social psychology and demographical studies. Statistics was an independent discipline in the 19th century, but it became an integrated part of nearly all branches of social science. Economists and statisticians played a dominant role in the planning systems of many countries in the 20th century. The huge capacity of storage and computational capacity of modern computers, has assisted greatly with the study of human behavior by social science. This can be illustrated by the Program of Engineering Social Systems of Harvard University.

(2) *Developmentalism*

The theory of social evolution was once dominant in Western Europe in the 19th century. Scholars of social science had once considered that the evolutionary theory of society could be applied to mankind in various regions, and follow the same stages of development. However, defects with this development of a single pattern were proven in many actual cases. People in different regions have different cultures. This is proven by the theory of "same end with different paths" of GST. Therefore, this trend of evolutional theory of society stopped in the beginning of 20th century.

There was a re-emergence of the concept of development through a derivative of evolutionary theory by study of nearly all social science in the later part 1950s. There are many emerging economies and new civilizations. Due to the dominance of the study of social science by the West, Western scholars of social science, especially the economists have many studies on development economics. It has become nearly a new branch of economics. Economists work at the international level, such as those working for the OECD and the World Bank, Asian Development Bank etc. OECD works on development issues, both for developed and developing countries. The latter two, especially the World Bank, have many projects related to assistance of developing countries. They produce country reports and suggestions for issues of development for a large part of developing countries, including transitional economies. The economist shad focused their studies on the development of Latin America in the mid of 20th century. However, by the later part of 20th century, due to the emergence of Asia, some international organizations began to focus on the Asian experience of development. For example, there are publications such as *The East Asia Miracle: Economic Growth and Public Policy* by the World Bank in 1993 and *Emerging Asia: Change and Challenges* by Asian Development Bank in 1997 and others, but all of these studies of development focus more from economic perspective, and the major researchers are Western scholars, there are more or less some gaps in the understanding of Asian society and culture.

(3) *Emergence of Social Systems Theory and Method*

One of the major theoretical trends of study of various branches of social science post WWII is the awareness of social system. There is a system feature of mutual interdependence of individual action and group action. There is also the application of a biological model in the study of social system. It seems that the concept of social system in the social science post WWII may be effected mainly by the new discipline of Systems theory, cybernetics, and some social scientists tried to apply that theory to effect the function and behavior of mankind. There are also many concepts from electrical and mechanical engineering which have been integrated into the study of social system, and the authors of this book consider that emergence of theory of social system, a very important theory of social science in the 20th century. The creator, Talcott Parsons was affected by systems science of the natural science with his background of structuralism and functionalism. Social systems theory should cover more than the area of cybernetics.

The study of social system includes mainly the actions of individuals, social groups and nations and their effects to society. The mutual interdependencies between them are considered sufficiently unified to warrant use of the word "system". Although the study of social system was first initiated by Talcott Parsons, and later by Jay Wright Forrester and Niklas Luhmann, this is debated among sociologists. A more detailed description of Parsons' thoughts will be given in 3.4 of this book. In fact, institutions played a very important role in social systems engineering. They have many definitions, according to Rhodes (2005) "Institutions there can be interpreted as reflecting habits and norms, more likely to be evolved than to be created. But institutions also may be seen as architecture and as rules that determine opportunities and incentives for behavior, inclusion and exclusion of potential players, and structuring the relative ease or difficulty of inducing change, and the mechanisms through which changes may be facilitated or denied" (p. xiii).

(4) *Structuralism and Functionalism*

Structuralism in its simplest definition is that we must understand things, such as phenomena and their structure. Structure consists of many elements. The relationships between these elements should be carefully analyzed. Furthermore, structures can be classified into surface structure and internal structure. It is the internal structure, the mode of connection forming the phenomena. It requires an in depth analysis to understand the details. While the surface structure shows the connection of visible external parts, it can be easily understood through perception by sense.

Functionalism is a theory of contemporary philosophy, its core idea is that mental states (beliefs, desires etc.) are constituted solely by their functional role, i.e., they have causal relations to other mental states, numerous sensory inputs and behavioral outputs. It is a theoretical level between the physical implementation and behavioral output. It is applied to explain the social institution through its function. Modern functionalism considers society as a system with its nature being the whole system, with structurally mutual interactive elements. Thereby, modern social systems theory is formed. This school of thought was led by Talcott Parsons and had once been dominant in the field of sociology in North America around 1950–1960s. However, this school of thought declined around the 1970–80s decade in the U.S.in the 20th century, but Parsons was still welcomed in Germany. This phenomenon has a complex background, there were imperfections in the social systems theory itself. This is not strange since mankind developed the general systems theory and systems engineering through nearly 2000 years of progress of natural science and engineering, while modern social science has a history of no more than 300 years since Adam Ferguson. Therefore, more studies are required to improve and perfect this new discipline. There may be also some ideological prejudice. For example, the U.S. takes the lead in the world in the development of systems theory and systems engineering, but only started the program of social systems in a very limited number of universities recently.

(5) *Conflict Theory*

Conflict theory is one school of thought of sociology, it focuses on the analysis of social, political or material inequality of social groups. It was created by Marx, from the perspective of theoretical system of sociology, there are three independent systems of theory, Conflict theory based upon Marx, Fuctionalism based upon Durkeim, and study of social action through interpretative means by Max Weber. It can be seen that Parsons' social systems theory is not only effected by Durkheim's functionalism, but also by Webers' influence of social action.

According to Chanbliss and Ryther (1975), there are six principles of Conflict perspective: Economic forms shape other institutional forms, usable economic strategies are limited in number; each economic form has inherent contradictions which produce classes, and class conflict, and which may produce its own destruction; Class-shaped institutions: class conflict will shape noneconomic institutions, and the more powerful classes will be better able to manipulate these institutions to serve their own interests; when the needs of the weaker classes are meet less and less through the dominant institutional patterns, force of social change arise, contradictions lead to radical transformation (pp. 67–68). According to *Contradiction Theory* of Mao (1937), (his paper had a philosophical perspective in the explanation of conflict). There were five major principles in his paper.

(i) There are two kinds of world views, one is metaphysics, which holds the view that everything is static, change is caused by external influences; the other is materialist dialectics, which holds the view that everything is dynamic, argues that external cause become operative through internal causes.
(ii) Contradiction is universal, it exists in all events and all processes, and the law of the unity of opposites, is the fundamental law of the universe.
(iii) Although contradiction is universal, it is necessary to recognize the particularity of contradiction, it is necessary to study it carefully and treat it appropriately.
(iv) There are two stages of process of recognition: the first is particularity of contradiction; then the universality of contradiction. It is precisely in the particularity of contradiction that universality of contradiction resides.
(v) There are identity and struggle of opposites of contradictories, one side of the opposites will be transformed to its opposite side under certain conditions. ([10]pp. 274–311).

[10]Note: The above content in English is abbreviated and translated by authors of this book.

3.3 Behavior of Individual and Groups, and Their Relations with the Social Environment

3.3.1 Personality of Individual and the Human Behavior

3.3.1.1 Necessity to Study Personality

The major area of application of social systems engineering is to study and improve the overall design of a social system. A social system is generally composed of individuals, families, various social groups, business units and a country with structure of a hierarchical level. The behavior of structural elements of a society and the behavioral relationship between these structural elements to perform appropriate functions are based on the personality and the behavior of individuals. Therefore, to explore the personality of individuals is meaningful and important.

3.3.1.2 Definition of Personality and Theories of its Development

(1) *Definition*

Personality is an organic combination of status of psychology and physiology specific to individual, it includes the components of heredity and those acquired in experience of living. Personality makes one distinguished from others. Personality can be shown from one's behavior and relationship with the environment (natural and social), with others and social groups.

The Chinese philosopher Mengzi (Mencius) and Xunzi in the Warring State Period had different perspectives on the nature of people, the former kept the view that the basic human nature was good, while the latter's view considered that the basic human nature was bad, therefore education and legal system are important to keep the society in order.

(2) *The Studies of the Theory of Personality in the Hellenistic Period in the Western Society*

Plato analyzed the nature of a person from two aspects, the soul (sprit) and the body (flesh). Aristotle integrated the study of psychological activity of a person with one's organism. He recognized that the person's inward activity was due to the transfer of information and knowledge by sensory organs (ears and eyes) to the heart and caused inward activities of the person. He had a major treatise *On the Soul* which discussed the kind of souls possessed by different kind of living things, distinguished by their different operations. Ancient Greeks had also given explanations to the difference of personalities from physiological perspective. The Greek doctor Galeno (AD 129–216) recognized that the sentiments and healthy of the people depended upon the balance of four types of humors in the human body.

(3) *Development of Various Schools of Psychology Post the Renaissance Period*

In the period of scientific revolution and afterward, social science and psychology were gradually entering into stage of development. English philosopher and physician John Locke (1632–1705) and David Hume (1711–1776), started to study the theory of mind with empiricism. Development of physiology in the period of scientific revolution had greatly impacted the study of psychology in 18th and 19th century. Five schools of psychology have been established since the 19th century.

(i) *Experimental Psychology or Psychology of Structuralism by Wilheim Maximilan Wundt (1832–1920)*

Wilheim Maximilan Wundt was a German doctor, psychologist, physiologist and philosopher. He recognized that psychology is a kind of science, he established it to be an independent discipline, separate from physiology and philosophy. Wundt is credited to be "Father of Experimental Psychology." He established the first formal laboratory in University of Leipzig to proceed the study of psychology. It was possible for him to explore the nature of religious belief, identify the abnormal consciousness and behavior, and discover the damaged part of the brain. He classified the consciousness of the people into sense (visual and auditory sense, etc.), emotion (good or vicious etc.) and image (concept) in the brain, and the process of self-observation of inner things, thereby leads to the analysis of consciousness. This approach is called 'The method of introspection', because it explores structural elements of consciousness. This psychology of structuralism had been dispersed to the U.S.

(ii) *Behavioral Psychology by John Watson (1878–1958)*

American psychologist John Watson observed the behavior of the rat. He held the view that although people have no way to understand the internal consciousness of the rat, it is possible to master its behavior, because behavior is the explicit expression of internal consciousness. Therefore he suggested that there was no need to observe the consciousness which cannot be mastered objectively. There was a need for personality measurement and personality test through scientific means. John Watson became the founder of the school of behavioral psychology.

Behavioral psychologists recognize that the purpose of psychology is "to predict and control the behavior." The process is to collect information and data and determine two objectives of research: "predict which kind of response will happen after stimulation" and "which kind of stimulation is effective in the production of certain responsive process."

In the definition of social system and social systems engineering given in this book, the term "behavior" plays a very important role. Therefore, "behavioral psychology" will be an important theory of micro-foundation in studying social systems engineering.

(iii) *Gestalt Psychology Founded by Max Wertheimer (1880–1943)*

Wertheimer founded his Gestalt psychology in 1925 in Germany during the period that Watson founded his behavioral psychology. Gestalt theory recognized: the perception of the people to various external stimulations is not recognized individually, it is processed to be a mutually interrelated whole with all stimulations to be recognized. There are similar features with respect to the memory of the people. Therefore, Gestalt psychology of Wertheimer is a type of psychology with the perspective of systems. He added new systems perceptions to psychology on the basis of analysis and inheritance of both Wundt's Empirical Psychology and Watson's Behavioral Psychology.

(iv) *Psychoanalytical School by Sigmund Freud (1856–1939)*

Sigmund Freud was an Austrian doctor of new biologist, known as the founding father of psychoanalysis, this term psychoanalysis was first used by him in 1896. His structure of the consciousness of the people is shown in Fig. 3.3.

Based upon Freud's view (Fig. 3.3), the consciousness of the people includes three elements, one is ego, which is a general term of psychology concerned with the conception about one's self including values and attitude. In Freud's psychology, the term refers specifically to the elements of human mind that represents the conscious process concerned with reality which is in conflict with id and the superego.

The second element Superego in Freud's psychology is the element of the human mind concerned with the ideal, responsible for ethics and self-imposed standard of behavior. It is characterized as a form of conscience, restraining the ego, and responsible for feeling of guilty when the moral code is broken.

The third element Id is the instinctual element of the human mind concerned with pleasure which demand immediate satisfaction. It is the unconscious element of the human psychology and it is in conflict with ego and superego.

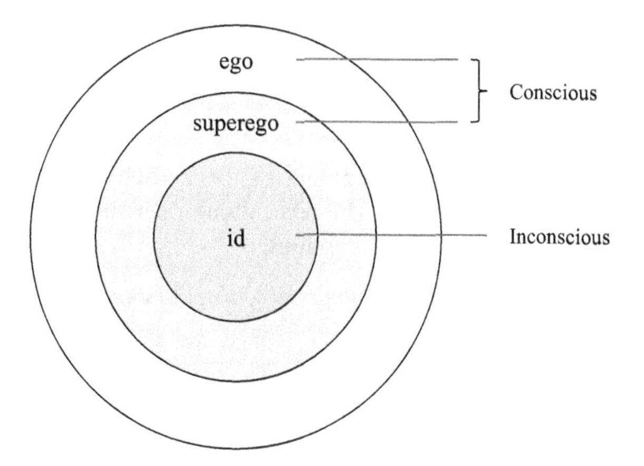

Fig. 3.3 Structure of consciousness of the people based on Freud's theory

(v)-1 *Humanistic Approach (or person centered) Psychology by Carl Rogers (1902–1987)*

He was an American psychologist and among the founders of the humanistic approach to psychology. He is also widely considered to be one of the founding fathers of psychotherapy research. He wrote 16 books. His theory was based on 19 propositions. Several selected propositions are listed below to explain partially 'person centered'.

- All individuals (organisms) exist in a continually changing world of experience, of which they are the center.
- The organism reacts to the field as it is experienced and perceived. This perceptual field is "reality" for the individual.
- A portion of the total perceptual field gradually becomes differentiated as the self.
- The result of interaction with the environment especially a result of evolutional interaction with others, the structure of the self is formed as an organized consistent conceptual pattern of perceptions of characteristics and relationships of one with others, together with values attached to these concepts.
- The organism has one basic tendency to actualize, maintain and enhance the experiencing organism.
- Behavior is basically the goal directed attempt of the organism to satisfy its needs as experienced. [Note: Needs will be explored in (v)-2]
- Any experience which is inconsistent of the organization of the structure of the self may be perceived as a threat, and the more of these perceptions there are, the more rigidly the self-structure is organized to maintain itself.

The above seven selected propositions of Rogers can provide some perceptions of formation of personality and the rigidity of the 'self' structure. And the following Maslow's theory will give the detail of changing human needs.

(v)-2 *Hierarchy of Human needs by Abraham H. Maslow (1908–1970)*

Maslow was an American psychologist, and he was also a humanistic psychologist who believed that every person has a strong desire to realize his or her full potential to reach a level of self-actualization. He established a hierarchical level of human needs which is shown in Fig. 3.4.

3.3.1.3 Two Schools of Thoughts of Personality and Their Core Concepts

The process of development of theory of personality and psychology has been described previously. The perception of personality can be classified into two types based on different assumptions.

(1) *The 'Rigidity' Assumption of the Personality*

The assumption of one type of theory does recognize that the personality of the people is rigid, unchangeable. This type of theory stresses the view that the people should recognize themselves, and obey this situation. This is so called

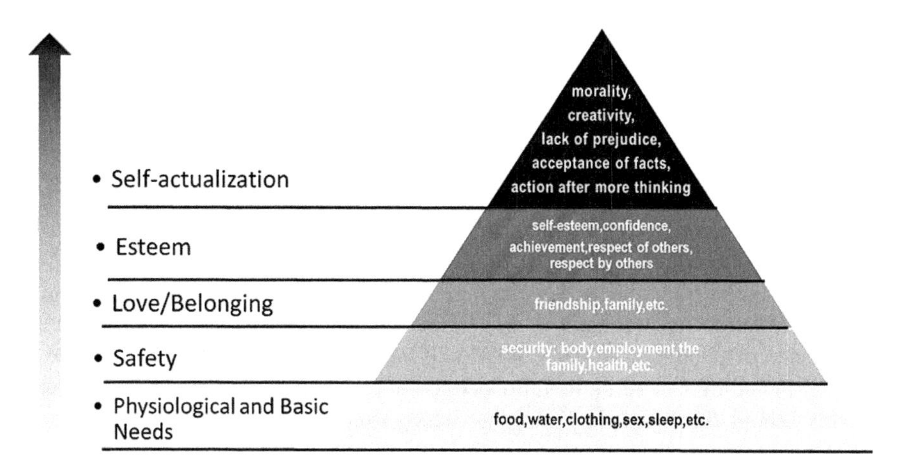

Fig. 3.4 Maslow' s hierarchical level of human needs

existentialism. This theory reflects a passive and pessimistic view to life, i.e. the source of troubles to human life is too broad to overcome, people can control, at most, a part of the trouble. While the other type of theory created by Rogers and Maslow is relatively optimistic, they consider that human nature is malleable and flexible, and can be changed and developed. Basic human nature is born good (which is similar to the view of Chinese philosopher Mencius), a bad society is the source of all troubles of the mankind, therefore, there are needs to reform society to improve the well-being of people. But the authors of this book recognize that it is possible that there are bad natures of a part of people due to inheritance and genetic reasons, but a good social environment can suppress the bad aspects.

(2) *The 'Flexibility' Assumption of the Personality*

The other type of assumption does recognize the personality is changeable with a system of five core concepts of theory of personality.

(i) *Motivation Theory*

This theory explains the reason of action of people is due to their motivation. Motivation is an important endogenous variable of theory of personality. There are also two types of theory related to the concept of motivation: one is based on the view of biology and physiology, such as the desire to achieve goals or to satisfy needs, passion, impulse etc. Physiological theory applies a few fixed motives to explain the motivation of human beings, nearly every person has similar motivation. These motivations belong to the scope of the unconscious mind, they exceed the scope to be controlled rationally, the other type of theory is based on the psychological perspective: such as value, expectation, attitude, preference and objectives etc. Psychological theory recognizes that human beings have a relatively broad individual specific motivation. These motivations are the choice of rationality, which can be modified through education or self-examination. In short, biological motivation drives the behavior, psychological motivation guides the behavior.

(ii) *Development Theory of Personality*

There are also two schools of theory related to development of personality. One is the Classic School, which recognizes that the development of personality can be divided into stages based upon growth of age. When people develop from one stage to another stage, the structure of their personality will have significant qualitative changes. Classic development theory of personality also assumes that the living experience of people at an early stage will have the most significant influence to people, including extending its influence to the personality in later stages. Scholars of this school, for example Freud, etc. recognized that the origin of neurosis or mental neurosis of adults comes from their infant period and childhood. The other is the Learning School, which recognizes that the development of personality is linear in nature. Scholars of this school, for example, the behavioral scholars, recognize: there are differences between adults and children, the former learn more methods in dealing with their external world. There is no structural change in the process of development. The more experiences acquired, the more development in breadth and in depth. Scholars of this school deny that the experience of development in an early stage is more important than that in later stages. The origin of mental neurosis of adults comes not from their childhood, it is due to the current status of the people.

(iii) *Theory of Development of Morality*

Due to the importance of development of morality with respect to the personality, and its important impact to social development, it has become an independent area of research. This area is sometimes called 'socialization'. Four aspects will be discussed for moral development. The first is the necessity of morality. Theory of psychological analysis recognizes the extreme importance of moral development with respect to the survival and development of human civilization. The process of establishment of morality is the process of transformation of an individual from a child of anti-society to become a member of social civilization. Children must suppress their inherent nature in order to survive in society. Classic sociology also recognizes that children must internalize the moral factor of culture, only then will they develop to become real people. The second is the exploration of causes in forming morality. There are two different views of exploration. One view recognizes that the moral development of people is a natural process of development of mankind from the children to develop into people, and adults should not interfere in this process. The other view held by Freud etc. recognizes that parents and society have the responsibility of training in the process of development of the morality of children. The moral quality of adults is the reflection of the quality of training. The third is the explanation of process of socialization. There are three different views.

The first view recognizes that development of personal morality closely depends upon the amount of internalization of social role and social value of that person, and also the sense of guilt of that person to destroy those rules. Generally criminals have less internalization of social rule and value, and they are also lacking the guilty feeling in destroying those rules. The second view is related to the attitude of people towards authority, those people holding a positive attitude to authority are easily in compliance with the requirements and orders of authorities, while those people with insufficient socialization are generally in opposition to authorities. The third view an explanation of socialization, recognizes that those people who can reflect positively on the needs and expectations of others are more socialized. The fourth aspect of moral development is related to personal autonomous consideration, many theoretical workers recognize that moral development depends upon the willingness of people to accept the rules and regulations set up by families and culture.

(iv) *Self Perception Theory (SPT)*

This theory was developed by psychologist Daryl Ben, it asserts that people develop their attitude by observing their own behavior and concluding what attitudes have caused it. This theory is in contradiction from conventional thinking of attitudes determining behavior. There are four different views with respect to this theory of elements that constitute the theory of personality. One view is to minimize its nature of importance; the second view recognizes that self-perception is the precondition of one's existence; the third view recognizes that self-perception is equivalent to personality; the fourth view acknowledges that self-perception is one of the elements which constitutes the personality. In fact, the reality of a person compared with one's ideal self-perception is one of the indicators in the measurement of psychological health of a person.

(v) *Process of Unconsciousness*

The process of unconsciousness has been an important concept of psychiatry and psychology, which is shown in Fig. 3.3 (the part id). There are different views and debates in the academic field. For example, the School of Existentialism recognizes that people in general lie to themselves for some unhappy facts (such as too fat, lazy etc.). These lies, including certain memories or events of what was said or done, are 'forgotten'. The contents of these lies are contained within the unconscious mind. No matter how many different theories exist, the development of the theory of personality has different views as to the importance of the theory of 'unconsciousness.' The older the theory was, the more the importance of 'unconsciousness' was emphasized. But relatively modern theories focus more on creativity, self-actualization, personal sound development, etc. except for the issue of insanity.

The above five: motivation, development of personality, development of morality, self-perception and unconscious mind constitute an overall system of personality theory.

3.3.1.4 Group Dynamics

(1) *Definition of Group Dynamics and its Application*

(i) *Definition*

Group Dynamics is a discipline which studies the system of behaviors and psychological process occurring within a social group (intragroup dynamics), or between social groups (intergroup dynamics).

(ii) *Application*

Mankind has existed in groups since the beginning history of its existence. Individual behavior is different from the behavior of social groups. This study can be applied to better understand decision-making behavior, better implement certain action plans for groups, and also follow the emergence of new ideas and technologies. It can also be applied to sociology, anthropology, political science, economy, education and business, etc.

(2) *Brief History of Development*

The development of Group Dynamics originated from the principle of system science, "The whole is greater than the sum of its parts." The social group should be studied as a whole, its nature cannot be understood by studying the nature of an individual. Study of Group Dynamics also originated from the development of psychology and sociology. Wundt, the founder of experimental psychology had a special interest in the psychology of groups. He recognized that some phenomena, human language, customs and religion could not be described through the study of individuals. The sociologist Durkheim also recognized collective phenomena, such as public knowledge. William McDougall, a psychologist believed that a "group mind" had a distinct existence born from the interaction of individuals. Gestalt psychologist, Max Wertheimer stated in 1924 "There are entities where the behavior of the whole cannot be derived from its individual elements, nor from the way these elements fit together; rather the opposite is true, the property of any of the parts are determined by the intrinsic structural laws of the whole" (Wertheimer, 1924, p. 7).

The social psychologist Kurt Lewin (1890–1947) had developed this discipline relatively systematically, and is credited as the founder of this discipline. He explained 'Group Dynamics' to be action and reaction of groups and individuals with respect to the changing environment. He defined the role of Group Dynamics as exploring the laws of development of nature of a group and their mutual relationship with individuals, other groups and large institutions. The term Group Dynamics can be understood from two sides. One is group, and group is constituted by two, or more than two persons with common belief and value, members of the group have goals and tasks in common; the other is dynamics, which means "coordinated continuous action to achieve pre-set goals." Lewin established The Group Dynamics Research Center at the Massachusetts Institute of Technology in 1945.

Wilfred Bion (1897–1989), an American psychologist, studied Group Dynamics from a psychoanalytic perspective. He established the Tavistock Institute in London, the institute has developed and further applied Bion's theory and practice. Bruce Tuckman, a professor of psychology at Ohio State University in the U.S., proposed a four stage model called Tuckman's Stages for a group. His model states that the ideal group decision making process should occur in four stages. Forming; Storming; Norming and Performing.

(3) *Types of Group Dynamics*

Study of Group Dynamics can be divided into two types: Intragroup Dynamics and Intergroup Dynamics.

(i) *Intragroup Dynamics*

The objects of study of Intragroup Dynamics are specific social groups, such as religious groups, political groups, military groups, environmental groups, work groups, etc. Scope of study includes: study of the cause of formation of a group, either through mutual attraction due to psychological effect or formation of social cohesion through attachment to social stratum, etc.; study of the specification to form the group, identification of the role of the members and their mutual relations and the set-up of common goals, etc.

(ii) *Intergroup Dynamics*

Intergroup Dynamics studies the behavioral and psychological relation between two, or more than two groups, which includes the recognition, understanding, attitude, opinion and behavior to one's group and to other groups. Intergroup Dynamics may be favorable to society in some cases, for example, organization of different research groups to accomplish a project with common goals. However Intergroup Dynamics may create conflicts in some cases, for example, it is common that there are conflicts between different countries or different social groups. There is the social identity theory in explanation of the intergroup conflict, which starts with a process of comparison between individuals in one group to those of other group. If this comparison is biased and not objective, then this process will become a mechanism of enhancing one's self esteem. Generally, the individual tends to favor its own members rather than the individuals of another group, and therefore will exaggerate and even overgeneralize the differences. Conflicts will occur in this comparison.

3.3.2 *Human Behavior and Social Environment*

All events in the history of mankind are marked by interaction between the people and the nature and also between people themselves. Correct behavior of the people will result positive effect to the progress of human society no matter how large or small it may be. Therefore, study of human behavior is important. Although several studies related to human behavior have been described in previous sections, the rapid explosion of knowledge requires a further supplementary discussion of this theme.

3.3.2.1 Limits of One Dimensional or Single View Approach

Although several schools of thoughts related to the human behavior, theory of development of personality and moral development had been described, there are relatively correct recognition of human behavior to certain extent due to progress of natural science and social science. But our knowledge to human behavior has not reached the stage of physical science, there are many laws of physics and chemistry, that the degree of accuracy can reach several places of decimals. But in the study of human behavior, there are only a few laws that can be established. For example, in the *World Development Report 2015*, with the theme of "Mind, Society and Behavior." Three principles of thinking and behavior are identified: First, people make most judgements and most choices automatically, not deliberately. The second, how people act and think often depends on what others around them do and think. The third, individuals in a given society share a common perspective on making sense of the world around them and understanding themselves. From the authors' perspective of working for policy research and gave advises to the leaders and governmental officials, for more than 30 years the above two principles can be termed to be "laws" rather than to be "principles." It can be seen from later two principles, all of them have close relations with the environment. Even the first relation has also the relations with the growth and development of the person with the environment state of interaction with other living systems and with other non-living components of the system's physical environment.

3.3.2.2 A Multidimensional Perspective of Human Behavior in the Social Environment

Ashord and Lecroy (2010) had established a multidimensional framework to study the human behavior and the social environment. They also have the perception that the social systems will guide this multidimensional framework. Only several major concepts will be briefed below:

(1) *Emergence of Multidimensional Perspective*

Due to rapid development of behavioral science, biology, genetic engineering and social science, the academic field recognized that the development of human behavior should not be observed from single dimension, a framework of multidimensional perspective should be used in the analysis. There is further recognition that the impact of biological, psychological or social factors to human behavior is not isolated, human behavior is a part of a system, which is formed by a network system of feedback from process of various causes. It must be emphasized that human beings are social and cultural animals.

According to Ashford and Lecroy (2010), there are four fundamental assumptions of framework of multidimensional perspective.

(i) There are three basic dimensions for assessing human behavior and the social environment: biophysical, psychological, and social.
(ii) These three dimensions are conceptualized as a system of biopsy chosocial function.
(iii) This system involves multiple systems that are organized in a hierarchy of levels from the smallest (cellular)to the largest (social).
(iv) This ascending hierarchy of system is in a constant state of action with other living system and with other non-living component of its physical environment (p. 17).

(2) *The Dimension of Biophysical System*

The biophysical system is an important subsystem to understand human growth and development. Genetics is the core theory in the study of biophysical system, it is the study of genes, genetic variation, and heredity in living organism. Greg or Mendel who is the father of genetics who studied "trait inheritance" patterns in the way traits are handed down from parents to offspring. He observed that organisms inherit traits by way of discrete "units of inheritance", so called "gene". The progress of studies continued from cytogenetics (around 1910–1940) to microbial genetics (around 1940–1960) and further to molecular genetics (after 1953). Scientists begin to understand that the basic function of the cell is controlled by DNA and RNA. These nucleic acids determine the human characteristics in heredity. There is a further understanding of the capacity of biophysical system of the human beings in response to changes through study of structure of brains, biochemical process in the nervous system, and growth and development of neurons.

(3) *The Dimension of Psychology*

Five schools of psychology had been described briefly in Sect. 3.3.1 previously. Some other theories will be supplemented here. In Freud's psychodynamic theory, the experience of people in the period of infant and childhood is emphasized and it is considered that those experiences will have a deterministic impact to the character of the people entering into adult time. Freud' s psychological analysis also recognized that the spiritual motive force of the people comes mainly from sex instinct. But some contemporary psychologists held different views. Several theories will be described. Erik Erikson (1902–1994), an American German development psychologist, known for his theory on psychosocial development. He believed that personality evolves across the life span. The sequence of development stages include tasks from which the results from both biological force and age related social or cultural expectations. He had classified eight psychological stages based on ages: the infancy, the early child hold, play age, school age, adolescence, young adulthood, maturity and old age. There are psychological conflicts in each stage.

In the stage of infancy, there are conflict of trust vs. mistrust. According to his view, people must grapple with the conflicts of one stage before they can move on to a higher one. Chinese philosopher Confucius had a more or less similar view in age related social or cultural expectations. In *The Analects,* Book II (1999) The Master said, "At fifteen, I set my heart upon learning. At thirty, I had planted my feet firm upon the ground. At forty, I no longer suffered from perplexities. At fifty, I knew what were the biddings of Heaven. At sixty, I heard them with docile ear. At seventy, I could follow the dictates of my own heart; for what I desired no longer overstepped the boundaries of right." (p. 11)

It had been described previously that John B. Watson is recognized as the father of *behaviorism.* He believed firmly that the development of personality and behavior depended upon learning. Burrhus Frederic Skinner built on Watson's theory and experiment studied a form of learning occurred under controlled conditions. He observed that behavior is repeated when followed by positive consequence and that it will not be repeated when followed by negative or neutral consequences.

Skinner recognized that the people change their behavior through learning due to passive impact by the environment. Later on, Albert Bandures (1925-), a Canadian/American psychologist and professor of social science in Psychology at Stanford University, was influential in the transition between behaviorism and cognitive psychology. He is credited to be the originator of social learning theory. He recognized that people can learn to develop themselves by observing the pattern of development of others. People can process the information, actively effect the environment which controls them rather than to be controlled by the environment surrounding them. The people should have the core element "Self-efficacy", that is, the adults shall believe their capability to achieve the expected goal of their behavior of action.

(4) *The Social Dimension*

Social systems theory is generally applied in the social dimension of multi-dimensional perspective.

General System Theory has been briefed in relative detail in Chapter 2. It had been accepted by social scientists and they recognized. According to Ashford and Le Croy (2010) that a social system is a system in which the components are people. These people participate in a number of social systems that influence their development. Brim (1975) and Bronfenbrenner (1979) had identified four levels of social systems: Microsystems, meso-systems, exo-systems and macro-systems. Microsystems involve face to face or direct contact among system participants; meso-systems are the network of personal settings in which the people spend their social lives; exo-systems in Bronfenbrenners's scheme are the large institutions of society that influence the personal system; and macro-systems are the larger sub-cultural and cultural contexts in which the micro-systems, meso-systems, and exo-systems are located (pp. 134–135). All these concepts are useful to study in impact of external environment to the behavior of individual or group in addition to various theories of development of personality as well as psychologies.

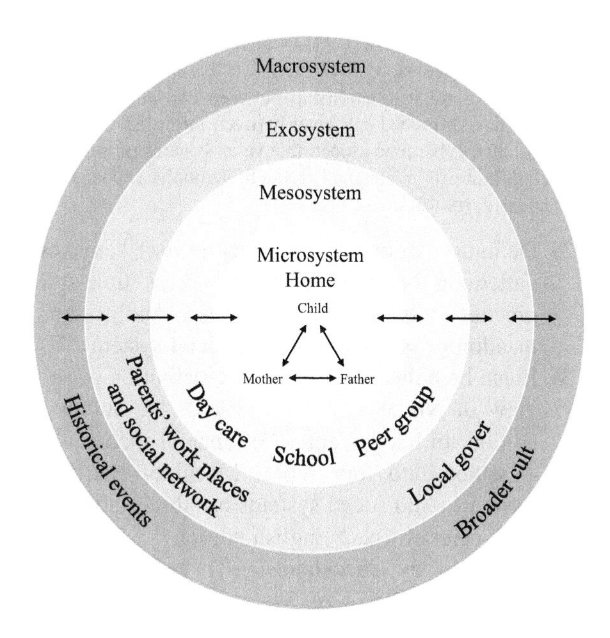

Fig. 3.5 Bronfenbrenner's ecological model *Source* p. 135 of Human Behavior in the Social Environment

Bronfenbrenner had established an ecological model to describe the environmental contexts of human development, which is shown in Fig. 3.5.

3.4 Social System and Parsons' AGIL Framework of System of Action

3.4.1 Discussion of Definition of Social Systems and Theory of Its Development of Social Systems Theory

3.4.1.1 Definition of Social System

Although the definition of "Social Systems" had been described in the very beginning of Chap. 1 of this book, it is better to have some further discussions on its definitions in order to have an in depth understanding. We shall compare several definitions from different sources.

(1) Definition of Social Systems from Marshall, G (1994) in his *Dictionary of Sociology* as follows:

> The concept of system appears throughout the social and natural sciences and has generated a body of literature of its own (general systems theory). A system is any pattern of relationship between elements, and is regarded as having emergent properties of its own, over and above the properties of its elements. The system is seen as possessing and inherent tendency towards equilibrium and analysis of systems is the analysis of the mechanism which maintain equilibrium both internally and externally, in relation to other systems.

The functionalism of Talcott Parsons offers the fullest employment of systems theory in sociology (see especially *The Social System* 1951). In Parsonsian terms social system can refer to a stable relationship between two actors, to society as a whole, to systems of societies, or indeed any level between these. All are analyzed principally in terms of their so-called cybernetic aspect: that is, as systems of information exchange and control, where equilibrium is maintained through, symbolic exchange with other systems across boundaries…" (p. 494).

(2) Definition from Jary and Jary (1991) "any, especially a relatively persistent patterning of social relations across time-space, understood as reproduced practice" (Giddens, 2001). Thus in the general sense, a society, or any organization or group constitutes social system…" (p. 598).
(3) It can be pointed out the above definition of social system is different from the view of Niklas Luhmann (1984), the author of 'Social System' in German edition. In his perception of general theory of social systems, systems can be classified into four types: Machines; Organisms; Social Systems; Psychic Systems. And social systems has three subsystems; Interactions; Organizations and societies. (p. 2 English edition).
(4) Definition by Checkland (1981), who is a British management scientist and emeritus professor of Systems at Lancaster University. He presents a comprehensive characteristic of social system, which calls human activity system. He writes in his book (1981).

Systems Thinking, Systems Practice, and states as follows:

Natural and engineered systems cannot be other than what they are Human activity systems, on the other hand, are manifested through the perceptions of human beings who are free to attribute a variety of meanings to what they perceive. There will never be a single (testable) account of human activity systems, only a set of possible accounts, all valid according to particular Weltans chauunge. (p. 14)

Given man and his abilities, we have the huge range of activities systems, from the one-man-with-a-hammer at one extreme to the international political systems (p. 21).

It can be seen from above quotations that there are different views on the definition of 'social system'. Even there are different views between the sociologists because there are only two available publications related to "Social Systems" by sociologists. Therefore a brief discussion of development of social systems theory will result a better understanding of 'Social System.'

3.4.1.2 Development of Social Systems Theory and Brief Introduction of Two Publications on "Social System"

(1) *Talcott Parsons* (1951) *The Social System*

Parsons' The Social System is the first publication to study the social system based upon the concept of 'systems'. He is influenced by Professor Henderson' Pareto's

General Sociology, to notice the extreme importance of the concept of system theory. His first publication is The Structure of Social Action in 1937 which may be related to the volume Toward a General Theory of Action by members of the Harvard University Department of Social Relations and their collaborators.

Therefore, in his book The Social System in 1951, it is the exposition and illustration of a conceptual scheme for the analysis of social systems in terms of the action frame of reference. The starting point of his analysis is system of social action which includes three aspects: a social system; the personality systems of the individual actors and the cultural system which is built into their action.

He also described in his book the units of social systems is the act, (p. 24) and a social system is a system of processes of interaction between actors, it is the structure of the relations between the actors as involved in the interactive process which is essentially the structure of the social system. The system is a network of such relationships (p. 25).

There are also functional prerequisites for a successful system of actions: the biological prerequisite; adequate motivation and adequate cultural resources and organization for the maintenance of the social system. In the aspect of adequate motivation, Parsons has divided further into two aspects: a negative and positive. The negative is that of a minimum of control over disruptive behavior. This means actions which interferes with the action of others in their roles in the social system.

Parsons' book contains 12 chapters with in-depth theoretical discussion of related aspects of systems of action and social systems. It has three chapters dealing with structure of social system, and one chapter of case of Modern Medical Practice, two chapters related to 'Roles', etc. Chapter 12, the concluding chapter with a subtitle "The Place of Social Systems in the General Theory of Action."

(2) *Niklas Luhmann (German Ed.* 1984, *English version 1995) Social Systems*

He studied under Talcott Parsons around the 1960s. In later years, he dismissed Parsons' theory, developing a social systems theory of his own. His book also contains 12 chapters, they are: System and Function; Meaning; Double Contingence; Communication and Action; System and Environment; Interpretation; The Individuality of Psychic Systems; Structure and Time; Contradiction and Conflict; Society and Interaction; Self-Reference and Rationality; Consequences for Epistemology. He challenged Parsons' theory of action through his concept of Double Contingency of Chapter 3 which may be too theoretical to be explained in simple words. Therefore, authors of this book will not discuss the academic debate between Luhmann and Parsons between Parsons and Giddens.

Several main concept of Luhmann's book:

(i) He emphasizes the basic element of the society is communication while individuals are socially meaningless.
(ii) The environment is a system-relative situation. Therefore, the environment for each system is different (p. 181).

(iii) He also emphasizes a determinate relation among elements is only realized under some conditions. The connections among relations must somehow regulated (p. 23).
It can be seen from the above limited quotations, we not only see the difference of Luhmann's concepts in opposing Parsons, but also the complementary aspects.

3.4.1.3 Discussions

Although there are many good studies of sociology, but there are only two studies of 'social systems' from sociologists. To study the society from the perspective of systems has a very short history in Western countries. It started with Talcott Parsons around the 1930s. It is also a macro-study containing many aspects and can be analyzed from various perspectives. Challenges to Parsons' concepts by Luhmann are based on the latter's perception that focuses more on communication in the age that the global society is in transition from an age of industrial society to an age of information society or knowledge economy. No theory can be perfect, there must be missing points so that there are many schools of economics and there are always debates among economists on certain policy. The social system is a complex and sophisticated system. Luhmann's social system can be complementary study to Parsons. It can also be seen from above quoted definitions, most of them are in consensus on the features of "relationship" and "activity" suggested by Parsons, therefore, Parsons' AGIL framework of analysis of action system will be discussed and explored in next section.

3.4.2 Parsons Framework of a System of Action—The AGIL Framework

3.4.2.1 Original Framework of Parsons

(1) *Parsons' Purpose to Establish the AGIL Framework*

The AGIL framework of a system of action is contained in the book *Economy and Society* published in 1956, which is written jointly by Talcott Parsons & Neil J. Smelser. The purpose of this book is for the "Integration of Economic and Sociological Theory." In fact, Parsons tried to create "The General Theory of Social Systems", which can be seen from the title of chapter six of this book, which is titled "Conclusion: Economic Theory and the General Theory of Social System."

This book is also a continuation of Parsons' past books, *Structure of Social Action* (1937) and *The Social System* (1951), many special terms used in this book had been fully explained in *The Social System*, but they have not been explained further in this book. It may cause readers to have some difficulties to have an in-depth understanding in reading this book.

(2) *Main Contents of Parsons' AGIL Framework*

The main contents are shown in Fig. 3.6 which is reproduced from Fig. 1 in his book *Economy and Society* (p. 19).

(3) *Explanation of AGIL Framework of Parsons*

A large part of original sentences are quoted from different pages of *Economy and Society* in order to avoid distortion of Parsons' purpose.

"According to the general theory: process in any social system is subject to four independent functional imperatives or "problems" which must be met adequately if equilibrium and/or continuing existence of the system is to be maintained.

A social system is always characterized by an institutionalized value system. The social system's first functional imperative is to maintain the integrity of that value system and its institutionalization. This process of maintenance means stabilization against pressure to change the value system, pressures which spring from two primary sources: (1) Cultural sources of change." (pp. 6–17) and (2) Motivational sources of change. "The first functional imperative, therefore, is "pattern maintenance and tension management" relative to the stability of the institutionalized value system.

Every social system functions in a situation external to it. The process of interchange between system and situation forms the second and third major functional imperatives.

"The first interchange concerns the situation's significance as a source of consummatory goal gratification or attainment. A goal state, for an individual actor or for a social system, is a relation between the system of reference and one or more situation objects which (given the value system and its institutionalization) maximizes the stability of the system." (p. 17) The system must "seek" goal states by controlling elements of the situation.

The second interchange deals with the problem of controlling the environment for purposes of attaining goals." (pp. 17–18) When a social system or the "adaptive" functions is simply an undifferentiated aspect of the process of goal attainment. But in complex systems with a plurality of goals and sub-goals the

Fig. 3.6 The functional imperatives of a system of action *Note* This figure is adapted from other figure in Parsons' other past working papers (abbreviated from Parson's note)

A Adaptive Instrumental Object Manipulation	**G** Instrument-Expressive Consummatory Performance and Gratification
L Latent Receptive meaning Integration and Energy Regulation Tension built-up and drain-off	**I** Integrative-Expressive Sign Manipulation

1.A—Adaptation 2.G—Goal Gratification 3.I—Integration
4.L—Latent-Pattern maintenance and Tension Management

differentiation between goal attainment and adaptive process is often very clear." (p. 18)

"The fourth functional imperative for a social system is to 'maintain solidarity' in the relation between the units in the interest of effective functioning this is the imperative of system integration."

"The four fundamental systems problems under which a system of action, in particular a social system, operates are this (latent) pattern maintenance (including tension management), goal attainment, adaptation, and integration. Their gross relation to each other are schematically represented in Fig. 1." (p. 18)

(4) *Briefing of the Book Economy and Society*

The authors had applied AGIL framework to prove "Economic theory is a special case of the general theory of social systems and hence of the general theory of action" and "An economy, as the concept is usually formulated by economists is a special type of social system." This application can be seen from the following figures. It can be seen that Parsons had applied AGIL framework to analyze the economic and society respectively in Fig. 3.7a, b, and he put further economy to be A (adaptive) in the AGIL framework of society and put polity to be the goal. This is true for every nation. He had also studied the second level of economic action systems in Fig. 3.7c, d.

The authors have also devoted substantial efforts to study institutional structure of the economy (Chapter 3, with more than 80 pages) and economic process within the social system (Chapter 4, with more than 60 pages) to prove their purpose that to integrate economic and sociological theory is academically rigorous.

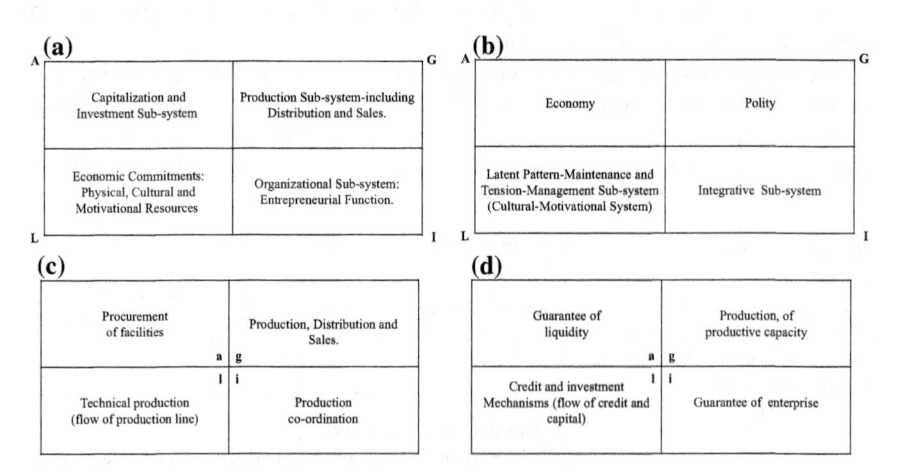

Fig. 3.7 **a** Functional and differentiation of the economy as a system (Fig. 2)* (Source: p. 44 of *Economy and Society*) **b** The differentiated sub-systems of society (Fig. 3)* (*Source* p. 53 of *Economy and Society*) **c** Production sub-system (Fig. 15)* (*Source* p. 199 of Economy and Society) **d** Investment-capitalization sub-system (*Source* p. 200 of Economy) (Fig. 16)*. *Note* No. of figure bracketed is the original number of figures in the book of *Economy and Society*

3.4.2.2 Revised AGIL Framework of Parsons' A System of Action

(1) From the perspective of authors of this book t Parsons' AGIL framework is a useful framework of system of action, through his continuous study of social action over 16 years, but his efforts in trying to integrate the theory of economy and social system seems to be debatable in spite of the fact that economic system is a subsystem of social system in a broader sense. His use of adaptation to be one of the functional imperatives may be inappropriate, therefore his concept as commented on by Bertalanffy "overemphasizes maintenance, equilibrium, adjustment,…etc." (Bertalanffy 1968) which we discussed previously. If he keeps the view of system of action, his AGIL framework can be simply revised to be a general framework of systems of action.
(2) Revised AGIL framework of systems of action. The revised framework of AGIL is shown in Fig. 3.8.

The rational of this revised framework of action is explained as follows:

(i) Based upon Parsons' original sentences, "…the 'adaptive' functions are simply undifferentiated aspects of the process of goal attainment." A process must involve actions. Therefore, we simply delete the word 'adaptive, and replace it with 'action'. Actions can be of different types depending upon the object to be studied.
(ii) All actions must be adapted to the environment, in Parsons' words, adaptation is 'relations to the situation.' In fact, adaptation is a separate subject of study. It can be passive or active. A U.S. scientist and Professor of psychology, electrical engineering and computer science John H. Holland (1929–2015) had published several books related to 'Adaptation', for example *Hidden Order-How Adaptation Builds Complexity* in 1995 and *Adaptation in Nature and Artificial Systems* in 2008.
(iii) The authors of *Economy and Society* may have spent too much effort to prove the subject 'The Economy as a Social System' due to pursuit of academic rigor, but based upon traditional classification of social science, economics is a branch of the social science. Economic system is a subsystem of social system in broad sense. We can also define social system in narrow sense to exclude economic activity.
(iv) The authors the *Economy and Society* also spent substantial effort to study the 'Boundaries' and the interchange between the subsystems (from pp. 52–100). This study is important not only with its academic rigor, but it is very important in the real world. There are some lessons in China's transition

Fig. 3.8 AGIL framework of system of action

A Action System	G Goal System
L Cultural System Motivational System	I Integrative System

from a former central planned economy to a socialist market economy. There is a need to avoid the confusion between the role of market and the role of public administration through a relatively perfect legal system. A detailed study of boundary between subsystem is critical. It is also true that there are overlapping areas between subsystems, or a so called grey area. This grey area needs special treatment to avoid rent-seeking behavior.

(v) The revised AGIL framework of System of Action can be applied to analyze economic system with 'A' to be economic action system and 'G' to be economic goal system. There are many available economic indicators devised by various international organizations for example, the UN has 14 indicators for economy, they can be used for 'G', the goal system. The action system on the supply side can be action system of production of primary, secondary and tertiary sectors, or a system of factors of production, i.e. the labor, capital and total factor of productivity, the action system on the demand side can be investment, consumption, export and import, etc.

(vi) In the revised AGIL framework, we use cultural system and motivational system in L to replace pattern maintenance which is passive in nature. Both cultural system and motivational system are used which is in consistent with the content of page 16–17 of *Economy and Society* that the social system's first functional imperative is to maintain the integrity of the value system while the pressure of change may come from cultural source and motivational source. Motivational source can also be a promotional source of change to a better future. Therefore, use of both of these two can adapt to external environment in a positive sense.

(vii) based upon revised AGIL framework of action system, Figs. 3.9 and 3.10 can also be established as follows:

(viii) This revised framework of AGIL of System of Action can be applied to any stage of the process of lifecycle of a system, i.e. planning or design, construction, operation and maintenance (or reform for social or economic system), etc.

Fig. 3.9 AGIL framework of S&T system of action

A S&T Action System	G S&T Goal System
L S&T Cultural System Motivational System	I S&T Integrative System

Fig. 3.10 AGIL framework of social system of action

A Social Action System	G Social Goal System
L Social Cultural System Motivational System	I Social Integrative System

3.5 Summary Points

1. This chapter 'Social Science and the Social System' is divided into four parts: (1) Definition of social science and its development. (2) Development of social science in 20th century and emergence of the concept of a social system. (3) Behavior of individual and groups, and their relations with the social environment. (4) Social system and Parsons' AGIL framework of systems of action.

The basic approach is similar to Chapter 2, (dealing with natural science and technology) it is necessary through the historical approach (vertical) and comparative approach (horizontal, or cross spatial) that it is possible to understand an object of study with relatively in-depth and in breadth.

2. Section 3.1 dealing with definition of social science and its history of development up to 20th century.

(1) Social science is also termed behavioral science post 1950s. There is also debate among sociologists that whether social science is scientific.
(2) Both social studies in China in the Spring and Autumn period and the Warring State period and the Hellenistic period are described. In fact, the social studies of China in that period were well developed. Some Western scholars, or experts of China's study had translated many social thoughts of Chinese philosophers in that period. Social studies in the Hellenistic period are described, which is also sources of Western civilization. Comparative study of these two is favorable to understand the two civilizations better.

3. Section 3.2 dealing with the development of social science in Western countries in the 20th century. This is period of rapid development of social science within the context of democratic and industrial revolution. The rapid and turbulent changes of society since the 19th century emerged many social issues. Three major thoughts effect the development of social science are described: (1) Impact of Marxism and others; (2) Influence of Freud, who promoted further the study of behavior and enriched the social science greatly; (3) Further specialization and cross-disciplinary approaches were developed in parallel. Social systems theory is the natural output of growth of social science and the trend of cross disciplinary studies.
4. Section 3.3 deals with behavior of individual and their relations with the environment. The feature and basic theories of social systems engineering are general system theory of Bertalanffy and theory of the social system of Parsons. The basic units of a social system are the individual and groups. Therefore, to understand Parsons' theory one must understand the personality of the individual and the human behavior, various schools of psychology are introduced, behavioral psychology by John Watson, Gestalt psychology by Max Wertheimer, psycho analytic school by Sigmund Freud, the humanistic approach psychology by Carl Rogers and Maslow are introduced.

Two schools of thoughts of personality and two types of group dynamics are also briefed.

Human behavior is also effected greatly by social environment, if academic rigor is pursued, human behavior is effected also by natural environment. In order to simplify the discussion, only social environment is emphasized with multidimensional perspective.

5. Section 3.4 deals with Social System and Parsons' AGIL Framework of System of Action.

(1) Four definitions of social system are given for comparative purpose. Parsons' definition of social system had been given in Chapter 1. Briefing of Niklas Luhmann's Social System is also given. But Luhmann has not given definition of social system in his book.
(2) Parsons' Framework of A System of Action-the AGIL framework is the core part of this chapter. Parsons' original AGIL framework, titled "the Function Imperatives of A System of Action" was published in the book *Economy and Society* jointly written by Parsons and his follower, Neil J. Smelser, their original purpose is "A study is the integration of Economic and Social theory", but on the cover of the book, it is written "A major contribution of economic and sociological theory". From the perspective of authors of our book, economics and sociology are separate branches of social science, each branch has its many theories and features. But if based upon actions and relations, they are universal in sense. The authors of our book have replaced "Adaptive" with "Action" to form a revised Parsons' AGIL framework of systems of actions. There will be relatively detail description of goal system (G), cultural system (L), and integrative system (I) in remaining parts of our book.

References

Ashord, J. B. & Lecroy, C. W. (2010). *Human behavior in the social environment—A multi-dimensional perspective* (4th ed., pp. 17, 134–135). International Education USA: Wadsworth, Cengage Learning.

Atkinson, S. (Senior Ed.). (2015). *The sociology book* (pp. 18, 31). Great Britain: Dorling Kindersley limited.

Bronfenbrenner, U. (1979). *The ecology of human development: Experiments by nature and design*. Cambridge, MA: Harvard University Press.

Brown, L. T., & Weiner, E. A. (1979). *Introduction to psychology*. Winthrop, MA: Winthrop Publisher Inc.

Chambliss, W. J., & Ryther, T. E. (1975). *Sociology: The discipline and its direction* (pp. 67–68). New York City: McGraw-Hill, Inc.

Checkland, P. (1981). *Systems thinking, systems practice*. New York: Wiley.

Confucius. *The doctrine of the mean* in *collection of four books and note*. Hong Kong: ChenXiangji Publisher (Chinese Version).

Duo, D. (Consultant Ed.). (1999). *The encyclopedia of Chinese medicine*. Spain: Calton Books Limited.

Fan, W. L. (1954). *General history of China* (Vol. I). Beijing: People's Publisher (Chinese version).
Fernandez-Armesto, F. (2009). *The World, a history* (2nd ed.) USA Pearson Education, Inc. (J. J. Ye, X. X. Qing, L. I. Song, S. M. Wang, Q. H. Wu, & W. F. Lu, Trans.) (2010). *The world, a history* (p. 185). Beijing: Beijing University Press (Chinese version).
Fu, H. S. (1999). *Introduction*. In A. Waley, & H. S. Fu *Laozi* (Chinese-English Trans.) in *Library of Chinese classics* (pp. 37, 40, 57). Hunan: Hunan Peoples Publishing House.
Fukahori, M. (2003). *Zukai De Wakaru Shinrigaku no Subete*. Japan: Nippon Jitsugyo Publishing Co, Ltd. (Hou, D. Trans.) (2007) *Yuan Jie Xin li Xue*. Tianjin: Tianjin Education Publisher (Chinese version).
Giddens, A. (2001). *Sociology* (4th ed.). Cambridge: Polity Press.
Goetz, R. W. (Eds.). (1986). *Concise Encyclopedia Britannica* (15th ed., Vol. 6). Beijing: Chinese Encyclopedia Publisher (Chinese version).
Guo, H. R. (Ed. & Trans.). (1962). *The art of war by Sunzi-contemporary translation and new edition* (p. 40). Shanghai: Zhonghua Publisher (Chinese version).
Jary, D., & Jary, J. (1991). *Collins dictionary of sociology*. London, Great Britain: HarperCollins Publisher.
Knoblock, J. (1988–1994). Jiang, J. (1999). *Xunzi* (Chinese-English Trans.) (pp. 267, 271, 295–271). *Library of Chinese classics*. China: Hunan People's Publishing House.
Luhmann, N. (1984). *Soziale Systeme: GrundrißeinerallgemeinenTheorie*. Frankfurt: Suhrkamp Veilay. (Bednarz, Jr, J. & Baecker, D. Trans.) (1995) *Social Systems* (pp. 23, 181). Palo Alto, CA: Stanford University Press.
Mao, Z. D. (1937). *Contradiction theory in committee on publication of Mao's selected works CPCCC (ED) (1951)* (pp. 274–331). Beijing: People's Publisher.
Mao, Z. D. (1951). *Collection of selected papers by Mao Ze Dong*. Beijing: People's Publisher (Chinese version).
Marshall, G. (Ed.). (1994). *The concise oxford dictionary of sociology* (p. 494). New York: Oxford University Press.
Mohun, J. (Senior Ec.). (2012). *The economics book*. London, Great Britain: Dorling Kin dersley limited.
Parsons, T. (1951). *The Social System* (pp. 24–25). Illinois: The Free Press.
Parsons, T., & Smelser, N. J. (1956). *Economy and society*. New York: The Free Press.
Rhodes, R. A. W., Binder, S. A., & Rockman, B. A. (Eds.). (2008). *The oxford handbook of political institutions* (p. xiii). New York: Oxford University Press.
Social Science (2012). *Encyclopedia Britannica-Encyclopedia Britannica Ultimate Reference Suite*. Chicago: Encyclopedia Britannica.
Stevenson, J. (1998). *The complete idiot's guide to philosophy* (pp. 4, 20–23, 39–41, 49–53, 59–62). New York: Macmillan Company.
UN. (2001). *Indicators of sustainable development: Guidelines and methodology* (2nd ed.). New York: UN.
Van Meerhaeghe, M. A. G. (1986). *Economic theory-a Cutic's companion*. Leiden: Stenfert Kroesse Publishing Company.
Waley, A. (1934). Fu, H. S. (1999). *Laozi* (Chinese-English Trans.) in *Library of Chinese Classics* (pp. 5, 87). China: Hunnan People's Publishing House.
Waley, A. & Yang, B. J. (1999). *The Analects* (Chinese-English Trans.) in *Library of Chinese classics* (pp. 91, 115, 117, 131). Hunan: Hunan People's Publishing House.
Wundt, W. M. (1973). *An introduction to psychology*. New York: Arno Press.
World Bank Group. (2015). *World development report 2015-mind, society, and behavior*. Washington, DC: The World Bank.
XunZi (Around 313–238 AD). *Enrich the country*. In Ke, M. C. & Zhang, F. X. (1993) *Explanation and translation of famous economic writings in Ancient China* (Chinese Version). Beijing: Economic Management Publisher.

Yang, B. J. (1999). *The Analects* (Chinese-English Trans.) in *Library of Chinese Classics*. Hunan: Hunan People's Publishing House (Translated into English by Author Waley).
(Note: China had established *Library of Chinese Classics* in 1999 with selected works of ancient Chinese philosophers and scholars and published them from Chinese into English to facilitate readers from abroad. An editorial committee was established headed by Yang, Muzhi. The editorial committee found that these were available very good translations from Chinese to English of those works in Western countries, such as translations of *The Analects* and *Laozi* by author Waley, A. of UK, who had been worked in the museum of Great Britain, and *Xunzi* by professor Knoblock, J. of U.S.A. The editorial committee of this *Library* simply chose available Chinese translations from Western publications and appointed one Chinese scholar to be responsible of modern Chinese and write introduction both in Chinese and English of those selected works. That's the general background of *Library of Chinese Classics*
Therefore, nearly most of the thoughts of ancient Chinese philosophers quoted by authors of this book was derived mainly from this *Library of Chinese Classics* if it is available. Of those no publication of related subjects in this *Library*, then all English translations were done by authors of this book.)

Chapter 4
Development of Social Systems Engineering

4.1 Development of "Social Systems Engineering" in International Society

Development of the discipline of "social systems engineering" is far slower than that of systems engineering in academic sense. It had been described in Chap. 2 that the United States published the first Handbook of Systems Engineering in the 1960s; it established the International Council on Systems Engineering (INCOSE) in the 1990s, and that Council had published several versions of Systems Engineering Handbook. The first edition of process framework of Systems Engineering was issued in 2002, this standard was adopted as IEEE 15288 in 2004. There was a former military standard of systems engineering MIL STD 499A in 1969, and it was updated several times, the most recent update is ISO/IEC 15288 (2015). However, there are large disparities in the development and practice of social systems engineering from a pragmatic perspective. There are some fragmented studies in Western Europe, such as Germany and the U.K. Japan started the academic study relatively earlier, while study by the United States is more or less a recent activity. The development in Japan and the United States will be briefly discussed below.

4.1.1 Academic Development of Social Systems Engineering in the United States and Japan

4.1.1.1 Development of Social Systems Engineering in the United States

(1) *Exploration of Social Systems Engineering in Early Stages*

In the field of systems engineering, many pioneering scholars and engineers had electrical or electronic engineering backgrounds, for example Harold (Hall)

H. Wang and S. Li, *Introduction to Social Systems Engineering*,
https://doi.org/10.1007/978-981-10-7040-2_4

Chestnut of General Electric and A. D. Hall of Bell Laboratory. Therefore, in publications and Journals of Institute of Electrical and Electronics Engineers (IEEE), there were several papers titled *social systems engineering* published in the 1960s, but most of them were written from the perspective of control engineering. In 1996, there was a change from this perspective, and one of the first book from this different perspective was *Designing Social Systems in a Changing World.* It was stated by the author Banathy (1996) in the Introduction that "In the design literature, there is today a back of attention to the fields of social work, health and other help services; to education and human development; to community and volunteer agencies; and to the field of public policy." (p. 2) His study was an approach to social systems engineering because 'design' is generally an important part of engineering.

(2) *In Recent Years, Several Universities in the United States have Launched Programs with Slightly Different Terms*:

[1]University of Pennsylvania, School of Engineering and Applied Science has a program 'Networked and Social Systems Engineering' for undergraduate students. The purpose of this program emphasizes that: It is necessary to understand people, systems and incentives and how the structure and properties of networks affects interactions in order to understand people and predict behavior and to design new capabilities and services."

The courses offered in the part of 'networked' of this program includes structural property of network, contagion in network, navigation in network, how do real network work, model of network formation, incentive and collective behavior, introduction to networked game theory, networked games: coloring, consensus and voting, trading in networks, internet economics etc.

[2]Harvard University, School of Public Health launched a program "Engineering Social System" ("ESS").

(i) *Purpose of this Program*

The purpose of this program is exploration of solutions for important global problems and to inform leadership and policymakers so that they can transform great ideas into changes of reality to make a better world.

(ii) *Criteria for Measurement of the Achievement of the Program*

There are four aspects of criteria for the measurement of success or failure of the program, which are shown in Fig. 4.1.

(iii) *Background of ESS and its Contents*

With the popularization of information and communication technology, huge amounts of data produced through the actions, communications and transactions by

[1]Source: www.nets.upenn.edu/.

[2]Source: www.hsph.harvard.edu/engineering/social/systems.

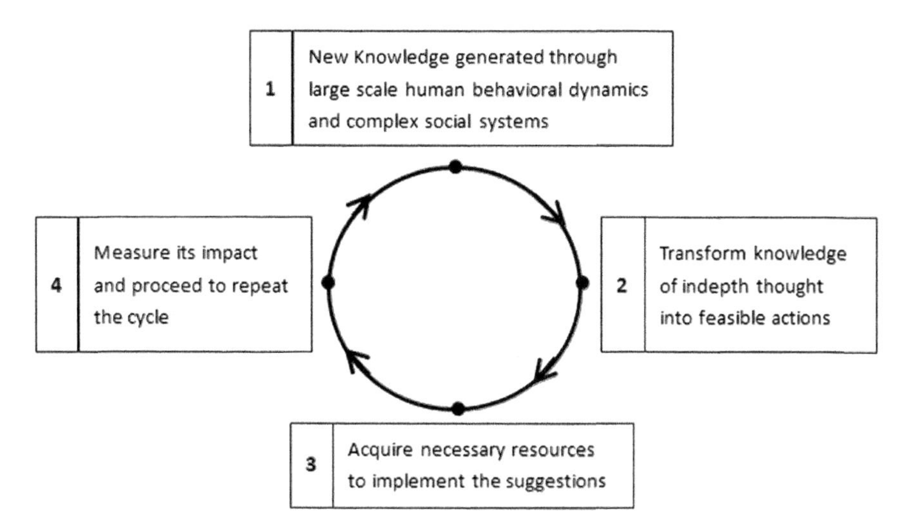

Fig. 4.1 Four aspects of criteria for measurement of the achievement of the program

mobile phones and credit cards are measured in Petabytes or PB,[3] this has formed the major background of ESS. This program will analyze through topology and dynamics of social networks of both developed and developing countries, it will couple the data of communication and financial transactions of millions of people with names unknown with vertical data of health care and socio-economic conditions. The ESS program is developing machine learning and computational rule of network analysis to better understand human society.

The meaning of research of the ESS program is to apply these new understandings to improve the society in which the billions of people producing the PB's of data live.

(iv) *Organized Research Groups*

Harvard University has cooperated with industrial organizations and other universities such as MIT and Northeastern University to enhance research capability.

(v) *Cross Disciplinary Study*

Cross disciplinary study is now being done by two new mutually complementary academic teams. One team involves social physics, while the other involves computational social science, an emerging discipline that studies the behavioral pattern of personal and groups through data in large scale of certain sets. The researchers of this team come from computer science and social science.

[3] 1 Petabyte = 10^{15} bytes.

4.1.1.2 Development of Academic Education in Japan on Social System Engineering[4]

(1) *Introduction*

Early in the 1970s, the Japanese academic field launched a program of social engineering in undergraduate studies of universities. Colleges specialized in the study of public policy and the making of plans. The Japanese also offered graduate studies of social engineering.

For example, the University of Tsukuba, set up a five year doctoral curriculum of social engineering in 1978, the major courses offered included formulation of institutions and plans, management engineering and formulation of cities' and regional planning, etc. Subsequently, there was a reorganization in the Graduate School of Social Engineering, four aspects of courses were offered: socioeconomic system; systems and mathematics of information; urban and environmental system; and, financial institutions and management.

In 2000, there was a further reorganization of the Graduate School of Social Engineering, which became the Graduate School of Information Technology, and a five year doctoral education program was offered. The Graduate School was further divided into two fields: social systems engineering; and, financial institutions and management. The courses for social systems engineering were based on three fields: socioeconomic system; systems and mathematics of information; and urban and environment systems. Further changes were made in 2005.

In 2014, University of Tsukuba announced, in view of the further progress of the cross disciplinary research and practice of business and public administration, three programs of graduate studies—social system and management, social system engineering, business management and public policy were to be unified into the Department of Policy and Planning Science (PSS). Education in the PSS strategically focuses on enhancing advanced capacity to design the society through organic integration of essential social aspects of assets and resource, space and environment, organization and behavior, PSS provides graduate programs at both the Master's and Doctoral levels.

(2) *Goal of Education and Qualification of Students in the PSS*

The goal is to train the student with three types of skills: skill in the analysis of socioeconomic systems; skill in organizational management; and, skill in the formulation of policy and planning of urban and environmental systems.

PSS seeks four qualities in the students for its Master's and Doctoral programs. Chinese have an old proverb: "Teach students in accordance with aptitude." Therefore, defining the qualities of students for this graduate study seems appropriate. The four qualities for students of social systems engineering are: (1) the

[4]Source: www.sk.tsukuba.ac.jp/sse/english.

students should be interested in continuous evolutionary society, and have the willingness to proceed with innovation; (2) have the willingness to explore theory of social systems of 21st century with originality; (3) possess global perspective and sensitivity to problems and issues, and solve them from a global perspective; and, (4) have interest and consider issues relating to morality and moral engineering, and possess the willingness to improve social welfare.

(3) *Content of Curriculum of the Master's Program of Social Systems Engineering*

The academic development of social systems engineering of University of Tsukuba has occurred for more than 35 years. Development can be seen from the general description in 2-(1) that there were several reorganizations of the educational program in this area. Table 4.1 is an abbreviation of the curriculum for illustrative purposes, and this may be a useful reference for other countries (Tables 4.1 and 4.2).

Table 4.1 Abbreviation of content of curriculum of master's program of social systems engineering

Socioeconomic system	Management science	Formulation of urban planning
1. 21 courses offered	1. 20 courses offered	1. 19 Courses offered
2. 10 courses in the name of economics, or related to economics. Eg.: Public Economics; Banks and Finance	2. 10 courses in the name of management, or related to management. Eg.: Strategic management; organizational behavior	2. 7 courses in the name of urban. Eg. Urban risk management; local survey of urban space
3. 3 courses common for socioeconomic, management science and formulation of urban planning. Eg.: Discussion and method of management science	3. 3 courses common for socioeconomic, management science and formulation of urban planning. Eg.: Discussion and method of issues of formulation of urban planning	3. 3 courses common for socioeconomic, management science and formulation of urban planning. Eg.: Discussion and method of issues of socioeconomics
4. 3 courses in the form of seminar. Eg.: Seminar of method of social survey	4. 4 courses of mathematics Eg.: Optimization theory and method	4. 3 Seminars. Eg.: International exchange seminar
5. 2 courses of Analysis. Ex: Data analysis	5. 3 other course: economic psychology; Information network; data exploration and treatment	5. 2 courses in the name of region. Eg.: Regional science
6. 3 other courses: Game theory; evaluation of systems; senior decision theory		6. 3 courses in the name of environment. Eg.: Environmental information science
		7. Others: Formulation of planning of advanced transport

4.1.2 Discussion of Development of Theory and Practice of Social Systems Engineering

4.1.2.1 Comparison of Academic Studies Between the United States and Japan

Global society was in competition with the development of science and technology post World War II. The United States had a program SDI (Strategic Defense Initiative) during the Reagan administration, and the European Commission had a program "Eureka" for funding research and coordination of science and technology activities in Europe. Japan was strong on technology, in that its manufacturing sector had once taken the lead among the world since the later part of the 1980s. The academic and industrial fields in the United States organized a joint research project to study how to revitalize their manufacturing sector and published a book *Made in America-Regaining the Productive Edge* in 1989. Therefore, generally speaking, the United States and certain European countries took the lead in science and education, and Japan took the lead in technology in the later part of the 20th century. This broad context of historical development illustrates the differences in the education of social systems engineering between the United States and Japan.

The academic training of ESS program is biased more towards the scientific side, it has explored and promoted a new discipline 'social physics' in this program. It has further promoted the application of mathematics to the study of social science so that social science may become a true 'science', similar to the natural sciences, to avoid the debate among social scientists. Computational social science was further developed by the ESS program of Harvard. In fact, the United States already established the Computational Social Science Society of the America. This society may also take the lead globally.

The development of social systems engineering in Japan is more pragmatic. For example, its three blocks of content of education, focus more on the linkage between economy and society, its application is focused on the formulation of urban planning. The study of urban planning will be simpler than that of a national planning, but they should have some features in common.

In fact, the United States published many excellent studies related to national planning in the 1960s, for example, Jan Tinbergen's *Development Planning* (Translated from the Dutch by N. D. Smith) published in 1967, W. Arthur Lewis's *Development Planning: The Essentials of Economic Policy* published in 1966, both were Nobel Prize winners. However, due to the ideology of politicians and a part of main stream economics, the United States has not implemented planning at the national level. Yet, at the township and city levels in several states, small planning offices are existed.

4.1.2.2 Development of a National Planning System is Effected by Politics in Western Countries

(1) *Planning Practice in Western Developed Countries*

Planning practice in Western developed countries is also affected by ideology of parties and it is biased to economic side. This can be seen in a paper *The Central Planning Bureau of the Netherland* written by Don and van den Berg (1990). The related part of their paper is abstracted below, which shows the impact of politics of parties and the role of leadership on planning institutions.

Case Study of Establishment of the Central Planning Bureau of the Netherland. Refer to Box 4.1 below.

Box 4.1 [5]Establishment of The Central Planning Bureau of the Netherland

1. General

The Central Planning Bureau of the Netherland is an independent Dutch government agency founded at Sept. 1945 by Nobel laureate Jan Tinbergen. It is well known for its performance of short term (up to two years ahead) and medium term forecasting and study of economic problems and analysis of policy proposals in medium term as well as study of structural problems and policy options in long term through scenario analysis.

2. The difficulties met in the establishment due to politics of parties

The Dutch government had drafted a law on "preparing the assessment of a National Economic Plan" post the second world war. The Minister of Economic Affairs by that time was the socialist Hein Vos, who intended to found a Central Planning Bureau to prepare a national economic plan to coordinate economic, social and financial policy on a regular basis and help the government in planning and guiding the economic process.

The Minister met with strong opposition from the non-socialist parties in parliament who eschewed central economic planning. Huysmans of the catholic party succeeded Vos as Minister of Economic Affairs in July 1946 after general elections. He held on to the text of the law as drafted, but in a memorandum to parliament explained the Central Planning Bureau would, be a strictly advisory body and only serve to compile information on the current economic situation and prospects.

[5]Note: Most of descriptions in Box 4.1 is from a conference paper contributed by Don and Berg in Nov. 1990.

3. The critical role of the Leadership of the new institution played

In Sept. 1945, a Central Planning Bureau (under formation) had been started in anticipation of a legal basis. With Tinbergen as its first director, it offered an assessment of the economic situation and suggested a number of policy measures in Feb. 1946. The usefulness of the information collected in this document, which was published in May has helped to convince parliament of the merits of the institution.

The law passed both houses of parliament unchanged in early 1947. It still mentions that the Bureau should regularly prepare a Central Economic Plan containing estimates and guidelines for the national economy, the final contents of the Plan to be decided by the government. Tinbergen contributed much to establishing the role of the Bureau While inspired by current questions of economic policy, he took a scientific and independent approach to help answering them. The open-mindedness and quality of the work done by him and the Bureau's staff in the early day soon commanded respect, a respect indispensable for the Bureau to work in the way it does to this day.

It can be seen from Box 4.1 that planning in Western countries is also affected greatly by the politics of parties and parliament, and the development of an institution is heavily dependent upon the leadership.

(2) *Planning Practices in the United Nations*

The United Nations, and its related organizations, have also engaged in planning and are similarly concerned about social issues, and provides a demonstration of the importance of planning in the process of development.

The United Nations was established on October 24, 1945, after World War II. The UN is a supra-national organization with 193 member states and 2 observer states, currently. It has launched four strategic international development plans since 1961, the so called "[6]*Decade of Development of the UN*." It was replaced by Millennium Development Goals from 2000–2015 and 2030 Agenda for Sustainable Development from 2015–2030. The following is abstract of *International Development Strategy and the Fourth Decade of Development of UN* for illustrative purpose.

(i) *Contents*

This document is composed of 6 parts with 112 articles. They are Preface (articles 1–12), objectives and Targets (articles 13–19), Policy and Measures (articles 98–

[6]UN had prepared several *Decade of Development* plans since 1945. The authors of this book had a copy of *Third Decade of Development*. It seems that UN had stopped this practice, and replaced with Millennium Development Goals since 2000. And Sustainable Development Goals since 2015.

102), Exceptional circumstances (articles 98–102), The Role of UN system (articles 108–112), examining process and Evaluation (articles 108–112) respectively. The previous 4 parts will be abstracted.

(ii) *Preface*

It gives a retrospect of objectives and targets set up in the Third Decade of Development of UN had not been achieved, basis of expected growth had been vanished due to appearance of unexpected unfavorable situation.

Projections of the situations of 1990s by various organizations of UN show that there will be no large difference between the previous decade and next decade if there are no important changes of policy.

Projects show that the growth of some Asian Countries will be relatively faster, but the growth of other countries will be continuously stagnated, especially.

(iii) *Objectives and Targets*

The main purpose of the strategy is to guarantee the 1990s should be a decade of accelerated development of developing countries and consolidation of international cooperation. There will be great improvement of developing countries and the reduction of disparity between rich and poor countries, and also the global society should find means to satisfy the needs but the environment will not be worsened in this decade. Strategy should guarantee the participation of economic and political life of all people, protect the cultural feature, and guarantee the measures of necessary survival of all people. Strategy should provide an environment, which support various regions to implement gradually a political system based on approval and respect of human rights and rights of society and economy, and legal system to protect all citizens.

Every country has the responsibility to formulate economic policy based upon concrete conditions of its country to promote its national development, every country should be responsible to the life and welfare of all citizens.

To achieve these main purposes, it is necessary to achieve six mutually related targets. These targets are: Accelerate the steps of economic growth of developing countries; Push forward the progress of sustainable development which can take care the social needs, reduce the large amount of worst poverty, promote the development and utilization of human resources and skills without detrimental effect to environment; Improve international monetary, financial and trading system to support the process of development; Create a strong and stable environment of the global economy, implement a sound macro-economic management at national and international level; Strengthen international development cooperation decisively; Solve the issues of least developed countries, i.e. the weakest countries in developing countries with special effort.

(iv) *Policy and Measures*

It is divided into 2 parts.

Part A is vitalizing development. It includes a framework of economic policy, foreign debt, fund for development, international trade and commodities, and also dealing with important domestic policies, such as science and technology policy agricultural policy and industrial policy.

Part B is priorities of development, which includes elimination of poverty and hunger; human resources and institutional development, population and environment.

(3) *Planning Techniques for a better Future*

This is a book written by Pyatt and Thorbecke, published in 1976 by International Labor Office of UN, and is the result of a three year research project on planning for growth, redistribution and employment supported by ILO (International Labor Organization). These two authors helped to further develop the Social Accounting Matrix which has becomes an important framework of planning and it also forms the backbone of the Computable General Equilibrium (CGE) model. Both CGE models and SAM (Social Accounting Matrix) have been heavily relied upon by the World Bank for development analysis. Pyatt moved to the World Bank in the 1980s.

4.1.3 Current Status of Development of Social Systems Engineering

4.1.3.1 Social Systems Engineering Is in the Process of Development, but There Are Barriers that Exist at the Country Level

As described in the beginning of this book "The social system exists ever since the existence of human beings." Mankind has the responsibility to improve the social system, that is the role of social systems engineering. It can be seen from Sect. 4.1.1 that the United States is well known for its advance in science, natural and social. It has taken the lead in the development of systems engineering, but its study of social systems engineering seems to be lagged behind that of Japan. Japan has an established Department of Policy and Planning Science in its University since 2014. While "planning" is a concept unacceptable by some political parties, even a part of scholars of economics, although the global society had been changed greatly compared to the period of Adam Smith. The development of planning practice of UN and a part of countries together with the social systems engineering seems to be on the right track. Theoretical studies of the United States are useful tools for development of social systems engineering, e.g. the development of social[7] physics and computational social science, which will promote further the development of

[7]Note: Social physics is a new discipline to be explored. It is defined by Pentland (2014) in his book *Social Physics* that "Social physics is a quantitative social science that describes reliable, mathematical connections between information and idea flow on the one hand and people's behavior on the other." Social physics helps us understand how ideas flow from person to person through the mechanism of social learning and how this flow of ideas ends up shaping the norms, productivity, and creative output of our companies, cities and societies.

social systems engineering. Some system engineers have also done work related to national development, which will be described to be a case study in Sect. 4.3.

4.1.3.2 Planning in the UN and Its Related Organizations Have Provided Conditions of Development of Social Systems Engineering

United Nations launched four *International Development Strategy, Decade of Development of UN* since 1960s. This document of development provides a very good example of planning with concepts of integration of science and technology, economy and society. UN recognized openly the failure to achieve the objective and targets of the *Third Decade of Development of UN*, but it is not strange due to the complexity of the global society. And also due to the incomplete statistics of various countries of the world. The UN had promulgated the Millennium Development Goals (8 Goals), since the beginning of 21st century. The UN General Assembly further adopted the "2030 Agenda for Sustainable Development" with a set of 17 development goals. The goals are to be implemented and achieved in every country from the year 2016 to 2030. The World Bank cooperated with many institutions and published *World Development Indications 2012* with more than 1000 indicators of 216 economies. Special focus to 8 goals of MDG[8] was given. It has also published *World Development Indicators 2016* with a special focus on the statistics of 17 goals set up in "2030 Agenda for Sustainable Development" of U.N.

It can be seen that the UN and its related organizations have done substantial work, including compiling statistics, to provide an opportunity and environment to improve national conditions through development and application of social systems engineering.

4.2 Planning and Design of Social Systems Engineering

4.2.1 General Discussion of Planning and Design of Social Systems Engineering

Although in the English-Chinese dictionary, there is little difference between the meaning of these two words, design and planning, there are differences in their application in daily affairs.

[8]Note: Millennium Development Goals were the eight international development goals for the year 2015 adopted by United Nation Millennium Declaration in 2000.

4.2.1.1 Planning

Planning is a set of steps that one takes towards a goal. In a national plan, there are a system of goals. The product of planning is a description of a sequence of activities (or action system in Revised AGIL Framework of Parsons' Systems of Actions) to be accomplished in a time schedule, (for example, five years' planning currently adopted by China and India, or planning of three or four years' period formerly adopted by France, etc.).

4.2.1.2 Design

Design is generally used in micro sense. It is a process which creates a description of a product or a system that has the capacity or capability to attain a set of purposes. The product of design is the model,[9] or the specification of the system which consists of four parts: description of inputs; specification of output; description of environment and location, and measure of value (Gosling, 1962). Once the specification is set up, then a plan can be prepared and one can proceed to bring the design to life.

4.2.1.3 The Difference Between Planning and Design

In an engineering sense, design and planning may be different in their details of study and content. It can be seen from the phases of the coarse structure of morphology of systems engineering and its activity matrix, in the phases of an engineering project, the first phase is program planning, and the second phase is project planning (and preliminary design), i.e. project planning is equivalent to preliminary design. In the practice of engineering design by Chinese design institutions, they generally classify design into three stages: preliminary design, technical design and drawings for construction details. This classification of stages of design is for hard systems engineering. This can also be useful for reference of design for soft systems engineering.

Since design is in more detail than that of planning, from the application of general systems theory, "the systems design process" and "planning" are treated in separate chapters in *Applied General Systems Theory*, and emphasize "Planning in the context of system design" (p. 533). Therefore, it is necessary for planners to have some understandings of design, which will be discussed in Sect. 4.2.2.

---

[9]Note: Model is a qualitative or quantitative representation of a process, or behavior or real substance. It can be classified into several types: its degree of semblance, it can be a scale model, analogue model, symbolic model or mathematical model. [Wang, 1980 Introduction to Systems Engineering (Chinese version, pp. 59–60)].

4.2.2 *Design in a Changing World*

The process of design is as old as the existence of human history, e.g. the design of stone tools for survival by early human ancestors. Also, there is the design of basic social systems by primitive tribes for basic survival in fighting against nature and competing tribes. However, design became a part of scientific and technological process during the scientific revolution. It is further developed quickly during the industrial revolution. It is more or less a matured process in the area of hardware engineering and hardware systems, but in the area of software or social systems, they have become an emerging new discipline, while design of software has progressed faster than the design of social system. Here under, a brief discussion of design, especially the design of hardware system will be given to show its relevance to social system.

4.2.2.1 Definitions of Design

There are many definitions of design because it is practiced by many professions in many different ways. Several definitions defined by different authors will be listed below to broaden the concept.

(1) Design is the initiation of change in man-made things (Johns, 1966). This definition is better adapted to design of hardware.
(2) Design is the use of scientific principles, technical information, and imagination in the definition of a system to perform specific functions with maximum economy and efficiency (Archer, 1966). This is a very general definition of design, which distinguished the process of design before the scientific and industrial revolution. It can be applied to engineering system or social systems engineering if the wording of "maximum economy and efficiency" after "with" is replaced by the appropriate phrase adapted to the objective or system to be designed.
(3) Design is the translation of information in the form of requirements, constraints, and experience into potential solutions which are considered by the designer to meet required performance characteristics (Luckman, 1984).
(4) Design generates, organizes, and evaluates a large number of alternatives; keeping focused on the best possible or most ideal solution, rather than on collecting and analyzing data about the problem (Nadler & Hibno, 1990).

Both (3) and (4), above, should be considered together as the general process of design. In fact, it is one of the principles of design that 'there must be an alternative', then the designer can identify a satisfactory solution from all alternatives. The later wording "rather than" of (4), above, should be replaced by 'in addition', because collecting and analyzing data are important steps of design. The wording of "rather than" may cause a misunderstand of the importance of "collection and analysis of data."

(5) A purposeful activity, design is directed toward the goal of fulfilling human needs (Asimow, 1962).
(6) Design is concerned with how things ought to be. The designer devises a course of action aimed at changing an existing situation into a preferred one (Simon, 1969).

These two definitions of design, (5) and (6), above, linked together with revised AGIL[10] frame of systems of action can be used appropriately for design of social systems.

(7) Reengineering is the fundamental rethinking and radical redesign of processes to achieve dramatic improvement in measures of performance (Hammer & Champy, 1993).

This definition is very important in engineering and social systems engineering. In the engineering field, there existed many engineering facilities and systems, but due to change of material, energy and information and also due to the changing demand, reengineering is an appropriate process to adapt to change, which is a universal phenomenon in the daily life of the human beings. It is especially important for social systems. There exist a great variety of social systems globally, although a large part of them may not be adapted to contemporary global environment. The author cautions that designing a new social system to replace the old one is generally not feasible due to cultural rigidity, and also, there may be larger unacceptable social disturbance. Reengineering or reform of a part of the existing social system is however feasible, and this can be illustrated from the relatively successful story of China's reform of gradualist approach, and the relatively unsatisfactory reform of the former Soviet Union due to a shock therapy approach.

4.2.2.2 Evolution of Design

(1) *Design of a Single Product for Sales*

The process of engineering design is well established due to the process of industrialization of more than 200 years. Nearly all large manufacturing companies in both developed and developing countries, have accumulated manuals for the design of equipment. With the manufacture of boilers, for example, the former Soviet Union transferred all of those design manuals to Chinese manufacturers of boilers, *manual for thermal calculation, manual for calculation of circulation of water, calculation of preparation of pulverized fuels*, etc. The design of a boiler is completed and standardized drawings are produced with modification of certain dimensions or components. China established her manufacturing capability quickly in her first and second five year planning period due to transfer of those designing and processing documents from the former Soviet Union.

[10]Note: Revised AGIL framework of systems of action, please refer to Chap. 3.

There were weaknesses in the manufacturing sector of China in her preliminary stage of manufacturing development. For example, the designers who worked in the manufacturing sector of power plant equipment, had no experience with the operation and maintenance work of the power plant, and this became a weakness of the manufacturing sector in China, which needed to be improved for the product in its initial stage of development. International experience has shown that the matured manufacturing companies have set up an application department to collect information from users, and studied the operational problem of their manufactured equipment. This organizational issue is also an issue of system concept. Figure 4.1 is the process of design of a new product for sales. This figure is adapted from a textbook *Introduction to Engineering* (Mayne & Margolis, 1982). Authors of this book have simplified the original figure to correspond more with conditions of the real world. A line of feedback information is added from block of sales to customers.

(2) *Design of a Process Plant and Engineering System in Western Countries*

The evolution of design from a single product to the design of a whole process plant developed quickly in Western countries due to accumulated experiences of designing and the construction of many industrial projects. Donald Beeman, the manager of Industrial Engineering Section of Industrial Power Engineering edited the *Industrial Power System Handbook* together with around 15 members who worked at General Electric, which had given an overall design of power system in a manufacturing plant, or commercial building, including generation, distribution, selection of equipment and protective devise, etc. in fact, this is also a book of systems engineering in nature. There is no shortage of books in dealing with the design of a process plant, for example, the publication of *Project Engineering of Process Plants* (Rase & Barrow, 1957), which dealt with the role of the project engineer, included major steps in plant design; business and legal procedures; details of engineering design and equipment selection. The actual process is quite complicated. This book was jointly written by a university associate professor of chemical engineering (Rase) and a project engineer Barrow who worked in the Foster Wheeler Corporation. Rase and Barrow's book shows clearly the fact that cooperation between the academic field and the practical engineering field is beneficial for the development of new knowledge with practical applications. The project engineer is also called chief designing engineer in China, or systems engineer from the perspective of authors of this book. At this time, there was also the publication of The Design of Engineering System (Gosling, 1962), etc. (Fig. 4.2).

(3) *Design of Engineering System in China*

Since the First Five Year Planning Period (1953–1957), China has established many engineering designing institutes supported by the former Soviet Union, such as electric designing institutes, transportation designing institutes, communication designing institutes, all of which are systems engineering in nature. A project engineer in Western countries is called chief designing engineer in China. Due to

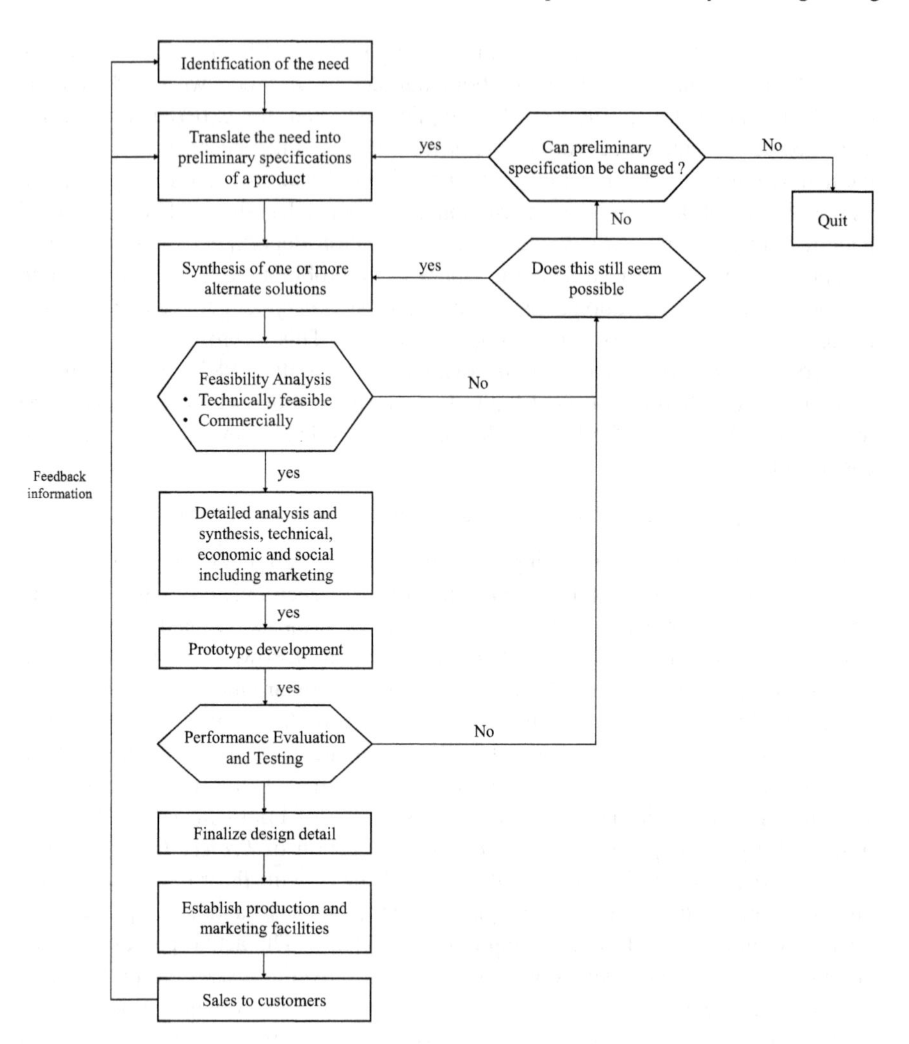

Fig. 4.2 Design process of new product (*Source* Adapted from Mayne and Margolis 1982 (Fig. 6.1, p. 328).

the many engineering projects for design and construction during China's industrialization process, even before its reform and opening, the Chinese chief designers have cumulated experience through doing and learning. The project organization of an electric power plant is shown in Fig. 4.3.

Table 4.3 shows the crucial points of design of engineering system (hard engineering system) summarized by the Chinese Electric Designing Institute. This reference is also useful for the design of social systems, shown in Table 4.3.

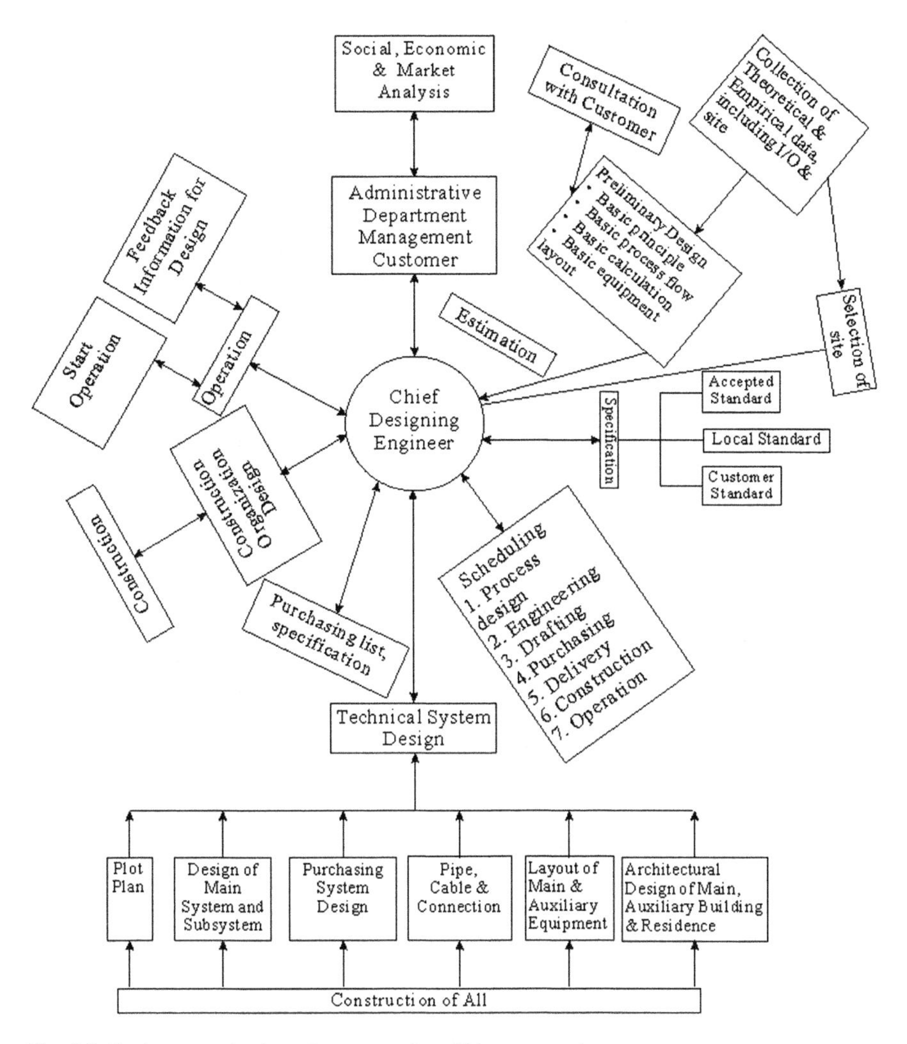

Fig. 4.3 Project organization of a power plant-Chinese experience

Table 4.2 Crucial points in engineering system design, the Chinese experience and its relevance to social system

Serial no.	Crucial points	Importance to engineering system design	Relevant reference for social system
1	Selection of sites	Site of plant is important to get access of all inputs and facilitate the transfer of outputs. (useful output and various wastes)	A country is situated in a global environment, collection and analysis of data of social (including political), economic, science and technology, etc. of the external environment of a country is crucial
2	Original data	All kinds of data, relevant to the performance of plant are important, especially, the precision of data is crucial. This is the basic procedure of engineering system design. The principle of selection of production process and choice of alternatives	All kinds of data relevant to the projection of supply and demand of the economy, budgetary income and public expenditure of social needs and security issues etc. should be collected, and carefully evaluated.
3	Principle of design	Choice of capacity of main equipment to adapt to current and future needs. Choice of matching capacity of all auxiliary systems. Layout of all plants including main and auxiliary plant etc.	Basic goals of socio-economic development should be set up. Projection of the feasibility of achievements of goals of system and all subsystems. Set up structure and function to perform the goal system, they should be carefully designed, improved and coordinated in consistency
4	Calculation	All principles of design should be based upon calculation with appropriate method. There is one principle: The best engineers do not use the most sophisticated mathematics, but instead use the most appropriate mathematics	All goals and performance of system and subsystems should be based upon quantitative analysis as much as possible. There are available many types of models. The choice of model should follow the same principle of calculation of design of engineering system
5	Construction organizational design	It is important that the process of construction can be implemented in order and in time	In planning or design of social system, it should be supported with qualitative judgement, such as historical or cross country comparison and logic. The process (phases of actions) of implementation should be carefully monitored. And behavior of various targeted groups should be analyzed and watched

(continued)

Table 4.2 (continued)

Serial no.	Crucial points	Importance to engineering system design	Relevant reference for social system
6	Connection and cooperation	For a large project, there is the involvement of many specialized disciplines. In the phase of construction, there are quite a lot of activities of cooperation and connection among different special fields, for example, buried parts or open holes in the structure of civil engineering required by mechanical, chemical or electric sectors, etc	In Revised Parsons' AGIL framework of Action System, integration and cooperation is the difficult and tedious part, integration of principle of design of various subsystem, coordination of major or minor action in the implementation of policy of different subsystems, must be analyzed in detail and coordinated properly with necessary trade-off among interested group but insisting on basic principle and goal

Table 4.3 Quality of life model for judging the performance of a nation

5 Major category with weight based upon Chestnut's paper	[a]Sustainable development goals 2030 UN
Social (24)	(1) No poverty (2) Zero hunger (3) Gender equality
Economy (20)	(4) Affordable and clean energy (5) Productive employment (6) Industry, Innovation and Infrastructure (7) Reduced inequalities (8) Responsible production and consumption
Health and education (22)	(9) Good health and well being (10) Quality education (11) Clean water and sanitation
Environment (15)	(12) Climate action (13) Sustainable cities and communities (14) Life below water (15) Life on land
National vitality and security (19)	(16) Peace, justice and strong institutions (17) Partnership for global development

[a]*Note* UN has 17 development goals and 169 associated targets. UN has not classified 17 goals into 5 categories. Authors of this book have classified these 17 goals primarily in correspondence to five major categories

4.3 Nation as Large Scale Social Systems Engineering—Three Case Studies

Planning or designing of a nation is an important field of study in social systems engineering. Three case studies will be provided here, which will assist the readers to better understand this theme.

4.3.1 Case Study 1 of [11]Nations as Large Scale Systems Studied by Harold Chestnut

4.3.1.1 General

The Case Institute of Technology, Case Western Reserve University organized the Case Centennial Symposium on *Large Scale Systems* to address the art and science of modeling, optimization, and decision making of large scale systems in the 1980s. Harold Chestnut of General Electric presented a paper *Nations as Large Scale Systems* in this Symposium. His paper was collected to be Chapter 7 of the book *Large Scale Systems* edited by Haimes, Y. Y. and published by North-Holland Publishing Co. in 1982.

4.3.1.2 Features and Outline of the Paper

Although it is a short paper with only 29 pages and 9 sections, it covers essential aspects of planning and designing a nation within the global context. The Paper has 12 diagrams which clearly explain the essential structure and function of a nation, and also trade relationship with other countries and analysis and suggestions of a world security system. The content of the Paper can roughly, be divided into two parts: Part 1 contains Sections 1–6, and deals with the basic concept of nations as large scale systems; Part 2 deals with national systems at the international level, including trade and security issues. These two parts will be abstracted in Sections 3 and 4 in the following to be backbone of the theme "Nation as Large Scale System", authors of this book have supplemented them in several aspects to adapt more to current situation.

4.3.1.3 Abstracts of Part 1 of the Paper

(1) *Descriptions and Comments*

Part 1 of the Paper includes six sections: introduction; systems definitions applied to nations; science, technology, and change; bases for the performance of a nation;

[11]Note: Chestnut (1982).

hierarchical levels of national large scale systems and models of large scale systems at national level.

In the introduction, Chestnut emphasizes that systems people should use their skill and expertise to understand nations better in terms of how the nation can be considered as a large scale system. He also pointed out that nations should set up objectives (goal system) to meet the needs of people. The planner must have the awareness that many people in a nation with their different needs, and a nation has many stakeholders for whom the nation must concern.

(2) [12]System Definitions Applied to Nations

The core part in this section is that Chestnut has classified the people of a nation into four groups: persons benefiting directly from the system, i.e. the users; suppliers of the nation, i.e., the owners', government (integrating system) and financiers; persons who plan and build the system; persons operating, performing and maintaining the necessary activities (action system) to enable the goods and services to be provided, i.e. the workers and managers who make the system operate, but they are paid to perform their parts so that the suppliers and users can both feel the benefits to them exceed their cost.

In considering the nation as a large scale system, it is necessary to consider the effects of the activities (action system) of the nation and their results from the view of above four groups. System function should be optimized with objective and subjective criteria to be kept in mind. There is also a need to have a dynamic perspective, because the benefits and costs to these different groups may change with time so that some sort of a "life cycle cost and benefit" consideration should be employed to be part of the judgement process.

The system's concept of looking at a nation as a system, which interacts with other nations in a world system, and interacts with the various subsystem with the nation that makes it, will provide useful insights in the contemporary world in the process of globalization and regionalization. Experiences, and concepts with cooperative systems and competitive systems in other contexts have useful counterparts as they affect behavior within and among nations.

(3) *Science, Technology and Change*

With the experience of a system engineer, Chestnut has a full awareness that science and technology have become a major driving force on society, but it is not a technological determinism Chestnut claimed that the number and size of many large scale systems, through application of technology, are frequently affected by the knowledge, wealth, and other resources of the nation, which is shown in Fig. 4.4.

[12]Note: The words in the bracket of this section are added by authors of this book to link Chestnut's concept to revised Parsons' AGIL framework of action systems.

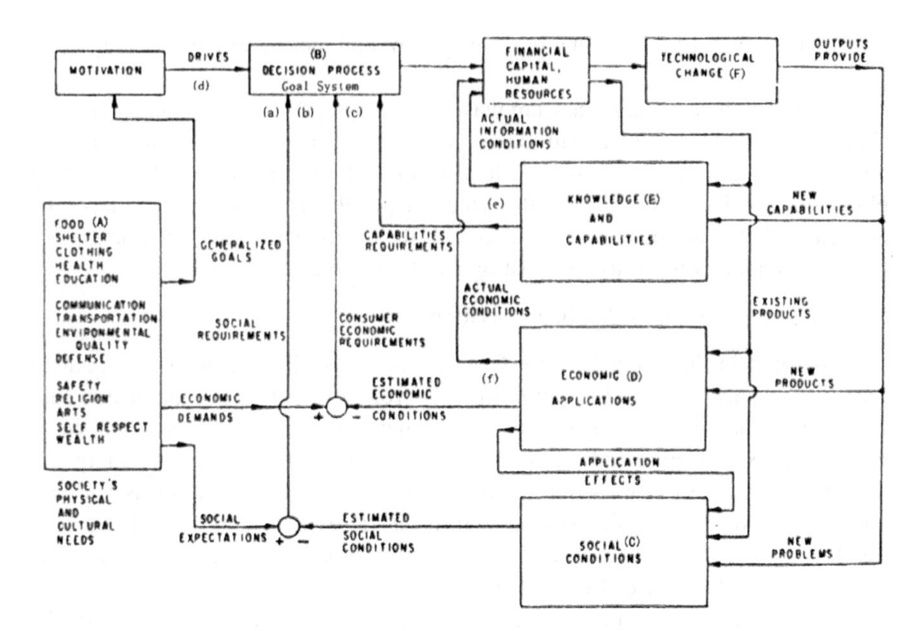

Fig. 4.4 Influence of technological change on society (*Source* Chestnut, 1982, Fig. 1, Chapter 7, p. 158)

Figure 4.4 shows that the physical and cultural needs of Block (A) form the driving function to the decision-making process of Block (B), which establishes an appropriate and feasible goal based upon (a) (b) (c) (d), and allocate financial and capital resources to existing social conditions, economic applications and technological change [Block (C) (D) (E)]. The outputs of technological change, Block (F) flows through to Block (D) and (E) providing new products for economic applications, and new capabilities for new knowledge. Chestnut also saw the issues that flow from technology change will cause new problems or new situations for the social conditions, for example, the change from an agricultural society to industrial society, there were many lessons and experiences to be learned from the history of development of many developed countries as well as developing countries. Currently, a part of countries especially the developed countries are in transition from the industrial society to an information or knowledge society, challenges to social conditions should be well analyzed.

It is also pointed out correctly by Chestnut (1982) "there is just an insatiable desire on part of people to have more and different things than heretofore, thus there is this drive to provide technological change."

(4) *Bases for Judging the Performance of a Nation*

During the period that Chestnut prepared his paper, the EPRI Journal presented quality of life indicators as a basis for judging the performance of a nation. There were five major categories with proper weighting, the five major categories and

Table 4.4 Share of Swedish Budget of different ministries and others 1990/91[a]

Serial No.	Item	Share of budget Ex.%
1	Ministry of Justice	1.42
2	Ministry for Foreign Affairs	3.67
3	Ministry of Defense	8.68
4	Ministry of Health and Social Affairs	27.88
5	Ministry of Transport and Communications	3.94
6	Ministry of Finance	6.43
7	Ministry of Agriculture	13.64
8	Ministry of Education and Cultural Affairs	1.2
9	Ministry of Labor	6.68
10	Ministry of Housing and Physical Planning	5.14
11	Ministry of Industry	0.6
12	Ministry of Public Administration	3.84
13	Ministry of Environment and Energy	0.4
14	Sum of Royal household and residences, Parliament and its Agencies, unforeseen expenditure	0.15
15	Interest on National Debt, etc.	2.4
16	Other expedited estimated	13.7

[a]*Note* Derived from the Swedish Budget 1990/1991

weighting were: social (24[13]), economic (20), health and education (22), environmental (15), and national vitality and security (19). Chestnut emphasized that "this idea of identifying a number of categories and weighting them and their sub-categories with appropriate weights does represent a way in which nations can be compared and in which optimizations can be used…" (Chestnut, 1982). Yet, there are also many kinds of indicator systems established by the UN, various international organizations, and the private sectors or World Forum, after the publication of Chestnut's Paper. Table 4.4 is prepared based upon Table 1 of the paper with five major categories supplemented with 17 sustainable development goals, set up in the 2030 Agenda for sustainable development adopted by the UN General Assembly on Sept. 25th, 2015.

The authors of this book are in agreement with the weight given to social, health care and education as well as environment. Experience of global development in past two centuries has yielded many lessons to be learned. Generally, many countries focus too much on economic goals and create numerous issues on social development with detrimental results to the environment.

[13]Note: Number in these brackets are weighting of the categories concerned.

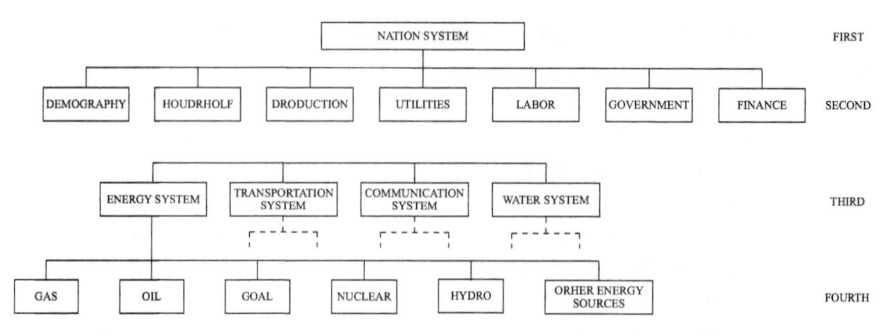

Fig. 4.5 Hierarchical structure of a nation system

(5) *Hierarchical Levels of National Large Scale System*

Chestnut studied the function and structure through hierarchical analysis of general system theory. He classified the national system into four hierarchical levels (Fig. 4.5). Nation system is the first level with seven sectors; demography, household, production, utilities, labor, government, finance is the second level, and under utility, there are four subsystems which are the third level, i.e. energy, transportation, communication and water, etc. He presented three diagrams to illustrate the hierarchical level of a national system, and one diagram of flow process of three large scale subsystem of fourth level, i.e. telephone communication system under the communication system (third level subsystem) which is under the scope and control of utility system (second level). A communication system can also be classified into several types of subsystems depending upon the criteria, for example, optical communication system, radio communication system, power line communication system etc., the above three subsystems chosen are based upon the media used in transmission. Similarly, the other two diagrams of flow process are the electric utility generating system, which is also at fourth level, and it is a subsystem of energy (third level); and jet air transportation system (fourth level) which is a subsystem of transportation.

Therefore, application of hierarchical analysis of general systems theory will facilitate the understanding of the structure of the system.

In the Chestnut's original paper, he ranked the system's structure from bottom to top, he ranked the lowest one to be the first level, for example, he ranked gas, oil, coal etc. to be the first level and the national system to be the fourth level. In our discussion, the labeling level is top-down, the national system is ranked to be first level.

In Chestnut's paper, he made hierarchical arrangement of various levels, he only made relatively detailed analysis of the utility system and energy system for illustrative purposes. His focus is biased to the relationship between sectors from economic perspectives, which can be seen from Section 6 of his Paper.

In the application of social systems engineering to national planning, it is necessary to focus on the structure and function of the national administration.

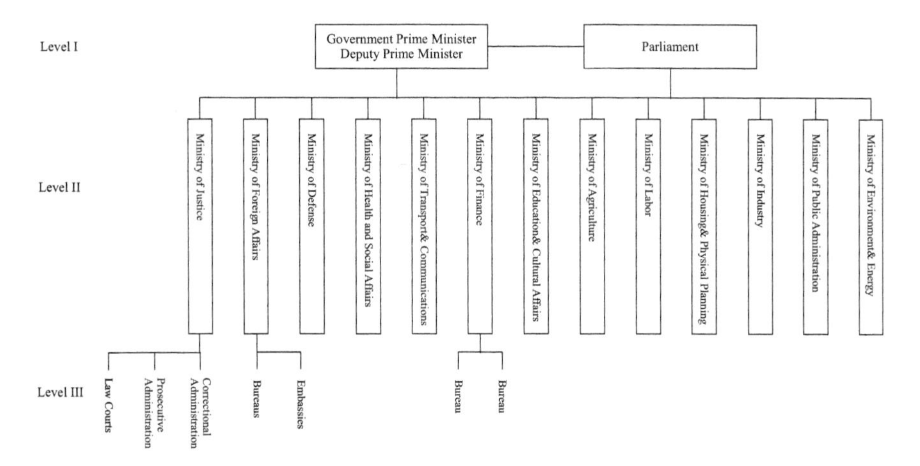

Fig. 4.6 Hierarchical administrative structure of a national system (*Note*: The first and second levels of Fig. 4.5 is drawn based upon *A Summary Published by the Ministry of Finance* (1990) *The Swedish Budget* (1990/91) Stockholm: Norstedts Tryckeri The third and fourth level are drawn arbitrarily based upon common administrative practice)

Figure 4.6 is prepared by authors of this book based upon the structure and function of the Swedish Government, functions of selected ministries are described to have some recognitions of function and structure of a government in the sense of administration.

According to the *Swedish Budget* (1990) descriptions of selected Ministries are quoted for illustrative purpose.

(i) Function of Ministry of Justice.

Principal responsibilities: Constitutional legislation, panel legislation and much of the other legislation regulating the position of individuals, vis-à-vis the state and other bodies, as well as laws regulating the affairs of private individuals, for instance family law, commercial law and copyright law. Authorities in the field of this Ministry includes the law courts, the prosecution administration and the correctional administration, as well as the Office of the Chancellor of Justice, the Data Inspection Board, Crime Prevention Board, the Crime Injury Compensation Board, the Accounting Standard Board and other bodies. (p. 24)

(ii) Function of Ministry of Health and Social Affairs.

Principal responsibilities: Matters concerning social insurance in the form of health insurance (including dental insurance), basic pensions, national supplementary pensions (ATP), semi-retirement pension insurance and occupational injury insurance; also health and medical services, social welfare and treatment, alcohol and narcotics policy.

> Also matters concerning child allowances and other financial support for families (including parental insurance), child welfare, child environments, care of the aged, service for the disabled, support from the State Inheritance Fund and social research. (p. 47)

(iii) Function of Ministry of Housing and Physical Planning.

Principal responsibilities: Housing provision, including the construction and financing of housing as well as housing allowances, management of natural resources, energy management in buildings, building administration, building research, surveying, mapping and real estate data, rent policy, housing cooperatives and the management of rental housing and issues. (p. 116)

The principal of the hierarchical level of a nation should be as simple as possible, but it depends upon the size and population of a nation. The function and responsibility of every line ministries or bureaus should be clearly identified, announced and transparent, and should avoid duplication or overlapping of responsibility as much as possible. The role of Ministry of Finance is important, because fiscal policy and monetary policy are two major policies to affect the national development. For developed countries, the fiscal expenditure is focused mainly on the social side, the growth of national economy depends on guidance of policy and create a sound and legal environment to create a normal environment for the activities of the market force. Table 4.6 presents the share of budget expenditure of Swedish government in 1990 for the activities of different ministries.

(6) *Models of Large Scale System at the National Level*

Model is a qualitative or quantitative representation of a physical objective, process and behavior, it should express the deterministic features of the above three under study. (Wang, 1980).

Chestnut discussed the development of national economic modeling, and there were probably from 20 to 100 or more economic models developed in the United States by the year 1982. Analysts and decision makers make forecasts and estimate the economic environment within the nation, or a geographical region, or industrial sector of nation.

Chestnut quoted a flow chart of the Wharton Annual and Industrial Forecasting Model to be an example to explain major variables and policies. He also constructed Fig. 4.7 which shows economic relationship among national seven sectors.

4.3.1.4 Abstract of Part 2 of the Paper

Large scale system at the international level is dealt with in Section 7 of his paper, which discusses both the issues of international trade and searching for world-security domains; Section 8 deals with barriers to success to establish a cooperation world system, and Section 9 is a conclusion, focused mainly on the world cooperative system.

As a system engineer, Chestnut had a clear concept that a national system is a subsystem of international system, and has a close relationship with other countries on flow of goods, services and funds, and also it is necessary to identify the actions

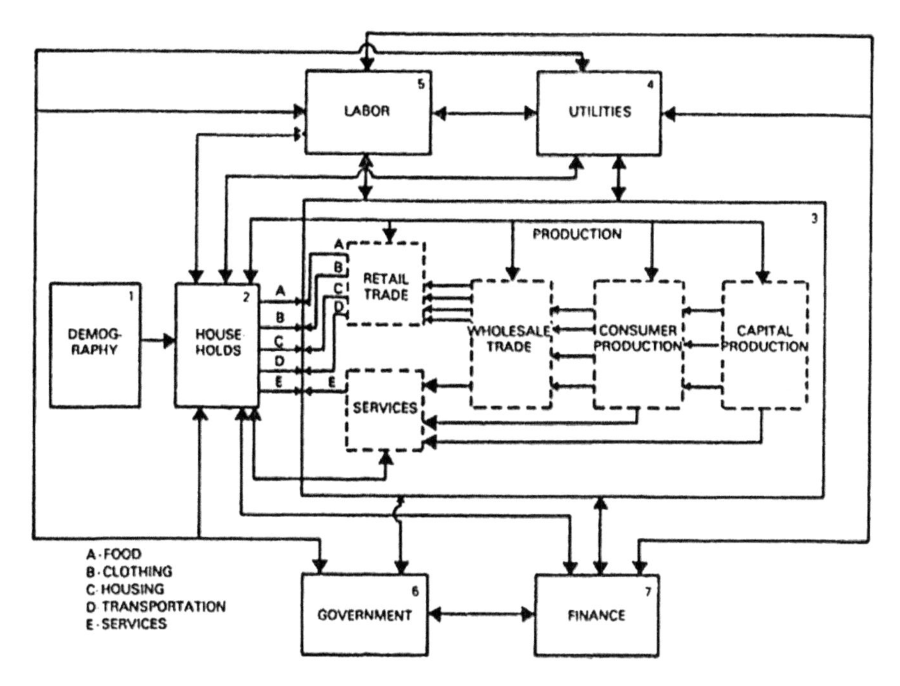

Fig. 4.7 Simplified description of relationships among national sectors (*Note* Chestnut, 1982, Fig. 9, p. 172)

needed to improve world stability and security. Several of his observations will be quoted respectively in (1) and (2) in the following, some supplementary remarks are also given to adapt Chestnut's discussion more to the contemporary world.

(1) *Large Scale System at the International Level*

Chestnut has a concept that in some particular specialized field, the world does act as a single large scale international information system. He studied the national system at international level from two aspects: the first is from the aspect of trade which is shown in Fig. 4.8 (quoted from his paper) with transfer of goods, services, funds and people, Chestnut also pointed, out that there are flows of information to the national decision makers who in turn may influence the operation of national sectors involved.

He also pointed out the possibility exists for instabilities among the nations involved to occur. Not only there should be some stabilizing means to overcome the tendencies for instability, but also there should be means for sharing information among decision makers in different countries to improve the possibility of international stability. His suggestions remind us of the two large global financial crisis post WWI.

In fact, the situation of trade had been changed greatly by the process of globalization. There is one chapter in Gidens's *Sociology* (2001) in which he states:

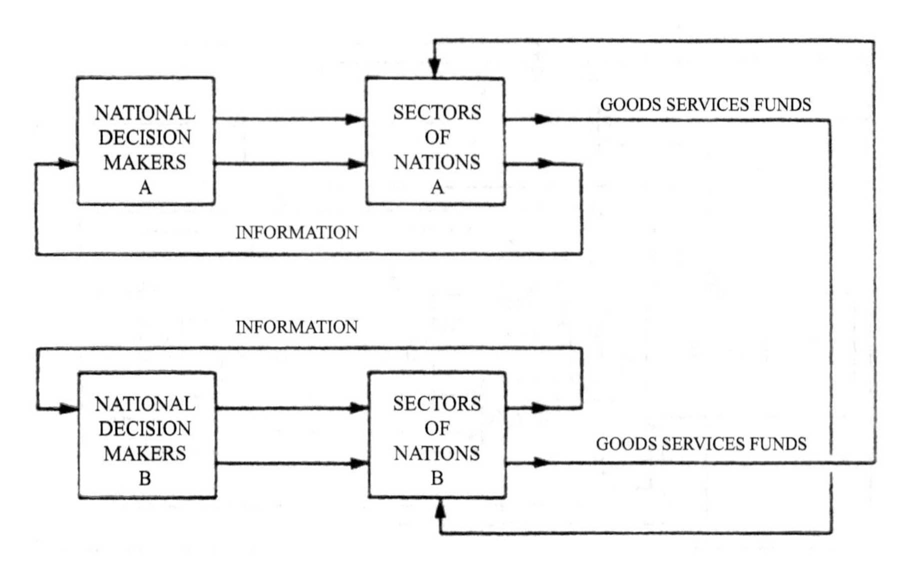

Fig. 4.8 Interaction of nations with transfer of goods, services, funds and people (*Note* Chestnut, 1982, Fig. 11, p. 175)

"Globalization is often portrayed solely as an economic phenomenon. Much is made of the role of transnational corporations stretch across national borders, influencing global production processes and the international distribution of labor. Others point to the electronic integration of global financial markets and the enormous volume of global capital flows." (p. 52). Economically, production overseas to serve local markets to lower production costs has long been a feature of FDI (Foreign Direct Investment) by the United States, a similar pattern has also taken place by Japanese TNCs (Transnational Corporation) Global production networks were developed, and there were reorganization of production at international level, In the 1990s, a new framework, called Global Commodity Chains (GCC) and the concept of value-added chain has also been used to the global organization of industries. Although these global production and trade networks are generally beneficial to the global economy, except for the issue of distribution of income, for example, China is responsible for final assembly of I-Phones, but China receives around only 1.5% of the total amount of the cost of sales. The global capital flow had caused two great financial crisis in recent years, the Asian financial crisis in 1997 and global financial crisis in 2008. Stability of the global economy due to financial flows is a real issue to be studied seriously.

The second aspect concerning Chestnut is from the perspective of searching for world security. He has identified nine aspects to be searched, which is shown in Table 4.5.

(2) *Barriers to Success*

Chestnut (1982) stated in his paper that "…in solving the larger scale system problems, in the part, has been to identify what the obstacles or constraints are,

Table 4.5 Search for world-security domain[a]

Serial No.	Domain to be searched	Chestnut's awareness
1	Search for common values and standards of conduct	Difficult to achieve now
2	Evaluation of principal world-security options	
3	Organization for global and regional security	
4	Peace keeping force	Realized, have been talked in early days of UN
5	Inspection and verification	Difficult to accomplish up to present
6	International dispute settlement	
7	Controlling new weapons-usable technologies	
8	Building international economic infrastructure	
9	Phasing the world-security process	Challenge is to develop a cooperative Security System

[a]*Source* Chestnut (1982, p. 176)

what are the barriers to success, and gradually eliminate them." This statement is true for engineering systems, but may only be true partially for social systems engineering especially for the international system, because there are many different cultural systems in different countries, and too many stakeholders with different interests. But Chestnut's analysis of major barriers to success is still meaningful as shown in Table 4.5, unfortunately, there is no knowledge as to how long it will take to overcome these barriers.

4.3.2 *Case Study 2 of China Towards the Year 2000*

4.3.2.1 General

China started its opening and reform process in late 1970s. It had been targeted by the 12th Congress of the Party to quadruple[14] the gross value of industrial and agricultural output by the year 2000 compared to the year 1980. There was also a trend of future studies by that time. For example, there was the publication of *The Global 2000 Report to the President Entering the Twenty-First Century* by the United States, focused on the intensifying stresses of environment, resource and

[14]Note: China had adopted the MPS (material Production System) of former Soviet Model by that time.

population to the quality of human life on our planet. The purpose of China's study was to serve as the "the foundation of long term planning". The Secretary General of CCP Hu Yaobang had high awareness both domestically and internationally, and requested Ma-Hong in 1981, the Vice President of Chinese Academy of Social Science (CASS) and also Director General of newly established Technical-Economic Research Center (TERC) of the State Council, to implement the study of China towards the year 2000, to give a relatively clear vision of China twenty years or even fifty years later." This study was listed to be key project of planning of social science in China's Sixth Five Year Plan. The former TERC (it was emerged into Development Research Center in later period) started this project on May, 1983, and completed this project on May 1985.

4.3.2.2 [15]An Overall Introduction to the Study "China Towards the Year 2000"

(1) *Background of the Study*

After ten years of turbulence during the Cultural Revolution, the Eleventh Central Committee of the Communist Party of China (CCCPC) at its Third Plenary Session had brought order out of chaos. The Party and the nation had successfully shifted the focus of national strategies to four modernizations. A series of guidelines of economic and social development had been formulated by the Party and the Government and achieved tremendous success that has attracted the attention of the whole world. On this basis, it is necessary for China to study its prospects to the end of the century and after. Since the Autumn of 1982, the former Technical Economic Research Centre (TERC will be used in later part) of the State Council[16] had carried out a systematic research project with the contents and organization of this theme. This subject was accepted in 1982 as a key research project for the philosophy and social sciences during the Sixth Five-Year-Plan period.

The research project was carried out with the objective of presenting a relatively clear, concrete and vivid scenario of the development of China's socialist economic, cultural, scientific and technological, people's living and spiritual civilization through an overall and comprehensive study, through synthesis and analysis of the subjective and objective conditions, domestic and international environmental studies under the guidelines set up by the Twelfth National Congress of the Communist Party of China. The research goal was to find out a relatively

[15]Note: Nearly the whole part of this case study from 2 to 4 are from the paper Wang and Li (1986).

[16]Note: The Technical Economic Research Center is emerged into Research Center for Economic, Technological and Social Development of the State Council in 1985. Later, the Center is further called to be Development Research Center of the State Council. (Abbreviated to be DRC).

satisfactory development pattern through comparative study of different feasible ways in achieving the strategic objective of socialist construction and also to study the basis and decision for the realization of these objectives as well as the necessary policy system for its realization. An analytical study and evaluation of this policy system had been carried out. In short, the project of the research "China Towards the Year 2000" is to realize better the strategic goal set up by the Twelfth National Congress of the Communist Party of China; to supply a systematic reference with scientific basis for the decision making and policy making for the Party and the State Council; to provide an overall scenario for the development planning of sectors and regions. It is hoped that through this concrete scenario Chinese people will be further encouraged to concert their efforts towards realizing the ambitious guidelines of the Party. In other words, the study "China Towards the Year 2000" is to find a way to achieve the socialist modernization process of the Chinese pattern. This project was highly recommended by former governmental leadership, his message to the annual plan submitted by the Research Centre in 1983: "'China Towards the Year 2000' was an important project, if this study was well researched it would be of a great significance, it was necessary that efforts should be made to do a good job of it."

Headed by Ma Hong, a number of leaders from the State Planning Commission, State Economic Commission, State Science and Technology Commission, the Chinese Academy of Social Science and the former TERC of the State Council formed a major leading team. They established a research team including personnel from the Research Centre and China Scientific and Technical Information Institute responsible for the organizational and research work. The research received cooperation from scholars and administrative officials from various scientific research institutes, economic management departments, planning institutes and institutes of higher learning. This research project was divided into three hierarchical levels. More than 400 research fellows were involved in the first and second level study. 108 science and technology associations were involved in the third level study. Through more than two years of hard work including collection of data, field survey, qualitative and quantitative analysis, synthesis and analysis, research results of the first level study were completed in May 1985.

The main results include one main report, thirteen sub reports and a collection of projection data. The titles of these are: *Main Report-China Towards the Year 2000*, *Population and Employment of China Towards the Year 2000*, *Economy of China Towards the Year 2000*, *Consumption of the People of China Towards the Year 2000*, *Natural Resources of China Towards the Year 2000*, *Energy of China Towards the Year 2000*, *Education of China Towards the Year 2000*, *Science and Technology of China Towards the Year 2000*, *Environment of China Towards the Year 2000*, *Agriculture of China Towards the Year 2000*, *Transportation System of China Towards the Year 2000*, *International Environment of China Towards the Year 2000*, *Macro-Economic Model Projections of China Towards the Year 2000*, and *Summary of Data of Projections of China Towards the Year 2000*. Besides the above mentioned main report and sub reports that were the direct responsibility of each group leader, the second level reports of special topics were done by various ministries i.e., social life, culture, athletic and sports, broadcasting and television, communication, foreign trade, construction materials, oceanic development strategy, natural resources, railway transportation, highway transportation, ocean transportation, domestic waterway transportation, urban passenger transportation, civil aviation, pipeline transportation and health care system.

(2) *Research Methodology*

The object of the study "China Towards the Year 2000" covers a large and complex socio-economic system. The research work must ensure that a proper coverage is given to the selection of the contents. Methodologically, it is necessary to sum up the experience of the past, to stand up in the present and to face the future. In organizational aspects, it is a multi-level interdisciplinary task oriented transient organization (ToTo). A system design for the research contents and organization was prepared.

In undertaking the research, attention was also given to the proper trade off between the long term goal and the current national status, with the latter as the starting point; to the combination of conceptual framework with the concrete scenario and started with the latter; to stress the theoretical research with the practical study and to emphasize the latter. Emphasis was given to combine the qualitative and quantitative analysis properly; the routine and the modern method in combination; an in depth survey in combination with the comprehensive research; special research combined with coordinated study. In short, mutual complementary research methods have been used throughout the study.

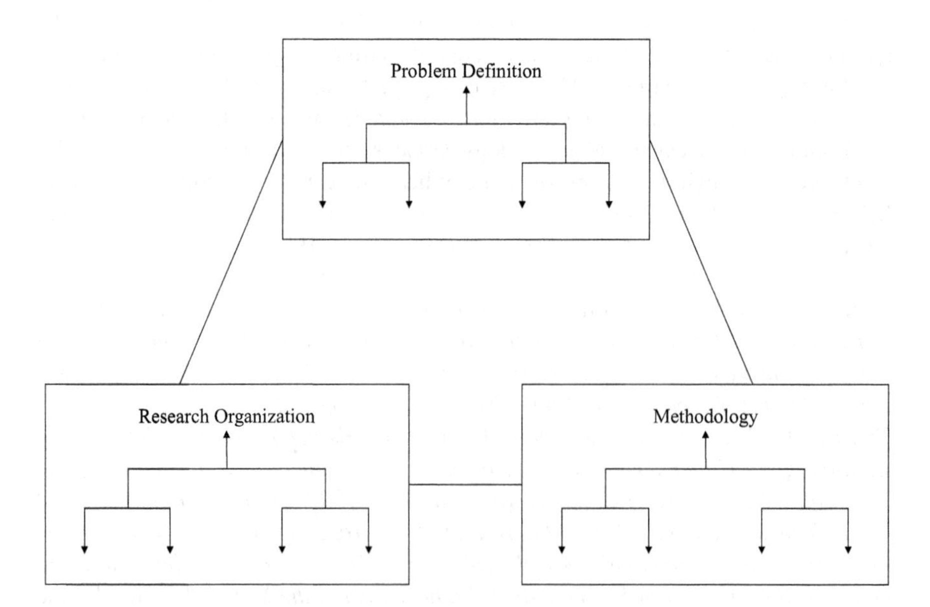

Fig. 4.9 System approach for research of complex problem

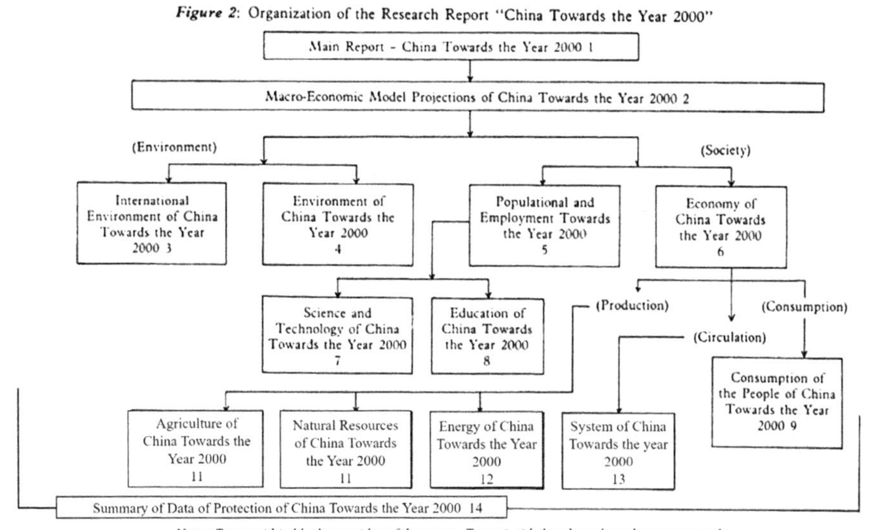

Fig. 4.10 Organization of the research report "China towards the year 2000"

(i) *A System Approach to the Subject of Study*

A system approach of this subject includes a comprehensive study of related aspects, i.e. the problems involved in the research object, the research methodology and the research organizational system design. These three are interrelated aspects, their relationships is shown in Fig. 4.9.

(ii) *The Organization of the Research Contents*

The research contents to be included have been carefully studied, the final topics have been chosen through screening from a series of topics of socio-economic activities including production, consumption and circulation done mainly by human resource and this socio-economic system is studied within an international and ecological environment. This conception of organization of the research contents is shown in Fig. 4.10.

(iii) *Inter-related Elements of Research*

Figure 4.11 also shows the dynamic concept and the connection between related elements. The research object includes a number of fields and socio-economic activities. It is necessary to get a clear picture of the current status and problems of all these fields and the possible state of their development towards the year 2000 in a dynamic pattern. Meanwhile the major concern is the development pattern under proper policy control for the coordinated development of economy, society, science and technology to achieve the strategic goal set up by the Twelfth Congress.

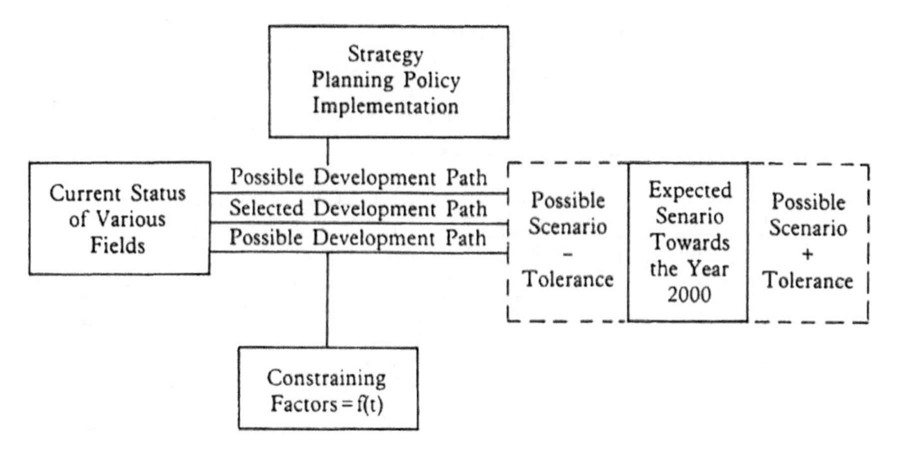

Fig. 4.11 The interrelated elements in the research objective

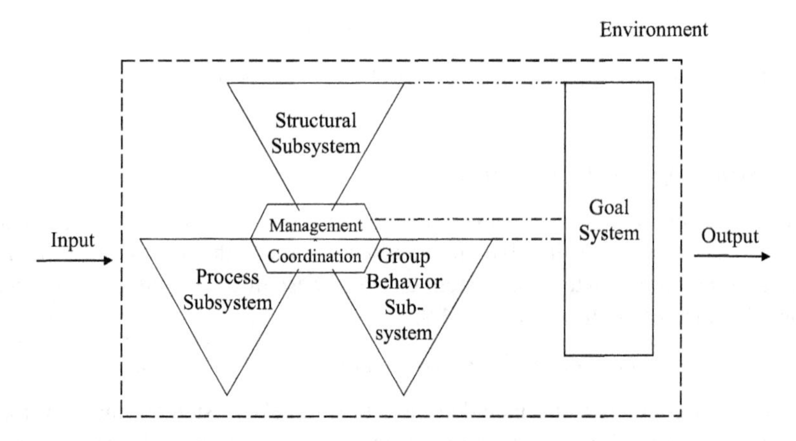

Fig. 4.12 An integrated system view of organization

Therefore, it is also necessary to study the design of overall strategy and policy system at different levels. These relationships are shown in Fig. 4.11.

China Toward the Year 2000 study covers very broad areas. The problems of these areas are mutually relative and interactive, such as how the change of the internal political and economic environment will affect the development of China and the policy to be adopted. Certain constraining factors are changing constantly with the changing environment. For example, in the First Five-Year planning period in China, there was a sufficient amount of land and water resources, and they did not impose constraints on the site selection for industrial plants but today, land and water have become serious constraints in the selection of plant site in certain regions. Therefore, care needs to be taken on the overall and regional restraining factors and their change in the future development process. Development strategies and policies need to be adopted to the changing situations.

(iv) *An Integrated Systems View of Organization*

Whether a project can be carried out effectively or not depends upon an effective organization. Organization theory and management practice are evolving continually, our organization for this task is also based upon the system approach, i.e. to view the transient organization in interaction with its environment and relationships among subsystems. The conception of organization is shown in Fig. 4.11. This point is important for the "Management Coordination" function performed by TERC and the Project Coordinator. In Fig. 4.12, the structural subsystem refers to the choice of a proper institution or personnel for this study. The process subsystem means a proper organization of the process for this project, such as preliminary system design by the project coordinator, preliminary group exploration to define the problem ... up to the final publication of the reports. The group behavior subsystem is also a subject to be taken care of by the management coordination, because scholars from research institutes, professors from universities and officials from the ministries may have different views and behave differently in thinking in this interdisciplinary study. Only when all of these aspects are taken care of, then can the project be completed in time and in accordance with the goal system set up.

(3) *Scenario of China Towards the Year 2000*

Eight scenarios were developed in this study.

(i) *Population will be 1.2 Billion or Slightly Larger*

The "Population" problem is the starting point and the end point of studying the development of society and economy. If the estimated annual rate of natural increase 0.95% is realized as from 1983 to 2000, the projected population size will be approximately 1.2 billion by the year 2000; if it is 1.34% the total size will be 1.28 billion. The data shows that from 1949 to 1982, with the only exception in 1959, 1960 and 1961, the average annual rate of natural increase was higher than 1.34%. China had two periods of baby boomers, i.e. from 1950 to 1957 and from 1962 to 1973, the annual rates of natural increase in those periods were all higher than 2%. The children born during these years have entered or will enter their reproductive age before or shortly after the year 2000. Therefore, despite the family planning efforts through the national programs, which have been indeed very successful for the last few years, to achieve the earlier targeted population size of 1.2 billion will be very difficult, if not impossible to be realized. The possibility is that if the 1.25 billion projection is realized, China's population will constitute 20.4% of total population of the world, which will be 6.12 billion approximately.

The other phenomena related to population include the infant mortality rate dropping from 2.1% in 1981 to 2% in 2000; the life expectancy increasing from 68 in 1982 to 72 in 2000; the age group of 65 and above occupying 4.9% in 1982 to 6.9% in 2000; a population transition from young age type to adult type; urban population rising from 20.8% in 1982 to 38.0% in 2000.

(ii) *"Relatively Comfortable Living Standard" of Different Levels*

The living standards of the people by the year 2000 will be up to the standards internationally termed as "satisfaction of basic needs" for the developing world or higher. Nation-wide consumption will be higher than present middle-income families in cities, i.e. 700 yuan (RMB) as annual consumption per capita averaged for both urban and rural population, three times higher than 227 yuan in 1980. Urban residents will have a consumption of annual 1200 yuan per capita, small town residents of annual 600 yuan per capita. The difference between the urban and rural sectors will be diminished by the end of the century, i.e. 2.7:1 in 1980 to 1.7–2:1 in 2000. Nevertheless, regional inequality will still exist in consumption, the more developed areas around Beijing, Tianjin, Shanghai and Guangdong (Canton), as compared with remote areas that are economically less developed. However, it is necessary to tackle this problem of regional inequality in the process of development. In that respect, there are different opinions among the Chinese scholars on the strategy of development. If one sticks to the latitudinal gradient theory, i.e. the coastal area of China is relatively developed, so the development process will be a gradient process from East to West. Others are advocates of the point-axis development process, i.e. with selected developed cities as points and these cities are connected by certain transportation system such as natural navigable waterway or railway etc. (axis), and then the development process should be a diffusion process from the point and the axis. The project coordinator is biased to the opinion that the development process should be a diffusion process. It is necessary to build up certain developed areas to form a gradient relative to its surroundings by utilization of the comparative advantage properly, then a proper diffusion process can be taken into consideration. There must be a proper trade-off between the development from the cost-benefit analysis on the basis of pure economic consideration, and the consideration of social equity on the other.

(iii) *The Fifth or Sixth Economic Power in the World*

The gross value of industrial and agricultural output of China by the year 2000 will be 2953.5 billion yuan, 4.1 times that of 720.7 billion yuan in 1980. This growth rate if realized will have exceeded the expected rate for the target of "quadrupling the GVIAO[17]". The average growth rate will be 7.3%. GNP by the year 2000 will be in US dollars, 1184 billion, 4.4 times that in 1980, and the average growth rate will be 7.7%. Taking 1.25 billion as the population size of that time, GNP per capita will reach approximately US$950. According to this projection, China's economic status will rank the fifth or sixth place in the world. Although the GNP per capita by that time is still very low, China viewed from an angle of total strength will be one of the leading economic powers in the world. By the year 2000, China still will not be able to stand among the first eighty countries as far as national income per capita

[17]Note: GVIAO is the abbreviation of gross value of industrial and agricultural output, which is unit of measurement of economic growth of MPS system adopted by former Soviet Union.

is concerned, China will still be lower-middle income country by the end of this century.

(iv) *The Gross Value of Industrial Output will be as high as the United States in the early part of the 1980s*

By the year 2000 China's gross value of industrial output will be 2420 billion yuan, 4.7 times that in 1980 (519.4 billion). This volume is as high as the United States in the early part of the 1980s.

With regard to industrial infrastructure, transport for instance, by the year 2000, the total length of railway will be 70,000–75,000 km; the total length of road transport system range from 1.3 million to 1.5 million kilometers; the airline 400,000–450,000 km.

Projection for the energy requirement will be 1.56–1.7 billion tons of coal, but the availability envisaged will be only 1.3–1.48 billion tons.

(v) *Agriculture will meet the needs of the people and country's economic development*

By the year 2000, the gross value of agricultural output (including township enterprises) will be 1010 billion yuan with an average annual growth rate of 7.2%.

By the end of this century, grain output envisaged will be 500–535 billion kg, cotton 4.8–5.25 million tons; rapeseed 21.3–21.8 million tons; meat 27.8–30 million tons. Both staple and cash crops will be sufficient to meet the needs of the people and the economic development.

(vi) *Remarkable Achievements in Science and Technology, but Still Lagging by Ten to Twenty Years Behind Advanced Countries*

China's science and technology by the year 2000 will be comparable to the level of the most developed countries in mid-latter part of the 1980s or in the beginning of the 1990s; as far as the overall technological capability is considered, China will compare herself to the world standard in the latter part of the 1970s or the early part of the 1980s.

By the year 2000, China will have a group of outstanding scientists in basic research, e.g. mathematics, astronomy, life sciences, earth sciences, probably leading the world.

China's technology will be less developed than basic sciences, but in petroleum, petrochemicals, machinery and some other fields the gap between the advancement level of China and the world will probably be very small. This is not only because China has already built up her technical capability in these areas, but also the rate of technical progress of these in advanced countries will be slower in the future. But the possible widened gap may be seen in such areas as nuclear fusion, artificial intelligence and oceanographic research. The impact of combining high technology and traditional technology forming a technological complex will become increasingly greater on the country's economic development.

(vii) *Rapid Development of People's Education of Three Orientation*

By the end of the century, the world-oriented, future-oriented and modernization-oriented education will develop rapidly and bear the following features: stress will be made on both formal education prior to job taking and on-job training, thus forming a continuous system of life education; the philosophy of education will be altered from knowledge-imparting to intelligence-developing.

By the end of the century, the rural population will get a primary school compulsory education, the urban population will get a junior middle school compulsory education, of which a segment living in big cities will get senior middle school education. The proportion of those having higher education attainment in the year 2000 will be 2.0% compared to the present percentage of 0.6. Efforts to achieve basic literacy will be further strengthened and the projected illiterate rate in the year 2000 will be 8% in comparison to 23.5% in 1982, the fact that more than 50% of the age group of 45 years old and above are illiterate makes the whole endeavor difficult.

By the year 2000 the projected demand of personnel with specialized education will be 49 million, three times greater than the 14 million available in 1983, hence from the fall of 1983 to 2000 China will need to train at least 14 million personnel.

(viii) *Development Prospect*

The whole national economy will become a vigorous one of the opening-to-the-outside-world type and a socialist economic system with Chinese characteristics will be basically formed. Traditional mentality will be greatly changed and socialist spiritual civilization considerably strengthened.

It is expected that by the end of this century, China will become a powerful socialist country with political stability, prosperous economy, strong national feelings and with peace-loving and happy people. China will display to the whole world a comparatively perfect, creative and vigorous socialist model with Chinese characteristics.

(4) *Basic national condition and major difficulties to achieve the goals*

(i) *Background for Formulation the General Strategy and Policy*

The recognition of the present natural conditions and the development goal of China is the foundation for the formulation of the general strategy and the full analysis of the constraints possibly encountered in the development process provides the major basis for working out the policy system.

The basic national conditions of China are as follows: China has a vast territory but a very unbalanced economic development. She has a large population and rich human and natural resources but the cultural and technological quality of the people are low which causes hardship to the society and economy; natural resources are abundant as a whole but on per capita basis it is small, economic and cultural constructions are considerably large in size but economic development is at a low level; science and technology lags behind those of the developed countries; in particular, infrastructures are insufficient; advanced socialist political and economic systems have been established but still need perfecting; the malady of economic system and the closed-door status of economy formed in past years are retained to a large extent; the 'open door' policy to the outside world has made it possible for China to use the experience of other countries for reference and utilize foreign capital, techniques and markets, but at the same time China shall stick to the establishment and perfection of her socialism.

The outlook may reflect various possibilities, but not realities. Between "possibility" and "reality" there is a gap which requires the full analysis of goal, present status, difficulties and various changes to be possibly encountered during the development. Studies show that in realizing the magnificent goal of the year 2000, China shall face difficulties as follows.

(ii) *There are Six Major Difficulties*

(a) There will be a net increase of more than two hundred million in population, causing heavy burdens on employment and social overhead cost;
(b) Education, science and technology are far from being adaptable to the development of the national economy;
(c) Transportation, energy, communication etc. will form the bottlenecks of development. These infrastructures are far from being satisfactory when compared to international standards but it is very costly to expand these infrastructures in greater scale.
(d) Some problems related to natural resources. China is short of surface resources and per capita resources are small. For instance, the per capita cultivated land, forests, grasslands and water resources are 36, 11, 46, and 25% of the average of those of the world, respectively. Although reserves of some of the ground minerals such as tungsten, bismuth, titanium, rare earth metals, pyrite, asbestos ranked first in the world, some of them are rarely used in world market and some cannot be fully processed by China at present. Furthermore, there are problems in natural resources such as the unbalanced distribution of resources, the low quality of resources etc., the abundance of low grade ore and scarcity of high grade ore and the large quantity of associated and Paragenic ore deposits and rareness of deposits with a single mineral or ore with a single composition. The surface resource problem is further exemplified by distribution of water resources and cultivated land. The area south of the Yangtze River domain possesses 80% of the total water resources of China and only 36% of the total cultivated land, whereas the area north of the domain has 18% of the total water resources and 64% of the total cultivated land. In particular, the Huanghe River, Huaihe River and Haihe River domains have only 7% of the total water resources and 40% of the total cultivated land. This situation is also the case with coal resources (Table 4.7).
(e) Ecological environment tends to be deteriorated. China is a forest poor country. The percentage of world forest is 31.3% while China has only 12%. There is also serious soil erosion, which is estimated around five billion metric tons per year.

(f) Insufficiency of capital is a major factor restricting the development of national economy and the reform of economic system.

A correct strategy and policy should be worked out from a long-term and overall point of view. The key point is to coordinate well the controlling factors between the long term and various stages as well as between the whole and individual parts so as to achieve the proposed goal. The macroscopic study of China in 2000 should of course be made from a long-term and overall point of view but the development strategy and policy of each stage and every local area must be coordinated with those of the long term and the whole country.

(5) *General Strategy of Development*

On the basis of the above background, this study has formulated, through research, the general strategy, which is described as follows:

(i) *The Strategy Arising from the Relationship between Systems and Subsystems*

(a) A strategy of coordinated development of the ecological environment, social, economic and scientific and technological system should be adopted. Compared with the growth of population, land and forests are limited and atmospheric environment is of limited capacity. Thus, great attention must be paid to this. This problem should be appropriately handled through comprehensive consideration from the long and short term conditions. China is a country with a low percentage of forest covering and hence afforestation is a very pressing problem. Vice Premier Wan Li has instructed that restrictions on afforesting barren hills and mountains should be relaxed. He said: "Covering the motherland with trees is a great cause that benefits our children and grandchildren. And it is one of the basic national policy". As for atmospheric pollution, the annual average of grained substances from waste gases discharged by factories in China is now 0.9 mg/m^3 in northern cities and 0.4 mg/m^3 in southern cities and that of sulphur dioxide is more than 15 million tons. However, the degree of urbanization of China is not high at present, and to process the sulphur dioxide would be very costly. Considering both the urgency of short-term economic growth and insufficient financial resource at present, and the importance of protection of ecological and physical environment in the long term, studies have been set up for the environment strategic goal of "controlling the deterioration of environment and emphatically improving environment quality and status" towards the year 2000.

(b) The social development strategy studies address three aspects. The first, is the relation between the social and economic system. The Central Committee of the Chinese Communist Party announced the Decision on Economic System Reform, the Decision of the Reform of Science and Technology System and the Decision on Educational System Reform in 1984 and 1985 respectively. In these decisions, the interaction between the social system and economic and science-technological developments had been considered to a certain extent. The second is to correctly handle the contradiction between the short and long-term social developments. For example, human resource development is of great importance to social and economic development. From a short-term policy, the investment in the education system is contradictory to the financial resource required in material and economic construction, but their relationship must be well handled. The third is to deal with the long and short-term relation of regional development, properly bring the comparative advantage of various regions into full play and promote the achievement of common prosperity. China has a vast territory and an uneven distribution of resources and exhibits a very imbalanced development of economic and production structure, it is necessary to a "gradient structure" in the level of development of the regions. The comparative advantage of various regions should be realized fully, dynamically promoting the dispersion of the gradient of economic potential and at the same time encouraging the constant formation of a new gradient structure. However, it should also be noted that the horizontal gradient of economy between the regions would not be increased uncontrollably, there are needs to set different development goals and adopt corresponding different policies in different regions or areas. These regions or areas should be developed from their comparative advantage in factors of production, and it is the role of the government to organize inter-regional professional coordination to realize the optimum inter-regional combination of factors of production to achieve the maximum social efficiency. At the same time, attention should be paid to the construction of modern industrial patterns and the enhancement of scientific and technological levels in less developed regions in order to transform them into developed ones, to promote the common prosperity and gradually reduce the degree of imbalance between them.

(c) In the formulation of the strategy for economic development, the project has studied through a system of mathematical models, simulated and calculated different consumption patterns, industrial structural patterns and accumulation rate and coupling with qualitative analysis and judgement, determined the scheme that under the premise of achieving the "quadruple" goal, the national economic strength will be considerably increased and the living standard of the people fairly raised and more material benefit obtained. In addition, studies have taken notice of restraint factors mentioned above such as infrastructures (energy, transportation and communication) and proposed the policies to give attention to the economic structure with energy saving kept in mind, and to strengthening these infrastructures in the development of economy. Based on the characteristics of China that the per capita quantity of resources is low and the quality and distribution of resources are uneven, studies have also decided to adopt the policy of economizing on resources and relying on science and technology to strengthen comprehensive exploitation and multi-purpose utilization of resources in the development of China's economy.

(d) China has now more than 0.40 million large, medium-sized and small enterprises and possesses a considerable production potential. For a period of time, in the past, the growth of China's economy mainly relied on the "extensive type", not the "intensive type", i.e. on the increase of investment, equipment and manpower, but little progress was made in the improvement of work quality, products, technology, organization and management. For a period of time in the future, the policy for economic development will be focused on the technological rehabilitation of existing enterprises, with the purpose of raising the quality of products and the economic efficiency of production.

(e) In the strategy of scientific and technological development, studies have noticed the international trend of rapidity in the development of science and technology and, in the light of the concrete conditions of China, proposed a development strategy of combining new technology and traditional industries. At the same time, China has emphasized a policy of accelerating the transfer of science and technology from foreign countries to China, military to civil utilization, coastal areas to the interior of China, and laboratory to production fields.

(ii) *The Strategy of Persistent, Stable and Coordinated Development*

A correct development strategy and a policy system ensuring its success requires the consideration of persistent, stable and coordinated development of science and technology, economy and society. The experience in development strategy acquired by developing countries since World War II has proved that to pursue a short-term super high speed growth of economy, if violating the objective law of economic and social development, would not generally lead to the effect of persistent and stable growth. In the strategy of economic development, therefore, studies have determined, through prediction and comparison, to reject the scheme of heavy industry-type structure and super high speed development and chosen the scheme of persistent, stable and coordinated development of national economy. The main point is to combine organically the growth of the economy and materially benefit the people. Summing up the experience in the past and international development, study has taken note of the role of consumption. The purpose of the socialist production is to meet to the maximum, the increasing needs of the people in material and cultural respects. This is the starting point, as well as the end result, of the research in development strategy and policy. Study indicates that there are different consumption patterns for suitably selected consumption levels. Production elements and resource conditions existed may decide a corresponding consumption structure. This structure should be taken as a Chinese long-term goal, to guide economic development so that various production factors are utilized to the highest possible degree in order to achieve a relatively high level of consumption, and provide more material benefit for the people.

(6) *Policy System*

The realization of the strategic goal of economic and social development requires a scientific and coordinated policy system that can promote the development of socialist commodity economy. The policy is a tool used by the state to macroscopically instruct, control and adjust the development of economy, society and science and technology, a lever to handle the relation between state, collective and individual, and a pivot for controlling major respects strictly and relaxing restriction on minor ones. It should possess a larger coverage, sufficient to control the social and economic activities of the whole state. In particular, under the conditions of the socialist[18] planned commodity economy, a policy system which is effective in adjusting economic activities, and considering social factors is needed. The whole society and national economy form a very large and complex system. Various links in the economic field are interconnected and interdependent. In adjusting the movement of their mechanism, both strategic and tactical policies, capable of solving both macroscopic and microscopic problems, are needed. For different economic activities, corresponding policies are required to control production, circulation, consumption, and domestic and foreign trades. These policies should form an organic whole.

(i) *Three Aspects should be Considered in the Formulation of a Policy System*

(a) The function of a policy is that it can affect the action of people under different conditions.

Generally, it can lead people to take some actions, adjust the action to the required pattern and check or forbid the undesirable action and its diffusion. A policy may have all these three functions but a particular one may be emphasized in the formulation of it.

(b) A policy should be selected based on the comparison of various policy alternatives.

When selecting a certain scheme, evaluation of the degree of possible effectiveness of this policy should be done.

(c) The consequences of the policy should be predicted.

A policy often induces secondary unexpected consequences. If the consequences are highly uncertain, collective exploration techniques should be adopted during the formulation of a policy, in order to expose various aspects of the uncertainty. Seminars are a possible form of solution and pilot schemes are another acceptable form.

The policy system suggested by the research is mainly as follows:

(iii) *Structure of the Policy System*

The structure of the policy system is diagrammatically shown in Fig. 4.13. The letters in the bracket designate the corresponding policy in the next paragraph.

There are many background considerations for each policy, but explanations will be only given to (n), (o), (u) below for illustrative purposes.

(a) The “quadrupling” strategy must be selected that makes both the nation and the people rich.
(b) Practical effects should be strived for full effort in the implementation of environment protection measures.

[18]Note: In the period of implementation of the study “China towards the year 2000”, the political economists advocated a system of “Socialist Planned Commodity Economy”.

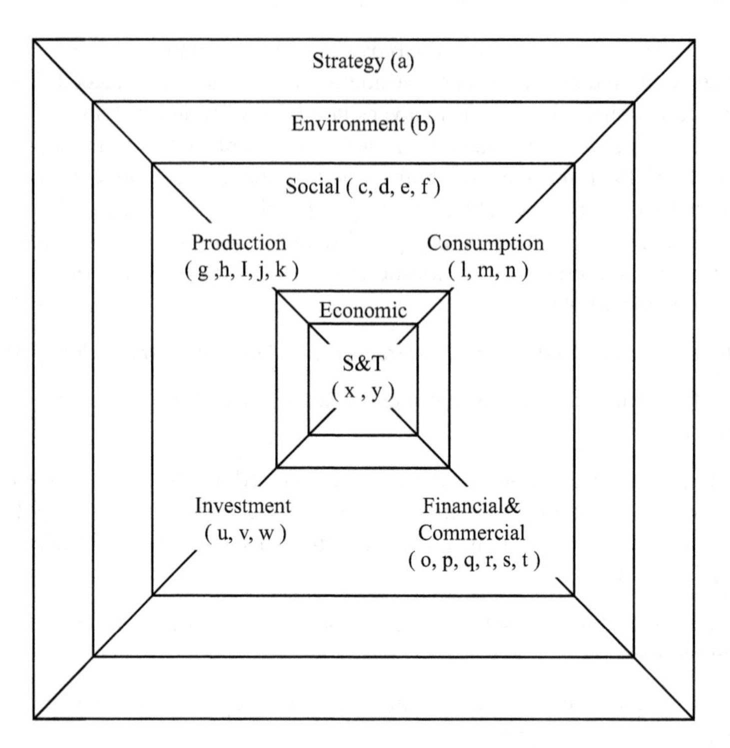

Fig. 4.13 Structure of the policy system

(c) The growth of population must be strictly controlled in order to realize a comfortable living standard for the people.
(d) A large variety of ways should be opened up for the employment of 250 million people, which is the number of projected growth of the labor force towards the year 2000.
(e) Educational concepts should be updated and creative-type personnel should be trained.
(f) The comparative advantage of different regions should be brought into full play and improve the economic gradient structure.
(g) The adjustment or industrial structure should be based on a guided consumption pattern of consumption.
(h) Technological rehabilitation of existing enterprises should be placed with priority.
(i) [19]An enterprise should be established as a community closely bound up with the destinies of the state, collectives and individuals.

[19]Note: In the period (1983–1985) of studying the project "China Towards the year 2000", the national environment has not matured to raise the policy on reform of State Owned Enterprise. This policy recommendation has only been adopted since the year 2016.

(j) The agricultural structure should be adjusted on the order of the trade-industry-agriculture framework.
(k) The bottlenecks of the transportation and communication system, energy system and the raw materials must be broken through.
(l) The growth of consumption by the people and that of per capita national income should be in synchronism.
(m) The consumption structure of the people should conform to the national conditions of China.
(n) [20]The fundamental way to solve the housing problem in cities and towns is to sell them as commodities. The housing problem is a social problem in general, but we classify this policy statement as one component of consumption policy. Because P.R.C., executed a very low rent housing policy, the rent paid for housing by ordinary staff and workers may be only three or four percent of the monthly wage. Housing was not considered as a type of commodity in the past, it forms part of the social welfare. The Chinese people have a very peculiar consumption pattern. The urban household has a high percentage of color TV sets, refrigerators and other consumer durables, but they paid a very small amount of their share on housing rent. The intention of this policy is to diversify the consumption pattern on one hand, and to promote the development of construction and the construction material industry on the other.
(o) In price adjustment, a "small step-by-step" method should be used. The price distortion problem in China has been serious in the past, and prices of nearly all types of goods were official designated prices. There are significant distortions in levels of relative prices between and within sectors. The prices of many commodities reflect neither their value, nor the relation or supply to demand. It had been pointed out in the "Decision on reform of economic structure" that this irrational price system has to be reformed. There are different opinions on how to carry out the price reform. The study foresees difficulties in implementing a comprehensive price reform in a country like China. This process will inevitably be a slow and difficult one. The solution recommended is to implement the price reform in a small "step-by-step" process.
(p) The issue of currency and the extension of credit must be strictly under macroscopic control.
(q) The system of criteria for judging the development of a national economy should be perfected.
(r) Accounting prices should be used to provide a basis for macroscopic decision making.
(s) The utilization of foreign funds should be appropriate in quantity and studied well in advance.

[20]Note: This policy recommendation is correct in the period of study of the project. But in the later period, especially the years post 2000, the housing rental market is not well established. The housing market has become too much commodity oriented, its basic social nature is distorted due to speculation.

(t) The optimum structure of foreign trade should be established on the basis of economic level; creation of a worldwide market channel should be examined.
(u) The reform should be carried out to enable state funds to be paid back after being used; (In the past, nearly all enterprises were state-owned and collectively-owned and a large percentage of investment was paid by the government. The suggestion for such a policy is to improve that situation. Now the state had changed the policy from direct allocation of funds for investment to new construction projects through credit.)
(v) The construction of key projects should be ensured by economic means.
(w) Investments in "non-production construction" should not be reduced.
(x) A new type of "technological complex" should be established.
(y) Resource-saving and energy-saving type technique should be placed with certain priority.

4.3.2.3 Foreign Trade of China and International Comparison of Similar Projects

(1) *Development Trend of World Trade and Foreign Trade of China Towards the End of this Century*

It is predicted that the growth of the economy in the Western countries will be still low in the second half of the 1980s. The growth of the world economy is expected to pick up speed in the 1990s, but instability and imbalance will still exist. Due to the low growth rate of the world economy, there will be unstable factors in the international monetary system and the development of world trade may be not fully optimistic. Currently, the trade protectionism is gaining momentum, prices of primary products in international market are dropping, the imbalance of world trade is sharpening and the burdens of international debts are heavy. These are all unfavorable factors to the development of world trade. The present low oil price, low exchange rate of the U.S. dollar, although favorable to the expansion of world trade, have caused different economic effects on various countries because of the difference in their position. Since there is a considerable uncertainty in the prospects of world trade, the figures or the development of world trade given below are estimates that can only be used for reference. According to our estimates, the average annual growth of world trade will be about four percent in the second half of the 1980s and about five percent in the 1990s, whereas the value of world trade will be nearly 2500 billion U.S. dollars in 1990. It is estimated that the world trade will be up to 4000 billion U.S. dollars by the year 2000. From the structure of different countries and regions, at the end of this century the developed, developing and centrally planned economy countries will account for 69, 25, and 10%, respectively of the total of world trade. The trade of the Circum-Pacific region will be considerably developed. With the experience of other countries coupled with the concrete conditions of China, it is necessary to adopt a mixed strategy of export orientation and import substitution in foreign trade and assign different proportions to different industries, and development stages in an attempt to obtain a suitable balance between the opening of economy and the protection of China's own industry. The prediction obtained by linear regression shows that the total export value of China will be US$100 billion (87 billion at minimum) in the year 2000, making up about 2.3% of the total export value of the world in the same year, and the total import and export value will be about US$180 billion by the year 2000. In the export commodity structure, the values of primary and finished products account for 49.9 and 50.1%, respectively in 1984, and 34.5 and 65.5%, respectively in the year 2000. In the import commodity structure, the values of capital goods and subsistence make up 81 and 19%, respectively in 1984 and 85 and 15%, respectively in the year 2000.

China's present import and export markets are relatively concentrated. China's imports from and exports to the markets of developed countries account for more than 60% of the total of its imports and exports, but China's trade with developing countries makes up less than 30% of the total, and that with the Soviet Union and Eastern European countries about 6%. The trade balances of China's imports and exports also are characterized by a regional disequilibrium. An unfavorable balance occurs in the trade with developed countries such as United States, Canada and Japan, whereas a trade surplus is achieved in the trade with Hong Kong, Aomen (Macao) and Southeast Asia. With the development of world trade and the shift of comparative advantage, in addition to the continued consolidation and expansion of presently available markets, a diversified market strategy should be adopted in the future in order to reduce the influence on the overall situation by events suddenly occurring in a certain key market to relieve the pressure caused by trade protectionism in some import countries. Besides maintaining the key market in North America, Japan and Western Europe and the export markets in Hong Kong and Aomen (Macao), the trade with developing countries will be expanded and that with the Soviet Union, Eastern Europe and the Circum-Pacific region will be remarkably increased.

(2) *International Comparison of Similar Projects and Prospects of the Asian-Pacific Region*

(i) After World War II, a new economic upsurge appeared around the globe. As people were intoxicated with optimism, the club of Rome issued a warning in 1972: if the world population, industry, pollution and food production continue to grow and the natural resources to decrease at the current rate, there will be physical limits to growth resulting from the limited availability of the world's natural resources; or the absorptive capacity of the ecosystem, stagnation occurs leading to sudden decline in population and industrial production with the probable consequence of loss of control and recession. This warning drew the attention of all countries. People thought it was necessary to begin research into the future in order to provide guidelines for the present. The World Futures Studies Federation was founded by scholars in 1973. Many projects of study or future had been initiated at government level with different emphasis of concern. For example, "*Interfutures*", a project initiated by the Government of Japan in May 1975 and organized within the frame work of OECD, was aimed "to provide OECD Member *Governments* with an assessment of alternative patterns of a longer-term world economic development in order to classify their implications for the strategic policy choices open to them in the management of their own economies, in relationship among them, and in their relationships with developing countries," and in *The Global 2000 Report to the President* it was proposed by former United States President Jimmy Carter to study the "probable changes in the world's population, natural resources and environment through the end of the century" to serve as the foundation of long term planning. "*Japan in the year 2000*" was to study the Japanese economy facing the challenges of the twenty-first century, the progress of internationalization, design for a Japanese type aging society and invigoration of a matured economic society. These types of studies change with the changing world and domestic environment. The concept of the "Pacific Shift", i.e. the transfer of economic wealth, cultural myth and idea innovation from the Atlantic Center to the Pacific Center, is one of these. The new post- industrial society and social impact of advanced computer technology are other subjects concerned by many nations and regions.

(ii) Since the industrialization process of' China is still ongoing, and the research objective has been mentioned in the background previously, economic development perspective is the practical objective of this study, although, the mutual effects between social, economic, and S&T systems should not be ignored. The World Bank carried out "a study of some of the key development issues that China faces in the next twenty years", the project of the World Bank initiated in 1983, was completed in May of 1985. The time period of the World Bank project coincidentally matched the Chinese study. The World Bank Study *"China-Long-Term Development issues and Options"* and Chinese study, shall be compared briefly below.

(iii) The World Bank Report studied mainly the economic system. The content of the report is: one main report with eleven chapters and supplemented by six volumes. *China Issues and Prospects in Education, Agriculture to the year 2000, the Energy Sector, Economic Model and Projections, Economic Structure and International Perspective, the Transport Sector.* It is also supported by many *background* papers such as *"The Asian Experience in Rural Non-agricultural Development and its Relevance for China", "Productivity Growth and Technological Change in Chinese Industry", "International Experience in Urbanization and its Relevance for China, "International Experience in Budgetary Trends during Economic Development and their Relevance to China"* etc.

The content of Chinese study has been mentioned previously. Chinese study carried down to the third hierarchical level into very detailed sectoral studies with

Table 4.6 Barriers to success[a]

Serial No.	Sources of barriers
1	National security distrust of other countries
2	Economic instability caused by disarmament due to shift of labor force from defense industry
3	Competition for needs of energy and material resources
4	Inadequate national leadership support for peace effort
5	Congress-executive leadership rivalry
6	Ideological differences between nations
7	Lack of financial support for R&D for world security and world stability

[a]*Note* Chestnut (1982, p. 179)

Table 4.7 International comparison of certain infrastructure

Country	Year	Transportation			Final energy consumption		Rate of popularization telephone set	
		Density per thousand inhabitants			Commercial (kgce/capita)	Total incl. biomass (kgce/capita)		
		Rail	Road	Total				
China (Taiwan not included)	(1981)	0.05	0.91	0.96	105	329	(1980)	0.43
India	(1979)	0.09	2.43	2.52	25	175		0.43
Japan	(1980)	0.21	9.52	9.73	1169	1169		46.00
U.S.	(1979)	1.60	28.02	29.62	4007	4007		78.90

seventy volumes responsible by the CAST (Chinese Association of Science and Technology) such as: *Electronics, Petroleum Industry, Iron and Steel, Light Industry, Standardization, Meteorology, Psychology, Geology, Library Industrial Design*, etc.

From the comparison mentioned above, it can be seen that the Chinese project covers a broader area while the World Bank project is concerned mainly with economic perspective. The Chinese team has the comparative advantage to carry out an in-depth study of domestic sectors while the World Bank team has the comparative advantage in making international cross country comparative study.

(iv) Both have used qualitative and quantitative analysis. Chinese projection by the year 2000 has been mentioned previously in this case study, and in greater detail in following section of this case study "General Quantitative Analysis of China's Economy Towards the Year 2000." While the projection of several aggregate indicators of the World Bank is shown in Table 4.6 (Table 4.7).

Table 4.8 Average annual growth of national income alternative projections, 1981–2000 (percent, at 1982 prices)[a]

Measure	Quadruple	Moderate	Balance
National income			
GDP	6.6	5.4	6.6
NMP	6.3	5.1	6.2
National income per capita			
GDP	5.5	4.3	5.5
NMP	5.2	4.1	5.1
Gross value of industrial and agricultural output	7.2	6.0	6.4

[a]*Source* The World Bank (1985), *China*, p. 36
Note A Western measure of national income, GDP (gross domestic product) is the net output of all sectors, including all services, plus depreciation. The Chinese measure, NMP (net material product) is the materially productive sectors.

Table 4.9 Prospects growth rate of GDP for Asian-Pacific region towards the year 2000

	1982	1985	198–1985 growth rate (%)	1990	1980–1990 growth rate (%)	2000	1990–2000 growth rate (%)
U.S.A.	3010	3460	4.8	4011	3	5391	3
Canada	290	323	3.7	373	3	503	3
Japan	1062	1210	4.5	1473	4	2180	4
Australia	164	183	3.6	212	3	285	3
New Zealand	24	25	1.7	29	3	39	3
Republic of Korea	68	87	8.2	122	7	239	7
Taiwan	49	63	9.1	88	7	174	7
Hong Kong	24	30	6.8	42	7	82	7
Singapore	15	18	8.0	26	7	51	7
Malaysia	26	31	6.6	44	7	104	9
Thailand	37	43	5.7	61	7	145	9
Indonesia	90	103	4.7	145	7	343	9
Philippines	40	29	1.0	47	4	112	9
China	260	351	10.4	492	7	967	7

(v) By comparison between the World Bank study and Chinese research, many conclusions are similar, for example, in the projection of composition of final demand, the shares of household consumption for food is 49% in the "Balance" scenario, this is exactly the same as Chinese prediction for the urban citizen in the year 2000. There are certain different opinions on human development. But generally speaking, the point of view is very similar in these two research results. The projection of growth of Chinese economy done by the World Bank is shown in Table 4.8.

(3) *The Pacific Shift*

In recent years, there has been chaos on "the pacific shift", i.e. the region of the Pacific Rim now has the possibility of running the world economy. The economic expansion of this region is very rapid and while we are not going to discuss this problem, a table of the projection of related countries and regions is quoted in Table 4.9.

4.3.2.4 [21]Overall Quantitative Analysis of China's Economy Towards 2000

(1) *Introduction*

In a large and complex research project such as "China Towards the Year 2000", it is inadequate to merely analyze the law of development of the main factors and the relationship among them by using only traditional methods. The system analysis and quantitative method should be integrated to enable an in-depth study to be made and scientific conclusions to be reached.

The macroeconomic study of China by the year 2000 is research of a long-term, strategic and comprehensive nature. The objects to be studied are the strategy of coordinated development and policy analysis in aspects of Chinese economy, technology and society. The national economy is a large and complex system which should be studied by combining qualitative analysis with quantitative analysis, as well as combining traditional methods with modern approaches, including establishing mathematical models and systems engineering.

The qualitative analysis is the prerequisite and basis of quantitative analysis. The former may help grasp the crux and essence, whereas the latter provides the deepening and the basis for conclusions of the former. Hence, both should be combined to supplement each other. As compared with the qualitative analysis, the quantitative analysis is a weaker link in the economic research in China. In the macroeconomic study of "China Towards the Year 2000", efforts are made to strengthen the research of quantitative analysis. Not only mathematical models are used in most of the sub-reports, but also establish a set of model systems for the overall quantitative analysis, The part of the research work is the basis of quantitative analysis for the main report and the means of the quantitative coordination among sub-reports as well.

(2) *Overall Quantitative Model System of Macro economy*

The system consists of seven models and their titles and designers are as follows:

[21]Note: This part comes from Li, Xia, Li, and Pan (1986).

(i) The development strategy and policy analysis model by the Technical-Economic Research Centre under the State Council of China;
(ii) The macroeconomic model by Fudan University;
(iii) The coordinated development of' population and economy model by Jiaotong University in Xi'an;
(iv) The model of quantitative analysis of economic structure by Tsinghua University;
(v) The model of expanded reproduction of two main categories by Nankai University;
(vi) The long-term development trend models by Jiaotong University in Shanghai and Shanghai Mechanical College; (System Dynamics Model)
(vii) The medium and long-term planning macroeconomic model by the Forecast Centre of the State Planning Commission.

By use of the model system made up of the above-mentioned seven models, the development trends of Chinese economy, technology and society are studied from various angles and with different approaches, the future pictures are depicted and the development strategies and policy problems concerning the long-range goals are described. Viewing from the nature and use of the models, there are the comprehensive balance model, the overall quantity analysis model, the structural analysis model, the goal system model, the scientific and technological advancement analysis model, the model of' coordinated development for population and economy, etc. The analysis periods of most models are from 1980 to 2000, whereas those of system dynamics models are from 1980 to 2080. The set of models includes both the theoretical mathematical model which describes socio-economic operations and the development mechanism in China with emphasis on the expounding of the interrelation among the variables and the practical qualitative calculation model in which analyses are carried out laying stress on the support of practical data and practical planning and prediction. The theoretical mathematical model or the principle model lays particular emphasis on the comprehension of the economic system and operation mechanism, while the practical quantitative calculation model pays particular attention to the development forecasting and policy testing as well as the acquisition of conclusion data. Thus, it can be seen that the former is the basis of the latter, which in turn is the judging factor of the former. There are also individual models which study the development trend of Chinese economy by use of the international comparison method and the structural analysis method and those which analyze the development prospect of China in the next 100 years. In the set of model systems, the coordinated development of economy, science and technology and society is taken into consideration, with the economic analysis as the main task. The problems studied by the models are partly overlapped so as to facilitate the horizontal checking. With the set of models, calculations have carried out and policy analysis of many development plans is done and strategic alternatives which lie ahead in the long-term development process are identified. The contents predicted include: the consumption level and structure of China by the year 2000, industrial structure, investment scale and structure, the import and export structure and rate of economic growth. The corresponding policies and measures are put forward based on the various figures predicted. With these models, the positive effects of technological progress to the economic growth in China and the constraint forces of population, resources, energy and transport and communication to Chinese economic development as well as foreign capital utilization scale and benefit plans are analyzed and discussed. In addition, problems concerning the reform of price, finance and banking are also studied. The model system as a whole has very rich and wide research content and strongly supports the research of the main report "China Towards the Year 2000", thus helping form the overall quantitative analysis report.

Only one of the models will be described as follows for illustrative purpose.

Model 1: The Development Strategy and Policy Analysis Model.

The model consists of three parts; the long-term planning and development law model, the annual or short-term optimal comprehensive balanced development model and the short-term model of the integration of plans and markets.

The long-term planning and development law model takes dynamic input-output equation as the kernel, expanding the distribution equation of the national income. It includes both the goal system model and the optimal structural development path model and the equation of productive capital construction investment, the equation representing the economic goals and the equation indicating the steadily balanced growth after the year 2000. The goal system model takes certain kind of long-term consumption structure plan as the goal for assessment under the presupposition that the gross annual value of industrial and agricultural production is to be quadrupled and works out the goal system suited to the consumption structure while meeting the requirements of the steadily balanced development after the year 2000. The goal system consists of the total social products, the national income, the sector output, the accumulation rate, the investment structure, the balanced development rate and other economic indicators, The optimal structural development path model consists or dynamic input and output model, the equation of consumption pattern, the equation of annual capital construction investment, the equation of national income and consumption level, the model of continual economic development after the targeted year and the equation of rough locus for controlling development. The optimal structural development path is solved by the annual dynamic simultaneous equations. With the long-term planning model, various long-range consumption structures, accumulation rate plans and different development strategies of economic structures are assessed and their natures are compared from the angle of the possibility for realization. The calculation results thus obtained of optimal annual distribution of accumulation rate and optimal investment structure time series can be input into the annual or short-term optimal overall balance development model to play the guiding role for the operation.

The annual or short-term optimal overall balance development model is one kind of model system of recursive programming and may be divided into three sub-models. Sub-model I is about the dividable decision and annual optimal condition. It divides, guiding by the long-term plan and taking into consideration various expected behaviors, as far as possible the plan decision or social decision

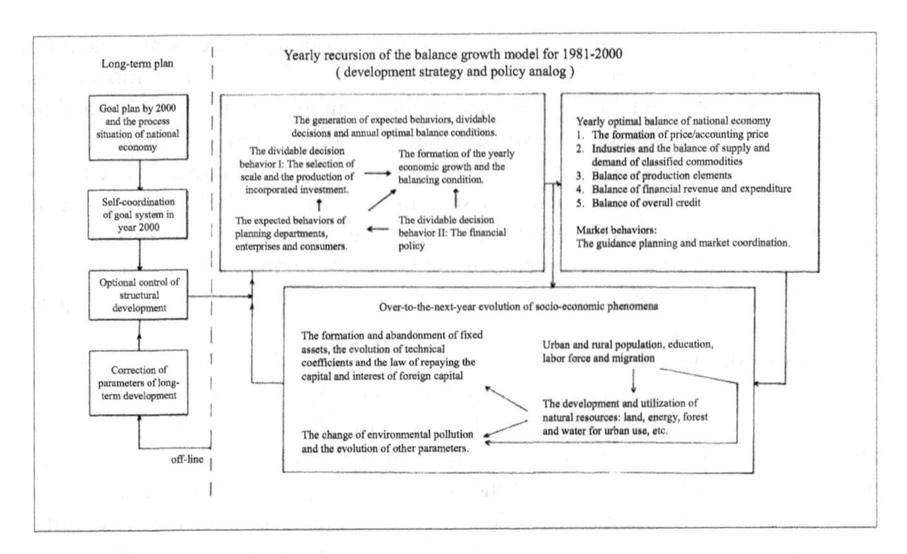

Fig. 4.14 Yearly recursion of the balance growth model for 1981–2000 (development strategy and policy analog)

into smaller problems, separating part of the decision problems from the annual overall balance for the reason that certain relevant data are hard to obtain. The dividable decision will include the selection of the share contributed by the industrial investment and environmental protection investment, the proportional relationship of sectoral investment which is used for expanding the reproduction of extensive or intensive type, the shares used in large, medium and small-sized enterprises, the proportion of investment in the simple reproduction in mining sector to the geological prospecting and the financial policies related to the overall economic scale and the overall investment scale. By summing up these factors, all data needed for the optimal balance problem can be obtained. Sub-model II is an annual optimal balance model of larger scale. It includes a total comprehensive balance relationship which may be used for determining the annual output value of sectors, the total amount and the structure of import and export, the utilization rate of foreign capital, the employment rate, the actual accumulation rate and the situation of financing and credit. Sub-model III describes the going beyond the year evolution of socio-economic phenomena and based on the data of the first two parts and using the system dynamics model and econometrics model, indicates the formation and abandonment of fixed assets, the change of natural resources, the change of population and cultural level, the index of pollution and the sum of repaying capital with interest of foreign capital.

The short-term model of interaction of plan and market is designed to describe the law and interaction of mandatory planning, guidance planning and market regulation. The state will increase the use of such economic levers as taxes, interest rates, exchange rates, credit, tariff and subsidies to guide the supply-demand behavior of the market. There exists between planning and market an inter-conditional and interdependent relationship. The model is designed to, through analogue study, assess economic levers and make the trend forecast so as to provide the basis for analysis and reference for the reform of the economic structure, see Fig. 4.14.

4.3.2.5 Features of This Case Study

This study has four unique features:

(1) *It is a Massive Project of Organizational Study*

At the top level it involves senior officials and scholars. A leading team with Mr. Ma Hong to be the head, who is a famous Chinese economist, as well as the former deputy secretary general of the State Council, vice president of Chinese Academy of Social Science and the head of TERC of the State Council. By that time there are also eight members in this team, for example, Zhangshou, the vice president of the State Planning Commission and Zhu Rongji, the vice president of the State Economic Commission at that time, and also other six members.

A working group is also established, led by its chief Madam Li Boxi from the Technical Economic Research Center (TERC) and two deputy chiefs, Li Jinchang from TERC and Bao Jinzhang from China's S&T Information Research Institute (CSTIRI). There are fifteen members in this working group, six of them come from

Table 4.10 International comparison of popularization rate of telephone set (the number in this table is end of 1980)

Name of country	Total no. of telephone set	Unit	Popularization rate (%)
America	19,160	10^4 set	83.7
Japan	5801	10^4 set	49.5
Britain	2778	10^4 set	49.7
West Germany	2855	10^4 set	46.3
France	2487	10^4 set	45.8
Former Soviet Union	2371	10^4 set	8.8
China	419	10^4 set	0.43
World total	50,829	10^4 set	11.5

TERC and nine of them come from CSTIRI. The junior author of this book Madam Li Shantong was one of the members. There are also two persons responsible for edition and publication.

There are also thirteen groups of people responsible separately for thirteen reports listed in Fig. 4.10. All groups are composed by members from production, academic field, research institutes and governmental officials. For example, the study of energy sector, is led by Lin Hanxiong, who is the head of the Bureau of Material of Construction and the Chairman of Energy Association, this group has forty-five members, five of them are responsible for writing the report of study, they come respectively from the Research Academy of Science of Electric Power, S&T Information Research Institute of Ministry of Coal Industry, Industrial technological Bureau of State S&T Commission, China's S&T Information Research Institute and S&T Information Research Institute of Ministry of Nuclear Industry.

(2) *Well Organized Process*

The process was well organized in the preparation stage and intermediate stage.

(i) *Preparation Stage*

In the preparation stage, Chinese State S&T Information Research Institute collects all relevant information for long term planning studies. Officials or professionals wrote background papers for this project. For example, the former vice minister Zhu BoLu of Ministry of Post and Communication had presented his background paper *Chinese Communication System Towards the Year 2000*, which described current status of Chinese communication system, gave international comparison of development of Chinese Communication System with other countries (which is shown in Table 4.10), forecasting the target of communication and investment required, etc.

Large scale conference and preliminary guidance at the top level was done in the preparation stage. TERC of the State Council and China Association for Science and Technology (with 108 member associations) jointly organized a large scale conference on the project "China Towards the year 2000" to mobilize the participants (professionals of natural science, social science and engineering). Mr. Ma Hong gave a speech including two aspects:

(a) Purpose of this study;
(b) Content of this study. He set forth twelve points to be studied, in fact, he gave his views to guide this study. For example, he emphasized an integrated study of China's trend of development of society, economy, culture, science and technology etc., and general rules under them; various schemes of projection and forecasting of different countries; case study of pattern of development of various countries, experiences of success and lessons of failure of them; to establish the goal system of economic and social development of China and other points.

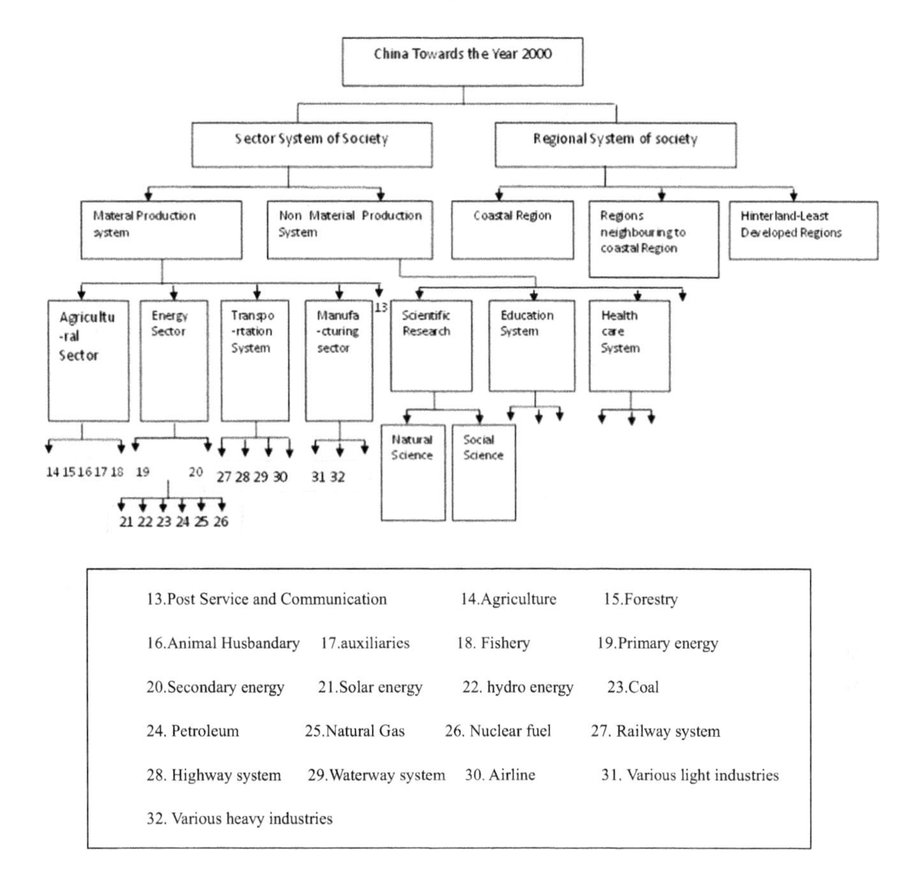

Fig. 4.15 Preliminary design of the research (first version)

(ii) *Intermediate Stage*

Preliminary design of this study had been done and revised several times. Figure 4.14 was the first version of structure of research content (prepared in early 1983) which differs greatly from the final version shown in Fig. 4.15.

Outline of the main report (2 schemes) was prepared by three responsible people, Zhang Pan, Lin Zixin, Wang Huijiong and the outline of several sub-reports was also prepared on January, 1984 by responsible people of each of the sub-reports.

Important research results were sent to the decision makers (for example, the secretary general, the prime minister etc.) in the process of study, so that essential information is accessible to the decision makers in time.

(3) *Huge Impact Domestically*

The studies wide and significant impact is due to strong organization at the team leading level (especially Mr. Ma Hong, who contributed great effort) and the working level (around four hundred people who participated in the first level research, and the contribution of thousands of people from the China Association of Science and Technology to the third level research with seventy reports. The influence of this study was huge domestically. It provided a demonstrative effect to all provinces and autonomous regions of China. For example, Lioning province, Jiangsu province etc. had prepared their provincial studies towards the year 2000, this trend of study even down to municipal level. It also impacted China's planning system, for example, the Seventh Five Year Plan of China. In the Sixth Conference of Sino-US Joint Economic Commission, it was also part of the speech given in the macro-economic management of that conference. China's State Economic Commission and the Ministry of Foreign Affairs hosted a Beijing Conference on the Asian and Pacific Economy Toward, the year 2000 jointly sponsored by the Asian and Pacific Development Center, the Chinese Academy of Social Sciences and the former TERC, and there were seven papers which are part of study of China Toward the year 2000 presented to that conference, the former Premier Zhao Ziyang had received all participants and had a photograph together with all participants.

H. J. Wang had been invited to give a keynote speech of "China towards the year 2000" in the opening session of a large scale international conference "Euro prospective" jointly organized by the Centre National de la Recherche Scientifigue/ CNRS, the Center for Long-term Forecasting and Evaluation/CPE, the French Planning Commission/CGP, the commission of the European Communities on 23rd, April 1987 in Paris. This speech was translated into French and published on the Journal "*futuribles*" April, 1988.

(4) *Attention from Abroad*

Wang also participated in the 15th International Conference of the Korean Institute of International Studies, and gave the speech China Towards the year 2000 and its impact on the Development of Asian Pacific hosted by University of Hong Kong on Aug. 1987.

He was also invited by Dr. Gerald U. Barney the study director of the *Global 2000 Report to the President*, to speak on behalf of Mr. Ma Hong on July 18th in the World Future Society's Sixth General Assembly and Exposition from July 16–20, 1989 in Washington D.C. This was a conference of significant scale with over 3000 participants and more than 200 sessions. 21st Century Studies at Future View is a special program of the 1989 General Assembly. There were around 46 "21st century studies" presented at that conference. They covered a broad range of future studies, and only a part of them dealt with whole country, for example, there were presentations on "Britain in 2010" and 'Reflections on the Project "Portugal 2000"'. The presentation by Dr. Wang on behalf of Mr. Ma Hong was "The Research Program on China's Future in the year 2000 and Beyond".

4.3.2.6 Conclusions from Theoretical Side

This massive organizational study of national long term planning is a tedious process that requires a disciplined inquiry. This process can be expressed in Fig. 4.16 as follows.

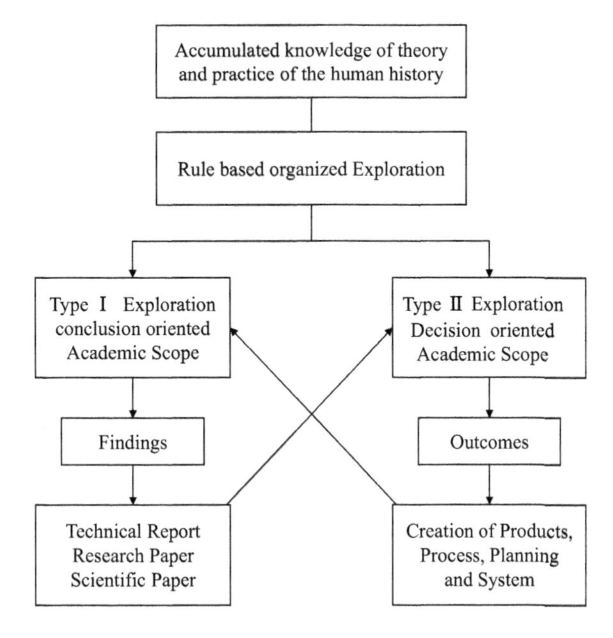

Fig. 4.16 Process of organized disciplined inquiry

The effective performance of rule based large scale organized inquiry depends upon several factors. If its purpose is national planning for the top decision makers, then a strong leadership team with significant influence is crucial, the successful results of the project "*China towards the year 2000*" illustrate this point. This can also be shown when compared with the next case study, "Integrated Economic Development Policies and Planning". One more point, is that documentary process and record keeping are also important, and this aspect is weak in many developing counties. Mankind has created a vast amount of knowledge and practice in history, but only part is kept and recorded. Yet, the progress of mankind depends upon both inheritance and innovation.

4.3.3 Case Study 3 Integrated Economic Development Policies and Planning

4.3.3.1 General

(1) *Initiation of the Project*

The launch of the concept for this project was in early 1988, the former Vice President Wu Mingyu of the Development Research Center of the State Council submitted a research proposal titled " A system of policy recommendations to the Chinese government to accelerate the process of industrialization of China" to

UNDP. The purpose of this proposal had two aspects: One is, there is an urgent need to explore emerging issues in the transitional period of China from a central planned economy to new economic system, to study new development strategy, targets of economic reform and related policies of opening and reform adopted since the third Plenary Session of the 13th Party's Congress; the other is that there are many policy research institutions established in China since 1980, there is a necessity for those institutions to exchange information with research organizations abroad and to establish a worldwide information network to promote research capability and upgrade the research level. This initial concept was supported by the former UNDP representative Mr. Kulessa. A preliminary research proposal was drafted by DRC. It had been discussed several times with the officials and staff of UNDP Beijing Residence Mission and Head Office of New York. The final title of the project "Integrated Economic Development Policies and Planning" was suggested by Mr. Sagakuchi from New York UNDP Head Office.

A large amount of preliminary work such as project design and organizational aspects were done before 1989. DRC appointed Mr. Wu Mingyu to be the director of the project. A project office was established based upon Wu's suggestion. Huijiong Wang was appointed to be director of the project office and also responsible one of three task forces. The other two task forces were led by Madam Li [22]Boxi and Mr. Wu Jinglian. The final English research proposal was evaluated and completed through collaborative effort by Professor Nathan Rosenberg of Stanford University. Dr. Oldham, the former director of Science Policy Research Unit (SPRC) and staff of DRC. The UNDP Representatives of Beijing Residence Mission involved in this project proposal were Mr. Kulessa, Mr. D. Morey, and Mr. H. Behrstork.

(2) *Sign of Project Document and Organization of Steering Committee*

The project document number CPR/88/029/B/01/99 was signed officially on Sept. 11, 1989 by three parties. Wang Huijiong signed on behalf of implementing agency (Mr. Wu Mingyu could not perform his duty by that time due to certain unexpected reasons); Long Yongtu the deputy director of CICETE on behalf of the executing agency; and Mr. H. Behrstock on behalf of UNDP. This UNDP project is a large scale policy oriented research project. The nominal inputs of UNDP are US$2,963,000. Based upon the project document, which is divided into five immediate objectives with 55 outputs and 212 activities, DRC regrouped these 55 outputs into 3 sub-projects.

Sub-project I deals with China's industrialization and policies, but it is further divided into 9 task forces to study 24 outputs of the singed document.

Sub-project II deals with issues of reform, it covers the study of major policy recommendations to demand management and China's evolution to a socialist market. It covers 6 outputs of the signed document.

[22]Note: Li Boxi is the same person Li Poxi in "China Towards the Year 2000."

Sub-project III deals with regional development issues and prospects. It covers 25 outputs of the signed document.

A Steering Committee was established for this project to monitor the progress and give necessary guidance to the director of project office. The Steering Committee consisted of eleven members, seven Chinese members and four international members. There were three members from DRC, headed by Mr. Ma Hong, the four other Chinese member were Gui Shiyong from the former State Planning Commission, Mr. Gao Shangquan from the former State Commission for Restructuring Economic System, Madam Zhu Lilan from the former State S&T Commission, and Mr. Long Yongtu, from the China International Center for Economic and Technical Exchanges, attached to the former Ministry of Foreign Economic Relations and Trade. The four international members were Roy D. Morey and Jan Mattson from Beijing Office UNDP, and also Professors Rosenburg and Oldham.

4.3.3.2 Research Content and Reports

(1) *Research level*

This research not only covered the national planning at macro-level as Case 2. It also studied in detail to the meso-level industrial policies, regional development etc. and select topics relevant to market economy and society. This research content has an important component for improvement of institutional building and improvement of the capacity of Chinese "think tanks". The earliest "think tank" in Western countries was the Institute for Defense and Security Studies founded in 1831 in London. Such organizations in China were only recently formed, having been established in the beginning of China's reform, and opening in late 1970s. DRC had a comprehensive national study of China from 1983–1985 from the project "China Towards the Year 2000", there is need to have an in-depth study at the level of sectors and regions to improve its learning curve, because China is too large to be understood only at a macroscopic level. Therefore, selected provinces in Eastern, Central and Western region and one municipality were chosen to be studied.

(2) *The Content of Research of Sub-project 1 includes Eleven Parts*

Economic development and policy; Restructuring of Chinese industrial structure-overall analysis, expected prospects and policy options; Comprehensive study of key industrial clusters; Study of Chinese industrial organization; Study of Chinese industrial technological policy: Issues and policy options; Study and design Chinese macro-economic multi-sector dynamic model; Current status of income distribution of China and policy options; The role of government in China's economic development and her macro-economic policy; Population, employment and urbanization in the industrialization process of China; Analysis of current status of utilization of foreign investment of China and policy recommendations; Study of optimization of structure of trade of China and projections.

(3) *The Content of Research of Sub-project II includes Five Parts*

Toward the evolution and the role of socialist market; Chinese financial policy and reform of financial system; Chinese fiscal policy and reform of fiscal system; Chinese investment system and some issues relevant to investment policy; Pricing and anti-inflation in the process of reform of China's economic system.

(4) *The Content of Research of Sub-project III includes Eight Parts*

Current status of regional development planning of China and explorations; Exploration of classification of economic regions of China; Comparative study of economic development policy and planning of Sichuan, Hubei, Shandong province and Tianjin municipality; Study of industrial policy and planning of Shandong province; Comprehensive study of economic development policy and planning of Hubei province; Comprehensive study of economic development policy and planning of Tianjin municipality; Quantitative analysis and study of economic development of Chinese regions.

(5) *Research Reports*

There are a large number of reports, in addition to the interim reports, there are five volumes of reports in Chinese, published by Social Science Literature Press in 1994 with different titles *Economic Development Reform and Policies*. In order to meet the needs of appraisal by experts from abroad, five volumes of *Integrated Economic Development Policies and Planning* in English are also printed in limited numbers.

In the above five volumes, Volume IA and IB were translated from the reports of sub project I, Volume II was translated from the reports of sub-project II, Volume III was translated from sub-project III. "Although DRC has accumulated certain experiences through purposefully exercising learning by doing principle in TOTOS (Task Oriented Transient Organization) for large policy oriented project by organization of domestic experts since its establishment in 1981. But this CPR/88/029 is the first large project including the organization of international experts. It also involved the invitation of selected qualified international experts to provide complementary study on the same subject and supplement new subjects which were not included in the signed document".[23] These sub-contracted studies are published in Volume IV of the English version of this study, which includes: *Macro-Economic Management in Development' Planning; Chinese Economic Development and Industrialization Policies; Comparative Advantage in Chinese Trade* to complement the study of sub-project I. *Strategy and Policy Issues for China's Economic Reform and Development From an International Perspective* to complement both sub-project I and II. *Social Security-A comparative Analysis* to complement sub-project II, and *Towards the New Strategies for regional*

[23]Note: Wang (1994) Preface to the English Version of Final Reports.

Development to complement Sub-project III. All of these contract research papers have provided significant concept and policy suggestions to China's national and regional economic development and reform, they also provide a basis for comparison between domestic study and expert views from an international perspective.

It was commented by Christin Wong, professor of University of California in her last sentence of her appraisal (1994) "Indeed, it would be very surprising if in the course of reading over 2000 pages of reports, I could not find some faults to note!"

4.3.3.3 Appraisal of the Project from Experts Abroad

A concluding conference was held in June 1994. Mr. Lu Baifu, the Vice President of DRC and Jan Mattson, the Chief Representative, Beijing Office, UNDP attended. There were several days of small meetings arranged in the conference. Four international experts were invited to appraise the final reports. The international experts invited were Lawrence Lau (professor of Stanford University), Shelley M. Mark (professor of University of Hawaii), Christin Wong (professor of University of California) and K. C. Yeh (Senior Researcher and professor of Hong Kong University of S&T and The Rand Corporation). Three of the invited experts gave their report of appraisal. They studied the final version of English reports in detail, and provided objective comments. All of the experts confirmed the reports. Selected comments from them are quoted which may be useful references for the study of social systems engineering.

(1) *Comments on Subprojects I (Mainly by Professor Mark and Yeh)*

> The unique feature of the study is its integrated approach.... It treats explicitly development and reform as two sets of interrelated issues to be analyzed within a single framework, in contrast to other approaches that consider them as separate issues or one as the core problem and the other peripheral. (by Yeh 1994)

(i) *Organization of the Study*

> A study of such breath in scope and such depth in analysis definitely needs a summary. Presumably, Vol. I, Part A, this part is clearly too long for the decision-makers to read and absorb. In my opinion, the report should be so organized as to anticipate the needs of three groups of audience: First are the decision-makers who will read only the policy options, the trade-offs between these options with respect to specific objectives and the recommendations. The second will be the large group of readers concerned with problems of economic development and reform in China. The third will be some specialists interested in the technical details of the analysis. Accordingly, the study should have three components: an executive summary of the sort one finds in a World Bank Report, a core study such as the one presented, and a set of technical appendices. As it is, the first and the third component are lacking. While the third may perhaps be skipped, the first is essential. (by Yeh 1994)

(ii) *Policy Recommendations and Alternative Options*

A basic objective of this study is to propose policy measures to the decision-makers. Not surprisingly, the study has come up with many specific recommendations that are both useful and timely. However, give the wide range of proposals it seems clear that not all of them can be carried out simultaneously, if only because of resource limitations. This means that setting priorities will be desirable. Yet no priorities for the large number of proposals have been established.

A related issue is that some measures complement each other, (e.g., defining property rights and corporatization) so that they need to be implemented together either simultaneously or sequentially. Identifying and grouping these policies might be useful. A third and perhaps more important question relates to the single set of recommendations for each issue. The decision-maker thus faces a take-it-or-leave-it situation. No options are offered other than the one proposed…. In general, when specific views of the current and future leaders as well as other exogenous factors are uncertain, it seems more useful to offer, first, alternative measures with their corresponding positive and negative impacts on the various objectives, and then the recommendations. (by Yeh 1994)

(iii) *Models for Projection and Simulation*

Several models have been developed or adopted from other studies for purposes of projection and simulation. However, not much information other than the results have been presented. It appears that some additional information on these models could have made the study even more useful to the readers, for two reasons. First, some of the results presented in the study are markedly different….

A second reason why such information will be useful to many readers is that the economy is in rapid transition and will remain so in the foreseeable future. This means that the coefficients in these models are bound to change because of differential rates of growth of the individual sectors, technological changes, and continuous institutional restructuring. Any projections into the future must allow for that (by Yeh 1994).

(iv) *Agriculture, Investment in Human Capital, and Interregional Resource Flows: Some Neglected Issues*

It is generally unfair to criticize a study for omitting issues that the reviewer believes important, However, in the present case, there are three possible exceptions. The first is agriculture. The importance of agriculture has been clearly recognized by the authors. (See I, pp. 971, 1043) Yet, the issue has been left virtually untouched in a 1271-page study on China's key development problems.

Another notable omission is the issue of investment in human capital. Empirical studies have clearly shown that educational attainment of a country's adult population is strongly positively related to that country's GDP growth rate. For China this is especially significant because at present the educational level of the population is rather low (by Yeh 1994).

(v) *Overall Comments on the Reports*

I observed that each report very carefully followed a procedure of identifying the key issues, providing necessary background, discussing the important factors involved, and then coming forward with recommendations. This appears to be a well structured and useful approach. However, in a few cases the analytical section was mainly an extended restatement of the previous background material, true economic analysis requires the writer not only to stipulate what took place and when, but also why it happened and how the situation might be changed before policy recommendations are made. My suggestion for future be further research or presentation of findings and recommendations to policymakers, is that the analytical portions be strengthened and some indication of priorities along with the underlying rationale for your choices be given. Presenting the economic leadership with five or six different options for the same task at the same time without some order of preference can be unproductive and quite confusing. (by: Skelly M. Mark)

(vi) *Changing Role of Government*

In my oral comments on the task force report, I have stressed a continued strong, though revised role for state planning…. Market reform is a complex process and cannot take place overnight. There has to be a long-term vision as to where the economy is going and how best to get there…market failure is prevalent even in the most industrialized countries… centralized oversight, though more indirect in approach, more open to alternative views, and much less rigid in implementation. Also warned against the prospect of government failure, where bureaucrats are more interested in seeking rents or maintaining power, thus jeopardizing the achievement of basic reform objectives…. Thus centralized planning and macro management must confront the challenge of how to coordinate and integrate the different policies to ensure macro stability and keep the economy on the path of sustained growth…. Extensive planning will be required as much to relinquish the old as to introduce then new. Establishment and maintenance of a nationwide social safety net to supplant the welfare functions of state enterprises, which have been a major reason for their continued budgetary support, would be a case in point…and other contingencies support the case for a revived revised role for Chinese state planning. (by Shelley M. Mark)

(2) *Comments on Subprojects II*

In sum, this is a world-class piece that demonstrates exceptionally clear thinking. It is comprehensive and boldly prescriptive. This is the kind of writing that should stimulate debate, further thinks and further work on the reform program.

One area of further work that is clearly indicated by the Wu Jinglian piece is the exploration of the relationship between the state (government) and state-owned enterprises (SOEs). I was disappointed that the Subproject did not include such a section. Although some of the other Subprojects dealt with enterprise reforms, none examined the topic in depth.

This omission was all the more unfortunate because of the paradigmatic shift in thinking about enterprise reform in China that took place in the early 1990s…. In fact, what to do with SOEs is the single most important problem of reform in China-the transition from a planned economy to a market economy cannot be successfully completed until the government can find a solution to the problem of state-owned enterprise. For this “problem” is at crux of all the “difficult” sectors of reform: the fiscal and monetary/financial sectors, in investment policy, and even in legal reform (by Christine Wong).

Finally, I want to make three criticisms of the overall project:

The first is the omission of politics and the distribution of political interests in the studies this is so critical to how the reform agenda is shaped … why some measures are implemented and others are not, or what separates a feasible set of reforms from an infeasible ones. While a discussion of elite politics and leadership changes may be too speculative and sensitive to broach, a discussion of the real politics of distribution of power among

> ministries and departments, between the central and local governments would have provided a useful guide to the prospects and projection of reform in the future. Especially neglected are the role and interests of local governments, and how they might be different between coastal areas and inland regions.
>
> The second is that the volumes do not make much effort to explain the methodology to their readers…the third is that the project did not develop its own data base for answering some basic questions, but instead, seemed to rely on official statistics (by Christine Wong).

(3) *Comments on Subproject III*

> A very important contribution of the study is the analysis of regional differences in China's Development. The study would have been even more useful if interregional resource flows were also analyzed. Apparently data for such flows were lacking. But it may be worthwhile to compile them, for they could provide the empirical basis for indentifying the changing pattern of interregional interdependence and developing regional link models. In this connection, one may also want explore the policy option of reducing regional income gaps through expanding interregional trade rather than simply transferring aid to the underdeveloped areas. (by Yeh 1994)

There are also some comments from International Consultant. Professor Nagamine' Haruo of Graduate School of International Development Nagoya University was not invited to appraise the final report. But he was invited to be international consultant to present a paper related to regional studies. Because there are few comment on sub-project III, but he had read the mid-Term Report of *Integrated Economic Development Policies and Planning* of Shandong, Sichuan, Hubei Province and Tianjin, professor Haruo's paper is titled "*Towards the New Strategies for Regional Development*", which will be abstracted here to supplement some views on subproject III.

His paper is divided into four sections and two appendixes. The four sections are: introduction: The problems; Systems Building for Development: Identification of needs and actions at the grass root; New horizons for rural development; and Conclusions. His paper has not only introduced his concept of regional planning, but it has also discussed the behavior of the bureaucratic system and its impact on development planning which is meaningful for social systems engineering. His paper has 44 pages. Only very limited aspects of his paper will be briefed in the following.

Professor Haruo pointed out correctly that a province in China often has both demographic and geographical scale comparable to countries in the Asian region such as Malaysia, Japan etc. The province of Sichuan is about 50% greater than Japan, and the population size of the two are quite similar (Sichuan: 108 million, Japan: 120 million), therefore, a provincial development in China should introduce sub-provincial territorial as one of framework of planning. Nonetheless, except the report on Shandong, other reports are totally silent on the sub-provincial dimension.[24]

---

[24]Note: This is an issue of information exchange and language barrier. In fact, Sichuan had published a study "Sichuan Towards the year 2000" in four volumes in Chinese on 1989. One of

Professor Harus also pointed out correctly that the nature of changes the policy makers in China confront today. "There can be no dependable textbooks whatsoever. Under the circumstances, the best approach is that of steering the path of development through trials and errors." It is necessary to establish a two-way feedback mechanism through which the central policy makers can accurately monitoring progress and problems encountered in the development process.

Haruo has noted the complexity of choice of policies in the development planning of a bureaucratic system. "In a bureaucracy, a failure in achieving expected results often means that the career of the incumbent is furnished. As a result, bureaucrats tends to prefer inaction to committing themselves in risking decisions, no matter how meaningful these might be in certain crucial respects. Consequently, it becomes extremely important to introduce a career promotion system that can distinguish the irresponsible inaction from sincere and risk-taking attitudes, even if the final result of the later might be a failure."

(4) *System Building for Development Management: Identification of Needs and Actions at the Grass Root*

Professor Haruo had a view that a development plan should be formulated with three fundamental objectives:

Objective A: To upgrade economic viability of the territorial unit in question;
Objective B: To meet the basic needs of the people in the unit;
Objective C: To ensure sustainability of development of the unit.

He also proposes a two level approach to regional development: the high level approach of Regional Planning Unit and the lower level approach by Local Development Unit, which are briefly shown below.

(i) *Regional Development Planning at RPU Level (Level-I)*

Main Purpose: Top-down disaggregation of national development objectives (i.a. Guidelines for resource allocation)

(a) Divide the entire country into 20 or less number of regions;
(b) Articulate development policies/strategies with regard to each of them;
(c) Determine the framework of development targets in terms of key development variables such as population, production (by sectors), land use, essential physical infrastructures environmental control, provision of basic social services pertaining to RPU, and guide lines relative to institutional reforms;
(d) Spell out principles of resource allocation, particularly the multi-level public fund allocation;
(e) Add other basic policies relative to problem/potentials specific to each region; and
(f) Relevant maps of appropriate scales

---

the four volumes focused on sub-provincial study. This is also true for development of other provinces.

(ii) *Local-level Development Planning at LDU Level (Level-II)*

Main Purpose: Bottom-up integration of essential needs as perceived by people, and organization of arrangements for development management at the grass root.

(g) Identify programmes/projects to be undertaken within each LDU in a given time frame, based upon a carefully conducted needs identification survey;
(h) In the above programmes/projects package, the training activities should constitute an essential component;
(i) Identify responsibilities to be borne by each agency/committee/community group individual concerned for ensuring successful implementation;

Spell out those institutional arrangements which are needed for ensuring close dialogue between government functionaries.

In Section 3 "New horizon for rural development", professor Haruo has introduced the experience of Agricultural Cooperative Corporation, a successful experiment of Yamagishism Commune (TGC) which may be a useful reference for further rural reform of China. And in the conclusion, professor Haruo emphasized correctly that "Development strategies merely aspiring for accomplishing economic targets cover only a small part of the game. It is the values of the people that ultimately matters."

4.3.3.4 Lessons and Experiences

(1) Although this project is massive in scale, it has passed the triparty evaluation with favorable comment, but its influence on policy makers and social influence are far smaller than Case 2. This is due to two reasons: the first, in Case Study 2 "China Towards the Year 2000", the theme of study was assigned by the top decision makers, the former secretary general Hu Yaobang and the former premier Zhao Ziyang. It is a top down process. It is also similar to the study "*The Global* 2000 Report to the President" by the United States, this study came from the President's environmental message to the Congress of May 23, 1971, "directed the council on Environmental Quality and the Department of State, working with other federal agencies, to study the 'probable changes, in the world's population, natural resources, and environment through the end of the century." This endeavor was to serve as 'The foundation of our long term planning.' Therefore, this project was initiated and had influenced the global society widely. Although this is also due to the high level of quality of study of the United States but intention of the top decision maker played a dominant role of its influence; the second, there was a strong organization of leading team composed of high ranking officials in the study of Case 2, the contribution and organizational efforts, of Ma Hong is important, while in the implementation of study of Case 3, the theme of study was raised by DRC, and Mr. Ma Hong was too old in age to take actual responsibility, the project was led by project office. This is the second reason for its lack of influence.

(2) The second lesson and experience learned in this project is the relationship between the subject of research and time span of the project. It is necessary to pay serious attention to the 'timing effect' of the applicability of the policy oriented research. In the implementation of this project in a rapidly changing global and domestic socio-political and socio-economic environment. Although, this project had been designed very carefully in the beginning, and although DRC is in a privileged position to submit the research results to the decision makers, a part of research faced 'Timing Effect Failure', i.e. a part of study cannot catch up with the changes. This is also an objective of policy oriented research. Recognition of this is important in design of a large policy oriented project. It is necessary to balance and coordinate properly the relationship between applied and basic research, long term strategy and short term tactics in the design of the project. And flexibility should be kept in implementation.

(3) The third point is adaptation of organization to deal with international experts and institutions. Although DRC has certain experience in organization of large domestic projects, this project is the first one for DRC involving the organization of relatively large number of foreign experts from different institutions. Although English is a common language applied internationally, due to historical reasons, a large part of Chinese scholars, although qualified in the domain of certain discipline, cannot write English reports by themselves, the final reports had to be translated, but all translators were only trained in English language with nearly no background in professional knowledge. Therefore, the English version of the final report cannot transmit the information of research in a perfect sense. Hence the English version of the final reports are printed in limited copies to be used by foreign experts who participated the project, and also to be the final product to be sent to UNDP. It is expected that this weakness will be overcome in the future.

(4) Implementation of this project has greatly improved the capability of Chinese researchers and also greatly improved the institutional building of China's "think tanks".

4.4 Summary Points

1. This chapter consists of three parts: Trend of development of social systems engineering in global society; design and planning of social systems engineering; three case studies of social systems engineering.
2. Trend of development of social systems engineering is far slower than systems engineering. The Institute of Electric and Electronic Engineers published several papers titled "Social Systems Engineering" from control engineering perspective. The United States started its academic study only in recent years. Harvard has a program "Engineering Social System" composed of two teams: "Social

Physics"; and, computational social science, which is science oriented. The program may be beneficial for development of social systems engineering in the future based on a sound foundation of scientific research.
3. Japan initiated the program of social engineering in undergraduate studies of universities quite early, and graduate studies of social systems engineering more than 35 years ago. It is engineering oriented, and its areas of application are planning and policy.
4. Difference of development of social systems engineering between the United States and Japan may be effected by an ideological barrier, party scholars and politicians in the United States and other developed countries think that national planning is in conflict with free market economy.
5. The United Nations prepared four plans of "Decade of Development" since its establishment. Its "4th Decade of Development" is abstracted for illustrative purpose. The global society is very complicated and the development of global society is very unbalanced, and certain planning is unavoidable to coordinate a relatively balanced development. The U.N. has provided a good example of the need to develop the discipline of social systems engineering.
6. The concept of planning and design is clarified, and the evolution of design from a single product, to project and to engineering system in Western Countries are summarized to illustrate the principle of process of recognition and thought.
7. Some of the Chinese experience of engineering design are summarized, and several points relevant to social systems engineering are discussed.
8. Three case studies to study nation as a large scale system are listed. One study came from Chestnut, an expert of systems engineering from the United States, two other case studies from organized research of DRC. One study is "*China Towards the Year 2000*", It is a national study of long term development, which was highly influential both domestically and abroad, the other is "Integrated Economic Development Policies and Planning" funded by UNDP, which was organized with participation from domestic and international experts. This project studied China's development policies down to meso level, i.e., it has focused on sectors and regions within the context of national development. Selected aspects of appraisal report from international experts are also quoted, which are pragmatic and objective. It has improved the research capacity of DRC, a national think tank.

References

Archer, L. B. (1966). *Systematic method for designers*. London: Council of Industrial Design.
Asimow, M. (1962). *Introduction to design*. NJ: Prantice Hall.
Babathy, B. H. (1996). *Designing social system in a changing world* (p. 2). New York: Plenum Press.
Chestnut, H. (1982). *Nation as large scale system*. In Y. Y. Haimes (Ed.), *Large scale system* (pp. 155–183).

Don, F. J. H., & van den Berg, P. J. C. M. (1990). *The Central Planning Bureau of the Netherlands: Its role in the preparation of economic policy.* Conference paper.
Gidens, A. (2001). *Sociology* (4th ed.). Cambridge: Polity Press.
Gosling, W. (1962). *The design of engineering systems.* London: Hey & Co Ltd.
Hammer, M., & Champy, J. (1993). *Reengineeirng the corporation.* New York: Harper Collins.
Johns, C. J. (1966). Design method revisited. In S. A. Gregory (Ed.), *The design method* (pp. 295–310). New York: Plenum Press.
Lewis, W. A. (1966). *Development planning: The essentials of economic policy.* New York: Harper & Row, Publishers.
Li, P. X., Xia, S. W., Li, S. T., & Pan, B. X. (1986). *Overall quantitative analysis of China's economy towards the year 2000.* In F. C. Lo (Ed.) (1987), *Asian and Pacific economy towards the year 2000* (pp. 269–273). Kuala Lumpur: Percetakan Jiwabaru Sdn. Bhd.
Luckman, J. (1984). An approach to the management of design. In N. T. Cross (Ed.), *Developments in design methodology* (pp. 83–98). New York: Wiley.
Ma, H. (Ed.) (1989). *China towards the year 2000.* Beijing: China's Social Science Publisher, Shanghai People's Publisher, Economic Dairy Publisher (Chinese Version).
Mayne, R., & Margolis, S. (1982). *Introduction to engineering* (p. 328). New York: McGraw-Hill Book Co.
Nadler, G., & Hibno, S. (1990). *Breakthrogh thinking.* California: Prime Publishing.
Pentland, A. (2014). *Social physics.* New York: Penguin Press.
Pyatt, G., & Thorbecke, E. (1976). *Planning techniques for a better future.* Geneva: International Labor Office.
Rase, H. F., & Barrow, M. H. (1957). *Project engineering of process plants.* New York: Wiley.
Simon, H. A. (1969). *The science of the artificial.* Cambridge: The MIT Press.
Swedish Ministry of Finance. (1990). *The Swedish budget 1990/91* (pp. 24, 47, 116). Stockholm: Norstedts Trycher.
The World Bank. (1985). *China long-term development issues and options.* Baltimore: The John Hopkins University Press.
Tinbergen, I. (1967). *Development planning* (N. D. Smith, Trans.). New York: McGraw-Hill company.
UN. (1990). *The fourth international strategy of decade of development of UN.* New York: UN (Chinese Version).
UNDP Project Office. (Ed.). (1993). *Integrated economic development policies and planning.* (4 Volumes Printed, not Published).
UNDP Project Task Force of DRC. (1994). *Economic development, reform and policies* (5 Vols.) Beijing: Social Science Literature Press (Chinese version).
Wang, H. (1980). *Introduction to systems engineering* (Chinese version) (Vol. 1, p. 59) Shanghai: Shanghai S&T Publisher.
Wang, H. J., & Li, P. X. (1986). *China by the year 2000.* In F. C. Lo (Ed.) (1987), *Asian and Pacific economy towards the year 2000* (pp. 245–266). Kuals Lumpur: Percetakan Jiwabaru Sdn. Bhd.
Wong, C. (1994). Appraised of DRC/UNDP project on integrated economic development policies and panning. www.sk.tsukuba.ac.Jp.
Yeh, K. C. (1994). Economic development, reform and policies: A commentary.

Part II
Outline of Social Systems Engineering (SSE)

Chapter 5
Methodology and Principle of Planning of Social Systems Engineering

5.1 Introduction

Social Systems Engineering (SSE) is a new discipline with coverage of natural science, social science and engineering. Its main area of application is social groups. The social group can be roughly classified into three groups hierarchy, country (nation), various organizations under a nation to perform proper functional role, and the family unit. Qualitative and quantitative approach are two major methodologies to be included in the topic of study of SSE. This chapter will focus on two major themes, methodology of qualitative approach and several basic framework of national planning and three models for studying social systems.

5.2 Definition of Methodology and Meaning of Its Study

5.2.1 Definition of Methodology

In its simple definition, methodology means a set of methods. Different disciplines may define the word "methodology" differently. For example, it is described by Box-Steffensmeier and Jones (2008) that "Political methodology provides the practicing political scientist with tools for attaching all these questions, although it leaves to normative political theory the question of what is ultimately good or bad. Methodology provides technique for clarifying the theoretical meaning of concepts…and for developing definitions…. It offers descriptive indicators for comparing…, and sample surveys for gauging… And it offers and arrays of methods for making causal inferences…." It can be seen from above quotation that the methodology for a single discipline has a broad coverage of scope. (p. 3).

H. Wang and S. Li, *Introduction to Social Systems Engineering*,
https://doi.org/10.1007/978-981-10-7040-2_5

Social systems engineering is a new emerging discipline with even broader coverage of scope than a single discipline. Therefore, the definition of methodology should cover a broad sense to adapt to its features. Wang (2015) had defined methodology as follows:

[1]"The methodology of social systems engineering is a combination of the following aspects: a set of procedures for planning and design; a set of logical steps for solving problems; a set of tools (diagrams, models, mathematical models, indicators) used by professionals of social systems engineering; and a set of relations between tools and measures taken." It is different from 'method' which is narrowly limited to the application of a single action, measure or tools.

5.2.2 Meaning of Study of Methodology

Mankind as a whole is in the process of entering into a postindustrial society, or so called information society and knowledge society. Although the global development of this postindustrial society is very unbalanced, certain countries and regions have already entered into an information or knowledge society. However, there are also countries or regions that are still in the developmental stage dominated by the primary sectors, i.e. agriculture, husbandry and fishery dependence.

Due to the accumulated knowledge and experience of mankind through more than several millennia, especially the development of computer and ICT, the gradual formation of a global networked society, the production, accumulation and diffusion of knowledge and practice of mankind have reached an advanced developmental status with extreme depth, breadth and speed. While social systems engineering is an interdisciplinary study of natural and social science, as well as engineering, the professionals of social systems engineering must have a broad base of knowledge. What is more important, is that they must master the correct methodology so that they can get access to various fields of knowledge quickly and apply them correctly. The Science Journal of China has organized discussions about the way of thought and scientific thinking. A paper written by H. C. Zen, the founder of the Journal titled *Reasons for No Science in China*, has been republished. Zen acknowledges the importance of methodology: "In short, the basic nature of science is not shown in its impact on the progress of physical aspects, it is mainly methodologically. There is no difference materially today compared to several thousand years ago. We do have science today which was nonexistent several thousand years ago, which is caused mainly by methodology. If we can acquire scientific method, then every reality we see is science in nature. Otherwise, if we buy all advanced theory and technology from others, it can only be a process of imitation, we can only be slaves and servants of others, when we will have independence, progress and knowledge, of ourselves? Those people bent on learning should know how to do!"

[1]Note: Translated from Wang (2015).

5.2.3 Exploration and Comments of Hall's "Three Dimensional Morphology of Systems Engineering"

5.2.3.1 A. D. Hall of Bell Labs

A. D. Hall of Bell Labs is a founding member of the Institute of Electrical and Electronics Engineer, published *"A Methodology for Systems Engineering* in 1962. He also published a paper, *Three Dimensional Morphology of Systems Engineering* in IEEE Transaction. This is an important paper regarding methodology. Although the paper was presented to the academic and professional fields, it seems that there was little attention or application of his methodology in either field. From the perspective of the authors of this book, the framework established by Hall's *Three Dimensional Morphology of Systems Engineering* is applicable to systems engineering, as well as social systems engineering and national planning with appropriate modifications.

5.2.3.2 Dimensions of Systems Engineering

Hall's (1969) *Three Dimensional Morphology of Systems Engineering* provides that morphology refers to the study of structure and form, morphological analysis means to decompose a general system into its basic variables, each variable becoming a dimension on a morphological box. There are at least three fundamental dimensions of systems engineering:

The first is a time dimension, which is segmented by major decisions. The interval between these decisions can be called phases, and they depict a sequence of activities in the life of a project (plan) from inception to retirement. (Refer to Fig. 5.2 "phases").

The second dimension models a problem solving procedure, the steps of which may be performed in any order and may be repeated in succession, but each of them must be performed. The flow of logic, is the essential feature of this dimension. (Refer to Fig. 5.2 "steps").

The third dimension refers to the body of facts, models, and procedures which define a discipline, profession, or technology. The measure for this dimension is the degree of formal and mathematical structure. The intervals along this scale are in decreasing order of formal structure which is shown in Fig. 5.2. (Refer to Fig. 5.2 "professions"). An activity matrix between first and second dimensions of his morphology is shown in Fig. 5.1.

5.2.3.3 Comments

(1) Hall's *"Three Dimensional Morphology of Systems Engineering"* is well adapted to development of a new product or project of a corporation. Although,

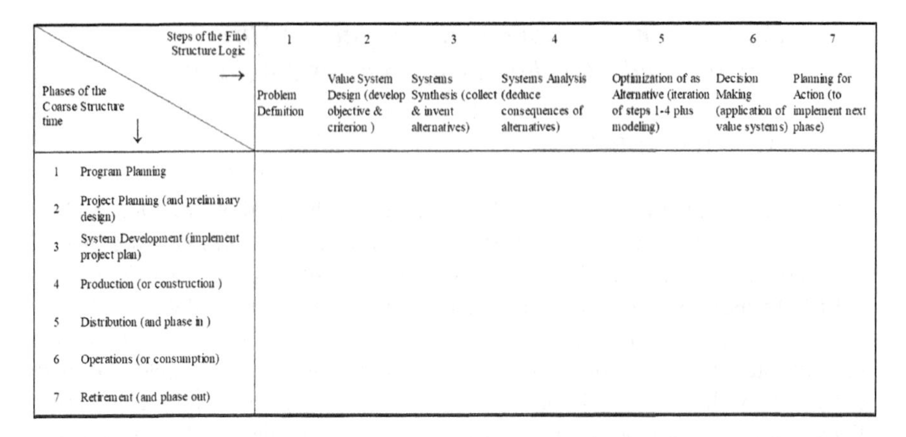

Steps of the Fine Structure Logic → / Phases of the Coarse Structure time ↓	1 Problem Definition	2 Value System Design (develop objective & criterion)	3 Systems Synthesis (collect & invent alternatives)	4 Systems Analysis (deduce consequences of alternatives)	5 Optimization of as Alternative (iteration of steps 1-4 plus modeling)	6 Decision Making (application of value systems)	7 Planning for Action (to implement next phase)
1 Program Planning							
2 Project Planning (and preliminary design)							
3 System Development (implement project plan)							
4 Production (or construction)							
5 Distribution (and phase in)							
6 Operations (or consumption)							
7 Retirement (and phase out)							

Fig. 5.1 Morphology of systems engineering and its activity matrix (*Source* Hall 1969)

Hall (1969) provided the explanation that "By program planning is meant conscious activity in which an organization strives to discover the kinds activities and projects it wants to purse into more detailed levels of planning." (p. 18). To simplify the expression, the item "Program Planning" can be simply expressed to be "planning". It can be corporate planning. It can also be a national planning. It even can be used as part of a project planning process, because the logic is well defined. If every project is carefully implemented through the logic steps, there will be no wasteful duplication of the construction of projects in the real world.

(2) Logic is important for many types of studies. The logic steps are well designed, and can be applied to other types of studies with only minor changes (adding or deleting certain steps, changes of wording etc.). Box 5.1 provides an example of policy analysis. Comparison of Hall's seven logic steps with the steps of policy analysis can be more clearly understood by the importance of logic steps and its application.

Box 5.1 [2]Eightfold Path of Policy Analysis

1. ***As stated by Professor Eugene Bardach*** **(2005)*, in***
   A Practical Gaide for Policy Analysis-The Eightfold Path to more Effective Problem Solving.
2. ***The Eightfold Path***

Policy analysis is more art than science. It draws on intuition as much as method. Nevertheless, given the choice between advice that imposes too much structure on the problem solving process and advice that is often too

[2]Source: Eugene Bardach (2005).

little, most beginning practitioners quite reasonably prefer too much. I have therefore developed the following approach, which I call the Eightfold Path:

- Define the Problem
- Assemble Some Evidence
- Construct the Alternatives
- Select the Criteria
- Project the Outcomes
- Confront the Trade-offs
- Decide
- Tell your story (p. xiv).

In fact, in Hall's paper, he explained the step "System synthesis refers to all means of compiling a set of contending alternatives, whose consequences are systematically deduced during the systems analysis step." (p. 18). Comparing Professor Bardach's Eightfold Path with Hall's Seven Steps, they are quite similar, but there is difference in making a decision. In the traditional practice of systems engineering, generally optimization is done by using a model. Yet, in the field of social systems engineering, such as analysis of public policy, usually trade-offs (or compromise) must be done for a scheme or policy between long term and short term goal, or among various interested groups (or stakeholders) to find a relatively satisfactory solution to meet reality rather than an optimal solution in theoretical sense. Therefore, Box 5.1, is complementary to Hall's Seven Logic Steps.

(3) In Fig. 5.2 the relation to the dimension of profession, ecology should also be added, and in the domain of social systems engineering, psychology should be added. Therefore, Fig. 5.2 will be modified to adapt to social systems engineering.
(4) With regard to the phases of coarse structure, planners of different countries may have different practice in dividing the phases.

5.2.4 Modified Hall's Three Dimensional Morphology of Social Systems Engineering

In brief, from the essence of Hall's (1969) *Three Dimensional Morphology of Systems Engineering* in 5.1.1.-2, there is a[3] bracket with the word "plan" inside. If "plan" is used to replace its preceding word "project'", the whole statement can also

[3]Note: The bracket with 'plan' inside is added by authors of this book.

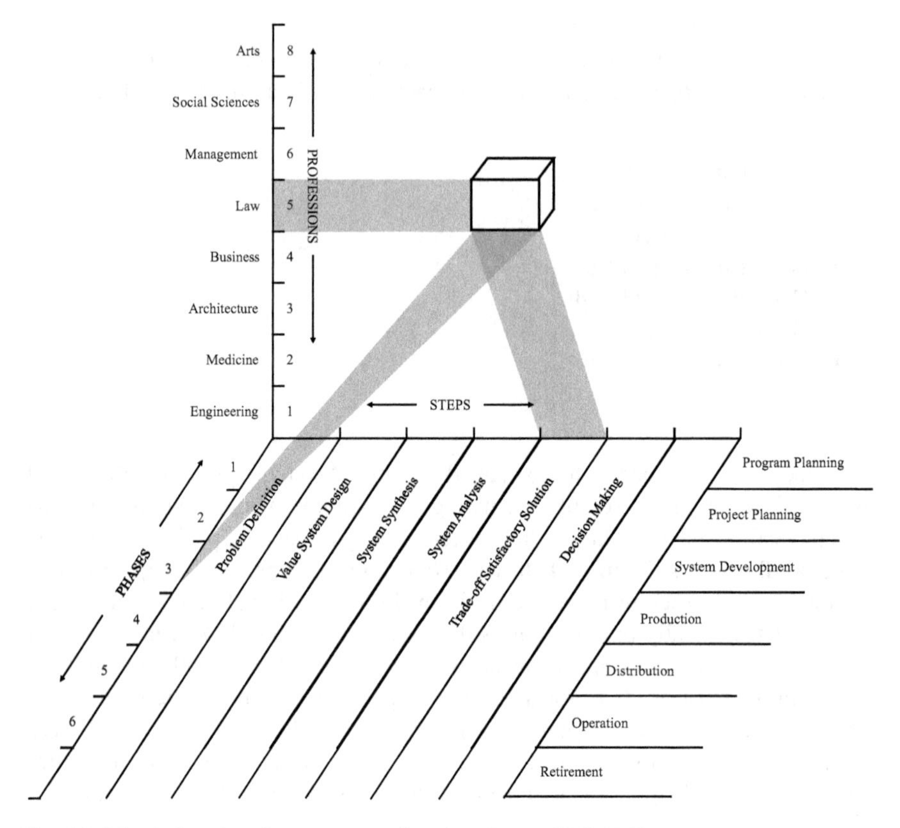

Fig. 5.2 Morphology box for systems engineering (*Source* Hall 1969)

be applied to social systems engineering. Figure 5.1 should be modified to Fig. 5.3, as shown. The symbol "√" in the bracket means activity should be taken for this designated logical step.

5.2.4.1 Morphology of Social Systems Engineering and its Activity Matrix

See Fig. 5.3.

5.2.4.2 Morphology Box for Social Systems Engineering

Figure 5.4 is the morphology box for social systems engineering. Its x-y plane is based on Fig. 5.4. Figure 5.3 has further classified the phases of the core structure in sequence of time and applied to national planning in the form of activity matrix. But there are modifications of professions along the Z-axis. Systems engineering

Phases of the Coarse Structure time ↓ \ Steps of the Fine Structure Logic →	1 Problem Definition	2 Value System Design (develop objective & criterion)	3 Systems Synthesis (collect & invent alternatives	4 Systems Analysis (deduce consequences of alternatives)	5 Optimization of as Alternative (iteration of steps 1-4 plus modeling)	6 Decision Making (application of value systems)
1 Initiation of new plan (Review past achievements and failures)	√					
2 Clarifying domestic and international environment (Information Collection)	√	√	√	√		
3 Preliminary Planning (Set up goal system, national strategy, coordination of sector and regions etc.)	√	√	√	√	√	√
4 Assessment and evaluation of preliminary planning	√	√	√	√	√	
5 Preparation of final plan	√	√	√	√	√	√
6 Implementation of plan(collect feedback information and adjust in time)			√	√		

Fig. 5.3 Morphology of social systems engineering applied to national planning activity matrix

including a broad base of different engineering activities. Social science is further classified into psychology, law, politics, sociology, and economics. because in the national planning system, they are studied separately. History is also classified to be social science, it is also listed separately and put on the top. From our perspective, the history of mankind recorded the interaction between mankind and nature, and also interaction between themselves. Progress of the interaction between mankind and nature is relatively fast because it is related to the basic survival of mankind against nature, it is more or less a common interest for all. However progress of the interaction amongst mankind is relatively slow, there were two World Wars, and moreover there is no shortage of prior wars throughout history. Can we learn lessons from our history? Therefore, history is ranked at the top of diagram Fig. 5.4.

5.2.5 *Methodology Provided by John N. Warfield*

5.2.5.1 John N. Warfield

John N. Warfield (1925–2009) was an American system scientist. He was elected President of the Systems, Man, and Cybernetics Society of IEEE and also the International Society for the System Sciences. He published a book *Societal Systems: Planning, Policy and Complexity* in 1976.

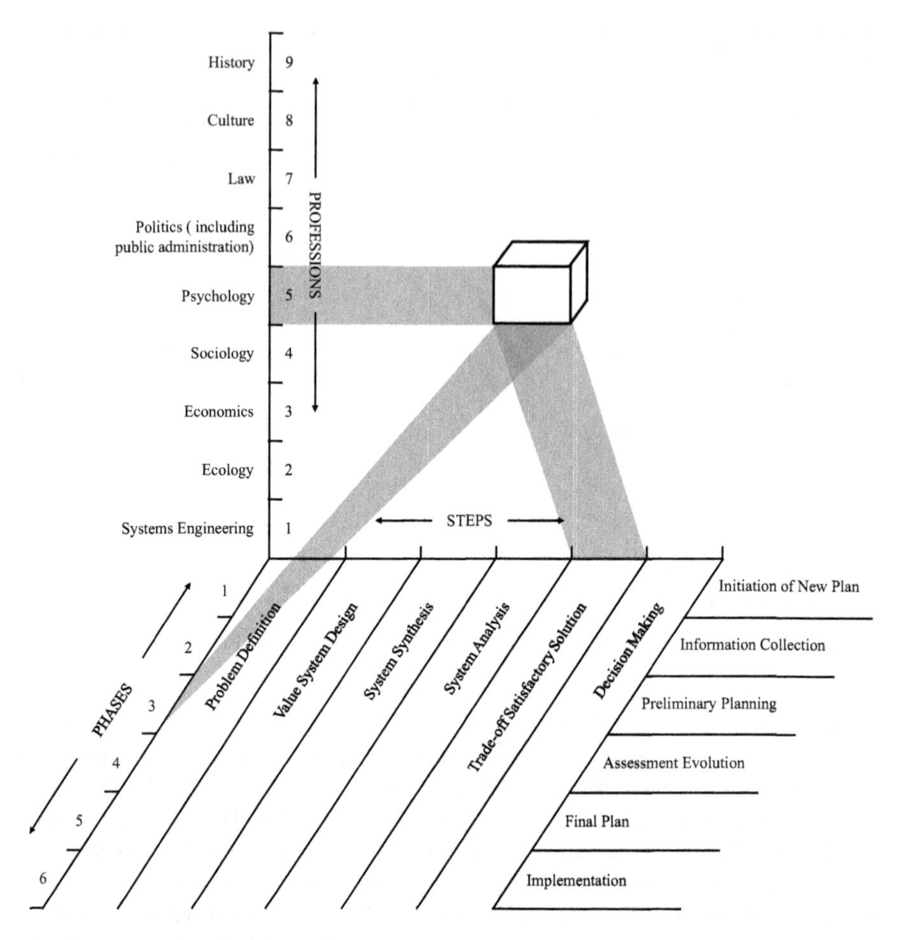

Fig. 5.4 Morphology box for social systems engineering

5.2.5.2 Societal Systems: Planning, Policy and Complexity

In *Societal Systems: Planning, Policy and Complexity,* Warfield (1976) raises in the beginning of preface of his book that "This book is primarily about method of coping with complexity." (p. ix). Warfield also raises concern for societal systems characterized by complexity.

This book contains 17 chapters. It can be divided into two parts: Part one is composed of six chapters that are qualitative in nature. It also raises consideration of the methodology of solving social issues, including the process of organized inquiry and giving means to make the process effective; Part two is composed of eleven chapters, many of those chapters, deal with Boolean algebra, linear matrix models and their application in interpretative structural modeling (ISM), which was proposed by the author in 1973. ISM is an effective methodology for dealing with complex issues. It is a computer assisted learning process that enables individuals or groups to develop a graphical representation of complex systems, and it can provide a fundamental understanding and course of action to solve problems. The last three chapters of this book (Chapters 15–17) discuss special types of interpretive structural model, such as intent structure, impact structure, preference structures, decision trees. All of these models are based on ISM and are available in software.

5.2.5.3 Several selected Points of this Book

It can be seen from above, that part two of this Book is mathematically centered on the development of interpretive structural modeling. Several selected points of Part One which are meaningful to social systems engineering will now be discussed.

(1) *The State of the Art of Studying Societal Problems*

It is expressed by Warfield that the state of the art of solving complex societal problems is dismal. There are several reasons for this. One reason lies in the fact that social and behavioral science generally lack established linkages for the translation of ideas from conception to implementation. Research information is generally limited to publication in scholarly journals, and has not been promoted to impact or influence decisions of governments and political leaders and educators. A second reason was quoted by Warfield in a report of the Special Commission on the Social Sciences of the National Science Board[4] (1969):

> Well organized obstacles to the utilization of social science knowledge lies within the social sciences themselves, and social scientists must share the blame for the failure to apply social science more broadly.....even when social science work is directed to application, it often produces fragments of knowledge that need to be joined with other fragments to present a program of action.

The Official Report quoted by Warfield was printed in 1969. And yet the situation has not improved in more than forty years, this can be seen from the conclusion given by Piketty (2014) in his book *Capital in the Twenty-First Century* that "Conversely, social scientists in other disciplines should not leave the study of economic facts to economists and must not flee in horror the minute a number rears its head, or content themselves with saying that every statistic is a social construct, which of course is true but insufficient. At bottom, both response are the same, because they abandon the terrain to others." *As long as the incomes of the various classes of contemporary society remain beyond the reach of scientific inquiry, there can be no hope of producing a useful economic and social history".* (p. 575)

Political society wants things simple. Social science needs a strategy for dealing with the condition.

[4]Note: National Science Foundation, Special Commission on the Social Sciences of the National Science Board Knowledge into Action: Improving the Nation's Use of the Sociences, NS1.28:63-3, Government Printing Office, Washington, D.C., 1969.

(2) *Synthesis is very much Emphasized by Warfield*

In the methodological framework established by Hall, synthesis is one of the Logical Steps. Hall simply explained it to be design and choice of alternatives. Honurs synthesis means more than that. Warfield's view of synthesis is broader than Hall's, which is quoted below.

> Synthesis 'means putting the parts together within various types of systems.
>
> The synthesis we need involves a better integration of the elements of our thought, policies and institutions in order to solve national problems effectively, via the efficient achievements of national goals through the use, where appropriate, of such means as science and technology (economic policy or social policy). At least two types of synthesis are important.
>
> First, we need synthesis within and between our perceptions of the problems in the real world, our goals for solving these problems and our means for attaining the goals, including our policies to support (the development of economy and society) *and use science and technology.*
>
> Second, we need to create synthesis within and between the elements of our thought, our policies and our institutions. (p. 17).

The above descriptions of meaning and types of synthesis are generally true, so that the authors of this book have added descriptions within brackets, i.e., the whole quoted statement can also be applied to study economy or society.

(3) *Methodology of Π-∑ Process*

Warfield had suggested the Π-∑ Process to solve societal problems. The Greek letter ∑ is used to indicate a summing action. It is used to indicate that things are being brought together. The Greek letter Π is used to indicate that things are being separated. The two letters Π and ∑ represent operations that have to be carried out in dealing with complex issues:

First, Warfield defined a general framework of solving societal problem shown in Fig. 5.5.

Figure 5.5 shows three basic connecting relations. First, there is the connection of the team to the issue. Establishing a proper team to deal with specific issues is no easy task. This can be seen from Case Study 2 and 3 of Chap. 4 of this book. Second, there is the connection of the issue to the methodology.

Figure 5.6a–c show the application methodology of Π-∑ process to the people and tools for illustrative purpose. In Warfield's book, there are also diagram of application of Π-∑ process to issue and others.

(4) *The Concept of TOTOS-the Organized Conduct of Inquiry*

In view of the complexity of societal problems, they are better addressed through the process of collective exploration. Warfield (1976) further raised the concept of TOTO (task oriented, transient organization) the organized conduct of inquiry to be one of methodology:

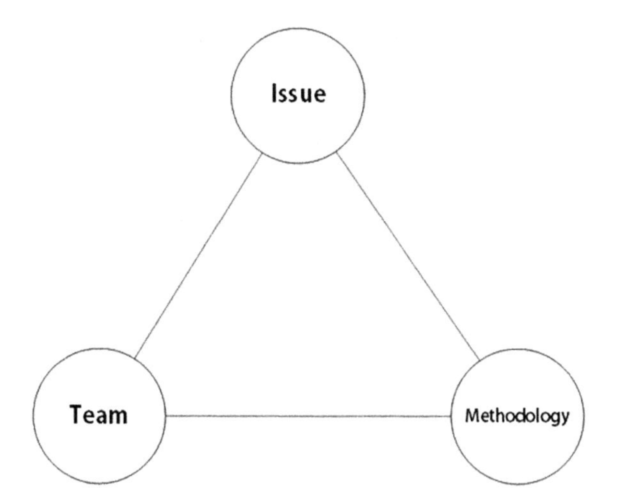

Fig. 5.5 The fundamental triangle of societal problem solving (*Source* Warfield 1976)

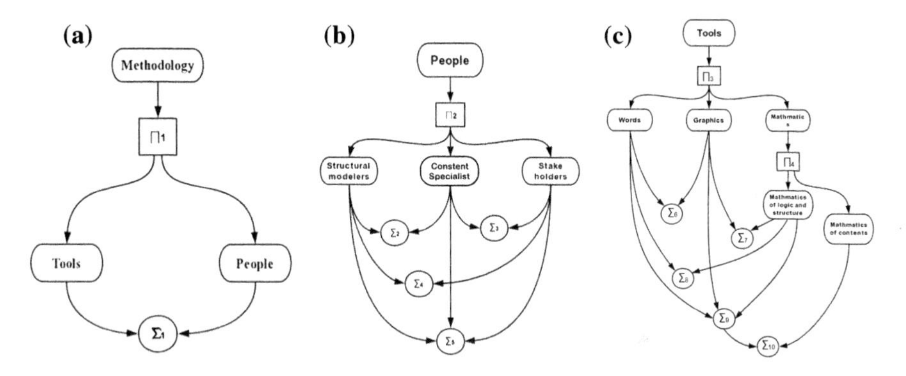

Fig. 5.6 Application of Π-$\sum$ process to methodology. **a** Overview of methodology [*Note* Warfield (1976) Societal System Fig. 1.4 (p. 28)]. **b** The people [*Note* Warfield (1976) Societal System Fig. 1.5 (p. 29)]. **C** The tools [*Note* Warfield (1976) Societal System Fig. 1.6 (p. 30)]

The task orientation arises out of some context that is sufficiently complex that individual effort is insufficiently broad, while group effort is insufficiently deep. The task then becomes the rationale for building an organization. But such an organization is considered to be transient, coming into existence for a time, which perhaps changing make up during the time as the issue becomes better clarified, and dissolving after its contribution has been made. (p. 73).

5.3 Further Exploration of TOTOS—Case Study: Expo 2000 OECD Forum for the Future

5.3.1 Mankind Has a Recorded History of More Than Several Millennia, with Many Complex Problems Faced by Policy and Decision Makers

Many complex problems cannot be solved by individuals, therefore the TOTOS concept has gained greater acceptance. There are many conferences and forums held worldwide, where TOTOS is a major point of discussion. TOTOS can perform several functions: First, it can perform the function of "Integration" (or coordination) in a modified Parsons' AGIL framework, which is a difficult part of the framework; Second, it can find solutions to specific issues; Third, TOTOS can be used to explore future trends (uncertainty in nature). Warfield had devoted exploration of this methodology with a full chapter, 9 principles and 11 conjectures are revised. Principles are capsule summaries of results of research or thoughts from other fields, which conjectures are related to the principles, but they are less well appointed by published research than the principles. TOTOS if well organized and performed, can be a useful methodology of social systems engineering and will assist greatly with national and corporate planning.

5.3.2 Case Study: Expo 2000 OECD Forum for Future

5.3.2.1 Expo 2000 in Hannover and OECD Forum for Future

At the turn of a new millennium, Hannover was awarded the privilege of hosting EXPO 2000 by the international world's fair organization B.I.E. in 1990. EXPO 2000 and four German Banks, supported the OECD Forum of Future to organize a series of four conferences to take place beforehand around the theme of "People, Nature and Technology: Sustainable Societies in the 21st century." Four international conferences were prearranged with four interrelated themes, the key areas were human activity: technology, economy, society and government. The process of organization and the methodology of study realized fully the function of exploration and qualitative coordination ("synthesis" to certain extent) of "TOTOS". Based upon the theme of each conference, related experts were invited to write specific papers, and participate the discussion. The organizers of these conferences summarized the concepts of specific papers and different views of discussions to form a final report.

5.3.2.2 Detail Process of Study of the First Theme-Technology

Whether effective achievement can be obtained depends upon leadership and proper procedure. EXPO 2000 OECD Forum for the Future was well organized and is a successful example of application of TOTOS. The process of study of the first theme "Technology" will be described in relatively detail for illustrative purpose.

(1) *Identify Proper Participants Beforehand*

Although the first conference was held at Schloss Krickenbeck near Düsseldorf, Germany on December 7–8, 1997. The theme was "21st Century Technologies: Balancing Economic, Social and Environmental Goals". The organizer of the Conference arranged a list of global experts to be invited, letters of invitation were sent to these experts on June 1997, 39 experts were invited (28 experts from Europe, 5 experts from North America, 1 from South America and 5 from Asia), 54 observers were also invited. Of the 30 experts that attended, 4 came from academic fields (including a Nobel prize winner), 4 were from Think Tanks, 10 from corporations, and 10 from governments.

(2) *Detail Design of Procedure, Content and Sessions in advance*

Although the Conference was held on December 7–8, a draft of concept and outline and arrangement of sessions was formed in early September. The draft was sent to participants in October, together with a Provisional Programme. Abstract of Outline for Conference One "21st Century Technologies: Balancing Economic, Social and Environmental Goals" issued on 3rd, Sept. 1997 will be given in the following.

21st Century Technologies: Balancing Economic, Social and Environmental Goals

Shaping the future in order to realize economic and social goals is a fundamental challenge of human society. Technology will play an important role to meet this challenge. But there are both positive and negative sides regarding the transformation of technological potential into economic and social outcomes.

The Conference has three goals. (1) Take an Unrestrained, look at the kinds of the changes rapid technological innovation has on everyday life and the way society functions (Session I). (2) Examine the analytically and important interdependence between economic and social conditions, and the rate and extent to which technology is likely to advance and be diffused. (Session II). (3) Consider those policies that look most likely to encourage economically and socially optimal rates and the distribution of technological development. (Session III)

The goal of the conference was identified.

Session I: Envisioning Technology's Potential: Opportunities

Session I aims to scan the horizon for technological possibility, paying less attention to economic or social constraints or incentive and more to the potential application of unbridled ingenuity. The focus will be on pervasive technologies that are likely to enter into a wide range of new product and process over the next twenty years changing the way we live and work, for example, microprocessors and biotechnology.

Three background papers were prepared for this Session, intended to provide a wide ranging assessment of the risks and opportunities proceeded by technically plausible application of technology over the next 25 years.

Paper I: Technology Trends: an Overview of the Potential
Paper II: Information Technology Trends: Exploring Potential Uses
Paper III: Biotechnology Trends: Exploring Potential Uses

Session II: Reaching Technology's Potential: The Enabling Micro, Macro and Global Condition

The aim of this session is to explain the interdependence between socio-economic condition and technology in order to reveal plausible approaches to making the best of technology's potential. Discussion in this Session: how the socio-economic condition plays a pivotal role in determining the rate of technological development (discovery and innovation) and the distribution of new tools (across economic sectors, space geography and social strata); how technological advances make new rates and patterns of economic and social development feasible by reducing costs on opening of new possibilities.

Discussion in the Second Session will concentrate on how various socio-economic framework condition lead to differences in the rate and distribution of technological innovation and diffusion as well on the implication of new technologies for society.

Three papers presented in this session: *Microeconomic possibilities: Organizational change and New Technologies*; *Macroeconomic possibilities*; *Global Possibilities: Technology and Planet Wide Challenges.*

Session III: Policy Priorities for Shaping Technological and Socio-economic Change

The aim of the Third Session is to identify policies most likely to generate the economic, social and political condition conclusive to making the best of technological opportunity while minimizing the attendant transition costs and risk factors.

The Secretariat will offer a framework for this Third session with an introduction, based upon the discussion in the previous two sessions. A few conference participants, representative of the main economic regions, will then be invited to comment on the policy priorities likely to predominate in other part of the world.

Given the tremendous opportunities opened up by technological advances, one of the main policy priorities may focus directly on finding ways of reconciling technological change with the widespread desire of stability and security. This balance will differ according to the historical and cultural preferences that distinguish different regions of the world. And one of the policy priorities over the next two decades could involve the search for ways of overcoming systemic tensions that arise from significant contrasts in either technological capacity or ecological sensitivity.

(3) *Evolution of Conference and Theme of Background Papers and Discussion*

The Provisional Programme of Conference on 21st Century Technologies: Balancing Economic, Social and Environmental Goals was finalized on the 23rd October 1997. Five of the background papers were sent to participants on the 26th November 1997, in advance the Conference.

The Conference was held on the 7th and 8th December followed closely the Provisional Programme of the Conference set up on 23rd October 1997. The name of the papers presented in the Conference differed slightly from what is shown in Box 5.2, although the contents were the same. The Six papers presented on the first day of the Conference: the *Next Twenty-Five years of Technology: Opportunities and Risks*; *Faster, Connected, Smarter; Biotechnology for the 21st Century*; *Technological Development and Organizational Change: Different Patterns of Innovation*; *Enabling Macro Conditions for Realizing Technology Potential*; and *Global Possibilities: Technology and Planet Wide Challenges.*

Session III: Policy Priorities for Shaping Technological and Social Economic Change was held on the 8th December. The lead speaker was Wolfgang Michalski, who was the Director Advisory Unit to the Secretary, and a major organizer of EXPO 2000 OECD for the Future Study. Michalski prepared the main issue paper which was also sent to the participants of the conference. This main issue paper summarized essential points of the six background papers, the purpose of Michalski's paper was to provide an outline of the central issues at stake, guide the discussions, and propose questions to stimulate the debate.

Several participants from North America, Europe and the senior author of this book were invited to give comments on policies and their impact relevant to their regions and countries.

A book with the name *21st Century Technologies—Promises and Perils of a Dynamic Future* was published by OECD in 1998. The book is a collection of six papers presented in the Conference. Chapter One was titled "The Promises and Perils of 21st Century Technology: An Overview of the Issues," and is based on the Main Issues Paper distributed in Conference held at Schloss Krickenbeck, Germany, 7–8 December. Part III of that paper was rewritten, with the title "Making the Most of 21st Century Technologies: for Encouraging Socio-Technical Dynamism", which is a summary of the main points of the discussion.

5.3.2.3 Successive Conferences of EXPO 2000 OECD Forum for the Future and Outputs

The second conference with the theme "21st Century Economic Dynamics: Anatomy of a Long Boom" was held in Frankfurt, Germany, on December 2–3, 1998, there were 34 participants, with 5 papers presented in the Conference. A book with the name "*The Future of the Global Economy—Toward a Long Boom?*" was published in 1999 by OECD.

The Conference was organized into three sessions. "OECD (1999) stated: the First Session looked at the generic factors likely to determine whether or not long-run economic dynamism will continue in the future. The Second Session addressed specific driving forces likely to accompany economic dynamism in the next century." (p. 4). And the last session discussed the policy options which are likely to influence different scenarios for a 21st Century boom.

There were five papers presented at the Conference. Chapter 1 of the book was titled *Anatomy of a Long Boom,* written by the organizers of the conference to summarize the main points of discussion.

The Third Conference with the theme "21st Century Social Dynamics: Towards the Creative Society" was held in Berlin, Germany on December 6–7, 1999. There were 46 participants with 6 paper presented at the Conference. According to OECD (2000), the first session provided a comprehensive global picture of the long-run trend in both the conventional and less conventional description of social change, e.g. shifting patterns and structures of income, population, wealth, social statue, health and cultural identity. The second session turned to possible outcome of the interaction between society and the change likely to be wrought by two interdependent development: first, the rapid diffusion and deepening of the knowledge economy and society; and second, more global and regional integration including markets for goods, services, finance, technology and labor. The Third Session looked at the policy synergies and conflicts." (p. 4). A book with the name *The Creative Society of the 21st Century* was published by OECD in 2000. There were 6 papers presented at the Conference and one keynote address by Donna E. Shalala, the U.S Secretary of Health and Human Services. Chapter 1 of the book titled *Social Diversity and the Creative Society of the 21st Century* was written by the organizers of the Conference.

The fourth conference with the theme "21st Century Governance: Power in the Global Knowledge Economy and Society" was held in Hanover, Germany on March 25–26, 2000. There were 34 participants with 6 papers presented at the Conference. A book with the name *Governance in the 21st Century* was published by OECD in 2001. 6 papers were presented at the Conference. Chapter 1 of the book titled *Governance in the 21st Century: Power in the Global Knowledge Economy and Society was* written by the Conference organizers. From OECD (2001), 3 main messages emerged from the discussion and analyses. First, older forms of governance in both the public and private sector are becoming increasingly ineffective. Second, the new forms of governance that are likely to be needed over the next few decades will involve a much broader range of active players. Third, and perhaps most importantly, two of the primary attributes of power that are ingrained in the structure and constitutions of many organizations; the tendency to meet initiative exclusively in the hands of those in senior position in the hierarchy, are set to undergo fundamental changes. (p. 3)

5.3.2.4 What Lessons and Experiences can be Learnt from this Case Study

This case study not only illustrates the application of TOTOS and its detailed steps, also provides a good example of cross-disciplinary study. The organizers of the conferences emphasized the points to experts before[5] invitation, and in the discussion of the Conference "you must think and talk about the development of science and technology within the context of development of economy and society."

The four topics are well identified, there are also essential elements to be studied integrally to be good example of a national planning. Science and technology is a driving force for the growth of economy within the context of society and appropriate governance. The growth of the economy is to establish the welfare of the society within the context of appropriate governance. All of these elements are mutually interactive. Therefore, the topics identified and the process of study provide a good example for theoretical and practice of social systems engineering.

There are weaknesses in these studies. Since the organizer is OECD, which is composed of mostly developed countries, the result of the study is biased to the optimistic side. (This may be also due to the reason to support EXPO 2000) The complexities of the human societies, presented in the papers will be discussed below to compare the challenges faced by people's today.

Siebert and Klodt (1998) *Toward Global Competition: Catalysts and Constraints In OECD* (1999). In *The Future of the Global Economy*, they described:

> Demand for more redistribution and an expansion resistance to further globalization can be expected especially from those groups that fear they will be among the losers. In highly developed countries the main losers are low qualified workers, because they are increasingly exposed to direct and indirect factor price competition from low wage countries. There is still an ongoing debate among economists about the relative importance of globalization for the income and opportunities of skilled and less skilled workers (Siebet, 1999b), but there is no doubt that the integration of China, Eastern Europe and other labor rich regions into the world economy will put the wage levels in Western industrial countries under strain. In a globalized world, high wages can only be earned if they correspond to high labour productivity which stems from a high qualification level of workers.

There is also a paper with the title Toward Globalism: Social Causes and Social Consequences written by Mϕller (1999), who was the Danish Ambassor to Singapore at that time. His paper was published in OECD (2000) The Creative Society of the 21st Century. A part of his paper is quoted below which can illustrate the cause of waves of anti-globalization in recent years.

> "Since the end of the Second World War, there has been an uninterrupted movement towards more internationalism and more globalism. The trade system has been liberalized. Capital movements flow freely between the major countries. Information and entertainment

[5]Note: The senior author was invited to participate the first conference focused on the theme of "Technology".

do not recognized borders. National seclusion as we knew it in the 1930s belongs to the past. Today our economic well-being and our knowledge base are firmly anchored in an international context.

A network of international institutions has been built to cope with the problems created by the nation state's failure to exercise control over economic, monetary, industrial and technological movements in its shift from the national level to the international level.

And yet, behind this screen can be detected a growing discontent with globalism. The impact on individual societies and on groups inside individual societies does not seen to be as undeniably positive as was once thought. Widening social disparities lead many people inside the nation states to dispute whether internalism and globalism really benefit them…..

The might of the large multinational enterprises, creating havoc with individual currencies, raise the fundamental question of who is really in charge-who is governing the world.

There still seems to strong support for globalism and internationalism but the support is not as strong as it used to be, and it continue to wane. Policy makers seem to be unaware of this.

…

[6]Second, internationalization and globalism result in higher production and a higher gross national product per capita, leads to a growing social disparity. The latest UNDP brings to light the following facts: in 1820, the ratio between the richest and the poorest nation was 3:1. But it had increased to 11:1, 31:1, 44:1 and 72:1 respectively in 1913, 1950, 1973 and 1992. In today's world, the 200 richest people have a fortune equivalent to the annual income of the least developed nations. Of these 200 people, 65 are found in the U.S., 55 in Europe, 13 in other industrialized nations, 3 in Central and Eastern Europe, 30 in Asia Pacific, 16 in Africa and 17 in Latin America. These figures do not prove that internationalization is to be blamed per se, but it cannot be disputed that there is a correlation between this growing disparity and the growing internationalization of economies. One topic that occurs frequently in discussion about the Chinese economy, is that while it is steadily producing growth rates between 7% and 10%, observers highlight the growing social problems inside Chinese society.

…

For many decades the political and economic architecture erected by industrialized society and industrial technology served us well. It no longer does so. Social structure based upon the nation state system is breaking down, and often faster than is imaged. During the1990's, the market as an economic factor had no opposition, the impact upon our well established social structure has been almost devastating.

A new design might feature:

First, less social disparity inside as well as between nation states.

Second, another concept of social welfare focusing more upon the role (rights and obligation) of the individual.

Third, economic and political institutionalizing on the international level comparable to the domestic political and economic architecture that served the industrialized nations well. (pp. 113–114, 119, 130–131)

[6]Source: The author had raised three pictures, only the second is quoted.

5.4 Systems Engineering Logic

5.4.1 General

In systems engineering, the system engineer should be a generalist rather than a specialist. Logic and mathematical formula are essential tools for cross disciplinary study. Because a mechanical, or electrical, or hydraulic phenomenon may be expressed in the same form of mathematical formula. The senior author of this book has written a chapter titled *Engineering Logic* in his *Introduction*[7] *to Systems Engineering* (Chinese Version) in 1980.

In social systems engineering, the social systems engineer may require a broader capability of cross disciplinary study between social science, economy and engineering, therefore, systemic application of relevant aspects of logic, integrated with certain mathematics, especially statistics can become essential tool for cross disciplinary study, the title "Systems Engineering Logic" is given in this section.

5.4.2 Induction and Deduction

5.4.2.1 Induction

Induction begins from individual or particular observation to achieve certain empirical generalizations, while deduction is a method to predict individual or particular observation based upon certain general theory known or hypothesis. Basically, the nature of these two processes of recognition of the external real world are in opposition to each other, but mankind recognizes the external world through these two in succession and in cycle. Every cycle of application of these two processes in succession will improve the depth of understanding of mankind to certain objective or phenomenon to be studied. This pattern of thinking is shown as follows:

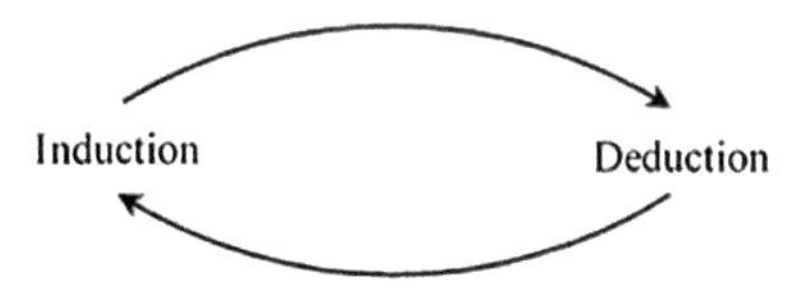

5.4.2.2 Induction and Classification

The general method derived from induction commonly used in social systems engineering is classification, since the subject and object to be studied in social

[7]Note: Wang (1980).

Table 5.1 Classification of countries in the world (based upon per capita GNP) [Date of 1987 from: The World Bank (1989). Data of 2008 from: The World Bank (2010)]

Type of countries	Year 1987 GNP/capita	Number of countries	Year 2008 GNP/capita	Number of countries
Low income country	≤ U.S.$480	42	≤ U.S.$975	21
Middle income country	$481–$5999	53	$976–$11,905	67
High income country	≥ $6000	25	≥ $11,986	44

systems engineering number in the tens of thousands. It is possible to reduce the amount of quantities to be studied though classification based upon certain criteria. If we try to study the development or policy options for certain countries, there are currently more than 220 countries in the world, within which, there are 195 sovereign states. In order to facilitate the study and comparison, the World Bank classified countries of the world into three major groups based upon per capita value of GNP in U.S. dollars, the low income, middle income and high income groups are is shown in Table 5.1.

GNP is the abbreviation of gross nation product which is the sum of all factor income adjusted for stock appreciation (gross domestic product) plus net property income from abroad. Through this classification of countries based upon inductive method, it is easier to study more than one hundred countries in number by grouping them into three categories of classifications. Classification based upon inductive method should follow four basic principles:

(1) The sum of numbers for every part is divided from the total number of objects to be studied, and should equal to the total number of objects. For example, in the year 1987, the total number of counties studied in 120, then the sum of number of low, middle and high income counties in the year 1987 should equal to 120. While in the year 2008, the total number of countries studied is 132, and the sum of number of countries in three groups should equal to 132.
(2) Classification of objects to be studied should be based on consistent basic quality. For example, the basic quality in measuring the state of development of countries of the world is GNP. It is one of the basic qualities of a country. For example, although China's gross domestic product in 2013 reached 56.88 trillion yuan, ranked the second among the world, but its GDP/capita was only 6852 U.S. dollars. China is still in the category of an upper middle income country. It can be found from the First World Development Report launched by the World Bank in 1978, in which the number of country groups classified by the World Bank[8] were five: low income country, middle income country, industrialized countries, capital rich oil exporting countries, and central planned

[8]Note: The World Bank World Development Report 1978–1981.

economic countries. It is evident that this classification of five groups had inconsistent nature of quality, this type of classification is in conflict with basic principles. Therefore, the World Bank modified its basis of classification of countries post 1980.

(3) Every part divided from the whole through classification should not have overlap or in duplication.

(4) Process of classification should follow the principle of hierarchical structure of general systems theory. This principle is used nearly universally in qualitative and quantitative analysis. For example, in the establishment of systems of indicators for measurement of progress of sustainable development, the United Nations had set up four major indicators at the first level, i.e. society, economy, environment and institution. Under the theme 'society'. There are six indicators of second hierarchical levels, i.e. equality, health, education, housing, security and population. There are also third and fourth level indicators, for example, under the theme equality, there are further classification into two indicators, poverty and equality of genders(third level), and under the theme of poverty, it is further classified into three indicators: percentage of population lived below poverty line; Gini coefficient of income equality and unemployment rate, etc.

5.4.2.3 Classification

Classification is an important part of social systems engineering. The objective social system studied by social systems engineering is its structure, function and behavior, if we can collect sufficient information related to behavior of the actors of the social system and their various groups, and analyze those information and classify into categories through appropriate criteria, it will form a very important foundation for the promotion and development of social systems engineering.

5.4.3 Deduction and Analogy

5.4.3.1 Deduction

Deduction is a type of reasoning from general to specific, while induction is a type of reasoning from specific to general. There is a third type of reasoning in logic, i.e. a type of reasoning from specific to specific, so called analogism. Deduction was described in the foreword of the book *Engineering Analogies* by Murphy, Shippy and Luo (1963) "The general concept of utilizing observations on one system to predict how a second system (different in some respects and similar in some respects to the first) will perform or behave has been known for many years, and the technique has been used in the solution of important problems". (p. v). The English word "analogy" originated from Greek meant "ratio", it was extended to mean the relationship of similarity between two events. Computers in their early stage were analog computers, which used the changeable aspects of physical phenomena such as electrical, mechanical, or hydraulic quantities to model the problem being solved. These were replaced by digital electronic computers post World War II. Two examples are described below as follows for illustrative purpose.

5.4.3.2 Principle of Analogy to Be Applied in Transport System of Goods in Analogy with Kirchhoff's Circuit Law

Transport system of goods is a system of process of flow of materials. In the theoretical analysis of a node (point of junction, it can be a port or railway station in reality of the world), based upon the principle of electricity or hydraulics, the total quantities of inflow and outflow should be equal at the node. This principle can be applied for the design and operation with respect to the size of the warehouse of a railway station or a port, or for the analysis and adjustment of flow of goods in each branch line connected at the station. Because in real world, it is impossible to have total amount of inflow equals to the total amount of outflow, a warehouse is needed in the railway station or port. Figure 5.7 shows this application of principle of analogy.

5.4.3.3 Principle of Analogy to be Applied in Explanation of Reform of Economic System

A large number of former central planned economies began to implement reform of their economic system to market oriented systems. Although the principles of economics, are correct, whether they can be applied to achieve effective results

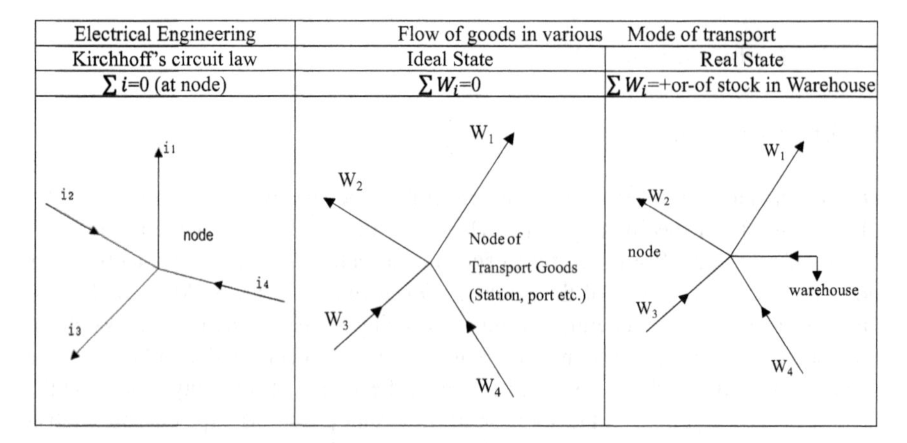

Fig. 5.7 Application of principle of analogy from electric system to transport system

depends upon their linkage with concrete conditions of a country. It is also related to the approach of social systems engineering. For example, Bolivia successfully tackle its hyperinflation in 1985 through a shock therapy approach. However, Bolivia is a small country with a population of around 10 million at the time, and an area of approximately one million km^2. Bolivia's population density was 9 persons/km^2 by the year 1985. Therefore, the shock therapy approach became successful. While this approach was adopted to economic reform of the Russia Federation in 1992, it resulted a serious failure, because Russia is a large country both in population and land area. Its population exceeds one hundred and forty million and its land area exceeds seventeen million km^2. Russia's political, social, cultural and economic conditions are very complicated. The application of analog of basic mechanics in physics, a large country with large number of population is equivalent to a huge mass (M), its traditional way of administration and operation equivalent to velocity (V). A huge mass also has a huge inertia and huge momentum (MV). It is very difficult to change the inertia or momentum in a short period of time without a huge force. It can be approximately known from mechanics that $F = M\frac{\Delta v}{\Delta t}$, if Δt is small compared with Δv, then the force will be huge and hardly bearable by a country. This analogy can explain the successful economic reform of China through the gradualist approach.

5.4.4 *Synthesis and Analysis*

5.4.4.1 Discussion of Relationship Between Synthesis and Analysis

Synthesis and analysis are the core part of qualitative and quantitative approach of social systems engineering. Based upon perception of logic, analysis is to separate an object or event through thinking, then to analyze every part and every feature of them. Synthesis means the composition or combination of parts to form a whole. In the process of engineering logic, generally it is necessary to have a whole (rough) concept of an object or phenomenon then separate the object and phenomenon and analyze them and recombine these into a whole, to have a better understanding of the whole. Then separate certain parts of an object or phenomenon through analysis of their features with in depth for certain purpose, and proceeds further the synthesis. Although the concepts of synthesis and analysis are in opposition to each other, but they follow the law of the unity in opposition, they must be carried out in pair, and it is generally an iterative process of thinking shown as follows:

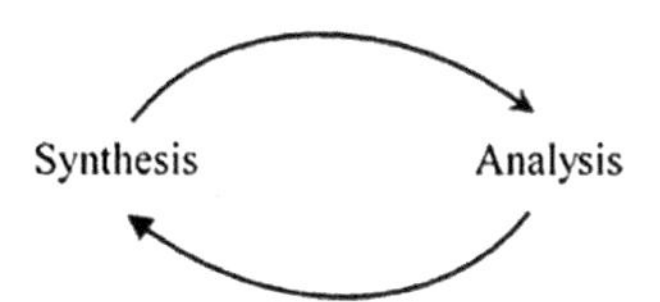

5.4.4.2 Some Discussions of Synthesis

The word synthesis comes from the Greek meaning "to place with". In the systems age, post 1950s, it became a dominant pattern of systems thinking in similarity with the thinking dominated by analysis during the age of determinism by mechanistic view. A system must be studied as a whole. If it is divided, it losses its property of synergism or coordination. The character of a nation is formed by many factors, population, geography, history, culture, political system, government, economy, social life etc. Studying only a single factor cannot lead to an understanding of the status of a country. Synthesis should be processed in three steps:

Identify the whole, in the first step of analysis one individual unit of the system must be considered to be the part of composition of the whole system;

Study the overall performance, features and behavior of the individual related to its subsystems and components;

Study the performance, features and behavior of subsystems and components within the context of overall system.

5.4.4.3 Analysis

The word analysis also comes from the Greek, although its original meaning was "a breaking up" or "a loosening", but since the period of scientific revolution, it becomes an important method of scientific discovery by Descartes, Galileo and Newton. It means more than "breaking up "or separating into parts in physical sense. It is a type of logic to separate an object or phenomenon through thinking, to analyze properties, features and relationship between various parts.

Traditional analytic method will proceed along the following three steps:

Proceed to divide or refine a certain concept or material substance;
Study in detail the properties and behavior of the parts separated from Step (1);

Based upon the properties and behavior studied from Step 2, predict the overall property and behavior of the object to be studied, or put the property and behavior of the parts separated within the context of the property and behavior of the whole, if they cannot satisfy the function or performance of the whole, then readjust and restudy the parts to be divided or redefined them from the whole.

The above process of analysis from Steps 1–3 is the system of scientific method, based upon inductive analysis, its basis is observation and experimentation, through analysis and inductive reasoning to understand the whole.

The methodology in the age of systems, sets up goals and functions of the whole system in the first, and then through analysis to obtain the structure of subsystems in hierarchy, synthesis of subsystems in hierarchy should meet the needs of the goals and functions.

5.4.4.4 Several Principles Related to Analysis and Synthesis

Analysis can proceed as follows:

Analysis of a system based upon its function, for example, in the economic action system, the economic system can be analyzed from the function of different products from the supply side, i.e. analysis can be done to the subsystem of primary (extraction of minerals and agricultural sector) secondary (manufacturing and construction) and tertiary sector (service). Analysis can also be done by the function of demand, i.e., investment, consumption, export and import.

Analysis of a system can also be done based upon hierarchical level of the system. With agricultural sector for example, it can be further classified into farming, forestry, husbandry, fishery etc.

Analysis can also be done based upon its sequence of flow of materials in the process of production, such as production, circulation and consumption. For the product of equipment for processing plant, then the sequence will be: R&D, design, material inspection and testing, production, quality control, package and transport, installation and testing, operation and maintenance, etc.

It has been described in the overall national design or planning, the first step of synthesis is set up a goal system. It should not be a single goal of economic growth. The experience of development of many countries as well as China have learned that "Growth of GDP alone" will create many social and environmental issues. China set up a goal system of seven priorities in its current Thirteenth Five Year Plan. But in setting up the goal system, a strong capability of synthesis and analysis is required. Because different goal systems may have different schemes to achieve them, they may have different input and output as well as different effects and risks.

5.4.5 Cause and Outcome

Features of Cause and Outcome of social systems engineering

The history of civilization of mankind is also the historical process of the recognition, adaptation and regulation of mankind to the objective laws of nature and society. Since the industrial revolution, mankind has a relatively in depth knowledge of the natural world, which enabled many laws to be established. These laws include the knowledge of relationship between cause and outcome. With Newton's Second Law of Motion for example, "the acceleration of a body is proportional to the force acting on it, and is in the direction of the force", the relationship between cause and outcome is simple, there is only one "cause"—the external force, and also only one "outcome"—the acceleration. However, the relationship of the object studied by social systems engineering is very complicated. There are four features of cause and outcome of social systems engineering.

(1) *Cause and Outcome generally occur together in nature rather than alone*

Certain social phenomenon are generally not caused by single cause. Also, a single cause may result in not only one desired outcome, sometimes, undesirable secondary or tertiary outcomes may be resulted. It can be expressed mathematically of 5.1.

$$\underset{\text{Set of causes}}{(\mathrm{C1,c2,c3}\ldots.)} \rightarrow \underset{\text{Set of outcomes}}{(\mathrm{O1,o2,o3}\ldots.)} \tag{5.1}$$

Capital letter C1 and O1, represent the major cause and outcome respectively.

Equation (5.1) can be applied conceptually, several people group together invest to manufacturing of car industry. C1, c2 etc. can be a set of input of capital, technology etc., while O1, o2…etc. may represent the outcome of output of number of cars, sales income, and the discharge of gases which causes undesirable outcome of environmental pollution.

The above example illustrates that it is necessary to consider cause and outcome in a broader way.

(2) *Nature of Succession of Outcome and Cause*

This means the outcome formed by cause in the first round can become the cause of the second round, which is shown in Eq. (5.2) as follows:

$$\text{Cause 1} \begin{Bmatrix} \text{Outcome 1} \\ \text{Cause 2} \end{Bmatrix} \longrightarrow \begin{Bmatrix} \text{Outcome 2} \\ \text{Cause 3} \end{Bmatrix} \longrightarrow \begin{matrix} \text{Outcome 1= Cause 2} \\ \text{Outcome 2= Cause 3} \end{matrix} \tag{5.2}$$

The application of this logic can be explained with discharge issue of environmental protection. For example, if a policy is implemented to allow the discharge of pollutant of a firm, with small amount of fine to control the discharge. This outcome resulted that the firm prefers to pay the small amount of the fine rather than stop the discharge. This has also induced an effect of demonstration that more firms will participate in the process of discharge of pollutant, causing further worsening of the environment. This is the principle of Total Quality Control to control from the beginning. This is also necessary for anti-corruption.

(3) *Time Effect and Cumulative Effect of Cause and Outcome*

The phenomena of cause and outcome of social system are not shown immediately. The outcome of low quality of education is not shown explicitly in the schools or universities. It is shown with time lag, reflected in low competitive strength and weakening of a nation. A country must respond and take action in time to various detrimental social phenomena, such as lack of honesty and credibility, corruption etc., otherwise the disruptive effect to whole society will be difficult to correct. For example, China has achieved good economic performance and external trade since its reform and opening in late 1970s. One of the causes which China devoted substantial efforts was to improve human resources. This was done around thirty years before the launch of reform and opening. The reforms include nearly free education at all levels, nearly free health care infrastructure, transportation, and an energy system. Then, in the implementation of reform and opening to outside, the cumulative impact of improved human resource is liberalized, the quality of China's labor force ranked in the front among south-east Asian countries, in combination with the import of foreign capital and technology, there is very rapid development of its manufacturing sector and becomes the number one large manufacturing country in the world. The cumulative effect of positive factors established in past (cause) is shown in later period due to change of external environment.

(4) *Dialectic Relationship between Cause and Outcome*

The relationship between cause and outcome is not absolute, there will be conversion of cause and outcome under certain conditions. With civil engineering for example, when a building is constructed on certain soil, the nature of the soil (cause) must be considered fully because it will affect the design of foundation and superstructure of the building (outcome). But post the construction of the building (cause), the nature of the soil (such as void ratio, compression ratio) under and nearby the foundation of the building will be changed (outcome). In the area of social systems engineering, it is the large multi-national corporations who drive the process of globalization and with many of their subsidiaries distributed globally. The mode of operation and organizational structure will follow the political, social, economic and cultural environment of the host country, in order to be successful. But the corporation culture, organization, management and mode of operation of parent company possessed by its subsidiary, will in turn, affect the economy, society and culture of the local environment where the subsidiary is placed. Some literatures explain this phenomenon by the theory of co-evolution.

5.4.6 Other Engineering Logic

5.4.6.1 Occasionality and Inevitability

Occasionality and inevitability are two aspects with the properties of mutually in opposition, as well as mutually in linkage, that existed in the process of development of objective events and phenomena. The development of event or phenomenon will include the aspect of occasionality, and will also include the aspect of inevitability. Emergence of parameters and variables with the nature of ocassionality and inevitability is a normal phenomenon of the complex object studied by social systems engineering. Workers of social systems must have sufficient recognition of this.

(1) There shall be existed factors of occasionality in the research and study any complex system.
(2) Development of modern science and technology that people can master regularity of occasionality. Systems scientists had raised the opinion quite earlier that these were too much mathematics dealing with solutions of finite values (inevitability); but not enough mathematics dealing with random variables and stochastic process (occasionality). Along with the development of applied mathematics, people can find certain inevitable regularity from some occasionalities. This is the probability distribution of random variables. If an appropriate model is chosen, it is possible to have the best estimation of value of unknown variables through statistical inference. For example, the book Social Systems Engineering (Japanese Version) by Oosawa Hikaru, and the curriculum of social systems engineering of University of Tsukuba. To sum up, both qualitative and quantitative analysis are two aspects of social systems engineering. Both aspects should be developed, weakness of both of them should be improved.

5.4.6.2 Qualitative Analysis Through Comparison

Comparison is a method of logic to understand the objective events or world quickly. It is also the basis to identify alterative scheme for decision makers. In the process of comparison of two objects, the similarity and difference of major subsystem or components, were compared first, then the similarity and difference of minor subsystem and components to be compared later.

Table 5.2 is the change of structure of GDP of global various economies from 1990–2008. It can be found from structural change of GDP of supply (economic activity) and demand side (expenditure) of China compared to other economies. There were nearly continuous decline of share of consumption of GDP and nearly continuous rising of share of investment of GDP from 1990–2008. This unbalanced pattern of growth far exceeded the average of the world and other economies. There is also a low share of service of GDP on the supply side. This is more or less a common pattern of growth of former central planned economies. The Comparison will provide important information to decision makers.

5.5 Principle of Planning Based upon Social Systems Engineering

5.5.1 General

From the academic perception, the overall design of a national system should study the national political system, structure and function of government, its role and

Table 5.2 International comparison of structures of GDP (1990–2008) (*Source* UNCTAD (2011)

Economy or country	Year	Structure of GDP based on expenditure (%)				Structure of GDP based on economic activity (%)			
		Final consumption	Formation of total capital	Export	Import	Agriculture	Industry		Service
							Industry total	Manufacturing	
World	1990	59.4	23.5	19.8	20.1	5.6	33.3	21.7	61.1
	1995	60	22.7	21.7	21.3	4.3	30.5	20.3	65.2
	2000	61.2	22.4	25	24.9	3.6	29.1	19.2	67.3
	2008	58.8	23.7	32.6	32.1	4	30.1	18.1	65.9
Developing economies	1990	59.4	25	25.8	24.6	14.8	36.8	22.4	48.4
	1995	59.3	27.3	29.5	30.1	11.9	35.1	22.5	53
	2000	58.5	24.7	34.9	32.3	10.3	36.3	22.6	53.4
	2008	52.7	31.2	42.1	38.6	9.8	40.2	23.7	50
Developed economies	1990	59.7	22.9	18.2	18.7	2.8	31.9	21.4	65.3
	1995	60.3	21.5	19.4	18.8	2.2	29.3	19.8	68.5
	2000	62.1	21.7	21.9	22.7	1.7	27	18.2	71.3
	2008	61.8	20.4	28.4	29.6	1.5	25.5	15.8	73
China	1990	48.8	34.9	18.4	15	26	39.7	35.5	34.3
	1995	44.9	40.3	22.2	20.1	19.7	46.6	40.6	33.7
	2000	46.4	35.3	23.4	21	15.2	46.4	40.7	38.4
	2008	37.3	49	37.8	33.2	11.6	48.3	42.8	40.1
Transitional economies	1990	53.6	28.9	24.9	27.2	19.9	44.8	24.8	35.3
	1995	55.9	24.6	32	31.9	10.7	36.3	19.2	53.1
	2000	52.6	19.3	45	33	10.3	36.3	19.5	53.4
	2008	51.3	26.3	35.8	30.2	6.1	36.4	17.4	57.5

(continued)

Table 5.2 (continued)

Economy or country	Year	Structure of GDP based on expenditure (%)				Structure of GDP based on economic activity (%)			
		Final consumption	Formation of total capital	Export	Import	Agriculture	Industry		Service
							Industry total	Manufacturing	
Russia	1995	52.1	25.4	29.3	25.9	7.6	37	17.1	55.5
	2000	46.2	18.7	44.1	24	6.7	37.9	19.4	55.4
	2008	48.6	25.5	31	21.9	4.9	36.1	17.5	59

relationship with various functional groups, families and individuals as well as its operation and maintenance. In the real world, however, National Systems are already in operation in more than 190 countries. So, there are no needs of overall design of a national system.

On the other side, due to changes of external and internal environment, it becomes a common issue in many countries that it is necessary to reform (maintenance in engineering sense) the structure, function and the mood of operation to adapt to changes. This is the element of "Adapt" of Parsons' original AGIL paradigm. For example, China had adopted the policy of openness and reform in late 1970s, and had set up the goals to accelerate the establishment of a socialist market economy, and ecological civilization. China has also summarized the core value of socialism system in 24 Chinese words: patriotic, devote to work, credibility and integrity, amiability, prosperity and strong nation, democracy, civilization and harmony. These are also goals pursued by most of the nations. Besides China, many other countries are in the process of reform. For example, the reindustrialization and reform of the medical system raised by the former president of U.S.A. Barack Obama. The European Union also faces the challenge of an aging society, fierce external competition, and has adopted a strategy of "Smart, sustainable and inclusive growth" in *EU 2020.* Therefore, planning of a national system is important to reform the existed system and to adapt better the changes of external environment, natural and social.

5.5.2 Three Basic Frameworks of National Planning

It can be seem from 5.2 that in spite of the growth and improvement of technology, economy, society and governance in general in the 20th centuries, there are still many remaining problems. The major issue are social disparity, aging society on the one side and addition of 80 million of new born people on the already crowded earth on the other. There were two World Wars in the 20th century and one great depression. The degradation of environment becomes increasingly serious. History of mankind shows the complexity of the issues involved. The global issue and prosperity cannot be solved by a single country focusing on its own development. A part of this depends upon the decision of the policy makers of major power players, which is outside the scope of this discussion. Improvement of the national planning system, however, can be one of the means to benefit the people and the society of one country within the global context. There are no shortages of national plans, some relevant experiences will be summed up in coming chapters. The perceptions of the authors will supplement the current knowledge pool.

5.5.2.1 Basic Framework of National Planning

Revised AGIL Framework of Systems of Actions can be used to be a basic framework of national planning. Figure 5.8 is quoted once more below and explained further in depth. (It is designated to be Framework 1 of National Planning).

(1) *Set Up a Rational Goal*

In the national planning system, the first step is to set up a rational goal system to adapt to the national conditions. For example, in the 13th Five-year Plan of China (2016–2020), seven goals are identified as follows:

- Maintain a medium-high rate of growth,
- Achieve significant results in innovation-driven development,
- Improve standards of living and quality of life,
- Improve the overall caliber of the population and level of civility in society,
- Strengthening further the coordinated development,
- Achieve an overall improvement in the quality of the environment and ecosystems,
- Ensure all institutions to become more matured and better established. pp. 16–18.

(2) *Economic Action System*

The action system can be broadly classified into S&T action system, social action system which had described in Figs. 3.9 and 3.10 in Chap. 3. In fact, there is also economic action system. Parsons' had studied the second level of hierarchy of economic action system in his "*Economy and Society*", specially, Parsons' had studied investment action and consumption action etc., which are also part of the economic action system. But these systems belonged to the first level of hierarchy of economic action system, i.e., demand action system, there is also supply action system at the first level of hierarchy.

(3) *The cultural system and the motivational system*

Based upon the simple definition given by Atkinson (2014), culture are the languages, customs, knowledge, belief, values and norms that combine to make up the way of life of any society. Culture may also refers to the arts (such as music, theatre, literature, and so on). p. 340.

Fig. 5.8 Revised AGIL framework of system of action (Framework I)

A Action System	G Goal System
L Cultural System Motivational System	I Integrative System

A relatively more detail definition of culture given by Marshall (1994) described:

> Culture is thus a general term for the symbolic and learned aspects of human society....... Social anthropological ideas of culture are based to a great extent on the definitions given by Edward Tylor in 1871, in which he referred to a learned complex of knowledge, belief, art, morals, law, and custom. This definition implies that culture and civilization are one and the same........In America, it is sometimes argue that the concept of culture can provide ways of explaining and understanding human behavior, belief systems, values, and ideologies, as well as particurally specific personality types...... In cultural anthpology, analysis of culture may proceed at three levels: learned pattern of behavior; aspects of culture that act below conscious level; and patterns of thought and perception, which are also cultural determined. pp. 104–105

The above quotation of definition explain the role of culture in the systems of action. When certain action systems are taken to achieve the goal system, the cultural system will perform the role to regulate the action system, for example, the law. The cultural system can promote the action system to achieve the goal system. For example, some Western scholar tried to explain the high growth rate of East Asia (*The East Asian Miracle* published by the World Bank in 1993) is due to the culture of Confucianism. But culture is formed from past, and may be not adaptable to contemporary or forward looking society. Therefore, an appropriate motivational system is necessary.

(4) *The Integrative (or Coordinative) System is of Crucial Importance and Difficult*

The people who are responsible for integrative function must have a concept of system thinking, i.e., he or she should not think in terms of a specific system, such as the economic system, the relevant connection and relationship with other systems should be thought in integration or in coordination. It had been described previously that there were rapid growth of science, technology and economy in the past, but there was also growth of social disparity in the past century. In the process of national planning, the integrative system can be performed through well designed TOTOS, which had been explained in the case Study of 5.2 or it can be solved through appropriate quantitative approach, which will be discussed later.

5.5.2.2 An Integrated Framework of Systems of National Planning (Framework II)

Figure 5.9a[9] is an integrated framework of systems of national planning, "The S&T system, economic system, social system and international system were studied as interrelated parts of an integrated whole. The system is a dynamic one, which changes with geographical location and time." Wang and Li (1987) had written *China* In Chamarik, S. & Goonatilake, S. (Ed) (1994) *Technological Independence.* (pp. 80–81). The social system, in a broad sense, includes the political system, and

[9]Note: Its original title was systems in time and space phase Sentences in quotation mark was explanation for the figure.

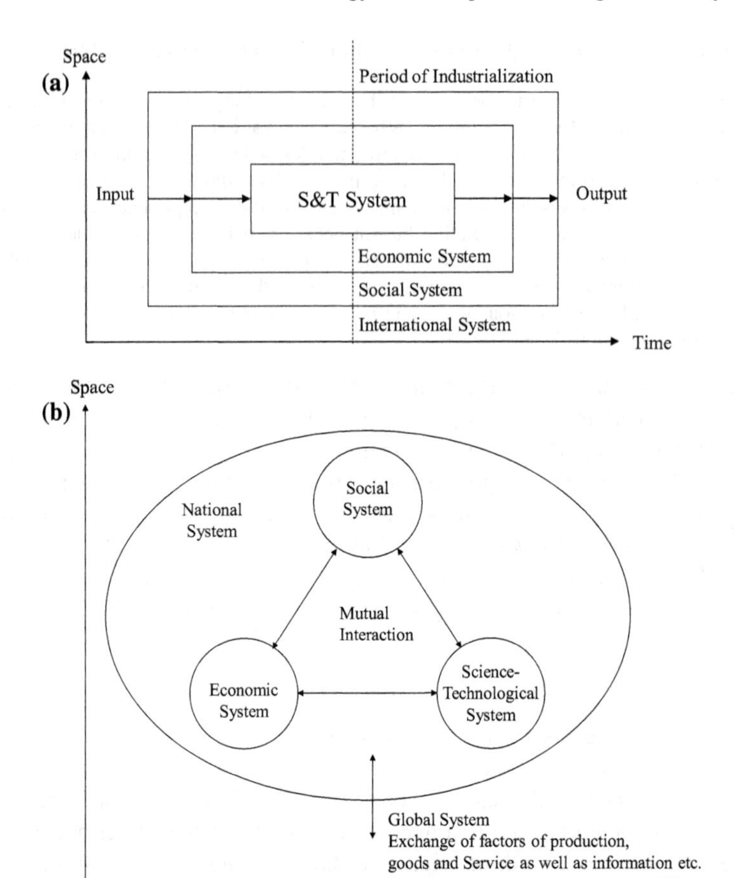

Fig. 5.9 **a** Integrated Framework of systems of national Planning (*Source* Fig. 5.1 System in time and space phase p. 81 Technological Independence). **b** Integrated Framework II of national planning system

economic system in the national planning. From the perception of authors of this book, Parsons' theory of social system focused very much on the action system, personality system and cultural system. The action system includes economic action (or behavior), and scientific research action. Therefore social system based upon the concept of action may cover all types of human action or behavior. In the synthesis analysis of a national system, the three subsystems should be studied separately but with linkages to other subsystems to be kept in mind. And synthesize three sub-systems through iterative process described in Sect. 5.4.4.1.

This framework can be redrawn in Fig. 5.9b to express the concepts more clearly.

A national system can roughly be classified into three sub-systems: social system, economic system, science and technological system. To simplify the discussion, [10]political system is included in the social system. These three subsystems develop individually based upon their own relevant theories, principles, or laws. These three subsystems are mutually interactive. For example, growth of science and technology will promote the economic growth and social development. But insufficient resource of a national economy will constrain the growth of science and technology. If the society does not pay sufficient attention to education, or protection of intellectual property right, it will also constrain the development of science and technology. In general, a balanced growth of three subsystems should be pursued. But a nation can put priority in development of certain sub-system based upon the stages of development. Generally many countries put priority to develop the economic sub-system first. For example, Japan had prepared Eleven medium-term economic plans before 1992. The Economic Planning Agency of Japan had prepared "Five-year Plan for Economic Self Support" since 1955, followed by its "New Long-range Economic Plan", "National Income Doubling Plan" and "The Medium-term Economic Plan". Japan also launched "The Economic and Social Development Plan (1967–1971), since 1967, in order to remedy various imbalance due to rapid economic growth in past ten years, many problem emerged, such as rises in consumer prices, delay in improvement of the living environment including housing and increase of pollution. Therefore, this plan tried to promote a balanced development of economy and society. Later on, social development was included to be the title of national planning of Japan, the New Economic and Social Development Plan (1970–1975), the basic Economic and Social Plan (1973–1977). Gini index in Japan was 24.9 in 1993, but raised to 37.9 in 2011. Based upon OECD data, 30 OECD[11] countries had Gini index from around 0.25 of Iceland to around 0.40 of U.S.A., with Chile, being the highest, a Gini Index of around 0.46. Income equality in developing countries is generally higher than developed countries. It can be seen from "social disparity" described in Sect. 5.3. Therefore, the integrative system (or coordinative system) of Revised AGIL Framework of Action is important. Figure 5.11b is designated to show the Framework 2 of National Planning.

The essential components to be considered in each system of Framework 2 are shown roughly in Tables 5.3 and 5.4.

5.5.2.3 Third Integrated Framework of Systems of National Planning (III)

Figure 5.10 shows the third integrated Framework of Systems of National Planning.

(1) The third integrated framework of systems of national planning, consists of three systems, macro-system, meso-system, and micro-systems. In general

[10]Note: Although political system is included in the social system, but it will not be discussed in this book. Because the authors have no experience in this urea.

[11]Current member states of OECD are 35.

Table 5.3 Essential components to be considered in each system

System component	Social system	Economic system	Science and technology system
1	Population and age structure	GDP and its growth rate	Number of higher educations
2	Race, minorities and migration	Component of GDP (supply side, i.e. primary, secondary and tertiary sectors)	Number of research institutes
3	Employment	Component of GDP (Demand side, i.e. investment, consumption external trade)	Number of S&T personnel and their distribution (Higher learning, business etc.)
4	Heath care	Public finance	S&T funding from various resources (gov., firms etc.)
5	Education	Structure of financial sector (i.e., banking, equity, bond, insurance etc.)	Mobility of S&T personnel (domestic and international)
6	Families	Balance of payments and reserves	Patents authorized
7	Income distribution of families	Physical infrastructure (Energy, transport, communication)	Years of compulsory education
8	Poverty	National debt	
9	Urbanization		
10	Structure of organizations		
11	Mass media		
12	Religion		

Table 5.4 Components of environmental system

System component	Domestic environment	International environment
1	Land	Global economy
2	Water	Geography of the world
3	Air	People, language and religion of major countries and neighboring countries
4	Pollutant discharged	Legal system, international laws
5	Ecological footprint	Communication and transportation of the world
6	Legal environment (Criminal rate, business environment)	Global energy system
7		Military and security
8		U.N. and its major organizations

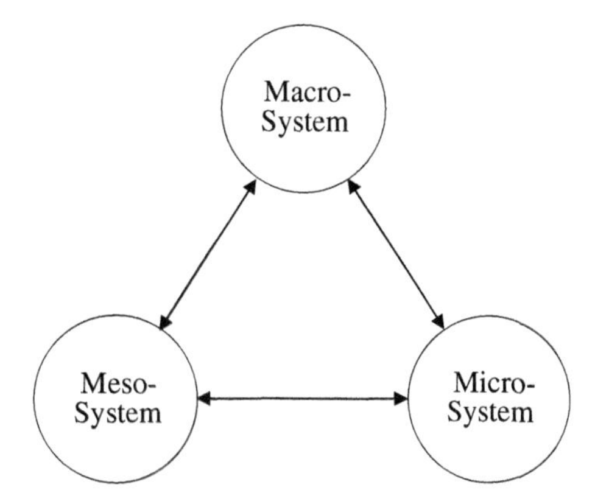

Fig. 5.10 Integrated Framework III of systems of national planning

systems theory, these three systems can be arranged into hierarchical level with macro-system on the top and micro-system at the bottom. But Fig. 5.12 has arranged them in the form of a triangle with an arrowhead of connecting line points to both sides. This expression emphasize the importance of mutual action among the three systems and a proper functional imperative must be performed by each system.

(2) The role of the macro-system refers to the role of the central government. It must create a right values system (moral value and value of action of its meso and micro-system). It must have the correct information and knowledge in time at meso and micro-level so that a relatively correct goal systems, action system and regulating system can be designed and effective integration can be implemented. The functional imperative of the macro-system is to create a sound and legal environment for the development of all economic, social, and scientific activity which will be favorable to a prosperous and fair society for all. The government must clarify the role of the market and market failure, and establish necessary social infrastructure (education, health care, unemployment insurance, pension system, care for infants, pregnant, disables and elders) within its fiscal capacity and proper policy for public private partnership. Try to satisfy and improve the basic needs of its people (cloth, food, housing and transport) within the context of national development.

(3) The role of the meso-system refers to the planning activities at different sectors and regions. Generally the principle and framework of planning at a regional level are in similarity of national planning. Planning the meso-systems should follow the value system closely with the macro-system and its goal system, generally in consistent with the macro-system with certain adjustment based upon its local concrete conditions, for example, the economic growth rate or priority of production activity. But in reality, a part of regions followed blindly the industrial policy set up by the central agencies without full analysis the feasibility of a project which has shown in steps of Fig. 5.2, and there were available many available publications by UNIDO and World Bank on the analysis and feasibility of industrial projects. It was pointed out by a World

Table 5.5 Dispersion of production activities among and within Provinces 1983 (*Source* World Bank 1985)

Production activity	Number of provinces involved in activity (out of 29)	Number of prefectures of municipalities involved in activity		
		Jiangsu (out of 14)	Hubei (out of 14)	Gansu (out of 13)
Foodgrains	29	14	14	13
Cotton	21	11	12	3
Coal	27	9	7	…
Cement	29	14	14	13
Pig iron	27	9	3	2
Steel	28	13	8	4
Steel products	28	14	7	…

Bank Report of China: "Long Term Development Issues and Options (1985)", China's past emphasis on local self-sufficiency rather than specialization has resulted in all twenty-nine provinces and hundreds of prefectures, being involved in a wide range of productivities (see Table 5.5). In fact, almost all provinces and many areas within provinces produce not only basic foodstuffs and materials such as cement whose production is dispersed in most countries, but also iron and steel products and consumer durables, whose production in other countries tends to be much more concentrated because of large economies of scale. In many areas of China, production is small scale, at high cost…". Table 5.5 is quoted partly from Table 5.3 of the World Bank to illustrate the issue of similarity of production activity among provinces. This can also explain the surplus production capacity of steel and cement of China currently and an adjustment policy of supply side is needed.

(4) It can be seen from the issues described in meso-system that they are closely related to the microsystems which include business and families. These business and families live in a macro and meso-system but also form the basis of a sound meso-system and macro-system. For some basic knowledge of business, please refer to Atkinson (2014). *The Business Book.* Family is also an important basic unit of micro-system and the foundation of a national system. The authors of this book suggest that there shall be a course on 'family' to teach young people of their role and responsibility in setting a family to supplement the driven by emotion and natural sexuality.

5.5.3 *Three Systems Models*

5.5.3.1 General

Up to the present, publications related to theoretical design of social system are relatively scarce, but there are no shortage of models. Based upon Banathy (1996) *Designing Social Systems in a changing world*, he had suggested "Three Systems Models that Portray Design Outcome."

(1) *The Systems-Environment Model*

This is an important approach. It had been described previously that every system exists in certain environment and with exchange of energy, material and information with its external environment. The system itself can be described by choosing appropriately from three integrated framework described in Sect. 5.5.2.

(2) *The Function-Structural Model*

This is the most common model used for national planning system. Different model builders with different professional background may have different views of sub-system of a nation. It can be seen from Chap. 4 Fig. 4.3, Chestnut emphasized that science and technology is a driving force of economic and social change. The authors of this book also put emphasis on the point that there should be a balanced development of three sub-systems, social, economic S&T, and there are mutually interactive. Figure 5.11 in the coming section, put emphasis only on society, economy and environment. Therefore, it should be understood that modeling is a useful tool for national planning, i.e., a system of tools and means are required to face the complexity of the reality of the world, in the application of certain model, it is necessary to understand the base of theory and assumption of the model, and if possible, an alternative model should be used to provide checking purpose.

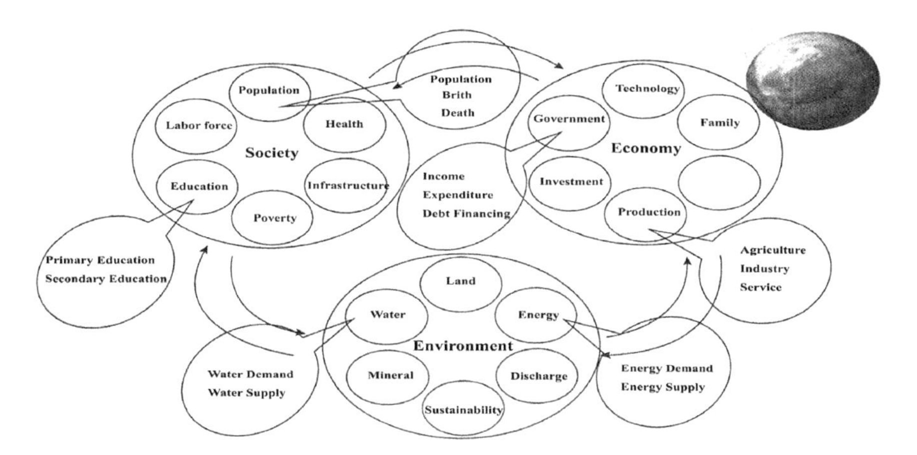

Fig. 5.11 Major concept of Structure of T21 Model (*Source* Andrea Bassi, Matter Pedercini (Oct. 2007) An Overview Threshold 21 (T21) Framework @ Millennium Institute org)

(3) *The Process/Behavioral model*

This model is only one of the key concepts of general systems theory, i.e. "input-transformation-output" model. Although there are behavioral equations in econometric model, human behavior is a very complicated phenomena requiring further collection of huge amount of data and classification. It is expected that joint efforts of international society are required to solve this basic issue.

5.5.3.2 Information related to T-21 World Model and National Model

(1) *General*

There are no shortage of various national plans. However, there are scholars and institutes focuses mainly in developing mathematical tools to study national planning and management. The following description is to provide some information relevant to the state of the art of this study.

(2) *Dr. Gerald Barney and his research group*

Dr. Gerald Barney, who had led the first project of national future study, *The Global 2000 Report to the President,* which included a large amount of quantitative study of a nation. Dr. Barney has established Global Studies Center and edited *Managing a nation-The Software Source Book* in 1986. Barney and Wilkins described in the preface that "*Managing a Nation: The Software Source Book* contains information on currently available software that can applied in the management of any nation. Far from being exhaustive, this first edition only hints at the range of software already available or under development. Future editions will extend substantially the amount of software reviewed."

That may be a trend of planning and managing a nation. There are already software available on multisector models, economy, rural and urban development, environment, ecology, politics and global model etc.

(3) *Millennium Institute and T21 Model*

Millennium Institute was established in 1983 at Washington D.C., and it is a non-profit organization devoted to develop new tools, concepts and models to assist decision makers to understand the mutual linkages among economy, environment and security. Dr. Barney had been involved in this institute. This institute had created T-21 Global Model based upon the concept of system dynamics. Creation of this model can be roughly divided into two periods: The first period, 1983–1993, was focused mainly to develop national modeling; the second period, started from 1995 to present, is to perfect and further apply this model to 25 countries. It had established a starting framework of a world model around 12 years, and a national planning system can be derived from this framework. Figures 5.11 and 5.12 show the major concept of structure of T21 model and linkage among major variables.

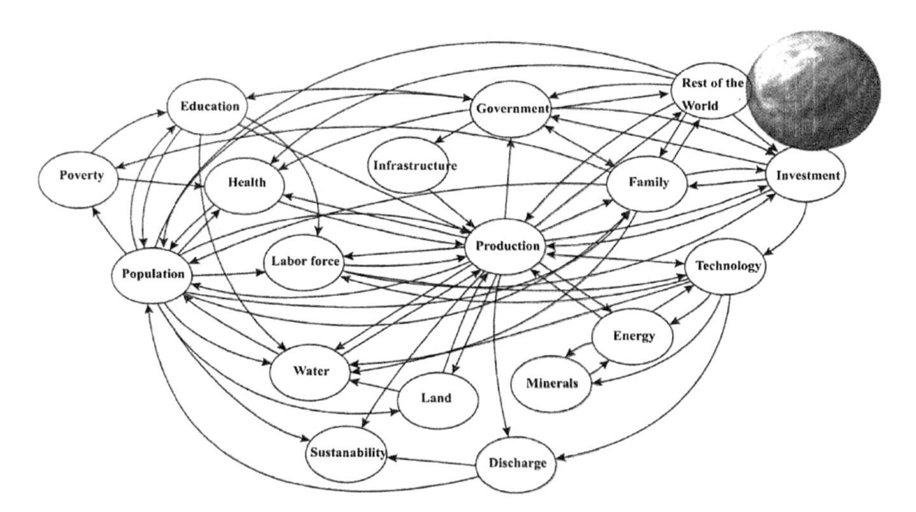

Fig. 5.12 Relationship of linkage among Major variables of T21 Model (*Source* Andrea Bassi, Matter Pedercini (Oct 2007) An Overview Threshold 21 (T21) Framework)

It is necessary to raise the point that Figs. 5.11 and 5.12 were version of 2007. Currently, it is evolved to be iSDG model with version of Jan. 15th, 2017. It is described in the introduction of very recent iSDG-Doc (2017) that "Starting January 2016. Country governments globally will be facing a dramatic challenge: how to turn the aspirations embedded in the 17 Sustainable Development Goals (SDGs) into actual development plans? Meeting the SDGs requires different strategies in different countries, and policy makers will have to address, among others, questions such as: How much resources are needed to achieve the SDGs? How to distribute investment across different areas? Where to invest first? How to finance such investment? The iSDG model has been designed and developed to support decision makers addressing such questions."

But Figs. 5.11 and 5.12 can show the basic concepts of the components of economic, social and environment system and their interrelationships, so the authors kept the old version in this book.

The above description illustrates that the quantitative approach of a national planning requiring a long term serious scientific research. Currently, there are many software available distributed in various institutions.

5.6 Summary Points

1. This chapter is consisted of four parts: Definition of methodology and meaning of its study; Further exploration of TOTOS-Case Study: Expo 2000 OECD Forum for the Future; Systems engineering logic and principle of planning based upon social systems engineering.

2. Professional of social systems engineering needs to solve issues of broad fields of disciplines, it is necessary for social systems engineering to master an appropriate methodology.
3. One of the pioneer of systems engineering Hall had created *Three Dimensional Morphology* of Systems Engineering in 1969. The three dimensional morphology provides an universal methodology for all types of engineering projects. But this methodology was derived on the basis of development of a new product and physical project. It is modified in a minor part to adapt to be methodology of social systems engineering.
4. The book *Societal Systems: Planning, Policy and Complexity* written by other eminent systems engineer John Warfield is briefed, his methodology Π-∑ process and TOTOS (Task oriented Transient Organizations) are introduced.
5. A case study to illustrate the application of TOTOS is described in relatively detail. This case study outlines a programme responsible by OECD Forum for the Future supported by EXPO 2000 in Hanover, Germany. There are four international conferences held each year around the theme of "People, Nature and Technology Sustainable Societies in the 21st Century." The first conference was held in Germany on December 1997. The theme was "21st Century Technologies: Balancing Economic, Social and Environmental Goals." The senior author of this book had been invited to participate the first conference of this programme. A relatively detailed description of the process of organization of this conference his given which can illustrate fully the implementation of TOTOS.
6. Important aspects of four publications of OECD related to development of technology, economy, society and governance are recommended. It can represent the core concept of social systems engineering in planning (or improving) a national system, a balanced development, i.e. balanced development of technology, economy, society and governance.
7. A system engineer must be trained to think in logic and in mathematics to get access to all disciplines. This is also the necessary capability of social systems engineer. The senior author of this book had published a book *Introduction to systems engineering* (in Chinese) in 1980 with a chapter of *Engineering logic* with examples of physical engineering in explanation. This approach is improved and revised several times. It was written to be part of qualitative and quantitative system in the book *Methodology of Social Systems Engineering* (Chinese version) published in 2014. This theme is presented in this chapter and titled *Systems Engineering Logic*. It includes a system of thinking in pairs of unity of opposites, such as induction and deduction, synthesis and analysis, cause and outcome, ocassionality and intentability and some derivatives of them, such as analogy and comparison. Examples from economic and social phenomena are quoted to illustrate its applications.
8. Three basic frameworks of national planning based upon social systems engineering are explained:

(1) a revised AGIL Framework of Systems of Action;
(2) an integrated framework of three mutually interactive systems, scientific and technological system, economic system and social system with dynamic change in space and time and with external environment;
(3) an integrated framework of systems (macro system, meso system and micro system) of National Planning.

9. Modeling is also an important methodology of social systems engineering. Three types of modeling are introduced.
   Every system is existed in certain external environment. The first type of modeling is The Systems-Environmental Modeling.
   Every system should be studied from two aspects: The Function-structural model and the Process-behavioral model.
   The pre-requisite of a mathematical model is exactness of the data. There are insufficient data related to human behavior. It is one of the basic task to engage collection of data and its classification.
10. T-21 World Model and national model are introduced briefly due to its relatively long history and with feedback loop based upon systems dynamics. From the perception of the authors, feedback loop is important means to link several systems together and to study their mutual interactions.

References

Atkinson, S. (Ed). (2014). *The business book* (pp. 104–105, 340). London: Dorling Kindersley Ltd.
Banathy, B. H. (1996). *Designing social systems in a changing world*. New York: Plenum Press.
Bardach, E. (2005). *A practical guide for policy analysis* (p. 14). Washington, D.C.: CQ Press.
Barney, G. O., & Wilkins, S. (Eds.). (1986). *Managing a nation-the software source book*. Arlington: Global Studies Center.
Box-Steffensmeier, J. M., Brady, H. E., & Collier, D. (2008). Political science methodology. In J. M. Box-Steffensmeier, H. E. Brady, & D. Collier (Eds.), *The Oxford handbook of political methodology*. New York: Oxford University Press Inc.
Child, J., Tse, K. K. T., & Rodrigues, S. B. (2013). *The dynamics of corporation co-evulution-A case study of port development in China*. UK: Edward Elgar.
Chinese Government. (2016). *The 13th five-year plan for economic and social development of the People's Republic of China (2016–2020)* (pp. 16–18). Beijing: Central Compilation and Translation Press.
Hall, A. D. (1969). Three-dimensional morphology of systems engineering. *IEEE Transactions on Systems Science and Cybernectics*, *SSC-5*, 156–160 (April 1969).
Marshall, G. (Ed.). (1994). *The concise Oxford dictionary of sociology*. Oxford: Oxford University Press.
Mϕller, J. ϕ. (1999). *Towards globalism: Social causes and social consequences*. In OECD (2000), *The creative society of the 21st century* (pp. 113–115, 118–119, 130–131). Paris: OECD Publications.
Murphy, G., Shippy, D. J., & Luo, H. L. (1963). *Engineering analogies* (p. 5). USA.: Iowa State University Press.

OECD. (1998). *21st century technologies-promises and perils of a dynamic future.* Paris: OECD Publications.
OECD. (1999). *The future of the global economy-towards a long boom?* (p. 4). Paris: OECD Publications.
OECD. (2000). *The creative society of the 21st century* (pp. 113–114, 119, 130–131). Paris: OECD Publications.
OECD. (2001). *Governance in the 21st century.* Paris: OECD Publications.
Okita, S. (1980). *The developing economies and Japan-Lessons in growth.* Tokyo: University of Tokyo Press.
Piketty, T. (2014). *Capital in the twenty-first century* (by A. Goldhammer). Cambridge: The Belknap Press of Harvard University Press.
Sibert, H., & Klodt, H. (1998). *Toward global competition; catalysts and constraints.* In OECD (1999), *The future of the global economy.* (p. 127). Paris: OECD publications.
The World Bank. (1989). *World development report* 1989.
The World Bank. (2010). *World development report* 2010.
UNCTAD. (2011). *Handbook of statistics.* New York: UN.
United Nations. (2001). *Indicators of sustainable development-guideline and methodologies.* New York: UN.
Wang, H. J.(1980). *Introduction to systems engineering vol.* (Chinese version) Shanghai: Shanghai Science and Technology Publisher.
Wang, H. J. (2004). *Integrated study of China's development and reform-preliminary explorations* (Chinese Version). Beijing: WuZhou Chuanbo Press.
Wang, H. J. (2015). *Methodology of social systems engineering.* Beijing: China Development Press.
Wang, H. J., & Li, B. X. (1987). China. In S. Chamark & S. Goonatilake (Eds.) (1994), *Technological independence-the Asian experience.* Tokyo: United Nations University Press.
Warfield, J. N. (1976). *Societal systems: Planning, policy and complexity.* New York: Wiley Interscience.
World Bank. (1985). *China: Long-term development issues and options* (p. 73). Baltimore: The Johns Hopkins University Press.
World Bank Group. (2015). World Development Report 2015.

Chapter 6
Indicators, Models and Mathematical Modeling

6.1 Introduction

One of the major difficulties of social systems engineering is measurement and quantification. Traditional systems engineering is mainly based upon the natural science. Mankind has mastered a relatively large part of the natural phenomena, and there are many laws, theorems and principles which can be applied to study certain objective or objective phenomena through models. For example, Newton's Law of motion, Kirchoff's circuit laws, Law of thermodynamics and many other laws of natural phenomena can be used to build mathematical models to understand, predict and control the objective and objective phenomena. But there is no laws at all in social science. There had once been the Say's Law of markets, but this argument was severely criticized by Keynes in the later period. Economics is one of the social science with application of various mathematics and with relatively reliable indicators to express quantitative results. But a large part of social phenomena have the difficulties in measurement, a shortage of relatively perfect indicator system as well as the issue of quantification. It was described clearly by Sellin & Wolfgan in 1964 that "the relatively underdeveloped state of measuring instruments for such purposes in the social sciences has been an obstacle to efficient and economic research. If measurements of time, temperature, length or weight had to be invented for each research analysis in the physical sciences, progress in those fields would indeed have been slow. Yet in criminology and in the social sciences generally, there has been little or no systematic theory stipulating how to select fundamental dimensions of conduct in order to measure certain social events." (Sellin & Wolfgang, 1964).

This situation is gradually improved since the U.S. Department of Health, Education and Welfare published *Toward a Social Report* in 1969. The book *Social Indicators 1973* was published with a collection of statistics of eight areas: health, public safety, education, employment, income, housing, leisure and recreation and population.

This chapter will be presented in four sections. Section one will present several new systems of indicators which can be used to promote the quantification of social

H. Wang and S. Li, *Introduction to Social Systems Engineering*,
https://doi.org/10.1007/978-981-10-7040-2_6

system engineering, and a brief explanation of preparation of these indicators will be given. Section two presents a general discussion of 'model', the purpose is to broaden the concept of 'model' and the process of modelling, rather than focusing on mathematical modelling blindly. Section three will describe the history of development of modelling. Currently, economic mathematical model played a dominant role in social systems engineering. Development of economic mathematical modelling from Western countries will be briefed. As a late comer of mathematical modelling, China's experience of development of mathematical modelling will also be briefed. The experience of Development Research Center of the State Council, (one of the state think tanks that both authors of this book are attached) will be briefed separately. Development of models in the social science especially the behavioral modeling abroad will also be briefed. China's Computable General Equilibrium Model-DRC CGE Model will be presented in the last section to illustrate what we have learned and achieved in mathematical modelling-an essential tool of social systems engineering.

6.2 Indicators

6.2.1 Definition of Indicator and Its Formulation

6.2.1.1 Definitions of Indicator

There are many definitions available for indicators, but the following definition may be more appropriate for public policy issue which will facilitate the process of communicating information to the decision makers and the public.

"Indicators provide information in more quantitative form than words or pictures alone; they imply a metric against which some aspects of public policy issues, such as policy performance can be measured. Indicators also provide information in a simpler, more readily understood form than complex statistics or other kinds of economic and scientific data; they imply a model or set of assumptions that relates the indicator to more complex phenomena." (Hammond et al., 1995, p. 1).

6.2.1.2 Formulation of Indicators

Formulation of indicators and indices are shown in Fig. 6.1.

Figure 6.1 shows the process of formation of indicators and indices. The process are explained as follows:

(1) Appropriate primary data shall be available, the correctness and applicability of indicators and indices depend on the length of the history of primary data. The longer the history is, the more applicable the indicators and indices are. For any primary data, there must be a permanent measuring method;

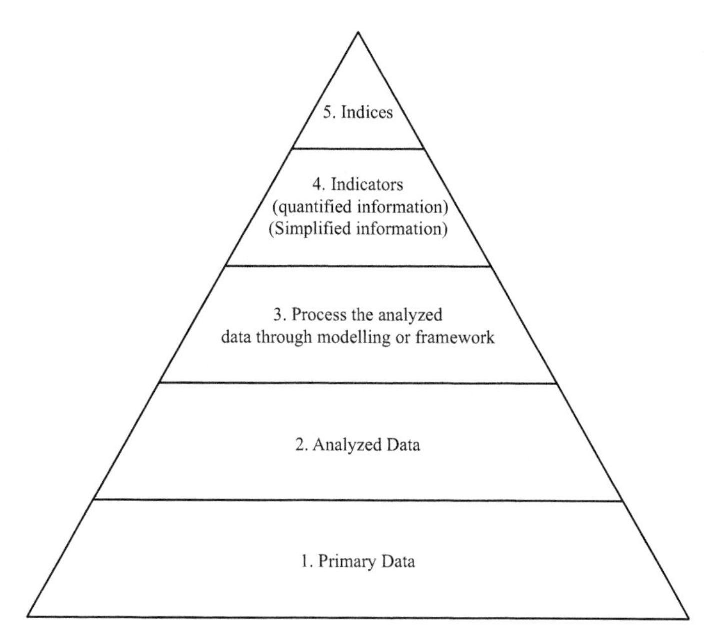

Fig. 6.1 Process of formation of indicators and indices

(2) Primary data must be analyzed before being applied, because some factors, like the sampling time and place of primary data, affect the applicability of indicators and indices;
(3) The analyzed data should be further processed into some kinds of information through a framework and a theoretical model;
(4) The information abstracted shown should be quantified and simplified, which is the basic nature of indicator. Indicators or sub-indicators shall have definite meanings and they shall be comparable.

It can be seen from the above definition of indicator that the process of indicator formation is complicated. This can be explained by a case study of United Nations in the preparation of Indicators of sustainable development.

6.2.2 Case Study Preparation of Indicators of Sustainable Development by United Nations

6.2.2.1 Background of This Case

The United Nations Conference on Environment and Development (UNCED), also known as the "Earth Summit" was held at Rio de Janeiro, Brazil, from 3 to 14 June

1992. The theme of this conference dated back to the 1972 UN Stockholm Conference which focused international attention on environmental issues, especially those relating to environmental and "transboundary pollution." In the years that followed, it also came to be recognized that regional or local environmental problems, such as extensive urbanization, deforestation, desertification, and general natural resource scarcity, can spread to pose serious repercussions for broader international security. International recognition of the fact that environmental protection and natural resources management must be integrated with socio-economic issues of poverty and underdevelopment culminated in the definition of "sustainable development" as defined by the World Commission on Environment and Development, 1987, as "...development that meets the needs of the present without compromising the ability of future generations to meet their own needs.". *Agenda 21* was accepted, which is a non-binding, voluntary implemented action plan of the United Nations with regard to sustainable development.

The following two points are described in Chapter 40 of *Agenda 21*:

(1) While considerable data already exist...more and different types of data need to be collected, at the local, provincial, national and international levels.
(2) Commonly used indicators such as GNP and measurements of individual resource or pollution flows do not provide adequate indicators of sustainability. Indicators of sustainable development need to be developed to provide solid basis for decision making at all levels.

6.2.2.2 Programme of the Commission on Sustainable Development (CSD) of UN

The Commission on Sustainable Development had approved the Programme of work on indicators of Sustainable Development in response to above calls with the involvement of organizations of the UN system, intergovernmental and non-governmental organizations.

The period of this programme was from 1995 to 2000, only the consideration of framework and modeling of indicators of sustainable development will be briefed in this chapter.

Two approaches had been applied in the preparation of indicators:

(1) The first framework adopted from the beginning of programme up to April 1999 was based on Pressure, State, and Response framework;
(2) The revised framework was based on themes and subthemes, this revised framework had achieved the final results of indicators for sustainable development.

6.2.2.3 Framework Based on Pressure, State and Response

(1) *DSR Framework for Sustainable Development Indicator*

The early work performed under CSD organized the chapters of *Agenda 21* under the four primary dimensions of sustainable development-social, economic, environmental, and institutional. Within these categories, indicators were classified according to their driving force, state, and response characteristics; which was a conceptual approach widely used for environmental indicator development. The term driving force represents human activities, processes, and patterns that impact on sustainable development either positively or negatively. State indicators provide a reading on the condition of sustainable development, while response indicators represent societal actions aimed at moving towards sustainable development. The initial framework established in the Programme for sustainable development indicators is shown in Table 6.1.

(2) *Original Source of Pressure State and Response frame*

The pressure-state response framework is not an innovation of this Programme, it was developed by the OECD from earlier work by the Canadian government. Following a cause effect social response logic, Fig. 6.2 is derived. The figure shows a matrix of Environmental Indicators done by OECD and UNEP.

This Pressure-State-Response Framework for indicators can also be expressed in a diagram shown in Fig. 6.3.

6.2.2.4 Revised Framework Based on Themes, and Subthemes

(1) The initial framework shown above and methodology sheets (sample of form methodological sheets is shown in Fig. 6.4) for 134 indicators were sent to 22 countries engaged in the testing process as a preliminary working list between 1996 and 1999.
(2) Some countries tested concluded that the driving force-state-response framework, although suitable in an environmental context, was not appropriate for the social, economic and institutional dimensions of sustainable development. With national testing experience and the overall orientation to decision making needs, the Expert Group of the Programme recommended the indicator framework to be policy and theme oriented related to sustainable development. Table 6.2 are key themes suggested by CSD testing countries.

Table 6.1 DSR framework for sustainable development indicators

SD dimension	Chapter of Agenda 21	Driving force indicators	State indicators	Response indicators
Social				
Economic				
Environmental				
Institutional				

Source Economic & Social Affairs UN (2001, Table 2, p. 26)

Issues	Pressure	State	Response
Climate Change	(GHG) emissions	Concentrations	Energy intensity; env. measures
Ozone Depletion	(Halocarbon) emissions; production	(Chlorine) concentrations; O_3 column	Protocol sign; CFC recovery; Fund contrib'n
Eutrophication	(N,P water, soil) emissions	(N, P, BOD) concentrations	Treatm. Connect.; investments/costs
Acidifcation	(SO_X, NO_X, NH_3) emissions	Deposition; concentrations	Investments; sign. agreements
Toxic Contamination	(POC, heavy metal) emissions	(POC, heavy metal) concentrations	Recovery hazardous waste; investments/costs
Urban Env. Quality	(VOC, NO_X, SO_X) emissions	(VOC, NO_X, SO_X) concentrations	Expenditures; transp. policy
Biodiversity	Land conversion; land fragmentation	Species abundance comp. to virgin area	Protected areas
Waste	Waste generation mun'pal, ind. Agric.	Soil/groundwater quality	Collection rate; recycling investments/cost
Water Resources	Demand/use intensity resid./ind./agric/	Demand/supply ratio; quality	Expenditures; water pricing; savings policy
Forest Resources	Use intensity	Area degr. forest; use/sustain, growth ratio	Protected area forest, sustain, logging
Fish Resources	Fish catches	Sustainable stocks	Quotas
Soil Degradation	Land use changes	Top soil loss	Rehabilitation/protection
Oceans/Coastal Zones	Emissions; oil spills; depositions	Water quality	Coastal zone management; ocean protection
Environmental Index	Pressure index	State index	Response index

Fig. 6.2 Matrix of environmental indicators (*Source* Hammond et al. 1995, p. 13)

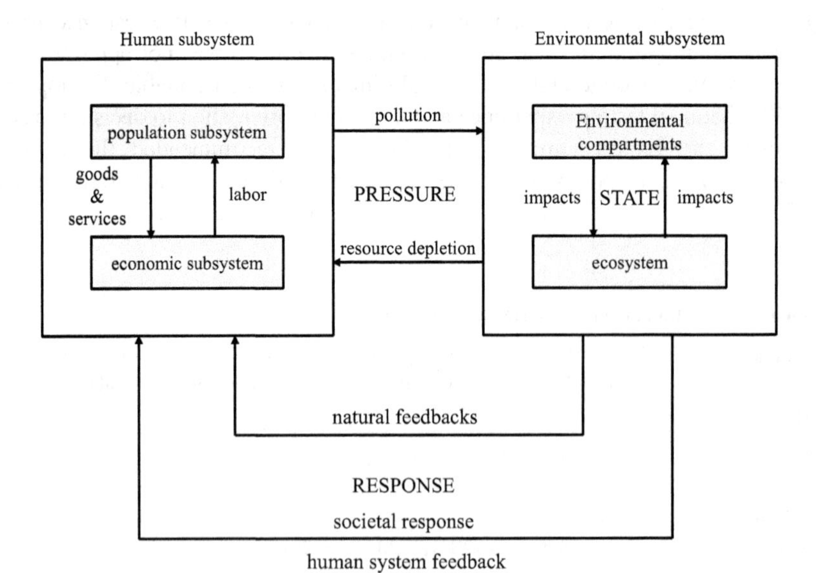

Fig. 6.3 Pressure-state-response framework for indicators (*Source* Hammond et al. 1995, p. 11)

Percent of Population Living Below Poverty Line		
Social	Equity	Poverty

1.Indicator

(a) Name: Percent of Population Living Below Poverty Line

(b) Brief Definition: The proportion of the population with a standard of living below the poverty line

(c) Unit of Measurement:%

(d) Placement in the CSD Indicator Set:

2.Policy Relevance

(a) Purpose:

(b) Relevance to Sustainable/Unsustainable Development (theme/sub-theme):

(c) International Conventions and Agreements:

(d) International Targets/ Recommended Standards.:

(e) Linkage to other indicators:

3.Methodological Description

(a) Underlying Definitions and Concepts:

(b) Measurement method:

(c) Limitations of the indicator:

(d) Status of the methodology:

(e) Alternative Definitions/Indicators

4.Assessment of Data

(a) Data needed to compile the indicator:

(b) National and International Data Availability and Sources:

(c) Data References :

5.Agencies Involved in the Development of the Indicator

(a) Lead Agency:

(b) Other Contributing Organizations

6.References

(a) Readings:

(b) Internet Site:

Fig. 6.4 Sample form of methodological sheet (*Source* Economic & Social Affairs UN 2001, pp. 65–69)

(3) The final framework of 15 themes and 38 sub-themes has been developed to guide national indicator development beyond the year 2001 based upon feedback information suggested by testing countries shown in Table 6.2 and also specific benchmarks or targets for many of the themes and subthemes by the international community (which is shown in Annex 1 of this UN publication). Only two of CSD theme indicator framework of social and environmental are quoted in Table 6.3 to illustrate two of the final results.

Table 6.2 Key themes suggested by CSD testing country priorities

Social	Environmental
Education	Freshwater/groundwater
Employment	Agriculture/secure food supply
Health/water supply/sanitation	Urban
Housing	Coastal Zone
Welfare and quality of life	Marine environment/coral reef protection
Cultural heritage	Fisheries
Poverty/Income distribution	Biodiversity/biotechnology
Crime	Sustainable forest management
Population	Air pollution and ozone depletion
Social and ethical values	Global climate change/sea level rise
Role of women	Sustainable use of natural resources
Access to land and resources	Sustainable tourism
Community structure	Restricted carrying capacity
Equity/social exclusion	Land use change
Economic	Institutional
Economic dependency/Indebtedness/ODA	Integrated decision-making
Energy	Capacity building
Consumption and production patterns	Science and technology
Waste management	Public awareness and information
Transportation	International conventions and cooperation
Mining	Governance/role of civic society
Economic structure and development	Institutional and legislative frameworks
Trade	Disaster preparedness
Productivity	Public participation

Source Economic & Social Affairs UN (2001, p. 38)

6.2.2.5 Later Work

The Department of International Affairs had continued its activities to monitor, evaluation the indicators of sustainable development; there were Expert-Groups Meetings respectively in Dec. 2005 and 2008.

United Nations at the Rio+20 conference, had passed a resolution, known as The Future We Want-Paragraph 246 of the Future We Want outcome document forms the link between the Rio+20 agreement and the MDG (Millennium Development Goals which are eight goals set up in Millennium Declaration of UN in September 2000. (The eight goals are: eradicate extreme poverty and hunger; achieve universal primary education; promote gender equality and empower women; reduce child mortality rates; improve maternal health; combat HIV/AIDS, malaria and other diseases; ensure environmental sustainability and develop a global partnership for

Table 6.3 CSD theme indicator framework

Social		
Theme	Sub-theme	Indicator
Equity	Poverty (3)	Percent of population living below poverty line Gini index of income inequality Unemployment rate
	Gender equality (24)	Ratio of average female wage to male wage
Health (6)	Nutritional status	Nutritional status of children
	Mortality	Mortality rate under 5 years old
		Life expectancy at birth
	Sanitation	Percent of population with adequate sewage disposal facilities
	Drinking water	Population with access to safe drinking water
	Healthcare Delivery	Percent of population with access to primary healthcare facilities
		Immunization against infectious childhood diseases
		Contraceptive prevalence rate
Education (36)	Education level	Children reaching grade 5 of primary education
		Adult secondary education achievement level
	Literacy	Adult literacy rate
Housing (7)	Living conditions	Floor area per person
Security	Crime (36, 24)	Number of recorded crimes per 100,000 population
Population (5)	Population change	Population growth rate
		Population of urban formal and informal settlements
Environmental		
Theme	Sub-theme	Indicator
Atmosphere (9)	Climate change	Emissions of greenhouse gases
	Ozone layer depletion	Consumption of ozone depleting substances
	Air quality	Ambient concentration of air pollutants in urban areas
Land (10)	Agriculture (14)	Arable and permanent crop land area
		Use of fertilizers
		Use of agricultural pesticides
	Forests (11)	Forest area as a percent of land area
		Wood harvesting intensity
	Desertification (12)	Land affected by desertification
	Urbanization (7)	Area of urban formal and informal settlements

(continued)

Table 6.3 (continued)

Social		
Theme	Sub-theme	Indicator
Oceans, seas and coasts (17)	Coastal Zone	Algae concentration in coastal waters
		Percent of total population living in coastal area
	Fisheries	Annual catch by major species
Fresh water (18)	Water Quantity	Annual withdrawal of ground and surface water as a percentage of total available water
	Water Quality	BOD in water bodies
		Concentration of faecal coliform in freshwater
Biodiversity (15)	Ecosystem	Area of selected key ecosystems
		Protected area as a % of total area
	Species	Abundance of selected key species

Source Economic & Social Affairs UN (2001, p. 30)
Note Numbers in brackets indicate relevant Agenda 2 chapters

development). It is described in paragraph 246 "the goals should address and incorporate in a balanced way all three dimensions of sustainable development (environment, economics and society) and their interlinkages. The Sustainable Development Goals (SDGs), officially known as *Transforming our World: the 2030 Agenda for Sustainable Development* with 17 "Global Goals" and 169 targets between them, were spearheaded by the United Nations through a deliberative process involving its 193 Members States and global society. The goals are contained in paragraph 54 United Resolution A/RES/70/1 of 25 September 2015.

6.2.2.6 World Development Indicators 2016

The World Bank Group published *World Development Indicators 2016* based upon sustainable development goals of the *2030 Agenda.* It is stated by Director Haishan Fu in the preface of above publication that "implementing the Sustainable Development Goals and measuring and monitoring progress toward them will require much more data than are currently available, with more accuracy, better timeliness, greater disaggregation, and higher frequency. The institutions fundamental to this effort should be supported through strong and renewed global partnerships. A new Global Partnership for Sustainable Development Data was launched alongside Agenda 2030 in September 2015 (See http://data4sdgs.org)." (p. 7)

41 international and governmental agencies and 11 private and nongovernmental organizations become renewed partnership of the above publication. The statement by Director Fu quoted here can illustrate fully the importance of primary data in research studies.

6.2.3 The National Competiveness Indicators-Developed by the Business School IMD

6.2.3.1 Introduction

IMD is the abbreviation of the International Institute for Management Development which was established in 1990 (through the merger of two institutes) and is located at Lausanne, Switzerland. It has a World Competitive Center which conducted its mission in cooperation with other partner institutes and published the *World Competitiveness Yearbook* since 1989. Its recent publication is *World Competitiveness yearbook 2016* which benchmarks the performance of 60 countries based on 333 criteria. The number of countries and criteria are increased along with the increase of years of this publication. For example, there were 47 countries and 290 criteria in the yearbook 2000. This case study shows only the design framework and the indicator system.

6.2.3.2 The Design Framework of National Competitiveness

The design framework of national competitiveness is shown in Fig. 6.5.

Figure 6.5 expresses the design framework of the competitiveness study in this yearbook. What are marked in four corners outside the ellipse can be regarded as four kinds of driving force, forming a competitive environment.

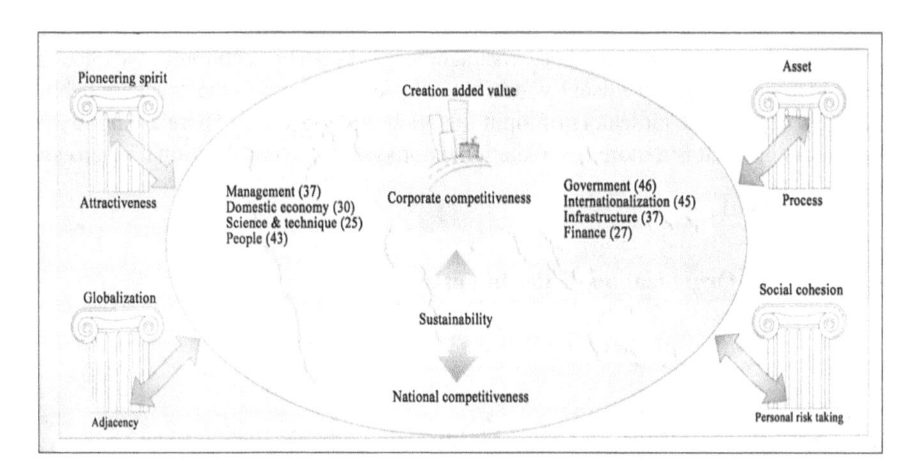

Fig. 6.5 Design framework of national competitiveness (*Source* Garelli 2000, p. 49)

(1) *Attractiveness and Pioneering Spirit*

They reflect a relationship between country-managed economy and world economy. The latter one refers to that a country encourages export and direct investment abroad, e.g. strategies adopted by Japan, Germany and South Korea. China also encourages export since economic reform, and in recent years, encourages enterprises to invest directly abroad. The former one refers to that a country attracts foreign investments through incentive policies, e.g. Singapore and China.

(2) *Adjacency and Globalization*

The former one refers to a country focusing more on making economic value adding activities adjacent to end users thereof. The latter one refers to developing international-operation-oriented transnational corporations, obtaining comparative advantages from the global market and configuring resources.

(3) *Asset and Process*

A country can handle the competitive environment in two ways, of which one depends on land, labor, natural resource and other assets, and the other depends on management and change of production process to form the competitive environment.

(4) *Personal Risk Taking and Social Cohesion*

The fourth one to form the competitive environment refers to establish a national economic system or model, or encourage individuals to take risks, i.e. international Anglo–Saxon Model, or depending on social consensus and cohesion, i.e., Europe Continent Model.
Under the four kinds of driving force, eight aspects, i.e. government, internationalization, infrastructure, finance, management, domestic economy, science & technique and people are listed within the ellipse. Numbers in the brackets behind the eight aspects are indicators of input factors in the yearbook. There are up to 290 indicators in all. It is a more complete indicator system covering national macro and micro states.

6.2.3.3 The Organization of the Indicator

The organization of the indicator system is shown in Fig. 6.6.

6.2.3.4 Activities Done by Other Institutions with Similar Nature

The World Economic Forum had published Global Competitiveness Report annually since 2004 with a design framework of 12 pillars.

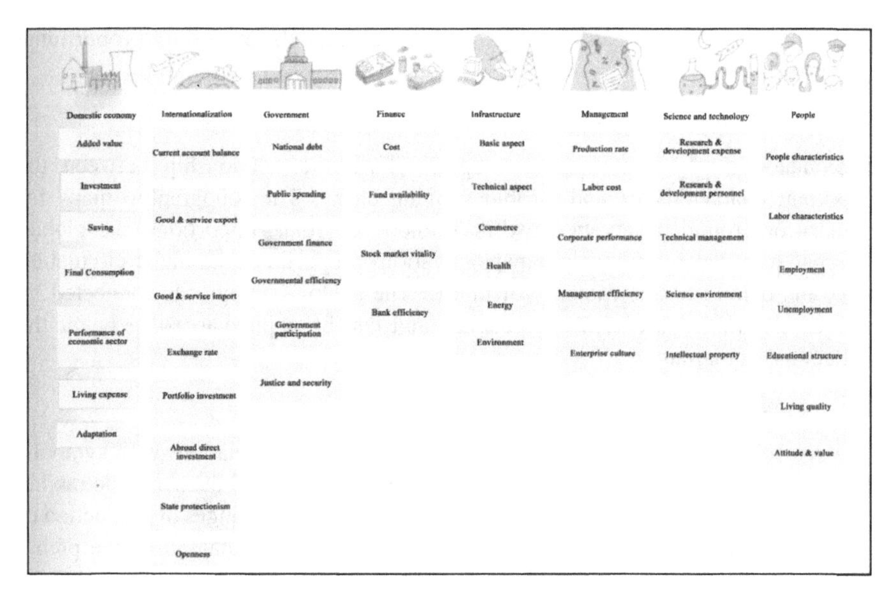

Fig. 6.6 Organization of national competitiveness state indicator (*Source* Garelli 2000, p. 57)

6.3 Models

6.3.1 General Discussion of Model

6.3.1.1 Basic Nature of the Model

The model is a tool which is a simplification of reality for analytical purpose, with a reduction in time and space that allows for a better understanding of reality. The representation can be expressed in words, numbers, diagrams or material substance shaped the same of the reality but in small scale, or a single equation or systems of equations showing the relationship among many elements and results of their interaction.

Model is a qualitative or quantitative representation of a process or actions of behavior, it should express the decisive or crucial relationship of the objectives to be studied.

6.3.1.2 Classification of Models

Models can be classified in many ways.

(1) *Traditional Classification of Model*

(i) *Image Model*

The image model is a material model, which is similar to actual objects in appearance, and can also be made in a certain proportion, e.g., the real estate models shown for sales, and professional fashion models are a fully proportional image model.

(ii) *Analog Model*

The analog model expresses in a special form the relationship between the important characteristics and attributes of an entity. The topographic maps for military or ground observation use in 1/50,000 or other proportions, the global, national, provincial or municipal maps in various proportions, the traffic-circulation map, etc. belong to this category of model. The analog computer is connected by various electronic parts and components, and can be applied to solve better the differential equations.

(iii) *Symbolic Model*

The symbolic model indicates various actual phenomena or relations with symbols. The digit, mathematical formula, language, chart, etc. belong to the symbolic model. For instance, the histogram can be used to indicate the GDP, changes of production of various products or foreign trade history, etc. The flow chart diagram in the preparation of the computer program also belongs to this category.

(iv) *Language Model*

The language model is a concept or the relation between various attributes, which is expressed by language. The language model is of great importance. In particular, it is required to clearly express each equation or functional relationship with language. The language model played a very important role in modeling. It is known that "A. Marshall himself formulated (not without humor) the following rules for the use of mathematics: (1) Use mathematics as a shorthand language, rather than as an engine of inquiry. (2) Keep to them till you have done. (3) Translate into English. (4) Then illustrate by examples that are important in real life. (5) Burn the mathematics. (6) If you can't succeed in 4, burn 3. This last I did often."

(v) *Mathematical Model*

The mathematical model describes and explains the represented system with mathematical symbols. The mathematical model is widely applied in the natural science and engineering field. Yet in the social sciences field, it hadn't been widely applied before 1930s. The application of mathematical model will be discussed further in Sect. 6.3 of this chapter.

(2) *Functional Model*

A model can also be classified based upon its function in the structure. For the social systems engineering, various functional models should be applied, the national planning model, policy model, prediction model, etc.

(i) *Rational Planning Model*

There are many theories and practices on the planning model. For instance, in Arthur Louis's *Development Planning*, he made the whole national planning model system with series of simple arithmetic.

(ii) *Projection Model*

The projection model can be a mathematical model, or a prediction trend formed by expert consultation.

(3) *Other Categories*

A model can also be classified based upon its application. For instance, there are economic models, agricultural models, global models, regional models, etc. According to the time criterion, there are the long-term projection models, long-term planning models or middle- or short-term forecasting models, mid-term and annual planning models, etc.

6.3.2 Difficulties of Modeling Social Systems Engineering

Although there are no shortage of supply of mathematical models in economics, a branch of social science, there are insufficient supply of mathematical models in sociology, because sociology is closely related to human behavior.

We will treat models of human behavior as a form of art, and their development as a kind of studio exercise. Like all art, model building requires a combination of discipline and playfulness. It is an art that is learnable. It has explicit techniques, and practice leads to improvement.

Fortunately, there are improvements on this issue. The National Science Foundation had supported the study of modeling of behavior. The problem can be solved through joint effort of academic studies.

6.4 Mathematic Modeling

6.4.1 Application of Mathematics in Social Science

6.4.1.1 Mathematics and Social Science

Mathematics is widely applied in the physics and engineering fields. However, its history of application in the social sciences is relatively short. From the perspective of science, the degree of mathematics application will further increase when the social sciences develops. This is the mark of any scientific maturity. Any developing discipline needs to become more accurate. The degree of accuracy can be realized through a certain branch of mathematics field. Of the social sciences, the application of mathematics in economics is more widespread than other branch of social sciences. For instance, the econometrics in economics has been widely applied and developed. Ragnar Frisch, a Norwegian economist, created the Mathematical Economics Society in 1930.[1] He said "statistics, economic theory and mathematics are indispensable for the development of economics. Yet only one of them cannot constitute the necessary condition to understand the economic life in modern times. Only the combination of three of them is powerful. Constitutes the econometrics." Up to now, the econometrics has been widely applied in the governmental planning and policy. Other widely applied mathematical model is Leontief's input-output model. One of the achievements of Brookings model in the 1960s was the matching of econometric model and input-output model. In the economic and mathematical models later on, the combination of them became a common practice. Yet the frequency of application of mathematical model in sociology and other social sciences is much lower than that of economics. This is because many events in the social sciences cannot be measured and quantified, or contain many random variables.

In 1970s, some electrical engineers tried to study the social systems with the control theory. Jay W. Forrester, creator of system dynamics, stated as follows in the 1971 and 1995 revised editions of his Counter-intuitive Behavior of Social System:

"[2] The human mind is not adapted to interpreting how social systems behave. Social systems belong to the class called multi-loop nonlinear feedback systems. In the long history of evolution it has not been necessary until very recent historical times for people to understand complex feedback systems.

Evolutionary processes have not given us the mental ability to interpret properly the dynamic behavior of those complex systems in which we are now imbedded.

The social sciences: which should be dealing with the great challenges of society, have instead retreated into small corners of research...." (Forrester, 1995, p. 2).

The quotations above describe the general situations and limitations of application of mathematical model in the research and development of social sciences. They can also help us to understand better the scope of applicability and potential of mathematical model in the social systems engineering.

[1]Note: Kemeny (1969).

[2]Source: Jay W. Forrester (updated March 1995) Counter-intuitive behavior of Social System. (Original text. Jan 1971)

6.4.1.2 Development and Application of Mathematical Model in the Economic Field of Western Countries[3]

French mathematician and economist Antoine Augustin Cournot (1801–1877) started to apply the economic model in the field of economics. In 1838, he studied some balance issues in the market with the mathematical model, and published the result in Researches on the Mathematical Principles of the Theory of Wealth in 1838. Later on, Leon Walras (1834–1910), a French economist, put forward the comparatively complete "general equilibrium theory" for the first time in 1874, and the abstract model developed by him was later extended by Viferedo Parets (1848–1923), an Italian representative for the social system theory. They became the first source of quantitative economics. The second source of development of macro econometric model were two mathematical models about the economic cycle in the early 1930s, which were respectively structured by Ragnar Frisch and Michal Kalecki (1898–1970), a Poland economist. These two models disclose the econometric method. The third main source is Keynes' *The General Theory of Employment, Interest and Money* and classic literatures on macroeconomic concepts that were published by him later on, esp. those on the consumption function. The fourth source affecting the development of macro model econometric is the thoughts and writings of German historical school. The fifth source is the study on classical statistics conducted by a batch of scholars in early 20th century, in particular, Trygv Haavelmo introduced the method based on the probability theory (realized by the opportunity rate through random disturbance) in early 1940s. On the basis of the above mentioned theories, a concept was brought forward by Walras in the Elements of Pure Economics: Economy is a system, the general behavior cannot be predicted from the properties of all constituent parts of the whole system. The argument is also the reflection of the relationship between sum and parts in the general system theory.

Jay Tinbergen (1903–1994) published a Netherlandish economic model consisting of 24 equations in 1935. He constructed further an annual American economic model consisted of 48 equations in 1939. Jay Tinbergen acted as Director General of Central Planning Bureau, Government of Netherland from 1945 to 1955, and energetically promoted the development of Netherlandish economic mathematical model for the purpose of predicting and formulating the policy. Wassily Leontief (1906–1999), a Russian economist, brought forward the input-output mathematical model which had also become an important constituent part of macro economy and microeconomic mathematical models. Klein published the Economic Fluctuations in the United States 1921–1941 in 1950, in which the result from the econometric modeling achieved by him was reflected. The United States National Research Council convened the American Economy Stability Meeting and determined to establish a detailed large-scale model in 1959. As a result, the SSRC/Brookings model was developed. The model included more than 200 equations and

[3]Reference: Authored by Klein et al. (1993).

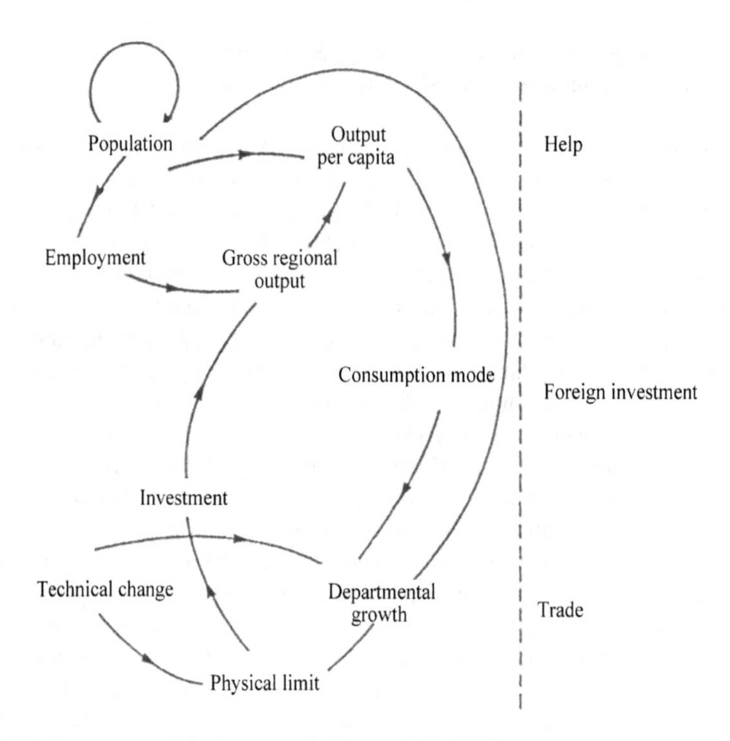

Fig. 6.7 Basic structure of limit to growth (*Source* Cole 1988)

realized the matching of input-output system and macro-econometric method. Ragnar Frisch, Jay Tinbergen, Klein and Leontief all were Nobel Prize winners. Throughout the common efforts of them, the economic theory and economic mathematical model have been developed simultaneously since 1930s. Other developed and developing countries also developed the study on the mathematical model. The scope of study ranges from the sector to the country and then to the globe, and in the academic field, the limits to Growth of Club of Rome is generally considered to be the first generation of global model. The world economic model compiled by Leontief, the British Government's[4] SARUM and the World Bank's model are the global model of second generation, etc. The United Nations'[5] UNITAD and Globus as well as the American 2000 global research models are called the global model of third generation, etc. Now, there are about 25 global models. The scope of study and structures of models become very complicated. Figure 6.7 indicates the basic structure and feedback circuit of limit to growth.

[4]Note: SARUM is compilation of System Analysis Research Unit Model. It was hosted by Englishman Peter Roberts in 1974. It is a multi-area economic model. It was firstly applied in the OECO's *Enter the Future*.

[5]Note: UNITAD is a combined research project of UNIDO and UNCTAD, used to research the growth mode and the relation between trade structure and industrial structure.

6.4.1.3 Promotion of Development of Mathematical Model by International Organizations

In the development of mathematical model, the United Nations and relevant international institutions played important driving roles. The United Nations held the "Seminar on Application of Planning Model System" in Moscow, former USSR in 1974[6]. The participants came from former Soviet Union, Norway, France, England, Hungary, Bulgaria, East Germany, etc. They put forward the economic policies and national planning models of the seven countries respectively. The International Labor Organization of the United Nations also published the *Planning Technology for a Better Future*.[7] The effort of the World Bank in the development of mathematical model also deserves special attention.

The World Bank established after the Second World War has huge manpower, material and financial resources. Chapter 10 will describe it in detail. It has a great influence on the trend and theory of policy studies. The World Bank had carried out much fundamental work for the development of computable general equilibrium model (CGE).

[8]Hollis B. Chenery (1918–1994) of the World Bank compiled a national mathematical model with the CGE principle since 1971 of the last century. Dervis published the actual guideline complied with the CGE model in 1982. At the same time, the World Bank supported such tools as data collection, working method, etc. The Social Accounting Matrix: Foundation for Formulating Plans was published in 1985. On the basis of the above mentioned theories and data development, staff of the bank also focused on the model software, the GAMS (General Algebraic Modeling System) was developed. The GAMS simplified the realization of CGE model in computer and promoted the analytical capability in transferring theory into practice and policy. The World Bank also developed the global model by cooperating with some institutions of higher learning in Belgium when participating in the Tokyo round negotiation, which was then used in its annual publications *World Development Report, Global Economy Prospect*, etc. In general, the improvements of analytical tool and hardware have been increasing in the exponential form since 1980s, yet the development of global data is obviously backward. The standard difference between different countries is quite large, since all global modeling teams need to collect the data of national income, product, tax rate and world trade in the early stage. The development of GTAP (Global Trade Analysis Project) data greatly improves the work of modeling.

Before and after the 1990s, under the leadership of some international organizations and the academic circle, the computable general equilibrium model has been

[6]Note: Reference: UN (1975).

[7]Note: Reference: Pyatt and Thorbecke (1976).

[8]Note: Dixon and Jorgenson (2013, p. 935).

widely developed and applied. The very recent publication of CGE modeling was *Handbook of Computable General Equilibrium Modeling* in two volumes edited by Peter B. Dixon and Dale W. Jorgenson in 2013.

6.4.2 Development of Mathematical Modeling of Social Since in China

6.4.2.1 General

China is a later comer of development and application of mathematical modeling in social science. Since the establishment of People's Republic of China in 1949, studies and application of mathematical modeling and statistics were more or less neglected in the economic and statistical field due to the consideration that the economic relationship was determinative under a central planned economy, and the MPS (Material Production System) was established in following the former Soviet model. Post the reform and opening in late 1970s, there was a training course of quantitative economics in Beijing, from June to August 1980, led by professor Lawrence R. Klein and other economists, most of them were American experts famous in mathematical modeling.

6.4.2.2 Development of Research Institutions

China began to modify its statistical system from MPS to SNA system through the assistance from the World Bank and other UN organizations. Chinese Association of Quantitative Economics was established in 1979. Many institutions were established to study the application of mathematics in modeling, for example research Institute of System Science was established under Chinese Academy of Science (CAS) (It was changed into Research Academy of Mathematics and System Science of CAS recently). Institute of Quantitative & Technical Economics of Chinese Academy of Social Science, State Information Center, Technical Economic Research Center (It was emerged into Development Research Center of the State Council in later period) etc., all of these institutions engaged in the work of application of mathematical modeling.

Since the establishment of State Information Center in 1987, it had constructed the Database, Method Base, etc. It tried to compile the national economy model, quarterly model, regional model, etc. It has also participated the Global Link project.

Institute of Quantitative and Technical Economics of Chinese Academy of Social Science issues a *Chinese Economic Blue Book* annually, and hosts the academic annual meeting for quantitative economics research. The meeting has been constantly expanding and improved in quality, since the first one in 1982.

6.5 Several Decades of Experience of Policy Modeling of DRC

6.5.1 Retrospect and Experience of Development of Policy Model

6.5.1.1 Introduction

Development Research Center of the State Council was established in the early 80s. It was one of the products of economic reform and opening to the outside world of China in the late 70s. The role of this center is to provide policy consultative service for the government. Therefore, the study and application of mathematical model in our work focused very much on policy analysis. Since China is a late comer in modern economic science as well as in themes of economic models for policy analysis, our center had tried with full effort to catch up and bridge the gap in this field with the contributions from all our partners in this development process. It should also be emphasized that there are technical difficulties to get exact results from economic mathematical modeling. This issue can be seen clearly from the criticism given by Keynes to the pioneering research of macroeconomic model of Tinbergen in that there were problems of misclarification, multi-collinearity functional forms, dynamic specification, structural stability, as well as difficulties associated with the measurement of theoretical variables, etc. The crucial issue in China is that mathematical economic models should be based on certain economic theory, while the transitional economics of the former central planned economy to a market-oriented economy is a new field in economic study. There may be also the issues of primary data and processed data.

In spite of these facts, it is recognized that mathematical models are an important tool of forecasting and policy analysis, and that it is unlikely that they will be discarded in the future. The challenge is to recognize their limitations and to turn them into a more reliable and effective tool. There seems to be no alternatives.

This description will be divided into two parts: the first part will give a brief retrospect of the policy modeling projects that had been done by us within the context of the economic situation of China at different stages of economic reform. This will give a general picture of evolution of policy modeling in China. And the second part will summarize the lessons and experiences.

6.5.1.2 A Retrospect of Policy Modeling of DRC

Policy modeling work of DRC can be roughly divided into two periods based on the broad context of progress of economic reform as well as the experience of our staffs.

(1) *Initial Period (1982–1990)*

In this period, China was at the initial stage of economic reform and opening-up, the planning system was undergoing changes and reform.

(i) *First Period of Application of Mathematical Modeling in Policy Analysis*

The first pioneering effort to the application of mathematical modeling in policy analysis was incorporated in the project of "Comprehensive Planning of Shanxi Province'. This project was assigned by the central government, it was to study the long-term development planning of Shanxi Province. This represented a natural process of development of policy analysis. As noted by Chenery, the earliest form of policy analysis in underdeveloped countries was typically described as "development planning", since one of its purposes was to assess the consistency of policy instruments and objectives. The term "development planning" is perhaps unfortunate, since it implies greater government control of economic activity. In fact, it is well known globally that a gradual approach had been adopted by China in the process of reform and opening-up. Therefore, in the project of "Comprehensive Planning of Shanxi Province', China hadn't implemented the reform of the planning system in the initial period. The concept of planning was still traditional in sense, and the only available policy instrument was "investment allocation".

But this pioneering project implemented in Shanxi Province from Feb. 1982 to June 1983 has two unique features: First, the planning period was extended to cover 20 years (1981–2000) compared to the past practice of 5-year planning period. Second, our center had the privilege to organize various government organizations, research institutions and academic people to work together in such a national key project. There were around 1400 people involved in this project. We tried to experiment policy analysis with new techniques in this project. A wide variety of modeling exercises were experimented in this project. There were around 100 people involved in the exercise of policy modeling, A total of 26 models were prepared. The scope of study covered very broad areas: the comprehensive planning of Shanxi Province, investment on coal sector, electric industrial planning, water resource utilization, optimal plantation, population model, environmental projection and planning, education projection and planning, investment on science and research, input and output of light industry, output projection of electronic industry, etc. The techniques used include input-output, econometrics, linear programming, multiple goal programming, decision analysis, etc.

The result was finally edited into a book by Wang and Tian entitled *Compilation of Economic Mathematical Models of Comprehensive Planning of Shanxi Province*. This pioneering project represented a very nascent stage of policy modeling of DRC. But we had achieved the purpose to get acquaintance to the application of various mathematical tools to the economic policy analysis. A team and organizational relationship had also been created among the researchers. An appendix with the brief introduction of seven mathematical tools were included in the above publication. In spite of the traditional concept of planning, the important types of interaction among the policy variables (objectives and instruments) and the constraints on the economic system of Shanxi Province had been correctly identified in the specification of the planning model.

Although it was a nascent stage in economic mathematical modeling, we came to understand that the key of economic mathematical modeling is an interdisciplinary study between economics, mathematics, statistics and the real world. The scholars involved in the Task-Oriented Transient Organization have different backgrounds, automation, economics, mathematics and others. Therefore, it is commented in the "Preface" of the above-mentioned book that "Economic modeling and mathematical 'model' are two fashionable terms in China. Model is designed through abstraction and simplification of the system of real world. Economic modeling is a simplified abstraction of economic activity. It can be in the form of a flow chart, statistical table, bar chart, hydraulic model to simulate economic system that had once been used abroad; but they were short-lived. Mathematical models for the economic system became widely adopted both domestically and abroad. It should be emphasized that our 'model' is a combination of economic activity and mathematical means. The precondition to design a relatively useful model is a detailed objective observation of economic phenomena with appropriate analysis and synthesis to understand clearly the interrelationship among the variables of various economic activities, to compare this with established theory and express these relationships with appropriate mathematics."

The question of how to allocate investment and other scarce resources among sectors and projects was also an urgent issue of development policy of China where governmental investment allocation playing a dominant role. Our center had proposed to the State Planning Commission the application of "Feasibility Study of Industrial Projects". We had organized a conference of national scale, and a symposium had also been published.

(ii) *Second Period of Policy Modeling of DRC*

(a) China Toward the Year 2000 was a pioneering project of strategic planning of China. With the experience gained in the development 'planning at provincial level, we had got aware of the weakness of traditional Soviet model of the planning system which became inappropriate in current stage of China's development. This point was also correctly summarized in a recent IMF report: in the case of Russia and China, at the outset predominantly agrarian economies where the majority of citizens were illiterate, the transformation to an industrialized and educated society was achieved roughly within a generation. However, once these economies entered intermediate or higher stages of development and resource allocation choices became more complicated was unable to cope. Therefore, in the national priority project of "China Toward the Year 2000", launched in 1983 and completed in 1985, a strategic development planning was developed which is totally different from the traditional model of Soviet planning. This study includes a main report and 13 sub-reports. Two of these sub-reports are: "Macro-Economic Model Projections of China Toward the Year 2000" and "Summary of Data of China Toward the Year 2000".

(b) *China: Economic Development and Modeling*[9] included fourteen models, which covered the study of development strategy and policy analysis, macro-economic model with application of econometrics, macro-economic model based upon production function and analysis of TFP, population and coordinated economic development planning model, quantitative analysis of economic structures, reproduction of two major sectors, long-term trend of development model, application of system dynamics, China's social economic development model, medium- and long-term macro-economic model, education planning mathematical model, energy system planning and decision model, energy demand model, China's environmental projection model, production structure plantation model, etc. The number of people involved in this sub-report was around 100. The collection of these models were published in 1990. This project represented the policy modeling with collaboration of people of different organizations at the national level. The mathematical tools used were the same as the first project described previously. In addition, system dynamics and recursive programming had been added. This effort of modeling exercise got the first-class national award of application of mathematical modeling in the national exhibition of computer application.

Through learning and practicing on policy analysis, 25 policy recommendations had been made to decision makers through this research. All these policy recommendations, reviewed today, seem to be correct. But not all of them had been adopted by the decision makers. Just quote one example, we had recommended that "an enterprise should be established to be entity closely bound up with the destinies of the state, collectives and individuals." This recommendation implicitly suggested that a new system of share company can be implemented to fit the interests of various actors of the economy. But it was not the right time to adopt this policy recommendation by the government in 1985. It should also be emphasized that not all policy recommendations were derived from the modeling system. These policy recommendations were derived through a systematic analysis and abstract from all sub-reports. For example, housing reform had also been recommended in that time, but implementation of this policy was launched around 1990s.

There is improvement in the consideration of the policy planning in this project compared to the initial period. As noted by Chenery "The logic of policy planning leads to distinctions among social objectives, constraints and policy instruments. A feasible program is then defined by a set of values for the policy instruments that satisfies the specified objectives and does not exceed the predetermined constraints." Broad policy instruments have been considered in this modeling exercise, for example, several conditions of balance had been incorporated in the model of Development Strategy and Policy Analysis, i.e., optimum balance between supply and demand, the factors of production, balance of payments, fiscal balance, and balance of credit. Although all these are theoretical in sense, this is nevertheless an

[9]Note: This book published in 1990 is based on the sub-report "Macro-Economic Model Projection of China Toward the Year 2000". Which described in previous paragraph.

improvement with sole consideration on investment allocation. Also, in the model of Population and Coordinated Economic Development Planning Model, four major systems with sets of criteria were established. These major systems are: the economic system, the ecological system, the population and resource system and the social system. This model had implicitly a broad perspective of sustainable development, which became the theme of the World Summit in 1992 in Rio De Janeiro.

(c) Regional Development Strategy and Regional Industrial Policy. Through the experience of policy modeling in a decade, we felt that China is too large that study at the national level result in insufficient consideration of the diversity of different regions. Therefore, we launched another national project, "Regional Development Strategy and Regional Industrial Policy," in late 80s. Nearly every province had included a chapter of I/O analysis in their study. Therefore, we have pushed forward further the development planning supported by mathematical modeling in every province. It should also be emphasized that a decade had been passed since the implementation of reform and opening-up policy, therefore, there bound to be gradual change in concept of planning as compared to the initial period.

(2) *Policy Modelling of DRC in the 90s*

Through the gradual approach of economic reform and opening-up, China had already created certain features and factors of market mechanism in the 80s: the establishment of four SEZs in 1980, the opening of 14 coastal cities in the mid of 80s and the further opening of the delta area of Yangtze River, Pearl River, etc. The decentralization of decision making process to different levels of government, the booming of the TVEs also greatly promoted the formation of market mechanism. The official announcement that "The target of economic restructuring is to establish a socialist market economy" in the 90s had accelerated further the formation and improvement of the market mechanism. Therefore, there was certain shift of focus of the policy modelling of the DRC in the 90s.

(i) *To Keep the Trait of Studying the Macro-economic Modelling Through Organized Research*

Our center had carried out the project through organized research. In view of the experience gained in the 80s, we kept on this trial and published "collection of models" in 1993 and 1999.

In the publication *China: Applied Macro-economic Models 1993*, which was edited by Wang, Li and Li, 19 models were collected. The techniques applied was generally the same as that used in the 80s. Integer programming was added in this exercise. In consideration that the application of model in the past generally focused on medium- and long-term goals, a model with quarterly forecasting was also added. There were three models which covered new areas of application in this publication, i.e., the quarterly macro-econometric model of the People's Bank of China, China's macro-fiscal model and a model of scheme of reform of pension system.

Some new concepts from abroad were also described in chapter 1 of that publication.

It is seen that through a decade of efforts in the application of mathematical modelling, some scholars and professionals have mastered a variety of sophisticated techniques, especially so for many scholars from fields other than applied policy research. It seemed that there was the need to have some institutional organizations to work specifically on model evaluation, so that there would be continuity to keep improvement to a specific type of model rather than to work on application of various mathematics with several feasible models in application to proper issue. China did not establish a planning or other policy model recognized domestically for continuous application. These ideas had also emerged in the West that: Models must meet multiple criteria which are often in conflict. They should be relevant in the sense that they ought to be capable of answering the questions for which they are constructed. They should be consistent with the accounting and/or theoretical structure within which they operate. They should also provide adequate representations of the aspects of reality with which they are concerned.

These criteria of model evaluation can be found in Hendry and Richard (1982) and McAleer and others (1985), although our concern on model evaluation is slightly different from them.

In the publication *China: Applied Macro-Economic Models 1999*, which was also edited by Wang, Li and Li, 13 models were collected. The new area of application includes "Model of International Balance of Payments and Money Supply", "Demand of Money, Modelling and Projection of Money Supply", "Modelling of Analysis and Projection of Chinese Agricultural Policy", "System Dynamics of infrastructure", and "Regional Water Resource Planning and Coordinated Economic Development". It can be seen from the titles of those papers the policy concern in this period. Indirect instruments, such as monetary policy and exchange rate policy, have become important policy instruments rather than direct control through mandatory planning.

There were three new types of models in the publication, i.e., "China Macro-econometric Modelling in Link Project of UN", "China: Multi-sectoral Development Analysis Model" and "DRC CGE Model". These latter two models were developed in our center through international assistance from Professor Clopper Almon and Professor David Roland-Holst of Inforum model, Dominique Vander Mensbrugghe of OECD Development Center, and Dr. Sebastien Dessus and Wang Zhi.

Emphasis in this publication is given once more to the continuity of improvement of the existing model with the evolution of IMF Multimode Mark III as an example. Attention is also given to the application of scenario analysis in explanation of long-term prospective.

(ii) *Micro Econometrics*

To be a governmental organization to provide policy consultation services, we have always kept our eyes open to the trend of the global society. We notice that emphasis on the use of micro-data in the analysis of the economic problems had been pioneered by Ruggles in 1956 on the development of a micro-based social accounting framework and by the work of Orcutt and his colleagues. We have learned the use of micro analytic simulation models of policy analysis from Urban Research Institute and experimented in the area of social security system reform in Shandong Province.

6.5.1.3 Recent Achievement

The very recent achievement of policy modelling is the application of CGE model to study the impact of China's accession to WTO. This project was carried out in 1997–1998. Two models are used in this analysis. The first one is a country model of China, including 41 sectors and 10 representative households (5 rural and 5 urban) with major focus on the evolution of the impact of China's domestic economy under different trade policies. The second model is a 17-region model, with 19 sectoral recursive dynamic models on World trade and production. The period of study covers 1995–2010. Useful and pragmatic conclusions and policy recommendations are derived from this project. The result of the study had attracted wide attention domestically and internationally. It has been quoted into reference in the recent IMF publication, China: Competing in the Global Economy. There are a few studies on this important theme of policy study, this project not only had an impact at the national and global levels, but also was a major concern at sectoral and even enterprise levels.

6.5.1.4 A Summary of Features of Policy Modelling of DRC in the Past Several Decades

See Table 6.4.

Table 6.4 Policy modelling of DRC in past several decades

Demand side	Supply side	
Economic background of major concern of decision makers	Policy researchers	Type of modeling
80s centralized planning economy Policy goal: Economic growth Elimination of bottlenecks Policy measures and instrument: Investment allocation Industrial policy	• To study the feasibility of quadrupling of GDP in 20 years • Major bottlenecks: Infrastructure Energy • Development Planning with high, medium and low scenario • Investment allocation • Preliminary study of strategy and policy • Understand with analysis of components of supply and demand	• I/O • Macroeconometric Model • System dynamics • Programming model etc.
90s socialist market economy Policy goal: Economic growth Control of inflation Provide more employment Opportunity Fiscal balance Balance of payments Income distribution and social issues Policy Instrument: Fiscal policy including taxation and tariff	• Improved focus on strategic planning • Long-term prospective exploration • Medium-term projection • Short-term forecasting • Policy analysis	• I/O • Macroeconometric model • Experimentation of microsimulation models • System dynamics • CGE • Programming model etc.

6.5.2 Lessons and Experiences

1. It is not the purpose of this paper to give theories and details of policy modeling, only a brief summary is made on the history of policy modeling of our center. It should be emphasized that there are many governmental and academic research institutions that contributed to the development of macro econometric modeling, mathematical modeling with various applications and theoretical study. It is also not our intention to be involved in theoretical debate either on economic theory, or on the feasibility of appropriate mathematical tools, etc. It is mentioned in the beginning of this paper that policy modeling requires an interdisciplinary study of economics, mathematics, and policy making process in a real world. It is not simply the application of mathematics to the real world, an in-depth knowledge is required of various disciplines of social science, the history of evolution, the knowledge of the real world, the features of various mathematical tools, etc. This brief history of policy modeling of DRC presented an overview of the learning

process of our center in working in the world of economy during transition of China. Some lessons and experiences will be briefed in the following items.

2. In policy modeling, we advocate the "combination of quantitative and qualitative analysis", it can be seen from the description in Part II of this paper that all 25 policy recommendations seem to be correct if reviewed today. But not all of them are derived from the policy modeling. Because the real world is too complicated to be explained completely with a single school of economic theory. It is correctly pointed out that "The ambiguity in testing Theories, known as the Duhem-Quine thesis, is not confined to econometrics and arises whenever theories are conjunctions of hypothesis (on this, see for example Cross, 1982). The problem is, however, especially serious in econometrics because theory is far less developed in economics than it is in the natural science."
3. International and domestic cooperation and exchange of information are crucial in an emerging knowledge based economy. Because the issues of the real world is so complex, while the number of researchers of China are very much limited, therefore, organized research or the form of ToTos (Task oriented, transient organizations) were adopted by our center through these several decades. The relatively successful modelling in the 90s was also due to international collaboration with OECD Development Center, Harvard University, the World Bank, ADB, modelling group of Inforum, ADBI and many others.
4. Although many models have experimented in our center, the critical role of the modeler is to select the appropriate technique best adapted to the issue under investigation. We should emphasize once more that the best choice is not the sophistication of mathematics.
5. Policy modelling researcher should also focus on the presentation of his results to the decision makers. A simple, straightforward expression should be designed to make the policy recommendations to be acceptable by the decision makers.
6. Through more than two decades of learning and experimenting of various types of models, it was determined that DRC must focus on one type of model due to limited manpower of our staffs, develop an applicable model, and improve the model continuously to make it a workable model. Professor Shantong Li, one of the authors of this book has worked on this continuously since 1990s. Section 6.6 will provide a case study of DRCCGE model and its analysis to Chine's 13th Five-Year Plan and Prospects of 2030 created by her team through the assistance and cooperation with many international experts, including professor Roland-Holst of Berkeley and Dr. Dominique of FAO.

6.6 Case Study of China's Computable General Equilibrium Model-DRCCGE Model

The DRCCGE model is developed and maintained by DRC, which is the abbreviation of Development Research Center of the State Council. At the same time DRCCGE model is a Dynamic Recursive Chinese Computable General Equilibrium model.

Since the 1990s, Development Research Center has begun to develop a computable general equilibrium model for China's economy. In 1993, Development Research Center developed a 10-sector Chinese computable general equilibrium model with the Shanghai Academy of Social Sciences and the Chinese Academy of Sciences jointly (Wang et al., 1993). However, the development and application of DRCCGE model began in 1995. In early 1995, Development Research Center, in cooperation with Huazhong University of Science and Technology, developed a 64-sector Chinese social accounting matrix based on the 1987 input-output table. This was the first national social accounting matrix developed by the Development Research Center of the State Council, which provides a basic data framework for construction of a DRCCGE model. From June to August 1995, Prof. Shantong Li and Dr. Fan Zhai worked with the OECD Development Center to develop a recursive dynamic Chinese computable general equilibrium model with 64 sectors and 10 groups of residents, which is the first version of the DRCCGE model. The benchmark year in this model was 1987. This version of the DRCCGE model is derived from the prototype of CGE model developed by the OECD Development Center for trade and environmental projects.

After that, Prof. Shantong Li led her team to improve the DRCCGE model in terms of taxation, trade, environment and health. Meanwhile her team also developed China's multi-regional DRCCGE model. So far, the DRCCGE model has been used to analyze economic growth, the adjustment of industrial structure, trade, labor market and income distribution, finance and taxation, energy, environment and health, water resources and other emergencies and major events (such as SARS, Olympic Games, etc.).

6.6.1 The Component of the DRCCGE Model

The latest version of the DRCCGE model is DRCCGE 2010. This model includes 34 production sectors; 2 representative households by area; and 4 primary production factors (capital, agricultural labor, productive workers, and professionals). As most of other computable general equilibrium model, the DRCCGE model contains the following modules.

6.6.1.1 Production and Factor Market

All sectors are assumed to operate under constant returns to scale and cost optimization. Production technology is represented by a nesting of constant elasticity of substitution (CES) functions. At the first level, output results from two composite goods: a composite of primary factors plus energy inputs, i.e., value-added plus the energy bundle, and aggregate non-energy intermediate input. At the second level, the split of non-energy intermediate aggregate into intermediate demand is assumed to follow the Leontief specification, i.e. there is no substitution among non-energy intermediate input. Value-added plus energy component is decomposed into aggregate labor and energy-capital bundle. Aggregate labor is further split into 3 types of labor force. And energy-capital bundles are decomposed into energy bundle and capital. Finally, the energy bundle is made up of 5 types of base fuel components.

The Model distinguishes two types of capital—new and old. This assumption of capital vintage structure allows the elasticity of substitution in the production function to vary for different capital vintage. This Model also reflects the adjustment rigidity in the capital market. The Model assumes that new capital goods are homogeneous while old capital goods are provided by the secondhand market. In dynamic simulation, when one sector is in shrinkage, the capital deployed in this sector can be taken out partially. The supply curve of such old capital is the constant elasticity function of the relative return of old capital. The higher return of old capital in relation to new capital, the more supply will be available. But the return rate of old capital shall not exceed that of new capital. In this Model, the supplies of three types of labor are fixed exogenously.

6.6.1.2 Foreign Trade

The rest of the world supplies imports and demands exports. Given China's small trade share in the world, import prices are exogenous in foreign currency (an infinite price elasticity). Exports are demanded according to constant-elasticity demand curves, the price-elasticities of which are high but less than infinite.

6.6.1.3 Income Distribution and Demand

Resident income comes from capital, and labor, as well as profit distributed in enterprises, and transfer payment from government and overseas. Rural residents get labor income through agricultural labor and productive workers while urban residents get labor income from productive workers and professional technicians. Capital income is distributed to residents and enterprises. The disposable income of residents is used in savings and consumption for commodities and services. Under the condition of meeting their budgets, the residents will maximize their own utility. Its effect function can be described with Stone–Geary utility function, and the derived resident demand function is the extended linear expenditure system (ELES).

Capital revenues are distributed among households and enterprises. Enterprise earnings equal a share of gross capital revenue minus corporate income taxes. A part of enterprise earnings is allocated to households as distributed profits based on fixed shares, which are the assumed shares of capital ownership by households. Retained earnings, i.e. corporate savings for new investment and capital depreciation replacement, equals a residual of after-tax enterprise income minus the distributed profits.

The government collects tax from producers, residents and foreign sectors, transfer it to residents and buy public products. Government income includes value–added tax, business tax, other indirect production tax, resident income tax and corporate income tax and import tariff. Subsidy is dealt as negative income of government. Total societal consumption and investment demand is described by the function of fixed expenditure ratio. The model assumes that all increased stock is the demand in domestic products. The composite commodity part of medium input, resident consumption and other final demands compose the total demand in the same category of Armington composite commodity.

6.6.1.4 Macro Closure

Macro closure determines the manner in which the following three accounts are brought into balance: (i) the government budget; (ii) aggregate savings and investment; and (iii) the balance of payments. Real government spending is exogenous in the model. All tax rates and transfers are fixed, while real government savings is endogenous. The total value of investment expenditure must equals total resources allocated to the investment sector: retained corporate earnings, total household savings, government savings, and foreign capital flows. In this model, the aggregate investment is the endogenous sum of the separate saving components. This specification corresponds to the "neoclassical" macroeconomic closure in CGE literature. Meanwhile exchange rate is chosen as the model numéraire.

6.6.1.5 Recursive and Dynamic

The DRCCGE model has a simple recursive dynamic structure as agents are assumed to be myopic and to base their decision on static expectations about prices and quantities. The dynamic characteristics of the model are reflected through the following factors: (1) quantitative growth in production factors; (2) TFP improvement and partiality of tech progress; (3) Vintage structure of capital. In such model structure, the basic factor driving structural change is the income demand elasticity of residents for different commodities (Engel Effect), structural change of intermediate input demand resulted from technology change, and factor composition change resulted from different factor accumulation speeds.

In this model, the growth rate of population, labor and productivity is exogenous. The growth rate of capital is determined endogenously by the savings/

investment relationship of this Model. On the level of total quantity, the current capital stock equals the capital stock in the previous period minus depreciation plus total investment. But on the level of sectors, because the demand in capital (including new capital and old capital) for some sectors is less than the depreciated old capital of the sector, their capital accumulation functions may vary. We assume that the producers use an optimal method to decide the Vintage structure of production. When the demand in product of certain sector exceeds its production capability of existing capital, the producer of the sector will need new capital input. Otherwise part of its existing capital will be shifted to other sectors.

6.6.1.6 Data

The model is calibrated to the 2010 Chinese Social Accounting Matrix (SAM) developed from the 2010 national Input–Output table. The SAM provides a consistent framework to organize the relevant flow of value statistics for China's economy to satisfy the requirements of a benchmark data set for CGE modeling. Some key parameters of the model essentially substitution elasticity and income elasticity were derived from a literature search. All other parameters mainly shift and share parameters are calibrated in the base year using the key parameters and the base data.

6.6.2 *The Application of the DRCCGE Model*

China has acquired huge achievements in its socio-economic development since the reform and opening to outside world around 30 years. The comprehensive national strength has increased greatly, China became the second largest economy of the world with its national GDP surpassed Japan by the year 2010; there is rapid rise of the level of people's income, per capita national income of the Chinese people has reached the standard of upper middle income countries. In entering the new historical stage of socio-economic development, it is necessary to have the awareness on several shortcomings existed in the pattern of economic development that supported the rapid growth of China in long term past. These shortcomings became serious increasingly in recent years, the most serious of them is low quality of economic growth. This low quality is shown concretely in three aspects: The first is low efficiency of utilization of factors of production resulting low efficiency of economic growth. In spite of the high growth rate of Chinese economy, it relies mainly on the large amount of input of low cost factors of productions (including labor, capital and land etc.), the level of utilization of agglomeration of factors of production is not high and also the allocation of them is not sufficiently optimized. The second, various disparities are too large, causing relatively low inclusiveness of economic development. The disparities among regions, urban and rural, and among various groups of people are increased continuously. Although there are some declines of them in recent years, they are still at a higher level. Furthermore, these disparities are not only shown in the aspect of income distribution, but also shown in the aspects of enjoyment of level of basic public service and equal opportunity in participation of development etc. In addition, there are continuous increase of degree of solidification and decrease of fluidity in continuity, these will not only affect the stability of society, but also affect the rise of efficiency of economic development. The third, the ecological environment is worsened rapidly and its pressure is increasing continuously. In spite of the fact that there are some improvements of efficiency of utilization of resources, the externality of resource-environment has not been well internalized, high energy intensive sectors and heavy pollutant sectors have developed rapidly, the level of economic structure with low pollutant is not high while the amount of discharge of pollutants is increased rapidly. Therefore, the level of destruction of ecological environment becomes increasingly severe.

The phenomenon of "competition at bottom line" with neglect of resource-environmental constraints during the process of regional development occurs constantly. The result is that part of the regional development has approached, or even surpassed, the upper limit of bearing capacity of resource and environment. This will not only effect the sustainability of development, but may also cause social conflicts. From the perspective of history, whether Chinese economy can be transformed successfully to an ideal new normal state over a long period of time, depends crucially upon whether the quality of economic growth can be improved in reality.

The current and future economic growth of China will face important changes from the domestic and external environment. Firstly, supporting conditions of economic quantitative expansion becomes increasingly weaker, it is necessary to rely more on the rise of efficiency in order to keep up certain economic growth. The change of age structure of the population results in the coming of peak value of supply of total amount of labor force. The coming of "Lewis turning point" illustrates that provision of rural surplus labor force will change its past pattern of "unlimited" supply. The rapid increase of aging population will push out the current state of "high saving rate and high investment rate." A continuous sluggish global economy, the rise of domestic factors of production, and the rise of price of currency have further limited the space for rapid expansion of quantity of export. Secondly, the society increases its demand on inclusive development along with the progress of China's socio-economic development. Meanwhile, the issue of income disparity becomes more evident with the decline of economic growth. It is possible to that the excessive social disparities may cause social conflict and limit the realization of potential of economic growth if the issue of cannot be solved properly. Lastly, the issue of environmental pollution has approached the tolerance limit of the people. Pressure from resource-environment will be developed to such an extent to form a real threat that cannot be ignored. It is necessary to note that the disruption of ecological environment has the nature of cumulative effect on the one side, and the increase of level of income of the people will result in the continuous decline of their enduring capacity toward environmental disruption on the other side. Therefore, in view of the domestic and external environment faced by China's socio-economic development in the near future, the increase of quality of economic development will be crucial to perform higher

Table 6.5 Design of scenarios for prospects of China's economic growth

Types of scenarios	Assumptions of scenario
Baseline scenario (scenario 1)	Related assumptions of baseline scenario are set in the following: (1) Trend of change of total amount of population is exogenous, using the projection data of UN (2) Level of urbanization and population of urban and rural, exogenous; annual average rise of rate of urbanization is 0.9 percentage point from 2016 to 2020. Annual average rise of rate of urbanization is 0.7 percentage point from 2021 to 2030 (3) Growth of total amount of labor force, exogenous. Change of supply of rural land, exogenous (4) Various domestic tax rate are kept without change. Various transfer payment, Exogenous (5) International balance of payments. It is kept in balance in the period of 2016–2030 (6) Growth of governmental consumption, exogenous (7) Total factor productivity (TFP), exogenous. Assume the growth rate of TFP from 2016 to 2020 will be lower increasingly, it is assumed to be lower than the average level of past 30 years, it will be maintained around the level of 1–2% (8) Bias of technical progress and the changes of rate of intermediate input, exogenous
Scenario of faster rise of quality of economic development (scenario 2)	(1) Accelerate the steps of technological innovation, promote free flow of factors of production and optimum allocation of them to raise the efficiency of growth (i) Reform and perfect the mechanism of institution of innovation, create a good ecological environment of innovation; push forward reform of institution of education and medical health work, accelerate the accumulation of human capital. Assume the growth rate of technical progress derived thereby will be 0.1 percentage point faster than the baseline scenario, the growth rate of technical workers will be 0.4 percentage point faster than the baseline scenario (scenario 1) (ii) Accelerate to push forward the construction of an integrated domestic market, promote free flow of factors of production among regions, sectors and enterprises. Abolish various explicit and implicit barriers which constrain free flow of factors of production, promote optimal allocations of them; reform and perfect related legal and policy system for free entry and exit of firms to create a better environment for survival of the fittest. Average annual transfer of rural labor force is assumed to be 1000–2000 thousand persons more than scenario 1, annual rate of urbanization from 2016 to 2030 will be 0.2 percentage point higher than the baseline scenario. Average annual rate of improvement of assumed technical efficiency will be 0.1 percentage point faster than the baseline scenario (iii) Perfect the reform of regulations of service sector, promote the opening of service sector to raise its efficiency. TFP of service sector will be 0.5 percentage point higher than scenario 1 from 2016 to 2030 (iv) Further expand the opening of domestic to external world to raise the position of China in global industrial chain, accelerate the process of upgrading Chinese industry

(continued)

Table 6.5 (continued)

Types of scenarios	Assumptions of scenario
	(2) Promote fair share of fruitful development, raise the inclusiveness of development (v) Promote the equalization of basic public service among regions, urban and rural and groups of people. Process of residentialization of migrant workers will be accelerated, annual increase of new residentialization of migrant workers will be 1–2 million persons, the growth rate of governmental expenditure on public service will be 1–2 percentage points higher than scenario 1 (vi) Reform the regime of income distribution to reduce income disparity and promote the growth of mid-income group. The share of mid-income group will be 0.3–0.5 percentage point higher than scenario 1, average propensity to consume accumulated will be raised by 5 percentage points. vii Adjust income distribution system of state owned enterprise and monopolistic enterprise, raise their share of profit delivering. Increase the share of governmental public expenditure: raise the transfer payment of government from the year 2016–2030 to poverty region and poverty people, it will be 10–15% higher than scenario 1 (3) Promote green growth, upgrade the sustainability of development (viii) Continue to accelerate the price reform of product of resource and energy (ix) Impose carbon tax to raise the efficiency of utilization of energy and reduce the intensity of discharge of pollutant. Collection of carbon tax can be started since 2016, the tax rate can be gradually increased from 50 yuan per ton of CO_2. The efficiency of utilization of energy in the period of 2016–2030 can be raised in average around 0.5 percentage point than scenario 1 (x) Push forward reform of institutions of central and local government and also their assessment system of achievement, raise the control efforts of pollution discharge, raise the standard of target of pollutant discharge
Scenario of slower rise of quality of economic development (scenario 3)	1. Global economy is sluggish continuously, progress of cultivation of domestic new competitive advantage is slower 2. Progress of construction of domestic integrated market is slower. Urbanization rate in the period of 2016–2030 is 0.2 percentage point lower than scenario 1, rate of transfer of labor force between urban and rural is also slower than scenario 1 3. Progress of technological innovation becomes slower TFP is around 0.4 percentage point lower than that in scenario 1 4. Push forward reform measures of income distribution and equalization of basic public service becoming slower 5. Green growth cannot be pushed forward effectively

efficiency economic operations, develop economic system with high inclusiveness, and maintain a more sustainable resource-environment to deal successfully all risks faced currently and in coming future.

From the analysis given previously, it can be seen that the change of quality of development will become an important characteristic variable. It is also the direction to be focused on by policies in later period. Therefore, in this paper, the theme of "Thirteenth Five-year Plan" and prospects of China's economic growth by the year 2030 will be described and centered around the change of quality of economic development with priority.

6.6.3 Scenario Design of Long Term Prospect of Chinese Economy

This study will apply the method of scenario to analyze the prospects of China's economic growth. This study will better reflect the uncertainty of socio-economic growth. This study will also describe the effect of various policies. Firstly, the baseline scenario (scenario 1) is designed based upon the historical trend of China's economic development in the past, as well as the changes of objective conditions to be faced in the future. Secondly, two contrasting scenarios (scenario with faster rise of quality of economic growth <scenario 2> and scenario with slower rise of quality of economic growth <scenario 3>) are designed based upon the main risks to be faced and adjustment of orientation of development strategy on the basis of baseline scenario. The purpose is to observe the priority of effects of various policies in promoting the quality of economic development. The concrete assumptions of various scenario are shown in Table 6.5.

6.6.3.1 Baseline Scenario (Scenario 1)

Growth in the baseline scenario is considered based upon development of the past and present. The most possible future changes, such as change in population, change of factors endowment, and general rules of technical changes, are also taken into consideration to derive the scenario through simulation. It is necessary to point out that scenario 1 does not consider much about policy changes, and therefore may be considered as a scenario without policy intervention. It thus serves as a reference base to be compared by other scenarios designed with various policy interventions. Under scenario 1, it is assumed that the supply constraint of labor force becoming stronger due to the effect of changes in population and population structure. Meanwhile, it is also assumed that there is a decline in the rate of transfer of labor force due to the approaching of the "Lewis turning point". There is a decline of the "saving ratio" along with the aging and rise of dependency ratio of the population. The growth rate of technical progress will become slower in more and more the frontier countries. Considering that the global economy will be in a state of sluggish growth for quite a long period, the growth rate of China's export will decline to certain extent. Under the above hypotheses of conditions, it is assumed in this modeling that average annual growth rate of TFP in the period of 2016–2030 will be in decline compared to past 30 years and will be kept at around 1–2%. Growth rate of urbanization will also be slower, the level of rate of urbanization will increase by 0.7–0.9 percentage point annually. Please refer to Table 6.5 for other important assumptions which will affect the medium and long term economic growth and structural changes in Scenario 1.

6.6.3.2 Scenario of Faster Rise of Quality of Economic Development (Scenario 2)

Based upon analysis given previously, three perspectives will be given here to consider the potential of upgrading quality of economic development and design of policies. There are two major aspects to upgrade the efficiency of economic growth, technical progress and upgrading of efficiency of allocation of factors.

Two aspects can be analyzed in technical progress. There are many disparities between China and developed countries in the aspect of technology. On the one side, the share of domestic value added of Chinese export is only 67%, while this value is 89[10], 85 and 73% for U.S.A., Germany and Japan, respectively. This illustrate clearly illustrates the technological disparity. It also shows that there are many "Non-frontier New Technology" held by developed countries, which can be imported and absorbed by China. On the other side, China's technology innovation capability itself is waiting to be strengthened further. The live patents[11] of China are less than one half of Japan or U.S.A.

With respect to the efficiency of allocation of factors, there are relatively large disparities of growth rate of TFP among three sectors. Firstly, the growth rate of TFP of the secondary sector of China is fastest in past 30 years, with an annual average growth rate approached around 4% growth rate of TFP of the tertiary sector is the slowest, with an average annual growth rate of only 1.2%. This means it is necessary to promote the re-allocation of factors, especially the labor force There is also a need to upgrade the efficiency of the service sector. Secondly, there are relatively large space of upgrading the efficiency of re-allocation in the internal part of manufacturing and service sector. Study shows that the coefficient of variation of capital value of marginal output among manufacturing sectors in the past 20 years was around 1.3 on average in China, while this value of is only around 0.3 in the U.S.A. This shows that the level of misallocation of capitals among Chinese manufacturing sectors is far higher than that of the U.S.A. If the value of the U.S.A. is taken to be the benchmark, there are larger space to upgrade the TFP of manufacturing sector through reduction of the level of misallocation in the coming future. Thirdly, further upgrading of efficiency within sectors can be achieved through "survival of the fittest." In fact, there are relatively large disparities of TFP among various types of enterprise within sectors. For example, TFP of export enterprise is 54.3% higher than enterprise engaged to non-export activities (Gong Guan etc. 2013). Furthermore, due to the imperfect mechanism of exit of Chinese

---

[10]Note: TIVA Data Bank of OECD-WTO: http://Stats.OECD.org/index.aspx? Data Set Code= TIVA 2015 C1

[11]Note: "World Intellectual Property Index 2011".

firms, there exist many "[12] zombie companies" that ineffectively held limited social resources and greatly reduce the efficiency of allocation of resource. Therefore, it will be an important channel of upgrading TFP in future through perfection of mechanism of competition and exit of firms to upgrade the efficiency of utilization of resources. Lastly, there is also space for optimum efficiency of allocation of factors. Study by Loren Brandt (2010) shows that there is a loss of 10% of TFP due to mis-allocation of labor force among regions.

Several aspects described above are important channels of upgrading growth efficiency in the future. The important points are acceleration of the process of technology innovation, promotion of free flow and optimum allocation of factors from perspective of policies, The concrete scenario of policies is assumed as follows:

(1) Reform and perfect the mechanism of technological institution to create a better innovative ecological environment; push forward institutional reform of education, medical and health, to accelerate the accumulation of human capital. Assumption of modelling on this basis will bring along a 0.1 percentage point faster of technical progress than scenario 1. The growth rate of technical workers will also be 0.4 percentage point faster than Scenario 1;
(2) Accelerate to push forward construction of integration of domestic markets, promote free flow of factors of production among regions, sectors and enterprises. Abolish various explicit and implicit barriers in restriction of free flow of factors to promote optimal allocation of factors; reform to perfect related laws, policy system for free entry and exit of enterprises to create a better environment of "survival for the fittest" of enterprises. The transfer of agricultural labor force will be 1–2 million persons more in average annually than scenario 1, the urbanization rate will be raised 0.2 percentage points annually than the baseline scenario. The rate of improvement of technical efficiency will be 0.1 percentage point faster in average than the baseline Scenario 1;
(3) Perfect the reform of regulation of service to promote the opening of service to upgrade the efficiency of service sector. TFP of service sector in the period of 2016–2030 will be raised cumulatively 0.5 percentage point higher than Scenario 1;
(4) Further expand the opening to both domestic and abroad, thereby to raise the rank of China in global industrial chain, accelerate the progress of upgrading of industry.

The insufficiently inclusive development of China is shown mainly in its unfair share of achievements of development. This can be expressed concretely in two aspects. The first is income disparity, which reflects to certain extent the opportunity in participation of development. There is difference of GDP/capita around three

[12]Note: Zombie company is a media term for a company that needs constant bailouts in order to operate, or an indebted company that is able to repay the interest on the debts, but not reduce its debts.

times between the income of province (or region) with highest level and that with lowest level in 2014. The ratio of income of urban and rural resident still approaches around three times. The Gini coefficient of income of national residents in China reached 0.469, which is far higher than the international picket line of 0.2. Based upon the study of NBER,[13] expansion of income disparity will not only be unfavorable to the expansion of consumption, but also affect the accumulation of whole social human capital and thereby affect the growth of the economy. The second is disparity of provision of basic public service. Currently, there exist large disparities among regions, urban rural and groups of people in enjoying basic public service. Based upon index of item of education and health reflecting quality of provincial development promulgated by "China's Human Development Index 2013", average years of education received by persons of Eastern region have approached around 9 years, while this is only 7.5 years for persons of Western region. The average life expectancy of Eastern region has surpassed 77 years, while this is only around 72 years for Western region. The background of these disparities in education and health is the disparity of provision of basic public service among regions.

The important aspect to upgrade inclusive development is to deal with unfairness of these two aspects. The crucial respect is reform and perfect the supply mechanism of income distribution and provision of basic public service to promote the fair share of achievements from the perspective of policies. Therefore, the design of scenario of policies is as follows:

(1) Promote the equalization of provision of basic public service among regions, urban and rural and groups of people. The rate of citizenization of rural migrant labor shall be accelerated, increase of citizenization of rural workers will be 1–2 million annually. The growth rate of governmental expenditure on public service will be 1–2 percentage points higher than Scenario 1;
(2) Reform income distribution system to reduce income disparity and promote the growth of mid-income group. The share of mid-income group will be 0.3–0.5 percentage point annually higher than scenario 1. Average propensity to consume will be raised around 5 percentage points cumulatively;
(3) Adjust the income distribution system of State Owned Enterprises and monopolistic enterprises, increase their share of profits turned over to the State. Increase the governmental public expenditure and increase the transfer payment of government to poverty regions and groups. This will be raised by 10–15% in the period of 2016–2030 than Scenario 1.

From the perspective in effecting the sustainability of development, the existing issues are shown mainly in two aspects: one is imperfection of the price system which cannot wholly reflect the supply and demand of the market, the level of scarcity of resources, the cost of destruction and the efficacy of restoration of ecological environment thereby extensive utilization of resources. The other is the

[13]Note: Oded (2011).

institutional arrangement between the central and local government and the assessing system of administrative performance, which focus more on economic growth rate of various levels of government, and neglect the quality of ecological environment constantly. Therefore, from the perspective of upgrading the sustainability of development, the crucial aspect is to induce the transformation of concepts of development at the institutional level, perfect the price formation through market and system of ecological recompense. The concrete design of scenarios of policy is as follows:

(1) Continue to accelerate the price reform of products of resource and energy;
(2) Collect carbon tax to upgrade the efficiency of utilization of energy and reduce the intensity of discharge of pollution. Carbon tax will be collected since 2016, the tax rate will be 50 yuan per ton carbon and raised to 150 yuan per ton carbon later. The efficiency of utilization of energy in the period of 2016–2030 will be 0.5 percentage point higher than Scenario 1 in average;
(3) Push forward reform of institutional arrangement between central and local government and system of assessment of administrative performance to upgrade the strength of control of pollution discharge and rise of rate of achieving target of discharge of pollution.

6.6.3.3 Scenario of Slower Progress of Upgrading the Quality of Economic Development (Scenario 3)

This scenario is to consider mainly various domestic and external risks and challenges faced by China's coming development, push forward various reform measures insufficiently for promotion of upgrading quality of economic development. In comparison with scenario 2, this scenario will consider mainly several changes as follows:

(1) The global economy is sluggish continuously, progress of cultivation of domestic new competitive advantage is slower. It can be seen currently that EU and Japan are still sunk deeply in the mud of financial crisis except there is trend of economic recovery of U.S.A. only. It can also be seen that the reform of the internal structure of developed countries have not been well implemented. The future prospects of recovery of global economy are still not clear. Under this scenario, it is assumed in this modelling that the global economy will keep a relatively low growth rate for quite a long period. And also in consideration of the domestic new competitive strength has not be well cultivated, the growth of export may be lowered further, upgrading of export structure will face larger difficulty.
(2) The growth rate of technical progress will become slower due to prospects of uncertainty and the slower rate of growth of input of innovation. The efficiency of utilization of resource cannot be upgraded.

Table 6.6 Economic growth and its sources during 2010–2030 (%, scenario 1)

	2010–2015	2015–2020	2020–2025	2025–2030
GDP	7.9	6.6	5.5	4.5
Source of growth				
Labor	0.3	−0.2	−0.2	−0.4
Capital	10.1	8.2	6.5	4.9
TFP	1.7	1.7	1.7	1.8

(3) Progress of integration of domestic market is slower because it is difficult to realize various policy measures to abolish local protectionism, and other aspects to promote free flow of capitals and free competition among major actors of the market.
(4) Progress of reform measures to be pushed forward is slower.
(5) Green development cannot be better pushed forward. There are distortions in the formation of mechanism of price of resources and energies, the market mechanism of reducing pollution discharge cannot be established, there is no consensus of new concepts to push forward green development.

6.6.4 Analysis of Trend of China's Economic Growth Under Various Scenarios

6.6.4.1 Baseline Scenarios (Scenario 1)

(1) There is continuous decline of growth of China's economy, the structure of the driving force of economic growth has not been better optimized. Table 6.6 presents the potential of growth rate of GDP in baseline scenario (Scenario 1) simulated by modelling.

It can be seen that the potential growth rate of China's economy in the period of "Thirteen Five Planning" will be slower further. The potential growth rate of GDP in the period of "Thirteen Five Planning" will be declined from 7 to 8% in recent years to 6–7%; the potential growth rate of GDP will be declined further to around 5% in the period of 2020–2030.

The background of decline of potential growth is the change of driving mechanism of economic growth. This change comes from the change of objective conditions mainly, and there is no change of trend of optimal adjustment expressed yet. The concrete conditions are shown as follows:

First, from the perspective of supply side, due to change of demographic age structure and the faster rise of dependency ratio, the rate of accumulation of capital

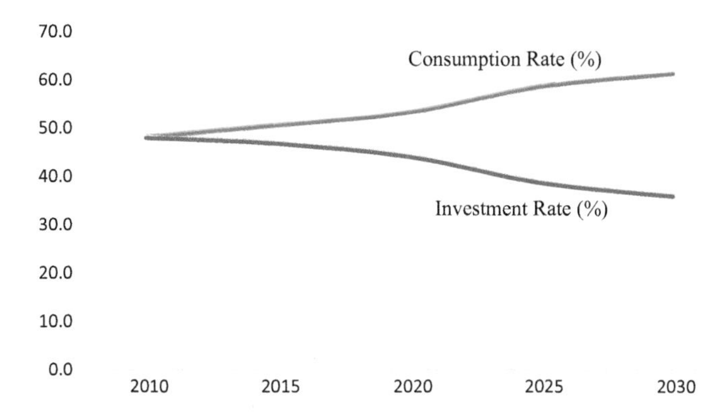

Fig. 6.8 Rate of consumption/rate of investment (%, scenario 1)

will decline continuously, while the driving force of economic growth will still rely upon still the input of capital. The contribution of technical progress on the economic growth is still not high enough. It can be seen from the results of simulation, in the period of "Thirteen Five Planning", the contribution to economic growth of input of quantity of labor force will become negative, and this trend will be in continuity. The share of aging population of China will be increased continuously, the share of aging people with age of more than 65 years in the whole population will be increased from around 10% currently to 17% by the year 2030. The saving capacity of whole society will decline along with the increase of aging population. The overall supply of investment will also decline in correspondence. From the result of simulation, the rate of accumulation of capital will decline rapidly from annual average around 10% currently to around 8% in the period of "Thirteen Five Planning," and it will be declined further to less than 6% in the period of 2020–2030. There will be decline of contribution to growth of economy from accumulation of capital in correspondence, but it is still the most important factor to economic growth. Anyhow its rate of contribution to economic growth will be declined from more than 70% currently to around 65% in the period of 2025–2030. Along with the rise of level of development, China's technological capability will approach the capacity of technological frontier countries closer and closer. In addition, due to the decline of efficiency of allocation of resources, the rate of technical progress will decline in correspondence, but its rate of contribution to economic growth will be raised from around more than 20% currently to around 35% in the period of 2025–2030 due to larger scale of decline of rate of accumulation of capital.

Second, from the perspective of demand side, the growth rate of export and the needs of investment will decline continuously. The efficiency of investment has not been improved yet. In comprehensive consideration of the trend of growth of the global economy and factors of changing cost of domestic factors. The growth rate of demand of export will be maintained around 6–8% in the period of "Thirteen

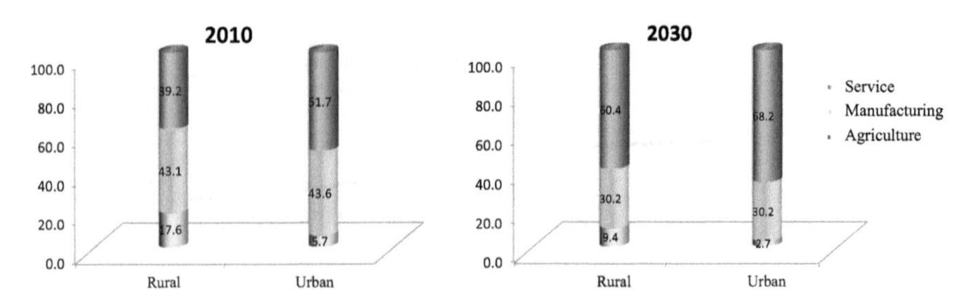

Fig. 6.9 Structure of urban and rural household consumption (2010–2030)

Table 6.7 Economic composition (GDP and employment) (%, scenario 1)

	GDP					Employment				
	2010	2015	2020	2025	2030	2010	2015	2020	2025	2030
Primary industry	10.0	8.1	6.7	5.5	4.2	36.7	31.0	25.9	21.6	17.1
Secondary industry	48.2	45.5	42.9	39.5	37.4	28.7	29.0	29.2	28.4	28.4
Tertiary industry	41.8	46.4	50.3	55.0	58.5	41.1	47.2	52.6	57.3	61.9

Five Planning"; this growth rate of export may be declined further around 5%[14] in the period of 2020–2030. In accompanying the near completion of the stage of industrialization, the process of urbanization tends to be slower, there will be decline of rate of investment of real estate, manufacturing sector and capital construction in coming 5–15 years period, furthermore these trends had been shown explicitly in recent years. Therefore, it is expected that the rate of investment on fixed assets will be declined to around 15%[15] in the period of "Thirteen Five Planning", it will be further declined to around 10% in the period of 2020–2030. Viewing comprehensively, due to decline of rate of export and investment, the role of demand pull of export and investment to economic growth will be declined continuously, the share of contribution of consumption to economic growth will be risen in correspondence, but the share of contribution of investment to economic growth is still relative large. Due to decline of rate of technical progress, the efficiency of allocation of factors of production has not been well optimized and the rate of deepening of capital is still relatively faster. All above factors have caused further worsening of efficacy of investment. Therefore, the incremental capital output ratio will be raised quickly from 5 currently to around 8 in coming 5–15 years (Fig. 6.8).

[14]Note: This is in similar compared to Japan and South Korea in the same stage of development: Japan had an average annual growth rate of export around 15% in 1960s, but it was declined to less than 5% in past twenty years; moreover, South Korea had its growth rate of export declined from 30% to less than 10%.

[15]Note: This is in consistent with the statistical data. Here, it refers to nominal growth.

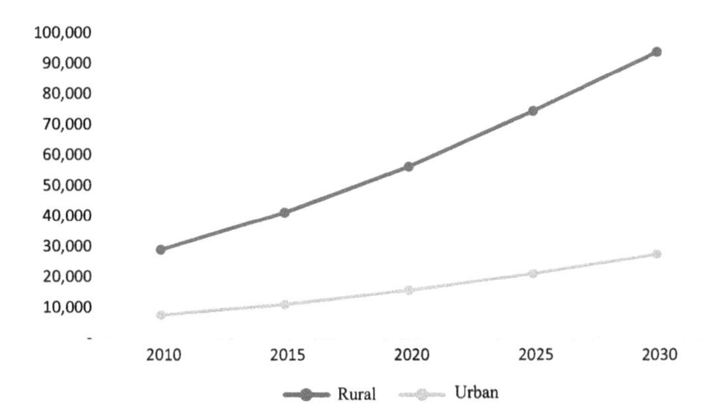

Fig. 6.10 Urban and rural household income (scenario 1)

Finally, from the perspective of industrial structure, the share of service will become increasingly higher than the secondary industry. In the period of "Thirteen Five Planning" continues to the year 2030 thereafter, along with the near completion of stage of industrialization, demand of investment will be declined continuously. The rate of demand of intermediate and capital goods will also be declined. Meanwhile, in accompanying the rise of level of income, the share of service consumption of the residents will rise continuously. With the pattern of consumption of rural residents to be an example, their share of service consumption in 2030 will be raised around 20 percentage points compared to the year of 2010, but the share of consumption of agricultural goods will be declined around 8 percentage points. Moreover, due to the rate of technical progress of service sector is far lower than that of manufacturing sector, the fast growth of demand of service will cause the faster rise of service price. Those factors act together will cause continuously faster rise of the share of service in the economy in coming 5–15 years. It can be seen from the structure of results of simulation, the share of service in the economy will be risen more than 10 percentage points, approaching 60% by the year 2030 (Fig. 6.9 and Table 6.7).

(2) The Level of Income of the Residents will rise continuously, but the income disparity is still relatively large. Keep pace with the growth of the economy, the level of income of the residents will rise continuously. It can be seen from the results of simulation that the disposable income of urban residents per capita will rise from around 30,000 yuan in 2010 to 56,000 yuan by the end of period of "Thirteen Five Planning" based upon price of the year 2010. It will rise further to around 90,000 yuan by the year 2030. The per capita net income of rural residents will rise from near 8000 yuan in the year 2010 to 16,000 yuan by the end of "Thirteen Five Planning." It will further be raised to the level around 27,000 yuan by the year of 2030. From the perspective of income disparity, although there is certain reduction of income disparity between the urban and rural residents, it is still at a relatively high level. The results of simulation show

Table 6.8 Energy consumption and CO_2 emission (scenario 1)

	2010	2015	2020	2025	2030
CO_2					
Emission (million ton)	7672.5	9600.2	11271.2	12304.5	12719.0
Intensity (Ton/10,000 yuan GDP)①	3.32	2.85	2.43	2.02	1.68
Energy					
Consumption (10,000 SCE)	324,939	412,222	487,265	534,351	554,109
Intensity (Ton S.C.E./10,000 yuan)①	1.41	1.22	1.05	0.88	0.73

Note GDP at 2010 constant price
Note ① Measurement of efficiency of utilization of energy in consumption is by its intensity of utilization. In Chinese practice, it is expressed by use of amount of standard coal equivalent (S. C. E 7000 calories/kg) in terms of Ton/GDP (in terms of 10^4 yuan)

that the ratio of income between the urban and rural residents is decreased from 3.7 currently to around 3.5 by the end of Thirteen Five Planning, and will be further reduced to 3.4 by the year 2030. The reduction of income disparity is closely related to the change of demographic structure and advance of urbanization on the one side, along with the coming of Lewis turning point. The cost of transfer of rural labor force rises rapidly, the income of migrant worker rises faster on the other side. In accompanying the continuous decline of surplus labor force in the agricultural sector, there is faster rise of marginal output of rural labor force. The share of rural labor force will decline from near 30% currently to around 25% by the end of Thirteen Five Planning, and further decline to around 17% by the year 2030 (Fig. 6.10).

(3) Although there is continuous improvement of efficiency of utilization of energy, but greenhouse gas will grow relatively faster as usual. There will be still marked increase of China's consumption of energy in scenario 1. The total energy consumption will be increased from less than 4 billion tons of coal to around 4.8 billion tons by the end of "Thirteen Five Planning", it will be further increased to 5.5 billion tons by the year 2030. There will be further improvement of efficiency of utilization of energy in later period due to the influence of factors of structural change of industry, the energy consumption per unit GDP will be reduced from 1.3 Ton S.C.E./10,000 yuan to 1 Ton S.C.E. [16]/10,000 yuan by the end of "Thirteen Five Planning", it will be further reduced to 0.7 Ton S.C.E./10,000 yuan by the year 2030. But it is necessary to see the assumption of scenario 1, the optimization of internal structure of manufacturing sector is insufficient, the share of high energy intensive sector of internal structure of manufacturing sector is reduced slightly from current 38.7 to 38% by the end of Thirteen Five Planning, it will be reduced only to less than 35% by the end of 2030.

[16]Note: S.C.E. = Standard Coal Equivalent.

Table 6.9 Economic growth and its sources during 2010–2030 (%, scenario 2)

	2010–2015	2015–2020	2020–2025	2025–2030
GDP	7.9	6.6	5.6	4.6
Source of growth				
Labor	0.3	−0.2	−0.2	−0.4
Capital	10.1	7.9	6.0	4.4
TFP	1.7	2.0	2.1	2.2

Along with the continuous expansion of total amount of consumption of energy, amount of discharge of greenhouse gas will also be increased continuously, the amount of discharge will be increased from around 9 billion tons currently to 11.3 billion tons by the end of "Thirteen Five Planning", it will be further increased to 12.7 billion tons by the year of 2030.

This means that under scenario 1, it is impossible to achieve the promise of the Chinese government that discharge of carbon will reach its peak value around the year 2030. Similar to the trend of change of energy intensity, discharge of intensity of greenhouse gas will be declined, it will be declined from current value of 2.9 Tons/10,000 yuan to 2.4 tons/10,000 yuan by the end of the period of "Thirteen Five Planning", it will be further declined to 1.7 tons/10,000 yuan by the year 2030 (Table 6.8).

6.6.4.2 Scenario 2

Simulation is done by application of modelling based upon assumptions set up in this scenario previously. Results of the simulation show that, although there is no explicit change of economic growth rate under this scenario, there is explicit improvement of quality of economic development.

(1) *Economic Growth will Rely More on Technology Innovation and Improvement of Technical Efficiency*

Under Scenario 2, although there is almost no change of the rate of economic growth compared to scenario 1, there is a larger change from the perspective of driving force of economic growth. It is shown mainly in that the rate of contribution of TFP growth to growth of economy is far higher than that of scenario 1. It can be seen from the results of simulation, the rate of contribution of TFP to economic growth in scenario 2 will be increased from less than 30% currently to 30% in the period of "Thirteen Five Planning", and will be further increased to near 50% in the period of 2025–2030. The rate of contribution of TFP to economic growth is increased by 4–10 percentage points compared to scenario 1, which means that the contribution of technical progress to economic growth is strengthened explicitly, and it becomes more and more a driving force that is equivalent to the force of capital in this scenario 2 (Table 6.9).

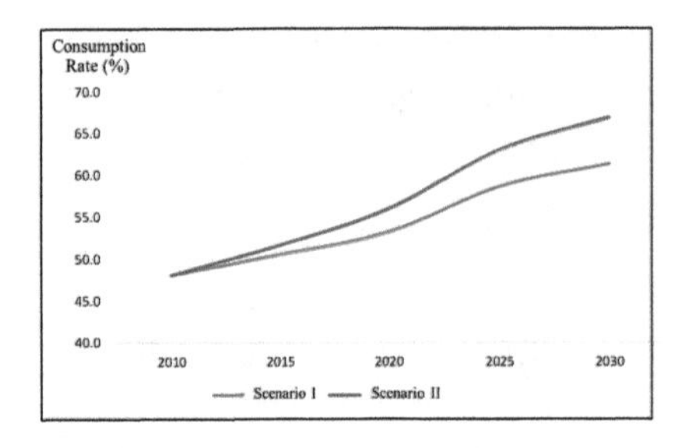

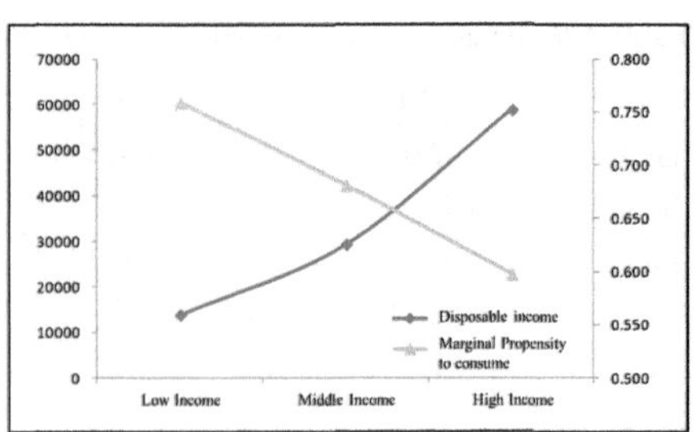

Fig. 6.11 Consumption rate and income and propensity to consume of the residents

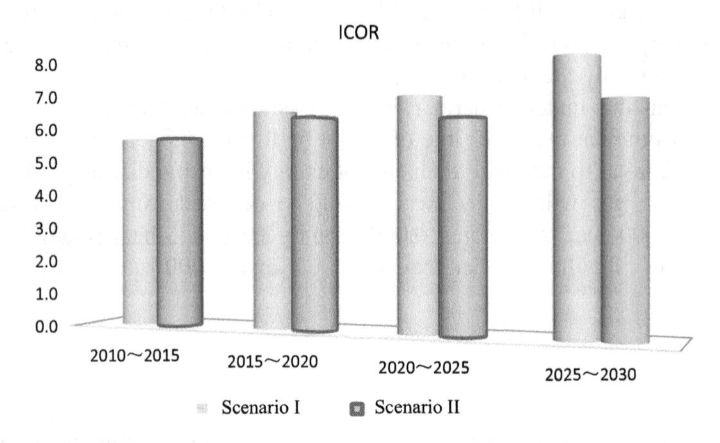

Fig. 6.12 Change of ICOR

In correspondence, the contribution of capital accumulation to economic growth of this scenario is lower explicitly than Scenario 1.

(2) *Economic Growth will Rely More on the Pull of Growth of Demand of Consumption, there is Certain Upgrade of Efficiency of Investment*

Growth rate of demand of consumption (including resident consumption and governmental consumption) of Scenario 2 is faster than scenario 1 explicitly from the perspective of demand side. Data of simulation show that the consumption rate of Scenario 2 will be increased fast around 50% currently to more than 50% by the end of "Thirteen Five Planning", and will be further increased to around 66% by the year 2030. This can be found from comparison that there are three main causes of this high consumption rate of Scenario 2, which is far higher explicitly than that of Scenario 1.

Firstly, in Scenario 2, there is acceleration of rate of increase of urbanization, which means there are more population lived in cities and towns. Meanwhile, the process of citizenization of migrant workers is accelerated, and the process of equalization of providing public service between urban and rural and among internal parts of cities is also accelerated. All these mean that the demand of governmental expenditure of public service will be increased explicitly. The

Table 6.10 Economic composition (GDP and Employment) (%, scenario 2)

	GDP					Employment				
	2010	2015	2020	2025	2030	2010	2015	2020	2025	2030
Primary industry	10.0	8.1	7.1	6.0	4.7	36.7	31.0	24.6	18.9	12.8
Secondary industry	48.2	45.5	41.7	37.4	34.7	28.7	29.0	29.1	28.3	28.4
Tertiary industry	41.8	46.4	51.2	56.6	60.6	41.1	47.2	53.8	59.9	65.8

structure of governmental expenditure needs adjustment. The expenditure of public service will be increased while the expenditure of investment construction needs to be reduced correspondently.

Secondly, along with the situation of improvement of income distribution and expansion of mid-income groups, the overall propensity to consume is upgraded continuously. It can be seen from family budget investigation, the higher the level of income, the lower the propensity to consume (refer to Fig. 6.11). Although the lower income group have higher propensity to consume, their income is too low to produce sufficient purchasing capacity. With respect to the high income groups, although they have very high income and strong purchasing capacity, their propensity to consume is insufficient. Therefore, the inverse relationship of the two groups described above shows the important role of promoting consumption through means of improving income distribution and upgrading the share of mid-income groups.

Thirdly, improvement of social security system will be favorable to the increase of propensity to consume of the residents.

In Scenario 2, the investment rate is lower than Scenario 1, which is in contrast with the trend of change of rate of consumption. Meanwhile, it is worthwhile to concern the point that there is improvement of investment efficiency explicitly in

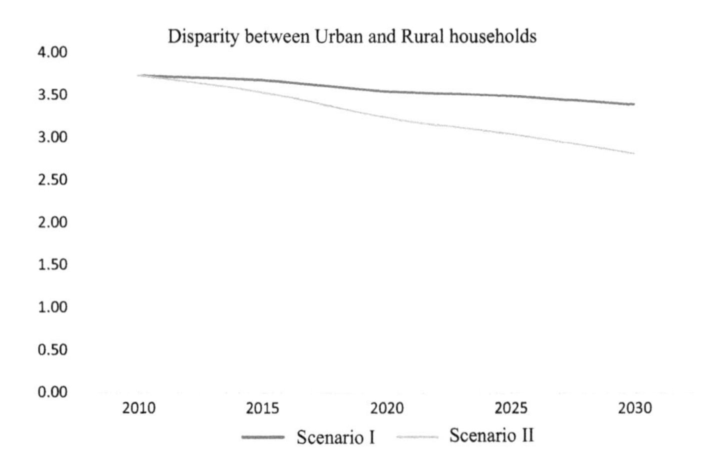

Fig. 6.13 Income disparity between urban and rural residents under scenarios 1 and scenario 2

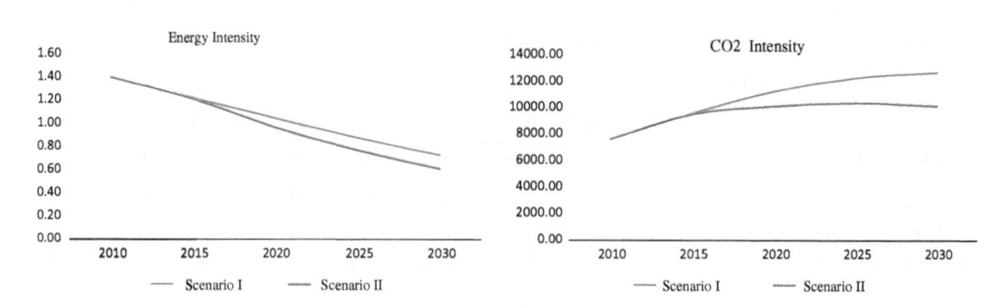

Fig. 6.14 Intensity of energy consumption and CO_2 emission under scenarios 1 and scenario 2

Scenario 2 compared to scenario 1. Figure 6.12 presents the ICOR under these two scenarios. Results of simulation show, although ICOR will be upgraded continuously in accompanying the deepening of capital in Scenario 2, but its rate of upgrading becomes slower explicitly compared to Scenario 1. The value of ICOR by the year 2030 in Scenario 2 will be lower around 15% compared to Scenario 1, which means the creation of the same output, the former will need to invest 15% less capital than the later. This can be seen clearly that there is upgrading of investment efficiency explicitly in Scenario 2 than Scenario 1.

(3) *Economic Growth will Rely More on the Growth of Service Sector and its Upgrading of Efficiency*

From the perspective of industrial structure, the share of growth of service sector is faster under Scenario 2. Results of simulation show that the share of service sector reached to 51.2% by the end of "Thirteen Five Planning" in Scenario 2, which is around 1 percentage point higher than Scenario 1. The share of service sector will reach to 60.6% by the year 2030, which is 2 percentage points higher than that of scenario 1. Meanwhile, the share of employment of service sector in Scenario 2 is also higher explicitly than Scenario 1. Therefore, the faster development of the service sector is not only favorable to the requirement of satisfying the faster upgrading of consumption structure of residents, but also favorable to the requirement of saving of resources and reduction of discharge of pollutions from the perspective of impact on resource and environment (Table 6.10).

(4) *Income Disparity between Urban and Rural Residents will be Declined Faster*

The level of per capita net income of rural residents in scenario 2 will be increased around 20% compared to scenario 1 based upon the result of simulation. From the perspective of comparison between urban and rural, along with the increase of level of income of rural residents, the income disparity between urban and rural residents will be declined in relatively large scale of scenario 2 than scenario 1. The result of Fig. 6.13 shows that the income disparity between urban and rural residents in scenario 2 will be reduced around 20% than scenario 1. There is small increase of fairness of sharing the fruits of development in scenario 2. This is derived mainly from two aspects: one aspect is that, along with the acceleration of pushing forward the process of urbanization and the rise of level of equalization in

Table 6.11 Economic growth and its sources during 2010-2030 (%, scenario 3)

	2010–2015	2015–2020	2020–2025	2025–2030
GDP	7.9	6.4	5.2	3.9
Source of growth				
Labor	0.3	−0.2	−0.2	−0.4
Capital	10.1	8.2	6.3	4.4
TFP	1.7	1.5	1.5	1.4

providing basic public service, the benefits received by migrant workers will be raised. Also, in accompanying with the transfer of more rural surplus labor force, the rise of marginal productivity of rural labor force will be faster. Therefore, there is increase of income level correspondently. The other aspect is that government will increase its transfer payments to poverty region and poverty groups in scenario 2. This is also favorable to the improvement of income situation of those crowd.

(5) *Efficiency of Utilization of Energy will be Risen Faster, Peak Value of Discharge of Greenhouse Gas will also be Emerged Faster*

In Scenario 2, there will be promotion of rational allocation of resources and improvement of its efficiency of utilization through series of policy measures in rationalizing the formation system of price of energy and resource goods and collection of carbon tax. Results of simulation show, there are explicit improvement either from the efficiency of utilization of energy or for the intensity of discharge of CO_2. By the end of "Thirteen Five Planning", the efficiency of utilization of energy will be increased around 7% in Scenario 2 than Scenario 1, it will be further increased around 16% by the year 2030. The more important point is, under Scenario 2, there will be coming of the peak value of discharge of CO_2 in the period of 2020–2030 along with the faster increase of intensity of energy consumption in scenario 2. The amount of discharge of CO_2 will be declined around 20% in Scenario 2 than the Scenario 1 (Fig. 6.14).

6.6.4.3 Scenario 3

Table 6.11 presents the future situation of economic growth of Scenario 3. Results of simulation show that the economic growth rate of this scenario will be lower around 0.2 percentage point than that of Scenario 1 in the period of "Thirteen Five Planning." The average annual economic growth rate will be lower around 0.5 percentage point than that of Scenario 1 in the period of 2021–2030. The major cause of decline of economic growth rate is due to the slow progress of upgrading quality of economic development thereby faster decline of both rate of technical progress and improvement of technical efficiency. This reflects the slow improvement of efficiency of economic growth to certain extent.

There is no explicit improvement of steps of adjustment of structure even it is in sluggish explicitly in the scenario 3. This is shown not only in the structure of demand and industrial structure of three components, but also in the internal structure of manufacturing sector. For example, it is shown in the result of simulation that the high energy intensive sector has a share around 35.93% of the manufacturing sector by the year 2030, which is 1.3 percentage point higher than Scenario 1.

Viewing from the energy consumption and discharge of pollution, the total amount of energy consumption grows faster as usual due to slow rate of upgrading efficiency of utilization of energy in this scenario 3, in spite of the relatively low rate of economic growth. Furthermore, although the growth rate of amount of discharge of CO_2 becomes slower, it cannot reach to the peak value by the year 2030, which is similar to Scenario 1.

6.6.5 Conclusion of this Case Study

This case study presents the application of a computable general equilibrium model to study and analyze China's "Thirteen Five Planning" and prospects of economic growth by the year 2030 through analysis by scenario method. The major conclusions and policy implications derived thereby are shown in the following:

1. Upgrading the quality of economic development is the necessary choice to deal with various risks. It is also an important safeguard to achieve the building up a well-off society (or Xiaokang society) in an all-round way and establishing a sound foundation for construction of modernization. This is inevitable either from the historical perspective or from the changes of future environment of development. It can be seen from the results of simulation that accelerate the process of upgrading the quality of economic development is the only means to deal with various risks brought from both changes of traditional mode of development and future environment, thereby to achieve the target in building up a well-off society, and to achieve sustainable development of economy and society.
2. Accelerate the process of upgrading the quality of economic development can promote the economy to achieve a "New Normal State of Virtuous Cycle". It can be seen from the result of analysis of simulation, along with a definite economic growth rate through acceleration of progress of upgrading quality of economic development, that there will be promotion of process of transformation from growth relying upon cumulative quantity of factors to growth relying increasingly on technical progress; from growth relying heavily on quantity of export and investment to growth based upon increase of demand of consumption and improvement of efficiency of investment; and from growth relying too much on industrial sectors to growth based more on coordinated development of various industries. At the same time, it can also promote the upgrading of

inclusive development and reduce the pressure of economic growth on resource and environment to achieve economic social and environmental development in harmony.

3. Push forward deepening of various reforms is the key to accelerate the upgrading of quality of economic development.

It can be seen from the results of analysis of simulation that the achievement of all objects, such as upgrading of efficiency of economic growth, reduction of social disparity, and lowering of discharge of pollution, depends upon the implementation of various reform measures, such as reform of conditions of innovation, factor markets, income distribution and ecological environment. The third plenary session of Eighteenth Party's Congress has made arrangement in an all-round way. It is necessary to accelerate the formation of consensus and common effort to push forward reform.

6.7 Summary Points

1. Measurement and quantification are two major difficulties to be solved in social system and social systems engineering.
2. The issue of measurement has been solved to a large part since 1970s. There was a collection of social indicators of eight areas: health, public safety, education, employment, income, housing, leisure and recreation, and population in 1973.
3. There are continuous work on refining the existed indicators and exploration of new indicators by many international organizations and non-governmental organizations.
4. Formulation of rational indicators must pass through a scientific process, collection of primary data, process of primary data, synthesize and analysis data through certain framework or theory to form applicable indicators.
5. A case study of preparation of system of indicators for sustainable development by UN and its related organizations is described in detail to illustrate the process described in '4'.
6. Framework of preparation of indicators of national competitiveness is all described in relatively detail. And the competitiveness indicator of World Economic Forum is also described briefly for comparative purpose on the one side, to promote in depth understanding of the complex process of preparation of indicators on the other side.
7. Model and modeling are not a new term in daily life and in engineering. Model is an abstract of real substance or process. Mathematical models had been very popular in natural science and engineering. The mathematical expression of Newton's Law of Motion seems to be a common sense in physical science. But the application of mathematical model in social science has a shorter history than the natural science. The application of mathematics in economics is

relatively well established, but its application to sociology is still lagging behind due to difficulty to quantify the action and behavior.

8. A general discussion of model and modeling is given to broaden the concept of them. Abstract, logic and observation of the object and objective process are crucial in the process of modeling. It is suggested to read this chapter together with Chap. 5 understand better the necessity of combination of qualitative and quantitative analysis in social systems engineering. The recent progress of modeling of behavior is described briefly to raise the awareness.
9. Development of mathematical model of social science, or more correctly, modeling of economic discipline both in Western countries and in China are given. These descriptions are a briefing of history of development of economic modeling. Understanding the history of certain discipline is beneficial to broaden the view of a scholar.
10. Four Decades of experience of policy modeling of DRC is given in Sect. 6.5. Because DRC is an governmental organization which provide policy consultative service for the government. Both authors of this book had served in this organization since its establishment. The authors think that although China is a late comer of mathematical modeling in social science, but China has shorten the gap relatively quickly more or less through learning and doing and collaborate with international experts in an environment of open system.
11. Section 6.5 are consisted of two parts: Sect. 6.5.1 is derived from a paper "Two Decade of Experience of Policy Modeling of DRC" presented in an international conference of modeling at Beijing 2001, with some modification, Sect. 6.5.2 is written more or less recently.
    Part 6.6 is a case study of DRC-CGE model, which is the continuous effort of Professor Li Shantong (the author of this book) and her team in modeling since late 1990s up to present.
12. In the case of DRC-CGE model, it has explained in simple words of the history of creation, the structure of the model of various module, and its application to the analysis of China's thirteenth Five-year Plan. Three scenarios are provided. This case study will also provide a concise and clear information of Chinese economy in general.

References

Boakin, R. G., Klein, L. R., & Marwah, K. (1991). *A history of macro econometric model building* (Chinese Trans.). Beijing: China Fiscal and Economic Publisher.

Bosworth, B., & Collins, S. M. (2008). Accounting for growth: Comparing China and India. *Journal of Economic Perspectives, 22*, 45–66.

Brandt, L., Biesebroeck, J. Van, & Zhang, Y. F. (2012). Creative accounting or creative destruction? Firm-level productivity growth in Chinese manufacturing. *Journal of Development Economics, 97*(2), 339–351.

Cole, S. (1988). *Global models and future studies* (p. 93). Herzog & Budtz AB.

Dixon, P. B., & Jorgenson, D. W. (2013). *Handbook of computable general equilibrium modeling*. USA: North-Holland.
Economic & Social Affairs UN. (2001). *Indicators of sustainable development: Guidelines and methodologies* (2nd ed.). New York: UN.
Forrester, J. W. (1995). *Counterintuitive behavior of social systems*. Technology Review. Retrieved from https://ocw.mit.edu/courses/sloan-school-of-management/15-988-system-dynamics-self-study-fall-1998-spring-1999/readings/behavior.pdf.
Garelli, S. (2000). *The world competitiveness yearbook 2000*. Switzerland: IMD.
Gigch, J. P. V. (1976). *Applied general systems theory* (2nd ed.). New York: Harper & Row Publishers.
Gong, G., & Hu, G. L. (2013). Efficiency of resource allocation and manufacturing total factor productivity in China. *Journal of Economic Research,* (4), 4–15, 29 (Chinese Version).
Guo, Q. W., & Jia, J. X. (2005). Estimating total factor productivity in China. *Journal of Economic Research*, (6), 51–60 (Chinese Version).
Hammond, A., Adriannse, A., Rodenburg, E., Brgant, D., & Woodward, R. (1995). *Environmental indicators: A systematic approach to measuring and reporting on environmental policy performance in the context of sustainable development*. New York: World Resource Institute.
Hsieh, C. T., & Klenow, P. J. (2009). Misallocation and manufacturing TFP in China and India. *The Quarterly Journal of Economics, 124*(4), 1403–1448.
Kemeny, J. G. (1969). *The social sciences call on mathematics in the mathematical sciences* (p. 22). Cambridge, MA: The MIT Press.
Klein et al. (1993). *Macro-econometric model history*. Beijing: China Financial & Economic Publishing House (Chinese version).
Lave, C. A., & March, J. G. (1975). *An introduction to models in the social science* (p. 4). New York: Harper & Row Publishers.
Li, S. T. (2002). *Productivity and the sustainability of China's economic growth*. Investigation Report No. 108 of Development Research Center of the State Council (Chinese Version).
Li, S. T., & Liu, Y. Z. (2011). *China's economy in 2030*. Beijing: Economic Science Press (Chinese Version).
Lin, Y. F., & Ren, R. E. (2007). East Asian miracle debate revisited. *Journal of Economic Research*, (8), 4–12, 57 (Chinese Version).
Liu, S. J. (2011). Traps or walls: Real challenges and strategic choices for China's economy. Beijing: CITIC Publishing House (Chinese Version).
Mao, Q. L., & Sheng, B. (2013). China's manufacturing firms' entry-exit and dynamic evolution of TFP. *Journal of Economic Research*, (4), 16–29 (Chinese Version).
Memories of Alfred Marshall, ed. A. C. Pigou. (1956). *Reprints of economic classics* (p. 427). New York: Kelly and Millman.
National Research Council. (2008). *Behavioral modeling and simulation-from individuals to society*. Washington, D.C.: National Academy of Sciences.
Oded, G. (2011). *Inequality, human capital formation and the process of development*. NBER Working Paper Series 17058. Retrieved from http://www.nber.org/papers/W17058.
OECD, & WTO. (2013). *OECD-WTO database on trade in value-added*. Retrieved from http://www.oecd.org/sti/ind/TIVA_stats%20flyer_ENG.pdf.
Pyatt, G., & Thorbecke, E. (1976). Planning technique for a better future ILO. Geneva.
Sellin, J. T., & Wolfgam, M. E. (1964). *The measurement of Delinquency*. New York: Wiley.
Song, Z., Storesletten, K., & Zilibotti, F. (2011). Growing like China. *Journal of American Economic Review, 101*(1), 196–233.
Van Meerhaeghe, M. A. G. (1980). *Economic theory: A critic's companion*. Deaden: Stenfert Kroese Pub. Co.
UN. (1975). *Use of systems of models in planning*. New York: UN.
Wang, H. J. (2001). Two decades of policy modeling of DRC. In H. Wang (2003), *Integrated study of China's development and reform-preliminary exploration of social system* (pp. 110–122). Beijing: Foreign Language Press.

Wang, H. J., Li, B. X., & Li, S. T. (Eds.). (1993). *A collection of China's macro-economic models 1993*. Beijing: China Financial & Economic Publishing House (Chinese Version).

Wang, H. J., Li, B. X., & Li, S. T. (2001). *Two decades of policy modeling of DRC*. Paper presented to the International Conference of Economic Modeling in Beijing, China, No. 2001.

Wang, H. J., & Tian, J. S. (1984). *Preface* (p. i) of *Technical Economic Center of the State Council & Planning Commission, Shanxi Province* (1984) (Eds). Compilation of Economic Models of Comprehensive Planning of ShanXi Province Taiyuan/Planning Commission, Shanxi Province (Chinese Version).

World Bank Group. (2016). *2016 world development indicators* (p. 7). Washington, D.C.: The World Bank.

World Bank & Development Research Center of the State Council. (2013). *China 2030: Building a modern, harmonious, and creative high-income society*. Beijing: China Financial and Economic Publishing House (Chinese Version).

Wu, H. X. (2014). *China's growth and productivity performance debate revisited-accounting for China's sources of growth with a New Data Set*. Economics Program Working Papers No. 14–01. New York: The Conference Board.

Wu, X. Y. (2013). Measuring and interpreting total factor productivity in Chinese industry. *Journal of Comparative Studies*, *69*(6) (Chinese Version).

Zhu, X. D. (2012). Understanding China's growth: Past, present, and future. *Journal of Economic Perspectives, 26*(4), 103–124.

Chapter 7
Planning System—The Major Aspect of Application of Social Systems Engineering

7.1 Introduction

7.1.1 Definition of Planning and Its Role

Definition 2 of Social Systems Engineering in Chap. 1 of this book states that "Social systems engineering guides the plan, design, construction and improvement of a social system and its process with certain value." Planning System is the major aspect of application of Social Systems Engineering.

Planning is a whole process of thinking about and organizing the methods and activities required to achieve the goals. Currently, there are many debates on national planning by academic field in both China and other countries. This is due to the basic understanding of meaning of "planning" and ideology of market-orientation in contrary with planning. From the practice of the mankind, there is no need to discuss whether the planning is required or not. The whole life of people is the process of doing something with goals, that is, planning must exist. There are no problems on whether planning should exist or not in the objective world where we lived, the major issue is the quality of planning. It was defined by Bowles and Whynes (1979), "We take the term 'planning' to refer to a purposive, means-ends process and we may define it as the deliberate manipulation of the parameters of a system to bring about a desired and specified alteration in the operation of the system." (pp. 1–2). Do people fully recognize the parameters effected the operation of the system? Do the people have deliberative thinking to manipulate the parameters of the system? These are the factors that determine the success and failure of a plan.

The history of the world tells us, although there were no overall national planning, planning of sectors, such as fiscal planning, diplomatic planning, especially military planning did exist. For example, the first emperor of Qin (259–210 BC) in the period of Warring State (475–221 BC) had consulted Wei Liao, the

H. Wang and S. Li, *Introduction to Social Systems Engineering*,
https://doi.org/10.1007/978-981-10-7040-2_7

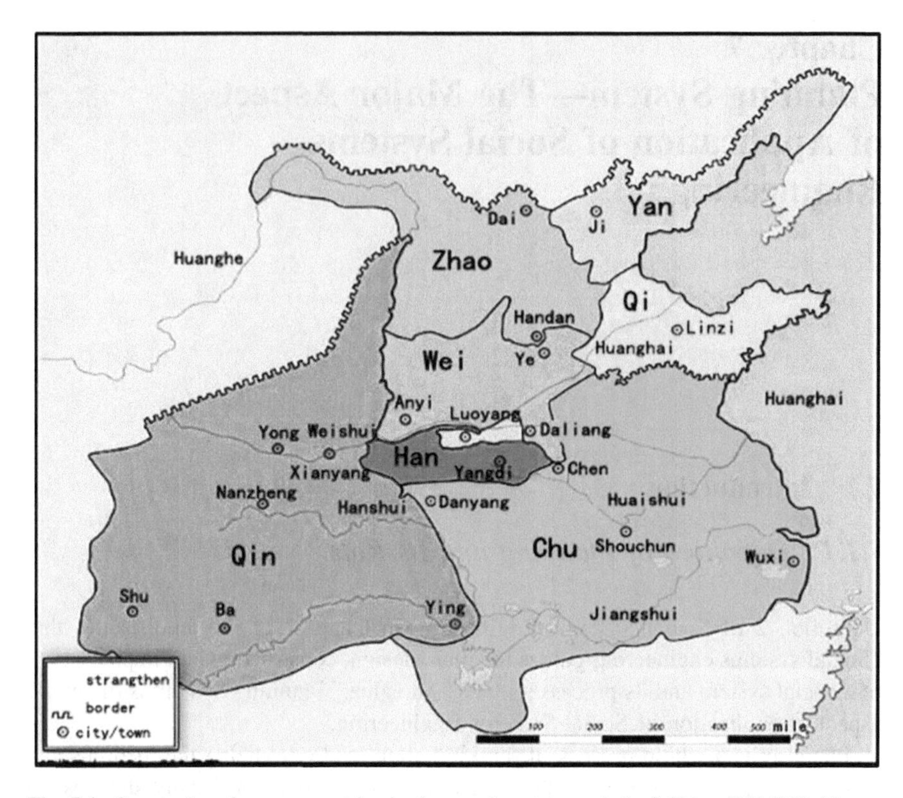

Fig. 7.1 Geography of seven countries in the warring state period of China (260 BC) (*Source* Seven strong countries 260 BC. Retrieved from https://www.google.com/search)

supreme government of official in charge of military affairs about the order to take off other six countries. Wei Liao replied: "Han was weak, it should be subjugated in the first; then the next would be Zhao and Wei, post the abolition of these three countries, then we raised full force to attack Chu. If Chu was subjugated: How Yan and Qi could go?" (Fig. 7.1).

This military plan had analyzed fully the parameters of national strength, geographical distance and also consideration with in depth the change of parameter of national strength through proper order of subjugation of other countries. Emperor Qin had followed the military plan of Wei Liao, and he unified China by that time. There are no shortage of various cases of planning in the global history. But there were also cases of planning failure, for example, the failure of the Second Five Year Plan of China due to Great Leap Forward with wrong conception to catch up the advanced developed countries within a short period of time. All these examples can explain the role of planning and factors related to a successful planning.

7.1.2 *Types of Planning*

Planning is a noun. If various adjectives are added before it, various types of plans will be formed.

(1) Based upon the nature of the actors who prepares the plan, then there are: national planning: corporate planning; family planning; individual planning etc.
(2) Based upon the time dimension of the planning period, there are: long term planning; medium term planning; annual planning, etc.
(3) Based upon the spatial administrative dimension of the object. Generally, different countries have different classification of administrative structure. With China for example; China has the level of administrative structure of province or autonomous region, prefecture, municipality and city, county etc. U.S.A. has the level of state, city, township and county. Although U.S.A. does not have a national planning, but it has planning commission at its lower level of administration. It can be seen from Daniels, Keller, and Lapping (1988), "Planning Commission: An official body appointed by the governing body of a city, township or county that is responsible for making the comprehensive plan. In addition, the planning commission makes recommendations to the governing body on the zoning ordinance and zoning decisions, on subdivisions, and on general planning matters." (p. 160). Academically, U.S. had many planning studies, it can be seen from the book related to strategic planning by Bryson and Einsweiler.
(4) Spatial Development Planning has a unique feature. It refers a planning by government to effect balanced and sustainable development of its territory of a country. Japan had established the first act on national spatial planning in 1949, its latest spatial planning was the *Sixth National Spatial Planning* launched in 2008. It includes urban planning, regional planning, planning for the use of land etc. South Korea has also its national spatial planning. European Commission had launched *European Spatial Development Perspective* in 1999, and it was agreed at the informal council of ministers responsible for spatial planning in Potsdam, May 1999. China had tried to prepare such plan in 1990, but it launched the first official formal national spatial planning in 2017.
(5) Based upon the nature of object of planning (functional), planning can be classified into social planning, economic planning, scientific and technological planning, etc.

    If the concept of hierarchical structure of general systems theory is applied, then social planning system can be further classified into several subsystems such as education planning, health care and medical planning, social security planning etc. The lower level of economic planning can be further classified into agricultural planning, industrial planning, service sector planning; or investment planning, consumption planning, etc.

Under industrial planning, the lower level manufacturing sector can be further classified as planning of 24 sectors based upon ISIC[1] Rev. 4.

7.1.3 *Classification of Planning in Economics*

7.1.3.1 Planning in Business

Business in other countries, especially the large corporations, have various types of planning long range plan with forecasting, marketing planning for sales, for quite a long time. Strategic planning grows in popularity since 1960s. According to Bryson and Einsweiler (1998).

> Strategic planning is a disciplined effort to produce fundamental decisions and actions that shape and guide what an organization (or other entity) is, what it does, and why it does it (Bryson 1998; cf. Olsen and Eadie 1982). Strategic planning is designed to help leaders and decision makers think and act strategically. The best examples of strategic planning-as is true of any good planning-demonstrate effective, focused information gathering; extensive communication among and participation by key decision makers and opinion leaders; the accommodation of divergent interests and values; the development and analysis of alternatives; an emphasis on the future implications of present decisions and actions; focused, reasonably analytic, and orderly decision making; and successful implementation.
>
> The public sector roots of strategic planning are deep, but few in number. Strategic planning in the public sector has been applied primarily to military purposes and the practice of statecraft on a grand scale (Quinn 1980; Bracker 1980) (p. 1)

The above is a brief description of planning in business. Planning at national level can be classified into mandatory planning (or centrally planned economy, as used in nearly all economic dictionaries) and indicative planning. The advantage and disadvantage of these two types of planning have been well summarized in economics through more than seventy years of practice of these two types of planning by various countries in the 20th century.

7.1.3.2 Mandatory Planning (Centrally Planned Economy)

Mandatory planning were applied to macro-economic planning of countries dominated by the state owned sectors, for example, the planning system implemented by former USSR, China and large part of Eastern European countries before the launch of economic reform and opening to outside world. The government of these

[1]Note: ISIC is a four letter acronym that stand for International Standard Industrial Classification. It is a United Nations' industry classification system. Revision 4 was published in 2008.

countries controlled all kinds of input and output of goods of state owned enterprises (SOEs), price of goods and wages and salaries of workers and staffs. Chief managers of SOEs were appointed by different levels of government, resulting in central owned SOEs or provincially owned SOEs etc. The activity of private sector was nearly in the margin of abolishment, or was very much restricted. All crucial economic activities were controlled by state planning department, they were not determined by market force. Most economic activities were determined by vertical information transmission through hierarchical levels of administrative structure, the target of input and output must be implemented. All above norm projects of all SOEs must be approved by different levels of government. Proponents of centrally planned economy claimed that planning enables wasteful duplication and unemployment to be avoided and goods to be distributed fairy.

Based upon personal perceptions of the authors who had lived in China in the stages of implementation of centrally planned economy, this type of planning is adapted to underdeveloped economies dominated by primary sector and is in the primary stage of industrialization on condition that there is nearly no corruption of the officials. The government can concentrate limited resources of factors of productions to the process of development and industrialization, this was a story that happened in China before 1980s. The corrupted officials were punished in the very beginning (two senior officials at prefecture level were sentenced to death on 1952) so that there were no followers. China had established nearly a national level of industrialization due to construction of key projects of various industries, and through its advocation of five small industries at rural area (small coal mine, small machine tool factory, small cement factory, small hydro power plant, small iron and steel plant), around 1960s and 1970s. Although the five small industries were poor in economy of scale, but this had created the necessary human resources and liberalized their capacity when China announced reform and opening in late 1970s.

The major difficulties and weakness of this types of planning is coordination and collection of feedback information in time. There are also distortion of transmission of information due to long chain of command through multiple linkage of bureaucratic administrative levels, which caused low efficiency of allocation of resources. When China became semi-industrialized, it was impossible to have a central planned economy to meet the variety of demand from its people. Therefore, economy of shortage were universal phenomena of former central planned economies.

7.1.3.3 Indicative Planning

In this type of planning, the central government will set up several targets to coordinate the investment planning and output of operation of both private and public sectors, the decision making of operation of economic activities is decentralized. The central government will influence the behavior of various actors of the society through laws, fiscal and monetary policy, social policy, science and technology policy, industrial policy, etc. to encourage the desirable actions and prevent the undesirable actions to achieve the targets set up by the government. The planning process can be either top down or bottom up. The major requirements of this type of planning is improvement of free flow, transparency and exactness of information to reduce the uncertainty of environment faced by decision makers, and a fair environment of competition of all players in various market within the legal context. There are many debates on economic planning among economists. This chapter will focus on discussion of national planning in a broader scope. Corporate planning will also be discussed briefly.

7.2 Evolution of Development of Modern National Planning System

7.2.1 Three Stages of Development of Evolution of Modern National Planning System

7.2.1.1 Period of Beginning of National Planning

The former Soviet Union had prepared its first Five Year Plan (1928–1932) in the second decade of 20th century. That planning document had three volumes with 1700 pages. The contents included light industry, heavy industry, agriculture, cooperative, finance, labor force, wages, education, transportation and communication, public health and social security, etc.

Preparation and implementation of the First Five Plan was very successful. According to Stavrianoo (1971) "By the end of the First Five Year Plan, in 1932, the Soviet Union had risen in industrial output from fifth to second place in the world." (p. 511). "The Five Year Plan attracted worldwide attention", particularly because of the concurrent breakdown of the West's economy…. Economic policies were influenced, consciously or unconsciously, by the Soviet success in setting priorities for the investment of national resources, which is the essence of planning. Some countries went so far as to launch Plans of their own, of varying duration, in the hope of alleviating their economic difficulties." (p. 512)

7.2.1.2 Age of Popularization of National Planning: 4th Decade–5th Decade of 20th Century

Post the WWII, many former colonized countries became independent sovereign states, all of them had pursued the objective of industrialization, and most of them, such as China, Hungary, Romania, Vietnam and Cuba engaged in the preparation of national planning demonstrated by the former achievements of Soviet model, These countries had also adopted socialism as the ideology of national development. Other countries, such as India, Bhutan, Nepal and Pakistan in South Asia; South

Korea, Malaysia and Indonesia in Southeast Asia; Brazil, Argentina in South America, also prepared medium term of four or five years plans. Even the developed countries, had set up planning institutions in their government in order to proceed reconstruction of their economy. For example, France and Netherland had developed the indicative planning system, supported by theories of development planning and economic models. These two countries took up a leading position in the exploration and preparation of national planning and their practice had effected Japan, South Korea and a large part of emerging economies in Southeast and East Asia. It is described in World Development Report (WDR) 1983 that, "some eighty countries indicated that four out of five have multiyear development plans; over the past ten years, approximately 200 plan documents have been prepared." (p. 66). Even United Nations had prepared several plans of *United Nations Decade of Development* before 2000.

7.2.1.3 Post 1960s up to Present: Decline of Preparation of Medium Term National Planning and Emergence of Long Term Strategic Planning and Scenario Planning

The indicative planning of some developed and developing countries depended upon heavily the forecasting models. Due to the data and modeling technique could not adapt to rapid political, economic and social changes, especially the two oil crisis happened in 1973 and 1978 had had serious impacts to global economy, many targets planned could not be achieved. On the other hand, it was described in WDR (1983), "At root, there is an inherent weakness in the 'blueprint' approach of planning agencies: available analytic techniques are just not able to cope with the complexity of economic change and to produce plans that are up-to-date, relevant, and comprehensive. Even the less ambitious forms of planning having their weakness. For example, investment planning based on input-output models has fallen foul of changing technical coefficients and demand patterns." (p. 67). By the late 1960s there was already widespread talk of a "crisis" in planning. In the seven decade of 20th century, there was a wave of privatization of SOEs in the Great Britain and U.S.A., and liberalization of trade and capital flow had become a dominant thoughts of main stream economists. The disintegration of former Soviet Union had pulled down totally the imperative planning. And the failure to meet the target of indicative planning had resulted the abandonment of mid-term planning in many developed countries, some of them simply claiming overall targets for macroeconomic policy, some of them turn to long term strategic planning such as European Union and Japan. With respect to method, scenarios method becomes popular. Although there is no national planning in the U.S.A., its annual fiscal budget, think tanks of Congress and two major parties, effected the national economic and social development implicitly. The national Intelligence Council of the U.S.A. had the Global Futures Project since 2010. It launched the first *Global Trends 2010*, and continued the launch of this report every five years. In its report

Global Trends 2030, launched on December 2012, four scenarios of alternative worlds had been identified. It is of interest to quote its four scenarios in Box 7.1 to compare the real global tend around 2017.

Box 7.1 [2]Four Scenarios in Global Trends 2030

1. Stalled Engines—a scenario in which the US and Europe turn inward and globalization stalls.
2. Fusion—a world in which the US and China cooperate, leading to worldwide cooperation on global challenges.
3. Gini-Out-of-the-Bottle—a world in which economic inequalities dominate.
4. Nonstate World—a scenario in which nonstate actors take the lead in solving global challenges.

Some countries, such as China, India still continue their practice in preparation of mid-term (five years) planning. China has been gradually transformed its planning system from planned economy to indicative planning, which will be described in Chaps. 11 and 12 in this book.

7.2.2 *Development of Planning Theory*

Planning requires a broad aspects of theory and knowledge of various disciplines. Narrowly speaking, the development of theory of planning has three related aspects.

7.2.2.1 Development of Planning Theory Itself

The former Soviet Union was the first country that practiced in the centrally planned economy. It had developed the contents, structure and procedures of planning. It had also developed the Material Production System (MPS) for the balanced development of the economy. Based upon the study by Bowles and Whynes (1979), "Gosplan had constructed a prototype input-output table in 1926, but the prevailing political climate was antagonistic towards mathematical techniques." (p. 178). Which can be seen from a speech given by Stalin in 1929 who contemptuously dismissed the Gosplan exercise. After the death of Stalin, input-output analysis re-emerged as a technical tool, and Leontief developed in its modern form in U.S.A.

[2]Source: NIC (2012).

France had developed the indicative economic planning as well as several quantitative approach of planning. Many Western economists, had learned experiences from the planning of former Soviet Union and France, and summarized the lessons and experiences to aid the development of developing countries. They had published system of planning theories and development planning. For example, Lewis, W. A. published *Development Planning: The Essentials of Economic Policy* in 1966, Lewis had worked in United Nations and the World Bank, and served as economic advisors of several developing countries, such as Ghana, Jamica, Nigeria etc. Therefore his book is not only theories of planning in nature, but also very practical to be applied in planning. Chapter 3 of his book is arithmetic in planning, which includes essential elements of economic planning with examples of calculations based upon arithmetic. Jan Tinbergen published *Development Planning* in 1960s. He is the founder of econometrics, as well as the first director of Bureau for Economic Policy Analysis of Netherland in 1945. Both of them are Nobel Prize Winners. There is also development of planning theory focused on *Planning Integrated Development: Methods Used in Asia*, which was a publication by Unesco in 1986. This book is a collection of papers of a sub-regional symposium organized by Unesco in collaboration with the Planning Commission of India by the end of 1984. This book has a special focus on integrating cultural and social dimensions into development planning. Other development of theory of planning is a book *Development and Planning* written by Sachs (1987), which is a collection of his papers. It was described in the preface that "The aim of this collection is to challenge a rigid, technocratic, economistic view of development planning and at the same time, more than ever, to claim a central role for a renovated and committed form of planning in the formulation of anti-crisis strategies." "The underlying epistemology is that of developmental social sciences: it rejects the domineering, scientist claim of economics, and at the very least marks a return to political economics and maybe even the beginning of an anthropological economics, but it refuses to beyond a heuristic function." (p. vii)

It can be seen from above quotation that Sachs kept a similar view with Piketty. And currently, the theory of planning is not dominated by economists alone. An integrated planning requires the coordinated efforts of all social scientists as well as natural scientists and engineers too.

7.2.2.2 Development of Mathematical Economics

Mathematical economics is the application of various mathematical method to express economic theory or for the analysis of economic issues or problems.

Mathematics includes a broad contents, arithmetic, algebra, geometry, calculus, differential equations etc. It had been described in Chapter 16 of Tinbergen's *Development Planning* (1967), Tinbergen raised the view that: Economists well trained in mathematics had spent their efforts three times to put the application of economics on more precise foundation in the history of science…. Only the third effort had been successful to enter the stage of practical application of economics… this effort started around the year of 1930, established the empirical relationship with economic life. Econometric Society was established in the same year. He had also mentioned the trend of development of Sociometrics. From perspective of authors of this book, there are needs to develop further the discipline of sociometry as well as its application in planning and social systems engineering. In Lewis's book Development Planning, Lewis had calculated economic growth rate, balance of payments, physical balance, input-output of industrial balance, demands of capital, and governmental budget with simple arithmetic, and put forward linear programming. This kind of treatment will also be acceptable by most decision makers, because a large part of decision makers are not trained in science and engineering. Planning became a controllable process, and a true science which was recognized by Tinbergen as so called scientific plan. France had continued to develop the application of mathematical models. In the seventh decade of the 20th century, the United National had organized an international conference on "Systems of planning Models" in former Soviet Union.

Application of mathematics in planning has promoted the preparation and development of scientific planning. But some eminent economists such John Maynard Keynes and Friedrich Hayek kept their attitude of criticism, They recognized that some human behaviors and their choice cannot be expressed by mathematics and probability. Mathematics, being a kind of method and tool, must be applied in accordance with objective regularity. It should reflect correctly the regularity and value existed objectively. But change of external environment can be very swift and volatile to cause the status represented by mathematical model noneffective totally. One of the causes of global financial crisis was complex formula, for calculation of derivative financial instruments were inconsistent with objective situation. Richard Brokstable (2007) had summarized his personal experience worked in financial market and hedge fund in his book *A Demon of Our Own Design* published in 2007. He pointed out in the introduction: "… [3]we had summarized many lessons in the development of other product and service in the 20th century, from the design of architecture to structure of bridge, from the operation of refinery to power plant, from the safety measure of automobile to airplane. The contrary, although the structural design of the financial market had improved greatly in past thirty years, but the results are more frequent and more serious breakdown."

Although design of product or engineering projects must be calculated by mathematical models, and design of structure of financial market has also developed to a new age in depending upon mathematical model and the new discipline of financial engineering. The object of design of engineering projects or product is

[3]Note: Description within the quotation marks are translated by authors of this book from the Chinese version of the book.

physical system of natural world. Current achievements in science can master a large extent of them. While the design of structure in the financial market involves the participants counted in hundred millions, their behaviors cannot be expressed in mathematics, which had been criticized correctly by Keynes. The issue is not the mathematical model itself, but that it has been applied wrongly with the object concerned.

7.2.2.3 Emergence of Future Studies and Scenario Planning

Internationally, the emergence of a new discipline, Future Studies, also has its impact on national planning. History looks back to the past, while Future Study looks forward to the future. There were some regularities of events, as well as some inevitability of events in the historical process of the development of human society. This will also be true for the future. Mankind had proceeded the exploration of the future in its earlier history, for example, I-Ching (The Book of Change, a Chinese book published around late 9th century BC) and its Hexagrams was a combination of Yang Yao and Yin Yao to form certain projections of probability. There were also many other historical stories and natural phenomena to explain the indications of the 64 combinations (Eight Hang). This book has included very preliminary elements of contemporary Future Studies.

The international Future Study (Futurology) started from H. J. Wells in 1932, who suggested establishing a new department in Future Study. Future Study or Futurology is defined as "studies of the future". The term was coined by German professor Ossip K. Flechtheim in the mid-1940s, who proposed it as a new branch of knowledge that would include a new science of probability. Herman Kahn (1922–1983), who had been employed at the Rand Corporation, and had a background of military strategist and systems theorist, worked together with Anthony J. Wiener and published *The Year 2000: A Framework for Speculation on the Next Thirty-Three Years* in 1976. This book had established a system of thoughts of the discipline of Futurology. There were many future oriented studies in Seventh and Eighth decade of the 20th century that were influenced by the book written by Herman Kahn. Bureau of Studies and Programming of Unesco had set up Major Programme 1 Reflection on World Problems and Future oriented Studies. And there were many studies in the 20th century such as *The Global 2000 Report to the President of U.S.A.*, *China towards the year 2000* of China, and *Japan by the year 2000*, etc. Post 2015, the United Nation has set up a new agenda, known as "Transforming our world": the 2030 Agenda for Sustainable Development with 17 global goals and 169 targets between them, the agenda was approved by 194 members of UN on Sept. 2015. It is also popularly known as the Future We Want. To achieve these goals, it requires knowledge of Futurology. Some organizations attached to UN are setting up programs to achieve these goals. [4]UNESCO has set

[4]Note: UNESCO is the abbreviation of United Nations Education, Scientific and Cultural Organization.

up the Global Education 2030 Agenda in response to SDG4 (quality education). UNIDO has launched a series forums of ISID (Inclusive and Sustainable Industrial Development) (Which is more or less similar to the case study OECD EXPO 2000 described in Chapter 4) in response to SDG 9 (Industry, Innovation and Infrastructure). Futurology may be further perfected through the process of implementation in the 2030 Agenda for Sustainable Development.

Herman Kahn was also the founder of scenario planning. Due to rapid change of environment and objective world, traditional planning generally may have not been adaptable to these changes. Scenario planning and scenario method may be quite useful. It is described by Godet (1994) "The future is multiple and several potential futures are possible: the path lead to this or that future and the progression toward it comprises a scenario. The word scenario was introduced into futurology by Herman Kahn in his book *The Year 2000*, usage was primarily literary, imagination being used to produce rose-tinted or apocalyptic predictions previously attempted by authors such as Anole France (Island of the Penguins), or George Orwell (1984)." (p. 154).

Although there is no national planning in U.S.A., its National Intelligence Council had launched the Global Futures Project which had been described previously. A more detailed discussion will be given later.

7.2.2.4 National Plans and Business Plans

(1) *Enlightenment of Business Plans*

As for the effect of national plans on various countries of the world, the understanding and application of various countries are not the same due to the effect of ideological trends of politics and economy, as well as the imperfection of planning techniques. But at the enterprise level, the preparation of plans is always the first priority for the management. All enterprises prepare different types of plans: strategic plans, long-term plans, annual plans as well as the plans of various departments and specific plans, marketing plans or environmental plans. For example, Hitachi of Japan has prepared *Hitachi Environmental Innovation 2050* for the environmental plans. In addition, all basic units of organizations shall prepare the working plans. Although America does not prepare the national plans, there is continuous exploration of the planning techniques at enterprises level. For example, the SWOT Analysis (strengths, weaknesses, opportunities and threatens) and PEST analysis (political factor, economic factor, social factor and technological factor) are developed by academic fields applied to enterprise plans.

(2) *The Development of the Social Organizations Requires the Plans of a Certain Extent*

Countries and enterprises are all social organizations of different forms. The economic scale and the number of staff of the modern large multinational corporations exceed those of some countries. For example, the asset of Exxon Mobil Corp, which ranks Top 500 Companies of American Fortune Magazine in 2014, was 348.4 billion dollars. The staff number of Wal-Mart was 2.2 million. However, the population of Timor-Leste near Indonesia in 2014 was only 1.2 5 million, whereas the population of Capital Dili is only 166 thousand. Its GDP based on purchasing power parity (PPP) which is only 6.3 billion dollars.

The development of these large and complex multinational corporations requires the plans of a certain degree to arrange and deal with the development in varying international environment. Due to the variation of the objective environment as well as the inadaptability of the planning techniques, the preparation and implementation of national and enterprise development plans cannot ensure the success of the development of countries and enterprises. Success depends upon being well-prepared and without such preparation there is sure to be failure. As for the planned and the unplanned, the plan is still one of the essential development tools for the social organizations. How to prepare better plans is a topic that requires the continuous exploration and efforts of professional and academia relevant to this field. It is expected that the development of social systems engineering can contribute to part of the effort.

7.3 Case Study of Medium-Term National Plans

7.3.1 Introduction

After World War II, some developed and developing countries attempted to prepare the national medium-term plans to guide the development of the national economy and society. Among which the exploration carried out by France was relatively systematic and advanced. This includes the determination of the development objectives, establishment of the national economic accounting system, as well as the exploration of application of mathematical models in the planning, the consultative system for the preparation process of plan, etc. These became examples of preparing the indicative planning system. Some emerging countries learned the experiences of the indicative plans of France. For example, South Korea had invited a special delegation of France in 1979 to introduce the French experiences of indicative planning during the preparation of its Fifth National Medium-Term Plan (1982–1986). The Sixth Five-Year Plan of South Korea became more refined and strategic after absorbing the experiences of France. Japan also has its unique features of indicative planning through learning and absorbing experiences from abroad. Japan's industrial policies, which were the combination of the Japanese unique culture and enterprise management system, made Japan turn into a powerful state in manufacturing rapidly after war. The Japanese experiences had attracted attention of a wide range of international community. Japan has also prepared National Spatial Development Plan quite earlier. India had always continuous practice of preparation of plans since its First Five-Year Plan (1951–1956) in 1950. Currently, India has entered Twelfth Five-Year Plan (2012–2017). As the officials of the planning department of India generally accepted the western education, India had adopted the model developed by the statistician Prasanta Chandra Mahalanob in 1953. Since the beginning of Second Five-Year Plan (1956–1961) to carry out the optimal configuration of investment of production department. The text of Twelfth Five-Year Plan adopted the scenario analysis method to determine the development objectives and also the system analysis in the process of planning. This part will mainly introduce the cases of mid-term planning of France and India, while other types of planning will be discussed in later part of this chapter.

7.3.2 Cases Study of French Mid-Term Planning

7.3.2.1 Principles for Preparing Indicative Plans of France

Plans of France were based on two basic concepts:

(1) French planning seeks to be "the search for a middle road reconciling commitment to freedom and individual initiative with a communal orientation in development." (Masse, 1965, p. 144)
(2) Planning must "express in its objectives a collective will in regard to building the future and in regard to the goals of economic and social development." (Delors, 1965, p. 155)

7.3.2.2 Extensive Social Participation in Preparation of Plan

During the preparation of the First Medium-Term Plan (1947–1950), France was at the postwar construction, and there were plenty of public departments at that time. Therefore, the preparation of plans had been supported by the public departments. So the planning had a nature of social contract. The preparation of the plan had attracted the participation of a wide range of representatives from economy and society, including administrative staffs and civil servants, famous experts, representatives of various professional associations (workers, foremen and managers), etc.

Due to the extensive participation of all levels of society during the preparation of plans, the plans can reflect the trends of all aspects of society well and clarify the social dispute and the political choices. The defined objectives thereof can integrate the objectives and hopes of all kinds of stakeholders in a plural society.

7.3.2.3 Function of Plans

The preparation of plans was based on the economic interests of all the social groups and was used as a tool to guide the public actions. It attempted to maintain the principle of freedom in economy of France and combined the social control function to construct the future of society collectively.

7.3.2.4 Two Characteristics of Indicative Plans of France

The first director of the planning agency of France Monnet summarized the following two characteristics for the plans of France in 1977 after several planning periods of France.

(1) *The plans of France were harmonious*

The harmonious plans were the objectives for preparing plans "[5]it is necessary to reflect the expectation of collectivity for the objectives of constructing future economy and society". Therefore, during the preparation and the implementation of economic policies, it was necessary to carry out a joint consultation participated by all the social representatives to achieve coordination. The key points were: "establish an environment to gather government and enterprise representatives, exchange information with each other and compare the respective projections. Then make decisions or form suggestions on government."

The concepts and the behaviors of this economic coordination meant that the function of the traditional public authority is required to be changed, i.e. the country shall not become a compulsive authority for the problems of people's concern. In 1929 of post-crisis era and post-World War II era, it might be necessary for the government to issue the unilateral regulation at this time in order to restore economic growth and recombine economy. However, as the only authoritative specification after the restoring of prosperity, the government was required to be changed. "Harmonious economy" had put forward a new concept for the intervention of the government. It required the government to fully respect the principle of freedom while taking actions on economic and social lives. These should be solved through consultation and negotiation rather than adopting the imperative instructions.

(2) *The plans of France were Flexible*

The indicative plans of France might be called the flexible plans, so as to be distinguished from the imperative plans or the rigid plans. In the imperative planning system, the production and investment projects in charge of the economic actors must be in accordance with instructions of the central authorities. However,

[5]Source: Jean Monnet in 'Les Chhiers Francias No. 181 May–June 1977' French Planning Comment Translated by World Bank on Aug. 1984.

the indicative plans were freely selected and operated by most of the economic decision makers so as to achieve the objectives of their own. The indicative plans emphasized the complementarity of market and plan. The price signals released by the market directed the short-term decision and the daily adjustment. However, the function of the plans concerned more in the long-term. The projection made by the long-term decision was used to provide the information about the future environment. The role of the planning of adjustment of market was for the purpose in reducing uncertainty.

7.3.2.5 Organizations of French Planning Agency and Determination of Growth Rate of Economy of Planning

(1) *Organizations of French Planning Agency*[6]

The organizations of French planning agency introduced here were the circumstances acquired by the State Commission for Restructuring Economic System of China while surveying abroad in 1991. These circumstances were before 1990s and were used for reference and comparison. French planning agency was directly controlled by the Prime Minister of France and had 150 staffs, actually about 70–80 experts are researchers. The planning agency was consisted of Economic Department, Department of Social Affairs, Department of Agriculture, Department of Scientific Development, and Department of Regional and Urban Service.

The Department of Regional and Urban Service, was also responsible for the cooperation of regional plans. There were 22 large regions in France, and the government of each large region had its own planning agency. After the preparation of its plans of large regions, a contract was signed with the central government.

Except for the five departments described above, there were three research units: *quantitative analysis research unit* (DESQ), which cooperated with Economic Department, researched the international environment as well as the economic and financial relationships between France and foreign countries; *long-term research unit,* which was responsible for the projection of economy, society and international affairs; and *coordination and socio-economic research unit*, which was responsible for coordinating the research in the social and economic fields related to the preparation of plans.

The planning agency was basically a research institute rather than an administrative institute and only served as the mediator between various departments of

[6]Note: Reference 1. French Planning Comment by John Monnet. 2. Wrote by Fan Wenhui, 1991 *Management of French plans Assembly of data of survey abroad of State Commission for Restructuring Economic System*. Foreign Affairs Department of State Commission for Restructuring Economic System.

government and the enterprises and the planning agency of large regions. The planning agency did not have any administrative power.

(2) *Determination of Growth Rate of GDP*

The French economic departments had devoted themselves to the development of economic model to carry out projection as well as coordination between departments. However, except for relying on results of projection of model, the determination of the economic growth rate was generally finalized according to the actual circumstances. This point can be seen from the discussion of Monnet below.

"The growth rate suggested by the plans more or less contains some kind of voluntarism". After the first plan of France and in 1950s, people widely believed that the economy of France shall restore to a more steady growth rate. As a large number of populations of the post-war generation would return to work, the Forth Plan of France had adopted a relatively high economic growth rate. The Fifth Plan emphasized the stability so as to determine a growth rate of 5%. As the determination of the Sixth Plan had estimated the probability of materials as well as the objectives of full employment, the determined growth rate was extremely ambitious.

It was difficult to judge and select the actual results of the growth rate. The government had many behavioral tools (tax policy, government procurement, social transfer payment, subsidy....) that could greatly influence on the growth rate. The growth rate determined by the plan was used as the reference for companies and administrative departments and was also used to affect the public opinions. Therefore, the growth rate announced by the plan had a certain function of mobilization.

7.3.2.6 Brief Introduction of Eight Medium-Term Plans of France

France had prepared ten medium-term plans according to the analysis of information obtained. This section will give a brief introduction from the First Plan to the Eighth Plan. The next section will give a relatively comprehensive introduction for the text content of the Ninth Plan. The profiles from the First Plan to the Eighth Plan of France are shown in Table 7.1.

Table 7.1 Brief introduction of eight medium-term plans of France[a]

Name of plan	First plan	Second plan	Third plan
Planning period	The original planning period (1947–1950) was delayed to 1953 in order to link up the termination period of Marshall plan	1954–1957	The original planning period 1958–1961 had not reached the predicted objectives, an adjustment planning period 1960–1961 was added
Planning background	• In the circumstances of economic shortage of the post-war • Reserved a large number of public departments with the Warfare system • Preparation of the plan was carried out at the same time as the French national accounting system was being developed	• Reconstruction was finished and a large part of shortage disappeared • Expected changes: more young people would enter the labor market; promoted the construction of European community • Established the economic accounts and the national budget commission	• Economic techniques used as basis of plans • Prepared the national accounting report of 1949–1955, and published I/0 table of 1951 in 1957. France scientifically applied the national accounting system at first
Objective	• The planning period required France to be provided with enough production means to produce more products quickly	• Transferred from quantity to quality; emphasized the production quality and efficiency to	• Prepared the employment of more young people and prepared entering the common market. Aimed at a high

(continued)

Table 7.1 (continued)

Name of plan	First plan	Second plan	Third plan
	• The gross production was increased by 25% compared with that of 1929 • Restoration of balance of payments	prepare France entering free trade system • Compared with those of 1952, GDP should increased by 25%; industry had increased by 25–30%; agriculture had increased by 20%; construction industry had increased by 60%; export had increased by 40% and import had increased by 25%	economic growth rate and the balance of international payment • GDP increased by 27% compared with that of 1956; expected a surplus of 300 millions of dollars
Action plan	• Gave priority to the bottleneck of shortage and heavily invested to the basic sectors • In 1946, after the determination of the six sectors such as coal, electric power, steel, cement, transport agricultural equipment, and mechanization two sectors such as fuel and azotic fertilizer were added	• Paid attention to the overall balance of economic development as well as the improvement of sectoral structures • Proposed to promote "basic action". The action referred to: develop science and technology; improve the adaptability and the professional level	• The third plan proposed "Urgent Task", which referred to strengthen the professional training and improve the market organizations

(continued)

Table 7.1 (continued)

Name of plan	First plan	Second plan	Third plan
		of industries and enterprises; strengthen training; improve the market organizations	
Name of plan	Fourth plan	Fifth plan	Sixth plan
Planning period	1962–1965	The planning period was 1966–1970, but the actual implementation period was extended to 1971 so as to coordinate with the operation of the common market	1971–1975
Planning background	• Political stability and peace. France entered a Europe with a greater degree of opening and was entering the "consumer society" • Replaced the accounting table of 1951 with the national accounting of 1960. There were I/O tables of 65 sectors in 1956	• Most part of the economic sectors of France had been opened to the outside world, [b]and the economic integration of six countries had been formed • The national accounting report of 1962 was adopted. I/	• When compared with the international precursors, French industries had defects in quantity and quality and required improvement • I/O tables of 77 branches adopting the data of 1962

(continued)

Table 7.1 (continued)

Name of plan	Fourth plan	Fifth plan	Sixth plan
		O Table adopted the data of 1959. Made progress in prediction and long-term research	replaced the table of 1959
Objective	• Paid more attention to the distribution of results as well as the local and the regional development • The name of plan was changed from the previous "Modernization and equipment" into "Economic and social development plans" • The annual GDP growth rate was 5.5%; the collective consumption goods were increased by 50%; the economic growth of less developed areas was promoted	• Improved the industrial competitiveness • The annual growth rate of GDP was 5%; the labor market was expected for a moderate salary increase to maintain the price competitiveness of the international market	• Strengthened competitiveness and maintained a balanced growth • Gave priority to the industrial development • The annual growth rate of GDP was 5.8–6%; the annual growth rate of industry was 7.5%; the annual increase of commodity price was controlled below 3.2%
Action plan	• Strengthened the research on science and techniques of economic growth and export capability • Social initiative-Strengthened the consumption and supply of society and collectivity, such as medical care, education, urban	• Innovated and improved industrial and commercial structure • Gave priority to develop the leading sectors in the Common Market	• Created a favorable environment for industrial and enterprise competitiveness • Eliminated the enterprises with loss, strengthened the

(continued)

Table 7.1 (continued)

Name of plan	Fourth plan	Fifth plan	Sixth plan
	planning etc. Paid attention to low wage, senior citizen and family with a large population	(heavy equipment, aeronautical construction and electronic products) • Improved the quality of workers; shift the emphasis, in maintaining equilibriums, toward changes in the "in value" variables (revenues policy, alert indicators	medium and small sized enterprises and made them specialized; established oversea industries • The preferential fields of the infrastructure industries were telecommunication, road, port etc.

Name of plan	Seventh plan	Eighth plan
Planning period	1976–1980 A political gap was formed due to the election of president. 1975–1976 temporary plan was introduced in 1974. The new leader confirmed the preparation of the seventh plan in October, 1974	The outline of the Eighth Five-Year Plan (1981–1985) was published in 1979. However, the outline had never been executed. This was due to the political change that Mitterrand of Socialist Party was in power in 1981
Planning background	• France and other developed countries were all affected by the energy crisis in the autumn of 1973. In 1974–1975, economy declined, and commodity price and	The outline was classified into three parts: the first part discusses the external dependence of French energy and provides the methods of reducing dependence; the

(continued)

Table 7.1 (continued)

Name of plan	Seventh plan	Eighth plan
	unemployment rate increased • The new national accounting system of 1976 was adopted; the product created during the sixth planning period (i.e. FiFi model) was continuously adopted	second part attempts to continue the policies of the seventh plan, brings French economy to the front of the international competitiveness and decreases the unemployment; the third part was the preferential actions of the six aspects
Objective	• Reestablish full employment by best use; reduced excessive inequality etc.; improved the quality of living environment; promoted research etc. • The annual average GDP growth was 5.5–6%; the foreign trade surplus of 1980 reached 10 billions of francs; there were 1.1 million of new jobs in 1980 at the end of the planning period; the inflation rate was decreased below 6%	
Action plan	• Listed 25 preferential project groups (PAP). Each project group was in the charge of the designated departments and was	

(continued)

Table 7.1 (continued)

Name of plan	Seventh plan	Eighth plan
	accompanied with the fund allocation. • The planning joint committee was established. The committee was composed of planning agency, president, prime minister as well as the ministers in direct relation with the plans. Aimed at strengthening the relations between the daily works of ministers and the long-term plans	

[a]*Source* French Planning Les Cahiers Francais No. 181 May–June 1977

[b]Referred to European Economic Community composed of Belgium, France, German Federal Republic, Italy, Luxembourg and Netherlands in 1957 according to Treaty of Rome

7.3.2.7 French Ninth Plan (1984–1988)

This part will provide a brief descriptions of official document of French Ninth Plan to have a relative overview of features of French planning, the pioneer of indicative planning for illustration and comparison purpose.

(1) *The Content of the Official Document*

Box 7.2 gives the content of the official document.

Box 7.2 Content of French Ninth Plan[7]

Strategy—modemize France (from the first planning law)	Page
The objective	1
Ways and means	6
Conditions	9
Introduction of second planning law	13
The programmes for priority implementation (PPIs)	
Encourage company to face future	19
Mobilize human resources and share initiativeness	35
Improve daily life in a spirit of solidarity	49
The main objectives quantified	
Incorporation into the 9th plan of the defence programming act	61
Develop cooperation	62
Incorporation into the 9th plan of the technological research and development policy and programming A	67
Energy	68
The principle sectors	
Agriculture, agriculture-food as well as forestry activities	71
Small business and self-employed	77
Transport	81
Telecommunication and broadcasting	87
Service	88
Housing	93
Planning contracts	
Planning contract of central government and enterprises	97
The purpose and coverage of plan contracts between central government and the Regions	101
Financial measures	105

(continued)

[7]Source: Press Service of the Commissariat General Du Plan (July 1984) Resume of the French Ninth Plan Translated by S. Dyson and L. Janssens.

(continued)

Financial coherence	119
Implementation of the ninth plan	125

(2) *First and Second Planning law of France*

France passed the first national planning law on July 13, 1982. The law laid down the strategic options and priorities for France's 9th Five Year plan. The Second Planning Act is an implementing act whose legal, financial and administrative provisions translate into practical measures the effort France must now deploy in its preparation for the future in according with the main thrust of the 9th Plan, namely to modernize France in conditions of social justice and greater democracy.

(3) *Objectives Contained Six Aspects*

The first aspect was the authority of France in the world. It emphasizes that France shall assume its international responsibility on all aspects. Although the international trade and relationship between the currency and finance was in a very bad status, the international responsibility of France shall promote the restoration of global economy, stabilizing the exchange rate, increasing assistance for the third world, consolidate European Community and maintaining international independence through the national defense programming act.

The second aspect was employment. It was anticipated that there would be a high rate of unemployment (8.5–9%) at the beginning of the ninth plan, and 725,000 labor forces would be added at the end of the planning period. It was necessary to exert a relatively high growth potential as far as possible based on the adaptability to the external balance and to expand employment. Meanwhile, the impact of the modernization of the production system on the labor market should be considered. The short-term supporting measurement employment should be provided, including energy saving policies, and the major public infrastructure projects to support the construction of housing, and promote the local government to create employment. In addition, the companies were required to strengthen training and reduce the weekly working hours to 35 h.

The third aspect was to pay attention to "the future of the young people". The planning text pointed out that: the economic crisis and the socioeconomic changes have resulted in the differences between the ambition of the youth and their education, and their lives of beginning entering adult hood. The education system must carry out complete reform on cultural construction, social function, as well as their relationships with the productive activities. The reform of each level should be provided with a double basis; the necessing staff training policies shall transform into the education methods and organizations; the independence, the openness and responsibility for the external world shall also be improved.

The forth aspect was to reduce inequality. The planning text had pointed out: "The modernization effort will not be accepted unless it is shared by all social categories". Measures for reducing inequality, including the measures for increasing the salaries of the lower paid and solving the uneven distribution of wealth.

The fifth aspect was to improve the social life. The content included the reform of community services, the modernization of public transportation as well as the improvement of urban management.

The sixth aspect was the regional balance. The measures in the five aspects had been proposed: establish a new future for the areas affected by the industry, including reconstruction of traditional activities, diversification, development of research projects, development of educating and training abilities etc.; further promoting the development of economy, society and culture of the overseas department and areas where nearly 1.7 million of French people lived generally in conditions of economic underdevelopment; continue to balance the regional development of France, including the West and South West areas as well as the rural vulnerable regions and mountainous areas; encourage to decentralize the development of the tertiary industry to rectify the disadvantage of concentrated far too much in the Paris region; the fifth aspect is to protect and improve the natural heritage, including environmental threat, mountainous area, coast, water body, etc.

(4) *Ways and Means*

There are four aspects:

The first aspect was to decentralize the administrative power and clarify responsibilities. The 9th Plan puts forward five ways in which decentralization and local development would be furthered: the first was to share the administrative power and clarify conditions; the second was to improve the financing after improving the decentralization of power to facilitate the coordination of local taxation and central government's help to local demands; the third was to ensure the central government financing for its contractual commitments including the Plan contracts between the central government and the regions; the fourth was to promote the development of regional economic information system; the fifth is to adjust the central government's territorial organization in line with decentralization.

The second aspect was to modernize industries and transform production systems. This text emphasized two aspects: the first aspect was to stop the declination of traditional industries and pointed out that machine tool industry, textile, steel, shipbuilding and chemical industries must be modernized. Emphasized that "In the process it becomes clear-as illustrated by the countries in the van of industrial progress-that there are no doomed sectors, only outdated technologies"; on the other hand, it pointed out that the leading industries till the end of this century (except for the energy departments) was mainly the electronic industry, be the center of modernization of all the industries. In addition, industries also include biotechnology, new communication technology, transportation, space technology and cultural industry. This text also emphasized the independence of state enterprise management, the measures for helping private enterprise financing, small enterprises and individual craftsmen.

The third aspect was training, research and innovation. The plan emphasized that great efforts must be made in the aspects of research, innovation, equipment replacement and new educational skills in order to make the producer adopt new technologies. The national research and development expenditure of France in 1985 was accounted for 2.5% of GNP and the industrial research and development spending was accounted for 1.5% of GNP.

The forth aspect was cultural renewal. The culture was an important part of economic and social development. The diversity of cultural needs would be promoted to make the relationships between artistic creation and society closer. The production capacity of audio and visual facilities will be strengthened under the challenges of the new communication technologies. The scientific and technical industrial culture would be developed so as to understand the influence of technical changes better. The external cultural relationships shall be redefined and the cultural exchange shall be adaptable to the modern media.

(5) *The Conditions*

There were conditions in two aspects: one aspect was to rebalance foreign trade; the other aspect was to control the application of resources strictly.

(6) *Priority for Implementing Projects*

French government's implementation of the ninth plan mainly focused on the implementation of 12 programmes for Priority Implementation (PPIs). There were years of budget arrangement as well as the support from non-budgetary finance and legal and administrative tools for each project.

These 12 programmes for Priority Implementation are shown in Table 7.2.

7.3.2.8 Develop Quantized Macroeconomic Management Tools

(1) *French Planning and Economic Departments Paid Attention to the Development of Quantitative Tools*

Complex macro economy in modern time requires quantitative tools to facilitate management. France has attached great importance to this work. The national statistics service department of France proposed the report for economic conditions of the state in December 1945. The planning department began "preparation of national balance sheet" in 1949. The national accounting system was completed in 1960. The new technologies for the macro measurement model had been introduced for the sixth planning period of France. The National Institute of Statistics and Economic Studies (INSEE) of France had developed the FIFI model. The model was based on the competitive economic theory of M. Raymond Courbis. The theory had classified economy into a sheltering and a non-sheltering department. The price was determined by the international market price of the non-sheltering department. The theories also included the investment theory based on profit or the "automatic financing" theory. It was also emphasized that the measures were adopted to make the non-sheltering department stimulate supply through reducing cost.

Table 7.2 12 Programmes for priority of implementation (PPIs)[a]

Category, number and theme of PPIs	Budget of 1984 billion of Francs	Total for the 9th plan billion of Francs
I. Encourage companies to look to the future		
PPI No. 1 Modernizing industry with the add of new technology	F. 3.2 billion	F. 19.9 billion
PPI No. 3 Encouraging research and innovation	F. 106 million	F. 64.6 billion
PPI No. 5 Reducing energy dependence	F. 2.8 billion	F. 15.5 billion
PPI No. 7 Selling more effectively in France and abroad	F. 4.7 billion	F. 27.7 billion
II. Mobilizing human resources and sharing initiative		
PPI No. 6 Acting positively for employment	F. 5.4 billion	F. 36.3 billion
PPI No. 2 Continuing the renovation of the education and training system for young people	F. 16.6 billion	F. 91.5 billion
PPI No. 4 Developing the communication Industry	F. 3.6 billion	F. 21.2 billion
PPI No. 9 Making decentralization succeed	F. 3.3 billion	F. 21 billion
III. Improving daily life in a spirit of solidarity		
PPI No. 10 Improving urban life	F. 2.6 billion	F. 15.1 billion
PPI No. 8 Providing an environment favorable to the family and the birthrate	F. 0.2 billion	F. 1.3 billion
PPI No. 11 Modernizing the health system and improving its management	F. 5.1 billion	F. 28.7 billion
PPI No. 12 Improving the quality of life also means improving justice and personal security for all	F. 1.3 billion	F. 8.1 billion

[a]*Note* This table is prepared based on p. 19, p. 25, p. 28, p. 31, p. 35, p. 40, p. 43, p. 46, p. 49, p. 52, p. 54, p. 57 of *Resume of the French Ninth Plan* (1984–1988)

FIFI was a global macro model, but did not include the financial departments. FIFI was a static model, which can give the prediction of the last year of the five year period. It was also a simulation model, which did not have the optimization function. The design of FIFI was to help the government make decisions in its controlled aspect.

INSEE had introduced the DMS (dynamic multi-department) model in 1976 to replace the FIFI model serving as the core model of the planning process. DMS was a dynamic medium-term model, which could provide forecast year after year. The economy was classified into 11 branches, and the industry was classified into departments such as consumption, intermediate input, and capital goods. DMS was a large simulation model. which had about 2000 equations. out of the 2000 equations 400 equations were behavior equations. DMS was an annual model of eight years.

The theoretical basis of DMS was different from that of FIFI. The model designer did not agree that FIFI should treat the international environment of the non-sheltering department as the factors of determining price. However, the price was determined by adding the unit labor cost to the intermediate product cost, and the salary of each department was determined by Phillips-Lipsey relations. The total consumption was determined by the savings function and was classified into 21 departments. Therefore, the expression of the behaviors of economic man—industry and family in DMS model was more sufficient than that of the FIFI model. One of the important characteristics of the DMS model was the inclusion of the productive force cycle (due to the lagging of output changes and employment effect). The cycle had an influence on wage cost and price.

(2) *Difficulties and Disputes In Carrying Out Model Work*

The FIFI model did not fit smoothly into the planning work of France. The introducing of the FIFI model into the sixth plan was regarded as an important progress. The FIFI model and its medium-term projection system are shown in Figs. 7.2 and 7.3. However, this new technology had brought new problems.

The complexity of this new technology as well as the new problems brought by it can be seen from Fig. 7.3.

Firstly, it cannot be understood. The social partners did not believe this new technology, so such as they cannot use it effectively. Secondly, the major trade organizations, FDT and CGT, refused to adopt this model.

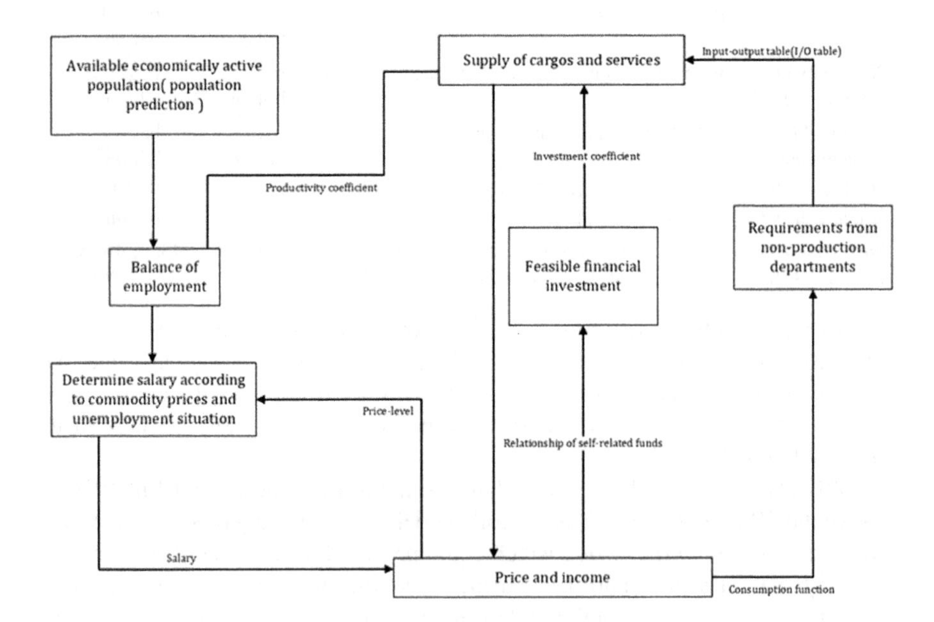

Fig. 7.2 FIFI model (version of the sixth plan) (*Source* UN 1975)

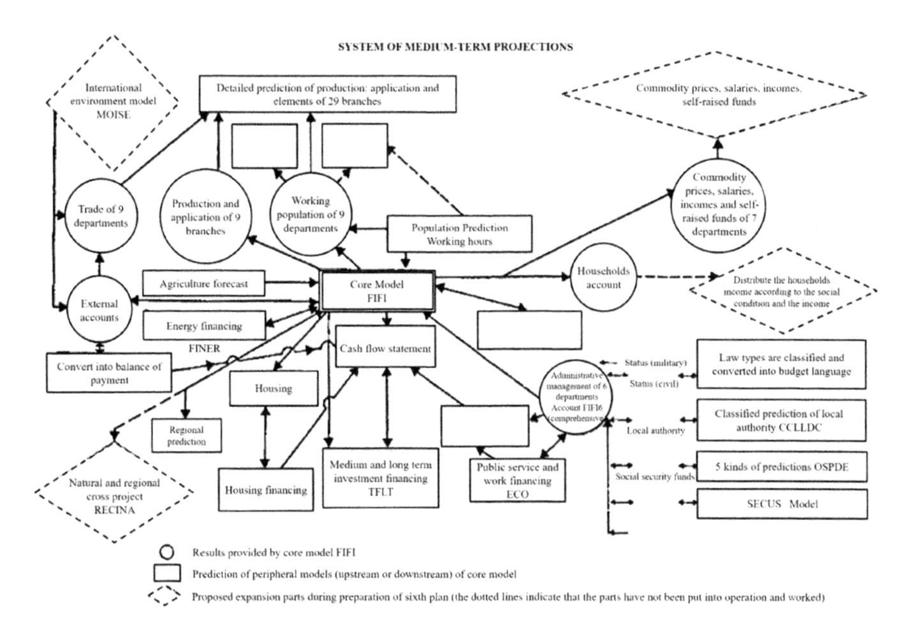

Fig. 7.3 Model system used for predicting the sixth plan of France (*Source* UN 1975)

The second problem was the mystery of the model. The model was a black box for many people. To use the model effectively, it shall be understood and the results thereof shall be illustrated accurately (extremely difficult).

The third problem of the model was that it cannot deal with some important structural problems effectively, such as the influence of income allocation or new technology.

The planning agency of France had published a report before the ninth plan and had proposed several aspects:

In recent years, the increasing of the complexity of model had concealed other problems in preparing plan to some extent.

The uncertainty in the model was not solved adequately.

The heed to recognize the internal and external uncertainty of French economy firstly and trace all kinds of possible values of uncertain variables so as to establish a detailed circumstance.

The hierarchical structure of the planning objectives shall be established.

One of the planning tasks was to deal with the different strategies of social partners in the structural changes. The tools used for preparing plans must be adapted to this task. The macro measurement model was not successful in this aspect.

(3) [8]*French Model Preparer's Views on Model*

During the planning process of France, there were academic controversies for the development and application of model. The opinions of critics related to the application of model for projection can be classified into three categories: the first category thought that the cost for preparing and maintaining model was too large; the second category thought that the complexity of model made the model become a black box, which was difficult in application; the third category thought that the measurement model was based on the previous data, where the results would be conserved and copy the previous results. The person in charge of the services of French INSEE project, Jean-Michel Charpin, thought it was expensive to prepare a high-quality and effective model, which usually required ten people and one year. The cost of maintenance of the model was not low, which included replacement of data as well as continuous research so as to find defects and redefine equations. The rationality of all these expensive cost depended on the intensive application of model. Therefore, as for the adoption of model and the organizations with great demands, the application of model for projection had actual meanings. However, as for the small organizations with small demands, the limited resources shall be used for understanding the previous changes and reflecting the future changes.

As for the complexity of model, Jean-Michel Charpin thought: many models used now were provided with publications for a detailed description. However, the problems were still not solved. As most of people worked relying on model did not have time or even have ability to understand and absorb plenty of complex data. Therefore, people would rather to adopt the models with simple and easy understanding. However, the complexity of economic phenomena did not permit experts to abandon professional knowledge.

Certain models were based on the previous data. This point can be discussed. People who was good at projection was not a fortune teller. People must research the past carefully, pay attention the structural and behavioral changes and distinguish from the previous circumstances.

From the perspectives of model preparation and user, Charpin thought that there were three advantages for the application of model: the first advantage was that the model can deal with plenty of interdependent relationships at the same time, including accounting, system, structure, logic, etc.; the second advantage was the calculation speed of model, which can rectify the secondary assumptions and test several changes during operation; the third advantage was that the model had a high storage level, and according to the equation, regulation and estimation of the historical changes, all the current production capacities formed by the previous investment decisions were important to the projection.

The conclusion was: the model shall be used rather than neglected, and the mechanical application of model shall be avoided.

---

[8]Source: Charpin (1984a, 1984b).

The work of four aspects requires to deal with the uncertainty.

The first aspect was to assemble the logical and different global circumstances relying on the projection of the oversea international organizations and the available tools. The second aspect was to finish all the works to draft the economic circumstances of France and propose more than two kinds of economic strategies to adapt to different international environment. The third aspect was to research the internal uncertainty by using other models or methods, and research on the uncertainty caused by the economic policies shall be given priority. The last aspect was to recognize the area of uncertainty compared with the known areas.

7.3.3 *Case Study of Indian Mid-Term Planning*

7.3.3.1 Overview

Although India is a developing country, it has a long history of preparing the national medium-term plan. India has become an independent republic country since 1947. During the administration of Nehru, India was inclined to adopt socialism and formulated Five-Year Plan for economic development by imitating the former Soviet Union. The Indian state planning commission was founded in March, 1950. In July, 1951, the committee proposed the First Five-Year Plan (1951–1956) draft of India, which was discussed in broad scope domestically. In December, 1951, Former Prime Minister Nehru submitted the draft to Parliament for approval. As most of Indian scholars and planning personnel have received western higher education, their planning theories and methods usually approach the frontier of Western social science. For example, the First Five-Year Plan had adopted the growth model of Harrod Domar. Although a part of countries had abandoned the preparation of mid-term national planning due to planning

Table 7.3 Indian Five-Year Plan and GDP growth[a]

Name of plan	Planning period	GDP annual growth rate %	
		Plan	Actual
First Five-Year Plan	1951–1956	2.1	3.6
Second Five-Year Plan	1956–1961	4.5	4
Third Five-Year Plan	1961–1966	5.6	2.2
Fourth Five-Year Plan	1969–1974	5.7	3.3
Fifth Five-Year Plan	1974–1979	4.4	5
Sixth Five-Year Plan	1980–1985	5.2	5.4
Seventh Five-Year Plan	1985–1990	5	5.7
Eighth Five-Year Plan	1992–1997	5.6	6.78
Ninth Five-Year Plan	1997–2002	6.5	5.35
Tenth Five-Year Plan	2002–2007	8.1	7.6
Eleventh Five-Year Plan	2007–2012	9	7.9
Twelfth Five-Year Plan	2012–2017	8	

[a]*Source* http://en.Wikipedia.org/wiki/FiveyearPlansofIndia

failure of two oil crisis in 1970s, and also a wave of strong belief of market forces, popularization of deregulation and privatization by the mean time. But India has continued its preparation of Five Year Plan which is shown in Table 7.3.

Although the new government led by prime minister Modi had scrapped the Planning Commission on August 13, 2014, and NITI (National Institution for Transforming India) was formed in Feb. 2015 to replace the former Planning Commission, but India to be one of the large developing country with a long history of Five Year Planning, its experience of mid-term planning is worthwhile to be described compared to the indicative planning experience of France, a developed country. Two documents of Indian Five Year Planning will be briefed below for study and comparative purpose, Sixth Five Year Plan (1980–1985) and Twelfth Five Year Plan (2012–2017) of India.

7.3.3.2 Briefing of Major Features of Indian Sixth Five Year Plan (1980–1985)

(1) *Overview*

Indian Sixth Five Year Plan has two documents. One is the main document with 463 pages, which presents the strategy, major programmes and policy issues involved in the pursuit of specified objectives. The other is *A Technical Note on The Sixth Plan of India (1980–85)*, with 238 pages, which presents, a description of the quantitative model that has been used in the formulation of the Six Five Year Plan of India. This kind of technical note was prepared on the approach to the Fifth Plan of India to meet the interest by economists and others in the detailed methodology of plan formulation.

(2) *The Main Document*

The main document is consisted of three parts:

Part I Presents the overall objectives, resources and policies of Sixth Five Year Plan. This part includes 8 chapters.

Chapter 1: Development Performance
This chapter gives a retrospective performance of previous five "Five Year Plan" in four aspects. Growth, modernization, self-reliance, and social justice. This part is supported with 12 annexures of statistical tables to show growth performance, resource mobilization, composition of GDP etc.

Chapter 2: Development Perceptive 1979–80 to 1994–95
This chapter gives choice of development strategy, alternative development scenario, perspective related to size of population, structure of demand, saving and investment, structure of output and long term sectoral profile and policies. A large number of social and economic indicators have been used in selecting the development strategy for the Sixth Plan and the decade thereafter;

Chapter 3: Objectives and Strategy of the Sixth Five Year Plan
Five subjects are explored in Chapter 3, objectives, macro-dimensions, sectoral growth profile, poverty and employment, and employment. Ten objective are identified in this plan. Marco-Dimensions are analyzed from aggregate saving and investment, rate and pattern of growth, pattern of public and private investment. Sectoral growth includes the study of supply and demand balance of cotton cloth and yarn, crude petroleum, manufactures mild steel, railway traffic and other five sectors. The pattern of growth is derived from a consistent system which is solved inter-temporally with an open economy model which consists of an 89 sector input-output model integrating the Sixth Plan period with the perspective period (1985–95) through a 14 sector investment planning model. Poverty estimation etc. are also done in *A Technical Note on the Sixth Plan of India (1980–85).*

Chapter 4: Public Sector Outlay
The Sixth Five Year Plan provides for a total output in the public sector of Rs. 97,500 crores[9]. This includes current outlay of 13,500 crores which is mainly the maintenance of service. The public sector outlay in the period of Sixth Plan will be Rs. 84,000 crores, an increase of 148% compared to the Fifth Plan. On the total of 97,500 crores, 47,250 crores are in the central sector, and the balance of Rs. 50,250 crores is in the States and Union Territories. This chapter includes three annexures of plan outlay to its 26 states, 90 other unclassified Union Territories and also public sector outlays.

Chapter 5: Resources for the Plan
This chapter gives an analysis of the financial resources likely to be generated and the funds needed for investment. These resources are derived from domestic and foreign resources. A detail analysis of gross domestic saving, investment and aggregate resources is done in this chapter through appropriate accounting model.

Chapter 6: Balance of Payments
The Indian economy, by and large, faced difficult balance of payment situation right from the Frist Plan, exports, imports, invisibles, foreign aid and borrowing, are analyzed in detail in this chapter, trade and payments policies are also identified.

Chapter 7: Policy Framework
The success of the Plan depends on many factors, but choice of correct policy framework is crucial requirements. Such a policy framework has to embrace mobilization of savings, supply and demand management, measures for improving the performance of the public sector, and adoption of specific steps which require for attainment of the objectives of the Plan. Also, such as in different sectors of the economy; accelerated growth in agriculture, pursuit of a well-coordinated energy policy to reduce dependence on imported oil, promotion of employment including self-employment, reduction of regional disparities, protection of environment and ecological balance, family planning and welfare and so on. An attempt is made in

[9]Note: Crore equal to 10,000,000 in Indian numbering, or 1 crore of Rupee means 10 million Rupees (Rs).

Chapter 7 to indicate the underlying assumptions of the Plan in respect of some of these matters and the kind of policy measure needed.

Growth with stability, mobilization of resources (fiscal policy, monetary and credit policies, income policy, pricing policy are discussed with this theme), public distribution system (infrastructure for public distribution, other improvements in the public distribution system, consumer protection and intelligence and information systems. Enforcement and training are discussed with the theme). Foreign trade policies, policies for accelerated growth in agriculture, policies toward industry in the private sector, regional disparities are described in detail.

Chapter 8: Plan Implementation, Monitoring and Evaluation

Successive five year Plans have stressed the importance of strengthening the implementation so that the projects and programmes in the Plans move according to time schedule and targets. Implementation needs to be supported by adequate monitoring and current and post-evaluation of major programmes to get lessons of experience to enable implements in the design of programmes. Topics of implementation of projects/schemes, Implementation of Rural Development and Employment Programme including employment, rural development, agriculture, irrigation, small and village industries, special component plan for scheduled castes, tribal sub-plan, antipoverty programmes, district administration monitoring of implementation, machinery for planning, people' s involvement in planning and implementation, personnel policies, programme evaluation; information system, data base for planning; training and national efficiency drive are explored.

Table 7.4 Theme of 20 programmes and its outlay

Serial No.	Theme of programme	Outlay Rs. crores
1	Agriculture and Allied Sectors	5693
2	Irrigation, Command Area Development and Flood Control	856.27
3	Rural Development and Co-operation	4723.07
4	Village and Small Industries	857.05
5	Manpower and Employment	–
6	Minimum Needs Programme	5807
7	Energy	26,465
8	Industry and Minerals	20407.05
9	Transport	12411.97
10	Communications, Information and Broadcasting	272.34
11	Science and Technology	3367.19
12	Environment	40
13	Education	2523.74
14	Health, Family Planning and Nutrition	1821.05
15	Housing, Urban Development and Water Supply	6410.42
16	Labor and Labor Welfare	161.9
17	Hill Area Development	900
18	Development of Backward Classes	2030.30
19	Women and Development	–
20	Social Welfare	272

Part II Programmes of Development

Part II includes Chapters 9–28, which have a detail description of objective and strategy, inputs and public policy, etc. Of 20 programmes and their outlay are listed in Table 7.4.

Part III This part includes "Foreword" by the prime Minister Indira Gandhi, prefaced by deputy chairman of planning commission Narayan Datt Tiwari and Appendix: *Sixth Five Year Plan 1980–85 A Framework.* This appendix includes a preface by Narayan Datt Tiwari and an attachment of the draft Plan-frames. The latter is consisted of six parts (127 items of description), they are introduction, objectives of the sixth plan, the growth rate, financial resources of the sixth plan, programme thrusts during the sixth plan, summary and issues. The draft Plan frames is a briefing of essential points of the main planning document. The terminology 'system approach' is used by Tiwari in item 56 'System approach to agricultural development' of part V of the Plan-frames.

The first paragraph foreword by the Prime Minister Indira Gandhi is quoted below to provide political context of India in the preparation of Indian Sixth Five Year Plan.

> Progress in a country of India's size and diversity depends on the participation and full involvement of all sections of the people. This is possible only in democracy. But for democracy to have meaning in our circumstances, it must be supported by socialism which promises economic justice and secularism which gives social equality. This is the frame for our planning. (Indira Gandhi, p.iii).

(3) *Contents of A Technical Note of The Sixth Plan of India (1980–85)*

(i) The contents of this "Technical Note" includes eight chapters, a preface, six annexures and tables, the tables have a high share of volume of contents (from pp. 95–237).

(ii) These eight chapters are: Introduction; Structure of the model; The sub-models; The core model: input-output, investment, private consumption, import and perspective; The core model: financial resource block; The core model: employment block; The material balance subsystem; Alternative scenarios and sensitivity analysis.

(iii) These six annexures are: Estimation of incremental capital output ratio for Sixth Plan; Mathematical formulation of private consumption blocks; Poverty estimates in the Six Plan; The model of the private sector investment and savings; Adjustment for changes in the terms of trade; Sector classification of Input-Output Table (89 sectors).

There are three salient features of the Sixth Plan Model[10]

First: A system of supply equations which is an extended and modified version of the Harrod Domar equation. The Hrrod Domar equation of the 1st and 2nd Plans can be presented in a simplified form shown in Eqs. 7.1 and 7.2.

[10]Note: Purpose of this section is simply to illustrate that modelling technique had been applied early in the First Five Year Plan of India and its successive improvement.

$$V_T = V_O + \sum_{i=0}^{4} I_t^* \, ICOR^{-1} \tag{7.1}$$

$$I_t = V_t(1 - B) \tag{7.2}$$

where

T	5th year (Terminal Period)
V_O	Value added, base year
I	New Investment
ICOR	Incremental capital-value added ratio
B	Average propensity to consume.

The above equations are modified in the sixth plan.

Second: A system of demand equations which is again an extended version of Leontief's input output system. The Leontief model used in the third, fourth and fifth Plan can be presented in a very simple form shown in Eqs. 7.3–7.5.

$$X_T = (I - A)^{-1}\{C + I + E - M + PC\}: \text{The third Plan} \tag{7.3}$$

$$X_T = B^{-1}\{C + I + E + PC\}: \text{The Fourth Plan} \tag{7.4}$$

$$X_T = B^{*-1}\{I + E + PC\}: \text{The Fifth Plan} \tag{7.5}$$

where

$(I - A)^{-1}$	Leontief inverse
B^{-1}	Extended Leontief inverse with import endogenized
B^{*-1}	Further extended Leontief inverse, with import and consumption both endogenized
PC	Public consumption

Lastly: A set of inequality relations.

7.3.3.3 Briefing of Indian Twelfth Five Year Plan (2012–2017)

(1) *Overview*

The theme of Indian Twelfth Five-Year Plan is "Faster, more inclusive and sustainable growth". This planning document has three volumes and was published by State Development Planning Commission Sage Publishing Company in 2013. Former Prime Minister Singh was also the director of Planning Commission in the period of preparation of this plan.

Contents of volume one include the foreword given by the prime minister and the preface given by the deputy chairman of Planning Commission. There are eleven parts in volume one. They are respectively Twelfth Plan: An overview;

Macroeconomic framework; Financing the plan; Sustainable development; Water; Land issues; Environment, forestry and wildlife; Science and technology; Innovation; Governance; Regional equality.

The second volume deals with economic sectors. There are eight parts, which are respectively Agriculture; Industry; Energy; Transportation; Communication; Rural development; Urban development; Other priority sectors.

The third volume deals with: social sector. There are five parts, which are respectively Health; Education; Employment and Skill development; Women's agency and Children's rights; Social inclusion.

(2) *Briefing of Forward by the Prime Minister Manmohan Singh*

During the Twelfth Five-Year Plan, there were both challenges and opportunities present at the same time. Because the plan began after the second global financial crisis, and the sovereign debt crisis occurred in Euro zone at the last year of "Eleventh Five-Year Plan". The crisis affected all countries including India. The estimated economic growth rate of the Fifth year of 12th Five Year Plan was only 5%.

This brought challenges for recovering economy to a higher growth path. Corrective measures must be adopted. Therefore, the primary task of "Twelfth Five-Year Plan" is to bring economy back to the path of rapid growth, and to guarantee "inclusive" and "sustainable" growth at the same time.

The 12th Plan has proposed a two pronged strategy focusing on rebalancing macro economy under control and also pushing for structural reforms in many area to maintain medium growth.

There are two aspects in rebalancing macro economy. The first aspect is to control financial deficit, and the financial deficit of central government at the end of planned period will not exceed 3%. The second is to correct the high current account deficit, both flow of foreign direct investment (FDI) and foreign institution investment (FII) should be encouraged. Caps on the permeated level in a number of sectors should be relaxed and a broad review of policies is being undertaken to facilitate easier flow of foreign investment.

The government has also acted on the supply side to tackle implementation problem holding up large infrastructure projects measures have been devised to revive the pace of investment.

The target growth rate for 2013–2014 and 2016–2017 in Twelfth Five-Year Plan is set as 8%.

Due to the low economic growth in the initial period of implementation of 12th Five Year Plan, higher economic growth rate in the later period of the plan will be expected. In order to achieve the target of 12th Five Year Plan, policy actions on many fronts must be adopted both by the central and state government. For instance, there are needs to accelerate the growth in agriculture, to address the challenge of managing the infrastructure sectors and to ensure its expansion to support growth: To face up the enormous challenges posed by urbanization. Also there is a need of much faster growth in manufacturing to provide employment to the young and increasingly educated population, which has high expectation and aspirations.

Growth must not only be rapid and it must be inclusive and sustainable. The benefits of growth must reached the scheduled castes (SCs)[11], scheduled tributes (STs)[12] and other disadvantaged groups in Indian society. These groups must get a fair share of the benefits of growth and must have a stake in society. Further, a growth process should be consistent with protection of environment.

One of the problems of the India plans in the past is that they have focused on the outlining an attractive future without, enough focus on what is needed to achieve the goal and the consequence of failing in this regard. Recognizing that outcome will be the result of actions. The 12th Plan for the first time, has resorted to scenarios to indicate the implications of different types of behaviors.

The objectives of the 12th Plan are ambitions and achieve them will be difficult. We have to prove that vigorously competitive politics in a democracy can achieve a sufficient consensus to implement policies we choose.

If India can continue the good performance for the next twenty years, we will have ensured the re-emergence of India as major economic player in the world, who is also one committed to its democratic secular and pluralistic tradition. It will inspire millions around the world, especially in Asia, Africa and Latin America, to seek their own destiny within the framework of a plural, secular and liberal democracy (pp. v–vii).

(3) *Briefing of Preface by Deputy Chairman of Planning Commission, Montek Sing Ahluwalia*

(i) *Basic Nature of National Planning*

National planning is a process of setting national targets, and preparing programmes and policies that will help achieve those targets. The policy and programmes must be consistent with each other, ensure optional use of national financial and real resource, and be based on an understanding of the response of the economy to this intervention.

(ii) *The Mulli-dimensional Nature of Plan Targets*

Indian plans always emphasize that the rapid growth of GDP is a necessary condition, but not a sufficient condition for improving the living standard of all the people.

Twelfth Five-Year Plan always complies with this tradition. Rapid growth can provide broad foundation for improving the living, and the rapid growth is considered as a necessary condition. However, it is recognized that faster growth is not

[11]Note: SCs and STs are variously officially designated of historically disadvantaged indigenous people in India.

[12]Note: SCs and STs are variously officially designated of historically disadvantaged indigenous people in India.

a sufficient condition if a growth process which may not be inclusive to ensure a spread of benefits to mass of population. Therefore, 12th Five-Year Plan accelerates the growth of GDP and also the growth of agriculture. The growth rate of agriculture for 12th Five-Year Plan is set as 4%.

As an agricultural population leaves rural areas, additional population will enter the labor market. Therefore, an inclusive growth must create employment. However, the highly capital intensive sectors (such as petrochemicals steel etc.) and depends on high end technical personnel, i.e. software development, information technology. These sectors cannot provide sufficient employment opportunities for plenty of personnel with medium technical level. Therefore, Twelfth Five-Year Plan required a rapid growth of the manufacturing industry, especially small and micro enterprises, in order to absorb the labor force currently employed in low productivity occupations.

Although the process bringing employment growth can provide a broad foundation for income growth, the process cannot ignore other important objectives, such as access to education, health care, sanitation as well as clean drinking water. They are not only a part of welfare, but also ensure a breathy and productive labor force. 12th Five-Year Plan emphasizes expanding the access to these services and views it as a critical role of government in the development process. The private sectors is now capable of investing in many economic fields that makes the investment from the public sectors unnecessary, and the government must take responsibilities for providing education, health care and primary service to the majority of people.

The inclusive growth also calls for pro-active intervention to bridge the many "divides" which segment our society. The plan return its tradition of reducing poverty. However, the Plan shall also focus on productive employment, upliftment of specific disadvantaged groups, ensure balanced development of all regions, and eliminating poverty. The multi-dimensional objectives are reflected in the adaptation of 25 monitoring objectives of 12th Five-Year Plan, wherein the GDP growth rate is only one of the monitoring objectives.

(iii) *Participation in Planning*

The process of preparing plan target and defining plan strategy is no longer a pure technical process, and it shall consider the extensive desire of plenty of stakeholders of participating in plan preparation. The planning commission has consulted central ministries and state governments as well as sector experts, economist, sociologist, scientist, civil society organizations during the preparation of plans. During the preparation of this plan, 146 work groups were established under the chairmanship of Secretary of the Ministry concerned, and included sector experts from within and outside the government.

The regional plan also adopts the same participating methods with the department. About 900 civil society organizations have participated in the preparation of the regional plan.

(iv) *The Role of Programmes*

The preparation of traditional plans usually includes the combination of many government programmes in order to achieve a certain objectives of the plan. However, the function of the policies for realizing the planned targets is usually not subjected to the sufficient consideration. This will be discussed in the next section.

Twelfth Five-Year Plan relies on the broad government programmes in order to realize an inclusive and sustainable growth. The programmes involve i.e. health, education, clean drinking water, sanitation, urban and rural critical infrastructures as well as the support for the living of vulnerable groups. Special programmes are also provided for the historically disadvantaged section of our population.

The investment in the public sector from central and state governments is about 12% of GDP, wherein a half belongs to the capital expenditure of investment and the other part belongs to the salary for providing services or belongs to the transfer payment. The economists usually focus on the investment of serving as the key driving force for growth, but shall also recognize that the access to services, such as health, education, children's nutrition etc., can make a key contribution to the medium-term growth. It is obvious that non-investment expenditures in the sectors are critical to achieve access. Therefore, the results of the project are subject to strict, independent and effective evaluation to determine the causes for any failure to achieve the expected results.

(v) *The Role of Policy Restructuring*

The content of the policies shall pay more attention to: with the increasing of the scale of the private sector in economy, decisions taken by individuals and other entities determined many critical economic outcomes. These decisions are therefore important. Especially since private investment by farmers, unorganized enterprises and the corporate sector, accounts for 75% of total investment in the economy.

Prime Minister has emphasized the importance of a sound macroeconomic framework in his Foreword. This essentially means having a reasonable fiscal deficit, a financeable current account deficit, a moderate rate of inflation and high rates of domestic savings. The 12th Plan has outlined a macro economic strategy. Except for the macroeconomic policies, the policies structuring must be carried out in other sectors in order to achieve the planned targets. The planning text has listed many ambitious agenda for policy change. The diversified development of agriculture as well as the sales of agricultural products by private sectors shall be supported. In thc field of energy it includes policies related to energy pricing and, greater involvement of the private sectors in the exploitation of primary energy. The policies of the industrial sectors are aimed at improving business, but should also encourage a flow of both risk capital and debt into the medium and small enterprises. Rationalization of tax policies is an another important area. The Plan emphasizes the needs to design policies regarding the efficient use of water through a combination of regulation and appropriate pricing. The policy also depends upon public investment for effective water conservation. In economic life, the policies for effective management of urbanization become more important day by day.

As the total resources available for government are limited. The sectors such as education and health have a great demand for resources. Therefore private resources shall be mobilized to realize the public objectives. This is especially important for the development of infrastructure, and the new policies shall be adopted. The new policies about Public Private Partnership (PPP) had been adopted for developing infrastructure during Eleventh Five-Year Plan.

Traditionally, infrastructure used to be created by the government. However, most countries of the world have increasingly experimented the policies about Public Private Partnership revenue, especially when the users would like to pay "a user charge" that will turn into a great income source. It needs to be emphasized that PPP strategy shall not be adopted blindly since the private investors are impossible to invest in the remote or less-developed areas of a country. These areas are still rely on the public investment.

Both the central government and the local government have ambitious programmes of developing infrastructure through PPP. Among the emerging market countries, India has the most PPP projects. Therefore, unexpected problems may appear in the course of the implementation of the projects. Both the central and the state governments shall explore the mechanism for supervising PPP project closely to ensure that they can provide services conforming to standard and recognize and solve the problems during the implementation of project in time.

(vi) *Implementation is the Key*

The success of any plan depends heavily on the quality of implementation. This is just the weakest in our planning process. It is often said that our plans are very good, but implementation is poor. This is actually a contraction. A plan can be regarded as a good plan only when it is based on the realistic assessment on the implementation with concrete proposals to increase implementation capacity as part of the strategy.

As for the projected status of the plan, the projected funding may not be sufficient on one hand, and the available funding is not put to good use on the other hand. The shortage of fund is usually unavoidable, because the available resources are usually calculated based on the income growth, which relies on the overall growth of economy. If the growth cannot be reached at the expected targets, the resources may be less than the predicted level. The solution to the above problems is better prioritization so that a shortage of resources will not result in a proportional funding shortage of all of the projects and eliminate the projects with the weakest production. A good sequence of priority must rely on the independent assessment on periodicity. The function of assessment shall be greatly strengthened in our planning process. The planning commission has established a new mechanism, i.e. establishing an independent assessment office that is under the charge of the members of planning commission.

Another cause for the poor implementation is lacking of executive capacity at the operation level. In Twelfth Five-Year Plan, therefore, provides for a portion of the funds in many programmes to be used to build implementation capacity, a new centrally sponsored scheme.

A major new initiative in Twelfth Five-Year Plan is to improve the implementation of the centrally sponsored schemes to rationalize the number of the CSS reducing them from 142 to 66 and permitting greater flexibility in the guideline. We have recognized that the national guideline of "one size fit all" cannot consider the different characteristics of various states. Therefore, differential treatment shall be adopted, and a new system shall be introduced. Firstly, each state can propose modifications in their national guidelines to adapt to the characteristics of the state. Secondly, each state should be allowed full flexibility for ten percent of its allocation under each scheme, which can be used for projects. The only requirement should be that the project must be within the brode objective of the scheme.

The midterm review of the 12th Plan should focus specifically on the success achievement in the dimension both at the level of the Center and each State.

(vii) *Scenario Analysis*

The important innovation of Twelfth Five-Year Plan is the recourse to scenario analysis. The plan no longer adopts a group of fixed objectives but the Plan outlines three scenarios. A more detail description of these scenarios will be given in (4) Twelfth Plan: An Overview.

(4) *Abstract from Overview*

Chapter one of the 12th Plan is an overview of the Plan, which gives the basic rationales of the Plan and the key area of intervention. It explains the need for faster growth, and the meaning of inclusiveness such as inclusiveness as poverty reduction, as group equality, inclusiveness as regional balance, as empowerment and inclusiveness through employment programme. The Plan also includes a brief discussion of environment sustainability with its relationship to energy, forest, transport policies and greenhouse gas emissions, the stated objective of reducing the emission of GDP by 20–25% between 2005 and 2020.

Developing capabilities is also briefed, which includes development of human capability (life and longevity, education, skill development, nutrition, health, drinking water and sanitation and enhancing human capability through information technology), development of institutional capabilities (implementation capability, delivery of public service and regulatory institutions) and development of infrastructure (power, telecommunication, road transport, railway, ports, financing infrastructure, the reach of banking and insurance, science and technology).

The other important topics reviewed include managing the natural resource and the environment, engaging with the external world and key policy initiatives.

Two aspects will be directly quoted from chapter one of the Plan.

(i) Alternative Scenarios

The Planning Commission has undertaken a systematic process of 'scenario planning' based on diverse views and disciplines to understand the interplay of the principal forces, internal and external, shaping India's progress. This analysis suggests three alternative scenarios of how India's economy might develop titled, 'Strong Inclusive Growth', 'Insufficient Action' and 'Policy Logjam'.

The first scenario 'Strong Inclusive Growth', describes the conditions that will emerge if a well-designed strategy is implemented, intervening at the key leverage points in the system. This in effect is the scenario underpinning the Twelfth Plan growth projections of 8%, starting from below 6% in the first year to reach 9 per cent in the last two years. The second scenario 'Insufficient Action' describes the consequences of halfhearted action in which the direction of policy is endorsed, but sufficient action is not taken. The growth in this scenario declines to around 6–6.5%. The third scenario 'Policy Logjam', projects the consequences of Policy Inaction persisting too long. The growth rate in this scenario can drift down to 5–5.5%.

Box 7.3 25 core indicators of monitorable targets of the Plan

Economic Growth

1. Real GDP Growth Rate of 8.0%.
2. Agriculture Growth Rate of 4.0%.
3. Manufacturing Growth Rate of 10.0%.
4. Every State must have an average growth rate in the Twelfth Plan preferably higher than that achieved in the Eleventh Plan.

Poverty and Employment

5. Head-count ratio of consumption poverty to be reduced by 10% points over the preceding estimates by the end of Twelfth Five Year Plan.
6. Generate 50 million new work opportunities in the non-farm sector and provide skill certification to equivalent numbers during the Twelfth Five Year Plan.

Education

7. Mean Years of Schooling to increase to seven years by the end of Twelfth Five Year Plan.
8. Enhance access to higher education by creating two million additional seats for each age cohort aligned to the skill needs of the economy.
9. Eliminate gender and social gap in school enrolment (that is, between girls and boys, and between SCs, STs, Muslims and the rest of the population) by the end of Twelfth Five Year Plan.

Health

10. Reduce IMR to 25 and MMR to 1 per 1000 live births, and improve Child Sex Ratio (0–6 years) to 950 by the end of the Twelfth Five Year Plan.
11. Reduce Total Fertility Rate to 2.1 by the end of Twelfth Five Year Plan.
12. Reduce under-nutrition among children aged 0–3 years to half of the NFHS-3 levels by the end of Twelfth Five Year Plan.

Infrastructure, Including Rural Infrastructure

13. Increase investment in infrastructure as a percentage of GDP to 9% by the end of Twelfth Five Year Plan.
14. Increase the Gross Irrigated Area from 90 million hectare to 103 million hectare by the end of Twelfth Five Year Plan.
15. Provide electricity to all villages and reduce AT&C losses to 20 per cent by the end of Twelfth Five Year Plan.
16. Connect all villages with all-weather roads by the end of Twelfth Five Year Plan.
17. Upgrade national and state highways to the minimum two-lane standard by the end of Twelfth Five Year Plan.
18. Complete Eastern and Western Dedicated Freight Corridors by the end of Twelfth Five Year Plan.
19. Increase rural tele-density to 70% by the end of Twelfth Five Year Plan.
20. Ensure 50% of rural population has access to 40 lpcd piped drinking water supply, and 50% gram panchayats achieve Nirmal Gram Status by the end of Twelfth Five Year Plan.

Environment and Sustainability

21. Increase green cover (as measured by satellite imagery) by 1 million hectare every year during the Twelfth Five Year Plan.
22. Add 30,000 MW of renewable energy capacity in the Twelfth Plan.
23. Reduce emission intensity of GDP in line with the target of 20–25% reduction over 2005 levels by 2020.

Service Delivery

24. Provide access to banking services to 90% Indian households by the end of Twelfth Five Year Plan.
25. Major subsidies and welfare related beneficiary payments to be shifted to a direct cash transfer.

(ii) Some Comments

Although India takes the lead in the planning technique among developing counties, and due to its effort of development since 1950s, its GDP (in terms of PPP), ranks number 5 among the world currently, only next to Japan. According to some projections, Indian economy will be as big or greater than the USA in the future.

A study in 2013 by *Corbsidge, S, Harrise, J., and Jeffrey, C.* found the following:

> India today is a place of great change, but it is hard at the same time not to be struck by how much has not changed. The inequalities of Indian society continue to erdre, and by the continuing prevalence of great poverty, in spite of such successful economic growth (as we discuss in Chapters 3 and 4). The persistence of the hierarchical values of Indian's ancient régime and of patriarchy, are reflected most clearly in the disabilities-In spite of years of affirmative action of the Scheduled Castes and the Scheduled Tribes who together make up more than a quarter of the population (see Chapter 12), and in the continuing deep disadvantage of Indian women (Chapter 13). (p. 304).

It remains to be seen whether these enduring characteristics of India's society and politics will come to limit India's economic growth (p. 304).

7.4 Strategic Planning and Scenarios Method

7.4.1 Introduction

The failure of mid-term planning in projection of two oil crisis in 1970s, and the wave of economic thoughts of liberalization of trade and capital flows around 1980–1990 had caused a part of developed countries to abandon mid-term planning such as France. Japan although had prepared at least eleven mid-term planning post WWII, had also started preparation of strategic planning. The Cabinet had a decision to issue a document *The New Growth Strategy-Blueprint for revitalizing Japan* on June 18, 2010 with provisional translation in English around 70 pages. This document had identified seven growth areas, "Green innovation"; "Life innovation", "the Asian economy" and "Tourism and the regions"; "Science and technology and information and communication technology"; "Employment and human resources"; and the "Financial sector". Targets of these seven areas to be reached by 2020 are described in detail.

Academically, there are differences between strategic planning and long range planning. According to Bryson and Einsweiler (1988a, 1988b). The differences are in the four fundamental ways: "strategic planning typically relies more on the identification and resolution of issues while long range planning focuses more on the specification of goals and objectives and their translation into current budgets and working programs"; "Strategic planning emphasizes assessment of the organization's internal and external environments far more than long-range planning does"; "Strategic planners are more likely than long-range planners to summon forth an idealized version of the organization—the organization's "Vision of success" while long-range plans typically are linear extrapolations"; "Strategic planning is much more action oriented." (p. 4). But Country's planning in real world, it seems that these two types of planning cannot be distinguished from each other.

Although U.S.A. has not practiced national planning, but its National Intelligence Agency had studied global trend every five years since 2010 through scenario method. In the following Sect. 7.4.2, a briefing of *Europe 2020: A New Economic Strategy* will be presented to be a case study of strategic planning and a summary of four scenarios of four documents of study of global trends (2015–2030) will be presented in Sect. 7.4.2 to be a case study of scenarios method.

7.4.2 Case Study of Europe 2020: A New Economic Strategy

7.4.2.1 Introduction

This is an official document with 32 pages. The contents include a Forward by José Manuel Barroso, an executive summary, six parts and two annexes. The six parts are: A moment of transformation; Smart, sustainable and inclusive growth; Missing links and bottlenecks; Exit from the crisis: First steps towards 2020; Delivering results, strong governance; Decision for the European Council. In fact, annex 1—Europe 2020: An Overview has given all core element of the document: The five headline targets, the three priorities of growth i.e. smart growth, sustainable growth and inclusive growth, and the related flagship unitive to achieve certain priority. For example, the three flagship initiative such as "Innovation Union", "Youth on the move" and "Digital society" are for the achievement of Smart Growth. This annex 1 together with the executive summary can explain the major concepts of Europe 2020. Therefore, simply the executive summary of the original document is briefed in 2–5 in the following and its annex 1 is presented in Sect. 7.4.2.5 in the case study. This document has covered all essential elements of a strategic planning as well as long range planning.

7.4.2.2 Executive Summary of Europe 2020: A New Economic Strategy

Europe faces the moment of transformation. The 2008 crisis has wiped out years of economic and social progress and exposed structural weakness in the Europe's economy. On the other hand, the entire world is making progress rapidly. However, various long-term challenges such as globalization, pressures on resources, aging intensify. European Union must now take charge of its future.

If Europe can act collectively as a Union, it can succeed. European Union needs a strategy to exit the crisis and to become more powerful. This strategy can turn European Union into a smart, sustainable and inclusive economy, delivering high levels of employment, productivity and social cohesion. *Europe 2020* sets out a vision of Europe's social market economy for the 21st century.

Europe 2020 puts forward three mutually reinforcing priorities:

Smart growth: developing an economy based on knowledge and innovation.

Sustainable growth: promoting a more resource efficient, greener and more competitive economy.

Inclusive growth: fostering a high employment economy delivering social and territorial cohesion.

European Union needs to define where it wants to be by 2020. Therefore, the Commission proposes the following EU headline targets following major objectives:

(1) 75% of the population aged 20–64 should be employed.
(2) 3% of the European Union's GDP should be invested in research and development.
(3) The climate/ energy targets of "20/20/20" should be met ("20/20/20" refers that the discharge of greenhouse gases shall be reduced at least 20% compared with that of 1990. If conditions are right, it should be reduced 30%. The percentage of renewable energy sources shall reach 20%. The energy efficiency shall be increased by 20%.).
(4) Share of early school leavers should be under 10% and at least 40% of the younger generation should have a tertiary degree.
(5) 20 million less people should be at risk of poverty.

The Commission suggests that EU goals are translate into national targets and trajectories.

The Commission is putting forward seven flagship initiative to catalyze progress under each priority theme. These seven flagship initiatives are "Innovation Union", "Youth on the mover", "A digital agenda for Europe", "Resource efficient Europe", "An Industrial policy for the globalization era", "An agenda for new skill and jobs" and "European platform against poverty" respectively.

7.4.2.3 A Moment of Transformation

Five aspects have been explained in transformation: the crisis has wiped out recent progress; Europe's structural weaknesses have been exposed; global challenges intensify; Europe must act to avoid decline and Europe can succeed. The later four aspects are briefed as follows:

(1) *Europe's Structural Weaknesses have been Exposed*

Before the crisis, there were many areas where Europe was not progressing fast enough relative to the rest of the world. The structural weaknesses in the following three aspects need to be noted.

(i) Europe's average growth rate has been lower than that of its main economic partners, which is largely due to a productivity gap that has widened over the last decade. Much of this is due to differences in business structures. And the structural weaknesses of it include lower levels of investment in R&D and innovation, insufficient use of information and communications technologies, reluctance in some parts of the societies to embrace innovation, barriers to market access and a less dynamic business environment.

(ii) In spite of progress, Europe's employment rates—at 69% on average for those aged 20–64—are still significantly lower than in other parts of the world. Only 63% of women are in work compared to 76% of men. Only 46% of older workers (55–64) are employed compared to over 62% in the US and Japan. Moreover, on average Europeans work 10% fewer hours than their US or Japanese counterparts.

(iii) Demographic ageing is accelerating. As the baby-boom generation retires, the EU's active population will start to shrink as from 2013/2014. The number of people aged over 60 is now increasing twice as fast as it did before 2007—by about two million every year compared to one million previously. The combination of a smaller working population and a higher share of retired people will place additional strains on the welfare system of EU.

(2) *Global Challenges Intensify*

While Europe needs to address its own structural weakness, the world is moving fast and will be very different by the end of the coming decade:

(i) Europe's economies are increasingly interlinked with global economy. Europe will continue to benefit from being one of the most open economies in the world but competition from developed and emerging economies is intensifying. Countries such as China and India are investing heavily in research and technology in order to move their industries up the value chain and "leapfrog" into the global economy. This puts pressure on some sectors of Europe's economy to remain competitive, but every threat is also an opportunity. As these countries develop, new markets will open up for many European companies.

(ii) Global finance still needs fixing. The availability of easy credit, short-termism and excessive risk-taking in financial markets around the world fuelled speculative behavior, giving rise to bubble-driven growth and important imbalances. Europe is engaged in finding global solutions to bring about an efficient and sustainable financial system.

(iii) Climate and resource challenges require drastic action. Strong dependence of fossil fuels such as oil and inefficient use of resources expose the consumers and businesses to harmful and costly price shocks, threatening the economic security and contributing to climate changes. The expansion of the world population from 6 to 9 billion will intensify global competition for natural resources, and put pressure on the environment. The EU must continue its outreach to other parts of the world in pursuit of a worldwide solution to the problems of climate change at the same time as it implements agreed climate and energy strategy across the territory of the Union.

(3) *Europe must Act to Avoid Decline*

There are several lessons we can learn from this crisis:

(i) The 27 EU economies are highly interdependent: the crisis underscored the close links and spillovers between European countries, particularly in the euro area. Reforms, or the lack of them, in one country affect the performance of all others, as recent events have shown; moreover, the crisis and severe constraints in public spending have made it more difficult for some Member States to provide sufficient funding for the basic infrastructure they need in areas such as transport and energy not only to develop their own economies but also help them participate fully in the internal market.

(ii) Coordination within the EU works: the response to the crisis showed that if the EU acts together, it is significantly more effect. It proves opportunity of taking common action to stabilize the banking system and through the adoption of a European Economic Recovery Plan. In a global world, no single country can effectively address the challenges by acting alone.

(iii) The EU adds value on the global scene. The EU will influence global policy decisions only if it acts jointly. Stronger external representation will need to go hand in hand with stronger internal coordination.

7.4.2.4 Three Scenarios for Europe by 2020

Three scenarios for Europe by 2020 are projected which is shown in Box 7.4.

Box 7.4 [13]Three Scenarios for Europe by 2020

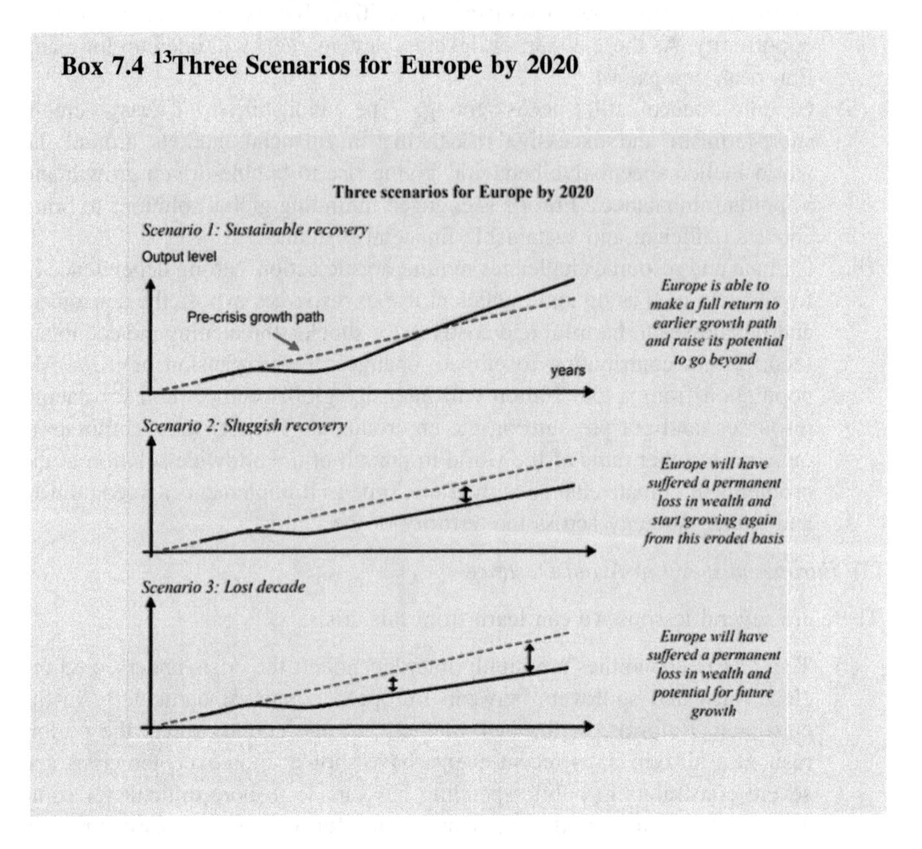

7.4.2.5 Europe Can Succeed

Europe has many strengths: it can count on the talent and creativity of its people, a strong industrial base, a vibrant service sector, a thriving and high-quality agricultural sector, strong maritime tradition, its single market and common currency, its position as the world's biggest trading bloc and leading destination for foreign direct investment. It can also count on its strong values, democratic institutions, its consideration for economic, social and territorial cohesion and solidarity, its respect

[13]Source: EC (2010), Europe 2020 COM (2010) 202, Brussels 3.3.2010. (p. 7).

for the environment, its cultural diversity, respect for gender equality—just to name a few. Many of its Member States are amongst the most innovative and developed economies in the world. But the best chance for Europe to succeed is if it acts collectively—as a Union.

When confronted with major events in the past, the EU and its Member States have risen to the challenge. In the 1990s, Europe launched the largest single market in the world backed by a common currency. Only a few years ago, the division of Europe ended as new Member States entered the Union and other states embarked on the road towards membership or a closer relation with the Union. Over the last two years common action taken at the height of the crisis through the European Recovery Plan helped prevent economic meltdown, whilst its welfare systems helped protect people from even greater hardship.

Facts have shown that Europe is able to act in times of crisis, to adapt its economies and societies. Today, Europeans again face a moment of transformation to cope with the impact of the crisis. Europe has structural weaknesses and intensifying global challenges.

In facing this moment of transformation, Europe exit from the crisis must be the point of entry into a new economy. For its own and future generations to continue to enjoy a high-quality of healthy life, underpinned by Europe's unique social models, it needs to take action now. What is needed is a strategy to turn the EU into a smart, sustainable and inclusive economy delivering high levels of employment, productivity and social cohesion. This is the Europe 2020 strategy. This is an agenda for all Member States, taking into account different needs, different starting points and national specificities so as to promote growth for all (pp. 3–17).

Annex 1[14] *Europe 2020: An overview*

Headline targets		
– Raise the employment rate of the population aged 20–64 from the current 69% to at least 75% – Achieve the target of investing 3% of GDP in R&D in particular by improving the conditions for R&D investment by the private sector, and develop a new indicator to track innovation – Reduce greenhouse gas emissions by at least 20% compared to 1990 levels or by 30% if the conditions are right, increase the share of renewable energy in our final energy consumption to 20%, and achieve a 20% increase in energy efficiency – Reduce the share of early school leavers to 10% from the current 15% and increase the share of the population aged 30–34 having completed tertiary education from 31% to at least 40% – Reduce the number of Europeans living below national poverty lines by 25%, lifting 20 million people out of poverty		
Smart growth	Sustainable growth	Inclusive growth
Innovation EU flagship initiative "Innovation Union" to improve framework	**Climate, energy and mobility** EU flagship initiative "Resource efficient Europe"	**Employment and skills** EU flagship initiative "An agenda for new skills and jobs" to modernize labour

(continued)

[14]Note: EC (2010), *Europe 2020* COM (2010) 202, *Brussels 3.3.2010.* (p.30).

(continued)

Smart growth	Sustainable growth	Inclusive growth
conditions and access to finance for research and innovation so as to ensure that innovative ideas can be turned into products and services that create growth and jobs, strengthen the innovation chain and boost levels of investment throughout the Union	to help decouple economic growth from the use of resources, by decarbonizing our economy, increasing the use of renewable sources, modernizing our transport sector and promoting energy efficiency	markets by facilitating labor mobility and the development of skills throughout the lifecycle with a view to increase labor participation and better match labor supply and demand
Education EU flagship initiative "Youth on the move" to enhance the performance of education systems to facilitate the entry of young people to the labor market and to reinforce the international attractiveness of Europe's higher education	**Competitiveness** EU flagship initiative "An industrial policy for the globalization era" to improve the business environment, especially for SMEs, and to support the development of a strong and sustainable industrial base able to compete globally	**Fighting poverty** EU flagship initiative "European platform against poverty" to ensure social and territorial cohesion such that the benefits of growth and jobs are widely shared and people experiencing poverty and social exclusion are enabled to live in dignity and take an active part in society
Digital society EU flagship initiative "A digital agenda for Europe" to speed up the roll-out of high-speed internet and reap the benefits of a digital single market for households and firms		

7.4.3 Case Study of Scenarios Method-Mapping the Global Future Project of National Intelligence Council of U.S.A.

7.4.3.1 Introduction

Although the United States has not practiced national planning, its National Intelligence Council launched a programme "Mapping the Global Future in 2010". The National Intelligence Council (NIC) undertakes a major assessment of the forces and choices shaping the world over the next two decades every four years. These studies are in the form of publications, and they are also accessible to its website. They use scenarios to illustrate some of the ways in which drivers examined in the study interact. This generates challenges and opportunities for future decision makers. The major drivers identified in *Global Trend 2015* are seven, they are demographics; natural resources and environment; science and technology; the global economy and globalization; national and international governance; future conflict and the role of the United States. These major drivers identified changed not very much, for example, they were named Game changers in *Global Trends 2030: Alternative World* published in December 2012. Six game changers were identified in that report, they are: crisis-prone global economy; governance gap; potential for increased conflict; wider scope of regional instability; impact of new technology and role of the United States. It can be seen in comparison the seven major drivers and the six game changers, although every item of them is different in wording, but they are the same in meaning. The very recent publication by NIC is *Global Trends-Paradox of Progress.* It was launched on January 2017. It is described by Treverton, G., in his letter from the NIC chairman that "This version, the sixth in the series, is titled "Global Trends: The Paradox of Progress".

7.4.3.2 Briefing of Global Trend 2015

(1) *Contents of the Report*

The report is consisted of three main parts and one appendix. The three main parts are Overviews; Discussion (seven major drivers and their trends) and Major regions (East and Southeast Asia, South Asia, Russia and Eurasia; Middle East and North Africa, Sub-Saharan Africa, Europe, Canada and Latin America). The appendix is the Four Alternative Global Futures. In the later publications of the series, "Four Alternative Global Futures" becomes the main theme of study. The four alternative global futures studied in later reports differed from the result of study in 2015.

(2) *Overview*

The NIC in close collaboration with U.S. Government specials and a wide range of experts in past 15 months has worked to identify major drivers and trends that will shape the world of 2015. Seven key drivers were identified. They are: Demographics; Natural resources and environment; Science and technology; The global economy and globalization; National and international governance;Future conflict and The role of the United States.

(i) *Methodology*[15]

It can be seen from the descriptions of the report series that the Global Trend 2010 and 2015 project depend upon mainly experts of NIC, U.S. government and non-governmental specialists to identify drivers, and to determine which ones matter

[15]Note: Discussion in this part will also take methodology of preparation of 2020 into consideration. In fact, methodology described in this part lasted to all other reports of Global Trend in later period.

most, to highlight key uncertainties, and to integrate analysis of these trends into a national security context. The result identifies issues for more rigorous analysis and quantification. NIC began the analysis with two workshops focusing drivers and alternative futures (four scenarios are identified, as "3" described). The Global Trend 2015 is also supported by many conferences cosponsored by the NIC with other government and private centers, just mention a few such as "Foreign Reactions to the Revolution of military Affairs" with Georgetown University, "The Global Course of the Information Revolution: Technological Trends" with Rand Corporation at Santa Monica, "Alternative Global Futures: 2000–2015" with Department of State/Bureau of Intelligence and Research and CIA's Global Futures Project.

To launch the *Global Trends 2020* report, NIC brought together some 25 leading outside experts of different disciplines to engage in a broad gauged discussion with intelligence community analysis. Three leading "futurists" are invited and they include Ged Davis, former head of Skell International's scenario project-to discuss their most recent work, and the methodologies they think about the future. Historian Haro James gave the keynote address, offering lessons from prior periods of "globalization". Various methodologies are surveyed and studied. Recent "futures" studies are reviewed. Six regional conferences in countries on four continents-one in the U.K., South Africa, Singapore and two in Hungary-to solicit the views of foreign experts from a variety of backgrounds. The regional experts contributed valuable insights on how rest of the worlds views U.S. The NIC 2020 Project Conferences and workshops reached 26 in number.

To jumpstart the global scenario development process, the NIC 2020 project staff create a Scenario Steering Group (SSG). SSG include a small aggregation of excellent members of the policy community, think tanks and analysis within the intelligence community, to examine summaries of the data collected and consider scenario concepts that take into account the interaction between key drivers of global change.

The NIC's 2020 Projects, applies information technology to expand the influence of participants. With the help of CENTRA Technologies, it created an interactive password-protected Website. This site also provided a link to massive quantities of basic data for reference and analysis. It contained interactive tools to keep foreign and domestic engaged and created "book on" computer simulations that allowed experts to develop their own scenarios.

(ii) *The Drivers and Trends*

The seven drivers and trends are analyzed in detail. Which is shown in Box 7.5.

Box 7.5 Drivers and Trends

(1) Demographics it is projected that the world population will be increased from 6.1 billion in 2000 to 7.2 billion in 2015, 95% of this increase will be in developing countries, nearly all in rapidly expanding areas.

Although people will live longer, but these will be divergent impacts both to developed and developing countries. Where political systems are brittle, the combination of population growth and urbanization will foster instability.

(2) Natural Resources and environment

It is projected by the report that overall production of food will be adequate to feed the world's growing population. But parts of Sub-Saharan Africa may be in malnourishment due to poor infrastructure and distribution, political instability, and chronic poverty. Energy resources will be sufficient to meet demand, because 80% of the world's available oil and 75% of its gas remain underground.

(3) Science and Technology

Looking ahead another 15 years, the world will encounter more quantum leaps in information technology and other area of science and technology such as biotechnology, material technology, nanotechnology will generate a dramatic increase in investment in technology. Older technologies will continue lateral "sidewise development" into new applications through 2015, benefiting U.S. allies and adversaries around the world.

(4) The Global Economy and Globalization

The networked global economy will be driven by rapid and largely unrestricted flows of information, ideas, cultural values, capital, goods and services, and people: that is, globalization. This globalized economy will be a net contributor to increased political stability in the world in 2015, although its reach and benefits will not be universal. Economic growth will be driven by political pressures for higher living standards, improved economic policies, rising foreign trade and investment, the diffusion of information technologies, and an increasingly dynamic private sector. Regions, countries, and groups feeling left behind will face deepening economic stagnation, political instability, and cultural alienation.

(5) National and International Governance

States will continue to be the dominant players on the world stage, but governments will have less and less control over flows of information technology, diseases, migrants, arms, and financial transactions, whether licit or illicit, across their borders. Nonstate actors ranging from business firms to nonprofit organizations will play increasingly larger roles in both national and international affairs. The quality of governance, both nationally and internationally, will substantially determine how well states and societies cope with these global forces. Shaping the complex, fast-moving world of 2015 will require reshaping traditional government structures. Effective governance will increasingly be determined by the ability and agility to form partnerships to exploit increased information flows, new technologies, migration, and the influence of nonstate actors.

(6) Future Conflict

Internal conflicts stemming from religious, ethnic, economic or political disputes will remain at current levels or even increase in number. Export control regimes and sanctions will be less effective because of the diffusion of technology, porous borders, defense industry consolidations.

(7) Role of the United States

The United States will continue to be a major force in the world community. US global economic, technological, military, and diplomatic influence will be unparalleled among nations as well as regional and international organizations in 2015. This power not only will ensure America's preeminence, but also will cast the United States as a key driver of the international system. Diplomacy will be more complicated. Washington will have greater difficulty harnessing its power to achieve specific foreign policy goals: the US Government will exercise a smaller and less powerful part of the overall economic and cultural influence abroad.

7.4.3.3 Global Scenarios Studied by Global Trends Report 2015–2030

(1) *Four Scenarios Studied by Global Trends 2015*

(i) *Scenario One: Inclusive Globalization*

A virtuous circle develops among technology, economic growth, demographic factors, and effective governance, which enables a majority of the world's people to benefit from globalization. Technological development and diffusion—in some cases triggered by severe environmental or health crises—are utilized to grapple effectively with some problems of the developing world. Robust global economic growth—spurred by a strong policy consensus on economic liberalization—diffuses wealth widely and mitigates many demographic and resource problems. Governance is effective at both the national and international levels. In many countries, the state's role shrinks, as its functions are privatized or performed by public-private partnerships, while global cooperation intensifies on many issues through a variety of international arrangements. Conflict is minimal within and among states benefiting from globalization. A minority of the world's people—in

Sub-Saharan Africa, the Middle East, Central and South Asia, and the Andean region—do not benefit from these positive changes, and internal conflicts persist in and around those countries left behind.

(ii) *Scenario Two: Pernicious Globalization*

Global elites thrive, but the majority of the world's population fails to benefit from globalization. Population growth and resource scarcities place heavy burdens on many developing countries, and migration becomes a major source of interstate tension. Technologies not only fail to address the problems of developing countries but also are exploited by negative and illicit networks and incorporated into destabilizing weapons. The global economy splits into three: growth continues in developed countries; many developing countries experience low or negative per capita growth, resulting in a growing gap with the developed world; and the illicit economy grows dramatically. Governance and political leadership are weak at both the national and international levels. Internal conflicts increase, fueled by frustrated expectations, inequities, and heightened communal tensions; WMD proliferate and are used in at least one internal conflict.

(iii) *Scenario Three: Regional Competition*

Regional identities sharpen in Europe, Asia, and the Americas, driven by growing political resistance in Europe and East Asia to US global preponderance, and US-driven globalization and each region's increasing preoccupation with its own economic and political priorities. There is an uneven diffusion of technologies, reflecting differing regional concepts of intellectual property and attitudes towards biotechnology. Regional economic integration in trade and finance increases, resulting in both fairly high levels of economic growth and rising regional competition. Both the state and institutions of regional governance thrive in major developed and emerging market countries, as governments recognize the need to resolve pressing regional problems and shift responsibilities from global to regional institutions. Given the preoccupation of the three major regions with their own concerns, countries outside these regions in Sub-Saharan Africa, the Middle East, and Central and South Asia have few places to turn for resources or political support. Military conflict among and within the three major regions does not materialize, but internal conflicts increase in and around other countries left behind.

(iv) *Scenario Four: Post-Polar World*

US domestic preoccupation increases as the US economy slows, then stagnates. Economic and political tensions with Europe grow, the US European alliance deteriorates as the United States withdraws its troops, and Europe turns inward, relying on its own regional institutions. At the same time, national governance crises create instability in Latin America, particularly in Colombia, Cuba, Mexico, and Panama, forcing the United States to concentrate on the region. Indonesia also faces internal crisis and risks disintegration, prompting China to provide the bulk of an ad hoc peacekeeping force. Otherwise, Asia is generally prosperous and stable, permitting the United States to focus elsewhere. Korea's normalization and de facto unification proceed, China and Japan provide the bulk of external financial support for Korean unification, and the United States begins withdrawing its troops from Korea and Japan. Over time, these geostrategic shifts ignite longstanding national rivalries among the Asian powers, triggering increased military preparations and hitherto dormant or covert WMD programs. Regional and global institutions prove irrelevant to the evolving conflict situation in Asia, as China issues an ultimatum to Japan to dismantle its nuclear program and Japan—invoking its bilateral treaty with the US—calls for US reengagement in Asia under adverse circumstances at the brink of a major war. Given the priorities of Asia, the Americas, and Europe, countries outside these regions are marginalized, with virtually no sources of political or financial support.

(2) *Four Scenarios Studied by Global Trends 2020*

(i) Davos World provides an illustration of how robust economic growth, led by China and India, over the next 15 years could reshape the globalization process—giving it a more non-Western face and transforming the political playing field as well.
(ii) Pax Americana takes a look at how US predominance may survive the radical changes to the global political landscape and serve to fashion a new and inclusive global order.
(iii) A New Caliphate provides an example of how a global movement fueled by radical religious identity politics could constitute a challenge to Western norms and values as the foundation of the global system.
(iv) Cycle of Fear provides an example of how concerns about proliferation might increase to the point that large-scale intrusive security measures are taken to prevent outbreaks of deadly attacks, possibly introducing an Orwellian world. (Executive Summary) (pp. 9–10).

(3) *Four Scenarios Studied by Global Trends 2025*

(i) *A World Without the West*

In this world, described in a fictional letter from a future head of the Shanghai Cooperation Organization (SCO), new powers supplant the West as the leaders on the world stage. The US feels overburdened and withdraws from Central Asia, including Afghanistan; Europe will not step up to the plate and take the lead. Russia, China, and others are forced to deal with the potential for spillover and instability in Central Asia. The SCO gains ascendance while NATO's status declines. Anti-China antagonism in the US and Europe reaches a crescendo; protectionist trade barriers are put in place. Russia and China enter a marriage of convenience; other countries—India and Iran—rally around them. The lack of any stable bloc—whether in the West or the non-Western world—adds to growing instability and disorder, potentially threatening globalization.

(ii) *October Surprise*

In this world, depicted in a diary entry of a future US President, many countries have been preoccupied with achieving economic growth at the expense of safeguarding the environment. The scientific community has not been able to issue specific warnings, but worries increase that a tipping point has been reached in which climate change has accelerated and possible impacts will be very destructive. New York City is hit by a major hurricane linked to global climate change; the NY Stock Exchange is severely damaged and, in the face of such destruction, world leaders must begin to think about taking drastic measures, such as relocating parts of coastal cities.

(iii) *BRICs' Bust-Up*

In this world, conflict breaks out between China and India over access to vital resources. Outside powers intervene before the conflict escalates and expands into a global configuration. The clash is triggered by Chinese suspicion of efforts by others to threaten Beijing's energy supplies. Misperceptions and miscalculations lead to the clash. The scenario highlights the importance of energy and other resources to continued growth and development as a great power. It shows the extent to which conflict in a multipolar world is just as likely to occur between rising states as between older and newer powers.

(iv) *Politics is Not Always Local*

In this world, outlined in an article by a fictional Financial Times reporter, various non-state networks—NGOs, religious groups, business leaders, and local activists—combine to set the international agenda on the environment and use their clout to elect the UN Secretary General. The global political coalition of non-state actors plays a crucial role in securing a new worldwide climate change agreement. In this new connected world of digital communications, growing middle classes, and transnational interest groups, politics is no longer local and domestic and international agendas become increasingly interchangeable. (Introduction) (p. 4).

(4) *Four Scenarios Studied by Global Trends 2030*

(i) *Stalled Engines*

In the most plausible worst-case scenario, the risks of interstate conflict increase. The US draws inward and globalization stalls.

(ii) *Fusion*

In the most plausible best-case outcome, China and the US collaborate on a range of issues, leading to broader global cooperation.

(iii) *Gini-Out-of-the-Bottle*

Inequalities explode as some countries become big winners and others fail. Inequalities within countries increase social tensions. Without completely disengaging, the US is no longer the "global policeman."

(iv) *Non-state World*

Driven by new technologies, non-state actors take the lead in confronting global challenges (Executive Summary) (p. ii).

7.4.3.4 Retrospect of Past Reports of Global Trends Works and the Three Scenarios in Global Trends Paradox of Progress Launched on Jan. 2017

(1) *Retrospect of Past Reports of Global Trends Projects*

Before launching report on Global Trends 2030, the NIC commissioned an academic study of the four previous reports beginning with the first edition in 1991–97. The reviewers examined them to highlight any persistent blind spots, bias and their strength. The key "looming" challenges that reviewers cited for GT2030 were to develop.

(i) A greater focus on the role of US in the international system. Past works assumed US centrality, leaving readers "vulnerable" to wonder about "critical dynamics" around the US role. One of the key looming issues for GT 2030 was "how other powers would respond to a decline or a decisive re-assertion of US power." The authors of the study thought that both outcomes were possible and needed to be addressed.

(ii) A clearer understanding of the central units in the international system. Previous works detailed the gradual ascendance of nonstate actors, but we did not clarify how we saw the role of states versus nonstate actors. The reviewers suggested that we delve more into the dynamics of governance and explore the complicated relationships among a diverse set of actors.

(iii) A better grasp of time and speed. Past Global Trends works "correctly foresaw the direction of the vectors: China up, Russia down. But China's power has consistently increased faster than expected… A comprehensive reading of the four reports leaves a strong impression that [we] tend toward underestimation of the rates of change…"

(iv) Greater discussion of crises and discontinuities. The reviewers felt that the use of the word "trends" in the titles suggests more continuity than change. GT 2025, however, "with its strongly worded attention to the likelihood of significant shocks and discontinuities, flirts with a radical revision of this viewpoint." The authors recommended developing a framework for understanding the relationships among trends, discontinuities, and crises.

(v) Greater attention to ideology. The authors of the study admitted that "ideology is a frustratingly fuzzy concept … difficult to define … and equally difficult to measure." They agreed that grand "isms" like fascism and communism might not be on the horizon. However, "smaller politico-psychosocial shifts that often don't go under the umbrella of ideology but drive behavior" should be a focus.

(vi) More understanding of second- and third-order consequences. Trying to identify looming disequilibria may be one approach. More war gaming or simulation exercises to understand possible dynamics among international actors at crucial tipping points was another suggestion (p. 4).

(2) *Three Scenarios of Global Trends Paradox of Progress*

There is very detailed explorations of three scenarios of part 4 of that report. "Three scenarios for the distant future: Islands, Orbits, Communities". Part 5 "What the scenarios teach us: fostering opportunities through resilience". Hereunder, simply descriptions of these three scenarios are quoted from the future summarized of the report.

Three stories or scenarios—"Islands," "Orbits," and "Communities"—explore how trends and choices of note might intersect to create different pathways to the future. These scenarios emphasize alternative responses to near-term volatility—at the national (Islands), regional (Orbits), and sub-state and transnational (Communities) levels.

(i) Islands investigates a restructuring of the global economy that leads to long periods of slow or no growth, challenging both traditional models of economic prosperity and the presumption that globalization will continue to expand. The scenario emphasizes the challenges to governments in meeting societies' demands for both economic and physical security as popular pushback to globalization increases, emerging technologies transform work and trade, and political instability grows. It underscores the choices governments will face in conditions that might tempt some to turn inward, reduce support for multilateral cooperation, and adopt protectionist policies, while others find ways to leverage new sources of economic growth and productivity.
(ii) Orbits explores a future of tensions created by competing major powers seeking their own spheres of influence while attempting to maintain stability at home. It examines how the trends of rising nationalism, changing conflict patterns, emerging disruptive technologies, and decreasing global cooperation might combine to increase the risk of interstate conflict. This scenario emphasizes the policy choices ahead for governments that would reinforce stability and peace or further exacerbate tensions. It features a nuclear weapon used in anger, which turns out to concentrate global minds so that it does not happen again.
(iii) Communities shows how growing public expectations but diminishing capacity of national governments open space for local governments and private actors, challenging traditional assumptions about what governing means. Information technology remains the key enabler, and companies, advocacy groups, charities, and local governments prove nimbler than national governments in delivering services to sway populations in support of their agendas. Most national governments resist, but others cede some

power to emerging networks. Everywhere, from the Middle East to Russia, control is harder.

7.4.3.5 Appendix

Mathematical approach of scenario methods.

The formal approach of scenarios method is quite complicated. Attached Fig. 7.4 is the diagram of mathematical approach of scenarios method.

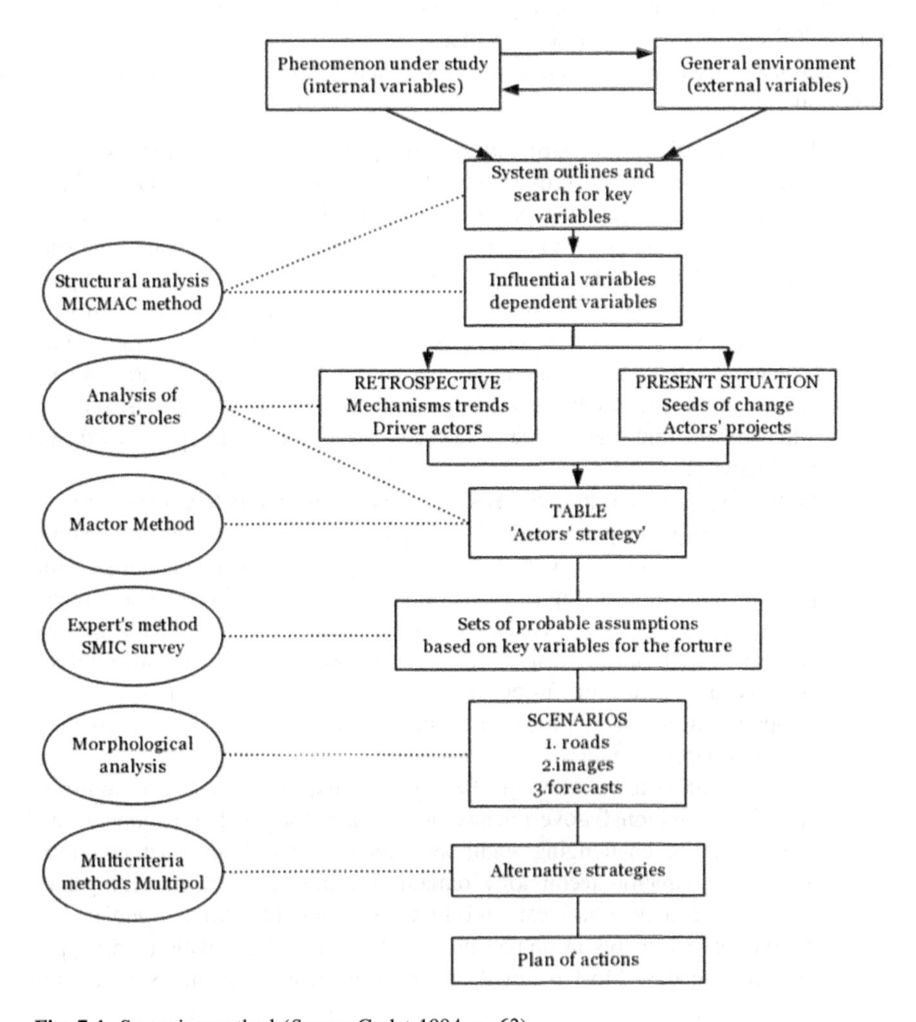

Fig. 7.4 Scenarios method (*Source* Godet 1994, p. 63)

7.5 Spatial Development Planning

7.5.1 Introduction

National territory and resources should be well planned in modern society. In most developed countries, except the US, have their own national legislation and spatial planning. The Federal Spatial Development of Germany, for example, defines the philosophies and principles of spatial development and fundamental policies for federal spatial development. Japan had taken the lead in the preparation of spatial development planning in Asian region Since the 1970s, South Korea has also successively enacted the spatial development planning for comprehensive territorial development. The European Commission had published *European Spatial Development Perspective* (ESDP) in 1999. This document was agreed at the Informal Council of Ministers responsible for Spatial Planning in Potsdam on May 1999. The spatial planning experience of Japan and the contents of ESDP will be briefed to be case study of spatial planning in Sects. 7.5.2 and 7.5.3.

7.5.2 Case Study 1 of Japanese Experience of Spatial Planning

7.5.2.1 Japan is a Latecomer Among the Developed Countries

Before the World War II, Japan had made full efforts to study and to introduce scientific technologies from the Western Europe and the United States, in particular, Germany. After the WWII, Japan studied scientific technologies of the US seriously and developed its own scientific and technological system based on its own cultural and national conditions. Around 1970s to 1990s, the Japanese manufacturing sector became dominant among the world. It was pointed out by Dertouzos, Lester, and Solow (1989) in *Made in America*, that "The survey of automobile assembly plant conducted by MIT's International Motor Vehicle Programs showed not only the cars designed by Japanese firms were of higher quality than U.S. designed vehicles but also that this quality advantage remained even when the Japanese designed cars were assembler in U.S. factories. In other words, the Japanese engineers had incorporated quality enhancing features into the design itself." (p. 70) The post-war Japan has created its spatial development planning system through several successive plans and with continuous improvement, a comprehensive system combining national economic planning, regional planning, and national territorial integrated development planning and legislation.

7.5.2.2 Experience of Japanese Spatial Development Planning[16]

(1) Since the formulation of the first comprehensive national development planning in 1962, Japan has successively enacted six comprehensive national development plans. (currently, it is also called spatial development plan in Japan) Comparisons of the first five territorial development planning are shown in Table 7.5.
(2) Major experiences of japanese spatial development planning

(i) It is started first at regional level, and also focused on establishment and perfection of legislative system. A gradualist approach is adopted from comprehensive regional development planning to comprehensive national development planning. In 1949 and 1950, it has released the Comprehensive National Development Act and the Hokkaido Integrated Development Act. Economic plans of Japan, such as the "Three-year Economic Plan" of 1952, and the "Five-year Plan for Economic Self-support" of 1955 were prepared with the goal of attaining a "self-supporting economy". A series of regional comprehensive development planning were also launched, Hokkaido comprehensive development plan was started in Oct. 1951. Promotional Development Acts of Capital region, Northeast region, Kyusha and Shikoku region were passed before 1962. Based on the experience of implementing comprehensive development plans in these regions, Japan commenced its first National Comprehensive Development Planning in Oct. 1962. Japan had passed the legislation of Planning Act of Formation of National Territory on 26th, May 1956, Act No. 205. The Act was revised and amended on 29th July 2005 to be Act. No. 89.
(ii) Japan has an excellent national planning system. At national level, it has two types of plans closely linked, the Economic Plan and the National Comprehensive Development Plan (Also called National Spatial Development Plan). Down to regional level, Japan had classified its country into 8 Broad Local Planning Regions, for example, the Northeast Region has seven counties.
(iii) Japan has also changed its concept of national spatial development planning from point-axis into promotion of self-support of broad regional development to adapt to change of social economic trends, such as reduction of Japanese population and its rapid aging society, development of globalization and East Asian region, and the development of information communication technology. Japan also has the awareness of that there are changes of the value system of her people, who focus more on diversification of life style, safety, security environment of the earth. It is necessary to promote the growth of major departments which realize the role of "public".

[16]Note: In the case study of Japanese experience, comprehensive national development planning and spatial development planning are used to be synonym. Because Japan used the term "Comprehensive national development planning" in its first planning in 1962. Japan followed the Western practice and change this term into "spatial development planning in later period.

Table 7.5 Comparisons of five national comprehensive development planning in Japan[a]

Comparisons of five national comprehensive development planning					
	National comprehensive development planning (1st NCDP)	New national comprehensive development planning (2nd NCDP)	3rd National comprehensive development planning (3rd NCDP)	4th National comprehensive development planning (4th NCDP)	21st Century territorial magnificent blueprint (5th NCDP)
Decided in cabinet meeting held on	05.10.62	30.05.69	04.11.77	30.06.87	31.03.98
Current cabinet during formation of rules	Cabinet of Ikeda	Cabinet of Sato	Cabinet of Fukuda	Cabinet of Nakasone	Cabinet of Hashimoto
Background	1. Start of rapid economic growth 2. Over-sized cities and enlargement of income gap 3. National income doubled (vision of Pacific region)	1. Rapid economic growth 2. Concentration of population and industry in large cities 3. Development of informationization, internationalization and technological innovation	1. Stable economic growth 2. Start of the trend of population and industry to diffuse to regions 3. Limitation of territorial resources and energies gradually stands out	1. Population and various functions totally centralized in Tokyo 2. Increasingly severe employment problems in some cities and surrounding areas due to sharp changes in industrial structure 3. All-round internationalization	1. Earth times (global environmental problems, mega-competition, and exchanges with Asian countries) 2. The era of population decrease and ageing 3. Highly information-based age

(continued)

Table 7.5 (continued)

Comparisons of five national comprehensive development planning					
Target year	1970	1985	About 10 years since 1977	Around 2000	2010–2015
Basic objectives	Realize inter-regional balanced development	Create abundant environment	Improve the comprehensive living environment of citizens	Realize multiplex distributed territorial structure	Lay foundation for development of multi-axial territorial structure
Development methods	Vision of growth pole development To realize the objectives, business needs to be diffused and development bases should be established to make them interconnected with existing large-scale geometries i.e. Tokyo. Traffic and communication facilities should be improved By organically combing the above to enable	Vision of large-scale projects Traffic networks including Shinkansen and highways should be improved, and the imbalanced territorial use conditions should be improved and the problems of overcrowding, undercrowding and regional differences.	Vision of settlement Concentration of population and industry in large cities should be controlled at the same time when local economies are revived so as to solve the problems of overcrowding and undercrowding; comprehensive living environment of citizens should be improved by realizing balanced use of land in Japan	Vision of exchange network To realize multiplex distributed territorial structure, 1. Make use of regional advantages to facilitate regional construction; 2. Nationally promote basic traffic, information and communication system improvements by the state or based on national guidelines; 3. strengthen the cooperation between the state, regional	Participation and cooperation Build beautiful homestead with participation of diverse subjects and cooperation among regions (Ideological strategy) 1. Creation of residential areas by making use of natural environment (small cities, villages, mountain villages, fishing villages, intermountain spaces, etc.)

(continued)

Table 7.5 (continued)

Comparisons of five national comprehensive development planning					
	mutual effects, characteristics of surrounding areas can be given full play to facilitate chain reaction development. Regional balanced development should be realized			autonomous mass, and various civic groups to create diverse exchange opportunities	2. Reconstruction of large cities (spatial improvement, update and effective use of large cities); 3. Enlargement of regional cooperation axis 4. Formation of wide-area international exchange circle (by establishing areas with global exchange functions)

[a]*Source* Ministry of Land, Infrastructure, Transport and Tourism in Japan, *New Territorial Planning in Japan*, Bureau of Territorial Planning (2008), published in the Seminar for Sino-Japan Territorial Planning held in Tsinghua University, Beijing (2008.3.10)

7.5.2.3 Contents of Sixth Spatial Development Planning of Japan

(1) *Four Strategic Objectives and Regional Construction Model*

To deal with globalization, decline of population, heritage and reconstruction, four strategic objectives are put forward in the sixth spatial development planning of Japan:

i. To carry out harmonious exchanges and cooperation with the East Asia;
ii. To create regions of sustainable development;
iii. To build a flexible territorial structure with strong power to defend disasters
iv. Heritage and management of the beautiful territory

Laterally, the new "public sector"-based regional construction should be promoted to implement the four strategic objectives.

(2) *Contents of the Sixth Territorial Planning*[17]

(i) *Chapter 1 Improvement of Broad region*

Improve living quality so as to ensure secured living (specific contents include: improve of housing quality to enable long service life; improvement of housing market; construction of medical cooperation mechanism; and reconstruction of traffic network)

Construction of peaceful living environment and energetic metropolis circle with full vitality (specific contents include: development of extensive living areas; mutual connection of urban functions; transformation to agglomerated urban structure)

Construct intergrowth and communications between beautiful and convenient villages, mountain villages and fishing villages

Promote interregional exchanges and cooperation and immigration of people, including the exchanges and co-existence between cities and villages, mountain villages & fishing villages

Support regions with severe natural, geographic and social conditions.

(ii) *Chapter 2 Industry*

Develop new technologies and upgrade productivity through reform (specific contents include: strengthening scientific and technological foundations; upgrading of industrial competitiveness; mobilization of regional resources to realize cooperation among the government, industry and university

Enliven the industries supporting regional development (specific contents include: improvement of environment of spatial location of prospective enterprises; support the construction of infrastructure of logistics for investment of private

[17]Source: Ministry of Land, Infrastructure, Transport and Tourism in Japan, *New Territorial Planning in Japan*, Bureau of Territorial Planning (2008), published in the Seminar for Sino-Japan Territorial Planning held in Tsinghua University, Beijing (2008.3.10) (pp. 8–9).

sector cooperated with the government; support the development of brands, high value-added industries and cross-domain industries)

Stable support of food products and new development of agriculture, forestry and fishery industries.

(iii) *Chapter 3 Culture and Tourism*

Build energetic regional society with rich cultural atmosphere (specific contents include: creation and introduction of new Japanese culture; full use of historical architectures to promote construction of blocks; popularization, heritage and introduction of traditional Japanese food culture to foreign countries; preservation, heritage and use of regional culture with rich characteristics)

Revive tourism and enliven regional economy (specific contents include: construction of scenic tourist landscape with international competitiveness for tourism; promotion of developing new tourism models of different regions; support and construction of landscape with long-term tourist attractions for tourism and support broad regional cooperation of tourism.

(iv) *Chapter 4 Traffic, Information and Communications*

Build the international traffic, information and communications system connecting the East Asia and the world (specific contents include: vision of high-speed channel of Asia ports; support the development of high-speed channels in ports of broad region; establishment of business circle of day return, goods delivery circle at the second day and the broadband environment of Asia)

Establish a traffic system covering national trunklines of land, sea and air to promote interregional exchanges and cooperation (specific contents include: connecting international bays, airports and the roads and rails accessible to regional tourism resources)

Support of broad regional traffic, information and communications system for living circles between cities and villages, mountain villages and fishing villages (specific contents include: improvement and popularization of broadband networks; improvement of regional traffic network for safe and secured living; activation and recovery of regional public traffic).

(v) *Chapter 5 Disaster Prevention*

Measures of integration of "software and hardware" (specific contents include: construction of facilities with sound disaster-prevention performance; promote the "software solution" aiming at disaster reduction and balancing "self-help", "joint assistance" and "public aid"; new tasks to carry out researches on coping with disasters brought about by global warming)

Construction of territorial spaces with sound disaster prevention performance (specific contents include: strengthening regional disaster prevention capabilities; territorial use that reduce disaster risks; expansion of path of circulation for traffic, information and communications).

(vi) *Chapter 6 Utilization and Protection of Territorial Resource and Maritime Spaces*

Focusing on territorial management of basin circles (specific contents include: construction and improvement of water recycling system; comprehensive management of soil and sand)

Improvement and protection of forests, farmlands, etc. (specific contents include: promote the farmland utilization by focusing on the opinion of "from possessing to using")

Use and protection of maritime spaces (specific contents include: construction of comprehensive planning of sea; comprehensive management of coastal areas based on perspective of integration of land and sea

Implement various policies to realize "national operation of national territory", namely, joint management of territory by people all.

(vii) *Chapter 7 Environmental Protection and Formation of Landscapes*

Build the material cycle in which human activities are coordinated with natural development (specific contents include: reduction of CO_2 emission; construction of "low-carbon society"; promotion of 3R (reduce, reuse and recycle))

Maintain and develop sound ecosystem (specific contents include: development of broad econetwork)

Construct landscapes with regional characteristics.

(viii) *Chapter 8 Realization of Regional Construction by New "Public Sectors"*

Ensure the person-in-charge of new "public sectors" so as to improve the environment of activities (specific contents include: promote of involvement of youth; cultivation of intermediate supporting organizations; full use of universities, experts and outsourced talents of regions)

Emphasize the regional construction of creativity and activities of the people (specific contents include: recovery and activation of geographical communities; support the village and mountainous areas to build new cooperation mechanism).

7.5.3 Case Study 2 of European Spatial Development Perspective (ESDP)

7.5.3.1 Introduction

European Spatial Development Perspective was published by the European Commission. The establishment of EU has a long history since the establishment of European Coal and Steel Community (ECSC) in 1951. It was evolved and called European Community in 1967. The term European Commission was introduced in

Nov. 1993, which was a simplification of the term "The Commission of European Communities." According to *Europe from A to Z* (1997), "The Commission is one of the main protagonists in the preparation, formulation, implementation and monitoring of binding decisions taken by The European Union." (p. 109).

This study was published in 1999. There were only 15 member countries by that time (including U.K.). Currently, EU has 27 member countries with exit of U.K. in 2017. In preparation of the ESDP, it includes the Chapter 5 in part A of this study, titled "The Enlargement of the EU: An Additional Challenge for European Spatial Development Policy". ESDP has considered additional 12 countries which had participated EU successively in 2004, 2007 (2 countries) and 2013 (one country). *European Spatial Development Perspective* is consisted of two parts. Part A deals with the theme "Achieving the balanced and sustainable development of the territory of the EU: The contribution of the spatial development policy." Part A is further divided into 5 chapters[18]. Part B is titled "The territory of the EU: Trends, Opportunities and Challenges" and it is further divided into 4 chapters.

7.5.3.2 Agreement by Ministers and Fundamental Goals of Policy

(1) This ESDP was agreed at the Informal Council of Ministers responsible for spatial planning in Potsdam, May, 1999. There is a page in the very beginning of ESDP, which is "Excerpt from the final conclusions issued at the close of the informal Council of EU Ministers responsible for Spatial Planning. The excerpt has five points, only part of point "3" and "5" are quoted in (2).
(2) Quotation of part of "Excerpt"

In the Ministers' view what is important is to ensure that the three fundamental goals of European policy are achieved equally in all the regions of the EU:

(i) Economic and social cohesion.
(ii) Conservation and management of natural resources and the cultural heritage.
(iii) More balanced competitiveness of the European territory.
(iv) The ESDP is a suitable policy framework for the sectoral policies of the Community and the Member States that have spatial impacts, as well as for regional and local authorities, aimed as it is at achieving a balanced and sustainable development of the European territory.
(v) It will serve as a policy framework for the Member States, their regions and local authorities and the European Commission in their own respective spheres of responsibility.

[18]Note: In the original document, it has not used the term 'chapter'. It is simply termed 1, 2, 3, and 1.1, 2.2, and 3.3.1 for notation.

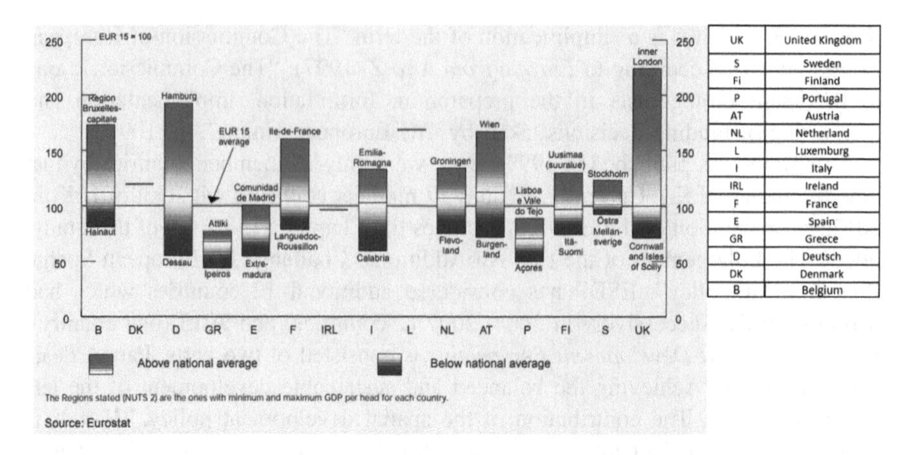

UK	United Kingdom
S	Sweden
Fi	Finland
P	Portugal
AT	Austria
NL	Netherland
L	Luxemburg
I	Italy
IRL	Ireland
F	France
E	Spain
GR	Greece
D	Deutsch
DK	Denmark
B	Belgium

Fig. 7.5 Regional disparities in GDP per capita (PPS) by member state, 1996 (*Source* Figure 1 of ESDP (p. 9))

7.5.3.3 Selected Aspects of Part A

(1) *Main Points*

Part A deals mainly with spatial development policy related to balancing sustainable development of the territory of the EU, and therefore, only selected aspects will be discussed.

(2) *Spatial Development Disparities*

A large organization of European Union with 28 members before the exit of U.K., has large income disparities among its member countries. EU per capita GDP in 2016 is estimated around #37800 [19](PPP) in 2016, but its lowest income member country is only #13000 [20]while its highest income member country reached #82000 [21](PPP) in the same year. While the 15 member countries studied in ESDP, Luxemburg had the highest per capita income of #99500 in 2015. While Portugal had the lowest per capital income of #26400 in 2015.

This spatial development disparity was also existed in 1999 in the publication of ESDP. There was not only income disparity among member countries, but there was also large disparity within one country. Figure 7.5 shows the situation.

(3) *Evolution of Concept of Spatial Development*

The issue of regional disparity, together with the influence of sustainable development raised by United Nations of the World Summit of 1992, has caused the

[19]Note: From World Fact Book. www.dni.gov. (Retrieved on 19th, Apr., 2017).

[20](See footnote 19)

[21](See footnote 19)

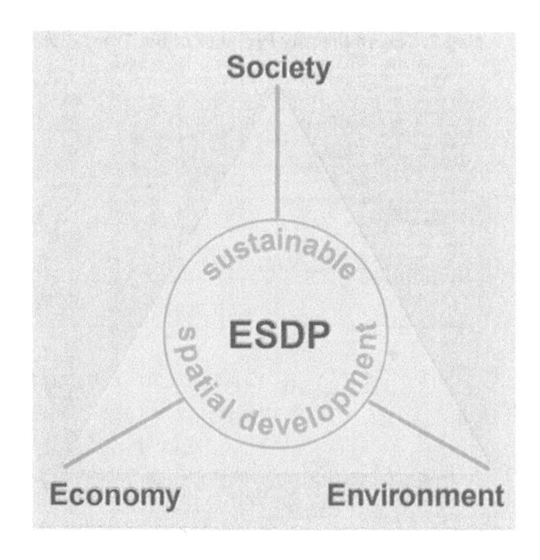

Fig. 7.6 Triangle of objectives: a balanced and sustainable spatial development (*Source* Figure 1 of ESDP (p. 10))

evolution of concept of spatial development of EU. It is described in 1.3 Underlying Objectives of the ESDP that "Considering the existing regional disparity of development.... The EU Spatial Development Perspective is based on the EU aim of achieving a balanced and sustainable development, in particular by strengthening economic and social cohesion.... The EU will therefore gradually develop, in line with safeguarding regional diversity from an Economic Union into an Environment Union and into a Social Union" (p. 10), which is shown in Fig. 7.6.

(4) *Selected Aspects Chapter 2 of part A of ESDP "Influence of Community Policies on the Territory of the EU*

Treaties of EU with implications for spatial development include Community competition policy; Trans-European Networks; Structural Funds; Common Agricultural Policy (CAP); Environmental policy; Research, Technology and development (RTD); Loan Activation of the European Investment Bank.

Trans-European Networks (TENS) In order to have better integration and operation of such large organization including 15 countries, Trans-European Networks is crucially important. It has been pointed out in ESDP that "The EU Treaty obliges the community to contribute to the organization and development of Trans. European Networks (TENs) in the areas of transport, telecommunications and energy supply infrastructure. This mandate should in particular, serve the Community objectives' of a smooth function of a Single Market as well as the strengthening of economic and social cohesion." (p. 14). The Priority Projects of TENS is shown in Fig. 7.7.

7.5.3.4 Contents of Part B

The contents of part B is shown in Box 7.6.

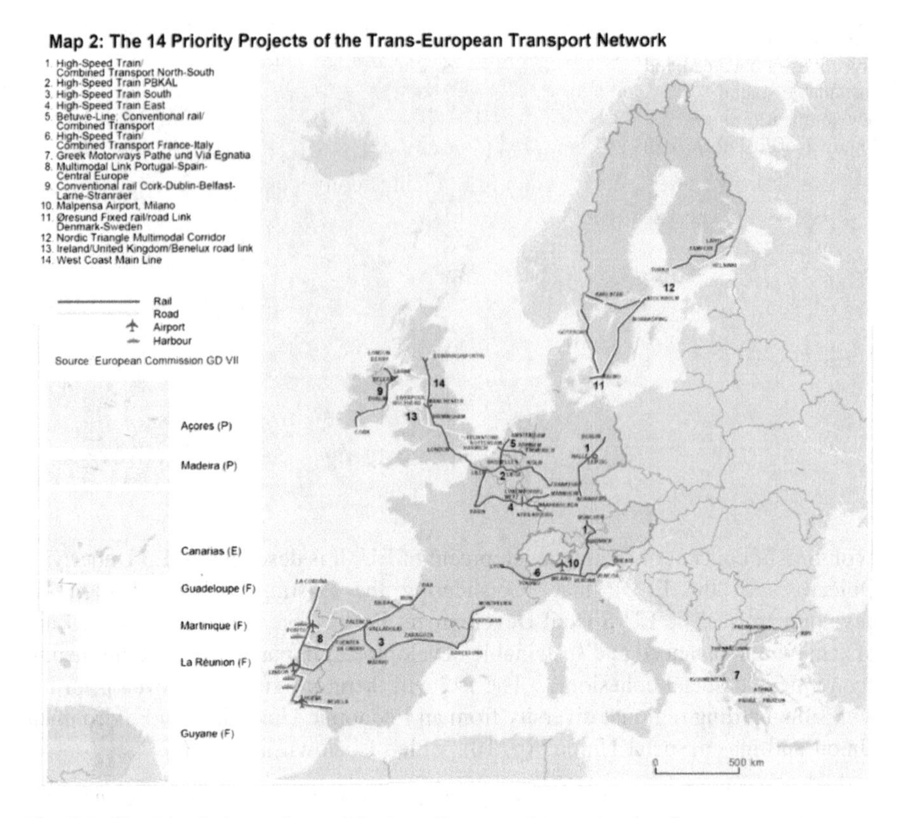

Fig. 7.7 The 14 priority projects of the trans-European transport network

Box 7.6 [22]Contents of the Territory of the EU: Trends, opportunities and challenge

1. **Spatial Development Conditions and Trends in the EU**
   - 1.1 Geographical Characteristics of the EU
   - 1.2 Demographic Trends
   - 1.3 Economic Trends
   - 1.4 Environmental Trends
2. **Spatial Development Issues of European Significance**
   - 2.1 Trends Towards Change in the European Urban System
     - 2.1.1 The Emergence of Urban Networks
     - 2.1.2 Changes in Urban Economic Opportunities
     - 2.1.3 Continuing Urban Sprawl

[22]Note: EC (1999).

2.1.4 Increasing Social Segregation in Cities
2.1.5 Improvements in the Quality of the Urban Environment

2.2 The Changing Role and Function of Rural Areas

2.2.1 Increasing Interdependence of Urban and Rural Areas
2.2.2 Different Lines of Development Trends in Rural Areas
2.2.3 Shifts in Agriculture and Forestry-Consequences for Economy and Land Use

2.3 Transport and Networking

2.3.1 Border and Integration Problems of the Networks
2.3.2 Increasing Transport Flows and Congestion
2.3.3 Inadequate Accessibility in the EU
2.3.4 Concentration and Development Corridors
2.3.5 Disparities in the Diffusion of Innovation and Knowledge
2.4 Natural and Cultural Heritage
2.4.1 Loss of Biological Diversity and Natural Areas
2.4.2 Risk to Water Resources
2.4.3 Increasing Pressure on the Cultural Landscapes
2.4.4 Increasing Pressure on Cultural Heritage

3. **Selected Programmes and Visions for Integrated Spatial Development**

3.1 EU Programmes with Spatial Impacts
3.2 Interreg II C Programmes
3.3 Pilot Actions for Transnational Spatial Development under ERDF Article 10
3.4 Spatial Visions

4. **Basic Data for the Accession Countries and Member States**

7.6 Summary Points

1. This chapter includes a comprehensive discussion of various types of planning systems. It is consisted of five parts: Introduction; Evolution of development of modern national planning system; Case study of medium-term national plans; Strategic planning and scenarios method and Spatial development planning.
2. There are many debates on national planning by academic field in China and others countries due to insufficient understanding of the meaning of "planning". Definition and roles of planning are clarified. Five types of planning are clarified in general sense. Classification of planning in economics is broadly discussed with three types. In spite of many types of planning, planning requires further improvement, it is a major area of application of social systems engineering.

3. Evolution of development of modern planning system, is discussed in four aspects.
   (a) There are three stages of evolution of development of modern national planning system: Period of beginning of preparing national planning (2nd–3rd Decade of 20th century; Age of popularization of national planning (4th–5th Decade of 20th century) and emergence of long term strategic planning and scenario analysis.
   (b) Development of the planning theory and application of models, especially mathematical models.
   (c) Emergence of Future Studies and scenario method.
   (d) National plans and business plans are discussed. National plans are discussed in detail in later parts of this chapter. A brief discussion of business plans are presented, planning method discovered in business plans has influenced also the national planning.

4. Case Study of Medium Term National Planning of France.

France is a pioneering country to develop an indicative planning system post the WWII followed by many countries in contrasting the imperative planning system adopted by the former Soviet Union and China. France has established the principle, process (with wider participation of the public in the planning process), planning theory and organization, national accounting system and quantitative approach of planning technique. Its experience and planning literatures have large influence globally.

Essential points of eight medium-term plans of France are briefed and relatively detailed briefing of official document of French Ninth Plan is given in this chapter.

5. Case Study of Medium Term National Planning of India.

India is one of the developing country who has developed its own Five Year Plan for economic development by imitating the former Soviet Union since 1951. But its plan is indicative in nature. Since the First Five Year Plan launched in 1951, Twelve Five Year Plan have been prepared. As most of Indian scholar and planning personnel have received Western higher education, their planning theories and methods usually approach the frontier of Western social science. These official documents of Indian Plan are briefed.

   (a) Contents of 8 chapters, 20 programmes and its outlays of the document Sixth Five Year Plan.
   (b) A Technical Note of the Sixth Plan of India (1980–85) which includes mathematical models and I/O Table.
   (c) Indian Twelfth Five Year Plan (2012–2017).

6. Strategic Planning and Scenario method

The failure of mid-term planning to meet the two oil crisis erupted in 1970s, and a global wave of liberalization of trade, capital flow and privatization erupted in the meantime, part of developed countries abandoned the preparation of mid-term planning and turn to strategic long term planning and scenarios method supported by development of Futurology.

Europe 2020: A New Economic Strategy which was launched by European Commission in 2010 is chosen to be case study of strategic long term planning, which has three mutually reinforcing priorities: Smart Growth; Sustainable Growth and Inclusive Growth. Five headline targets; employment; expenditure of R&D; climate/energy; education and poverty alleviation are set up by 2020.

There are also seven flagship initiative putting under each priority theme and three scenarios of EU 2020 are identified.

7. Case Study of Scenario Method-Mapping the Global Future Project of National Intelligence Council of U.S.A.

Although the United States has no practiced national planning, its National Intelligence Council (NIC) had launched a programme Mapping the Global Future since 2010. NIC undertakes a major assessment of the forces (drivers) and choices shaping the world over the next two decade every four years. There are six publications up to the present, from Global Trends 2010 to Global Trends 2030, and the recent publication Global Trends Paradox of Progress launched in Jan. 2017.

8. Spatial Development Planning-Case Study of Japan.

National territory and resources should be well planned in modern society. Japan takes the lead in spatial development planning in Asian region. Japan had a long tradition of urban planning before the WWII. Early in 1919, it had promulgated Urban Planning Act and Urban Architectural Standard Act. The planning professionals were dominated by civil engineers and architects. With this historical background, Japan had promulgated the Comprehensive National Development Act and the Hokkaido Integrated Development Act respectively in 1949 and 1950. It commenced its first National Comprehensive Development Plans in Oct. 1962. Four National Comprehensive Development Plans in succession were prepared respectively in 1969, 1977, 1987 and 1998 which are briefed in Table 7.5 of this chapter.

The very recent one is Sixth Spatial Development Plan launched in 2008. It contains eight chapters.

9. Case Study: European Spatial Development Perspective (ESDP).

This ESDP was published in 1999, there were only 15 members countries by that time (including U.K.), EU has 27 member countries in 2017. But in preparation of this ESDP, it had taken "The Enlargement of the EU" into consideration.

ESDP is consisted of two parts. Part A deals with the theme "Achieving the balanced and sustainable development of the territory of the EU: The contribution of the spatial development policy.", it has five chapters. Part B is titled "The territory of the EU: Trends, Opportunities and Challenges", it is divided into four chapters. They are briefed selectively in this chapter.

References

Black, J. (Ed.). (2002). *Dictionary of economics* (2nd ed.). Oxford: Oxford University Press.

Bowles, R. A., Whynes, D. K., Allen, G., & Unwin, G. (1979). *Macroeconomic planning.* London: Wiley.

Brokstable, R. (2007). *A demon of our own design: Markets, hedge funds, and the perils of financial innovation (Chinese version).* Beijing: Zhongxin Press.

Bryson, J. M., & Einsweiler, R. C. (Eds.). (1988a). *Strategic planning threats and opportunities for planners* (p. 1, 3–4). Chicago: Planners Press.

Bryson, J. M., & Einsweiler, R. C. (1988b). Introduction (Chap. 1, pp. 1–14). In J. M. Bryson, & R. C. Einsweiler, (Eds.), *Strategic planning threats and opportunities for planners* (pp. 3–4). Chicago: Planners Press.

Charpin, J. M. (1984a). Quantified macroeconomic projections. *Futures, 18*(2), 158–169 [Note: Reprint by Economic Development Institute of The World Bank to be EDI Training Materials 405/715 Aug. 1984].

Charpin, J.- M. (1984b). *Quantified macroeconomic projections* [Translated from Les Projections Macroeconomiques Quantifiers]. New York: The World Bank.

Corbridge, S., Harriss, J., & Jeffrey, C. (2012). *India today: Economy, politics and society* (2nd ed.). Cambridge, UK: Polity Press.

Daniels, T. L., Keller, J. W., & Lapping, M. B. (1988). *The small town planning handbook* (p. 160). Chicago: Planners Press.

Delors, J. (1965, March). *Planning and trade union reality* (p.155). Droit Social, No.3, March 1965.

Dertouzos, M. L., Lester, R. K., & Solow, R. M. (1990). *Made in America: Regaining the productive edge, New York* (1st Harper Perennial edition, p. 70). New York: HarpPeren.

European Commission. (1999). *ESDP European spatial development perspective* (pp. 0, 2–3, 9–10, 14–15). Italy: European Communities.

European Commission. (2010). *Communication from the commission Europe 2020 a Strategy for smart, sustainable and inclusive growth.* Brussels: European Commission. 3.3.2010. COM (2010) 2020.

Godet, M. (1994). *From anticipation to action—A handbook of strategic prospective.* Paris: UNESCO.

Government of India, Planning Commission. (1981). *Sixth Five Year Plan 1980–85* (pp. iii, 5, 431–463). New Delhi: The General Manager, Government of India Press.

Goudard, D. *Planning process and national accounts in France* (Refer to Note 31).

Indian Planning Commission, Perspective Planning Division. (1981). *A technical nate on the Sixth Plan of India (1980–1985)* (pp. iii–iv, 2). New Delhi: The General Manager, Government of Indian Press.

Institute für Europäische Politik. (1997). *Europe from A to Z-Guide to European integration.* Belgium: EC.

Lewis, W. A. (1966). *Development planning: The essentials of economic policy.* USA: Harper & Row.

Masse, P. (1965). *The plan or the anti-risk* (p. 114). Paris: Éditions Gallimard.

Monnet. J. (1984). *French planning* (P. Pascallon, Trans.). Washington D.C.: EDI [Note: EDI Economic Development Institute Training Materials 405/714] (Original work published 1977).

Monnet, J. (1989). French planning-comment (Caracteristiques de la Planification Francaise, Trans. J. Monnet), In Pascallon, P. (Ed.), *La Planification Trancaue* (pp. 21–27, 1977). New York: The World Bank [Note: Reprint by Economic Development Institute of the World Bank to be EDI Training Materials 405/714 August 1984].

N&R Planning Bureau of Japan. (2008). Japanese new spatial development plan (Original PPT in Chinese pp. 1–10, Wang, H. J. Trans.). In *Seminar of Spatial Development Planning of China and Japan Jointly Sponsored by N&R Planning Bureau, Ministry of Territorial Transport of*

Japan and Tsinghua Urban Planning and Designing Research Academy on 10th, March, 2008. Beijing: Tsinghua University.

National Intelligence Council. (2000). *Global trends 2015: A dialogue about the future with nongovernment experts*. Retrieved from https://www.dni.gov/files/documents/Global%20Trends_2015%20Report.pdf.

National Intelligence Council. (2004). *Mapping the global future: Report of the National Intelligence Council's 2020 Project*. Retrieved from https://www.dni.gov/files/documents/Global%20Trends_Mapping%20the%20Global%20Future%202020%20Project.pdf.

National Intelligence Council. (2008). *Global trends 2025: A transformed world*. USA: U.S. Government Printing Office. Retrieved from https://www.files.ethz.ch/isn/94769/2008_11_Global_Trends_2025.pdf.

National Intelligence Council. (2012). *Global trends 2030: Alternative worlds* (p. 108). Retrieved from https://globaltrends2030.files.wordpress.com/2012/11/global-trends-2030-november2012.pdf.

National Intelligence Council. (2017). *Global trends: Paradox of progress*. Retrieved from https://www.dni.gov/files/documents/nic/GT-Full-Report.pdf.

Planning Commission (Government of India). (2013). *Twelfth five year plan 2012–2017* (Vol. 1–3). New Delhi: Sage Publication India Pvt Ltd.

Sachs, I. (1987). *Development and planning* (P. Fawcett, Trans.). Cambridge: Cambridge University Press (Original work published in 1984).

Stavrianoo, L. S. (1971). *A global history* (2nd ed., pp. 511–512). USA: Prentice Hall.

The Commissart General Du Plan. (1984). *Resume of the French ninth plan* (1984–1988) (S. Dyson & L. Janssens, Trans.). Paris: Press Service.

The World Bank. (1983). *World development report 1983*. New York: Oxford University Press.

Tinbergen, J. (1967). *Development planning* (Dutch by N. D. Smith, Trans.). New York: McGraw-Hill Book Company.

Unesco. (1986). *Planning integrated development: Methods used in Asia*. Paris: United Nations.

United Nations. (1975). *Use of systems of models in planning* (pp. 66, 71). New York: United Nations.

Chapter 8
Boundary, Environment and Social Change of a Social System

8.1 Introduction

All systems are located in space and time of a certain environment. The system exchanges material, energy and information with the external environment. The performance of a system depends upon its internal structure and function on the one side, and also depends upon its active adaption to its external environment on the other side. Moreover, the system also has impact to the environment where it is existed. It should be emphasized that the relationship between system and environment is relative in sense from the nature of hierarchy of a system. For example, in the hierarchical system of social organization, it includes the effect of the international system, the national system, the enterprise system, the family system, etc. Meanwhile, the enterprise system and the national system can also become an environment of the family system. In study of a policy system, it is necessary to consider the broad context of preparation of certain policy. The broad context to be considered includes environments of social, political governmental administrative, economic and cultural, etc. But the above 'environment' in the study of policy system are 'systems' by themselves, i.e. they are social system, political system, governmental administrative system etc. Therefore, the word environment and system can be used interchangeably, and it depends upon thc issuc and thcme to be studied. And, the boundary between the environment and system is divided depended upon the theme under study.

This chapter discusses relationship between the environment and an engineering system as well as between the social system, the latter is much more complicated than the former. Section 8.2 of this chapter will also give a case study of interaction of a corporation, a subsidiary of a large multinational corporation with its external environment, Shenzhen, special economic zone of China. This case provides studies related social systems engineering at firm level which is also a part of study of this book. A general discussion of the levels of arrangement of social system and functional structure will be given in Sect. 8.3 of this chapter. Boundary, social

H. Wang and S. Li, *Introduction to Social Systems Engineering*,
https://doi.org/10.1007/978-981-10-7040-2_8

change and human development will be discussed briefly. A discussion of cultural system and its relationship to social system and social change will be given in Sect. 8.4. Globalization and regionalization will be given in last section of this chapter, as well as the mega trend of social change, a subject of debate in contemporary world.

8.2 Environment and System

8.2.1 General

8.2.1.1 Two Types of Systems

Since the emergence of general systems theory and systems engineering in last centuries, there were many man-made engineering established. For example, many large power stations, large petroleum refining factories, and various types of trans-continental transport projects etc. In planning, design and operating these large man-made engineering systems, there are many factors of natural system(s) to be considered, or narrowly speaking, a natural environment of the system(s) to be faced. For example, in the earlier publication of System Engineering Handbook edited by Machol et al. (1965), part II is titled System Environments. This part includes six chapters, which are "The Ocean", "Land Masses", "Urban Areas", "The Lower Atmosphere", "The Upper Atmosphere", and "Space and Astronomy". These natural environments are important factors to be considered in man-made engineering systems. The mankind has acquired relatively better knowledge of natural system due to progress of natural science, especially since the scientific revolution emerged in the seventeenth century. In recent years, the mankind also known further, such as the interactive effect of their activities to the natural system where the mankind is lived. For instance, the state of the air, the water and the soil affected by the human activities and by the man-made engineering system. In a broader sense, the natural systems are also effected by the other type of social systems defined by Parsons (see Chap. 1). These two types of systems interact with each other, and promote changes with each other.

The mankind has acquired a fair amount of knowledge of the natural system, i.e., to adapt to the nature of the natural system, or even actively manipulates the natural system, i.e. to regulate or to utilize the natural system to certain extent for the benefits of themselves. But the man-made system (the social system is also made by the man) is more complicated than the natural system. For example, one of the environmental problem (of the natural system), caused by the human activities (of the social system), is climate change due to the greenhouse effect. This phenomenon had been discovered by scientists and was recognized globally by the decision makers of many countries among the world. Some international organizations has taken the lead to study this environmental issue and its impact to the mankind, such as UNEP (United Nations Environmental Programmes) and also at the regional level. For example, the European Environment Agency had also given assessment of Europe's environment. A report was prepared and published which was requested at the DobříŠ Conference of European environment ministers in 1991. This report, titled Europe's Environment-The DobříŠ Assessment edited by David Stanners and Philippe Bourdeau, which was published in 1995 has the following statements related to climate change:

> Greenhouse gases in the atmosphere have increased since preindustrial time by an amount that is radiatively equivalent to about a 50 percent increase in carbon dioxide (CO_2), although CO_2 itself has risen by about 25 percent; other gases have made up the rest, the global emission of most greenhouse gases is expected to rise in the next decade: CO_2 at 0.5 percent/year, methane (CH_4) at 0.9 percent/year, and nitrous oxide (N_2O) at about 0.3 percent/year. In contrast, CFC emissions are expected to decrease to near zero around the year 2000 following internationally agreed phased out measure. As a result of these trace gas trends, an effective doubling of greenhouse gas concentrations (as commonly measured as CO_2-equivalents) is expected around 2030. (p. 513)

But the global social system has no consensus on the impact of climate change, the recent US withdrawal from Paris Climate Change Agreement can illustrate the complexity of the social system.

8.2.1.2 The Boundary

(1) *The Meaning of Study the Boundary*

The boundary between a physical system or man-made engineering system with respect to its external environment is relatively easy to be identified. It is also true to identify the boundary between the natural system and social system. But in the planning and design of a social system, the determination of the boundary line is relatively complex, and it depends upon the issue of the relationship of the social system to be studied. It may be physical, geopolitical, social, economic, psychological or technological. In physical engineering system, the system engineer needs to pay attention to the points of linkage between different disciplines and between different subsystems. Design and planning of each subsystem are responsible by experts or professionals specialized in the discipline of certain subsystem. Similarly, the role of system engineer, a generalist, who should also be responsible to coordinate the linkage among all subsystem and to make sure that the whole system is balanced with satisfactory performance. It is also true for the planning or design of a complex social system. For planning of a nation or a large firm, the economists, sociologists and engineers or scientists will be responsible for planning each subsystem, such as, economic system, social system or science and technological system. Economists, sociologists and engineers or scientists make plans based on their expertise. It is the role of the chief planner, or team of professionals who can perform the role of generalist (of social systems engineer) to look for the boundary among economy and society, among science and economy etc., so that the whole social system will have a balanced and satisfactory performance.

(2) *Broad Classification of Types of Boundary in Social Systems Engineering*

The boundary between social system is generally complicated since the boundary line is not physical, and it is identified through norms, institutions, and regulations. The boundary may be subjected to change when there is social changes. Possibly a broad classification can be of two types, physical and conceptual. The former types is easy to be understood, much can be discussed for the later type. For example, various established disciplines have relatively clear boundary of their scope of study. But there were emergence of interdisciplinary studies in recent decades, and it becomes a complex subject to be studied further.

8.2.1.3 Business Planning and Environment

(1) *Environment of Business*

As discussed in the introduction, that the higher order system in hierarchical order forms the environment of the system of lower order. Enterprises or business firms have to be existed or operated under certain environment. These environments, have been well studied by planners of business, which include political environment, economic environment, social environment and technological environment. These four together form PEST analysis in modern business planning. This book has focused on the formation of basic framework of social systems engineering and its applications. The application to a national planning has been discussed. Business firms to be one element of social system, and the subsystem of the national system, and their growth and development is also an important part of study of the national system. Therefore, a very brief discussion of business planning will be given here.

(2) *Planning in Business Firms*

Western business firms had a long tradition in preparation of various types of plans, for example, forecasting and long range plan, strategic plan, work plan, marketing plan etc. It is interesting to note that one of the title of *The Business Book,* edited by Atkinson, in 2014, "Plans are useless, but Planning is Indispensable" Follow the title, is the discussion of the scenario for planning, preparation for change. Contingency planning is also discussed in that book, because firms have to prepare for sudden changes to markets or the environment to ensure the continuous operation of the business (Atkinson, 2014, p. 211). In recent hundred years, there are relatively rapid development of multi-national corporations since the first Multinational Corporation, the East India Company arose in 1600. These MNCs either resource seeking type or market seeking type, and they must do environmental analysis with in-depth in order to achieve their strategic goal. In a dynamic and changing society, launching a new business requires careful planning. This business plan generally includes the vision, goals, strategy of the business and tactics employed to achieve them. The substance of the plan is crucial. The strategies and tactics employed should be the outputs through a logical and comprehensive business planning process. The process is generally begin by evaluating the environment.

(3) *PEST Analysis*

The PEST Analysis is commonly done together with the scenario planning in business's plan. In fact, PEST analysis and scenario analysis should also be applied in national planning. Some concrete contents of PEST analysis are shown below:

(i) *Political Factors or Political Environment to be Analyzed*

(a) Taxation System

The direct and indirect taxes influenced consumer spending and market demand; and corporate taxation has impact on profitability of business.

(b) Fiscal expenditure

Public spending by government has direct impact on demand of the economy.

(c) Monetary Policy

Monetary policy and interest rate will influence demand and debt service ability of business. Exchange rate policy will influent foreign trade organizations.

(d) Regional and industrial policy

These will affect the operation of business and regional preferential policies, and it may be a determining factor to locate a business.

(e) Change in international trade can create new export markets

For example, China becoming a member of WTO on 2001 had promoted greatly its external trade. Its recent 'One belt, one road' initiative will also affect the environment of trade.

(f) Education and training

It will have a long term impact on a business' ability to recruit qualified personnel.

(g) Regulation and deregulation

(ii) *Economic Factors or Economic Environment*

(a) Stage of economic development

The stage of economic development will affect the demand of the product and service supplied by the business. And also the stage of development will affect the level of infrastructure, such as electricity supply, availability of communication and transport service. All these are important factors to effect a successful operation of the business.

(b) Financial environment

This means easy access to credit and relatively low rate of inflation.

(c) Employment level

High levels of unemployment in a region will reduce demand of the market.

(d) Housing and land price

These will affect the cost of production and they are generally the price of factors of input of production function.

(e) Oil and commodity price

The dramatic increase of oil prices in 1970s had caused so called global oil crisis and put downward pressure on profits in many parts of the economy. This situation only recently has being changed due to the progress of technology of the exploration of shale oil in U.S.A.

(iii) *Social Factors or Social Environment*

(a) Size of population and its growth rate

The size of the population is a very important factor for business plans, since it effects the scale of the market. China has attracted Foreign Direct Investment rapidly since its reform and opening in late 1970s. Especially post its establishment of four special economic zones and coastal opening cities. This situation is also true for India and Africa.

(b) Demographic structure

Differences in the age structure have influences on savings and spending. Japan, for example, it can be adapted to its aging population domestically; it also develops product and service to adapt children and youth in the global market.

(c) Social and cultural pattern

Different countries and different regions differ greatly in social and cultural pattern. This will also affect the type of product and service supplied by business.

(d) Mobility and migration of the population and labor force

In a mega trend of globalization and regionalization at international level and rapid rate of urbanization in many developing countries, there is a trend of increasing mobility and migration of the population and labor. This trend will have implications for business sales and distribution strategy.

(iv) *Technological Factor or Technological Environment*

(a) Rate of investment on R&D

This depends upon the type of product and service supplied by the business. Generally, a high ratio of expenditure of R&D to total sales will be applied to high technology firms, this ratio declined following the order of medium technology and low technology firms. The level of expenditure should also be determined by comparison with competitors of the market.

(b) Buy and make decision analysis

In contemporary society, there are a lot of technological resources available. Developing a new product can be achieved by transfer of technology through license; import crucial parts of the equipment or develop by business firm itself. Meanwhile, resource of professionals of the firm itself should also be considered as a factor.

(c) New markets and innovation of production method

Does the adoption of new technology can create a new market and how the technology can be used to improve production methods with cost-effectiveness should be analyzed.

8.2.2 Case Study of Interaction Between Enterprise System and Its Environment-Growth of Yantian International Container Terminals (YICT) in Shenzhen Special Economic Zone (SEZ) of China

8.2.2.1 Background of This Case

China had launched its reform and opening to outside world in late 1970s, i.e. post the Third Plenary Session of the Eleventh Party's Congress opened in Dec. 1978. But it takes quite a long period to transform China from a former centrally planned economy into a socialist market economy with indicative planning. A gradual approach to regional opening was adopted by the Chinese government. Four Special Economic Zones (Shenzhen, Zhuhai, Shantou and Xiamen) were opened in 1980, fourteen coastal cities (Tianjin, Shanghai, Dalian, Yantai, Guangzhou, Zhanjiang etc.) were opened in April 1984, three coastal economic opening regions were launched successively on February 1985 and on March 1988 respectively. The three coasting economic opening regions including Yangtze River Delta, Pearl River Delta and Southern Fujian (Xiamen, Zhangzhou and Quanzhou). In later period, there were opening regions including opening of municipalities, counties and coastal cities of Eastern Liaoning Peninsula, Shandong Peninsula and Bohai rim. The Pudong New Zone in Shanghai was opened to overseas investment in 1990.

Evolution of Foreign Direct Investment policies can roughly by divided into three stages:

(1) *The Initial Stage (1979–1982)*

In the initial period of attracting FDIs, China had no or little experience in dealing with foreign companies. Also its legal system was unsuitable to facilitate business transactions with them. There were only 3841 contract projects with contracted FDI around 28.1 billion U.S.D. But total amount of U.S.D. actually utilized were 18.2 billion.

(2) *The Sustained Growth Stage (1985–1992)*

It can be seen from above that there were further opening of 14 coastal cities, and further improvement of FDI incentives was made in 1986 with the promulgation of regulations for encouragement of foreign investment. There were sustained growth of FDI in this period. The FDI actually utilized in 1985 was 4.76 billion U.S.D., it was increased to 19.2 billion U.S.D. in 1992.

(3) *High Growth Stage Post 1992*

FDI actually utilized was increased from 38.96 billion U.S.D. in 1993 to 49.6 billion U.S.D. in 2001. It was increased further from 55.0 billion U.S.D. in 2002 to 126.3 billion U.S.D. in 2015, i.e. the period post China's joining of WTO.

A series of laws and regulations were promulgated to promote various forms of FDI venture. Equity joint venture was the first type of foreign funded enterprises established in China. Its establishment had been made possible by the "Law of the People's Republic of China on Joint Venture Using Chinese and Foreign Investment" promulgated on July 1979 by the National People's Congress.

Hutchison Port[1] Holding (HPH) (a subsidiary of CK Hutchison Holding, the former Hutchison Whampoo Limited, which is an investment holding company based in Hong Kong with diverse array of holdings) and Shenzhen Dong Peng Industry Company (later renamed to be Yantian Port Group YPC) signed the joint venture agreement on Oct. 5th 1993 establishing Yantian International Container Terminals [YICT]. This is a joint venture with HPH as the majority owner and the Yantian Port Group (owned by the Shenzhen government) as the principal minority shareholder. This case study is in the context of interaction between the growth of a corporation with its parent company which is a large Multi-National Corporation (listed in the Global 500 of Fortune Magazine) and the meso and macro environment of China in transition from a former centrally controlled economy to a socialist market economy.

8.2.2.2 Purpose of This Case Study

This case study is a very brief abstraction of a book, *The Dynamics of Corporate co-Evolution.* This case can also illustrate authors' view of this book, use active adaptive action to replace "adapt" in Parsons' AGIL framework. That book can supplement many useful materials relevant to social systems engineering at enterprise level.

[1]Note: HPH had a network comprised 48 ports globally in 2016.

The *Dynamics of Corporate Co-Evolution* is briefed in the following:

(1) This book is the product of cooperation of three professors, professor Child, J., Emeritus Professor of Commerce, University of Birmingham, professor Tse, K.K.T. Adjunct Professor, the Chinese University of Hong Kong and professor Rodrigues, S.B., Professor of International Business and Organization, Erasmus University. Professor Tse was also the former general manager of YICT since its establishment. This book, a case study of port development in China is the product of Prof. Tse's experience in Shenzhen SEZ in close association with detailed academic research by two eminent professors from abroad.
    The major theme of this book is a historical retrospect of the growth of YICT, a port corporation which is an equity joint venture of Hutchison Port Holding (a multinational corporation based in Hong Kong) and Yantian Port Group, an enterprise owned by Shenzhen city government within a highly politicized and institutionalized environment.
(2) This book contains three parts and ten chapters. Part I is Introduction, Perspective and Method. Chapter 2 of part I has given a detail discussion of co-evolutionary perspective. In Fig. 1.1 of this book, not only it shows the interaction of the human social system with the ecosystem, but also illustrates the concept of co-evolution and co-adaption and the human social system within the ecological environment. While the co-evolution perspective described in this book is co-evolution between social systems, i.e. between an enterprise and different levels of governmental system. Chapter 3 of part I focused on relevant methodological issues, a research design is presented and a summary of research methods and sources are attached which can provide precious reference for sociological studies.

Part II of this book is titled Environment, Evolution and Managerial Initiative. It contains four chapters. Chapter 4 with the title "Yantian port and its changing environment." This chapter begins by putting Yantian on the map which is essential to study an enterprise to know the feature of its location and its environment, and then, describes the changing context in which it evolved. YICT's macro environment and meso environment are described in succession. A table titled "evolution of the environment and its relevance to YICT," and listed evolution of macro and meso environment in three periods: 1978–1989, 1990s and 2005. Chapters 5, 6 and 7 focus on the "'intentionality' of YICT's management in the process of the company's co-evolution with its environment." These three chapters also illustrate the rationality to use the concept of active adaptive action to replace "adapt" of Parsons' AGIL framework of action system. Chapter 5 gives detailed analysis of the stages through which YICT evolved during the first 15 years of operation and become a world-class port. Change of concept and management initiate in different periods is outlined. Chapter 6 describes areas of innovative management practice that YICT introduced to strengthen internal capability to realize its intent of strategy. The advanced practices introduced by the management of YICT also improves its influence of co-evolution of its external environment. Chapter 7 deals with relation management, a framework of relationship is created in dealing with the Chinese joint venture partner, in dealing with ministries of Chinese government over the expansion of port and in dealing with governmental regulatory authorities operating in the port.

Part III with the title Co-Evolution: Theory and Practice is to analyze the main themes from the Yantian story, and then, to advance the understanding of theory and practice of co-evolution. Part III contains Chapters 8, 9 and 10. Chapter 8 deals with forms of co-evolution. Two types of co-evolution are identified: the 'asymmetrical' type of co-evolution and a more 'symmetrical' kind of evolution. The former type of co-evolution is that changes in the environment are its primary driver, and the 'fittest' firms must adapt to environmental requirements in order to survive. This is in consistent with traditional evolutionary view. While the later type of co-evolution is in similar with active adaptive action raised in our book, that both changes of environment and active actions adopted by firms have mutual impact to each other. In Chapter 8, the authors of The Dynamics of Corporate Co-Evolution have adopted the concept of system components in the co-evolution of YICT and Its environment, which is shown in Fig. 8.1.

Chapter 9 "develops a systematic analysis of the political dynamics of co-evolution, illustrating these primarily by reference to interactions between the company and government agencies on the issue of port practices" (pp. 11–12). Chapter 9 has a Table 9.1 "*Bases for power and influence in YICT's Co-evolution*", five types of power resources are identified, material resources, coercion, legitimacy, reference and expertise. Chapter 10 is the closing chapter. Seven approaches are derived from the case study of YICT for the management practioner. They are: Read the environment holistically for first mover advantages; Manage key external relationships proactively; Innovate to survive and break the vicious cycle; Mobilize organizational resources to meet growth challenges: Develop collaboration based on 'Power With'; Build a sustainable business ecosystem and learning consistently in the crafting of strategies.

Figure 8.1 is composed of three blocks A, B and C which represent macro, meso and micro system respectively as well as the interactions. Co-evolution is represented by heavy black line with arrowhead at both ends. Descriptions in block A and B presenting features of the respective system. Block A represents the macro system or national system of China. This macro system is marked by features of developing countries, its imperfect market situation, immature legal framework etc. This national system A is also the macro environment of the meso system B and the micro system C, the corporation of YICT. Block B is meso environment of micro system C. The sectoral features of meso system B, the shipping and port industry are shown by their oligopolistic competition, barriers to entry, capital intensity etc. Block B also shows the unique ownership system, the international joint venture ownership especially the majority non-mainland Chinese ownership. Block C represents the micro system, YICT Corporation. It has set up strategic intent and evolution of visions and strategy as well as its international joint venture management highly influenced by its parent company and should also take due consideration of the mainland partner. Its organizational changes to adapt to macro and

Forms of co-evolution

Macro environment – Contextual variables

Developing country factors

- Imperfect market situation
- Immature legal framework
- Government control and intervention
- Bureaucratic administration with underdeveloped rules & practices
- China's historical and cultural legacy

(a)

Meso environment

Shipping and port industry dynamics

- Oligopolistic competition
- Barriers to entry
- Capital intensity
- Government involvement in project approval and implementation
- Control by government agencies in everyday port operations

International joint venture ownership

- Majority non-mainland Chinese ownership, an exception for the sector

(b)

Micro level –internal/intra-organizational dynamics

Strategic intent

- Evolution of vision and strategies

IJV management

- Partner relations and parent influences

Organizational adaptation and change

- Strategic positioning
- Sales and marketing
- Managing partner relationships
- Managing government relationships
- Establishing a favourable ecological system

(c)

Organizational performance

- Viability and contribution to the community
- Financial Performance

⟷ = Co-evolution

Fig. 8.1 Macro, meso and micro system components in the co-evolution of YICT and its environment (*Source* Adapted from Child, Tse, & Rodrigues, 2013)

meso environment are shown by its strategic positioning, sales and marketing, managing various relationship etc. And with its organizational performance, it can achieve viability and contribution to the community and also its sound financial performance, then it can influence macro and meso environment to achieve co-evolution.

8.2.2.3 Comment by Author of This Book

The author appreciates highly the contribution of general manager of YICT who has a working experience to launch YICT project in a difficult environment. Through the effort of the general manager as well as his team around 16 years, YICT becomes a world class port. The first author of the book has association between Kenneh Tse, around 10 years. Due to the contributions of both academics and practitioners, this book is backed by advanced theoretical exploration and validated with reality of objective world. It is an important part of social systems engineering relevant to enterprises. Wang, H. J. had written a paragraph of acclaim (2013) to this book that:

> The dynamics of corporate co-evolution provides an excellent exploration of co-evolution from the perspective of power relations within a hierarchical system. It is relevant not only to firms working within a political environment, but also useful for people working in think tanks and policy analysis. Its treatment of relation management has universal implications. (p.v)

8.3 Social Structure, Function and Social Change of Social System

8.3.1 Social Structure and Function of Social System

8.3.1.1 Clarification of Basic Theory Adopted

In our exploration of social systems engineering, we are biased more to Parsons' Functional Structural theory. Niklas Luhmann had also developed his social systems theory around 1970s and had got wide acceptance in the field of sociology. But Luhmann has focused more on communication rather than on individual action. Communication and information exchange are part of important activities of the human society. Luhmann correctly point out the integration of different systems is a difficult issue, which requires certain forms to connect separated systems, in which some systems producing translation communications, can be understood by other systems. But we feel Parsons' functional structural theory and action theory which approaches more the reality of the objective society. Its weakness can be complimented by other theories. The social system engineer shall adopt appropriate theories fitted most to his applications, not necessary a complex or advanced theory.

8.3.1.2 Social System Analysis

Social system can be analyzed from two perspectives: one perspective is to analyze the social system from its structural components in hierarchical level, which is shown in Fig. 8.2. This figure is a simplified structural functional diagram of a national social system. The left hand side represents the hierarchical structural relationship of a national society, which is composed of five levels of individuals to be the basic unit. The actions top down and the reactions bottoms up are shown by two parallel lines with arrowhead connecting the blocks representing the level of

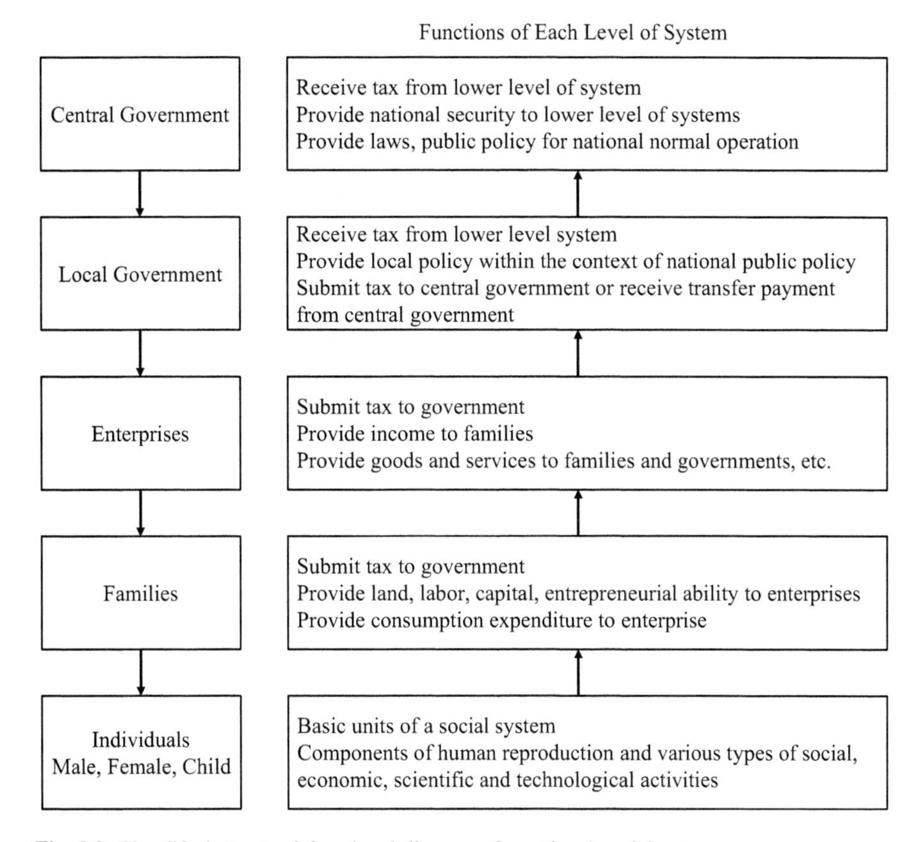

Fig. 8.2 Simplified structural functional diagram of a national social system

the system. The functions of each level of systems are shown in blocks at right hand. This is a very simple structural functional diagram of the national social system for illustrative purpose to show the concept of structural functional approach. The functions listed in blocks of right hand are mainly economic in sense[2], the main flow between different levels of system is fiscal income and expenditure and resource and product flow between family and enterprise, which is shown in a part of functions listed in blocks related to enterprise and family.

The second structural functional perspective of a society is an equilateral triangle, the ideal national social system preferable to have a balanced structural and functional approach of its three subsystems, the social system (including political system), the economic system and scientific and technological system, which is shown in Fig. 8.3. But the general growth pattern of many countries are not balanced, some may emphasize the social side etc., sometime may emphasize the

[2]Note: This function of social system and S&T system can be drawn separately.

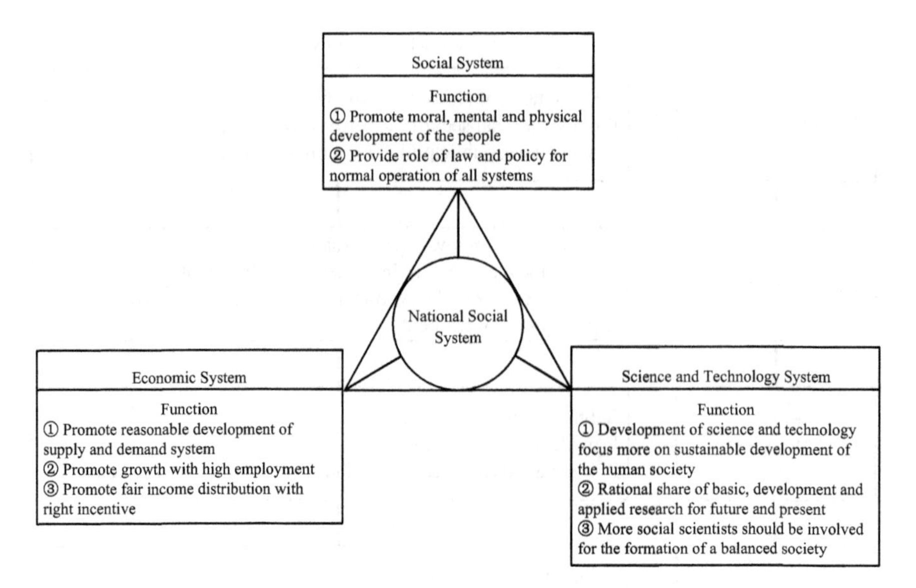

Fig. 8.3 Balanced approach of functional structural diagram of a national society

economic and science and technology side. The later one is very popular in many developing countries. But a part of developed countries have relatively balanced approach of its social system, which should be analyzed from historical, cultural and development perspective. This can be shown from the data of Table 8.1. The Nordic countries share in common the Nordic model of economy and social structure, market economy combined with strong labor union, a high universal welfare financed by heavy taxes. And there is also a high degree of income distribution with little social unrest. Table 8.1 shows that all the Nordic countries have lower Gini coefficient compared to other countries. But even the Nordic countries have also disadvantages in lacking of competitive advantage at several sectors of their economies due to excess welfare. Lessons and experiences should be learned to improve the balanced approach in the history and future prospects of development of the mankind.

In the balanced approach of functional structural analysis of a national society, three functional systems of first level are classified in studies of this book.

(1) *The Social System*

Political system is an important part of social system. It is separated listed to be an individual system in the Parsons' analysis. In our functional structural analysis, political system is included in the social system. In pure academic sense, the broad nature of social system should include the economic and S&T system. But in the analysis and synthesis of social systems engineering, it is also conceptually reasonable to set up boundary line between three systems.

Table 8.1 Comparison of economic performance of selected countries or union (data are mainly 2015, exception will be noted)

Country or group	GDP official exchange rate	GDP growth rate (%)	GDP/capita (PPP)	Gini coefficient
EU	16.27 trillion $	1.9	37,800	30.9 (2014 est.)
Nordic countries				
Denmark	295 billion $	1.2	45,700	24.8 (2011 est.)
Finland	229.7 billion $	0.4	41,100	26.8 (2008)
Norway	389.5 billion $	1.6	68,400	26.8 (2010)
Sweden	492.6 billion $	4.1	47,900	24.9 (2013)
Iceland	16.72 billion $	4	46,100	28 (2006)
U.S.A.	17.95 trillion $	2.4	55,800	45 (2007)
Russia	1.325 trillion $	−3.7	25,400	42 (2014)
Japan	4.123 trillion $	0.5	38,100	37.9 (2011)
China	10.98 trillion $	6.9	14,100	46.9 (2014)
India	2.091 trillion $	7.3	6200	33.6 (2012)

Note Data source from CIA (2017) The World Factbook. www.cia.gov/library/publicatoin

(2) *The Economic System*

From academic classification, economic system should be a part of social system. However, due to the global trend focusing development of the economic system to be priority, and its role and function had been well defined. Therefore, the economic system is considered individually to become an independent system in our functional structural analysis of a society.

(3) *Science and Technology System*

Generally, this system was neglected by functionalism of sociology since science and technology may be related more to natural science. But it is exactly the role and performance of science and technology, which drives the major force of the society. The mankind is evolved from stone age, bronze age, iron age, the polymer age and currently the information age. It is the application of those basic material to be tools and goods of a society, and therefore, those materials become the mark of stage of historical development of the mankind. The non-material information to be a mark of stage of the mankind is a turning point of human development, which shows the development of society of the mankind depending less on physical materials. Nonetheless, it is necessary to have the awareness that a large part of developing countries are still depended more on material wealth currently. From the perspective given above, science and technology should be an important system of constituents of social structure with their function to be a driving force of social progress.

8.3.2 Social Change

8.3.2.1 Basic Concept of Social Change and Its Types

It had been described in Chapter 1 that "The social system is dynamic in sense, it is subjected to change through internal and external force." The common social change is historical change, which is shaped by human beings in large and pervasive scale, for example, war and peace, trade and production, science and technology (breakthrough of S&T), religion, belief and thoughts, power struggle and politics etc. Changes can also be classified into subjective change and objective changes. The former is change of perspective of the individual of the actors, while the latter is changes of environment external to the individual or actors.

In Fig. 8.2, there may be change of individual consciousness. At the family level, there were changes from an extended type of family structure with co-living of several generations, to smaller unit of family of nuclear type in the transition of human producing activity from the agricultural society to an industrial society. There were also changes of organizational ownership from sole proprietorship to partnership to corporation and to multinational corporations, etc.

8.3.2.2 Meaning of Study of Social Changes

Social change is pervasive. There are different perspective to deal with social changes. The first one is adaptation, i.e. with personal subjective change to adapt to the objective changes of environment. The second one is active adaptation which had been used in revised Parsons' AGIL framework. Active adaptation means the actor who kept personal value and criteria to judge whether the external change is beneficial or detrimental to the human development. Adapt actively the change which is beneficial to the human development and adapt with reservation or resistance to the detrimental changes. It is also interesting to note, there are four styles of people in responding to social changes: The reactive orientation: "Back to the Future". The inactive style: "Don't rock the boot", the proactivity style: "Riding the tide" and The interactive. Style: "Shooting the rapids", (Banathy, 1996). This is a subject required broad studies.

There are also study of social change and human development by applying challenge-response model to study the social change and its impact to individual's conviction (Silbereisen & Chen, 2010).

8.3.2.3 Factors Effect Social Changes

The human being has been living on earth for approximately 500,000 years. The human way of life and social mechanism have undergone tremendous changes. The modern society is already radically different from the society several hundred years ago. Social changes are influenced by many factors, the top three of them are:

(1) *Physical Environment*

Physical environment exerts remarkable impacts on the development of social organizations of humans. People must organize their ways of life and work in order to meet the local conditions such as climate and geography. A social group and organization of people living in the cold and mountainous regions for example have different ways of work and life from a social group living in the tropical plain region.

(2) *Political Organizations*

Social shifts are impacted by the political organizations. The landlords, syndicates, conglomerates, kings, presidents and the various political parties have appeared in the agricultural and industrial societies after the primitive society have a significant extent influenced the ways and models of social development. This is still the major factor of social change. It is worthwhile to remember that many leaders and a part of followers had not got off from the trap of wars, i.e. two world wars happened in 20th century and caused loss of eighty millions of people. There were also happened twice the global financial crisis, which had caused loss of property severely of a large amount of people. The mankind, to be a rational animal, have to learn lessons from the history and promote the trend of beneficial social change for their development.

(3) *Cultural Factors*

Cultural factors may include religions, information and leaderships, among which information plays an essential role. The development of languages enabled people to record the accumulated knowledge, and therefore, increase the control of material resources and form organizations of major scales. Many of the leading figures in history have also brought positive or negative impacts on social changes. Religious leaders, political leaders and outstanding scientists and philosophers have all played a huge role spurring social shifts. While some leaders changed people's ways of thinking, others inspired distinct social transformations by overthrowing the established social orders.

8.4 Cultural System and Social Changes

8.4.1 Introduction

Cultural system existed as old as the existence of human society. And it is the culture which had effected the life of the people and the society more than several thousand years. But its systematic study may be around only 200 years. Currently there are more than several tenth of definitions of culture by sociologists or anthropologists. But the oldest definition given by Sir Tylor, E. B. (1832–1917) can be served to be a starting point, it is "that complex whole which includes knowledge, belief, art, morals, law, customs and any other capabilities and habits acquired by man as a member of society." (Taylor, 1871).

It can be seen from above primary definition of culture, it nearly includes all kinds of human activities. There are many studies related to culture. This book will focus more on culture and social change, i.e. cultural system and development. More definitions of culture will be given to broaden concepts. A brief discussion of formation of culture and its changes will be described, and development of various schools of cultural theory abroad will be briefed. Development of Chinese cultural system will be introduced further. It is suggested by authors of this book that this part should be read together with Chapter 3 to understand better China's cultural system, and its relation to development.

8.4.2 More Definitions on Culture

8.4.2.1 Definition of Culture Given by Robert Bocock (1992)

Definition of culture given by Bocock (1992), and then adapted by Schech and Haggis (2000), which is shown in Table 8.2.

The above definition of culture has defined the word "culture" in five times. This represents culture is dynamic in sense. It changes along with evolution of the society. In the primitive stage of society, the culture of the mankind is simply cultivating land to get crops or domesticate animals for basic survival. When the mankind can achieve basic survival, then they turn to cultivation of mind, arts and civilization. And along with the development of the mankind, they further go up to culture 3, 4, and 5 respectively.

8.4.2.2 Definition of Culture Commonly Used by Sociologists, and Supplemented by Authors of This Book

Culture is the summation of symbols (immaterial) and finished products which was created and applied by the mankind. It constitutes the ways of life and concepts of individuals that make up different societies. It encompasses the criteria and belief

Table 8.2 Definition of culture

1. Culture = Cultivating land, crops, animals
2. Culture = cultivation of mind, arts, civilization
3. Culture = Pross of social development
4. Culture = Meanings, values, way of life
5. Culture = Practices which produce meaning

systems concerned with clothing, food, housing, languages, attitudes, rules, rituals, marriage customs, family life, work, leisure and behaviors.

From perspectives of authors of this book, the culture of a nation is also affected very much by the political culture. The general culture of a nation should be considered within the context of the political culture.

Political culture is defined to be the "set"[3] of attitudes, beliefs and sentiments that give order and meaning to a political process and which provide the underlying assumptions and rules that govern behavior in the political system.

8.4.2.3 Definition and Interpretation of Culture Given by Anthropologists

The anthropologists in contemporary world give their interpretation of culture in following five aspects[4]:

(1) Culture provides the rules and norms, and can promote orders, regularity, familiarity and foreseeability. Otherwise, a world of unorganized people, events and conducts will result;
(2) Culture explains the meanings of human conducts and events in the material and non-material worlds, so that humans can construct their practical worlds and communicate among them;
(3) Culture contains values. A problem faced by the human race is the inability to scientifically construct and improve values. In the process of boosting cultural progress, humans face a common problem. That is, the emergence of harmful concepts in human societies, or even the concepts of superior and inferior cultures that discriminate against the human races. An accurate concept of culture is the practices, social relations, concepts and emotions favored by the human societies;
(4) Culture is a commonly shared phenomenon in which systematic concepts, interpretations of people and their behaviors, and phenomena of material and non-material worlds are jointly shared by members of the same cultural groups;
(5) The expressions of cultural models are highly impacted by the environment, though not determined by it. Factors in material and social environments can therefore both influence the expressions of culture. While the impacts of material environments may take several generations to show their effects, the influences of social environments due to feedback mechanism of social culture will immediately and continuously affect expressions of culture.

[3]Source: Sills and Merton 1968.

[4]Source: Whiltehead (2003).

8.4.3 Formation and Change of Culture

8.4.3.1 Formation and Change of Culture

Foreign sociologists and anthropologists have been probing into the nature of culture, its formation and change for centuries, and formulated understandings and concepts below.

Sociological theories generally believe: human behaviors are results of nurture (social determinism), rather than the products of nature (biological determinism that promotes natural endowment). The distinction between humans and other animals is that the humans possess the ability to collectively create and disseminate culture of symbols.[5] The production and dissemination of knowledge is realized through complex social processes. Humans, both influent and are influenced by cultures, and thus, producing new forms and contents of culture. The traits of different cultures are therefore determined by their historic properties, relativity and diversity. They change in line with the organizations of politics, as well as economic and social changes. Agricultural society, for example, has culture of agricultural mode of production and consumption, while industrial society has culture of industrial mode production, consumption, social organization and life style. The cultural properties are thus different between a feudal society and a modern democratic society.

In the process of cultural changes, the insides of the society are usually influenced by two forces, i.e. promoting changes and fighting against changes. These forces prompt the formation of cultural concepts and practices in the existing social structures, and are both subjected to the influences of the social structures, natural conditions and development of technology. Social conflicts, technological development and changes in economic structure can all cause changes within a society, and promote the formation of new cultural patterns, changes of concepts and other cultural changes. Changes in energy technology, for example, have driven the human evolution from an agricultural society to an industrial society, and the style of people's life and major mode of production of these two societies differed greatly. Development of information technology is driving the human society from an industrial to post-industrial society or information society. As the nature of the society changes, the material and non-material human cultures will undergo corresponding changes.

8.4.3.2 Diversity of Culture

Cultural belief varies among different ethnic groups. Human behaviors and practices also feature notable differences. The acceptable behaviors vary tremendously

[5]Note: Symbols comprise the capability of characters, pictures, human's postures, works, music notes, etc.

between different cultures. In the contemporary western society for instance, parents usually do not interfere with the study of their children, whilst in some eastern societies, the "Tiger Mother" is a main role model to boost children's study. On the other hand, cultural differences and diversities are also often the sources of conflicts between people. It was pointed in the *Report of the World Commission on Culture and Development* of UNESCO titled *Our Creative Diversity* (1995) that:

> In a world in which 10,000 distinct societies live in roughly 200 states, the protection and accommodation of minority rights is a principal concern. But minorities have also asserted the rule of the majorities such as British in India in the past, the Afrikaners in South Africa and the Central and Eastern European Communist Parties. Minority rights should not be at the expense of the rights of majorities. Nor should vocal bullies, pretending to speak for minorities, as the voice of their people. Democratic "voice" should also be heard much more at the international level than has been the case so far. The commitment to the pace resolution of conflicts and to fair negotiation, and equity both within and between generations, are other important principles of this global ethics. (p. 16)

The above quotation shows the complexity of cultural diversity. It shows that there are around 10,000 distinct societies with very different cultures. When acquiring or merging with companies in another country, the major multinational corporation nowadays for instance have to consider the cultural backgrounds of the staff in the countries where companies being acquired or merged with are located, for it will have significant impacts on the future operations of the companies being acquired or merged with.

8.4.4 Western Culture and Exploration of Cultural Theories Since the Scientific Revolution

8.4.4.1 Western Culture

It had been described previously that the existence of culture is as old as the existence of the mankind. And culture differed greatly between regions, countries, races etc. Because modern cultural theories were emerged in Western countries, a very brief discussion of Western culture and several schools of cultural theories emerged since the scientific revolution will be briefed here.

Western culture, which is quite different from the Chinese culture, is nearly equivalent to Christian culture throughout most of its history. Ancient Greeks, such as Plato, Socrates and Aristotle, had helped to establish Western philosophy which still has its influence on Western culture in the modern world.

(1) *Explorations of Cultural Theories in Western Countries before the 20th Century date back to the Period of Scientific Revolution following the Renaissance*

The German philosopher Kant (Immannel Kant 1724–1804) developed the concept of "enlightenment" which is similar to the German word "bildung" (culture). He argued that "enlightenment is man's emergence from his self-imposed nonage." He believed that the nonage which "lies not in lack of understanding but in indecision and lack of courage to use one's own mind without another's guidance." He urged people to "dare to know! (Sapere aude.)" German scholars after him further explained the creativity of mankind, arguing that human creativity features ways of unpredictability at different levels, and is as important as human reason. Due to the need of nationalism for the establishment of the German state, German scholars also proposed that "culture is a world view" and "different races have different world views" and also different theories. In 1860, the German scholar Adolf Bastian (1826–1905) suggested to conduct scientific comparisons of all human societies to discover the same set of "basic concepts" shared by different races in their different world views. Bastian reasoned that different cultures around the world developed by adapting those "basic concepts" to local conditions. This viewpoint paved the way for the formation of the modern cultural theory. The famous anthropologist Franz Boas (1858–1942) accepted this concept and later left Germany and took this concept to America.

The 19th century British humanist, poet and prose writer Mathew Arnold (1822–1888) used the word "culture" to imply the individual ideological culturing of the humanities, and referred culture to the "world's finest ideas and words". In his era, the word "culture" was a concept of the elite class and was associated with activities like arts, classic music and high-class cuisine which are in turn associated with city life (Throsby, 2001). The word "culture" almost became synonymous with civilization in the Great Britain of that age. Scholars also began to study the "culture" of the non-elite classes, that is, the so-called folk culture. The concepts of high culture and low culture thus appeared in the European societies in the 18th and 19th century. It reflected the inequality of the European society in that historic age. This may be more or less the misleading concept in a class society.

In 1870, anthropologist Edward Tylor (1832–1917) used the theory of high-low culture argument to found the theory of religious evolution, in which he contended that religion evolved from polytheism to monotheism. In the process, he also redefined culture as "the characteristics of different types of activities shared by all human societies", a concept that paved the path for the exploration of the contemporary culture. He later illustrated culture in 1874 as such, "culture, or civilization,[6] taken in its broad, ethnographic sense, is that complex whole which includes knowledge, belief, art, morals, law, custom, and any other capabilities and habits acquired by man as a member of society." He made this statement not as an attempt to provide a general theory of culture, but to explain his understanding of culture in a major discussion on the nature of religion. American anthropologists after Tylor either complemented or revised his understanding of culture to shape their definitions of culture.

---

[6]Note: Ethnography is a multi-discipline endeavor, which makes descriptive study on human specific society through social investigation and comparison of various racial societies.

(2) *Explorations of Cultural Theories in Western Countries in the 20th Century*

Both anthropologists and sociologists had continued the exploration of cultural theories in the 20th century. There is even a branch discipline of anthropology-Cultural anthropology established.

(i) *Explorations of Cultural Theories by Anthropologists*

In the 20th century, especially in the American anthropology community based on Tylor's theory, "Culture" became a central and uniformed theory. Culture referred to "the universal ability of mankind to classify and encode human experiences with symbols, and disseminate and communicate these encoded experiences in the social domain through symbols". In America, anthropology was divided into four branches: biological anthropology, linguistic anthropology, cultural anthropology and archeology. In Europe, however, archeology is often viewed as a discipline in its own right. The achievements made in these four branches also influenced the researches made by other countries.

Biological anthropology aimed at clarifying two questions in the research of culture: (a) is culture exclusively possessed by mankind or do other species, primate apes in particular, also claim this feature; and (b) how did culture evolve among the mankind? When studying culture as "learned behaviors" in the research of these questions, anthropologists further discovered that the learning of children is based on imitation, while the learning of other primate animals is of a competitive type. Anthropologists also searched for the relations between languages and cultural studies.

After arriving in America from Germany, Franz Boas started the first graduate course for the study of culture in Columbia University in 1896. He was also the founder of the modern anthropology in America. The focus of cultural studies at that time was regarding the cultural evolutionism as the dominant model. Franz Boas however disagreed with this focus, arguing that the mankind's knowledge of different cultures was very incomplete and attainment of the knowledge was usually based on unsystematic and unscientific researches, making it unable therefore to develop a scientific, effective model for general applications on human cultures. He established the principle of cultural relativity, in which he trained his students to conduct observations and onsite researches in different societies. He believed culture as a "state" is always in a flowing status. The cultural anthropology promoted by him and his students once played a dominant role in the era of the World War II. Their researches were focused on how different forms of cultures and individuals worldwide act creatively through their own cultures.

(ii) *Exploration of Culture by Structural Functionalists*

The French structural anthropologist Claude Levi-Strauss combined some of cultural anthropologist Franz Boas' and French sociologist Emile Durkheam's theories in his study of culture. He attempted to develop a universal structure of mind by systematically comparing the cultural structures of different societies. He believed that in the laws of chemistry, the finite number of elements can combine to form infinite compounds. So there might be a few fixed cultural elements that can be used by humanities to formulate the massive forms of culture observed by cultural anthropologists. Structuralism once held a dominant position in the anthropologic field in France and produced immense influences on the British and American anthropology in the 1960s and 1970s.

Structural functionalism is in essence one of the sociologic theories. Society is seen as an entity combined of "structural and cultural components" or "events". In his cultural anthropologic theory, Franz Boas regarded anthropology as a genre of natural science committed to the study of mankind. The structural functionalism however views anthropology as one branch of social science committed to the study of one particular aspect of mankind. As a result, structural functionalism redefined "culture" and narrowed its scope of research.

The theory of social behavior developed by the American social system promoter Talcott Parsons was closer to the British social anthropology. The model he used to interpret human behaviors is made up of four systems: "behavioral system" due to biological needs; "personality system" formed by characteristics of personal character which impacts the functions of the society and world; the "social system" of the models of interplays between units of the society, especially the identity and importance of social units; and "cultural system" of behavioral codes and values which regulates the social behaviors through symbols. According to his model, the "personality system" belongs to the research scope of psychology; "social system" to the research scope of sociology; and only the behavioral codes and values in the remaining "culture system" belongs to the research scope of cultural anthropology. The theory of Franz Boas, however, believes that all these systems fall within the research scope of anthropology, contending that Talcott Parsons' research of anthropology and culture is of a narrow concept. The anthropologic community thinks the cultural values of the structural functionalism resulted in the conceptual confusion of "society" and "culture". Anthropologists believe that a society means a group of people, and that a culture refers to "pan-human" ability and the summation of non-genetical phenomena. While a society features clear boundaries, a culture is fluid and its boundaries have openings and can be penetrated.

8.4.4.2 Cultural System Theory

There were debates of scope of study of the conceptual model in explaining human behaviors. The model adopted by sociologist Parsons to explain the human behavior consists of four systems, while Franz Boas had a different view. From the perspective of social systems engineering, theories which fitted the best reality will be adopted. Two working papers by Whitehead, T. L. in 2002 and 2003 studies healthcare system from the perspective of cultural ecology. Three models were established in Whitehead's papers: Cultural System Paradigm (CSP), Cultural System Approach to Change (CSAC), and Cultural Systems Approach to

Programme Planning, implementation and evaluation (CSAPPE). These three models provide a useful framework for social systems engineering, therefore, the major concepts of Whitehead's two papers are summarized below:

Culture is an "integrate" system, with coherence among interactive parts. The cultural system consists of the following 5 sub-systems or parts.

(1) *Behavioral Model System of Individual and the Norm*

There are various models of individual behaviors, standard or risky/sick. In CSP, the behavior model consists of behavioral actions and the social culture describing those behaviors, including: what are the behaviors (content), how are they conducted (method), where are they conducted (place), who conduct them (participants), when are they conducted and are they general behaviors (generalization).

(2) *Individual and Shared Concept Systems*

Individual and shared "concept" or "concept" structure (including knowledge, faith, attitude system, values, "meaning symbol") are the frame to explain the meanings of human behaviors, including sick/risky behaviors and other categories of CSP. The concept system is able to drive special behavior model, and affects other cultural ecological community. In CSP, concept system is an essential part of the cultural system, also important to the institutionalized patterns of behavior formed by the cultural community and social system.

(3) *Important Social System*

Domestic units (family or residential community), community binary relation outside the residential region (minority groups, social networking and kinship, volunteer association/organization, symmetrical binary relations such as friendship, ranch worker or actual/changed kinship binary relation, asymmetric binary relations such as employer-employee, patron—client, etc.), policy agencies and agents for broader community/society, etc. are included in the important social system. Important social systems are "engines" of cultural production and reproduction, driving the coherence of behavioral models, and their feedback system on the community concept and behavior, on the other hand, could affect the concept system.

(4) *Material Cultural System*

The material cultural system includes material art products and other artificial products, especially those of "symbol" or "meaning" values.

(5) *Expressive Cultural System*

The expressive cultural system includes expressions with cultural meaning and symbol form such as music, singing, dancing, conversation, literature, proverbs, preaching, artistic expression, etc.

8.4.4.3 Human Ecological System

Cultural system is part of the human ecological system, which is also consisted of other three system.

(1) *Material Environment System*

Human community lives in the material environment system. The community cultural system is one factor for successfully maintain life continuation, guards against potentially life-threatening events, and seeks for overcoming the factors restricting life continuation.

(2) *Actual and Need Felt Systems*

The actual and need felt systems are systems that human community and individuals adopt to reach material and psychosocial functions, including: biological needs (for example, reproduction, consumption, food, water and other energy resources, discharging waste, disease prevention and treatment; climate risk prevention, etc.), means needs (economic, educational/social, governing/political, legal, public, etc.); expression needs (cognitive, meaningful and orderly world views), emotional (social status and accepted, loved or liked, self and group categories, etc.), and love message (need to explain, communicate, etc.).

(3) *Important Historical Processes and Events*

The events can be physical (e.g. floods, waterlogging) or sociocultural (e.g. violent coup, war, new economic or marketing system). They make one cultural system or part of it institutionalized and sustainable, or cause culturally "revolutionary" mixture of various concepts, etc.

8.4.5 The Chinese Culture and Re-emergence of Asian Culture

8.4.5.1 Researches of East Asian Region

The rapid economic growth of many developing countries in East Asian region post the Second World War had attracted the attention of global society. There are several publications of international organizations focused on the experiences of development of those countries, for example, *The East Asian Miracle* and *Rethinking the East Asian Miracle* published by the World Bank, Emerging Asia by Asian Development Bank. In the search for explanations of high economic growth of Asian countries, "some analysts turned to culture, in particular to those cultural features of Chinese societies that are shaped by Confucian ideas (Schech & Haggis, 2000).

Although it is recognized by Western sociologists and a part of scholars that the influence of Confucianism has also been significant in Korea, Japan, Vietnam and Singapore as a source of learning and an ethical code (Marshall, 1994), these countries have shaped their own culture based upon their history and major ethnic group. Therefore, only the Chinese culture will be introduced in the follows.

8.4.5.2 The Confucianism and Its Impact to Chinese Culture

It had been described in Sect. 8.3.2.1 that there are many definitions of culture. But this book will limit the discussion of two major thoughts and philosophies of two schools, the Taoism and the Confucianism which have the greatest impact on Chinese culture, especially the Confucianism and the other aspects of culture will be described very briefly. It is necessary to emphasize that one of the major difference between the Western culture and the Chinese culture is that the Western culture had been dominated by the Christian culture throughout most of its history but religion had not played dominant role in most of the Chinese historical period. Although Taoism had once become a type of religion in China (for example, Daoist (or Taoist) were favored by some emperors in early Han dynasty, Tang dynasty and Song dynasty), it is less popular than Buddhism in China, and the Confucianism had a large influence on China's culture.

On the other side, Confucianism had played its role in China similar to Christian culture of Western countries. This is due to two reasons: the first is that Confucius had been responsible for the editing of *Five Classics* and his thoughts in the *Four Books*, and the *Analects* are collections of sayings and answers of Confucius to his students and followers. And Confucianism became state orthodoxy in the Western Han Dynasty since BC 134; the second, the dominant position of Confucianism in China around 2000 years is caused by political influence of empires. In addition the system of imperial examination to select candidates for the officials was launched since Sui dynasty (581–618 AD), Four Books and Five Classics were major contents of imperial examination. Therefore, nearly all officials and elites in later period of Sui dynasty had provided a demonstrative model of ethical code set up by Confucius for whole Chinese society around 1400 years up to 1911. There was New Cultural Movement of the mid 1910s and 1920s sprang from the disillusionment following the failure of the Chinese Republic, founded in 1912 to address China's problems. This New cultural Movement as a whole is good for the progress of China toward a modern society, but it has some negative effect to criticize the Confucianism indiscriminately, a part of Confucianism is outdated, for example, discrimination against female, a part may need to be reformed, for example, over emphasis on hierarchical ranks, but as a whole the Confucianism had set up a moral system which had maintained the stability of social order of China around 2000 years. Unfortunately, Confucianism was suffered once more a heavy blow during the period of Cultural Revolution of China.

A part of thoughts of traditional Chinese philosophers had been introduced in Chap. 3 of this book, including some thoughts of Laozi (who is widely considered the keystone work of the Taoist tradition) and Confucius, but *Analects* is the most important part of Confucianism, therefore a brief systemic introduction of *Analects* will be given below to be the foundation of understanding of Confucianism as well as its influence to the Chinese culture.

8.4.5.3 The Analects

(1) *Introduction and Table of Contents*

The *Analects*, pronounced "lunyu" in Chinese and meaning "Edited conversation", includes the social philosophy, political philosophy and concepts of education by Confucius and his major followers. Many Western experts had done notable translations of Analects. The translation by Arthur David Waley (1889–1996) was adopted by one of the book series of the Library of Chinese Classics (Chinese-English), and published in 1999. The Analects contains 20 chapters. The traditional titles given to each chapter are mostly an initial two or three incipits. Table 8.3 is contents of the Analects.

(2) *Briefing of preface of The Analects by Yang Fengbin*[7]

(i) *Importance of Analects*

The Chinese civilization has a long history of 5000 years. But now, in China and the Chinese communities overseas, any Chinese tradition, be the major tradition held by the social elite or the minor tradition understood by common citizens or countryside peasants, is just the same. Every Chinese, every act and every move of his or hers may be originated in one book—*The Confucian Analects*, a canon formed 2400 years ago in the so-called Axial Periods.

In China, without Confucius and his Analects, we can say nothing of "philosophical breakthrough". Chinese philosophy is rather ethical. The core of Chinese philosophy is the doctrines of ethics and morality. If we take it that the various forms of Western culture lay stress on the seeking for truth, we may say that the various forms of Chinese culture lay stress on the seeking for goodness. In the ancient times, the first lesson a disciple was taught by any teacher was to cultivate the virtue, be the latter a learned scholar or a master of martial arts. Teachers might let his disciples or students recite some maxims or mottoes, which, in most cases, are quoted from the Analects. In some "under developed" rural places, the moral ideas contained in the ancient classics still remain. A countryside couple, though they may be poorly educated, embody the Confucian moralities in either their speech or their behavior. And their children are brought up in such an atmosphere unconscious of the fact that this is a teaching of traditional morality. So we must read the Confucian Analects it is a textbook of life for all the Chinese people.

[7]Note: Yang Fengbin had written the preface of the book The Analects of the book series of the Library of Chinese Classics (Chinese, English).

Table 8.3 Contents of *Analects*[a]

Book	Title	Translation	Notes
1.	Xue Er 學而	Studying	
2.	Wei Zheng 爲政	The practice of government	This chapter explores the theme that political order is best gained through the non-coercive influence of moral self-cultivation rather than through force or excessive government regulation
3.	Ba Yi 八佾	Eight lines of eight dancers apiece	Ba Yi was a kind of ritual dance practiced in the court of the Zhou king. In Confucius' time, lesser nobles also began staging these dances for themselves. The main themes of this chapter are: criticism of ritual impropriety (especially among China's political leadership), and the need to combine learning with nature in the course moral self-cultivation. Chapters 3–9 may be the oldest in the *Analects*[35]
4.	Li Ren 里仁	Living in brotherliness	This chapter explores the theme of *ren*, its qualities, and the qualities of those who have it. A secondary theme is the virtue of filial piety
5.	Gongye Chang 公冶長	Gongye Chang	The main theme of this chapter is Confucius' examination of others' qualities and faults in order to illustrate the desirable course of moral self-cultivation. This chapter has traditionally been attributed to the disciples of Zigong, a student of Confucius. Gongye Chang was Confucius' son-in-law
6.	Yong Ye 雍也	There is Yong	Yong is Ran Yong, also called Zhou Gong, a student of Confucius
7.	Shu Er 述而	Transmission	Transmission, not invention [of learning]
8.	Taibo 泰伯	Taibo	Wu Taibo was the legendary founder of the state of Wu. He was the oldest son of King Tai and the great-grandfather of King Wu of the Zhou Dynasty
9.	Zi Han 子罕	The Master shunned	Confucius seldom spoke of advantage
10.	Xiang Dang 鄉黨	Among the Xiang and the Dang	A "xiang" was a group of 12,500 families; a "dang" a group of 500 families. This chapter is a collection of maxims related to ritual
11.	Xian Jin 先進	Those of former eras	The former generations. This chapter has traditionally been attributed to the disciples of Min Ziqian, also known as Min Sun, a student of Confucius
12.	Yan Yuan 顏淵	Yan Yuan	Yan Hui was a common name of Zi Yuan, a favorite disciple of Confucius
13.	Zilu 子路	Zilu	Zilu was a student of Confucius
14.	Xian Wen 憲問	Xian asked	This chapter has traditionally been attributed to the disciples of Yuan Xian, also called both Yuan Si and Zisi, a student of Confucius
15.	Wei Ling Gong 衛靈公	Duke Ling of Wey	Duke Ling ruled from 534–493 BC in the state of Wey

(continued)

Table 8.3 (continued)

Book	Title	Translation	Notes
16.	Ji Shi 季氏	Chief of the Ji Clan	Ji Sun was an official from one of the most important families in Lu. This chapter is generally believed to have been written relatively late;[35] possibly compiled from the extra chapters of the Qi version of the *Analects*
17.	Yang Huo 陽貨	Yang Huo	Yang was an official of the Ji clan, an important family in Lu
18.	Weizi 微子	Weizi	Weizi was the older half-brother of Zhou, the last king of the Shang dynasty, and was founder of the state of Song. The writer of this chapter was critical of Confucius
19.	Zizhang 子張	Zizhang	Zizhang was a student of Confucius. This chapter consists entirely of sayings by Confucius' disciples
20.	Yao Yue 堯曰	Yao spoke	Yao was one of the traditional Three Sovereigns and Five Emperors of ancient China. This chapter consists entirely of stray sentences resembling the style and content of the *Shujing*

[a]*Note* Data source from CIA (2017) The World Factbook. www.cia.gov/library/publicatoin

(ii) *What is the Confucian Analects about*

As for the fundamental thought of Confucian Analects, people have invariably insisted that there are two points of view. One is the ren (benevolence), the other is li (rite).

One famous scholar, Mr. Yang Bojun, thinks that the core of Confucian Analects is benevolence. He noted that in Zuo Qiuming's Chronicle, the character rite is used 462 times, while the character benevolence is used only 33 times. In the Confucian Analects, however, the character rite is used 75 times, while the character benevolence is used as many as 109 times.

We think that the core of Confucian Analects is, for a certainty, benevolence.

The benevolence in Confucian Analects does not refer to only one thing. Extensively speaking, it means to apply a policy of benevolence in all lands under heaven. Intensively speaking, it means to love people. Benevolence refers to loyalty and consideration, and the fundamental way to be a real human being, namely, filial piety to one's parents and love and respect to one's elder brothers. To be a benevolent man in true sense is very difficult, but everyone can do something to be benevolent, that is to say, one may keep the benevolent virtue by doing good things. Benevolence can be discussed further from four points.

First, Benevolence is the very nature by which the human distinguishes himself from the animal. If one is living, he should put the benevolent virtue in practice, Meanwhile, benevolence is the lofty goal one should try to achieve all his life. On one side, people with high ideals are those who exist with benevolence all the time. On the other side, they take it for a duty to apply a policy of benevolence in all lands under heaven.

Confucius said, "Benevolence is the humanity" ("The Doctrine of the Mean"). He also said, "Riches and honors are that men desire. If it cannot be obtained in the proper way, they should not be held. Poverty and meanness are what men dislike. If it cannot be obtained in the proper way, they should not be avoided."

Second, benevolence is an important procedure to realize the highest goal to the effect that "the whole world is one community."

In *The Book of Rites,* the chapter entitled "Li Yun" says as follows. When the great Tao prevailed, the whole world was one community. Then people might select the worthy and promote the capable, and have regard for credit and amity, so people would love not only their own kinsmen and children, but also do their duties to the others. As a result, the aged would live their full span, the able-bodied would be of some use, the young would grow up, and the spouse-bereaved, the orphan, the single, the invalid and the diseased would be supported, the male had a sufficient share, and the female had proper marriage. Goods seemed to dislike to lie on the ground, so there was no need to keep them in one's own house; the strength seemed to dislike not to come from the body, so there was no need to use it for one's own purpose. Therefore nobody would plot anymore; there were no robbers and bandits, thus it was un-necessary to close the gate. If so, we may say, the whole world has truly become one community.

Third, benevolence may be accumulated bit by bit in everyday life. Benevolence means to love people. The way to carry out the benevolent virtue is to put oneself in the place of another, and from the near to the distant.

Fourth, judging from the relationship of benevolence and rite, benevolence is the root, and rite is the branch; benevolence is the ins, rite is the outs; benevolence is the content, rite is the form. Benevolence is the ultimate aim, rite is a kind of binding force and norms and system which guarantee the achievement of the purpose.

According to Mr. Yang Bojun's *A Dictionary of Confucian Analects*, the word rite has been used 74 times in the Confucian Analects, meaning ritual intention, ceremony and propriety, system, and discipline. We think that ceremony and propriety, system, and discipline rites and law all serve the aim "all under heaven will converge at benevolence."

Zixia asked, "What is the meaning of the passage-The pretty dimples of her artful smile! The well-defined black and white of her eye! The plain ground for the colors?" The Master said, "The business of laying on the colors follows the preparation of the plain ground." Zixia said, "Are ceremonies then a subsequent thing?" The Master said, "It is Shang who can bring out my meaning! Now I can begin to tale about the Odes with him." ("Ba Yi"). Here, the word "subsequent" tells us that the rite should come after benevolence.

8.4.5.4 Other Aspects of Chinese Culture

From the definition of culture described previously, it should cover a broad aspects. Several selected aspects are described below:

(1) *The Chinese medicine*

China's *The Inner Classic of the Yellow Emperor,* published in late the Warring States Period, is the first comprehensive Chinese medicine book,. The feature of Chinese medicine is the system concept and dialectical view. The *Thousand Pieces of Gold Formulae, written* by Sun Simiao in Tang Dynasty, details the material selection and production of Chinese medicine. The *Compendium of Materia Medica,* which took Li Shizhen in Ming Dynasty 27 years to finish, comprehensively summarizes the research achievements of Chinese medicine before the 16th century, and introducing 1094 kinds of botanicals and 444 types of animal materials. Chinese acupuncture also attracted attention from foreign medicine communities.

(2) *Literature*

In terms of achievements of the Western literature, people will talk about Greek Homer's epics, Iliad and Odyssey, British Shakespeare's plays and Dickens' novels, French Dumas' historical novels, British Shelley's poems, etc. In China, there are *Book of Songs* and Qu Yuan's *The Sorrow of Separation* in Western Zhou Dynasty and the Spring and Autumn and the Warring States Periods,, the Tang and Song Poetry. the Legends in Tang Dynasty, such as *Biography of Ying-Ying*, Guan Hanqing's poetic dramas in Yuan Dynasty, such as *The Grievance of Dou E.* Yuan. Ming, and Qing Dynasties have witnessed a rich collection of novels, of which there are world-famous *Three Kingdoms, Chronicles of the Eastern Zhou Kingdom, Dream of Red Mansions, Journey to the West (Monkey), Water Margin*, etc.

(3) *Others*

China is the hometown of tea. Chinese cuisine such as Sichuan and Shandong cuisine have a history over 2000 years. Kun opera is listed as historical and cultural heritage by UNESCO (United Nations Educational, Scientific and Cultural Organization). Jingju developed in Qin Dynasty is well known to be Beijing Opera overseas. Pottery and Chinese cultures have a long history of several thousand years. China's lacquerware, jade, bronze, gold and silver articles, as well as Sichuan brocades and Suzhou's embroidery. are also well known around the world, etc.

8.5 Globalization and Regionalization—A Mega Trend of Social Change

8.5.1 Globalization—A Mega Trend of Social Change

In very recent years, there are debates of globalization in a part of countries. But globalization, although a modern term, represents the trend of development throughout human history. The basic principle of sociology (i.e., the systematic study of the development structure, interaction, and collective behavior of organized groups of human beings) recognizes that the trend of social change of human history is from small organized groups to larger groups. Man cannot live alone. Therefore, human society had progressed from primitive to agricultural, industrial, post-industrial and informational or bio-society. Every stage of social change links the organized groups of mankind to a larger scale. And globalization is a trend of social change.

In ancient China, we had a saying: The household can hear the barks and crows of the neighborhood, but they never communicate with each other, even down to the end of their lives. This saying was true for a society thousands of years ago, because there was no ties for them, based on economic aspects. In a modern global society, the people in different countries are deeply dependent on each other for their living. Chinese food and clothing, Japanese T.V. and household electric appliance. German cars, French fashion and U.S.A's Apple iPhones can be found everywhere throughout the whole world, just to mention a few examples. That is one aspect of globalization. We shall quote some sentences from Arnold Toynbee: "Our age will be remembered chiefly neither for its horrifying crimes nor for its astonishing inventions, but for its having been the first age since the dawn of civilization, some 5000–6000 years back, in which people dared to think it practicable to make the benefits of civilization available for the whole human race."

Discussion of globalization will not only focus on the tangible and intangible facts, but will also emphasize the fact that the sense of global consciousness and global responsibility are important, and should be recognized, along with such trends as regionalization. They will be further elaborated in coming parts.

8.5.2 Four Aspects of Globalization

8.5.2.1 Globalization of the Biosphere

Mankind shares the common biosphere for living, as well as the resultant problems, in a globalized world. Thousands of years of productive activity of mankind, especially centuries of industrialization have greatly influenced the relationship between the human being and the environment. The basic biosphere issues today are: Climate change, natural resources, energy, food and agriculture, waste disposal, health and population. Only a few of them will be explained below.

(1) *Climate Change and Environmental Problems*

The issue of climate change has been recognized worldwide, and this subject has been discussed in Sect. 8.2.1.1 previously. The theme of resources will be discussed briefly. There are already many studies on the limit of growth. One aspect is the shortage of resources. The common example of resource with major impact to daily life of most of the people is water. But it had been studied quite earlier that "In some situations, irrigated agriculture is threatened by falling water tables."

(Brown, 1984). This happened in the southern Great of U.S.A. as well as in the Soviet southwest. And with energy resource for example, based upon *Statistical Review of World Energy 2017*, the workable reserve for coal, petroleum and natural gas are 110 years, 53 and 54 years respectively. The shortage is also true for other resources. The first United Nations Conference on Environment and Development was held in Rio in 1992, and sustainable development is defined to be "meeting the needs of the present without compromising the ability of future generations to meet their needs." Therefore, it is necessary to have a sense of global consciousness and responsibility.

(2) *Food/Agriculture*

The agricultural society of mankind has lasted for thousands of years throughout human history; Currently, agricultural society still exists in a large part of developing countries, even in many industrialized countries and in countries that are moving into the post-industrial or information age. Although the percentage of population engaged in agricultural activities in the U.S.A. and Japan has declined to 0.7 and 2.9% respectively, a majority of the people of the world have to earn their living based upon this activity, especially the emerging economies and the whole of mankind still has to live on food and production from agriculture, however old a history it will be. Besides satisfying nutritional needs, agricultural operation still determine the use of global land and water resources, provide the bulk of the jobs in the world, and act as the primary links of commerce and culture between regions. Agricultural production techniques have changed dramatically during the twentieth century; the new millennium could even introduce more dramatic changes that would raise agricultural productivity, make improvements in food nutrition and taste, and preserve natural environments. Modern biotechnology is emerging; sustainable and alternative agricultural techniques will emerge. It is also a global responsibility to explore the potential for these new technologies and to study their impact on farmers, business, consumers and world commerce.

8.5.2.2 Globalization of Scientific Sphere

Science and technology provide human beings with both opportunities and problems. Three great revolutions in world history have changed the world from regional to global since the year 1500 BC. The scientific revolution, industrial revolution, political revolution, and development of modern transportation and communication techniques have linked a dispersed world to form an integrated whole. The scientific revolution took place primarily during the sixteenth and seventeenth centuries between the publication of Copernicus's De Revoutionibus Orbium Coelestuim (1543) and Newton's Principia. Science has the potential of indefinite advance. Furthermore, science is universal or global, being based on an objective methodology. The industrial revolution during the 1780s caused "a take-off to self-sustained growth": A breakthrough of productivity did occur. The lives of more and more people are being changed by the spread of products based on technological innovations. The protection of intellectual property rights also helped the development and distribution of new technologies. In pragmatic sense, the title of this theme can be "Globalization of Technosphere." Plastics, electric household appliances, synthetic textiles, audiovisual equipment, televisions, cars and improvements in new building materials, etc. have also promoted the process of globalization and the life style of global people in general. The achievements of science and technology have improved the living standard of the people globally. The breakthroughs achieved by the development of semiconductor technology and efficient micro-processors, at a progressively declining cost, signal even more sweeping changes. Communication and information industries are in some countries acquiring such economic importance that they are tending to become dominant over heavy industry and manufacturing as the chief component of the national product. Remote sensing and other techniques help greatly in regional planning. And modern communication systems have more or less cemented the world together. The hotline between the superpowers has saved the people from the possibility of a destructive global war. Advances in agronomic research may be of the great help in solving the hunger of the world. Appreciable progress can still be made in the rational utilization of soil and water resources to survive the economic activity. In the field of health, synthetic drugs are increasingly being supplemented by medicaments produced by living organisms, microbes and animal cells, which have an optimum curative impact but sharply reduced side effects. Improvement of the health care system has prolonged the life expectancy of the global population much longer than the generation of their grandfathers. The global society is in the process of transition from industrial society to postindustrial society or information society driven by internet and internet of things.

Despite the achievements of science and technology for the improvement of global living standard as a whole, they are often regarded as a menacing Pandora's box. Atomic power can destroy man, but it can also transform living conditions through the globe. Rockets can be used as intercontinental ballistic missiles, but are also convenient for transporting people around the globe. Although the global network of society provides the increase of overall wealth and innovative capacity of contemporary society, the "new achievements remain limited to certain classes and don't benefit all." (Fuchs, 2008 p. 353). We must perceive deeply this globalization effect and promote the positive contribution of technology in preventing their negative aspects. One more point should be emphasized: Science is universal but technology has traditionally belonged to firms, and intellectual property rights are legal rights globally. It is necessary to be aware of the trend of scientific globalism and technological nationalism.

8.5.2.3 Globalization of the Econosphere

The dominant feature of globalization is in the econosphere, because it is the basis of economic activity that has formed the demarcation of the history of mankind, the primitive, the agricultural, the industrial society, and currently, the transition to postindustrial society etc. The major issues of globalization of the econosphere include the evolving new economic relationship, the establishment of global markets, the development of multinational corporations with their unique role in the process of globalization with their capacity to move capital and technology rapidly from place to place, drawing opportunistically on resources, labor markets and consumer markets in different parts of the world. There is an abundance of literature in recent years dealing with "the New International Economic Order". But a retrospect of world history will better help us recognize the issues of globalization and regionalization. Although man and his hominid ancestors have existed on the globe for over 2 million years, man lived largely in regional isolation until 1500 BC which formed a major turning point in human history: The voyages of Columbus and Gama and Magellan broke the bonds of regional isolation. Direct sea contact was established among all regions, and the trend of globalization emerged as one of the historical processes. Subsequently, there were globally repercussions of such major events as the scientific revolution and the industrial revolution. In the twentieth century, There were two World Wars and the Great Depression, but there were also emergence of many developing economies, a large part of them were formerly belonged to a colonial world. In the twenty-first century, there was once more the global crisis erupted of around 2007–2008, but the emerging market and developing countries were less effected. There is change of economic landscape of the global econosphere. Based upon World Economic Outlook (2016). Oct of IMF, by the year end of 2015, 39 advanced economies had a share of 42.4% of the world's GDP, 63.4% of the exports of goods and service and 14.6% of the population of the world. while 152 emerging market and developing economies, had a share of 57.6% of the world's GDP, 36.6% of the exports of goods and services, and a 85.4% of the population of the world. Emerging and Developing Asia especially had a share of 30.8% of the GDP (in terms of PPP), 18.4% of the exports of goods and services, and 48.8% of the population of the world. This is the new macro landscape of the econosphere. Restructuring of the global economy is another feature of global interdependence. The mankind live in an ever-changing and dynamic world; the economic structure of every nation needs to be changed and upgraded. Upgrading the economic structure implies the utmost effort to change the existing pattern of economic structure and employment, i.e. new competitive products should be created by learning new technology and acquiring the skills required in production and marketing. The spurt of globalization in the manufacturing sector specifically in recent years has brought more competition among producers within and between countries and regions, spread new technologies in wider areas, and increased interdependence between them through greater trade and investment. At the same time, imbalances have developed and become major management problems. New approaches should be studied and created to meet the needs of today's and tomorrow's global econosphere.

8.5.2.4 Globalization of the Sociosphere

From our personal academic point of view, we generally include the politisphere within the sociosphere. Politics and the political system are one type of activity and system within a broader societal context. But they are mutually interactive; generally, the political system can shape the dominance of a certain religion or culture. This had been explained previously that why Confucianism is more dominant in China than Taoism.

(1) *Globalization of Culture and Life-Style*

Although culture and life-style are generally deeply rooted and shaped by history, they are subjected to change in an evolutionary world. Our life-style is far different from our predecessors, and the same is true for generations younger than ours. With the intensive circulation of products and growth of tourism, and the rapid strides made by direct satellite television broadcasting, the world has become a totally open field.

The means for storing data, sounds and images are being extended and improved with the aid of computer memories and digital means. Processing and computation capacities are being extended in space by way of communication satellites, and on the surface of the earth or under the sea, through the use of microwaves or optical fiber. Remote sensing techniques help data collection. People of different countries learn culture and life style from each other through a demonstration effect; that is, the modern trend of globalization of culture and life-style. Of course, preservation of cultural diversity may also persist for several centuries; identity and diversity are two major forms of our world.

(2) *Other Global Social Trends*

There are other social trends around the world, such as aging people in developed countries; maintenance of a stable family; changes in the conception of governance at government, community and corporate levels; and increased influence of international organizations, such as the United Nations and its related institutions.

8.5.3 A Retrospect of Global Industrialization and Trade from a Historical and Cross National/Regional Perspective

In studying globalization and regionalization from a development perspective, the basic method of approach may be a study of the past as well as of worldwide experience at the present time. Chinese have an old proverb: "the failure of the previous cart may be a good guide for the next cart." The econosphere and technosphere play important roles in globalization as well as regional development. A very brief description of the two aspects will now be given; the main lessons will be summarized.

8.5.3.1 The Three Periods of Global Industrialization

Although the process of globalization is driven by trade and investment of the multinationals and promoted by science and technology, the core factor is the national development through industrialization and diffused to global industrialization. The process of global industrialization can roughly be classified into three periods:

(1) *Industrial Revolution: First Period, 1770–1870*

The industrial revolution cannot be attributed merely to the genius of a small group of inventors. The combination of favorable forces operating in eighteenth century Britain played a more significant role. Before the industrial revolution, there was a commercial revolution that provided large and expanding markets for European industries, and to meet the demands of these new markets, industries had to improve their organization and technology. On the organizational side, Adam Smith emphasized specialization and the division of labor. On the technology side, there were James Watt's improvement of the Newcomen steam engine between 1776 and 1781, and the later transport revolutions. Industrialization centered on steel, railway and steamships from about 1820 to 1870. Railways integrated national markets, and bigger ships could be built with steel and steam power. Freight costs dropped, which enabled the emergence of a global market. Bulky agricultural products from the American Midwest could be sold to distant markets in Europe. Belgium, France, Germany and the United States began to industrialize in this period.

The spread of industrialization to the European continent was also facilitated by institutional reforms and the removal of many of the restrictions that had hindered domestic commerce. France, Germany, Belgium, the Netherlands, Switzerland, the United Kingdom, and the U.S. nearly all adopted more liberal external trade policies between 1820 and 1870. In 1870, more than three quarters of the world's industrial production was accounted for by four countries: U.K. (about 32%), U.S. (about 23%), Germany (about 13%) and France (about 10%).

(2) *Industrial Revolution: Second Period*

The second period of industrial revolution differed from the first in two aspects. First, technological advances became more dependent on scientific research that was systematically organized in companies and universities. Second, for the first time, industrial growth in the industrialized countries became partly dependent on supplies from elsewhere, such as raw materials needed by the new technologies and ingredients for new alloys. The growth of different countries became much more dependent on each other. Between 1870 and 1913, Russia and Japan joined the "Industrial League." Industrialization in Japan was paralleled by reform. The Meiji Emperor was restored in 1868 as head of a centralized state. Guilds were eliminated and private property rights to land were established. People were now free to choose their trade or occupation. Taxation was reformed and made uniform across the country. These experiences showed that successful industrialization is generally accompanied by certain institutional reforms.

From 1913 to 1959, there was the collapse of liberalism and of the global market. Two world wars happened in this period.

(3) *Global Industrialization Post the Second World War: the Third Period*

After the Second World War, the world economy entered a period of unprecedented output and growth. Manufacturing led the way in both output and growth. As in the nineteenth century, export grew faster than output, i.e. a deeper and broader global market was forming.

Postwar growth in manufacturing was promoted by an explosion of new products, new technologies, liberation of international trade, and increased integration of the global economy. Assembly line production, the automobile, electricity and consumer durables were given a push by the postwar release of postponed consumer demand. Some of the new technologies assisted the globalization of markets. The jet aircraft shortened travel time greatly. Telecommunications made it easier for MNCs to coordinate subsidiaries in different countries. The new electronic media helped to shape a global market with increasingly similar consumer taste. Trade liberalization among industrialized market economies under the GATT (later, it is evolved to become WTO) helped to create a global environment that was conducive to the development and distribution of the new technologies to a certain extent.

8.5.3.2 Two Major Actors of Global Industrialization

(1) *Roles of Multinational Corporations (MNCs)*

Although multinational corporations in manufacturing date back earlier to the history of the nineteenth century, it has been only since the 1960s that they have become major actors in the globalization of industrialization. By 2016, the FDI outward stock reached 26 trillion US dallors which was around 34.7% of the GDP of the world (UNCTAD, 2017 World Investment Report 2017 Geneva: UN). Beginning in the late 1960s, several MNCs started rationalizing their global production. In the past, most foreign subsidiaries had produced finished products. Now all the subsidiaries were increasingly linked into a unified production process, in which each subsidiary performed only those parts of the process in which it had a comparative advantage. Sometimes, the arrangements are between locally-owned and foreign-owned companies, known as international subcontracting. Once again, there emerges the effect of Adam Smith's specialization and division of labor in a global production process, and a global value chain is formed in this process.

(2) *Decolonization and Postwar Performance of Developing Countries*

Between 1500 and 1763, Europe had emerged from obscurity by gaining control of the oceans and the relatively empty spaces of Siberia and the Americas. But Europe's impact in Asia and the Pacific still remained negligible at the end of the eighteenth century. By 1914, this situation had changed tremendously. Vast territories, including nearly the entire continent of Africa and the greater part of Asia, had become outright colonial possessions of the European powers. Of the 16.819 million mile^2 comprising Asia, no less than 56% of the area was under European rule. Japan, the only truly independent Asian nation in 1914, accounted for a mere 161,000 mile^2 (Stavrianos, 2004).

Many of the countries emerging from colonialism after the period of the Second World War chose an industrial development strategy that emphasized import substitution behind high government protection. Many also believed that protection granted through government restriction on imports had been a successful experience in the early industrialization of Europe, North America and Japan. Some of the newly independent countries, inspired by the Soviet Union's rapid industrialization, combined import substitution with government ownership, planning and production. In these countries, government sought to restrict domestic trade, while in Europe and Japan early industrialization had been assisted by the easing of restrictions on domestic trade. But in the recent economic reform in these countries, they had learned through experience that trade and industrialization have the effect of reinforcing each other.

It had been commented by a study jointly sponsored by UNIDO and the World Bank in 1980 that "the past three decades of industrialization in developing countries have created a second industrial revolution that is transforming the world economy even more radically than did the changes that took place in Great Britain in the late eighteenth and early nineteenth century." (Cody, Hughs, & Wall, 1980) and currently (with another three decades). Their GDP have surpassed all developed economies.

8.5.3.3 Lessons from History

The above retrospect of global industrialization not only serves as a description of the trends and factors in forming a global economy; but also provides some lessons for regional development of different regions.

(1) The role of agriculture should be fully emphasized: Although industrialization is important for economic growth and is a catalyst for globalization, history also shows that as countries started industrializing, agriculture was important as a market and source of labor for industry. Germany's success in its initial period of industrialization was aided by a large home market for its many agricultural families, producing more than enough for their own consumption. Only when countries reached quite high levels of industrialization could urban growth replace agriculture as the major source of nonagricultural labor, and industry earn enough foreign exchange to permit heavy dependence on food imports.
(2) The role of institutions: Both agricultural and industrial growth depends heavily on institutions. historical experience suggests that regardless of the strategy, growth today will be widespread only if agrarian institutions provide a large number of cultivators with a significant share of a substantial marketable

surplus, and if industrial institutions also adapt to change in an ever-changing world, whether regional change or global change.

(3) Initial conditions: The initial conditions of a country or region should be carefully emphasized: A country with a large domestic market is in a better position to establish industrial plants that take advantage of economic scale.
(4) Domestic and foreign trade policies: The successful industrialize countries generally first acquired technology through imports, then rapidly moved to producing manufactured goods for export. Policies that allowed free communication between foreign markets and domestic producers, that allowed domestic resources to move freely in response to the opportunities, and that complemented existing resources through education, training and infrastructure, all contributed to their success.
(5) Education, skill formation, technology adaptation: Transportation and communication networks all contribute to a sustainable growth of a prosperous global society.
(6) The last and not the least is that every country should have a sense of global consciousness and responsibility.

8.5.4 Regional Development Within the Context of Globalization

In order to have an in-depth study of regional development, it is necessary to clarify the concept of "regions".

8.5.4.1 Typology of Regions

Regions can be classified based upon different concepts. There are several categories of regions:

(1) *Classification of Regional Pattern based upon Economic Geography*

This refers to the economic regional pattern based upon geography. For example, the term North and South is the broadest economic geographical region. It divides the economic regional pattern based upon the latitude of the earth, because many poor countries are located south of 30° north latitude.

China has classified its nation into Eastern, Western and Central regions, which is also an economic geographical classification in a relatively narrow sense.

The Yangtze delta economic region and the Nile delta region are other examples of this category.

There is one other type of economic region, related to type of administration. The Tennessee Valley Authority crosses seven states of Southern United States, but it does not cover all the areas of those states; only the regions of the Tennessee Valley Basin are included.

(2) *Classification of Regions based upon Hierarchical Administrative Level*

In a general sense, the nation is the first-rank administrative region; the second rank is the province or municipality, the third is the county, the fourth is the township, the fifth is the village. The above classification is based upon Chinese practice; different countries may have different terms. For example, the Chinese translation of the Japanese province, in Chinese characters, is county.

(3) *Classification of Region based upon Function*

Regional pattern can also be classified by function; the Free Trade Zone is an example. The urban-rural pattern classification is not only based upon function, but it is also a pattern of economic geography which reflects core-periphery cleavage. In recent years, China has also classified its regions based upon its ecological function, i.e. the bearing capacity of environment. There are four functional zones: the leading developing area; the key development area; Restricted development area (major agricultural production area); and Prohibited (key ecosystem service area) development area.

(4) *Regional Classification based purely upon Geography*

For example, this includes Eastern Africa, South Africa, East Asia, South Asia, Europe, Middle East, Asia Pacific Region, and the Caribbean countries.

(5) *Geographical and Economic Integrated Regions with Political or Economic Implications*

Such as U.S.-Canada-Mexico Free Trade Zone and European Community.

(6) *Other Types of Classification*

There are also many types of regional classification based upon topography, wealth, language, race, etc.

Regional development strategy and policy must take full consideration of such features of regional pattern.

8.5.4.2 Trends of Regionalization

Trends of regionalization should be studied; they appear in different forms based upon the typology of regions discussed above.

(1) The first type of regionalization, which most concerns the contemporary world, is the geographic and economic integrated region with political implications. This is because the increased scale, sophistication and interdependence of the modern global economy is formed through new technologies and new forms of organization which make it possible to overcome some frictions that tend to

operate against hierarchical flows of production and consumption. There are growing trends toward formalized transnational integration, such as the European Union. This form of integration or super-regionalization has the advantages of potential for the smallest nation states, as well as the potential for the creation of multiplier effects from the existence of an enlarged market and for strengthening regional interaction by easing the movement of labor, goods and capital. This trend started early in 1949 with the establishment of Council for Mutual Economic Assistance (Comecon), followed by the establishment of European Coal and Steel Community (ECSC) and the European Economic Community (EEC) successively in 1951 and 1957. European Free Trade Association (EFTA) was established in 1960, and finally the European Union was formed in 1993 through Maastricht Treaty. There are also many other multilateral Free Trade Area such as CISFTA (Commonwealth of Independent States FTA), GAFTA(Council of Aral Economic Unit, Greater Arab FTA), PAFTA (Pacific Alliance FTA), and NAFTA (North American FTA) etc.

(2) The second type of regionalization is administrative regionalization at different hierarchical levels—the nation, the province, the prefecture and county, etc. In this typology of regional classification, processes of economic growth and decline are not geographically neutral in their impact. In a trend toward global industrialization, the locational requirements of new profitable manufacturing production are likely to differ from those of existing production. If the companies were left to make their own decisions, there will be market failure and externalities. It is in this context that appeals for governmental action arise to "help" a particular region or set of regions (several provinces or countries in the Chinese sense), either to "adjust" to a new economic situation or to encourage compensating investment by fiscal policies or measures. Regional governments are viewed as agents for maintaining or attracting private investment in general. In China, the province is also a unit for the allocation of resources. In the U.S., the states compete with one another in establishing the most attractive "business climate". In some countries, such as Italy, Norway, Britain and Southeast Asian countries, regional authorities have been introduced to encourage regional economic planning and foster local industrial generation. Such regional governments generally demonstrate a trend of regionalization in a certain form. The national government must assure local and regional constituents that they represent their best interests. When impelled by geographically differentiated pattern of economic growth and decline, regional policies and regional devolution have been important responses.

(3) In functional classification of typology, the urban-rural relationship should not be underemphasized. The same holds true for the North-South relationship in a general sense. It must be noted that a rich North and a poor South (the poor majority living mostly south of the 30° N), and rich urban or poor rural areas will cause many social problems and unrest, and a large amount of immigration from poor to rich areas. The growing interdependence, which is characteristic of the second half of the twentieth century, has meant that although the world has enjoyed a rapid rate of economic growth, the share of the third world is

different from the developed countries. The disparity between the rich and the poor areas has thus increased. This is also true for the urban-rural relationship. Although China has done relatively well in its economic reform, especially in the rural areas, it has been pointed out by a research paper of a foreign institution that: "research in 1990 found that from 1978 to 1984, rural incomes in average and wealthier parts of the county grew rapidly.... the greatest progress among poor counties occurred in 1983. But the gap between the poorest countries and the rest of rural China increased dramatically." That trend in functional regionalization should be fully taken into consideration in regional planning.

8.5.4.3 Regionalization—Other Megatrend

The first type of regionalization described above is another Megatrend complemented to globalization. With WTO for example, it has more than 160 member countries, but it is very difficult to get consensus for such a large number of member countries with different situations and national interests, this can be fully illustrated from the negotiations of the Doha round. And it was reported there are several hundred regional trade agreements (RTA) established. Therefore it can be concluded that regionalization is another megatrend of social change.

8.6 Summary Points

1. Environment is an important aspect to be studied in two types of engineering systems: Systems engineering and social systems engineering. The environment itself may be also one level in hierarchical level of system related to the lower level of system.
2. Study of boundary line is important, but it is relatively complex in the study of social system.
3. Modern business planning is generally consisted of two parts: study of environment of the business through PEST analysis, and scenario planning.
4. Growth of Yantian International Container Terminal (YICT) in Shenzhen Special Economic Zone of China is chosen to be a Case Study. This case study is an explanation of application of concept and major principle of social systems engineering at two levels, one level is for national planning, the other is for planning of enterprise.
5. Social struction and function are two essential elements of a social system. Although Niklan Luhman had constructed his social system theory focused more on communication, authors of this book inclined more to Parsons' structural functional theory. Because it seems more pragmatic to be basic theory of social systems engineering. But exchange of information (Communication),

energy and material within the social system and to external environment is still emphasized in this book.

6. Two types of structure of national social system from functional perspective are proposed: Vertical hierarchical type and triangular type, a balanced triangular approach is advocated. It should be emphasized that the 'social system' used in this book is academic in sense, because all kind of human actions, such as political action, economic action and scientific and technological action belonged to the activities of the mankind, or within the social system. But for the analysis of the social system from engineering perspective, it is convenient to classify the social system into three subsystem: social, economic, science and technology. The activities of the latter two types of structure are well defined, the boundary can be defined clearly, with all the rest activities of the mankind to be included in the social system.
7. Social change is an important part of the study of social systems engineering. The social system is dynamic in sense. From engineering perspective, this study tries to explore whether or how the mankind to regulate or control the social change in the direction toward a peaceful, prosperity and harmonious society for us all. Of course, this is an objective far from us, but it is necessary to strive for that.
8. Cultural system played a very important role of social change. Several important definitions given by sociologists and anthropologists are presented. Formation and change of culture are discussed. There are two crucial points: human behaviors are results of nurture rather than the products of nature; human behaviors and practices feature notable difference, i.e., diversity of culture.
9. Exploration of cultural theories of Western is explained briefly since the scientific revolution and period of renaissance, the concepts of Kant (German philosopher), Bastin (German scholar), Arnola (British humanist) and Tylor (English Victorian anthropologist) are briefed in succession. Tylor had redefined culture as "the characteristics of different types of activities share by all human societies" which paved the path for the exploration of contemporary culture.

   Exploration of cultural theory by both anthropologists and structural functionalists in Western Countries in the 20th century is briefed. The theory of social behavior of Parsons is consisted of four subsystems: "behavioral system" due to biological needs; "personality system" formed by personal characteristics; "social system" formed by the interplays between units and society; and "cultural system" of behavioral codes and values. But anthropologist Franz Baos held a different view with Parsons' perception of cultural system theory of contemporary anthropologist is quoted for comparative study.
10. The Chinese culture. The emergence of East Asia in recent several decades had attracted the attention of several international organizations and Western scholars to study the relationship between culture and development. One study by Schech and Haggis (2000) has concluded the Confucian cultural features in "explanations of East Asian economic development usually included: human

relations and social harmony based on the idea of filial piety; respect for authority and a strong identity with the organization; subjugation of individual rights in four of community obligations; and diligence." p. 41 Authors of this book give our view in Sect. 8.4.5 of this chapter. Especially the book *Analects* which represents the core concepts of Confucius is introduced.

11. Globalization is viewed by authors of this book to be a megatrends of social change, although there are some debates on that. This trend is explained from four aspects: globalization of the biosphere; globalization of the technosphere; globalization of the econosphere and globalization of the sociosphere.
12. Regionalization is the other megatrend of social change. It is a stage of globalization in perfect sense. These two process of social change are running in parallel, while the process of social change of regionalization has a shorter time span.
13. Regionalization is the other megatrend of social change. It is a stage of globalization in perfect sense. These two process of social change are running in parallel, while the process of social change of regionalization has a shorter time span.

References

Adorno, T. W. (1991). *The culture industry*. New York: Routledge.
Atkinson, S. (Ed.). (2014). *The business book* (p. 211). London: Dorling Kindersley Limited.
Banathy, B. H. (1996). *Design social systems in a changing world*. New York: Plenum Press.
Broun, L. R. (Ed.). (1984). *State of the world 1984*. New York: W.W. Norton & Company.
CIA. (Eds.). (2017). The World Factbook.
Child, J., Tse, K. K. T., & Rodrigues, S. B. (2013). *The dynamics of corporate co-evolution—a case study of port development in China*. UK: Edward Elgar Publishing.
Cody, J., Hughs, H., & Wall, D. (1980). *Policies for industrial progress in developing countries* (p. 12). New York: Oxford University Press.
Friend, G., & Zehle, S. (2009). *Guide to business planning* (2nd ed., pp. 31–35). London: The Economist Newspaper Ltd.
Fuchs, C. (2008). *Internet and society-social theory in the information age* (p. 353). New York: Routledge.
Garner, R. A. (1977). *Social change*. Chicago: Rand McNally College Publishing Company.
IMF. (2016). *World economic outlook* (p. 207). Washington, D.C.: IMF.
Knox, P., & Agnew, J. (1989). *The geography of the world economy*. Great Britain: Edward Arnold.
Machol, R. E. (Ed.). (1965). *Systems engineering handbook*. New York: Mc Grew Hill Book Company.
Marshall, G. (1994). *The concise Oxford dictionary of sociology* (p. 83). Oxford: Oxford University Press.
Sardar, Z., & Loon, B. V. (1998). *Introducing cultural studies*. New York: Totem Books.
Schech, S., & Haggis, J. (2000). *Culture and development* (pp. 10, 41). UK: Blackwell Publishing Company.
Silbereisen, R. K., & Chen, X. (2010). *Social change and human development-concept and results*. London: Sage Publications Ltd.

Sills, D. L., & Merton, R. K. (1968). In D. L. Sills (Ed.), *International encyclopedia of the social science* (Vol. 12, p. 218). New York: Macmillan and Free Press.
Stanners, D., & Bourdeau, P. (Eds.). (1995). *Europe's environment—The Dobris Assessment* (p. 513). Copenhagen: EEA.
Stavrianos, L. S. (1999). *A global history: From prehistory to the 21st century* (S. H. Dong, et al., Trans., 2004, p. 626). Beijing: Beijing University Publisher.
Throsby, D. (2001). *Economics and culture*. Cambridge, UK: Cambridge University Press.
Toynbee, A. J. (1961, October 21). 15 not the age of atoms but for welfare for all. New York Times Magazine.
UNCTAD. (2017). *World investment report 2017*. Geneva: UN.
Wang, H. J. (1991). Globalization and regional development. In P. Hall, R. D. Guzman, C. M. M. Bandara, & A. Kato (Guest Eds.). (1993), *Multilateral cooperation for development in the twenty-first century—UNCRD's twentieth anniversary commemorative volume*. Nagoya: UNCRD.
Wang, H. J. (1998). Foreign direct investment policies and related institution-building in China. In H. J. Wang (2003), *Integrated studies of China's development and reform* (Chapter 9, pp. 251–277). Beijing: Foreign Language Press.
Whitehead, T. L. (2003, September). *Assessment of cultural system* (A CEHC/EICCARS Working Paper).
World Commission on Culture and Development. (Ed.). *Our creative diversity*. France: Egoprim.
Yang, F. (1999). Introduction. In M. Yang (Chief Ed.), *The Analects* (one of Library of Chinese Classics Chinese English). Changsha: Human People's Publishing House.

Chapter 9
Regulation of Social System

9.1 Introduction

It is common sense that regulations are necessary to keep order, security, and development of a social system. From Chapter One, the social systems engineering definition of a social system includes that ".... The social system exists in certain physical and non-physical environment with the boundaries. With the aim towards optimal mutual satisfaction, individuals and groups within the external boundary of a social system interact with each other and with those outside the boundary via exchange of information and action. Their interactions within and outside the boundary are influenced by cultural systems, shared value systems, existing regulations and institutions."

Regulations will be studied in much more detail in this Chapter. There are several theories of regulations and certain concepts regarding the change of regulations. According to Yandle (2011), there are five theories of regulations. The capture theory was developed by political scientist Bernstein (1955) and economic historian Kolko (1963). The special interest of economic theory of regulation was developed by the late Nobel laureate Stigler (1971). In 1991, a law school professor McChesney (1991) developed the fourth theory of regulation, a theory of political wealth extraction; it is damage control with a twist. The bootlegger and Baptist theory of regulation (B&B), was proposed by the author Yandle when he was the Executive Director of the U.S. Federal Trade Commission (FTC) in 1983. The discussion in this Chapter will focus on the most appropriate concept, theory and practice rather than the most advanced. When new trends emerge, scholars must integrate appropriate knowledge based upon their own perspective and life lessons into these new trends.

The authors of this book believe that public policy, regulation, law etc., all perform the functions of regulation, and belong to the discipline of public administration. This Chapter will provide a brief discussion of public administration in the broader context relating to regulation, followed by a brief description of Western

H. Wang and S. Li, *Introduction to Social Systems Engineering*,
https://doi.org/10.1007/978-981-10-7040-2_9

beliefs and a discussion of China's legal system. A detailed discussion of public policy will be a major part of this Chapter, since both authors of this book worked in the Development Research Center (which provides policy consultation for China's government) for more than 30 years. Therefore, we wish to share our experience with the readers. Two case studies will be provided by the end of this Chapter.

9.2 General Discussion of Public Administration

9.2.1 Public Administration as a Discipline

Public Administration as a discipline developed largely in the Twentieth Century. It was defined in 1947 by Appleby, P.H. to be "public leadership of public affairs directly responsible for executive action", but it was not initially recognized as a discipline in the academic field because it covered a very broad area. We provide below both a suggested definition for Public Administration along with its scope of studies based upon available literature.

9.2.1.1 Suggested Definition for Public Administration

In its simplest form, Public Administration is a discipline that studies whatever government does. As an activity, it should reflect the cultural norms, beliefs and power of society, and yet account for the fact that similar administrative acts can be performed differently in different cultures. It is the totality of the working day activities of all the world's bureautics—whether they are legal or illegal, competent or incompetent, decent or despicable. Public Administration is inherent in the execution of public law as well as the formulation and implementation of the public policy. The regulation of enactments is also part of Public Administration, it is government telling citizens and businesses what they may and may not do, which is one of the oldest functions of government. Public Administration performs the executive function of government, implementing public interest and also organizing and managing people and other resources to achieve the goals of public interest, as well as other governmental goals. It is both the art and science of management applied to the public sector (Shafritz, 2004).

9.2.1.2 Scope of Study of Public Administration

The scope of study of Public Administration covers very broad areas, for example as studied by the Indian Council of Social Science, it includes politics and administration (including policy and administration, and the goals of national policy), union-state relations, research on union and state administration (the central-local relationship and administration in general), government headquarters, district administration, administration of urban areas, planning, budgetary and financial control, fiscal administration, public personnel administration, administration of development programs in agriculture and community development, administration of non-agricultural development in rural areas and development of small industries in urban areas, administration of science and technology, administration of law and order, administration of external affairs, administration of defense, administration of social services and welfare programs, industrial regulation, and administration of public enterprises (including project planning).

9.2.1.3 The Definition of Public Administration by Professor David H. Rosenbloom

Professor David H. Rosenbloom (1943–) is a well-known scholar in the field of Public Administration. He defined Public Administration as "the use of managerial, political, and legal theories and processes to fulfill legislative, executive, and judicial governmental mandates for the provision of regulatory and service functions for the society as a whole or some segments of it." *Public Administration,* 2d ed. (1989).

This definition of Public Administration is better understood after reading definitions 1 and 2, above.

9.2.1.4 The Definition of Public Administration from the Dictionary of Sociology

The bureaucratic systems and their procedures which serve the government and implement its policies, and the field of study which describes policy analysis and development, and the policy implementation processes (Marshall, 1994).

9.2.2 Trend of Change of Public Administration

9.2.2.1 Branch Theories of Public Administration

There are several branches of Public Administration: comparative public administration focused on the comparative analysis of the systems and politics of public administration in differing nation states; competitive public administration which emphasizes the public sector policies that force differing organizations to compete with each other for the opportunity to do work for the public; entrepreneurial public administration, which is a style of public management that reflects managerialism in stressing results, a search for innovation and calculating risk taking with a resulting reward for successful managers; and feminist theory of public administration, which is a discipline through various approaches to interpreting the aspects of public administration from a feminist perspective.

9.2.2.2 Emergence of New Public Administration

A general, mostly undefined, movement inspired by younger scholars who challenged several tenets of public administration. Dwight Waldo, noted that public administration was in a time of revolution at a conference held in 1968 at Syracuse University. Papers of this conference were edited by Frank Marini and published as *Toward a New Public Administration*: The *Minnowbrook Perspective (1971)*. The goal of that meeting was to identify what was relevant about public administration and how the discipline had to change to meet the challenge of the 1970s. H. George Frederickson, a professor at the University of Kansas, contributed a paper "Toward a New Public Administration", which called for social equity in the performance and delivery of public service. Professor Frederickson has led the movement from the 1970 to present, and it has had a positive impact, in that now the ethical and equitable treatment of citizens by administrations is at the forefront of concern in public agencies (Shafritz, 2004).

9.2.3 Response to Changes

9.2.3.1 A Changing World and Disputes Among Scholars

During the past several decades, there were debates among Western scholars related to contents of public administration, or more narrowly, the role of the government.

(1) According to Salamon (2002):

> A new era of public problem solving has dawned in the United States and many other parts of the world. Instead of relying exclusively on government to solve public problems, a host of other actors are being mobilized as well, sometimes on their own initiative, but often in complex partnership with the state. In this new setting, traditional notions of public and private responsibilities are being turned on their heads and traditional concepts of public administration rendered largely obsolete. (p. 41)

(2) In the *World Development Report* 1997 titled *The State in a Changing World,* it was stated that "The last fifty years have shown clearly both the benefits and the limitations, especially in the promotion of development." It also raised new worries and questions about the state's role of four recent developments:

(i) "The collapse of command-and-control economies" in the former Soviet Union and Central and Eastern Europe; (p.17)
(ii) "The fiscal crisis of the welfare state" in most of the established industrial countries; (p.17)
(iii) "The dramatic success of some East Asian countries in accelerating economic growth and reducing poverty" due to the role played by the state; (p.17)
(iv) "The crisis of failed states in parts of Africa and elsewhere all of these have challenged existing conceptions of the state's place in the world and its potential contribution to human welfare." (p.17)

9.2.3.2 Different Responses to Changes

(1) From the perspective of Professor Salamon (1943–), a new guide to the new governance is to improve the relative effectiveness of government and private action. Professor Salamon studied the development of a more systematic body of knowledge about the varied "tools" of public action in wide use in the book *The Tools of Government.* Many topics were studied in detail, including: government corporations and government sponsored enterprises; economic regulation and selection of tools; social regulation and patterns of tool use, tool selection, etc. For example, the tool of "government insurance", was studied through "defining the tool", "design features and major variants", "pattern of tool use", "basic mechanics", "tool selection", "management challenges and potential response", "overall assessment", and "future directions". Similar approaches were applied to other topics such as, "public information," "corrective taxes, changes, and tradable permits", "contracting", "purchase-of-service contracting", "grants", "loan and loan guarantees", "tax expenditures", "vouches", "tort liability", etc. This book has given a relatively detailed explanation and analysis of this new governance from both the U.S. and European experience.
(2) Response of the World Bank to the changing role of state, from the development and macro perspective of global society.

From the perspective of the World Bank, an effective state is indispensable to promote the economic, social and sustainable development of a country, but states should work to complement markets, not replace them.

The following four points are recommended by the World Bank in part two of its report titled Match Role to Capability:

(i) States at all levels of institutional capability should respect, nurture, and take advantage of private and voluntary initiative and competitive markets.

(ii) States with weak institutional capabilities should focus on providing the pure public goods and services that markets cannot provide (and that voluntary collective initiatives underprovide), as well as goods and services with large positive externalities, such as property rights, safe water, roads, and basic education.

(iii) Credibility is vital for success. States with weak institutional capabilities should also focus on the tools for policymaking and implementation that give firms and citizens confidence that state officials and organizations will not act arbitrarily and will live within their fiscal means.

(iv) Matching role to capability is a dynamic process. As institutional capability develops, states can take on more difficult collective initiatives (initiatives to foster markets, for example), and use efficient but difficult-to-manage tools for collective action, such as sophisticated regulatory instruments. (p. 40)

9.3 The Legal System—An Essential Means of Regulation of Social Systems

9.3.1 General Discussion

It is well known that a legal system is an essential system in the regulation of human behavior. The legal system of Western countries is well established with many theories and practices. From China, Shen Buha (400–337 BC), Shang Yang (390–338 BC) and Han Feizi (280–233 BC) are well know legalists not only in China, but also known by Western scholars who specialize in study of China. Therefore, both the Western legal system and the detail of Chinese legal system will not be discussed here. However, the contents of a book relevant to legal system of Western countries, with unique features of policy analysis, as well as a very brief discussion of China's legal system will be presented.

9.3.2 Legal Environment of Business, *a Contribution by Professor Butler, H. N. in 1987*

9.3.2.1 Contents of the Book

A Table of Contents from the book are shown below in Table 9.1.

9.3.2.2 Unique Feature of This Book

It can be seen from Table 9.1 that Chapter 1, is titled Introduction to the Law and Policy Analysis. The author of this book has applied policy analysis to selected chapters of this book.

(1) In Chapter 1 of this book, "A Note on the Role of Theory in Policy Analysis", the author emphasized that economists and other public policy analysts use numerous abstract models, such as supply and demand, to develop theories about real-world behavior. Theory helps to explain why; it helps to develop a broad framework that can be applied to a large number of phenomena. A theory must simplify and abstract from reality.

Table 9.1 Contents of *legal environment of business*[a]

No. of parts and chapters	Titles of parts and chapters
Chapter 1	Introduction to the Law and Policy Analysis
Part 1	**Business and the legal system**
Chapter 2	Business Litigation and the court system
Chapter 3	Business and the constitution
Chapter 4	Administrative agencies and procedures
Chapter 5	Business and Society: Political Activity, Ethics, and Social Responsibility
Chapter 6	The Legal environment of International Business
Part 2	**Private Law: A Summary of Selected Business Law Topics**
Chapter 7	Contracts and Commercial Transactions
Chapter 8	Torts and Business Activity
Chapter 9	The Law of Real, Personnel, and Intellectual Property
Chapter 10	Business Associations
Part 3	**Information Problems and Government Intervention**
Chapter 11	Securities Regulations: The 1933 Act and Issuance of Securities
Chapter 12	Regulation of the Securities Industry: The 1934 Act and the Trading of Securities
Chapter 13	Product Safety and Work Safety
Chapter 14	Consumer Law: Advertising and Debtor-Creditor Relations
Part 4	**Externalities: Environmental Laws and the Regulation of Business Pollution**
Chapter 15	Environmental Protection
Part 5	**Monopoly Power: Antitrust Law and the Protection of Competition**
Chapter 16	Statutory Introduction and the Goals of Antitrust Law
Chapter 17	Monopoly, Collation, and Horizontal Mergers
Chapter 18	Antitrust Issues in the Distribution of Goods and Services
Part 6	**Unequal Bargaining Power and the Distribution of Income: Government Intervention on Behalf of Employees**
Chapter 19	The Employment Contract: Wages, Hours, and Benefits
Chapter 20	Labor Management Relations
Chapter 21	Correcting and Deterring Discriminatory Employment Practices

[a]*Source* Butler (1987, pp. xi–xix)

The author also emphasized the four accepted theoretical justifications for government intervention in the free market, due to market failure caused by information problems, externalities, monopoly and unequal distribution of income. pp. 14–17.

(2) The authors also present “Principles of Policy Analysis” in Chapter 1. The main point of that Chapter is whether regulatory policies are achieving their goals. If policy analysis reveals that the goals are being achieved, the next issue to

explore is whether the goals may be achieved in a less restrictive or less costly manner. If policy analysis reveals that the goals are not achieved, the next issue to explore is whether the goals can be attained through other forms of regulation or whether the market should be left alone as the best, albeit imperfect, alternative.

The Principles of Policy Analysis raised by the author include:

(i) The first step in policy analysis is to describe the world as it is, engaging in positive analysis ("what is"), as opposed to normative analysis ("what ought to be").
(ii) Benchmark for the comparison of alternative policies. The typical benchmark is the hypothetical, perfectly competitive market.
(iii) The fact that the real world market deviates from the ideal results of the hypothetical competitive market, means that the ultimate policy question is whether the cost of correcting the market failures are greater than the costs of the continuance of the market failures?
(iv) The concept of efficiency should be addressed. The notion of change is central to understanding the concept of efficiency. The crucial question for policy analysis is to compare benefits for some with costs for others. The author discusses the topic of "Opportunity cost and the allocation of scarce resources", "Marginal analysis and Trade-offs", and "Efficiency as a normal concept" at pp. 18–21of his book.

9.3.3 China's Legal System

9.3.3.1 China's Legal System Before the 20th Century

As described previously, China had the School of Legalism represented by Shen Buhai (400–337 BC). China began to establish its own written code of laws in the period of Spring and Autumn and Warring State (770–221 BC). The Qin Dynasty had unified China in 221 BC, and the entire period from the Qin Dynasty to the Tang Dynasty was characterized by its centralizing tendencies and economic organization of the population. In the early period of the Tang Dynasty, the Minister Zhangsun Wuji and Li Shiji authored a book of 30 volumes, 12 chapters and 102 articles related to the Code of Feudal Laws, upon the command of Emperor Gaozong of the Tang Dynasty in 653 AD. The contents of this book integrated both criminal law of the Tang Dynasty *and its annotation*, it was also the chief source of law of the Song, Yuan, Ming and Qing Dynasties. The Code of Feudal Laws developed in feudal societies are different from modern legal culture.

9.3.3.2 Development of China's Legal System Since The 20th Century[1]

In the early 20th century, China passed through the critical period of the late Qing Dynasty. Under Shen Jiaben and others, the Qing government tried to revise the law along the lines of the western legal model. In 1908, the Qing government issued the Outline of Constitution granted by the Emperor, which was the first written constitutional document in Chinese history. Years later, the legal document became waste paper and the Qing Dynasty withdrew from the stage of history.

With the establishment of the People's Republic of China in 1949, the Chinese began the pioneering work of building socialism. A new era in the construction of Chinese law and rule of law was opened. The Chinese embarked on a path of building a socialist legal system from scratch.

In 1954, at the first meeting of the First National People's Congress, the Constitution of the People's Republic of China was enacted, marking the establishment of the Chinese legal system. The Constitution is now known as Constitution 1954 with 106 clauses. Shortly afterwards, there were periods of the Great Leap Forward and the Cultural Revolution stopped the process of legalization.

In the late 1970s, Chinese reforms had a far-reaching impact, the work of the State was shifted to modernization. The principle that law of rule should govern the Country was established.

In September 1982, the new Constitution of the Communist Party of China was drafted at the CPC's 12th National Congress. The Constitutional Amendment was deliberated and adopted on the first meeting of the Eighth National People's Congress in March 1993, which put the theory of constructing socialism with Chinese characteristics, reforms, opening up, and socialist market economy into the Constitution. China began to approve various economic laws and statutes on a large scale. Table 9.2 shows the valid and effective laws established before 2010.

"Maintain Commitment to the law-based governance of China" has been put into one of six guiding thoughts of *The 13th Five Year Plan for Economic and Social Development of the People's Republic of China (2016–2020).*

Table 9.2 provides the current situation of leave of the legal sector around the year 2010.

[1]Note: This part is composed mainly abstract from Reference 9.

Table 9.2 Valid and effective laws[a]

Legal sector	Amount of legislations (Case)	Percentage (%)
Administrative laws	81	34.9
Economic laws	54	23.3
Constitution and constitution-related legislation	38	16.4
Civil and commercial laws	34	14.7
Social laws	17	7.3
Procedural & non-procedural laws	7	3.0
Criminal laws	1	0.4

[a]*Source* Pan and Ma (2010, p. 35)

9.4 Public Policy

Policy and public policy seem to be synonyms in common sense. Yet, they are two distinguishable terms in global academia. Therefore, there is a need to clarify certain terminology relating to this theme.

9.4.1 *Clarification of Terminology*

9.4.1.1 Policy

Policy is a common noun in Western meaning. The Oxford Dictionary (1974) refers to it as a "plan of action, statement of aims and ideals, especially one made by a government, political party, business company, etc.", the word "Policy" was applied generally to the "one made by a government" in ancient times through to present. Yet, the *Dictionary of Public Policy and Administration,* refers to "A policy is a standing decision by an authoritative source such as a government, a corporation, or the head of a family" (Shafritz, 2004). Governmental decisions may include demanding corporations and individuals to pay income tax. Corporate decisions may define the working hours for employees. Family decisions may decide when dinner will be served. People mistake policy for governmental decisions, the proper name of which is "public policy". From a theoretical perspective, governments, corporations and families are equally entitled to make policies in the general sense. Whereas "public policy" is policy in the narrow sense. Another aspect of the definition of policy is that it is "a standing decision", which means a policy decision should be comparatively long term instead of transient in nature. However, in the practice of policymaking, policies need to be adjusted dynamically based on changing objective circumstances and evolving perceptions of practices.

9.4.1.2 Public Policy

Several definitions of Public Policy from different sources are given below for comparison purposes:

(1) Public Policy. Outputs of a political system, are usually in the form of rules, regulations, laws, ordinances, court decisions, and administrative decisions. Public policy may be perceived as a pattern of activity applied, in the words of political scientists Heinz Eulau and Kenneth Prewitt, consistently and repetitively (Eulau and Prewitt, 1973, 25). Eulau and Prewitt are known for their work in the 1960s on election studies and policy development. Policy may include both the ends and means; it may encompass both the objectives it aims to achieve and the actual procedures by which the objective(s) are sought. Public policy may be short or long-range in duration (Kruschke & Jackson, 1987).
(2) Public Policy. Certain objectives relating to health, morals, and the integrity of government that the law seeks to advance by declaring invalid any contract which conflicts with those objectives, even though there is no statute expressly declaring such a contract illegal (Butter, 1987).
(3) A policy made on behalf of the public by means of a public law or regulation that is put into effect by public administration (Shafritz, 2004).
(4) Decision-making by government. Governments are constantly concerned about what they should or should not do, and whatever they do or do not do is public policy (Shafritz, 2004).
(5) The implementation of a subset of a governing doctrine. The governing doctrine is necessarily an ideology, a comprehensive set of political beliefs about the nature of people and society. It is perhaps best thought of as an organized collection of ideas about the best way for people to live and about the most appropriate institutional arrangements for their societies (Shafritz, 2004).

Although five definitions of public policy are quoted above, it should be mentioned that definition 1 was given in an earlier publication of *The Public Policy Dictionary* in 1987, while definitions 3, 4 and 5 all come from Shafritz (2004). Definition 2 was from Butler (1987). By comparing these five definitions, it can be seen that there are different views of the definition of "Public Policy". Definition 1 is a concise definition, and one related to the very nature of public policy.

9.4.1.3 Public Policymaking

Refers to the totality of the decision process by which a government decides whether to deal with a particular problem. In order to explain the mechanisms that produce policy decisions or no decisions, there are several different theories, however the rational approach is commonly used in the real world. The rational decision making approach was first proposed by Harold Lasswell (a pioneer in policy sciences) in 1963, and posited seven significant phases for every decision:

(1) the intelligence phase, involving an influx of information;
(2) the promoting or recommending phase, involving activities designed to influence the outcome;
(3) the prescribing phase, involving the articulation of norms;
(4) the invoking phase, involving establishing correspondence between prescriptions and concrete circumstances;
(5) the application phase, in which the prescription is executed;
(6) the appraisal phase, assessing intent in relation to effect; and,
(7) the terminating phase, treating expectations established while the prescription was in force.

9.4.2 History and Schools of Public Policy and Policy Sciences Studies

9.4.2.1 History of Public Policy and Policy Sciences Studies

It is true that public policy studies only started to assume the status of an independent academic discipline during the latter half of the 20th century in Western countries. Yet, according to the previous definitions of public policy as "decision making by government" which "encompass both the objectives it aims to achieve and the actual procedures the objective(s) are sought", such policies have existed as long as there have been governments. Numerous scholars of history have committed themselves to conducting policy studies and advisory activities. As a country with five thousand years of civilization, China had its first centralized government—the Qin Dynasty-from 221 to 207 BC. Yet, even before the existence of a central government, "hundred schools of thought" were already flourished during "the Spring and Autumn Period" and "the Warring States Period", which included many schools of policy studies. The Chapter of "Li Yun" in *The Book of Rites* includes a passage describing the society during the primitive commune period before Yu the Great: "When the great Tao prevailed, the whole world was one community. Then people might select the worthy and promote the capable (leaders are chosen through public election and have regard for credit and amity policy goals of creditability and social harmony should be pursued), so people would love not only their own kinsmen and children, but also do their duties to others (Achieve "charity" in modern sense). As a result, the aged would live their full span, the able bodied would be of some use, the young would grow up, and the spouse-bereaved, the orphan, the single, the invalid and the diseased would be supported, the male had a sufficient share, and the female had proper marriage (Goals of social policy)." Were primitive commune societies without any written records really like what was described above? Or did it reflect the author's (Confucius was considered by many to be the compiler of *The Classic of Rites*) idea about what should be the best way for people to live and the most appropriate institutional arrangements for their societies? Yet it did illustrate the level of policy studies in ancient China. Another example would be Xunzi's "On

Enriching the State" written during the Warring States Period: "The ways to make a country self-sufficient include moderating the use of goods, letting the people make a generous living, and being good at storing up the harvest surplus. Moderate the use of goods by means of ritual principles, and let the people make a generous living through the exercise of government (through the execution of policies). Xunzi further explicated the exercise of government through specific policy measures and instruments in the following paragraph: "If one taxes lightly the cultivated fields and outlying districts, imposes excises uniformly at the border stations and in the marketplaces, reduce the number of merchants and traders, initiates only rarely projects requiring the labor of the people, and does not take the farmers from their fields except in the off-season, the state will be wealthy." These two examples illustrate the level of China's policy studies during the Warring States Period. Among the "hundred schools of thought" during the Spring and Autumn Period and the Warring States Period, many schools produced their various policy study ideas. The later history of China was filled with cases of policy studies, which once adopted by a government would become actual policies. Therefore, policy and policy studies have more than two thousand years of history in China. Yet, they failed to develop into modern disciplines by themselves. Overseas policy studies also has a long history. It is safe to conclude that public policies have existed ever since governments came into being. After the industrial revolution, driven by the development of the forces of production and natural and social sciences, as well as national demands, a number of policy institutions were founded in different countries during the earlier half of the 20th century. A list of policy institutions founded before 1950 includes: Germany—Hamburg Institute of International Economics (1908), Kiel Institute for the World Economy (1914), Friedrich Ebert Foundation (1925), and Ifo Institute for Economic Research (1949); United States—Hoover Institution of Stanford University (1919), National Bureau of Economic Research (1920), Council on Chicago Council on Global Affairs (1922), Social Science Research Council (1923), Brookings Institution (1927), Battelle Memorial Institute (1929) (focused on study of technical and scientific innovation and education), and the RAND Corporation (1948); Britain—the Royal Institute of International Affair (Chatham House) (1920); Switzerland—the Graduate Institute of International and Development Studies (1927); Canada—the Canadian International Council (1928); and, Japan—the Kyushu Economic Research Center (1946).

The foundation of the above policy institutions and their practices expanded policy studies into a cluster of theoretical systems including policy sciences, policy analysis, etc.

In 1951, Harold Lasswell edited *The Policy Sciences: Recent Developments in Scope and Method* and was thereby considered the founder of both the concept and methodology of policy sciences. From 1968 to 1971, Yehezkel Dror, a former employee of RAND Corporation, published a series of policy sciences books, including: *Public Policy Making Reexamined (date)*; *Design for Policy Sciences (date)*; and, *Ventures in Policy Sciences: Concepts and Applications (date).* Later, together with the development of behavioral sciences, systems science, statistics, and a number of new mathematical disciplines, many research institutions started to apply quantitative analysis to analyze the impacts and consequences of various programs and policies. Public policy gradually developed into an independent discipline.

During the 1960s, public policy studies and public administration were both developing very fast. While the rapid development of the latter became stabilized in the beginning of the 1980s, the discipline of policy studies kept its momentum. Different from public administration, the range of policy studies is very broad and yet to be defined. Biologists and chemists have emerged from their labs to assume the status of environmental policy experts; civil engineers and urban planners have become urban policy experts; municipal bond analysts and governmental accountants have become financial policy experts; social psychologists, whose original research interests were to stimulate people to act for public welfare, have now become social policy experts; the list goes on and on. Every discipline has a gap that policy studies can fill. Due to the interdisciplinary nature of policy studies, people from various disciplines seek to claim their share of research funds for policy studies. These factors together have caused the continuous development of policy studies. However, conflicts have existed between the studies of all of these specialized policies and the interdisciplinary requirement of "policy sciences". A number of additional policy institutions have also been founded since the 1950s. The United Nations and other international institutions launched their programs of global policy studies, including: the United Nations Economic and Social Council, as well as other Regional Commissions; the United Nations Industrial Development Organization; the United Nations Educational, Scientific and Cultural Organization; the World Health Organization; the World Bank; the International Monetary Fund; the Organization for Economic Co-operation and Development; and, a number of research branches founded under the European Union. The basic concepts, vocabulary and range of components of public policy have gradually taken form. American scholar Earl R. Kruschke compiled *The Public Policy Dictionary* in 1987 which provided a series of somewhat unified definitions of certain concepts related to public policy. There were numerous publications related to studies of public policy. One relatively recent publication was *The Oxford Handbook of Public Policy* edited by Moran, Rein, and Goodin, published in 2006, which is composed of nine parts with forty-four papers. The themes of nine parts are: Introduction; institutional and historical background; modes of policy analysis; producing public policy; instruments of policy; constraints on public policy; policy intervention: styles and rationales; commending and evaluating public polices; and public policy, old and new respectively. As an emerging discipline, public policy is still in the process of development.

9.4.2.2 Schools of Public Policy Studies

Throughout its process of development, public policy studies can be generally divided into two schools that intersect at certain points.

(1) *The School of Quantitative Analysis*

This school emphasizes the use of quantitative analysis. And its success is reflected generally in the micro-field, since most micro-issues are comparatively simple. The birth of this school was triggered by military mathematicians hired by corporations and research institutes after World War II, who started to apply operational studies and other mathematical disciplines to corporate production, storage, delivery, sales and logistics. This school also studies macro-policy issues. Represented by RAND Corporation in the US and International Institute for Applied Systems Analysis in Europe, this school depends on mathematical models for policy analysis. In the field of macro-economy, a number of economists have developed various types of models to conduct policy analysis. Subsequently, these models were also applied to the studies of social and environmental issues.

(2) *The School of Qualitative Analysis and Historical Studies*

This school mostly conducts analysis from qualitative and historical perspectives. Yehezkel Dror is still the academic leader of Israel's policy studies. In 2001, he was asked by the Club of Rome to write a book, *The Capacity to Govern*. Dror represents a way of thinking that emphasizes historical experience, as well as the function of political leaders. A number of United Nations institutions, such as the United Nations Industrial Development Organization and the World Bank, usually include chapters of historical review in the annual reports they published. For example, the theme of the 2005 *Industrial Development Report* published by the United Nations Industrial Development Organization was "Capacity Building for Catching-up", with the subtitle of "Historical, Empirical and Policy Dimensions".

Generally speaking, there are no rigid boundaries between these two schools. The division usually comes from various academic backgrounds of policy researchers and research orientations. Most policy studies generally adopt a methodology that combines qualitative and quantitative analysis.

9.5 A Social System Approach of Public Policymaking Process

9.5.1 Elements Involved in Social Systems Approach of Public Policymaking Process

The three boxes shown in Fig. 9.1:

1. Box A is the external environment, where many issues of policy originate. It is also the context of analysis of public policymaking, i.e. the environment related to the policy issue and policy making that must be analyzed.
   Only the first level of the hierarchy of environment is shown for illustrative purpose, there are also second and third levels of social environment, for example: cultural environment, educational environment, etc.

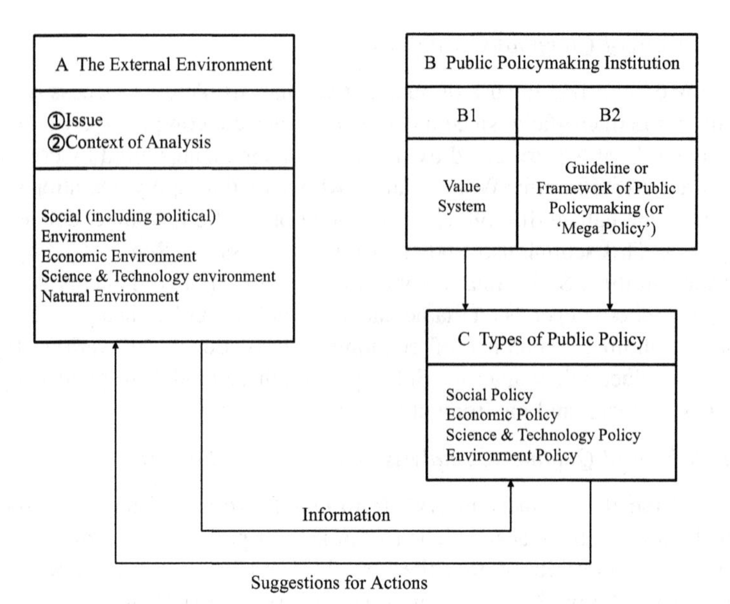

Fig. 9.1 Elements involved in Public policymaking process

2. Box B is public policymaking institution, i.e. it is responsible for studying policy issues and making appropriate policy recommendations to the decision makers. The real functional blocks are B1 and B2.

   (1) B1 is a value system. Every policy recommendation is based upon the value system of the policymaker. The value system is mostly determined by the historical, cultural background of a country or its organizational groups. For example, the core value system of China is currently composed of 12 items: prosperity, democracy, civilization, harmony, freedom, equality, justice, the rule of law, patriotic, dedicated, honesty and friendship.
   Different countries may have different core value systems. There are differences in subjective values and objective values. This is a complex subject, and is beyond the scope of this topic's discussion.
   (2) B2 is the guideline or principle of public policymaking, most of the considerations listed there are applicable to all types of policies.

3. Box C is types of policies to be applied to the relevant policy issues concerned. It is similar to the descriptions for Box A, only the first level of hierarchy of policy is listed for illustration, with economic policy for example, there are second level policies such as agricultural, industrial and service sector policy, etc.

9.5.2 *Guideline of Policy Making (B2)*

9.5.2.1 Setting up Appropriate Overall Goal System

One of the major issues that policy studies has been facing is "complexity" derived from the cluster of mixed driving forces of the discipline. To solve this "complexity" problem, a number of goals and concrete policies have been generated. The establishment of overall goals to serve as guidelines for large sets of concrete policies is a main requirement.

Different conflicts usually exist in concrete policies, such as the conflict between present goals and future goals, and the conflict between positive policy results and uncertain negative results. These conflicts shall be resolved on the level of overall goals.

Taking one project of the authors of this book as an example: we were asked by the State Development and Reform Commission of China to study the feasibility of planning regional development based on "Functional Zones." The overall goal of the project was "building a new pattern of regional development based on functional zoning". This overall goal was proposed in order to meet the following challenges: past economic layout had caused unbalanced distribution of population and economy, with increasing gaps between different regions; past urban-rural segmentation and regional segmentation systems caused the large-scale seasonal migration of population and huge pressure on transportation during holidays; the spatial separation between working-age and dependent population led to increasing gaps between the levels of public services in different regions; the spatial imbalance in terms of the distribution of economy and resources caused increasing pressure on trans-regional resource mobilization, for example the lasting issues of coal and electricity transportation among others; deteriorating ecological environment caused by administrative-division-driven development strategies.

With the overall goal of "building a new pattern of regional development based on functional zoning", the mega-policy aims at solving the five development challenges mentioned above and reflects the long term strategic mindset that respects natural laws and embraces the future. The studies and design of concrete investment, financial, industrial, land, and population policies need to conform to this overall goal.

9.5.2.2 Clarify Policy Boundaries

Policy boundaries have to be defined before any specific policy is made. Different goals require different systems and policy environments. For example, when designing basic public service policies such as compulsory education or basic health care policy, we need to decide whether to seek a market solution, a governmental solution, or a mixture of the two.

9.5.2.3 Setting up of Time Frame

The question of the policy target time—when does one wish the main results of policies to be produced?—is another mega-policy issue. For example, China had set a goal that by the end of 1957, it would surpass Britain in a relatively short period of time. The policy would have been a correct one if the target time had been set for "forty years". This example shows the importance of studies on policy target time. In other words, mega-policies need to consider when a policy (whatever it may be) should be proposed.

9.5.2.4 Estimate Fully the Secondary and Tertiary Impact of Policy

It is the accumulated experience of policy implementation of the real world that there are undesirable or expected secondary and tertiary impacts. With China's process of industrialization as an example, while it is successful as a whole, there are many secondary impacts, such as environmental issues and other issues of adjustment currently necessary.

9.5.2.5 Estimation and Determining Risk and Its Acceptability

All policies involve certain degrees of risk, not to mention those policies that require constant reform during the process of social and economic development. Mega-policies need to consider whether one is ready to accept the higher risks associated usually with more innovative policies, or whether one prefers the lower risks often associated with incremental policies. But it is important to bear in mind that the risks of "maintaining a contemporary situation" may sometimes be higher than those innovative ones.

9.5.2.6 Comprehensiveness Versus Narrowness

This megapolicy facet treats the extent to which a policy should deal with a broad range of components or should focus on a few components, or even a single one. The choice between the two is determined by the policy boundaries, on one side, and the number of applicable policy instruments within the range of research, on the other side. Thus, a policy dealing with social welfare will be relatively comprehensive, while a policy dealing only with retraining will be narrower. However, the range of the latter could also be broadened to include the lifelong learning of people living in the information society, an extremely broad public demand issue. It is important to point out that more comprehensive policies are not necessarily more important or more significant. Thus, the narrow policy decision of a state to develop its nuclear capability will nevertheless be a serious international issue.

9.5.2.7 Balance Oriented Versus Shock Oriented

The question faced by the guideline of policy is, should a policy be directed at achieving given goals through a shock approach, or should the goals be achieved through an effort to promote the change of components together in a mutually coordinated and balanced way. When a shock approach is adopted, usually the goal system would be transformed into a new state of existence, including increased openness to change and propensity to transformation. This megapolicy research dimension of this guideline is also closely related to the comprehensiveness versus narrowness facet described previously, in the sense that comprehensive policies are mostly balance oriented and narrow polices are easily shock oriented. This assumption is strongly influenced by general systems theory, systems analysis, and planning theory. Especially in planning theory, the tradition that favors comprehensiveness and balance may be detrimental to innovation. Therefore, the choice between balance and shock oriented strategies is very important in this guideline.

9.5.2.8 Relevant Assumptions on the Future

In the past, policymaking usually assumed that "what exists now will continue in the future" and avoided engaging in systematic efforts to study non continuous policymaking, which often lead to errors in policymaking.

As a facet of guideline, the establishment of explicit alternative futures assumptions is an important contribution to the improvement of policymaking. There are trends of paying close attention to "futurology" and "strategic" studies and their impact on public policy making. For example, policy studies concerning UN sustainable development goals 2030 are currently widely implemented by many UN organizations such as UNIDO, UNESCO etd. method adopted by many foreign economic forecasts, alternative futures assumptions have already played indispensable roles.

9.5.2.9 Theoretical Bases

Policymaking processes are usually influenced by the strategic mindsets of policymakers, which more than often come from the policymakers' own subjective ideas that have not yet been explicated by systematic theories. Policymakers shall correct their assumptions through comprehensive reasoning, and improve these assumptions with systematic knowledge, rational structures, and organized innovation until they master sound theoretic bases. It is worthwhile to be mentioned briefly that there are available many theories related to public policy, such as Elite Theory, Group Theory, Institutional Theory, Rational Choice Theory, game theory, incrementalism theory etc. All of them can be selected appropriately based upon concrete conditions.

9.5.2.10 Resources Availability

Insufficient research regarding resources availability often lead to failure of policies. The impracticality of many policies often comes from erroneous assumptions that all the resources demanded by a given policy do exist and are available. The reality is all but opposite. Many policies cannot be executed due to insufficient resources. There is need to have a comprehensive understanding of resources, which not only include money, but—sometimes even more important for policymaking—also qualified personnel, sufficient information, usable equipment, etc. When considering policymaking practices both home and abroad, there have been many cases of policy failures that are not caused by insufficient hard resources but by overlooking the availability of soft resources (human resources, system and mechanism resources, knowledge regarding a certain discipline, etc.).

9.5.2.11 The Range of Policy Instruments

(1) Policy instruments had be well studied and defined even quantitatively in the field of economic policy. The following quotations will illustrate this.
According to Acocella (1994), a variable can be defined as a policy instrument if the following three conditions are satisfied:

> (i) Policy makers can control the variable; that is, they can decide what value it should have and fix it directly with their own actions (controllability).
> (ii) The variable whose value has been fixed by policymakers has an influence on other variables, which are assigned the role of targets (effectiveness). In the simple case of one target and one instrument linked by a functional relation, effectiveness can be measured as the derivative of the target with respect to the instrument, $\frac{dy}{dx}$, where the targets is y and the instrument x. With multiple targets and instruments, measuring effectiveness with derivatives is more complicated (see Tinbergen, 1956).
> (iii) It must be possible to distinguish the variable from other instruments in terms of its degree of controllability and, above all, effectiveness: two instruments with the same effect on all targets are not really two separate instruments (separability or independence). (pp. 185–186)

(2) Table 9.3 is quoted partially from a table prepared by Chenery, H. B., the former vice president of the World Bank from 1972 to 1982. This table is quoted to illustrate there are a range of policy instruments adapted to various economic policy.

9.5.2.12 Balance Between Goals of Long Term and Short Term

This is a crucial issue of public policy of the politics of the real. For policymakers are generally not the decision makers who are in power to determine the

Table 9.3 Clarification of policy instruments (General)[a]

Area of policy	Price variables		Quantity variables	
	Instrument	Variables affected	Instrument	Variables affected
Monetary	Interest rate	(1) level of investment (2) cost of production	Open market operations	(1) money supply (2) prices
Fiscal	Personal income tax	(1) consumption and saving	Government expenditure	(1) national income (2) price level
	Corporate income tax	(1) profits (2) investment		
Foreign trade	Exchange rate General tariff level	(1) cost of imports (2) price of exports (3) balance of payments	Exchange auctions	Exchange rates
Foreign investment	Taxes on foreign profits	Level of foreign investment	Foreign loans and grants	(1) investment resources (2) exchange supply
Consumption	General sales tax	Consumption	Social insurance, relief, other transfers	(1) consumption (2) income distribution
Labor	Wage rates	(1) labor cost (2) profits and investment (3) labor income	Emigration and immigration	labor supply

implementation of polices recommended by the policy maker or public policy-making process. Generally, the decision makers have a definite time period of terms of service depending upon the voting process of election. It is not unusual that they will focus on policies favorable to his personal preference with his terms of service. For example, goals of high economic growth versus resource constraints, goals of high economic growth versus environmental degradation, or in concrete example, in the development of science and technology, how much share should be allocated to basic research which has no significant contribution to national socio-economic development in short term…. Balance between them depends upon option of the decision makers, usually compromise and trade-off will be made to achieve a satisfactory solution.

9.5.3 Context of Environment of Public Policymaking

In order to define the public issues to be dealt with by public policies and their respective urgency thereof, as well as to measure the success of public policy making and execution, the external environment need to be analyzed in detail.

9.5.3.1 Social Environment

(1) *Social Environment has big Impacts on Public Policymaking*

The major components of social environment include size, age composition, distribution and mobility of the human population. In developed countries, population aging has become a priority concern for social policy, which is also related to health care and social security issues. China is also facing problems of population aging and increasing aged dependency ratio. With an annual increase of working-age population amounting to 4 million, and a large-scale influx of rural workforce into the cities. The annual increase of employed persons in urban areas reached more than 11 million from 2011–2015, employment pressure (including the employment of both college graduates and migrant workers as well as the protection of their legitimate rights and interests) has become an urgent public policy issue in the present social context.

Fast urbanization will also cause a series of problems such as congested transportation, increasing crime rate, housing supply pressure, as well as urban air and water pollution. In 1992 the "World Summit on Sustainable Development" was held. During the presidency of Bill Clinton, the White House formed the Council on Sustainable Development to advise the President on "bold, new approaches to achieve our economic, environmental, and equity goals". In 1994 China passed its *National Agenda 21* and formed a corresponding office for the agenda. However, due to the development phase that China was undergoing during that time and the different understanding people had regarding sustainable development, the execution of *National Agenda 21* was far from satisfactory.

(2) *Political Environment*

Political environment is considered within the social environment in our study the impact of political environment upon policymaking is universal around the world descriptions of an American public policy studies are quoted below to illustrate this point.

"It is impossible to understand public policy without considering politics, which affects public policy choices at every step, from the selection of policymakers in elections to shaping how conflicts among different groups are resolved. To appreciate the political context, one must be aware of the relative strength of the two major parties; the influence of minor parties; ideological differences among the public, especially the more attentive publics such as committed liberals and conservatives; and the ability of organized interest groups to exert pressure. It is equally important to consider how much interest the public takes in political process, its expectations for what government ought to do, and the level of trust and confidence it has in government. For example, both in the United States and in other industrialized nations there has been a notable erosion of public trust in government in recent decades, typically because of historical events such as the Vietnam War and the Watergate scandal. This trend affects not only the way people are likely to judge government programs and what public officials do, but also the way the press covers public policy debates and actions."[2]

(3) *Cultural Environment*

Public policymaking is also influenced by cultural environments. Culture in the broad sense is an integral part of the sum total of social systems. It includes a series of social norms and belief systems such as attitudes, rules, clothing, language, ceremonies, behaviors, etc. Sociological theories generally hold that human behaviors are the results of nurture (social determinism) instead of nature (biological determinism). Due to the importance of behaviors, behavioral sciences is considered as part of the theoretical base of policy sciences. Since the aim of all public policymaking is to influence the behaviors of certain groups of people, culture is of vital importance to public policy making and execution. Culture in the narrow sense refers to political culture, i.e. values, beliefs and attitudes regarding government and political process. Different countries and regions have different cultural environments. For example, the cultural environments of coastal regions of China such as Guangdong and Shanghai are different from that of Xinjiang and Inner Mongolia. It is necessary to take regional cultural differences into consideration while making differentiated regional policies.

9.5.3.2 Economic Environment

Issues of economic environment is generally high on the international agenda, there were the following issues such as inflation, employment and maintenance of a balanced budget. In the meanwhile, China was facing quite a different set of major economic challenges that included maintaining a high rate of economic growth to guarantee employment as well as to raise the funds in order to solve a number of social problems, such as education, health care, elderly care and some other social security issues. The pressing economic issue for China was to keep up a high growth rate and to curb inflation at the same time while simultaneously adjusting its industrial structure to guarantee a high-speed growth of the service industry in order to create more jobs, reduce high energy-consumption and high polluting industries, and promote the development of low energy-consumption high-tech industries.

[2]Source: Kraft and Furlong (2007).

There are also needs to readjust the relationship between central and local financial departments to ensure that central financial departments could influence the economy through macroeconomic regulation while local financial departments could regulate local economies to satisfy local demands for social and economic growth.

9.5.3.3 Science and Technology (S&T) Environment

It is recognized world-wide that science and technology is a driving force of socio-economic development. Both the developed and developing countries pay attention to promote the development of science and technology an try to improve environment for their growth. Due to the historical background that scientific revolution and industrial revolution were initiated in Western countries, therefore, the Western developed countries have relatively better institutional environment for the development of science and technology, for example, the educational system, the funding system and the organizational system (professional society, project organization) which provide essential input to S&T system as well as the evaluation and incentive system.

OECD (Organization for Economic Cooperation and Development) founded in 1960, currently with 34 member countries, most of them are high income economies. It had focused to study S&T policy for quite a long period. It had launched a study and publication *Science, Technology, and Industry Scoreboard 1999-Benchmarking Knowledge-Base Economies.* In this publication, it focused only countries of OECD and EU. For example, with the indicator of investment in tangibles and in knowledge, this indicator is further classified into three aspects: public spending on education; R&D and software. Sweden ranked the highest in the year 1995, its investment in knowledge as a percentage of GDP was slightly over 10%, while this indicator averaged for EU and OECD were around 8% respectively, and Japan ranked the highest of the indicator of gross domestic expenditure on R&D as a percentage of GDP, which was 2.8% in 1997 while this indicator was around 2.2% for OECD.

In the 2015 publication, China was also to be taken in consideration. And in the executive summary of the report, it is described in the executive summary with the title "The research 'mix' matters":

> Since the mid-1980s, OECD spending on basic research has increased faster than applied research and experimental development, reflecting many governments' emphasis on funding scientific research. Basic research remains highly concentrated in universities and government research organizations. Significant share of R&D in such institutions in dedicated to development in Korea (35%) and China (43%). Overall in 2013, China invested relatively little (4%) in basic research compared to most OECD economies (17%), and its R&D spending is still heavily oriented towards developing S&T infrastructure, i.e. buildings and equipment. (p. 15)

It is expected that the above abstracts and quotation of *Science, Technology and industry scoreboard* of OECD will provide some macro international environment of development of S&T for public policymakers.

9.5.4 *Types of Public Policy*

9.5.4.1 Social Policy[3]

This is a complex subject. It will be discussed through identification of proper definition and illustrate this through a concrete example.

(1) *Proper Definition of Social Policy*

The available definition of social policy can be found in *Collins Dictionary of Sociology* edited by Jary and Jary (1991) that "a field of study which entails the economic, political, sociological and sociological examination of the ways in which central and local governmental policies affect the lives of individuals and communities." p. 590. It is also correctly pointed out in above Dictionary that Social policy is notoriously difficult to define and its use varies between authors. This can be seen from the description of social policy in Bardach, E.'s *Policy Dynamics (2006),* chapter 16 of *The Oxford Handbook of Public Policy.* "Social Policy: A quarter-century ago, Wildavky was writing about the social effects of policies, and sounding very much like Jay Forrester and his students in his concern over the sheer complexity of things. Today there is a second, if not third generation of problems that arise from the complexity of interactions, and these are the problems of making policy adjustments in an environment already dense with interconnected policies. In social policy, for instance, eligibility for one program is sometimes conditioned on eligibility for another, so that reasonable cutbacks (or expansions) in the later have unexpected and undesirable effects in the former." pp. 360–361. *Distributive and redistributive policy* of chapter 30 of above mentioned *Handbook* is considered to be an essential social policy, but in the case of EU, social and employment policies is considered to be the priority of social policies due to the work-force in the EU is over 150 million, while 8 million people are self-employed, and small and medium-sized firm (SMEs) account for about 70% of employment.

Although education policy and health care policy are generally considered to be part of social policies. But there is a trend of rapid process of urbanization in many developing countries and emerging economies. Housing is one of the basic needs of lives of the people, lessons should be learned from many slum areas in the developing countries, it[4] should be one of the priority of social policy for developing countries in general.

(2) *Cases of Social Policy of the European Union*[5]

Some EU countries have the tradition of focusing on social development. For example, the Swedish government's fiscal expenditure on society and health

[3]Source: Jary and Jary (1991).

[4]Note: Housing policy is put in the category of economic policy in Shafritz's (2004) *The Dictionary of Public Policy and Administration* because it is related to interest rate.

[5]Source: Roney (2000).

usually amounts to 28% of its total budget; on culture and education, about 14%. In 1989, the European Communities issued the Social Charter which were co-signed by its member countries (not including the UK). This charter reflected certain aspects of the EU's social policy. Box 9.1 shows headings of the Charter of Fundamental Social Rights for Workers.

Box 9.1 [6]The Charter of Fundamental Social Rights for Workers

1. The improvement of living and working conditions (including those of seasonal and part-time workers), and proposals relating to, for example, collective redundancy procedures.
2. The right to freedom of movement. Under the Charter this implied entitlement to equal treatment, including social and tax advantages, and that conditions of residence should be harmonized.
3. The right to exercise any trade or occupation on the same terms as those applied to nationals of the host state, and social protection as for nationals of the host state for Community citizens in gainful employment.
4. The right to fair remuneration for all employment and, to this end, that a fair wage should be established.
5. The right to social protection, together with a minimum income for those unable to find employment or no longer entitled to unemployment benefit.
6. The right to freedom of association and collective bargaining, i.e. the right of people to belong to the trade union of their choice (and also the right not to belong to a trade union), bargaining rights and the institution of conciliation and arbitration procedures.
7. The right to vocational training. Every worker should have the right to continue vocational training all through his or her working life.
8. The right of men and women to equal treatment.
9. The right to information, consultation and worker participation. This is aimed particularly at companies or groups having operations in more than one Member State.
10. The right to health, protection and safety at the work-place.
11. The protection of children and adolescents. The minimum working age should be 16, and there should be special vocational training for the these groups (see below).
12. Provisions for elderly people. Every citizen in retirement or early retirement should have an income that guarantees him or her a reasonable standard of living, with a specified minimum income, even if he or she has no entitlement to a pension.

[6]Source: Roney (2000, pp. 190–191).

13. Provisions for disabled people. There should be measures aimed at achieving the best possible integration into working life for this group, together with various related measures on mobility and so on.

9.5.4.2 Economic Policy

The content and boundary of economic policy are uncontroversial in all academic (economics or public policy) dictionaries. Economic policy refers to the processes by which the government regulates the economy. Basically, there are three subfields covered by economic policy: fiscal policy, monetary policy, investment policy, trade policy and other policies with economic significance, such as energy policy, agricultural policy, labor policy (in developed countries there are also trade union policy) and so on. Due to the fact that policies do not exist in vacuum, the interaction among different aspects of economic policy becomes a major research topic. For example, monetary policy mainly regulates the amount and cost (interest rate) of currency and credit; fiscal policy primarily deals with budget, deficit and tax. The execution of many projects is simultaneously influenced by fiscal and monetary policies. Academic studies can make all kinds of ideal theoretical combinations of different policies. For instance, macroeconomic regulation has four sets of combinations: tight fiscal policy, tight monetary policy; easy fiscal policy, easy monetary policy; tight fiscal policy, easy monetary policy; easy fiscal policy, tight monetary policy. In reality, however, these policies are hard to put into practice because they are restricted by different executive departments. Take the US as an example. Monetary policy is within the jurisdictions of the President and the Federal Reserve. And fiscal policy is determined by the President, the Fed, and the Congress. However, the taxation right belongs solely to the Congress. Therefore, the execution of combined policies is in fact restricted by political systems and political institutions.

9.5.4.3 Science and Technology Policy

Science and technology policies are key topics in developed and developing countries. Once upon a time, the difference between science policy and technology policy was a problem for intense controversies in foreign countries. Recently, controversies have become less because science is more and more considered as the foundation of technology. Western governments used to put more emphasis on science policy. This can be seen from the quotation of executive summary of *OECD Science Technology and Industrial Scoreboard 2015* that most OECD economies have a share of 17% on basic research of their R&D spending. Because technology policy is generally considered to be more effective when the market and the private sector play bigger roles. This would be instructive for the making and execution of science and technology policy for late comers in development of science and technology. Table 9.4. shows the major differences between National Science and Technology policies in objectives, reference criteria for permanence, main types and scope of activities etc.

Table 9.4 Differences between national science and technology policies[a]

Aspect	Science policy	Technology policy
Objectives	(A) To generate scientific (basic and potentially useful) knowledge that may eventually have social and economic uses, and will allow understanding and keeping up with the evolution of science (B) To produce a base of scientific activities and human resources linked to the growth of knowledge throughout the world	(A) To acquire the technology and the technical capabilities for the production of goods and the provision of services (B) To acquire a national capacity for autonomous decision-making in technological matters
Main types of activities covered	Basic and applied research that generates both basic and potentially useful knowledge	Development, adaptation, reverse engineering, technology transfer, and engineering design, which generate ready-to-use knowledge
Appropriation of results of activities covered	Results (in the form of basic and potentially useful knowledge) are appropriated by wide dissemination; publishing ensures ownership	Results (in the form of ready-to-use knowledge) remain largely in hands of those who generated them; patents, secret know-how, and human embodied knowledge ensure appropriation
Reference criteria for performance	Primarily internal to the scientific community. Evaluation of activities is based mainly on scientific merit and occasionally on possible applications	Primarily external to the technical and engineering community. Evaluation based mainly on contribution to social and economic objectives
Scope of activities	Universal: activities and results have worldwide validity	Localized (to firm, branch, sector, or national level): activities and results have validity in a specific context
Amenability to planning	Only broad areas and directives can be programmed. Results depend on the capacity of researchers (teams and individuals) to generate new ideas. Involves large uncertainties	Activities and sequences can be programmed more strictly. Little new knowledge generally required, and existing knowledge is used systematically. In valves less uncertainty
Dominant time frame	Long and medium term	Short and medium term

[a]*Source* IDRC. Science, Technology and Development: Planning in the STPI Countries, Ottawa: IDRC, STPI Project, 1979 (pp. 16–17)

9.5.4.4 Environmental Policy

Environmental policy is very complicated for policy-makers, and they even become focal topic of policy for international controversies, which can be seen in the debates on policies between the US and the EU, and between the Democratic and the Republican parties within the US. On the other hand, environmental policy is closely related to energy policy and resources policy. It is necessary to study environmental policy together with energy and resources policies.

Development of EU environmental policy and US environmental policy will be briefed. Because they are two major economies with relatively long history of development of environmental policy.

(1) *Development of Environmental Policy of the European Union*

Principles of Environment policy of European Communities were defined early in 1972, they were:

(i) Prevent pollution at source;
(ii) Environmental policy must be compatible with economic and social development;
(iii) All planning processes must consider environmental issues;
(iv) Polluter pay for pollutions.

According to the fifth Environment Action Program, "Towards Sustainability, 1992–2000", medium-to-long term environmental targets are based on five specific sectors: industry, agriculture, energy, transportation, and tourism, with five priorities: (i) Integration of the environmental dimension into other policies; (ii) Supporting the development of public transportation and use of clean technology, and paying attention to local as well as regional issues; (iii) Make use of different sectors, such as education, to promote environmental protection; (iv) Take ecological issues into consideration; (v) Strengthen the EU's role in international initiatives.

Since Nicholas Stern, the former Lord of the Treasury of U.K., put forward the Stern Review in 2006 with the support of the UK government, the EU began to take reducing greenhouse gases emissions as a major mission. The EU determined that by 2020, developed countries should reduce 30% of greenhouse gas emission. And by 2020, energy efficiency should increase by 20%, and the use of biofuels should reach 10% in the transportation sector.

(2) *The Development of Environmental Policy in the United States*

The development of environmental policy in the US fully manifests the political nature of public policy. In 1970, United States Environmental Protection Agency (EPA) was established. Their description of environmental policy was to "formulate guidelines and policies upon objectives and standards regarding the use and protection of material environment, including soil, water, air, wild animals and plants. Environmental policy covers fields including broad range of areas, such as water pollutions, air pollution, solid waste management, radiation control, pesticides and toxic chemical control and so on." The EPA, established in 1970, was in charge of the legislation and supervision in those areas.

From 1965 to 2005, more than 40 acts were passed in the US regarding environmental issues. Some of the important ones are listed in Table 9.5 for reference.

Although traditional American values include the inviolability of private property, individual rights, limited government and the emphasis upon economic growth, the new social forces rose in the early 1920s began to clash with them. This conflict brought about the government's new functions in environmental protection and natural resources management. In the second half of the 19th century, municipal authorities started to realize the importance of municipal services, such as clean water, waste management and wastewater treatment services. In the 1930s, a series of natural disasters happened. Therefore President Roosevelt's New Deal extended the scope of protective policy to include flood control and soil conservation. In 1933, Tennessee Valley Authority was set up, which proved that the government's management and planning of land was favorable to public interest. Before the 1970s, energy policy was not the focus of the US government. The only targets of energy policy at that time were to stabilize prices and to ensure supplies exception was the commercialization of nuclear power plants.

In 1970, the Environmental Protection Agency was established. Americans started to care about environmental and health risks, and strongly supported the

Table 9.5 Major environmental acts in the US, 1964–2005

Year	Acts
1964	Wilderness Act PL88-577
1969	National Environmental Policy Act (NEPA) PL91-190
1970	Clean Air Act (Extension) PL91-190
1972	Federal Water Pollution Control Amendments (Clean Water Act) PL92-500
1973	Endangered Species Act PL93-205
1974	Safe Drinking Water Act PL93-523
1976	Resource Conservation and Recovery Act (RCRA) PL94-500
	National Forest Management Act PL94-588
1980	Comprehensive Environmental Response, Compensation, and Liability Act (Superfund) PL96-510
1988	Ocean Dumping Ban Act PL100-688
1990	Clean Air Act Amendments PL101-549
	Pollution Prevention Act PL101-508
1992	Energy Policy Act PL102-275
1996	Food Quality Protection Act (amended FIFRA) PL104-120
2003	Healthy Forests Restoration Act PL108-148
2005	Energy Policy Act of 2005 PL109-58

promulgation of public environmental policy. However, the US polity prevented the Bush Administration from taking actions against climate change. In June 2002, the Bush Administration submitted the US Climate Action Report to the United Nations, showing that the U.S. would obey the United Nations Framework Convention on Climate Change passed during the Earth Summit in 1992. Since 1992. The U.S. government was facing the dual challenges of political acceptability and economic feasibility. So this report was secretly sent to the UN and no press briefings were held. When the news ran, President Bush was intensely criticized by the conservatives. Bush had to deny his support for the report and claimed that it did not reflect his administrative policy. Moreover, he emphasized that he opposed to the Kyoto Protocol on climate change.

9.5.4.5 Policy Process-Theories

There is one additional theory which had not been raised in 9.4.2–9 on available theories of public policy, i.e. policy process theories. There is a Politics-Policy System Model which is reproduced to be Fig. 9.2 to conclude this part without further discussion. It is clearly shown in the figure that the political system is a "black box" in sense. Scholars of public policy studies should understand the implication of this.

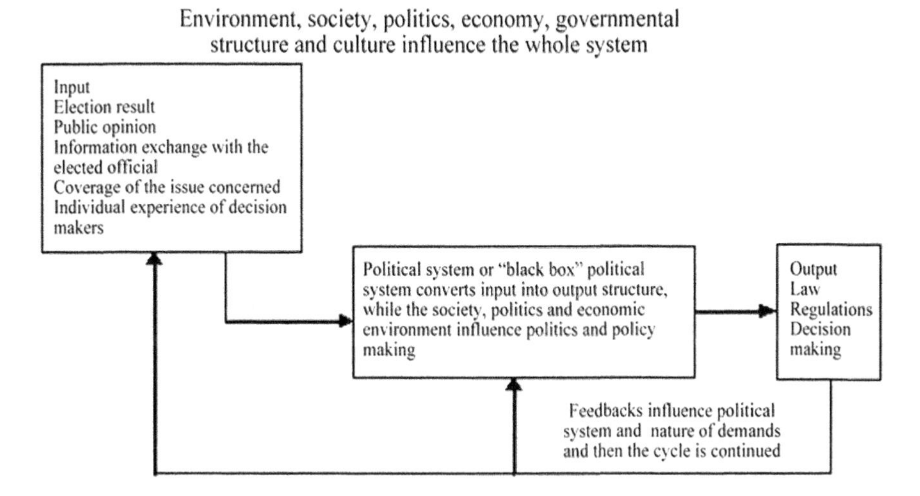

Fig. 9.2 Politics-policy system model (*Source* Birkland 2005)

9.6 Case Study 1 of Applying Analytic Hierarchy Process to Policy System

9.6.1 Background Information of This Case Study

In the national plans of China, the regional planning was generally based on the geographical regions, i.e., Eastern region, Central region and Western region etc. before 2005. Based upon the "Outline" of preparation of Eleventh Five-Year Plan, the concept of "Functional Zone" was raised, i.e., to clarify the regions based upon their ecological and bearing capacity. The relevant bureau of State Planning and Reform Commission had entrusted Research Center of China's Development Planning of Tsinghua University to implement a study on "Differentiated Policy to Construct the Functional Zone". This study was completed around December, 2007. Analytic Hierarchy Process is experimented to determine the weight coefficient of various policies in these four functional zone (i.e., leading development zone, key development zone, restricted development zone and prohibited development zone). Only the section of AHP applied to leading development zone is shown in this case study for illustrative purpose.

Due to the multiplexity of the development policy of functional zones, it is necessary to consider the relative importance of various policies in the four different functional zones. In order to generate a more quantitative-based descriptive analysis, this section adopts analytic hierarchy process to decide the relative weights of various policies in the four different functional zones—greater weights will be assigned to prioritized or key policies.

9.6.2 Brief Introduction to Analytic Hierarchy Process

Analytic hierarchy process (AHP) is a multi-objective decision-making method developed by US operational research expert Thomas L. Saaty in the 1970s which combines quantitative and qualitative approaches. The core of this method is the quantification of decision makers' judgments and the use thereof as the quantitative basis for decision-making. Users of AHP first quantify their decision problem into a hierarchy of elements that are correlated to each other. Once the hierarchy is built, the decision makers use their judgments systematically to set the priority of these elements. AHP method requires relatively smaller amount of data and shorter time to make decisions.

Procedure of AHP in decision making includes:

1. Analyze the correlations between and structure of the internal elements of a decision problem based on sufficient knowledge of the problem, and decompose the structure into a number of levels, such as measure level, principle level, goal level, etc. Then express the correlations between these elements on different levels through charts.

2. Once the hierarchy is built, various elements on the same level should be compared to each other "pairwise" based on their respective impact on another element on a higher level. A judgment matrix should be built to calculate the weighting factors of elements on the same level.
3. Conduct consistency check. If the results could pass the consistency check, the weighting factors will be accepted. Otherwise, the factors will not be accepted, which will trigger another comparison resulting in another judgment matrix.
4. Calculate the weighting factors of elements on different levels based on the weighting factors of same-level elements.

9.6.3 Relative Weight Analysis Regarding Policies for Optimized Development Regions

We will illustrate the concrete application process of AHP with the policy support system of optimized development regions. Similar analytic procedure will be used for key development regions, restricted development regions and no-development regions.

9.6.3.1 Decompose Internal Elements into Levels (Hierarchies)

First of all, a hierarchy of functional zone policy system is built as follows based upon policy analysis. Hierarchies of other functional zones are similar (Fig. 9.3).

9.6.3.2 Calculate the Weighting Factors of Elements on the Same Level

Build judgment matrixes based upon pairwise comparisons, calculate weighting factors and conduct consistency check. Judgment matrix is a matrix formed by the comparisons of various elements on the same level based upon their respective impact on another element on a higher level in the hierarchy.

If we are going to compare n elements $\{y_1, y_2, ..., y_n\}$ regarding their respective impact on a target element, we need to select two elements y_i and y_j each time. a_{ij} refers to the ratio of the impact that y_i has on the target to the impact that y_j has on the target. The value of a_{ij} is defined by Saaty's one to nine scaling method[7] (Table 9.6).

[7]Saaty et al. compared through experiments the accuracy of people's judgments with different scales. Results of these experiments showed that a scale of one to nine is the most appropriate.

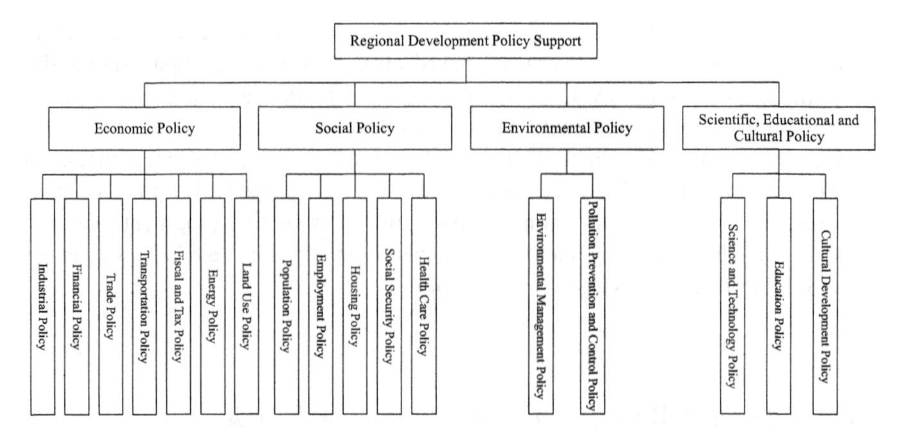

Fig. 9.3 Policy hierarchy framework

Table 9.6 Scales and their meanings

Scales	Meanings
1	Element y_i and y_j are of equal importance
3	Element y_i is slightly more important than Element y_j
5	Element y_i is significantly more important than Element y_j
7	Element y_i is strongly more important than Element y_j
9	Element y_i is of extreme importance compared to Element y_j
2, 4, 6, 8	Median values between any two adjacent judgments
Reciprocals of above	If the ratio of the importance of y_i to the importance of y_j is a_{ij}, then the ratio of the importance of y_j to the importance of y_i is $a_{ji} = 1/a_{ij}$

Judgment matrix A = $\left[a_{ij}\right]_{n\times n}$ calculated through pairwise comparisons has the following features:

$$①a_{ij} > 0;\ ②a_{ii} = 1;\ ③a_{ji} = 1/a_{ij}\ (\mathrm{i,j} = 1, 2, \ldots, \mathrm{n})$$

Feature ③ is true because: if the ratio of element 1 to element 2 is recorded as a_{ij}, then the ratio of element 2 to element 1 is $a_{ji} = 1/a_{ij}$, i.e. transposed corresponding elements are reciprocals of each other (Table 9.7).

The data in the above table can be explain in the following: 4.000 in boldface shows that in optimized regions social policy is significantly more important than environmental policy when these two elements are compared. The result of reverse comparison is 1/4 = 0.2500. Other pairwise comparisons are similar. Included in the right-most column are the weights calculated based on the judgment matrix. There are many methods to calculate the weights, among which the eigenvector method is selected by our project for its higher computational accuracy. The computational procedure includes: first calculate the maximum eigenvalue λ_{max} of the matrix; then

Table 9.7 Judgment matrix calculated through pairwise comparisons of second-level elements

Regional development policy support	Economic policy	Social policy	Environmental policy	Scientific, educational and cultural policy	Weight
Economic policy	1.0000	0.5000	2.0000	0.2500	0.1333
Social policy	2.0000	1.0000	4.0000	0.5000	0.2667
Environmental policy	0.5000	0.2500	1.0000	0.1250	0.0667
Scientific, educational and cultural policy	**4.0000**	2.0000	8.0000	1.0000	0.5333

calculate W, the corresponding eigenvector of λ_{max}; the weights included in the right-most column will be calculated through the normalization of W.

When people conduct pairwise comparisons of elements within complex issues, it's impossible to achieve complete consistency. Certain amount of errors have be tolerated. Therefore before deciding whether to accept the calculated weights, we must conduct consistency checks on judgment matrixes. The procedure of consistency check includes: first calculate consistency indicator CI[8]; then find its corresponding average random consistency indicator RI; Thomas L. Saaty has given the list of values of RI; finally calculate the consistency ratio CR = CI/RI. When CR < 0.1, the consistency of the matrix is considered acceptable.

In the following we will give a concrete example of how to compute the weights of the aforementioned judgment matrix. The computation processes of other matrixes are similar.

$$\text{Judgment Matrix A} = \begin{bmatrix} 1 & 0.5 & 2 & 1/4 \\ 2 & 1 & 4 & 1/2 \\ 0.5 & 0.25 & 1 & 1/8 \\ 4 & 2 & 8 & 1 \end{bmatrix}$$

Use matlab and input [V, D] = eig(A)

Then we can get the maximum eigenvalue and its corresponding eigenvector of Judgment Matrix A:

$\lambda_{max} = 4.000$, its corresponding eigenvector P = [0.2169, 0.4339, 0.1085, 0.8677]

Normalize P, i.e. W = P/sum (P); then we will get the weights vector

$$W = [0.1333, 0.2667, 0.0667, 0.5333]$$

[8]CI = (− n)/(n − 1), n refers to the number of order of the judgment matrix.

Table 9.8 Corresponding values of RI

N	1	2	3	4	5	6	7	8	9
RI	0	0	0.58	0.90	1.12	1.24	1.32	1.41	1.45

N = 4, therefore RI = 0.90
CR = CI/CR = 0.000/0.90 = 0.000 < 0.1

Computation process of the consistency check is (Table 9.8):

$$CI = (\lambda_{max} - n)/(n - 1) = (4.000 - 4)/(4 - 1) = 0.000$$

After calculation, the consistency ratio of the matrix is 0.000, which means it has passed the consistency check. The weights derived from this matrix are acceptable. In this matrix, the weight of economic policy regarding the overall goal is 0.1333, the weight of social policy is 0.2667, the weight of environmental policy is 0.0667, the weight of scientific, educational and cultural policy is 0.5333—all are acceptable (Table 9.9).

Included in the right-most column are the weights calculated through the judgment matrix. After calculation, the consistency ratio of the economic policy matrix is 0.0239, which means it has passed the consistency check. The weights derived from this matrix are acceptable (Table 9.10).

Included in the right-most column are the weights calculated through the judgment matrix. After calculation, the consistency ratio of the social policy matrix is 0.0102, which means it has passed the consistency check. The weights derived from this matrix are acceptable (Table 9.11).

Included in the right-most column are the weights calculated through the judgment matrix. After calculation, the consistency ratio of the environmental policy matrix is 0.0000, which means it has passed the consistency check. The weights derived from this matrix are acceptable (Table 9.12).

Included in the right-most column are the weights calculated through the judgment matrix. After calculation, the consistency ratio of the scientific, educational and cultural policy matrix is 0.0000, which means it has passed the consistency check. The weights derived from this matrix are acceptable.

9.6.3.3 Calculate the Weighting Factors of Elements on Different Levels

Based on the weights of different levels, the combined weights can be calculated as follows:

Taking the computation process of the combined weights of industrial policy as an example, the weight of industrial policy regarding economic policy is 0.1333. Therefore the combined (final) weight of industrial policy regarding regional development policy is 0.0322 = 0.2412 * 0.1333 (Fig. 9.4; Table 9.13).

Table 9.9 Judgment matrix calculated through pairwise comparisons of third-level elements under economic policy

Economic policy	Industrial policy	Financial policy	Trade policy	Transportation policy	Fiscal and tax policy	Energy policy	Land use policy	Weight
Industrial policy	1.0000	2.0000	5.0000	6.0000	4.0000	3.0000	0.5000	0.2412
Financial policy	0.5000	1.0000	4.0000	5.0000	3.0000	2.0000	0.3333	0.1596
Trade policy	0.2000	0.2500	1.0000	2.0000	0.5000	0.3333	0.1667	0.0449
Transportation policy	0.1667	0.2000	0.5000	1.0000	0.3333	0.2500	0.1429	0.0308
Fiscal and tax policy	0.2500	0.3333	2.0000	3.0000	1.0000	0.5000	0.2000	0.0678
Energy policy	0.3333	0.5000	3.0000	4.0000	2.0000	1.0000	0.2500	0.1040
Land use policy	2.0000	3.0000	6.0000	7.0000	5.0000	4.0000	1.0000	0.3517

Table 9.10 Judgment matrix calculated through pairwise comparisons of third-level elements under social policy

Social policy	Population policy	Employment policy	Housing policy	Social security policy	Health care policy	Weights
Population policy	1.0000	0.1667	0.2500	0.5000	0.1250	0.0434
Employment policy	6.0000	1.0000	2.0000	4.0000	0.5000	0.2694
Housing policy	4.0000	0.5000	1.0000	2.0000	0.2500	0.1427
Social security policy	2.0000	0.2500	0.5000	1.0000	0.1667	0.0756
Health care policy	8.0000	2.0000	4.0000	6.0000	1.0000	0.4690

Table 9.11 Judgment matrix calculated through pairwise comparisons of third-level elements under environmental policy

Environmental policy	Environmental management policy	Pollution prevention and control policy	Weight
Environmental management policy	1.0000	4.0000	0.8000
Pollution prevention and control policy	0.2500	1.0000	0.2000

Table 9.12 Judgment matrix calculated through pairwise comparisons of third-level elements under scientific, educational and cultural policy

Scientific, educational and cultural policy	Science and technology policy	Education policy	Cultural development policy	Weight
Science and technology policy	1.0000	2.0000	4.0000	0.5714
Education policy	0.5000	1.0000	2.0000	0.2857
Cultural development policy	0.2500	0.5000	1.0000	0.1429

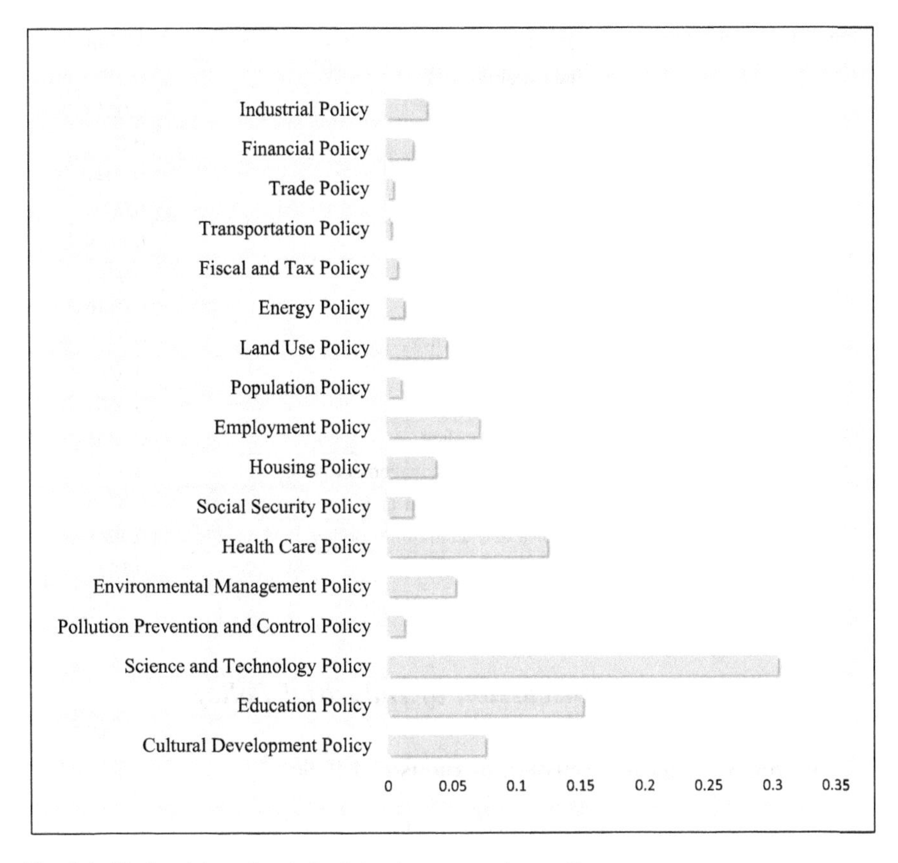

Fig. 9.4 Final weights of optimized development regions policy system

9.7 Case Study 2 of Industrial Policy of China

9.7.1 Introduction

Industrial policy is an important policy for national development, in a global trend of increasing integration of global and regional economic activities, an appropriate national industrial policy will not only be beneficial to the development of a single country, but it will also be favorable to the regional and global prosperity as well. This section will be divided into three parts:

1. Some general discussions of industrial policy;
2. China's industrial policy 1953–2013;
3. China's new situation and new industrial policy within the context of globalization and regionalization.

Table 9.13 Final weights

Options	Weight
Industrial policy	0.0322
Financial policy	0.0213
Trade policy	0.006
Transportation policy	0.0041
Fiscal and tax policy	0.009
Energy policy	0.0139
Land use policy	0.0469
Population policy	0.0116
Employment policy	0.0718
Housing policy	0.038
Social security policy	0.0201
Health care policy	0.1251
Environmental management policy	0.0533
Pollution prevention and control policy	0.0133
Science and technology policy	0.3048
Education policy	0.1524
Cultural development policy	0.0762

9.7.2 Some General Discussion of Industrial Policy

9.7.2.1 Some General Discussion of Industrial Policy

(1) *Brief history of industrial policy (IP)*

The concept of IP had been raised early in 18th century, the concept of selective protection of industries was contained in the *Report on the Subject of Manufactures* of U.S. economist and politician Alexander Hamiltion and the work of German economist Friedrich List. The first IP in the India was introduced by the British in 1923, it was officially raised by independent India in 1948. According to the study by Gerard and Robert (1983), they had studied French Industrial Policy from 1945–1981, Federal Republic Germany's Industrial Policy from 1967–1981, Italian's Industrial Policy from 1947–1981 and Japanese Industrial Policy from 1945–1980 etc. Therefore, there are rich historical experiences globally on this subject.

(2) *Definition of Industrial Policy*

Nearly all economic dictionaries have definitions of Industrial Policy. A recent study by OECD has given 17 definitions. Hereunder, two quotations from *Dictionary of Economics* will be quoted for comparative purposes and clarification of some ambiguities.

(i) *According to Black, J. in Oxford Dictionary of Economics (2nd Ed.)*

"Official policies concerning the direction of economic activity to particular parts of the economy. For many years governments were concerned with the division of economies between industry and primary production: the less industrialized countries tried to growth of local industry, while the more industrial countries protected their native agriculture and mines. In more recent years attention has centered on try to encourage the rise of high-tech sectors using advanced technology….." (p. 232)

The above quotation has not only provide the definition of IP, but also gives the different purpose of IP between industrial and less industrial countries.

(ii) *According to Pass and Lowes (1993) Collins: Dictionary of Economics (2nd Ed.)*

> A policy concerned with promoting industrial efficiency and competitiveness, industrial regeneration and expansion and the creation of employment opportunities. Industry policy can be broadly based, encompassing, for example, measures to increase competition (see competition policy) and promote regional development (see regional policy) as well as specific across the broad measure to stimulate efficiency and adoption of new technology; …. (p. 254)

Compare definition 2 and 1, it can be seen that definition 1 is more broadly based, it focused on economic activity while definition 2 focused on industry only. But definition 2 also described "Industry policy can be broadly based", a brief discussion of types of Industrial Policy may provide further clarification of those concepts.

9.7.2.2 Types of Industrial Policy

According to Gerald and Robert (1983), "The problem of defining what refers to IP and what does not is not a trivial one. This is so partially because this concept is rarely clearly defined in the literature and partially because its meaning has changed through the time and is perceived differently by various countries". p. 14. They have classified IP based upon scope, and distinguish the policy actions in descending order of generality as follows:

(1) *Industrial Policy in general.* This includes all types of IPs, both general and specific.
(2) *General or nonselective industrial policies (GIPs).* GIPs are intended to be available on equal terms to all industries of the economy. These may include broad policies aimed at improving, in general, the resource allocation mechanism and encourage investment or technology. They are also called "horizontal industrial policies" (Stoffaes ,1981).
(3) *Activity-specific policies (ASPs).* This concept is a subcategory of GIP.

(4) *Regional-specific policies (RSPs).* These may overlap with the previous two categories, for regional policies may not be targeted to a particular industry or activity.
(5) *Sector-specific policies (SSPs).* These are policies directed at specific but broad sectors of the economy, for example, manufacturing, export or import sector.
(6) *Industry-specific policies (ISPs).* Policies directed at specific industries, defined broadly or narrowly.
(7) *Firm-specific policies (FSPs) or project specific policies (PSPs).* These are policies designed to benefit particular firms.

9.7.2.3 Trend of Study of International Organizations

(1) *The Work of UNIDO*

UNIDO is the abbreviation of United Nations Industrial Development Organization which was established by UN in 1966 and converted to a specialized agency in 1985. Its primary objective is the promotion of development in developing countries and international industrial cooperation. It had studied Structural Changes in Industry World Industry in 1980 etc. It has also a flagship annual report. Before the year 1995, its name was Industry and Development-Global Report, its name changed to Industrial Development Report around 1995. It had also established a competitive Industrial Perform Index (CIP) since 1985 covered 80 economies, the number of economies increased to 141 in 2013. It had also organized several important forum to discuss global and regional IP, for example, it had organized Global Forum on Industry: Perspective for 2000 and Beyondat New Delhi, India in 1995, and Asia-Pacific Regional Forum on Industry post the Asian financial crisis at Bangkok, Thailand in 1999. UNIDO also published annually International year Book of Industrial Statistics more than 20 years, it has a detail classification of industrial activity set out in tables contained in the yearbook follows either Revision 2, Revision 3 or 4 of ISIC depending on the individual country data scheme. It covered up to 4-digit level of 24 manufacturing sectors around 200 countries in his latest edition.

(2) *The Work of OECD*

With OECD for example, it has many studies related to "Industry and Entrepreneurship", "SMEs and entrepreneurship" "Industry and Globalisation" and also provide statistics of Industrial Production for OECD countries and major non-member economies. It has also provided studies related to "Policy Evaluation of Industrial Polices". But study of industrial policy is broad based rather than focusing on manufacturing sector only as done by UNIDO.

9.7.3 China's Industrial Policy 1953–2013

9.7.3.1 General

With regard to China's Industrial Policy, it can roughly be divided into two periods: periods since 1953–1985, that is from the First Five Year Planning period up to the Sixth Five Year planning period, five of the Five Year Plan were in the pre-reform era. In the pre-reform era, China had followed the former Soviet model of central planning system which was mandatory in nature. MPS (Material Production System) was used for the measurement of growth of national economy, Gross Value of Industrial and Agricultural Value of Outputs (GVIAO) was used to measure the economic growth, service sector was considered to be nonproductive. There was no official term of IP, but implicitly, there were Industrial Policy in general. China began to reform its planning system gradually from mandatory planning to indicative planning gradually since its Sixth Five Year Plan and Industrial Policy was raised to be an official term since its Seventh Five Year Plan.

9.7.3.2 China's Industrial Policy in the Pre-reform Era

It can also be broadly classified into two periods.

(1) *"General Line of Transitional Period"-given by former president Mao on Sept. 8, 1953*

There was implicit general industrial policy in the beginning of part II of his paper.

"We shall determine the major tasks of the First Five Year Construction Plan under the guidance of the General Line.

The basic tasks of the First Five Year Construction Plan are: the first, concentrate major resources to develop heavy industry, establish the foundation of national industrialization and modernization of national defense; cultivate technological personnel, develop transportation, light industry, agriculture and expansion of commerce in accordance; promote the cooperation of agriculture and handicraft in steps and reform the private industrial and commercial sectors. Bringing into play the role of individual agriculture, handicraft and private industrial and commercial industry correctly….."

(2) *"Discussion of Ten Great Relations"—speech (paper) given by former president Mao on April 25, 1956*

After the implementation of the First Five Year Plan more than three years, the former president Mao had many new concepts of implicit IP. Selected important policies are quoted from his speeches (paper) in the following:

(i) It is appropriate to adjust the share of investment to heavy industry, agriculture and light industry, to develop more the agriculture and light industry…. (Targeting policy).

(ii) In order to balance the location of industrial development, industry in hinterland should be developed with full effort (Regional policy).

(iii) Technology transfer (Functional policy) was emphasized throughout the pre-reform ear. In the 1st Five Year Planning period, there were technology transfer with both hard and software mainly from the former Soviet Union. There were imported technology from the advanced economies in 1970s.

(iv) Growth of SMEs (Functional policy). This is a by-product of the people's commune movement around 1958. Under the slogan of "Mobilizing the whole nation to make great efforts to develop industry." Rural people's communes established a large number of commune and brigade enterprises, by the end of 1958, they employed 18 million people with a total value of output around 6 billion. This movement was stopped in the period of adjustment. But the growth of them resumed during the latter part of the period of Cultural Revolution (1966–79). There were so called five smalls, small coal mine, small iron nines small cement industry, small machine shops, small hydro-power plants. In 1978, there were more than 1.52 million of such enterprises, the value of output reached 49.3 billion yuan. These had created certain technological and managerial capacity in rural area. This has laid a foundation of high growth rate of Town and Village enterprises since China's reform and opening.

(v) Regional location. There were further transfer of the location of industrial projects in the hinterland in the 3rd and 4th Five Year Planning period.

(vi) There was once a period of wrong slogans, "Grain first" and "Iron first", i.e. targeting to products too much.

(3) *Lessons, Experiences and Achievement*

(i) *Lessons and Experience*

(a) It can be seen from above that China is successful in its implicit IP on the growth of primary and secondary industry, especially the growth of heavy industry. But China had failed in its IP for the underdevelopment of tertiary industry. But the underdevelopment of tertiary industry is a universal phenomenon for those countries effected by the former Soviet Union, even Russia by itself. It can be proved by the data of Russia in *Handbook of Statistics UNCTAD* before the year 2000.

(b) Its common practice for industrialization of developing countries to focus on light industry to be priority to accumulate capital in its initial stage of industrialization, then shifted to heavy industry to be priority. The IP adopted by China and the former Soviet Model was in contradiction to general practice, this type of industrialization will be at the cost of consumption of the people, may be a cause of low growth of GDP. But it is good for long term to establish a good foundation of industrialization and the capacity of competitiveness.

(ii) *Achievements*

China's achievements of implicit IP and industrialization can be shown by the quotation of description of China's Industry from the Main Report (1981) China: Socialist Economic Development of the World Bank.

In terms total output, however, China ranks among the world's major industrial countries. The net value of manufacturing production (again at Chinese prices and official exchange rate) is about one seventh that in the U.S.A. In 1979, in terms of quantity produced, China led the world in output of cotton yarn and fabric. It ranked third in output of cement, coal and sulfuric acid, fifth in steed production, and seventh in electric power generation.

9.7.3.3 IP of China Since Launch of Reform and Opening in Late 1970s to 2013

(1) *General*

China had learned the international practice of IP in this period. It had integrated industrial structure and IP to be a part of its development planning since its Seventh Five Year Plan. There are evolution of preparation and implementation of IP throughout the later period of Seventh Five Year Planning period.

(2) *Evolution of IP of China*

(i) Part II of Seventh Five Year Plan is industrial structure and IP. Adjustment of industrial structure is raised, especially promote the growth of tertiary sector (or service sector), seven sectoral policies are described in detail, they are agriculture, consumer goods industry, energy (with all sub-sectors), raw material (metals, chemicals and petro-chemicals), mechanical and electronic industry, consumption and material of construction, transport and communication.
(ii) The former State Economic Commission had launched large scale import of technology (around 3000 item 1979–1988) for technology re-vitalization of existing enterprises.
(iii) A joint research project "Industrial policy" was implemented between the World Bank and DRC.
(iv) Bureau of IP was established in the former State Planning Commission.
(v) The first national IP was promulgated on 1994.3.25. China had also promulgated industrial specific policy (ISP) of "Industrial Policy of Automotive Industry".
(vi) Later on, China and its State Planning Commission had promulgated a series of documents related to sectoral-specific policies (SSPs), such as *Catalog Listing of Products and Technology for Key Industries to be Encouraged for Development by Country* and *Guidance Catalog for Foreign Invested Industry* etc.

(vii) In the year of 2009 post the period of global financial crisis, the Adjustment and Revitalization Plans of ten key sectors were issued, the ten sectors were: automotive industry, iron and steel, textile, equipment manufacturing, shipping industry, electronic information, petrochemical, nonferrous metal, light industry, logistics industry etc.

(viii) In the later half year of 2010, the State Council had issued decisions to develop seven strategic emerging industries including accelerate to nurture and develop the high-end equipment, next-generation information network technology, energy conservation and environment protection, new energy, biotechnology and new materials thereby to occupy in advance the commanding frontier of future competition and push forward structural optimization of Chinese industry.

(ix) The State Council also announced the action plan of "Made in China 2025" and "Internet+" in 2015 to achieve the purpose to push forward the transformation of China from a large country to a strong country in its manufacturing industry and also reform and upgrade its traditional industries through internet.

(3) *Brief Conclusion*

China is currently ranked number one in GDP (in terms of PPP), its output of manufacturing sector and its international trade in goods among the world. A part of people both in domestic and abroad think all these achievements coming from China's reform and opening since late 1970s. President Xi Jinping had given a speech "Recognized correctly the two "thirty years"; before and post reform and opening." He had three aspects of in depth description, hereunder, only selected descriptions of two aspects (Central Propaganda Department, 2014) will be quoted:

(i) "Although there are large differences on ideological guidance, general and specific policy, and practical work in carrying out the construction of socialism in these two historical periods, but these two are by no means separated from each other furthermore, they are not in opposition fundamentally."

(ii) "Practice of exploration of socialism in the pre-reform era has accumulated conditions for practice of exploration post reform and opening, while practice of exploration of socialism post reform and opening is persistence, reform and development to the previous period…" (p. 19).

(iii) This quotation of President Xi's speech has not only pointed out correctly correct evaluation of two historical periods of China in its socialist market economic development, they are also true for China's development of industrialization and industrial policy.

9.7.4 *China's New Situation and New Industrial Policy Within the Context of Globalization and Regionalization*

The mankind is living in a dynamic and changing world. There are changes of organization of production and trade in the global economy led by the transnational corporations post WWII. China itself is also undergoing changes from a country dominated by agricultural sector to become the largest manufactures of the world. China's IP must be changed to adapt to concrete global and regional environment as well as the strength and weakness of its industry and the concrete conditions of its country. This part will be divided roughly into four topics: the new situation of the global and regional economy; the new situation of China; analysis of competitiveness of performance of China's export; and new IP recommended for China within the context of globalization and regionalization.

9.7.4.1 New Situation of the Global and Regional Economy

There are Reorganization of Production and Trade in the Global Economy through Innovation in Process of Production and the Business Model of Intraindustry Trade in Parts and Components.

(1) *The Transnational Corporations have become the Primary Movers of the Global Economy Post WWII*

The international economic order was defined by the U.S. and other core powers that supported it in terms of the ideology of free trade, but TNCs had performed the role to link the production of goods and services in cross-border, value-adding networks post WWII which made the global economy in the last half of the 20th century qualitatively distinct from what preceded it. Although the first transnational business organization, the East India Company arouse around 1600, later, there were growth of them in the natural resource (oil, mineral) sectors up to the early 20th century, but they did not play a central role in shaping a new global economic system until after the WWII. The large TNCs have the power to coordinate and control supply chain in more than one country, even if they do not own them.

(2) *The Emergence of International Trade and Production Networks*

There are new features of international trade and production in last half of 20th century. They are:

(i) the rise of intraindustry and intraproduct trade in intermediate inputs;
(ii) the ability of producers to "slice up the value chain". i.e., to break a production process into many geographically separated steps; and

(iii) the emergence of a global production networks that have altered governance structure, mode of competition and the distribution of gains in the global economy.

(i) *The Rise of Intraindustry and Intraproduct trade in Intermediate Inputs*

This form of fragmentation of production is the international division of labor that allows producers located in different countries with different ownership structure to form cross-border production networks for parts and components. These specialized "production blocks" are coordinated through service links, which include transportation, insurance, telecommunications, quality control and management. This forms of trade connect "integration of trade" in the world market with the "disintegration of production", the TNCs are more profitable to outsource (domestically or abroad) an increasing share of their noncore manufacturing and service activities. This represents a breakdown of the vertically integrated mode of production, so called Fordist model, originally by the automobile industry of U.S., replaced by Toyota automobiles' model of "lean production" which has to coordinate exceptionally complex interfirm trading networks of parts and components as a new source of competitive advantage in the global economy.

(ii) *Slicing up the Value Chain*

The concept of the value-added chain in the design of international business strategies are based upon the interplay between the comparative advantage of countries and the competitive advantage of firms. The concept of comparative advantage helps to determine where the VA chain should be broken across national borders, while the competitive (or firm specific) advantage influences the decision of a firm on what activities and technologies along the VA chain it should concentrate its resources in (This situation is shown in Figs. 9.5 and 9.6.).

Therefore, there are two distinctive shifts of contemporary global economy: the unparalleled fragmentation and reintegration of global production and trade patterns since the 1970s; and the recognition of the power of VA chain as the basis for formulating global strategies of TNCs that can integrate comparative (location specific) advantage and competitive (firm specific) advantage.

(iii) *Production Networks in the Global Economy*

Due to the dominance of production networks in the global economy, the GVC (global value chain) approach offers the possibility of understanding how countries and firms are linked in the global economy, it also acknowledges the broader institutional context of these linkages, including trade policy, regulation and standards. Because there are a variety of overlapping terms used to describe the complex network relationships that make up the global economy, it is necessary to clarify the concept of GVC based upon available literatures.

Global value chains are defined to be "Emphasis on the relative value of those economic activities that are required to bring a good or service from conception, through the different phases of production (involving a combination of physical

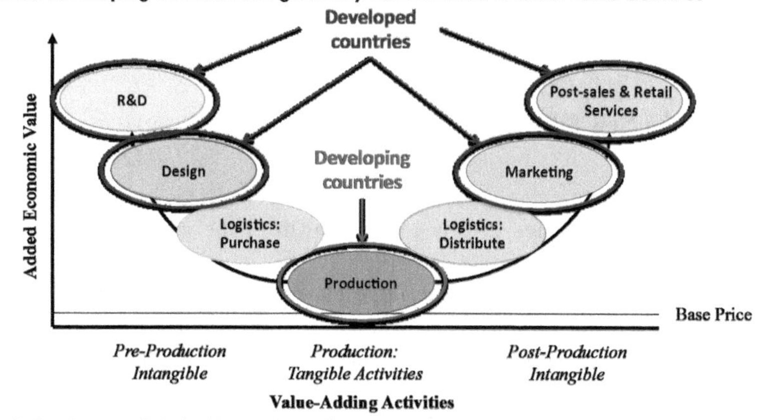

Fig. 9.5 Distribution of value in GVCS: apparel global value

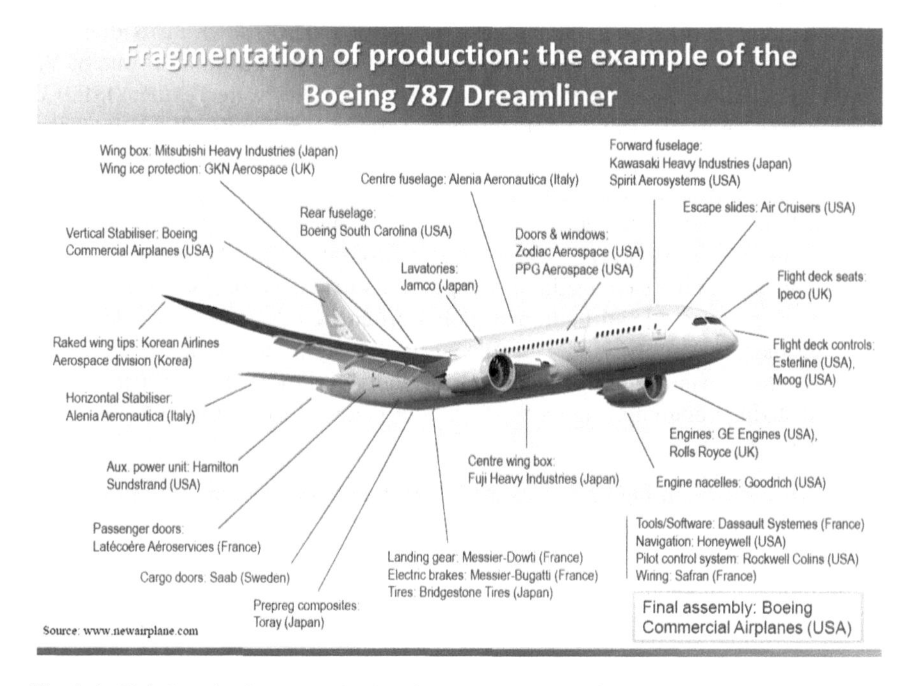

Fig. 9.6 Global production network of boeing 787 (*Source* Retrieved on September 8, 2017 from https://garywordpressblog.files.wordpress.com/2013/03/boeing787_.jpg)

transformation and the input of various product services), delivery to final consumers, and final disposal after use. (KAPLINSKY 2000; Gereffi and Kaplinsky 2001)." (Smelser & Swedberg, 2005).

9.7.4.2 The New Situation of China

(1) China has become the upper-middle income economies by the year 2010. And China has contributed greatly to the growth of the global economy which is shown in Fig. 9.7.
(2) China's has emerged to be the largest manufacturer of the world in the year 2015. This is due to its continuous effort of industrialization of more than sixty years since the establishment of PRC in 1949. On the other side, the surge in China's share in world MVA is also due to the sluggish growth of the global economy, especially the industrialized countries were suffered from serious downturns post the global financial crisis. Figure 9.8 shows change of share of leading manufacturers in world MVA in 2005 and 2015 (at 2010 constant price). China has an average annual growth rate around 10.6% of MVA between 2005 and 2015 compared to 1.0% in the U.S.
(3) [9]The strength of China's manufacturing sector is also shown in its dominant share of VA in various manufacturing sectors. China ranks number one of VA of eight sectors based upon ISIC division in 2014. They were Textiles (ISIC13), Wearing apparel (ISIC14), Leather and related products (ISIC15). Other non-metallic mineral products (ISIC23), Basic Metals (ISIC24), Computer, electronic and optical products (ISIC26), Electrical equipment (ISIC27), Machinery and equipment n.e.c. (ISIC28). It ranks number two of VA, only next to U.S. in other eight sectors of ISIC division in 2014.
(4) Insufficient China's competitive strength of its manufacturing sector in the new pattern of international trade and global production networks. In spite of the high share of VA of China's manufacturing sector, but the competitive strength of China's manufacturing sector is still relatively low in the new pattern of a globalizing economy. Figure 9.9 is a case study of formation of factory gate price.

The case shown in Fig. 9.9 gives the fact that, although China is responsible for the final assembly of i-phone4, but it contributes only 3.37% ($6.54) VA of the final production ($184.04).

In shortly, although China is the leading manufactures of the world MVA (Fig. 9.8) but its contribution to global value chain is relatively low (Fig. 9.9).

[9]Note: UNIDO (2016).

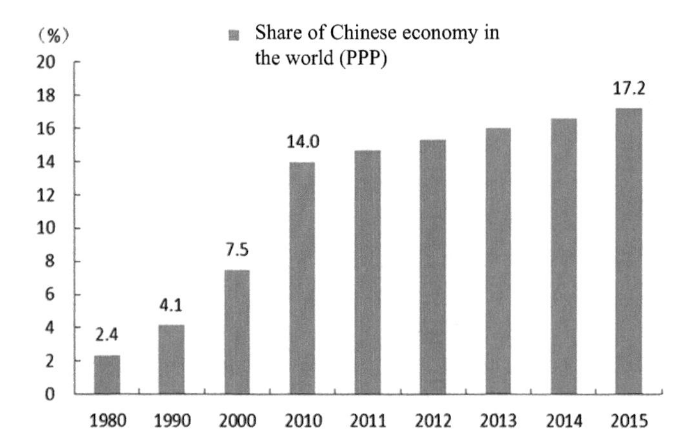

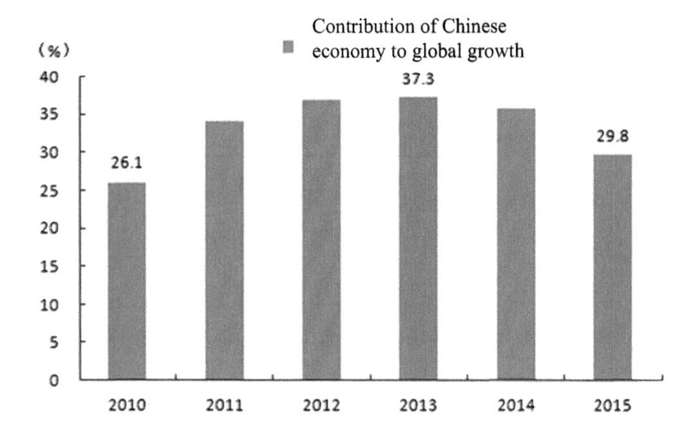

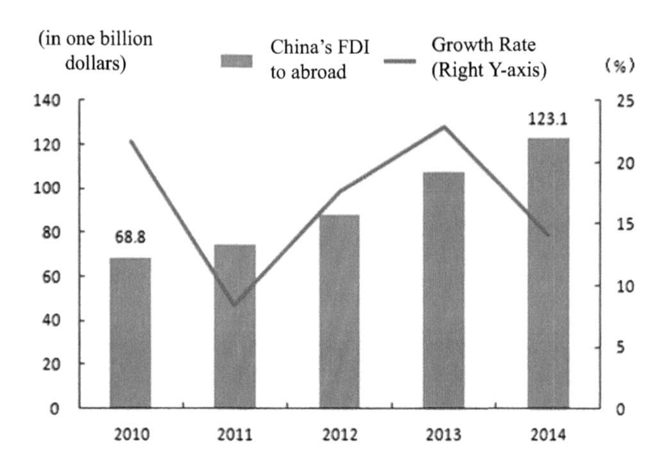

Fig. 9.7 The contribution of chinese economy

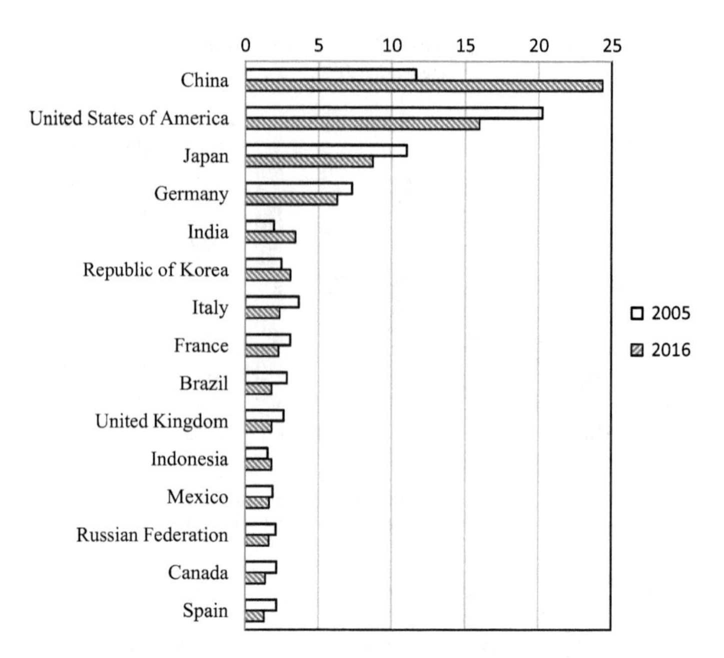

Fig. 9.8 Share of leading manufacturers in world MVA, at 2010 constant prices 2005 and 2015 (*Note* UNIDO (2016). *International yearbook of industrial statistics 2016* (p. 38). U.K.: Edward Elgar Publishing Limited)

Case Study: Formation of Factory Gate Price of iPhone 4 in Global Production and Trade Networks. (unit US$)

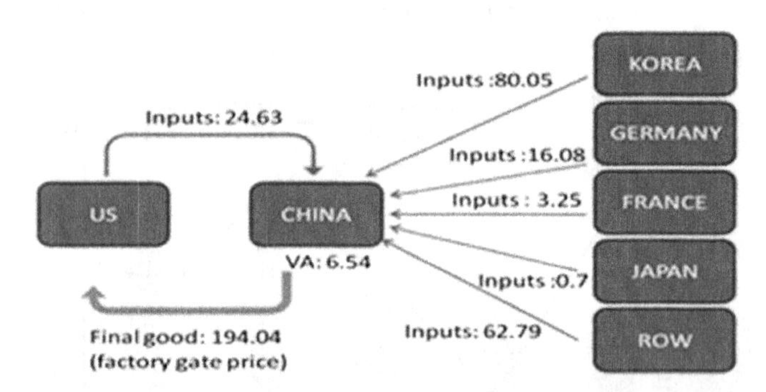

US trade balance with	CHINA	KOREA	GERMANY	FRANCE	JAPAN	ROW	WORLD
Gross	-169.41	0	0	0	0	0	-169.41
Value added	-6.54	-80.05	-16.08	-3.25	-0.7	-62.79	-169.41

Fig. 9.9 Case study of formation of factory gate price of a product in global production network

9.7.4.3 Analysis of Competitiveness of Performance of China's Export

Although China has ranked to be number one of the output of its manufacturing sector among the world in 2015, but its capacity to acquire value-added of global value Chain of its export is different for its different components. An analysis of competitiveness of performance of its export will give better guidance to its IP in an increasingly globalizing world.

(1) [10]*Brief explanation of Method of Analysis of Competitiveness of Performance of China's Export*

The analysis is done through KWW method and the data base of WIOD.

(i) *The KWW Method*

The KWW method classifies the total value of the export of a country into four parts:

DVA This is an important part of constituents of GDP of a country. It is the implicit value contained in domestic VA of export, i.e. it is domestic value added absorbed abroad;

RDV This is the domestic value added that is exported initially, but finally returned to home via imports from other countries and consumed at home;

FVA Foreign value added used in the production of domestic exports;

PDC Duplicated counted terms due to trading back and forth of intermediate goods that cross border at multiple times.

(ii) *WIOD*

This is the abbreviation of World Input-Output Database of EU. It contains I/O tables of 40 countries with data from 1995 to 2011. Its classification of manufacturing sector is shown in Table 9.14.

(2) *Several Features of China's Export of Manufactures*

(i) *International Comparison of Export Performance through Domestic and Foreign VA of Selected Countries*

Table 9.15 is comparison of export performance of selected countries in 2011 through KWW method and data of WIOD.

It can be seen from the international comparison of export performance of Table 2 that the share of domestic VA of China is 76%, which is higher than the average level of the world (69%) by seven percentage points. And it is higher than most other countries, but it is only lower than Japan, U.S.A. and Australia. This illustrates that China has established a stronger capability to acquire VA of its

[10]Note: All materials presented in 3.1.3 are based on Li and Wu (2015).

Table 9.14 Classification of manufacturing sectors of WIOD

Code of sector	Manufacturing sectors	Classification of types
C03	Food, beverage manufacturing and tobacco	Labor intensive
C04	Textile	
C05	Leather, fur, feather and footwear manufactures	
C06	Wood processing, and wood, bamboo, rattan, palm and straw manufactures	
C07	Paper and paper products printing, reproduction of recorded media	
C08	Refined petroleum products, coke, processing of nuclear fuels	Capital intensive
C09	Basic chemicals and chemical products manufacturing, chemical fiber manufactures	
C10	Rubber and plastics manufactures	
C11	Nonmetallic mineral manufactures	
C12	Metallic manufactures	
C13	General and special equipment manufactures	
C14	Electric and optical equipment manufactures	Technological intensive
C15	Transportation equipment manufactures	
C16	Other manufactures and recycling of waste and scrap	Labor intensive

Table 9.15 Comparison of export performance of selected countries 2011 (Unit 100 mil. U.S.D.)

Country/world	DVA	FVA	DVA share (%)	FVA share (%)
Japan	5679	968	81	14
U.S.A.	8820	1745	79	16
Australia	651	131	78	16
China	13,286	3249	76	18
Turkey	960	245	75	19
India	1673	567	72	24
Italy	3434	1120	70	23
Germany	9411	3007	69	22
Britain	2740	922	69	23
France	3719	1347	67	24
Spain	1845	767	64	27
Mexico	1482	822	61	34
World	85,344	28,778	69	23

export through her effort in industrialization around more than six decades. It is also interesting to note the lower share of DVA of export of members of EU. This may be caused by high integration of internal trade of EU. (UNCTAD).

(ii) *International Comparison of Export of Labor Intensive Product*

Table 9.16 gives international comparison of export of labor intensive product of selected countries. Table 9.17 gives changes of domestic VA of export of labor intensive product from 1995–2011 through calculations based upon KWW method and data of WIOD.

Table 9.16 shows that China has a higher capability of its labor intensive manufacturing sectors (C03–C07 of Table 9.14) to acquire VA in the division of labor of GVC. The domestic VA of export of labor intensive products reached 374.4 billion U.S.D. which ranked number one in the world in 2011. Its share of DVA is 86%, which is equal to U.S.A. and both of them are next to Japan and Australia with their share of DVA of 87%.

Table 9.17 shows that there is stable growth of domestic VA of China's export of labor intensive products in the period from 1995–2011. It was 55.2 billion U.S.D. in 1995 and reached 374.4 billion U.S.D. in 2011. There was unique feature of changes of DVA share since 1998. The value of DVA share was 87% in 1998, and it was declined gradually to 81% in 2004. And it was re-rising gradually to 86% in 2011. This change illustrates the upgrading capability to acquire VA in GVC of China's manufacturing sector. The reason of declining value of DVA share in later period of 1990s was due to the high growth of international sub-contracting of China's textile sector. Higher share of intermediate product from abroad was used, it caused the lowering of share of domestic VA of China's export. While in recent years, there is shift of international subcontracting to the creation of production of China's own brand, therefore, there was re-rising of DVA share.

Table 9.16 International compassion of export performance of labor intensive product of selected countries 2011 (Unit 100 mil. U.S.D.)

Country/ world	Implicit domestic VA in export (DVA)	Implicit foreign VA in export (FVA)	DVA share (%)	FVA share (%)
Australia	216	29	87	12
China	3744	540	86	12
Germany	1438	420	74	22
Spain	476	126	77	20
France	840	198	79	19
Britain	545	112	81	17
India	666	355	64	34
Italy	949	233	78	19
Japan	210	24	87	10
Turkey	342	89	77	20
U.S.A.	1517	201	86	11
Mexico	216	66	75	23
World	20,010	5406	76	21

Table 9.17 Changes of domestic and foreign VA of China's export of labor intensive product 1995–2011 (Unit 100 mil. U.S.D.)

Year	Implicit domestic VA of China's export (DVA)	Implicit foreign VA of China's export (FVA)	DVA share (%)	FVA share (%)
1995	552	97	84	15
1996	566	84	86	13
1997	603	87	86	12
1998	582	76	87	11
1999	552	97	84	15
2000	692	116	84	14
2001	737	116	85	13
2002	840	138	84	14
2003	1003	180	83	15
2004	1180	238	81	16
2005	1514	286	82	16
2006	1927	344	83	15
2007	2384	402	84	14
2008	2724	418	85	13
2009	2532	315	85	11
2010	3090	438	86	12
2011	3744	540	86	12

(iii) *International Comparison of Export of China's Capital Intensive Product*

Tables 9.18 and 9.19 show respectively "International comparison of capital intensive product of selected countries 2011" and "Changes of DVA and FVA of China's export of capital intensive product 1995–2011".

Capital intensive product contains manufacturing sectors with code number from C08–C13. China's capital intensive manufacturing sectors have lower capability to acquire VA in the division of labor of GVC. This fact can be shown clearly in Table 9.18, although China has a higher value of DVA in 2011 which reached 375.7 billion U.S.D., only next to Germany and U.S.A. in the same year, but its value of DVA share is only 75%, equal to Mexico in the same year. And this value is lower than Japan, U.S.A., Germany and even Turkey and India. In spite of the high growth of value of DVA from 28.3 billion U.S.D. in 1995 to 375.7 billion U.S.D. in 2011 (refer to Table 9.19), but the DVA share is declined from 84% in 1995 to 75% in 2011. This decline of DVA share can be explained from two causes: on the one side, China's capital intensive product has used a larger share of intermediate product and parts from abroad in integration into the division of labor of GVC; but it also shows that China has not upgraded its capability of capital intensive manufacturing sector to acquire VA in the division of labor of GVC on the other side.

Table 9.18 International comparison of export performance of capital intensive product of selected countries 2011 (Unit 100 mil. U.S.D.)

Country/ world	Implicit domestic VA in export (DVA)	Implicit foreign VA in export (FVA)	DVA share (%)	FVA share (%)
Japan	2572	515	87	11
U.S.A.	3914	930	85	13
Germany	4542	1277	79	15
India	599	132	77	17
Turkey	427	86	77	15
China	3757	911	75	18
Mexico	469	117	75	19
Australia	370	89	74	18
Britain	1306	433	67	22
France	1632	572	66	23
Italy	1813	678	65	24
Spain	862	391	61	27
World	36,903	12,359	67	22

Table 9.19 Changes of DVA and FVA of China's export of capital intensive product 1995–2011 (Unit 100 mil. U.S.D.)

Year	Implicit domestic VA of China's export (DVA)	Implicit foreign VA of China's export (FVA)	DVA share (%)	FVA share (%)
1995	283	44	84	13
1996	332	46	86	12
1997	426	61	85	12
1998	411	50	87	11
1999	552	97	84	15
2000	490	81	82	14
2001	516	79	83	13
2002	593	99	82	14
2003	776	158	79	16
2004	1081	269	75	19
2005	1311	339	74	19
2006	1676	422	74	19
2007	2308	573	74	18
2008	2939	679	75	17
2009	2342	455	80	15
2010	3025	669	77	17
2011	3757	911	75	18

(iv) *International Comparison of Export of China's Technological Intensive Product*

China has a relatively lower capability of its technological intensive manufacturing sectors to acquire VA in the GVC which is shown in Table 9.20. It can be seen from this table that in spite of the highest DVA of China.

In the year 2011, it reached 578.4 billion U.S.D. in that year, but its DVA share is relatively low. Its value is only 71%, not only it is lower than Japan (84%) and U.S.A. (82%), it is even lower than India (78%). Therefore, it is necessary to improve the performance of China's technological intensive sectors. In order to have an in depth understanding of this issue, further analysis is done in the next section.

(v) *Case Study of Bilateral Trade and Decomposition of Electrical and Optical Equipment Manufactures between China and U.S.A.*

In order to understand better the performance of China's technological intensive product manufactures in the world trade, a case study of bilateral trade of electrical and optical equipment manufactures between U.S.A. and China is done. Because this bilateral trade has the highest value of this type of trade of all sectors between two countries. It reached 212.0 billion U.S.D. in 2011, within which, the value of export from China to U.S.A. is 176.9 billion U.S.D., while this value is 35.1 billion U.S.D. from U.S.A. to China. The difference between them is more than four times. (Refer to Tables 9.21 and 9.22).

Tables 9.21 and 9.22 have also shown the components of decomposition of total value of export. It can be seen from Tables 8 and 9 that the export from China to U.S.A. is dominated by the final product, this value is 104.2 billion U.S.D. in 2011, a share of 58.9% of total value of exports while the value of export of intermediate

Table 9.20 International comparison of export performance of technological intensive product of selected countries 2011 (Unit 100 mil. U.S.D.)

Country/ world	Implicit domestic VA in export (DVA)	Implicit foreign VA in export (FVA)	DVA share (%)	FVA share (%)
Japan	2897	430	84	12
U.S.A.	3390	615	82	15
Australia	65	14	79	17
India	409	80	78	19
Germany	3431	1310	76	2
China	5784	1799	71	22
Italy	672	209	71	22
Turkey	190	71	68	25
Britain	890	377	65	28
France	1247	578	63	29
Spain	507	250	82	3

Table 9.21 China's export of electrical and optical equipment manufactures and their decomposition to U.S.A (Unit mil. U.S.D.)

Year	Item	Total value of export	Total value of export of final product	Total value of export of intermediate product	DVA	FVA	RDV	PDC
1995	Value	10,998	7634	3364	8544	2262	16	176
	Share (%)	100	69.4	30.6	77.7	21	0.1	1.6
2005	Value	87,608	53,492	34,116	53,784	29,997	341	3485
	Share (%)	100	61.1	38.9	61.4	34	0.4	4
2011	Value	176,924	104,156	72,769	123,187	46,496	1296	5946
	Share (%)	100	58.9	41.1	69.6	26	0.7	3.4

Table 9.22 U.S.A.'s export of electrical and optical equipment manufactures and their decomposition to China (Unit mil. U.S.D.)

Year	Item	Total value of export	Total value of export of final product	Total value of export of intermediate product	DVA	FVA	RDV	PDC
1995	Value	3400	1284	2116	2691	396	182	130
	Share (%)	100	37.8	62.2	79.2	12	5.4	3.8
2005	Value	16,402	3845	12,556	11,926	1482	1777	1216
	Share (%)	100	23.4	76.6	72.7	9	10.8	7.4
2011	Value	35,059	10,584	24,475	28,314	2762	2470	1513
	Share (%)	100	30.2	69.8	80.8	8	7	4.3

product is 72.8 billion U.S.D., with a share of 41.1% of total value of export; while the export from U.S.A. to China is dominated by intermediate product, its value is 24.5 billion U.S.D. with a share of 69.8% of total value of its export, the export of final product is 10.6 billion U.S.D., with a share of only 30.2%. These figures reflect different positions of electrical and optical equipment manufactures occupied by U.S.A. and China in the GVC. China produced the final product of electrical and optical equipment mainly through the process of assembly, while U.S.A. is responsible mainly the supply of intermediate product of electrical and optical equipment manufactures.

Tables 9.21 and 9.22 have also shown the value of DVA, FVA, RDV and PDC through further decomposition of total value of trade between China and U.S.A. in the manufacture of electrical and optical equipment. It can be seen from the values of components of exports of Chain and U.S.A. They have very different structures of VA. Firstly, the share of DVA of export from U.S.A. to China is 81%, which is higher than the share of DVA export of 70% from China to U.S.A.; Secondly, the share of FVA of China's export is 26% which is greater than the share of FVA of export from U.S.A. to China. This shows that there is higher share of foreign VA from the export of China's electrical and optical equipment manufactures to U.S.A., while the share of foreign VA of export from U.S.A. to China is lower; Lastly, the share of RDV of export of electrical and optical equipment manufactures from China to U.S.A. is very low, which is only 0.7% in the year 2011, while the share of RDV of export from U.S.A. to China is explicit, which is 7%. This reflects different positions occupied by China's and U.S.A.'s electrical and optical equipment manufactures in the division of labor of the global value chain. Because U.S.A. engaged mainly in the design of product and export of product of components and parts, therefore, it is positioned in the upstream of GVC, a fair part of VA of export of U.S.A. return back to home country through imports from other countries and used by consumers of U.S.A., and relatively speaking, China's product is positioned at the downstream end of the GVC, there is a few VA of China's intermediate export returns back domestically through import.

There is need to have a new concept of IP in international production networks.

It had be warned by UNCTAD in its Trade and Development Report 2014 that "The business model underlying international production networks is build on asymmetric governance relations, where lead firms shape the distribution of risks and profits in their favor" p. 105. It has also been pointed out in that report: the ability for China to climb up the value chain remains a challenge, with electronics sector for example, despite being the largest exporter, China accounts for just 3% of the share of profits derived from this sector (Starrs, 2014:91).

But it has also been pointed out by UNIDO in its Industrial Development Report 2016 that "Global value chain can be important assets to achieve structural transformation in the long run". (p. 181)

Industrial policy in a globalizing world should have a systems perspective. It must be get rid of past practice of IP. Although innovation policy is currently recommended nearly by all international organization, but it is necessary to consider both the positive and negative consequences of innovation. "A one-size-fits-all approach to economic policy is unlikely to bring structural change". (p. 184)

It is necessary to have trade-off, which is a crucial concept of systems engineering, i.e., tradeoff between economic versus social, social versus environmental, and environmental versus economic, understanding these trade-off is the pre-condition of an integrated industrial policy.

9.7.5 *Concluding Remarks*

In spite of there are set backs in several countries toward the process of globalization. But the process of globalization and regionalization is a major trend of the world economy in general, this trend is irreversible. Regionalization is a part of process of globalization. The lessons and experiences of industrial policy of China in the process of globalization (WTO works for global trade) and regionalization (There are too many RTAS) may provide some references for developing counties in general.

Case study 1 is an example of application pf quantitative method to the analysis of impact of various policies to the construction of functional zones in regional planning. This case study is a partial abstract of a study on "Differentiated Policy to Construct the Functional Zones" done by Research Center of China's Development Planning of Tsinghua University entrusted by relevant bureau of State development and Reform Commission before the preparation of China's Eleventh Five-Year Plan. A detail description of process of application of Analytic Hierarchy Process (AHP) to the study of one of the functional zones, the leading zone is described in detail for illustrative purpose of combination of quantitative and qualitative approach to policy analysis.

Case study 2 is Industrial Policy of China. This case is a summary of working experience of policy consultative service of Development Research Center of State Council where both authors of this book are attached. Industrial policy had once been a subject of academic debate due to different thought of ideology of economic liberalization against governmental intervention. But in the beginning of 21st century, most of international organizations relevant to policy research have confirmed the role of industrial policy, especially the East Asian countries. This case study if a briefing of industrial policy studies and practice of China including general discussion of industrial policy, types of industrial policy, study of industrial policy by UNIDO and OECD and China's industrial policy in the pre-reform era and its lessons and experience, China's industrial policy since the launch of reform and opening up to 2013., and China's current situation and new industrial policy within the context of globalization and regionalization.

9.8 Summary Points

1. Social system must be regulated in order to be developed in order and security, Regulation is a part of public administration.
2. Public administration is an evolving discipline due to change and evolution of cultural norms, belief and power reality of society. This subject is discussed by tentative definition (covered various views), scope of study in traditional sense. Then two definitions are introduced.

3. Trend of change of public administration is discussed, especially the emergence of new public administration since 1971.
4. Response to changes from academic perspective and international organization working with development issue of the global society are introduced. The former perspective represents trends in developed countries while the latter focused on changing role of the state from the development and macro perspective of the global society, especially focused on developing countries.
5. Legal system is one type of regulation. Because the legal system in Western countries is well established in theory and practice, therefore, this theme is not discussed for Western countries, but a *Legal Environment of Business* by Western professor is introduced briefly, because the author kept a view of "Policy Analysis" applied to his presentation of Western legal system.
6. China is a late comer in the establishment of a modern legal system. But China had a school of Legalism early in the Spring and Autumn, and Warring State period (770–221 BC). The history of legal system of the feudal societies is introduced briefly. And the development of China's legal system since 20th century is given.
7. Public policy is other important type of regulation, several terminologies such as policy, public policy, public policymaking are further clarified. It should be emphasized that public policymaking in Western usage means the totality of the decisional process of the government to deal with a particular problem. But based our understanding of policy process, the decision maker is not necessary the policy maker. The policy makers are generally the experts in studying the policy under certain environment and raise policy recommendation to the decision maker of the government.
8. History of public policy and public science studies are described, especially the public policy studies of China during "the Spring and Autumn period" and "the Warring State period" were introduced. The thoughts of Xunzi have attracted Western scholars in recent years specialized in studies of public policy.
9. Public policy had also existed in Western countries since government came into being. But public policy is becoming contemporary policy science driven by process of industrialization of the developed countries. Harold Lasswell edited *The Policy Science: Recent Development in Scope and Method* was considered the founder of concept and methodology of policy science.
10. During the 1960s, both the discipline of public administration and public policy were developing very fast. The development of discipline of public administration became stabilized around 1980s, but the discipline of public policy keeps its momentum of growth due to broad involvement of various disciplines.
11. There are two schools of public policy studies, the school of quantitative analysis focused more on application of mathematical models, the other focused more on qualitative analysis and historical studies.
12. A social system approach of public policymaking process of researchers is introduced. This process includes analysis of external environment where the policy issue is originated, it is also the context of analysis of application of

various types of policies. Study and application of various types of policies based upon guideline and value system of policy makers (refer to policy researcher).

13. Guideline (or framework, or mage policy) is necessary concepts applicable to the study of all specific policy applied to specific environment. They are: setting up appropriate goal system, clarify policy boundaries, setting up appropriate time frame, estimate fully the secondary and tertiary impact of implementation of designated policy, estimate and determining risks and its acceptability, comprehensiveness versus narrowness, balance oriented versus shock oriented, relevant assumption on the future theoretical base, resource availability, choice of range of policy instruments, balance of goal between long term and short term.
14. Contexts of Environment of public policymaking are discussed with respect to social environment (including political environment and cultural environment), economic environment, and S&T environment. Available studies of OECD are quoted for benchmark the study of S&T environment.
15. Social policy is a complex theme, available definition from dictionary of sociology and the recent study in the *Oxford Handbook of Public Policy* are quoted to illustrate different views. Housing policy is classified into social policy because it is one of the basic needs of daily life of the people.
The Charter of the Fundamental Social Rights for Workers of EU is quoted for illustrating a minor part of concrete social policy.
16. Macro-economic policy is discussed briefly. Science and technology policy is also discussed. Clarification of science policy and technology is given, the former is focused more by developed countries while the latter is focused more by the developing countries. The guideline of long term versus short term can be a frame of reference in the consideration of choice of public policy.
17. Environmental policy is a hot topic raised in recent several decades, both the experience of development of environment policy of EU and US are discussed in relatively detail. This section is concluded with a policy process-theories diagram, the essential meaning of this diagram is "political system is a black box". This point should be noted by all policy researchers.

References

Acocella, N. (1998). *The foundations of economic policy-values and techniques* (B. Jones, Trans. from the Italian) (pp. 185–186). Cambridge: Cambridge University Press.
Adams, F. G., & Klein, L. R. (1983). *Industrial policies for growth and competitiveness—an economic perspective* (pp. 14–15). Lexington: D.C. Health and Company.
Birkland, T. A. (2005). *Policy process-theories, concepts and models of public policy making* (2nd ed.). New York: M.E. Sharpe.
Black, J. (2002). *Oxford dictionary of economics* (2nd ed., p. 232). Great Britain: Clays Ltd.
Butler, H. N. (1987). *Legal environment of business* (pp. xi–xix, 14–21, 873). Cincinnati: South-Western Publishing Co.

Central Propaganda Department (PCCC2014). *Readings of series of import speeches* by Secretary General X. Jinping (pp. 18–19). Beijing: Learning Publisher & People's Publisher.

Dror, Y. (1971). *Design for policy science* (pp. 63–73). New York: American Elsevier Publishing Company, Inc.

Hornby, A. S. (1974). *Oxford advanced learner's dictionary of current english* (p. 656). London: Oxford University Press.

Jary, D., & Jary, J. (Eds.). (1991). *Collins dictionary of sociology.* Harper Collins Publisher.

Kirchivety, G. W. (1905). Civil Administration. In D. C. Gilman, H. T. Pech, & F. M. Colly (Eds.), *New international encyclopedia* (1st ed.). New York: Dodd, Mead and Company.

Kraft, M. E., & Furlong, S. R. (2007). *Public policy* (2nd ed.). Washington, D.C.: CQ Press.

Kruschke, E. R., & Jackson, B. M. (1987). *The public policy dictionary* (p. 35). Santa Barbara: ABC-Clio Inc.

Li, S. T., & Wu, S. M. (2015). *Analysis of capability to acquire va of china's manufactures in participation the division of labor of global value chain (research report of DRC).* Beijing: DRC.

Moran, M., Rein, M., & Goodin, R. E. (Eds.). (2006). *The oxford handbook of public policy.* New York: Oxford University Press.

OECD. (1999). *OECD science, technology and industry scoreboard 1999, benchmarking knowledge-based economies.* Paris: OECD.

OECD. (2015). *OECD science, technology and industry scoreboard 2015-innovation for growth and society* (p. 15). Paris: OECD.

Pan, G. P. & Ma, L. M. (2010). *China's Laws (*G. J. Chang, Trans.). Beijing: China International Press (pp. 6–11, 14, 35).

Paranjape, H. K., & Pai Panandikar, V. A. (1973). *A survey of research in public administration* (Vol. 1, pp. xiii–xvi). New Delhi: Allied Publisher Private Limited.

Paranjape, H. K., & Pai Panandikar, V. A. (1973). *A survey of research in public administration* (Vol. 2, pp. xiii–xix). New Delhi: Allied Publish Private Limited.

Pass, C., & Lowes, B. (1993). *Collins: Dictionary of economics* (2nd ed., p. 254). Glasgow: Harper Collins Publishers.

Roney, A. (2000). *EC/EU fact book* (6th ed.). London: Kogan Page Limited.

Salamon, L. M. (Ed.) (2000). *The tools of government-A guide to the new governance* (p.v, p. 41, pp. 186–213). New York: Oxford Press.

Shafritz, J. M. (2004). *The dictionary of public policy and administration* (pp. 236–245). Colorado: Westview Press.

Smelser, N. J., & Swedberg, R. (Eds.), (2005). *The handbook of economic sociology* (2nd ed., pp. 166–168). Princeton: Princeton University Press.

The World Bank (1981). *China: Socialist economic development the main report* (p. 87). Washington D.C.: The World Bank.

The World Bank. (1997). *World Development report 1997-the state in a changing world* (p. 1). New York: Oxford University Press.

UNCTAD. (2014). *Trade and development report 2014* (p. 181). New York: UN.

UNIDO (2016). *International yearbook of industrial statistics 2016* (p. 38. pp. 62–72). U.K.: Edward Elgar Publishing Limited.

Yandle, B. (2011). *Bootleggers and Baptists in the theory of regulation.* In D. Levi-Faur (Ed.), *Handbook on the politics of regulation.* U.K.: Edward Elger.

Zhong Hua Ren Min Gong He Guo Guo Shi Tong Jian (Vol. 1 1949–1956, p. 685). Beijing: Red Elag Publisher.

Chapter 10
The Development of Think Tanks

10.1 Introduction

Development of the global society has entered into a new millennium, although the material wellbeing of most of the global people has been improved through progress of science and technology, improvement of economic system and also improvement of social side in certain system aspects. But the growth of these systems are very much unbalanced. Although U.S.A. ranked number one on its economy and capacity of science and technology, but there are series divide between the blue collar class and the white collar class. China, although ranked number two in the global economy and number one of its manufacturing capacity, but there are no shortage of social issues in health care, education, housing and bottom line of morality. Officials of governments globally are generally occupied in busy routine and urgent short term works, they have no spare time to think about long term issue. And also due to the specialization of discipline, every department concerns development of its own without coordination with other department or no consideration of secondary or negative effect of certain development. And also, due to their position of administration, they have vested interests to pursue short term results. The emergence of modern think tanks concentrated with experts of various disciplines, if they are not funded by government or business sectors, they can perform relatively better results of issues faced by the country or global society. But how such kind of think tanks can be existed is an issue to be explored. This chapter will give a brief discussion of role of think tanks, type of think tanks, its process of development etc. Relative detail case studies are given to several selective thanks with description of their organizations, histories, work on education and training, programmes etc. The authors of the book expect that these materials may be helpful to assist most of the developing countries to have the information how to establish a think tank. The authors suggest that if a think tank is well established with researchers known the real world politics, the think tanks can perform the role of

H. Wang and S. Li, *Introduction to Social Systems Engineering*,
https://doi.org/10.1007/978-981-10-7040-2_10

national or other types of planning, while the planning organization of the government is responsible for implementation, coordination, monitoring feedback information in time and making corrective actions, the think tank is a consultative organization, while the governmental planning organization has administrative power in sense.

10.2 Definition, Function and Classification of Think Tanks

10.2.1 Definition and Function of Think Tanks

The term of think tank originated in the 1950s. However, the earliest think tank in the world is the British Royal United Services Institute (RUSI) established in 1831. Such institutions are consultancy service providers specialized in the collection, analysis and study of situations at home and abroad for national decision makers in various countries in their formulation of public policies. Meanwhile, think tanks also play their role in transmitting to the general public the results of selected relevant issues or policy alternatives under research. Abiding by such a definition, organizations established by companies to study their development strategy or planning are excluded from the main research scope of this book.

The creation and development of think tanks can be attributed to the evolution of human society from the traditional and simple agrarian society to industrialized and even post-industrial societies. Human survival and development are subjected to complex factors looming from the natural, societal and political environments, not to mention the explosive growth of knowledge generated and accumulated from economic and social undertakings, as well as in the fields of science and technology. The solution of complex problems now requires not only specialized expertise of respective fields, but also a wider range of knowledge compound, straining the knowledge, time and energy limitations of the overwhelmed national policy makers. Various think tank organizations thereby play their roles as bridges to or catalysts for decision makers. When function as a bridge, think tanks work to analyze, identify and highlight situations and research results from academic and industrial circles as well as all kinds of world problems to formulate concise and correct situation analysis and timely policy advice passed to decision makers for alternative consideration. When function as a catalytic agent, think tank organizations also explore new problems which may happen in the future and provide decision makers with early warning and countermeasures. For example, leaders in the world held a conference in 1992 on environment and development under the auspice of the United Nations, adopted the "*Agenda 21*" and reached a consensus on sustainable development. In effect, early in 1970s, the Club of Rome, a private think tank, issued its research report of "*The Limits to Growth*", putting forward early warnings on problems related to resources and environment. In 1987, the Brundtland Commission released its report "*Our Common Future*" that defined the concept of sustainable development. In 1984, the Worldwatch Institute, an American think tank, published its first report "*State of the World*" that provided early warning analysis and sound policy advice on population, resources, energy supply, reduction of oil dependence, soil and forest conservation, recycling of resources and materials, development of renewable energy, food supply security and automobile development review. Such are the catalytic functions of think tanks. Regrettably, however, not all good policy advice by think tanks are adopted by political decision makers in various countries due to their appreciative limitation or their realistic or political considerations. For example, the previous American government treated the United Nations environment conference held in South Africa in 2002 with indifference. Development of renewable energy became a focal point of American energy policy only under the Obama administration.

Through the release of its research report "*State of the World*", the Worldwatch Institute also disseminated its research results and policy advice to the general public to raise awareness of certain public sectors of about what is happening or may happen in future.

10.2.2 [1]Classification of Think Tank

Think tanks could be classified by various perspectives. For example, they could be categorized according to their affiliation (e.g. some countries have think tanks based on right-wing or left-wing beliefs), to their different types of funding, or to the nature of their respective policy specialty or field of research that they are engaged in, namely, categorized into economic, social, environment and foreign policy think tanks. Examples of economic think tanks include the Hamburg Institute for International Economics established in 1908, and the Academy of Macroeconomic Research under the State Reform and Development Commission of China. Example of diplomacy think tanks include the U.S. Chicago Council on Foreign Relations established in 1922. Classification by specialty had certain significance in the early 20th century. However, during the contemporary era, problems under research of think tanks have become more comprehensive. Therefore, the author refers to the method of classification by think-tank affiliation as described in *Global Think Tanks: Policy Networks and Governance* written by James G. McGann in 2011, with minor changes to his original classification in Table 10.1.

[1]Note: Most of this chapter were completed in 2012; therefore, the referenced statistics are only for illustration of problems and situations.

Table 10.1 Think tank classification by affiliation[a]

Serial No.	Categories of think tanks	Definitions and indicative examples
1	Independent think tank	Relatively or completely independent public policy research organizations. Their operation and research are free from influence of government or any interest group or contributor
2	Semi-independent think tank	Public policy research organizations free from influence of government. However, they are controlled by interest groups and main contributors, which may have major influence on think tank operation and research results
3	Think tank subordinate to colleges and universities	Policy research centers established by universities
4	Think tank subordinate to political party or in favor of a particular political party	Public policy research organizations formally subordinate to a political party or in favor of a particular political party. Such as the Center for American Progress in favor of the Democratic Party, the Heritage Foundation in favor of the Republican Party, etc.
5	Think tank subordinate to the government	Public policy research organizations subordinate to government departments, such as the U.S. National Bureau of Economic Research, The Academy of Social Sciences and the Development Research Center of the State Council in China. They are all staffed by government employees
6	Quasi-governmental think tank	Public policy research organizations not staffed by government employees but are funded by government or contract research
7	For-profit think tank	Public policy research organizations operated as per profit-making enterprise model

[a]*Source* For this table, reference is made to McGann and Sabatini (2011)

10.2.3 Development of Think Tank Research

10.2.3.1 Importance of Think Tank Research

Think tanks generate ideas, policies, strategies and tactics. Before public policy research organizations in modern society came into being, most leaders in the world depended on knowledge-based advices of expert teams to achieve military or political victories. For example, during the Battle of Guandu that took place in the Three Kingdoms Period, both Cao Cao and Yuan Shao employed groups of strategists.[2] Cao Cao adopted correct advice by his strategists and finally won the battle even when out-numbered by his opponent, and laid a foundation for future development. Reaching the contemporary society, those expert groups providing ideas and policy advices evolved into public policy research organizations. Establishment and development of such organizations are of vital importance to national development. Early examples of think tanks set up in developed countries may include the Kiel Institute for World Economy established in 1914 and the Royal Institute of International Affairs (now renamed as Chatham House) established in 1920.

The developed countries attached importance very early to researches on world economy and international affairs outside their countries to facilitate their own development. For various think tanks emerged after the World War II, some focused on domestic economic, social, scientific and technological development policies (mostly developing countries) and others also placed emphasis on global research. In the current age of economic globalization, there is need to investigate the development of think tanks of other countries in the world and the state of their research engagement in order to promote development and research quality of think tanks at home. Some developing countries are short of various intellectual resources in its early days. Knowledge on current standing of international think tanks may facilitate development of these countries by virtue of resources from such think tanks. That is because human society is entering the information and knowhow age and the quality and quantity of knowledge-based (including empirical knowledge) think tanks are an important indicator reflecting a country's international competitiveness.

10.2.3.2 Current State of International Think Tank Research

In 1980s, the [3]Nippon Institute for Research Advancement (NIRA) in Japan carried out an overall investigation of 72 major policy research organizations in the U.S., Europe and Japan, focusing on their respective purpose of establishment, scope of activities, organizational structure, source of funding and major research subjects. The United States takes the lead in theoretical research and statistics of global think tanks. Besides publication of many works (such as *Think Tank: Public Policy and Technopolis* written by Andrew Kich in 2006 and *Intellectual Competition: Washington's Think-tank World* written by Wedenbaum in 2011), "think tank and civil society" course is offered by Professor McGann of Pennsylvania State University in the same university. Data was collected from global think tanks.

[2]Note: Three Kingdoms, i.e. Wei, Shu and Wu, existed in history of China during 220–280 A.D., which is called the Three Kingdoms Period. Cao Cao, the creator and King of Wei, had been an official of the Eastern Han Dynasty. He fought with Yuan Shao, a noble in Eastern Han Dynasty at Guandu. Yuan Shao dispatched 700,000 forces and Cao Cao only had 70,000 soldiers. Cao Cao adopted advices given by Xun You, Xun Yu, etc., but Yuan Shao turned down the good advices of Tian Feng and Xun You and even made Xun You surrender himself and his advisory to Cao Cao. Yuan Shao lost the battle in the end. This was later known as the Battle of Guandu.

[3]Note: NIRA (1985, p. 224).

"Global Think Tanks Ranking" action was initiated in 2006. With the participation of nearly 250 experts, 150 medias and scholars, 100 think tank representatives and think-tank course directors as well as contributors around the world, the global think tanks ranking report (2010) was completed in 2011. This report ranks 5491 think tanks around the world and lists the top 75 global think tanks. The Academy of Social Sciences of China ranks 24th among the top 25 global think tanks. Global think tanks ranking report (2010) is the fourth report in its series. It is also a report with relatively complete and systematic data collection among current global think tanks researches.

10.2.3.3 Think Tank Research in China

Although in Chinese history, many leaders depended on advisories for decision-making and military and political victories at an early time, China lags behind developed countries in terms of establishment of public policy research organizations. Establishment of think tanks mainly dates from 1970s. However, the number of think tanks in China grows rapidly and China ranks second in the world after the United States in terms of the number of think tanks. In China, public policy is called soft science, since in the general sense, public policy research is within the scope of social science. In China, the Soft Science Research Association has been established. There are limited works related to statistics, analysis and systematic research on development of think tanks in China. Development and the number of think tanks in China are by and largely covered in [4]*Introduction to Soft Science Research in China* written by Gan Shijun in 1980s. In recent years, series of books such as *Think Tanks in China* and *National Think Tank*, as well as *China Think Tank Development Report* and ranking have also been published.

10.3 Development and Construction of Think Tanks

10.3.1 Development of Global Think Tanks

10.3.1.1 Overview of Global Think Tank Development

Although nearly all decision makers in various countries around the world need support of knowledge experts to pursue political, military, social and economic objectives, think tanks as an institutional component is originally a creation of

[4]Note: *Introduction to Soft Science Research in China* was published in *Soft Science Research and Decision Making* prepared by the Science and Technology Policy Bureau under the State Scientific and Technological Commission and the Educational, Scientific and Cultural Department under People's Daily in 1986.

industrialized and developed countries. For example, the United Kingdom as the global dominant in the 19th century, was the first to establish the British Royal United Services Institute (RUSI) in 1831 to fulfill the needs of a global power. With the gradual transfer of global economic center from Europe to North America, the Brookings Institution was established in 1916 in the U.S. As described in Annual Ranking Report of Global Think Tanks issued by Pennsylvania State University, the United States retains the largest number of think tanks around the world, with the Brookings Institution ranking first above all other top think tanks in the world.

Global think tanks began to emerge in the 19th century and developed and increased at a slow pace until 1940. During the World War II, growth of think tanks was witnessed due to the need for defense experts and technical experts. For example, the famous RAND Corporation founded in 1946 is derived from Douglas Aircraft Company and became an independent company in 1948. After World War II and to meet the need for post-war recovery and reconstruction of some developed countries, as well as during the period of the cold war, the need for research on global political, social and economic developments in industrialized countries prompted the rapid growth of think tank development and advancement in various countries worldwide. This development also corresponded to the new emergence of a large number of developing countries who were former colonies that are faced with the need for industrialization and development in general. Think tank growth peaked in 1992 and dropped off thereafter (refer to Figs. 10.1 and 10.2).

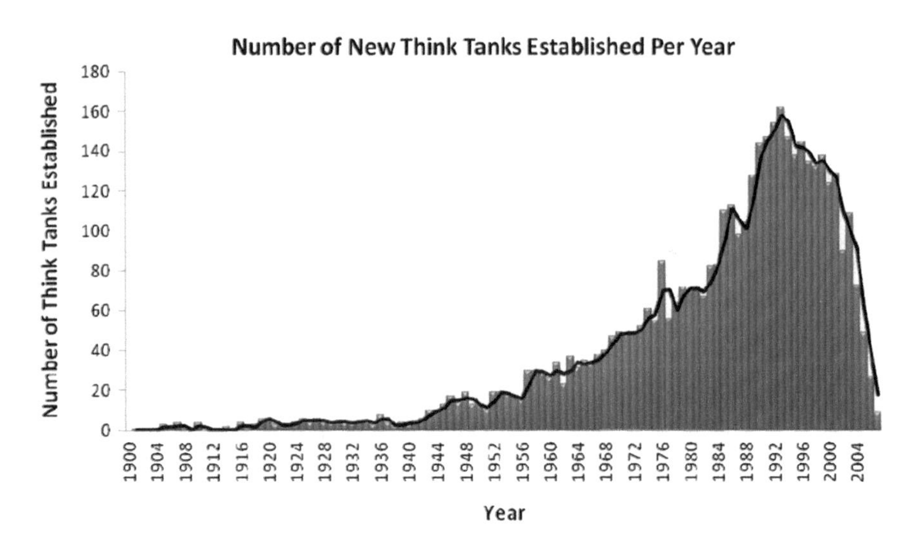

Fig. 10.1 Establishment and development of global think tanks (1900–2004) (*Source* McGann 2010)

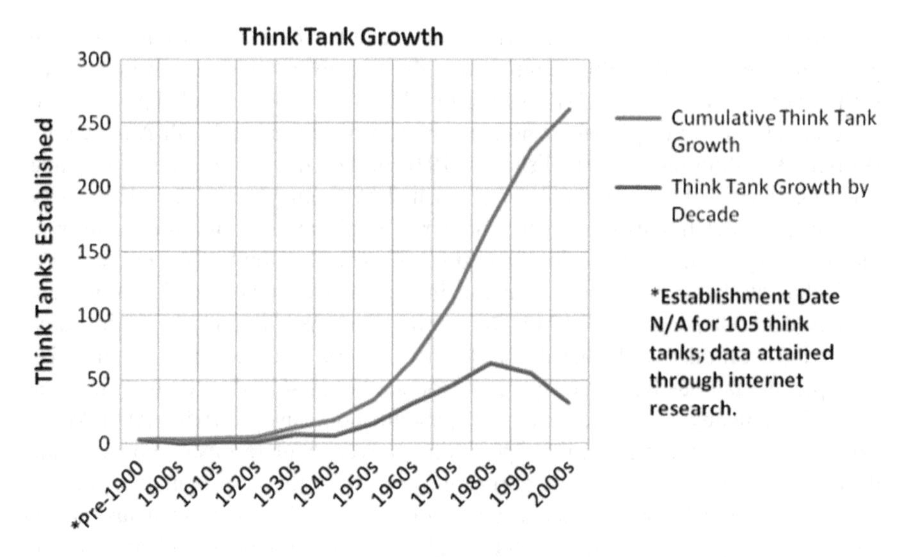

Fig. 10.2 Growth of think tanks (*Source* McGann 2010)

10.3.1.2 Status of Think Tank Development in Different Regions (2010)

Figure 10.3 and Table 10.2 show the regional distribution of think tank development around the world in 2010. As shown in the figure and table, North America has the largest number of think tanks on a global scale (having 30% of global think tanks). Europe ranks second (27%). Asia and Latin America have 18 and 11% of global think tanks respectively. Africa (8%), the Middle East and North Africa (5%) and Oceania (1%) take the last three places.

10.3.1.3 Global Think Tanks Ranking Report of Pennsylvania State University

Professor James McGann of Pennsylvania State University published the first global think tanks ranking report in 2007 through his "think tank and civil society" course in "international relations" curriculum. With improvements to the evaluation process, the fourth global think tanks ranking report (2010) was published in January 2011.

(1) *Main Report Contents*

The report (2010) mainly includes four categories of ranking subdivided into 29 tables. Ranking of primary categories and secondary categories is shown in Table 10.3.

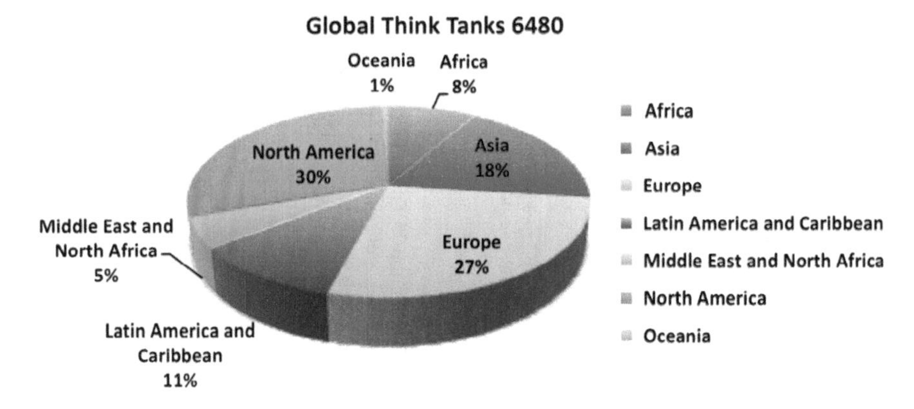

Fig. 10.3 Regional distribution of major global think tanks (2010) *Note* Same source as with Fig. 10.1

Table 10.2 Regional distribution of global think tanks (2010)[a]

Regions	Quantity of think tanks	Proportion in global think tanks (%)
Africa	548	8
Asia	1200	18
Europe	1757	27
Latin America and Caribbean Sea	690	11
Middle East and North Africa	333	5
North America	1913	30
Oceania	39	1
Total	**6480**	**100**

[a]*Source* McGann (2010, p. 19)

(2) *Criteria for Global Think Tanks Ranking*

Criteria for global think tanks ranking consist of four key indicators and 26 secondary indicators, as shown in Table 10.4.

(3) *Examples of Top 25 Think Tanks in the World (2010)*

Table 10.3 shows 29 ranking tables based on different categories in the global think tanks ranking report (2010). For illustrative purposes, the following Table 10.5 lists the top 25 think tanks (in and outside America) in the world covered by Secondary Category 3 out of the Primary Category entries for top think tanks ranking as shown in Table 10.5.

Table 10.3 Classification of top think tanks in the world[a]

Primary categories	Secondary categories
Top think tanks ranking	The best think tanks in the world (Table 1); the top 50 think tanks in the world (excluding America) (Table 2); the top 25 think tanks in the world (including America) (Table 3)
Top think tanks ranking by regions	The top 50 think tanks in America (Table 4) The top 25 think tanks in Latin America and Caribbean Sea, Middle East and North Africa, Sub-Saharan Africa, Western Europe, Central and Eastern Europe, Asia, etc. (Tables 5–10)
Top think tanks ranking by research area	The top 25 think tanks for international development (Table 11), the top 15 think tanks for health care (Table 12), the top 25 think tanks for environment, defense and international affairs, domestic economic policy, international economy, social policy, science and technology, etc. (Tables 13–18) The top 20 think tanks for transparent and good governance (Table 19)
Top think tanks ranking by special achievements	The top 25 think tanks with the most innovative policy ideas and suggestions (Table 20) The top 20 think tanks newly established (in the past year and a half) (Table 21), the top 25 think tanks with the most excellent policy research subjects (Table 22), the top 10 think tanks most capable of reaching the public through internet (Table 23), the top 10 think tanks Mst capable of disseminating their projects and researches through media (hardcopy or electronic) (Table 24) The top 25 think tanks with the best external relationship/public participation as well as the most influential public policy and subordinate to colleges and universities (Tables 25–27), the top 20 think tanks run by governments (Table 28) The top 25 think tanks run by political parties (Table 29)

[a]*Source* McGann (2010, pp. 32–52)

Table 10.4 Indicator system for evaluation of global think tanks ranking[a]

Key indicators	Secondary indicators
Resource indicator (7 secondary indicators)	The ability to attract and retain excellent scholars and analysts; level, quality and stability of obtained financial support; proximity to decision makers and other policy elites; availability of contingents capable of timely serious research and analysis; institutional flexibility; network quality and reliability; critical connection to policy groups, academic society and media
Utility indicator (7 secondary indicators)	Reputable willingness of media and national policy elites to visit; the quantity and quality of citation by media and website hits; involvement in legislative and administrative agencies for hearing; approaches by government officials for consultation, official appointment or press release; the number of publications sold; the number of reports issued; research and analysis results quoted as reference data in academic and public publications, as well as participation in organized conferences and seminars
Output indicator (5 secondary indicators)	The quantity and quality of policy advice and ideas generated; publications produced (such as books, essays in magazines, policy brief, etc.); news interviews performed; the number of press conferences and seminars organized; and the number of staff appointed with government positions or as consultants
Impact indicator (7 secondary indicators)	The number of recommendations considered or adopted by decision makers or civil society organizations; central position in network argumentation; the extent of consultancy for political parties, candidates and interim teams; the number of awards; academic journals, argumentation and media publications having influence on policy debate and decision making; prominent influence of website; success rate in challenging conventional wisdom and standard operating procedures of bureaucracy and elected officials

[a]*Source* Derived from McGann (2012)

10.3.2 Trend and Challenges Facing Global Think Tank Development

10.3.2.1 Trends of Global Think Tank Development

During the period from 1831 when the first think tank was established to 1940, global think tanks growth was rather slow, with the establishment of think tanks mainly in developed countries and the largest number of them found in the United States. The developed countries retained their lead during the growth peaks in think tank development amidst and after the World War II, as well as in the annual growth of new think tanks from 1964–1970 (Ref. Fig. 2.1). In developing countries, think tanks were mostly established until after 1970.

Table 10.5 Top 25 think tanks in the world (2010)

Serial No.	Think tanks
1	The Brookings Institution (U.S.)
2	Council on Foreign Relations (U.S.)
3	Carnegie Endowment for International Peace (U.S.)
4	Royal Institute for International Affairs (U.K.)
5	Amnesty International (U.K.)
6	RAND Corporation (U.S.)
7	Center for Strategic and International Studies (U.S.)
8	The Heritage Foundation (U.S.)
9	Transparency International (Germany)
10	Peterson Institute for International Economics (U.S.)
11	International Crisis Group (Belgium)
12	Cato Institute (U.S.)
13	American Enterprise Institute for Public Policy Research (U.S.)
14	The International Institute for Strategic Studies (U.K.)
15	Center for European Policy Studies (Belgium)
16	Human Rights Watch (U.K.)
17	Woodrow Wilson International Center for Scholars (U.S.)
18	Bruegel (Belgium)
19	Adam Smith Institute (U.K.)
20	Stockholm International Peace Research Institute (Sweden)
21	The Open Society Institute (Hungary)
22	The Urban Institute (U.S.)
23	Center for Global Development (U.S.)
24	Chinese Academy of Social Sciences (China)
25	Fraser Institute (Canada)

The advanced establishment of think tanks in developed countries prompted a trend of certain think tanks expanding their institutions beyond national borders. For example, the Brookings Institution, No. 1 top think tank based on 2010 ranking report, established a Brookings research center in Tsinghua University, Beijing of China in October 2006. The Japanese Sasakawa Peace Foundation also set up a branch in America and held a conference together with the Brookings Northeast Asia Policy Research Center in December 2011 to study Japan's gain and loss resulting from its participation in Trans Pacific Partnership (TPP).

Another trend in think tanks development is that following the creation of the United Nations after the WWII, a host of international, regional and collective think tanks under the UN auspice came into being. These include United Nations Industrial Development Organization, Food and Agriculture Organization of the United Nations, the United Nations Conference on Trade and Development, the World Bank, International Monetary Fund and some United Nations subsidiary regional organizations such as Economic and Social Commission for Asia-Pacific Region. Organizations independent of the United Nations such as the Organization for Economic Cooperation and Development (OECD), the Asian Development Bank, etc. were also established. All these global and regional organizations have professional expertise and most of them also are securely funded to organize experts from all around the world to engage in research.

The advancement of information technology and network technology further promoted the globalization and networking of think tanks, as the case of the International Center of Economic Growth founded in 1985 and the Global Development Network established by the World Bank in 1993. The latter was separated from the World Bank in 2001 and became an independent non-profit think tank.

10.3.2.2 American Experience in Think Tank Development

(1) *Significance of Research on American Experience in Think Tank Development*

Although the earliest think tank of modern significance emerged in the United Kingdom, America achieved the largest number of think tanks with the best quality during the process of global think tank development. Besides, America was the first to publish series of statistics, analyses and literature resulting from systematic research of global think tanks. Due to tremendous differences in political cultures of different countries, the political, industrial and academic circles have different appreciation of the value of think tanks. It is the belief of authors of the book, nonetheless, that think tank is an integral component of overall planning of a state and is complementary to planning departments of a national administrative system. The United States, however, while possessing the most and best think tanks worldwide rejects the concept of importance of overall national development planning due to the influence of dominant economic ideology. The American academics however do recognize that the sharp rise of think tanks was against the backdrop that government staff, especially senior staff began to realize that size and complexity of American government would increase day by day during and after the World War II, resulting in increasingly tough problems related to planning and adaptation. Emergence of think tanks was regarded as an important measure to solve this problem. Since they are small-scale and independent organizations and free of recessive administrative departmental benefits, think tanks may hold a [5]trans-department and cooperative view on problems. Besides, they have no time limit and are free from influence of divisional responsibilities; therefore, they may generate forward-looking thought. Due to the above-mentioned characteristics, American think tanks underwent a rapid growth after 1945 and also influenced think tank development in other regions.

[5]Source: McGann and Sabatini (2011).

(2) *Special Historical and political background of American Think Tank Development*

Special historical and political background of American think tank development falls into three aspects.

(i) *America, a Relatively Late Comer of Development of Think Tanks Compared with Western Europe*

Although the United Kingdom and Europe had the earliest think tanks, America is the creator of the earliest think tank system. North America has been in the lead in the global think tank development since the 20th century (Table 10.6).

(ii) *In Terms of Political Ecosystem, America is Different from other Countries*

In comparison with other countries, formulation of public policy in America is more susceptible to influence of non-government sectors. The American government has a very clear division of labor between its administrative and legislative departments which is different from a standard parliamentary system. In the latter case, the Prime Minister is concurrently the chief executive officer and the chief party representative on legislative issues. In the case of the America presidential system, the President is not only the head of state but also the head of government who is elected into the office rather than through parliamentary designation. Heads of ministries are by presidential appointments and are accountable to the President rather than the parliament. In America, legislative initiatives may come from the Congress and the White House in parallel.

Table 10.6 The number of think tanks newly established all around the world

Time period	Worldwide	Africa	Asia	Eastern Europe	Latin America	Middle East	North America	Western Europe
1700–1800	3	0	1	0	0	0	1	1
1801–1900	18	0	1	0	0	0	8	9
1901–1910	16	0	1	0	0	0	12	3
1911–1920	25	0	0	1	0	0	19	5
1921–1930	39	1	3	4	0	0	22	9
1941–1950	117	2	17	8	6	0	54	30
1951–1960	196	3	24	8	18	3	70	70
1961–1970	340	10	50	15	28	13	127	97
1971–1980	476	30	76	6	54	24	248	38
1981–1990	956	42	113	50	96	31	417	206
1991–2000	1248	110	135	249	85	63	348	258
2001–2010	414	30	48	61	34	24	89	128
Total	3883	229	473	402	322	158	1435	864

Budget appropriation and approval is a major issue during decision-making. In America, such an issue manifests as competition between government sectors rather than political partisan bickering seen in a parliamentary system. Given the multitude of government sectors in the United States, the government may be subjected to assorted external lobbies in the process of policy formulation. The legislative branches of the Congress themselves have significant research functions that reside in such offices as the Office of Technical Assessment, the Congressional Budget Office and the General Accounting Office, etc. These offices not only have strong research capacities but are also closely connected to external think tanks.

(iii) *Main Mode of Impact of American think tanks on American Political Culture*

The hierarchical selection process of government officials in America is not set in a closed system. A Congress representative may become a government minister. For example, Leslie Aspin, a U.S. Congressman, became the Secretary of Department of Defense. Local politicians may become national leaders, such as President Carter and President Reagan. President Reagan was also a famous actor. Strobe Talbott (a pressman and journalist), Robert Rubin (a businessman), Henry Kissinger (a scholar), etc. all became senior government officials.

Think tanks play an important role in the American political process. Certain think tanks are in effect government official reserves, as in the cases of the Brookings Institution and the American Enterprise Institute that are well-known shadow government reserves. When an opposition party is in power, other parties will return to think tanks. Think tank members from the academic circle also introduce former politicians, pressmen and cabinet officials into university life circle. For example, Professor Joseph Nye and Graham Allison of Harvard University were former senior government officials. Senior staff of American foundations also hold posts among government think tanks. The principal of the Ford Foundation had been the United States Secretary of State; Hillel Fradkin, a member of the Bradley Foundation had left the foundation for a short term and became a researcher of the American Enterprise Institute. All these suggests that American think tanks are different from those in other countries and has a "revolving door" relationship with politics and government. Bush depended on the experts from conservative American Enterprise Institute, while Obama relied on Center for American Progress. These think tanks help the ruling authority with idea generation and policy formulation. Many government officials migrate among the White House, the Congress and think tanks under the "revolving door" mechanism. Some American think tanks are even established by presidents or senior politicians. For example, the Hoover Institution of Stanford University was established by Hoover, the 31st President of the United States (1928–1932). Rice, the United States Secretary of State during Bush's government, now works in this institute. As for advantages of the "revolving door", P. J. Crowley who is a former senior researcher of the Center for American Progress and now holds a post in the Bureau of Public Affairs in U.S. State Department once pointed out that "think tanks are much more rational than government sectors and your work will not be so humdrum. (Government sectors) make you work on affairs within identified boundaries. While working for a think tank, you always start from scratch." His observation clearly explains the characteristics and advantages of American think tanks as well as their relationship with the American government.

(3) *Two Categories of American Think Tanks*

American think tanks could be divided into two categories from another perspective. One category of American think tanks is universities without students and had existed before the World War II. Demanded by advanced industrial economic management and mounting international commitments, it was required to introduce science and rationality into government at the beginning of the 20th century. Establishment of the Brookings Institution was such a response to this demand. The other category of American think tanks is mainly contract-based research organizations which carry out subject researches for specific government sectors. For example, the American RAND Corporation performed contractual researches for the United States Department of Defense all along till today and has now expanded the same contract-based research for other departments.

(4) *Experience in American think Tank Development*

Not all American think tanks perform successfully. Some think tanks without access to generous charitable funding and dependent on donations from companies or individuals tend to target for research results that serve the contributors due to their commercial motives. In general, successful experience in American think tank development could be referenced by newcomers in six aspects:

(i) Think tanks could be more future-oriented than governmental research organs. That is because staff working in governmental research divisions are not encouraged and rewarded for "creative destruction";
(ii) Think tanks are most likely to generate agenda for policy reshaping. Administration bureaucrats are usually accustomed to the security of standard operating procedures;
(iii) Since think tanks have no permanent recessive benefits deriving from any set field of research, they are more prone to promote closer cooperation among researchers from different groups for common goals;
(iv) They are more capable of overcoming bureaucratic impediments and promote integrated utilization of intellectual capacities;
(v) They may achieve better dissemination of relevant policy researches across government apparatus than can be done by government sectors themselves, since the latter is often subject to linear bureaucratic control or even "turf fight";
(vi) In contrast to government officials, think tanks could better fulfill the policy functions ranging from data collection, factual identification, idea formulation to means of implementation in a more comprehensive manner, because

policy functions of government officials are often impeded by internal division of duties that fragment the integral policy effectiveness of the government.

10.3.2.3 Challenges Faced by Think Tank Development

Challenges facing think tank development are generally different in developed and developing countries. Still, there may be problems common to all, like the following three.

The first issue is the independence of think tanks and their research activities relative to the source of funding. Think tanks in developed countries depend less on direct government grants but more on various funds. Their research activities appear relatively independent. In reality, except those think tanks subordinate to the United Nations organizations that could objectively provide policy consultancy on a global scale, research results of other think tanks are only relative in terms of objectiveness and independence. Research results of think tanks are swayed in varying degrees by fund contributors. Especially for some think tanks in developing countries, they usually depend on government grants and often act as justifiers of government policies or mouthpieces of government through their research work.

The second challenge facing think tank development lies in the types of policy makers and the average quality of the public. Since the research results of think tanks are for the consumption of policy makers and the public, the particular type of policy maker determines the extent of acceptance of think tank products. Some policy makers can be self-willed, or may have set their own mind with regard to a certain policy, or have formed their opinion under the influence of surroundings. Accordingly, some policy makers may only accept think tank research results conforming to their own opinion and reject those that are in conflict. Some policy makers may have their mind set on a certain policy but without prejudice. These policy makers may adopt the useful elements from think tank research results to improve, supplement or even modify their original ideas, so as to formulate correct policies. Still some policy makers may be neither assertive nor prejudiced and, instead, are capable of utilizing think tank research results. Of cause, there are those that do not have such capability. In short, the extent of impact of think tanks also depends on different types of policy makers.

Quality of the public is dependent on public culture traditions, state and maturity of economic and social development, age structure, occupation pattern, etc. Some correct research results may not be accepted by majority of the public at that time, while some superficial "results" or publicity stunts may spread like viral. The professional ethics of think tank leaders and staff here comes into play. They may either earnestly conduct their research work and release their results complying with professional ethics, or cater to a momentary popular trend with some emotional calls that do not fit the long-term interests of the country and people. This is a challenge facing think tanks during their development process—competence in research and professional ethics of think tank workers.

10.4 Case Study of International Think Tank

10.4.1 Case Study of Significance of International Think Tank

The establishment and development of modern think tanks originated from some western developed countries. For one and a half centuries of such effort, developed countries have accumulated some sound operation and organization experience that are of reference value for developing countries when establishing their own think tanks later. In this chapter, some think tanks from developed countries, like the Brookings Institution and RAND Corporation as well as International Development Center of Japan, will be selected for case study. The reasons for the selection are that the first two rank among the global top think tanks and that the third is engaged in planning research that is consistent with the subject of this book.

10.4.2 Case Study of Brookings Institution[6]

10.4.2.1 A Brief History of the Brookings Institution

The Brookings Institution was established in 1916 by a group of American leading reformers to conduct research for the government. It was the first private organization devoted to analyzing public policy issues at the national level. Since its founding, the Institution has contributed at all historical junctures of development of the United States. The contributions it made at each stage are as follows:

(1) *Pioneering Stage: Mission of the Institution*

In 1916, Robert S. Brookings, together with other governmental reformers, established the first private organization which was devoted to fact-based study into national public policy issues aimed at making the government become the key promoter for beneficial and efficient public service provider and exploring ways to introduce "science" into research fields of governmental organizations.

Meanwhile, Brookings established two sister institutes, i.e. the Institute of Economics in 1922 and the Robert Brookings Graduate School in 1924. In 1927, these three were merged into the Brookings Institution with the mission of conducting, promoting and consolidating research in "various fields such as economy, governance, politics and social science".

Harold Moulton, the first president of the Institution, is a professor of University of Chicago known for his study of war debts. The economists of the Institution played a critical role in the formulation of the act leading to the setting up of the first

---

[6]Note: The case of Brookings Institution is briefing of its various annual report from 1985 to 2010. 1985 is a printed Version Glodberg (1985). Remaining are derived from Brooking Website.

American Budget Office in 1921. Warren G. Harding, former US President, said that the founding of the Budget Office was the greatest reform in government operations since the setting up of the United States.

(2) *War and Peace Periods*

During the World War II, experts from the Institution helped the government to conduct wartime mobilization and subsequent management work. Leo Pasvolsky, an expert of the Institution, worked in the State Department and played an important role in improving the United Nations blueprint planned by President Roosevelt. He also contributed to the formulation of the Marshall Plan after the World War II. In 1948, Arthur Vandenberg, a Senator from the Council on Foreign Relations of the United States Senate, praised a report of the Institution as "a worksheet for solving complex and critical problems facing the Congress."

(3) *Contribution to Shaping the National Institution*

The year before the 1960 presidential election, Laurion Henry, a governmental research specialist of the Institution, published a book named *Presidential Transitions*. This book was designed to help future presidents (John F. Kennedy or Richard Nixon) to smoothly start their governance. After publication of this book, experts from the Institution prepared a series of recommendations on confidentiality policy.

The Institution also played a role in the Watergate Scandal happened in 1970s. President Nixon ordered his assistant to intrude into the office of Leslle Gelb who was a researcher of the Institution and was an analyst working for Daniel Ellsberg in United States Department of Defense before. The latter disclosed "Pentagon Papers" to New York Times and Washington Post. At a summer night of 1971, two military officers attempted to slip into the office but were stopped by Roderick Warrick, the guard of the Institution.

In the same year, the Institution conducted a series of researches on Federal budget and made in-depth analysis on various plans, giving the public a hand and making a centralized choice for Congressional spending. Three years later, the Congressional Budget Office was established under the impetus of the Institution. Alice Rivlin, an eminent economist of the Institution, served as the first director of the Office. In the succeeding years, she held alternating posts between the Institution and the government.

On September 29, 1966, President Lyndon Johnson made a speech on importance of public service and American metropolises on the 50th anniversary of the Institution.

In 1992, the Institution published [7]*Memorandum to the President* written by Charles L. Schultze, the economic department head of the Institution. Charles L.

[7]Note: This book was translated into Chinese in November 1993 by the UNDP "economic development, reform and policy" research group of the Development Research Center of the State Council with approval of the Institution and was named as *Macroeconomic Decision Orientation—Memorandum to the President* and published by Shanghai Far East Publishers in 1994.

Schultze previously served as the director of the Council of Economic Advisers. Base on his government experience, Charles L. Schultze deemed that government policy makers, pressmen and the public needed some basic knowledge on macro economy to help them make judgment and decision on relevant macroeconomic phenomena and issues. In Charles L. Schultze's opinion, all kinds of macroeconomic issues, inflation, unemployment, economic growth, interest rate, trade deficit, etc. are usually focuses of political debates. Policy makers, pressmen and the public should not be at the mercy of others and should make correct judgment by themselves on the basis of certain knowledge on mode of economic operation and various government policy measures. In the above-mentioned book, Charles L. Schultze explains economic operation and roles of various macroeconomic policies for busy policy makers and the public in an innovative manner by 29 memorandums. He envisages that the new president should spend an hour every week to learn critical economic principles under the guidance from the director of the Council of Economic Advisers. Contents of these 29 memorandums are special subjects which need to be understood by the President when making macro policy decisions.

(4) *Contribution to Economic Growth*

Joseph Pechman, Director of the Economic Research Planning Department of the Institution, energetically promoted reform of U.S. tax law at the beginning of 1980s. As a result of his research, the Tax Reform Act was enacted in 1986, having profound impact on U.S. economy.

In 1990s, the Federal government pushed to introduce series of social programs to states and cities. The Institution developed a new generation of urban policies to promote link between metropolises and suburban areas. When President Bill Clinton was prepared to sign the historic welfare reform legislation, Ron Haskin (former Republican Congressman) and Isabel Sawhill (former officer in President Clinton's Office of Management and Budget) established a group in the Institution specializing in policies related to child and family. In 2001, child tax credit was proposed by Isabel Sawhill and Adam Thomas and became a major component of taxation legislation.

Bill Gale, Mark Irory and Peter Orszag (economists of the Institution) continued their hard work on taxation system improvement. Combining financial incentives of the middle-income and low-paid workers as well as new situation faced by companies, these scholars pushed the research on easier savings for retirement for Americans into legislation, which made them become the most influential economists in the U.S.

(5) *Meeting Challenges from Global Threats*

The terrorist attack on September 11, 2001 increased the urgency to study and deal with threats to national development strategy. The United States is required to sustain its capacity to guarantee prosperity and stability overseas and maintain an open society internally. Experts from the Institution proposed recommendations on the maintenance of internal security and intelligence operation at an amazing speed.

They provided hearing in the Congress and made good use of their outreach capacities, including its in-house television studio, to illustrate the up-to-date global situation to the frightened public.

(6) *Governance and Update*

Faced with the Great Recession during 2008–2009, scholars from the Institution explored the cause and consequences and shaped the debate by a series of analyses and recommendations. When the era was faced with convergent influence of climate change, arms proliferation, weak and failing states and other multi-facet problems calling for complex remedial measures, experts of the Institution made speedy response by means of highly eye-catching events and comprehensive strategic research.

The Institution will celebrate its centenary anniversary in 2016. At that time, the five departments of the Institution will join hands under the research planning subject of "governance and update", so as to pool intellectual resources of the Institution together to study the prioritized subjects, such as growth through innovation, new opportunities and happiness, perfecting energy and climate policies and management of global changes.

10.4.2.2 Top-level Organization and Funding Source of the Institution

(1) *Top-level Organization*

The top-level organization of think tanks plays a critical role in think tank operation and development. An top-level organization comprised of top distinguished personages and highly competent candidates from the society will facilitate think tank branding, secure funding and guarantee research independence. The top-level organization of the Institution is comprised of the Board of Trustees, the President and Vice President. The Board is responsible for supervision of the organization, approval of research scope and safeguarding the Institution's operational independence. The President works as the chief administrator and is responsible for policy formulation and coordination, project recommendation and staff designation with the help of staff at all levels in the advisory committee.

Responsibilities of the Board are clearly described in the organization's enterprise law: "The Board of Trustees is responsible for conducting scientific researches and publishing achievements under the most favorable conditions and guaranteeing independence of researchers while conducting such researches and publishing relevant achievements. Decision, control or influence on specific research process or its achievements is beyond the duties of the Board."

The Board consists of a chairman, two vice chairmen and dozens of permanent and honorary members. The President of the Institution is a permanent member of the Board. The Board is constituted by election and serves three years in principle. Some successive permanent members become honorable members later.

Permanent members of the Board mainly consist of distinguished persons from the top American political, legal, academic, industrial and financial circles. Take the Board constituted in 1985 as an example, there were 36 permanent members and 19 honorable members at that time. Permanent members included three university presidents such as from University of Chicago, and two professors from MIT Sloan School of Management. Members from political and legal circle included Robert S. McNamara (former Defense Secretary and President of the World Bank) and the then incumbent President of the World Bank, as well as William T. Coleman (Principal of American Legal Adviser Company) and Lloyd N. Cutler (a White House lawyer during Carter and Clinton administrations). Besides, 7 members are presidents or executive officers from financial and banking circles. Permanent members from the industrial circle include presidents or vice presidents of major companies in America, such as executive presidents and vice presidents of AT&T and IBM ranking among the worldwide top 500 enterprises listed by Fortune Magazine.

The Board constituted in 2010 was comprised of 50 permanent members and 44 honorary members. The then incumbent president of the Institution was Strobe Talbot who worked as a pressman as well as a senior officer in American government before. Vice President Glenn Hutchins of the Board is a former vice president of American investment bank Golden Sachs. Incumbent permanent Board members consist of 14 presidents or senior managers from the financial circle; 3 presidents and vice presidents from various foundations; 4 members from the academic circle, including. President of the Energy Research Institute, President of American Rensselaer Institute of Technology, professor of Georgia Law School and senior investor of Yale University. Bart Friedman who comes from law profession and is a senior partner of the most profitable law firm in Wall Street, Cahill Gordon and Rendel LLP and Ann Fudge who comes from the political circle and was appointed as a member of Fiscal Responsibility and National Reform Committee by President Obama in early 2010. From the industrial circle, there were presidents and vice presidents of MGM International Leisure Company, Zeonic Company, Encana Corporation, etc.

The President of the Institution has the final say on research undertaking. The important and influential achievements of the Institution are presented as publications of research findings. The President applies the final clearance for such publications judged by competitiveness, accuracy and objectivity of the research findings quality. When making a final decision, the President will take into consideration both advice from relevant project leaders and private quality assessment results from external specialists through advance reading of research reports. Release of a research achievement indicates that such an achievement deserves attention from the public and does not mean that conclusions and suggestions contained in the publication are recommended.

(2) *Funding Source*

Funding source of the Institution mainly comes from charity, companies and private foundations or donation. The Institution also collects fees from its conference events, computer service as well as sales of publications. In addition, it also collects fees from its contract research conducted for the government. In this case, the Institution reserves its right on the release of their contract research findings. According to data shown on annual report of the Institution (2010), the annual gross operating income of the Institution is USD 102 million mostly from donation and contract research income of USD 52 million as fiscal revenue in 2010. This is followed by asset investment income which reached USD 12 million and income from sale of publications of USD 2.3 million and other revenues.

10.4.2.3 Research Organization of the Institution

(1) *General*

Research organization of the Institution evolved with development of times and mainly include research[8] departments (programmes) and centers.

(2) *Research Organization and Research Contents before 1990s*

Before 1990s, the Institution set three research departments and two research centers.

(i) The economic research department: It is responsible for increasing the understanding of economic operations so as to improve economic performance. Alice Rivlin who once served as the first director of the Congressional Budget Office was the principal of the economic research department at that time. The department focused mainly on three policy aspects:

Policies leading to sustainable development of American economy;

How to achieve more effective social planning and policies in time of resource constraints?

How to improve international economic relations in an increasingly interdependent world?

(ii) The government research department: The department is the main component of research on American political system—from election to institution Such research work is generally divided into three categories:

Political system: Researches on the presidency, the Congress and interaction between the two. These researches clearly pointed out that the President's office is usually funded by the public and the President often focuses on wider national interest. Political forces of the Congress works in a direction contrary to that of the President. Election of Congress representatives is mainly funded by interest groups and Congress representatives serve for their own states and regions.

[8]Note: The original text is "Programme" which has a literal meaning of planning or scheme; however, it actually means research department here.

Regulations, economic policies and social policies.

(iii) Foreign policy research department: The Institution attaches importance to foreign policy research, because formulation of foreign policy requires comprehensive understanding of political intention, institutional operation and historical evolution of various countries and regions. The Institution considers that in the real world, progress in reaching a consensus on disposal of complex issues can be very slow due to power division and different views among government sectors. Therefore, integration of various expertise and specialists is all the more necessary in foreign policy research. Researchers conducted before 1990s include defense analysis, Soviet Union studies, Middle Eastern studies, East Asian studies, as well as researches on international economy and trade.

All these researches received funding from MacArthur Foundation, Carnegie Foundation and Ford Foundation.

The above-mentioned two centers are public policy education center and social science computing center respectively

(3) *Present Research Organization (2012)*

The Institution set 5 research departments and 11 research centers in 2012.

(i) *5 Research Departments*

The economic and foreign policy research departments established in the last century are retained and the government research department is renamed as governance research department. Besides, global economy and development research department and urban policy research department are added.

The governance research department mainly works on improvement of government performance, economic security and social welfare as well as creation of opportunities for American people.

The global economy and development research department mainly focuses on opportunities and challenges arising from globalization, to which policy makers, the business community and the general public pay close attention. Researches were mainly conducted in three aspects, i.e. shaping of global economic driver, measures for poverty elimination and rise of new emerging economies.

The urban policy research department works on redefinition of challenges from American urbanization process and studies about the innovative approach to inclusive, competitive and sustainable development of constructive communities.

(ii) *11 Research Centers*

The eleven policy centers include Blanc education policy center, Northeast Asia policy research center, technology innovation center, universal education center, child and family center, dynamic social and policy center, America and Europe center, Engelberg Medical Reform Center, John L. Thornton China Center, Sabin Middle East Center, Brookings Urban-Tax Policy Center.

10.4.2.4 Determination of Research Subjects

(1) *Source of Subjects*

Proceeding from its mission objective, the ability of any given think tank to choose a research subject suited to the need both at home and abroad is a critical factor for institutional survival and development. Staff of the Institution persistently explore and conduct researches in areas of interest. They often exchange ideas with government officials, Congress representatives, corporate leaders, principal officers of foundations and other scholars to identify issues required to be studied. Subjects are selected based on their significance and timeliness, adequacy of information and available technological capability, availability of personnel and funds, as well as the relevance between the selected subjects and purpose of the Institution. All research subjects of the Institution have a high academic standard and deserve consideration by the public. The Institution does not take positions on nor bears any responsibility for policy issues.

(2) *Examples of Research Topics*

In 1980s, research subjects of the Institution in economic field paid attention to improvement of innovation and productivity. The Institution collected detailed data from major industries at that time, such as chemical, textile and steel industries, and made itemized analysis. It proposed policy recommendations on various new technologies to the government, such as what technologies should or should not be encouraged by the government for commercialization. During the Reagan administration (1981–1989), large scale deregulation was implemented in the U.S. Economic department of the Institution conducted researches and made recommendations on regulation or deregulation of banking, automobile and aircraft industries.

As for social studies, the Institution researched on aging of population and medical care, evaluation of public and private education in America, prevention of illegal immigrant as well as future development and employment policies.

Researches in foreign policy had been always focused on national security and defense. A completed research project in this includes *Strategic Command and Control: Redefinition of Nuclear Threat*, in which it analyzed the coordination of U.S. command, control and communication system for the purpose of dealing with strategic attacks in case the U.S. has to face, and the vulnerabilities in the U.S. system response. The work also analyzed deployment of armed forces and strategies in the Persian Gulf. On the topic of U.S.-Soviet relations before collapse of the U.S.S.R., the research work included *Relaxation and Conflict: the U.S.-U.S.S.R. Relations from Nixon to Reagan*, etc. As for researches on the Far East, besides studies on changes in the Japanese economy, the Institution also focused on Sino-American economy and security issues in the post-Mao era. In 1980s, the Institution held the third joint meetings together with the Chinese Academy of Social Sciences. Scholars and participating representatives from both parties conducted joint studies on economic policies and prospects of major countries in the Asian-Pacific region, impact respective internal developments of these countries on their trading partnerships, as well as the Sino-U.S. relations.

(3) *Earlier Research Topics*

Major research topics of the Institution before 2012 included American prosperity blueprint, election reform, global health care, Greater Washington, international volunteers and services, democracy and development in the Middle East, youth in the Middle East, opportunities in 2008, independent approaches of the next President, etc.

(4) *Current Research Topics (2012–)*

Current research topics of the Institution include defense actions in the 21st century, African growth actions, arms control scheme, Brookings-Rockefeller urban innovation projects, national budget priorities, election campaign in 2012, climate and energy economy projects, development aide and governance schemes, energy security projects, global urbanization studies, researches on Great Lakes economic zone, enterprise and public policy researches, Latin-American studies, global order management, retirement security projects and studies on U.S.-Islamic world relations.

10.4.2.5 Public Policy Education

(1) *Historical Evolution of Public Policy Education promoted by the Institution*

Public policy formulation and implementation have an significant impact on social, economic and cultural activities of the public, as well as national development in right directions. Public policy researches need wide participation of the public. Organizations conducting policy researches also bear responsibility for disseminating research achievements among relevant persons and the public to empower the latter to reach consensus and ensure better implementation of public policies. The Institution established its higher education research department around 1950, which served as a department for public policy education for nearly a quarter century. On the basis of the higher education department, the Institution established its public policy education center in 1985. The center extended the scope of the Institution's meetings and seminars to expand the influence of the Institution's achievements and provide resources for national public policy education and development. Goals of the education center are as follows.

Develop innovative curricula to fulfill various demands of incumbent leaders;
Assemble leaders of various departments to discuss policy issues and make choices;
Propagate research findings of the Institution's scholars to a wider public;
Organize public forum on public policy issues;
Enhance influence of the Institution's educational activities at the international level.

(2) *Development of Public Policy Education at present*

At present, public policy education activities of the Institution are further deepened on the basis of achievements of the past half-century and focus on diversification and partnerships. Public policy education consists the following categories:

(i) *Enterprise Leader Curriculum*

It is offered jointly by the Institution and University of Washington and assembles elites from all walks of life in Washington, such as scholars, consultants, legislators, business leaders, military leaders, media and diplomats, to discuss public polices, aimed to understand senior government officials' insider thinking and to explore talents for cross-departmental management and leadership. There are about 175 enterprise leaders and heads of various government sectors participating in the curriculum every year.

(ii) *Certificate Program*

The course takers are required to attend classroom studies for no less than 20 days and undertake two compulsory courses, i.e. ethnical behavior and integrative leadership and execution, lasting 8 days. Learners will have a free choice of any course offered by the Institution in the rest 12 days. At the end of the curriculum, learners are required to write 3–5 pages of study notes.

(3) *Customer Curriculum*

The American National Institutes of Health selected the Institution as its partner for policy training. Every year 20 leaders are chosen to undertake the curriculum for 6 months.

(4) *Degree Program for Master of Science Provided for Leaders*

It is co-sponsored by the Institution and Olin School of University of Washington and mainly aimed to train qualified talents to lead public sectors of the federal government.

(5) *Legis Congress Scholarship Program*

Learners of the Legis Congress scholarship program are government officials (above GS-13 level) or private sector leaders interested in legislation. Applicants need to receive and pass intensive training offered by the Institution at first and when approved for the scholarship grant the candidate will then work in the Congress members' office or serve at a Congressional committee for 7–12 months. Through this program, learners could learn about the internal procedure of policy formulation and legislative drafting and attend in hearings of committees, so as to develop high-level strategic planning skills, new leadership capabilities and political in sightedness.

(6) *Courses in 2012 and Examples*

The Institution offered 37 short courses in 2012, varying in length from one day to six days and covering a wide range of public policy topics. For example, highly practical political and diplomatic courses and training were provided in 2012, including the five-day course of *In the Congress—How the Congress Operates* in February and May, the three-day training of *Inside Washington—Knowledge on Formulation of Federal Policies* in June and the three-day training of *Inside the*

Middle East: Security, Politics and Developments in April. Besides, there were short-term courses related to academic policy issues, such as *Politics and Policy Formulation, Leadership Elasticity, Leadership Learning Organization, Finance for Non-financial Managers, Scientific and Technical Problems, Strategic Thinking —Promoting Long-term Success*, etc.

10.4.2.6 Publications and Publishing

Publications: Publication of research findings is an important measure for propagating achievements of a think tank. Publications of the Institution include books, documentation, news bulletins and articles published on various magazines. In 1980s, the Institution was a member of the Association of American University Presses and its publications could be bought from bookstores or ordered directly. Later, the Brookings Institution Press was founded. Annual fiscal revenue of the press reached USD 2,3 million from June 2010 to June 2011. Due to the rise of China, books related to China studies became the selling point of the press. For example, the book *Managing the China Challenge: How to Achieve Corporate Success in the People's Republic* written by Kenneth Lieberthal (director of the China Research Center of the Institution) was recommended by the press in 2011, adapting to the demand of transnational corporations. The press also published *Danger from Neighbor: Sino-Japan Security Relationship* written by the director of the Northeast Asian Policy Research Center of the Institution, *Emerging Middle Class in China: Beyond Economic Change* compiled by Cheng Li (a senior researcher for Chinese Thinkers Series), as well as *China in 2020: New Superpower* written by Hu Angang from Tsinghua University of China.

10.4.3 Case Study of RAND Corporation[9]

10.4.3.1 A Brief History of the RAND Corporation[10]

RAND Corporation was formally founded on May 14, 1948. It was both a post-WWII making and was an independent non-profit organization separated from Douglas Aircraft Company in California. In its 50 years of development, RAND assembled a group of natural scientists, engineers, economists and sociologists to conduct interdisciplinary cooperative study. RAND Corporation was thus known for its strict scientific studies on various social and economic issues from its empirical and non-partisan approaches. Before 1970s, it had made special

[9]Case study of Rand Corporation is also based on various annual reports of Rand Corporation.
[10]Source: Rand Corporation (2014).

contribution to American military and high-tech development. Besides, it also developed a series of theories and research methods, such as game theory, system analysis, etc.

(1) *The Origins of RAND*

During the World War II, academic scholar and scientists outside the military field applied successfully technological R&D to the need of battlefields, which significantly impacted on the outcome of the war. As the war drew to a close, the War Department, the American military establishment, the science R&D offices and relevant industries felt the need for a private research organization to connect military planning with research and development decisions.

In a report to Secretary of War, Commanding General of the Air Force Arnold wrote:

> During this War the army, air force and navy made unprecedented use of scientific and industrial resources. The conclusion is inescapable that we have not yet established the balance necessary to insure the continuance of teamwork among the military, other government agencies, industry, and the universities. Scientific planning must be years in advance of the actual research and development work.

Holders of the same view included Edward Bowles from Massachusetts Institute of Technology, General Lauris Norstad, Curtis Lemay, Donald Douglas (the President of Douglas Aircraft Company), Arthrur Ramond (the Chief Engineer of Douglas Aircraft Company) and their assistant Franklin Collbohm. The above-mentioned persons signed the contract to establish Project RAND with Douglas Aircraft Company on October 1, 1945. On March 2, 1946, Franklin Collbohm took charge for execution of Project RAND. Originally, Project RAND was designed to be executed at a separate place in Douglas aircraft plant.

In May 1946, Project RAND released its first research report, i.e. "Experimental World-Circling Spaceship", which in effect was a detailed engineering feasibility study on artificial satellites. This report illustrated the necessity of developing such a carrier, defined outer space as the future and suggested the air force consider space as its natural point of departure, since space provides an important base for sufficient reconnaissance, communication development and meteorological forecast. This report also pointed out that technologies for space launch and various space activities were already available. In 1947, Project RAND was separated from Douglas aircraft plant and set up its office at Santa Monica of California. It also attracted sociologists' participation in the research team in the same year. By early 1948, Project RAND had grown to 200 staff with expertise in a wide range of fields, including mathematics, logistics, economics, engineering, chemistry, psychology, and aerodynamics. In its second annual report, Project RAND stressed that "the complexity of the problems, and the rapid, if uneven, advances in the various fields call for coordination, balance, and cross-fertilization of effort. Coming from the laboratories of industry, the seminars of universities, and the offices of administration, the RAND staff is very conscious of this need for team work." Working modes of different professional research teams were reflected in RAND's history. RAND Corporation was formally founded on May 14, 1948. In February 1948, the command of American air force informed Douglas Aircraft Company in writing of its approval for RAND to be an independent non-profit company separated from Douglas Aircraft Company. At that time, RAND got an interest-free loan from Ford Foundation with the support of H. Rowan Gaither who served as a lawyer before and later became the chairman of council of the Foundation and also got loans from private banks with Ford Foundation as its guarantor, raising USD 1 million initial fund to start its business independently. In its institutional principle, RAND holds the goal of further improvement in science, education and charity for American public welfare and security. The first council was comprised of 11 persons, mostly leaders in science and engineering researches, including Franklin Collbohm (former principal of Project RAND), H. Rowan Gaither, Jr. (served as a lawyer before and later became chairman of Ford Foundation), L. J. Henderson Jr. (deputy head of RAND Corporation), Charles Dollard (President of Carnegie Corporation of New York), Lee A. Dubridge (President of California Institute of Technology), John A. Hutcheson (director of research laboratory of Westinghouse Electric Company), Alfred L. Loomis (scientist), Philip M. Morse (physicist from Massachusetts Institute of Technology), Frederick F. Stephan (professor of social statistics from Princeton University and director of the Office of Research and Statistics), George D. Stoddard (President of University of Illinois) and Clyde Williams (director of Battelle Memorial Institute).

Manning of the first board of directors determined RAND's research and development orientation as well as its achievements in quantitative analysis and systematic approach for researches on science policy and social economic studies. As early in 1947, Arthrur Ramond, the Chief Engineer of Douglas aircraft plant had remarked that "RAND is characterized by its research systems and method rather than specific devices, instruments or weapons. We not only focus on physical aspects of these systems but also pay close attention to human behavior in them". RAND's development in the succeeding years was an intermingled development process of natural science and social science.

In 1953, RAND was moved to a new office designed for easy face-to-face contact of experts in different disciplines.

RAND's major contributions to development of defense, science and technology as well as social economy in America in the next years include:

(2) *Contribution to American Intercontinental Missile*

Augenstein from Purdue University had made great contribution to the development of United States air force intercontinental missile. At the beginning of 1950s, atomic bombs were balky and rocket inaccurate. In Augenstein's opinion, with more accurate guidance, smaller rockets could be used to carry small warheads. Augenstein and his team made calculations on guidance accuracy, rocketry, reentry technology and strategic reconnaissance and proposed system balance and integration for feasibility and designs of different schemes. In 1954, he made a four-hour report on the convention held by United States Department of Defense with Mathematician Van Neuman as the President. Suggestions given by United States Department of Defense later were all based on the report of the Augenstein team. This report enabled the Department of Air Force to enlist development of intercontinental missiles into its main plans as early in 1955, to ensure that America would not lag behind other international competitors in space projects.

(3) *Mathematicians' Contribution and Computer Creation*

RAND's mathematics section assembled a large number of top mathematicians, such as Van Neuman (RAND's consultant) and Willis Ware from Princeton University in 1952. Van Neuman created the game theory and RAND applied this theory to nuclear confrontation under different scenarios. These mathematicians created the first computer and made contribution to development of peripheral equipment. Besides, they also laid the foundation for modern computer network by integration of communication technology.

(4) *Participation in Internet-Initiation Development*

Paul Baran, a young engineer worked for Hughes Aircraft Company before, joined RAND Corporation in 1959. He was strongly interested in improvement of "command and control" system of the military and devoted himself first to improving reliability of the communication network. He later shifted to application of digital network communication technology and developed the Packet Switching method. However, AT&T Company, a monopoly enterprise in America at that time, insisted on its analogue technology and rejected exploring digital network requested by the Air Force. By recommendation of RAND Corporation, the Air Force decided to carry out this work by itself. The U.S., Department of Defense started to intervene and defined the work to be within the scope of the 1969 ARPA Project. Later, this project gradually expanded from Internet for military use to universal Internet at present.

(5) *Development of System Analysis and Planning, Programming and Budgeting System (PPBS)*

RAND Corporation was the first to develop system analysis. In 1996, David Jardini, a historian from Carnegie Mellon University, commented that "system analysis is a social policy for planning and analyzing basic research methods in different fields, such as urban decline, poverty, health care, education, etc. It also plays a role in effective functioning of municipal departments, such as police strength composition and fire fighting, etc."

RAND Corporation also developed the planning, programming and budgeting system (PPBS), which was once used as the federal government budget standard in 1965 during Lyndon Johnson administration. At present, the system is still in use by South Korean defense departments.

RAND Corporation currently employs 800 experts from all walks of life. They are engaged in prospective researches, such as public obesity in America, review of the military response to Hurricane Katrina for future disaster reaction, research by some departments on evaluation indicators of the "No Child Left Behind" programme, and prevention and reduction of terrorist threats in regions with weak government control.

10.4.3.2 Top-level Organization and Funding Source of RAND Corporation

(1) *Top-level Organization*

(i) *Board of Trustees*

The Board of Trustees is the top-level management of RAND Corporation. The Board was comprised of 23 persons in 2010, including[11] James A. Thomson (former RAND President and CEO); four permanent members, i.e. former Secretary of the Navy, Secretary of the Air Force, a general and one officer in technology and procurement section of U.S. Department of Defense; another four members were former Congressmen, principals from the financial circle and presidents of universities of international affairs. From the enterprise circles, there were former vice president of General Electric and former President and CEO of Sony Pictures Family Entertainment Company. Most of the rest members were principals from the financial circle and some of them, such as Michael K. Powel, once worked as the director of Federal Communications Commission. In addition, there were two honorary members served by former U.S. Defense Secretaries.

(ii) *Administrative Management*

The 2011 RAND Annual Report had an entry of its leadership that falls into two categories. The first type is administrative management, that included Michael D. Rich, the fifth president and CEO of RAND Corporation replacing James A. Thomson in November 2011, other general secretaries, chief financial officers (CFO), and 11 heads of various departments, such as department of foreign affairs

---

[11]Note: James A. Thomson was a physicist before and worked for White House and Defense Department in affairs such as national defense and arms control. From 1989–2011, he acted as the fourth President and CEO of RAND Corporation.

research and analysis department and global research department. Michael D. Rich, the current President, is a doctor of law and comes from an avionics family. His father Ben Rich was an outstanding aviation engineer who took part in development of F-117A. After joining RAND, Michael D. Rich successively held the posts of researcher, member of the research group leaders and Vice President. He once assisted James A. Thomson in diversification of RAND in 1995, which brought changes to RAND's organization and operation. The second type is heads of eight research departments, such as RAND education, health care, labor and population, Project Air Force, etc. The third type is principals of RAND Europe institute, Cartel institute, etc.

(iii) *Advisory Committee*

In organizing interdisciplinary researches, in addition to its own researchers, RAND also established a series of advisory committees, such as graduate school advisory committee; advisory committee for practical networks for children, families and communities; Asia-Pacific center policy advisory committee (Zhuang Jianzhong from Shanghai Jiaotong University is a member); ethnics and governance advisory committee; global risk and security advisory committee and security center; Middle East public policy center advisory committee; Gulf region policy institute advisory committee; infrastructure, security and environment advisory committee; national defense research center advisory committee, etc.

(2) *Funding Source*

As for funding source, RAND Corporation relies more on contract research fees, which is different from the Brookings Institution. However, all top-level research institutes in America have income from donation and return of investment. As in RAND's 2011 annual report, its total income in 2010 fiscal year (September 2009–September 2010) was USD 250 million, with investment return of about 2.3%, general income of 4%, charities of 4 and 90% were from contracts and donations.

10.4.3.3 Organization Structure of RAND Corporation

(1) *Organization Chart (see Fig. 10.4)*

As shown in Fig. 10.4, organization of RAND Corporation is mainly composed by three parts: administration, research departments and graduate school.

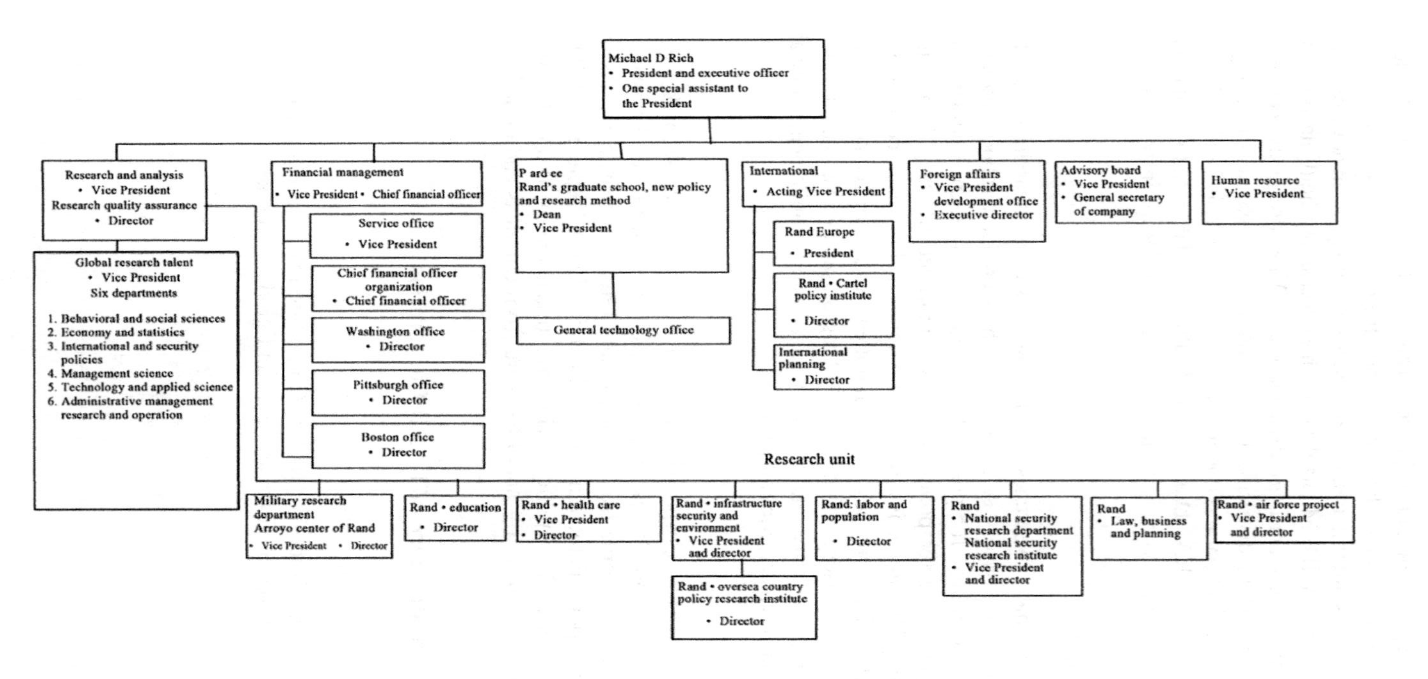

Fig. 10.4 Organization chart of RAND Corporation

(i) *Administration*

The President and Chief Executive Officer are responsible for administration and management of research, education, etc. As for administration, non-research departments include Financial Management Department, Human Resources Department, Advisory Board, Foreign Affairs and Development Office, etc. All think tanks exist in a particular political and economic society and their survival and development are dependent on their external connections. In RAND organization, special persons are assigned to contact with the Congress and media; besides, international divisions are specially established to conduct researches on Europe, Middle East (Cartel Policy Institute) and other regions.

(ii) *Research Departments*

RAND organization is of typical matrix structure. It consists of research and analysis service departments and research quality assurance departments under the leadership of Senior Vice President. The established Global Research Talent Department assembles talents in various professions, such as behavioral and social sciences, economy and statistics, international and security, management science, technology and applied science, supplementing the eight professional research departments, i.e. education, health care, labor and population, national defense security, etc. Three of eight research units of RAND are concerned with national defense, i.e. RAND Arroyo Center, National Defense Research Department and Project Air Force.

(iii) *Graduate School*

Pardee Graduate School of RAND Corporation was founded in 1970. It has a history of over 40 years and is mainly engaged in public policy research and confers doctor's degree (5 years) and master's degree. In 2010, the Graduate School had 110 students, coming from 26 countries and regions (including graduates from China) and different organizations, such as Israel's National Economy Commission, the Boeing Company, the Kennedy School of Harvard, etc. The Graduate School aims at cultivating a new generation of policy leaders able to lead and create local and global change. Such leaders should have distinct personality, broad vision and keen insight. RAND Graduate School puts up a prize for outstanding essays with financial support of various parties. Paper contents in 2010 involve a wide range of areas, such as decisive factors of agricultural productivity in China and India; can local control of schools expand learning opportunities, survey of New York City; fair sense of mileage-based road toll collected from users; sustaining vitality of symphonic orchestra; sustainable social security system and maintaining standard of living in aging society; long-time care of the aged; what can we learn from international experience; nuclear power generation; a sustainable option of Russia for reduction of greenhouse gas emission; understanding China's behavior in the South China Sea; What's on the menu, exploring the role of catering industry in obesity.

10.4.3.4 Examples of Research Topics in 2010

The research topics of Rand in 2010 laid particular emphasis on three aspects:

(1) Major policy issues in the U.S. in recent years. For example, drug legalization (California enacted a law to legalize the production and distribution of marijuana), ending of sex-oriented discrimination, medicare reform, fraud in financial crisis, study on improvement of company ethnics and governance and on education improvement (issues such as improvement of teaching quality and motivation mechanism which were proposed due to the provision that "to enhance children's reading and mathematical ability by 2014" stipulated in the "No Child Left Behind Act" issued by the U.S.).
(2) Study on the cost effectiveness for crime prevention and police force increase (Rand published an online calculator for this purpose), homeland terrorist problem and public security.
(3) Innovation research on new problems, such as financing of health care and welfare for the aged, and prevention of space debris danger for space flight, etc.

10.4.3.5 Composition of Rand Experts

The quality of researchers is critical to the growth and development of think tanks. Rand currently employs 1800 staff in total distributed research institutions in the U.S. and abroad. There are 735 experts specializing in public policy area. For number of experts in sub-areas of public policy, see Table 10.7.

Table 10.7 Number of experts in sub-areas of public policy in rand

Serial No.	Classification of public policy	Number of experts
1	Children and family	23
2	Education and art	24
3	Energy and environment	37
4	Public health and health policy	181
5	Infrastructure and transportation	23
6	International affairs	102
7	Law and commerce	49
8	National security	137
9	Aging of population	25
10	Public security	25
11	Science and technology	41
12	Terrorism study and homeland security	63

10.4.3.6 Unique Strengths of Rand and Its Public Policy Analysis

Since its establishment, the employees of Rand have been mainly technicians from military industries and part of its research work is to serve the needs of US national defense, which requires precise quantitative analysis supported by qualitative analysis. As a result, policy analysis becomes the unique strength of Rand. The policy analysis process develops from operational research to system analysis and then to policy analysis. Modeling is one feature of policy analysis. A model is formed by a group of mathematical formulae to reflect the correlation among main data. Application of mathematic model has become a universal tool for making policy studies in modern economics circle. However, the application of models and policy analysis by other institutions has not been as successful as that by Rand. One difficulty of modeling is in making correct long-term analysis. However, seeing from the predications in the *Long-Term Economic and military Trends 1950–2010* of 15 countries made by Wolf ([12]Charles Wolf Jr.) from Rand in a project concerning Rand's international economic policy planning in 1980s, the trend (not the absolute figure) is quite correct. As a case in point, this report also shows the necessity of applying quantitative tools correctly in social systems engineering. The report was published in April, 1989, attracting much attention from the domestic media and senior leaders of China. A brief summary of this report is given below. Due to the strict confidentiality rules of Rand concerning overseas distribution, this report was meant for the U.S. Department of Defense. The official release of the overall report by the Department of Defense was in effect earlier than that by Rand.

10.4.3.7 Report Summary of Long-Term Economic and Military Trends 1950–2010[13]

(1) *Background*

In October, 1986, a long-term strategy general committee was established jointly by US Defense Secretary and the special assistant to the President for national security affairs for the purpose of giving an overall consideration of the changing security environment, emerging technologies and national defense strategies under the condition of resource constraints in the next 20 years.

In order to accomplish the above assignment of the committee and the final report, four working teams are established to focus on future security environment, offence and defense forces the third world conflict and technology.

[12]Note: Charles Wolf, Jr. is present international economy chairman and senior economic adviser of Rand. He was also dean of Pardee graduate school of Rand during 1970–1997. He held a policy tutorial class in Tianjin, China invited by Institute of Industrial Economics of CASS in September, 1982.

[13]Source: Wolf, Hildebrandt, Kennedy, Henry, and Terasawa (1989).

The Rand report is mainly to provide the global economic and military trends during 1950–2010 for the use of the future security environment working team.

(2) *Report Structure*

The report consists of a summary and four parts: I Forward; II Gross National Product (GNP) trend (Different GNP estimates of the Soviet Union, the uncertainty of Chinese GNP estimates, Per Capita Gross National Product, significance of GNP); III Military spending trend (differences and comparison of US programs with Soviet programs); Military spending of other countries; IV Military asset stock. An appendix is included to explain the data sources and calculation methods.

(3) *Detailed Description of Part II, III and IV*

Parts II, III and IV of this report are quoted in the tables and figures. Figure 10.5a–c are achievements respectively about GNP trend, annual military spending trend and military asset stock. We can see from Fig. 10.5a that the achievement has made a correct forecast that China's economy growth would rank second in the world by 2010. However, there is a big gap between the forecast about the former Soviet Union in the report and the actual situation, because later collapse of the Soviet Union was not predicted in the report. We can see from this case that quantitative analysis has limitations once applied to the impossible prediction of contingencies in long-term projection.

10.4.4 Case Study of International Development Center of Japan[14]

10.4.4.1 Development of Japanese Think Tanks

Japan is a newcomer among the developed countries. However, Japan has strived to absorb experience from all developed countries and domesticate them during its development process. Before the World War II, Japan attached importance to the experience of Germany or Europe. It attached more importance to US experience after the war. The establishment and development of Japanese think tanks started relatively late among the developed countries. The Japanese National Institute for Research Advancement conducted in 1983 a study of several domestic research institutions under the research topic of *Trends of Research on Policies of the US, Western Europe and Japan*. The earliest think tank was the Research Institute of National Economy founded in December, 1945 which aims at investigating the economic conditions and policies of Japan and other countries to better contribute to world peace. Its main research content includes economic forecast, analysis of

---

[14]Source: Case study of International Development Center of Japan is written based upon various years of "IDCJ Report".

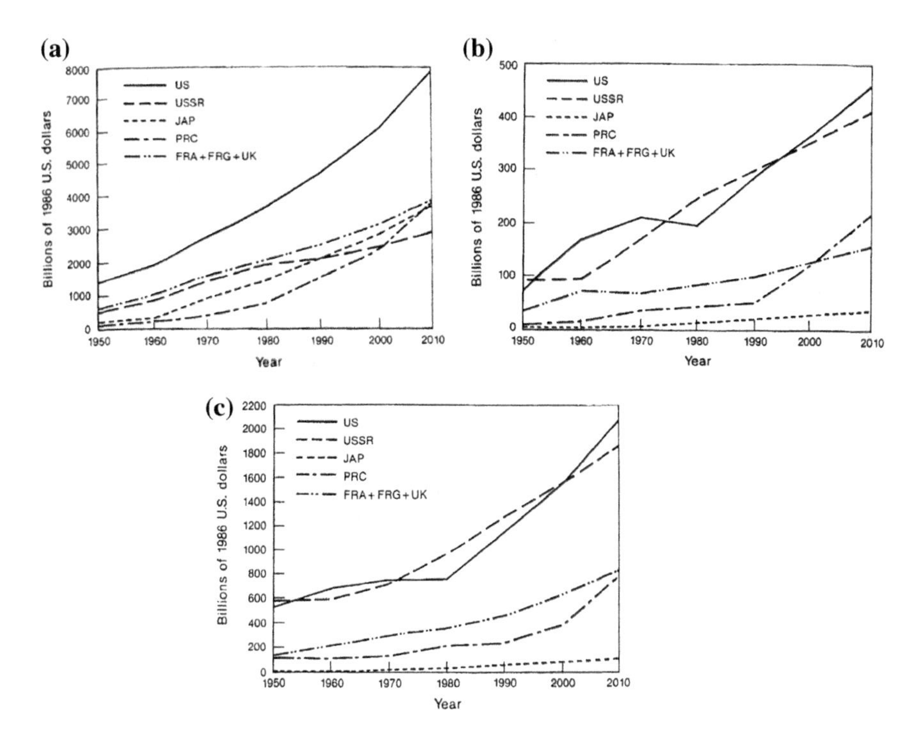

Fig. 10.5 **a** Gross national product: U.S., Soviet Union, Japan, China and three NATO countries (*Source* Wolf, 1989, p. 5). **b** Annual military expenditure: U.S., Soviet Union, Japan, China and three NATO countries (*Source* Wolf, 1989, p. 19). **c** Military asset stock: U.S., Soviet Union, Japan, China and three NATO countries (*Source* Wolf, 1989, p. 22).

industrial structure and policies, as well as regional economic analysis and business environment adaptation. Other think tanks were mostly founded during 1960s–1970s. The research system of Japan differs from other countries in terms of originality. For example, Japan set up such think tanks as Future Technology Research Institution, Policy Science Research Institution and Social Engineering Research Institution.[15] Social engineering was first proposed by Japan. Other developed countries may have advanced and mature system engineering research but not social systems engineering research. In Japan, even those research institutions attached to enterprises also conduct wide researches on the society, and economy. For example, Hitachi founded in 1910 established the Hitachi General Plan Research Institution in 1973. The purpose of this institution is to conduct independent and entrusted research on economy, society, environment and social science (including industrial research). It also carries out independent or entrusted

[15]Note: The Ministry of Education, Culture, Sports and Technology in Japan established a committee in 1994 for studying the future policy research and education in Japan, and in 1997, the Postgraduate Institute of National Policies of Japan was founded.

research and consultation on enterprise management. In 1980s, its research content included the "Enterprise Strategy in 21st Century as an International Enterprise".

10.4.4.2 The Case of International Development Center of Japan[16]

(1) *Overview*

The International Development Center of Japan was founded in 1971. It is a private non-profit think tank and also the first Japanese research institution engaged in development and cooperation. Its activities are mainly for three functions: first, as a training agency, it trains workers of development planning and economists. Second, it offers policy consulting and makes wide research on macro economy, development planning and programs. Third, as a research institution, it analyzes mainstream issues and forms conceptual frameworks for development strategies.

This research institution follows three principles when accomplishing activities for the three functions:

(i) *Cross-discipline/integration*

The Center often researches on development issues by way of cross-discipline/integration.

(ii) *Matching Theory with Practice*

The Center applies theoretical perspectives when analyzing practical problems and also attempts to enrich theories with truthful findings from practice.

(iii) *Remain Neutral*

The Center makes an effort to stay free of any influence of vested interest, government or private organizations so as to ensure neutrality of its research results.

(2) *Organization Structure (see Fig. 10.6)*

(3) *Top Level Organization and Funds*

(i) *The top level Organization consists of a Council and its Chairperson, the President and Executive Director*

In the 1990s, the Council had 41 distinguished persons from the political, financial, industrial and academic circles. The Chairperson at that time was Saburo Kawai who became the principal of the Asian Development Bank Institute at the turn of the century after leaving that Center. The Chairperson adviser then was Saburo Okita, the former Foreign Minister of Japan. Among the 41 members of the Council, four are outstanding scholars in Japan, e.g. Professor Shinichi Ichimura who served successively as professor of Kyoto University, Vice-Chancellor of

[16]Source: International Development Center of Japan (2010).

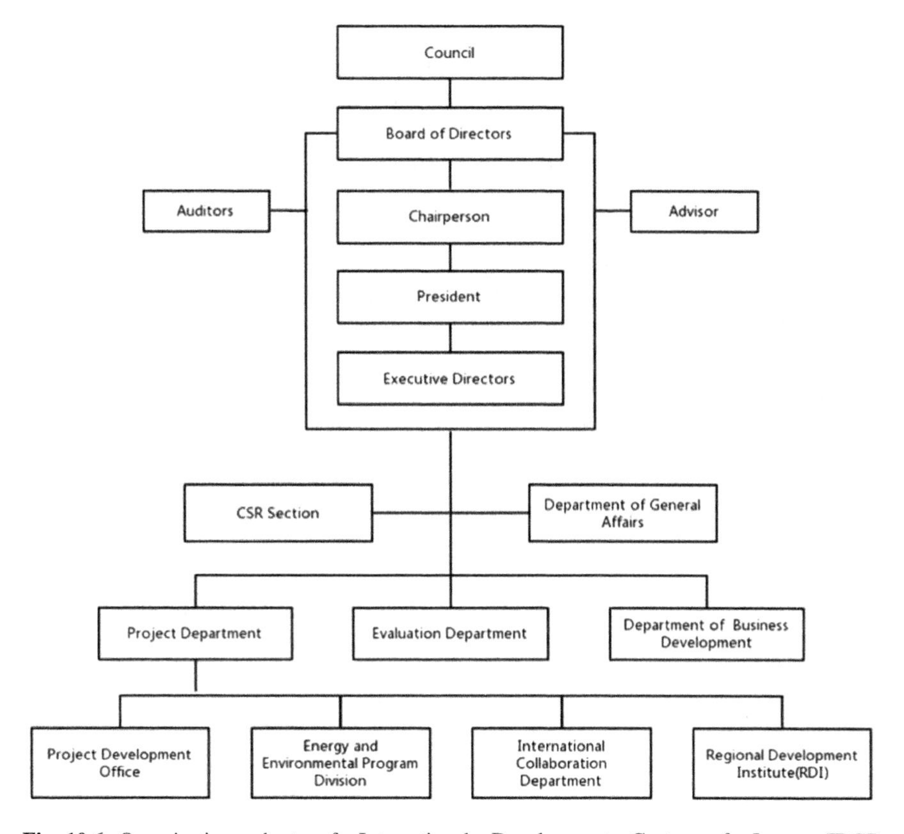

Fig. 10.6 Organization chart of International Development Center of Japan (IDCJ) (*Source* International Development Center of Japan 2010)

Nagoya International University and Director of Kitakyushu Institute. He is an internationally known economist with outstanding achievements in quantitative economics. Miyohei Shinohara is also another professor of economics well-known to the world.

The top level organization is supported by a large research advisory group that had 27 members in 1990s, 22 of them were professors or associate professors from universities in Japan. The other 5 members were foundation member, former government official, Japanese economist in the World Bank and investment adviser.

The Chairperson then Saburo Kawai was also the Executive Director concurrently. There were 8 Executive Directors in total. The President of the Council, Hideo Monden, was the chief executive.

(ii) *Funding Sources*

The funding sources of the Center are from two channels. One part is donations from associations, foundations and private sectors. Private sector donations take up 63% of the total. The other part is income from commissioned research, education and training activities of government departments concerned. The commissioned research income takes up 60% or so of the total income not from donation. The donation income is 8% more than non-donation one. By April 1, 2010, the research center held a fund of 1227 million Yen.

(4) *Specific Research Activities*

In 1990s, the Center had four main research offices: planning and research, development research, training and general affairs. There were 26 people in charge of administration in the four offices. Among them, 17 were famous senior researchers and 15 were general researchers. The research activities and organizational structure are adjustable to changes in the external environment.

Figure 10.6 is the latest organization structure of the Center. Compared with the organization in 1990s, CSR Section and Energy and Environment Program Division are newly added. The activities of the original three operation offices may have been delegated to the bottom level of organizations in Fig. 10.6. Main activities of the Center are divided into six aspects.

(i) *Research on Policy Making and Planning*

The scope of research on policy making and planning is extremely wide, which includes national and regional research, sector research, aid management and evaluation research as well as planning research. The national and regional research incorporates Japan's research on international relations and assistance, such as those projects in 1990s of bilateral technical cooperation between Japan and Singapore, and regional technical cooperation around Singapore. The Center had also recommended setting up economic policy data exchange networks in developing countries starting from Thailand and its neighboring countries. As Japan is a main foreign aid country in the world (Official Development Assistance), it also evaluates the water supply, maritime education, nursing education and fishing port construction under the joint assistance project to Fiji by Australia and Japan. As to planning research, it helped Sumatra Selatan of Indonesia in regional overall development research and formulated a 20-year regional development plan and sector development strategies for 300 investment projects.

(ii) *Training*

Training is an important activity of International Development Center of Japan for international cooperation. This activity is supported by the Ministry of Trade and Industry (MTI) and Economic Planning Department of the Japanese government, and by such international organizations as World Bank, UNIDO and so on. The primary training courses are practical, e.g. policy making and project management. The training program is mainly designed for policy planners and administrators of developing countries and also for Japanese workers involved with assistance.

(iii) *Promotion of Social Responsibility Enhancement*

Social responsibility of enterprise had become popular among many transnational corporations adopted as self-discipline rules in their business models during 1960s–

1970s. Nevertheless, following the social development conference held in Copenhagen, Denmark by the United Nations in 1995, the UN pushed forward a global convention in 2000, highlighting 10 principles such as human rights, labor, environment and anti-corruption. Some transnational companies also incorporated these principles into their social responsibilities. Japan reacted promptly to the UN calls. The social responsibility office was then upon added to the Center (see Fig. 10.6 CSR Section). The Center not only focuses on implementation of various projects in developing countries but also on promotion of enterprise social responsibility activities in these countries with its rich knowledge and experience in international cooperation. In addition, it also plays as a consultant and bridge between academic cooperation fields and private sectors at home and abroad.

(iv) *Theoretical Research*

The Center also conducts internally series of theoretical studies on international cooperation and some leading-edge theme explorations about international assistance. The research results are release on publications and website of the organization.

(v) *International Exchange*

The Center holds international forums and invites foreign experts on development issues to attend. It dispatches experts of the Center to the government departments of developing countries and cultivates overseas networks among relevant organizations for partnerships.

(5) *Recent Work Examples*[17]

(i) *Source of Projects*

The Center had 62 projects in 2010 from Japan International Cooperation Agency (JICA), Ministry of Foreign Affairs, Ministry of Trade and Industry (MTI), and from such international organizations as World Bank International Finance Corporation, European Bank of Reconstruction and Development (EBRD), the United Nations Development Programme (UNDP), Economic Research Institute for ASEAN and East Asia (ERIA) and so on.

(ii) *Example of Planning Project: Urban Development Master Plan Study for Vientiane Capital*

This project is undertaken by the International Development Center of Japan and other Japanese organizations such as Nippon Koei, Pacet and Oriental Consultants. The project lasted from January 7, 2010 to March 31, 2011.

Project content and background: Laos is a landlocked inland country. Its geographical constraints impede international economic activities in Laos. In recent years, it has improved the highway networks in the East-West and South-North

[17]Source: IDCJ Report 2010 its 60 projects, involving approximate 30 countries in Asia, Africa and South America.

Corridors, overcoming its disadvantages into advantages to become an important cross way for regional transportation from the former "isolate inland country".

Its capital Vientiane is expected to be the central hub of these economic corridors. With a current population of 700,000, Vientiane is also a destination for domestic and foreign investment. In addition, Vientiane also creates most employment opportunities for Laos. For the above reasons, the rapid urbanization of Vientiane will be one of the main issues of Laos. By 2030, the urban population proportion in Laos will take up to 29% of its total and the population in Vientiane will reach 1.4 million according to predication.

In 1991 and supported by UN Commission on Human Settlements, Vientiane once made an overall plan about future land use. However, the actual development activities in Vientiane now do not accord with the requirements in that plan, leading to a disorderly use of urban land. To address this situation, the Lao government invited the Japanese government to conduct a "Research on the Urban Development Master Plan of Vientiane, Capital of Laos".

The research targets are as follows:

To formulate the Urban Development Master Plan of Vientiane, Capital of Laos by 2030.
To explore effective measures to improve urban development.
To formulate basic strategies for infrastructural development for the city center and for urban design.
To form an implementation plan.
To form an implementation plan for developing planning capabilities.

(iii) *Training cases–Kyrgyzstan Republic-Japan Center (Phase II)–Business Administration Course*

This project lasted from June 6, 2008 to July 11, 2011.

Project background: the business courses about human resources development of "Kyrgyzstan Republic-Japan Center (KRJC)" have entered phase II for the purpose of promoting market economy orientation. The main target trainees are enterprise owners, mid-level managers and joint ventures (including candidates and local counselors directing entrepreneur management). This project was approved by Japan International Cooperation Agency (JICA). The project objective was to transfer techniques and know-how of business administration courses to local workers and instructors. The work in the third year included 16 topics of four courses and also 3-month training in KRJC.

(iv) *Case of Regional Cooperation in East Asia—industrial Development to Promote Economic Cooperation*

This project was supported by MTI of Japan. It lasted from February 18, 2011 to March 31, 2011. (The project might have been preparations for negotiations.)

Project background and conclusion: The International Development Center of Japan explored the influence of Comprehensive Economic Partnership in East Asia (ASEAN + 6, CEPEA) on Japanese domestic industries through the project. The research identified in detail the supply chains from Japan to ASEAN regions, China and South Korea and finally to India. It also calculated in detail the accumulative profitability based on the plan for countries of origin.

The target industries included automobile, electronics, chemical engineering and textile. The project not only conducted surveys of these relevant industries but also investigations of the main companies and associations (33 in total) of the above industries by way of hearing.

Research findings: The accumulative advantage of CEPEA will be greater if industries related to India such as automobile and electronics are taken into account. The research also confirms that there is equal accumulative profitability for developing automobile industry via Thailand and for developing electronics industry via Malaysia including when the supply chain from India is taken into account.

10.5 Development of International Think Tanks

10.5.1 Overview

The foreign think tanks introduced in Sect. 10.3 above are those founded from the beginning for the purpose of serving domestic economy and social development (except cases in Japan). With this given, against the backdrop of current day globalization of modern economy, these think tanks have to study the overall international trends to be integrated with domestic researches. In addition, they are also engaged in certain worldwide training services to expand their influence. Nevertheless, there are also some other types of think tanks seen in the development of foreign think tanks that do not serve specific countries. In this chapter, these types of think tanks are called international think tanks. They have not yet been listed in above-mentioned global think tanks sorting report system published by the University of Pennsylvania in the U.S. The authors think that they should be one type to be researched. They can be roughly classified into three classes.

One class includes organizations subordinate to the UN or their derivative organizations. For example, the UN is equipped with Economic and Social Council under which there are 9 functional commissions, including Commission on Science and Technology Development, Commission on Sustainable Development and Statistical Commission. This Council is also equipped with 5 regional commissions, e.g. Economic and Social Commission for Asia and the Pacific (ESCAP). The UN is also set with Programme and Funds under which there are 13 organizations such as United Nations Conference on Trade and Development (UNCTAD) and United Nations Environment Programme (UNEP). The UN is also set with 15 specialized agencies, including the Food and Agriculture Organization of the United Nations, International Labour Organization (ILO), United Nations Industrial Development Organization (UNIDO), United Nations Educational, Scientific and Cultural

Organization (UNESCO), International Monetary Fund (IMF) and the World Bank, etc. These organizations, besides undertaking certain administration, project investment and assistance to relevant countries, also boast a significant number of experts who not only conduct in situ researches but also organize experts across the world to conduct global, regional or national studies in all fields, establish various standards and publish all kinds of global annual reports or national reports. These research organizations have abundant expertise and impressive capacity to mobilize financial resources. Their research and promotion of achievements have significant impacts. Almost all global development agendas have been proposed by them after their synthesis. Such agendas include, for example, sustainable development, social development, millennium development goals and green growth.

The second class includes group or regional organizations. For instance, the Organization for Economic Cooperation and Development (OECD) composed of developed countries and African Economic Research Consortium. The former focuses on its members of developed countries but is also engaged in research of global issues as it has a long history and abundant human and financial resources. For example, it recently devoted itself to the initiative of "better life" as it feels that it is imperfect to take GDP as the single development goal. The African Economic Research Consortium mainly focuses on the development of African member countries.

The third type of international think tanks is world-wide organizations formed mainly by relying on modern communication and computer communication technology. For example, the "International Center for Economic Growth" and the "Global Development Network" which was founded by the World Bank in 1997 and later separated out as an independent organization. This chapter will make some introductions to the functions of OECD and the World Bank as think tanks. In particular, the research methodology of the World Bank can be a reference for the conventional think tank construction. Besides, it will also make brief case introductions to the International Center for Economic Growth and African Economic Research Consortium.

10.5.2 Case Study of the Organization for Economic Cooperation and Development (OECD)

10.5.2.1 Profile

The Organization for Economic Cooperation and Development (OECD), formerly the Organization for European Economic Cooperation (OEEC) founded in 1948, was set up to manage U.S. and Canadian assistance in the frame of the Marshall Plan for recovery of Europe after World War II. In 1949, OEEC set up its headquarters in Paris. After expiration of the Marshall Plan, OEEC switched to economic problems. In 1958, OEEC set up an subordinate as the European Atomic

Energy Agency and in 1960 it proposed the resolution for the founding of OECD to handle European and Atlantic problems and assist the less developed countries. The Organization for Economic Cooperation and Development (OECD) was officially founded in 1961 with the U.S. and Canada as its founding members, originally 20 in total. Later on, Japan, Finland, Australia and New Zealand joined in. At present OECD (2014) has 34 member countries in all.

10.5.2.2 Major Tasks

OECD provides solutions to common problems facing its member governments, shares experience and offers better policy alternatives to improve people's livelihood. In the past 50 years, OECD contributed in formulating global standards on governance, anti-corruption, corporate responsibility, development, international investment, taxation and environmental problems, and organized numerous international conferences to reach consensus and policy recommendations.

10.5.2.3 Organization and Fund Source

OECD has approximately 2500 workers. Its annual budget is about USD 510 million (EUR 342.9 million) financed by its member countries in proportion to their GNP. The assessed ratio of contributions by the U.S. is the highest followed by Japan (about 16%), Germany (about 9%) and Britain, France and other countries (about 7%) (Fig. 10.7).

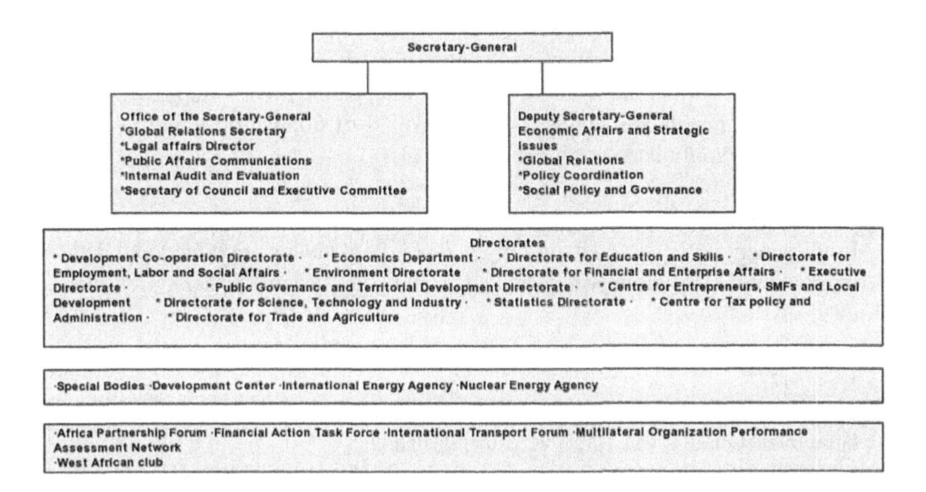

Fig. 10.7 Organizational chart of the Organization for Economic Cooperation and Development

10.5.2.4 Work Content Examples

(1) *Detailed descriptions of all work of each department are provided in the annual reports of OECD*

Box 10.1 extracts some highlights about the "Directorate for Employment, Labor and Social Affairs" from the 2014 annual report to illustrate its routine work content of this department.

Box 10.1 OECD Directorate for Employment, Labor and Social Affairs

(i) Presentation by Stefan Scarpetta, Deputy Secretary-General and Director of the Directorate.

"One of current difficulties and challenges is to offer income support to those families severely affected by crisis, or encourage or help the unemployed to find a job and take a positive attitude to explore the right balance between the two. In order to promote the recovery more intensively in an inclusive approach, policy makers need to advance the employment-friendly social policy integrated, thus compromisingly dealing with economic growth and promoting the relation between happiness to prevent the crisis."

(ii) Brief description of effectiveness of Employment Outlook 2013.

Give a brief description of effectiveness of the OECD publication of Employment Outlook 2013, explain the OECD Employment Outlook 2014 and introduce the OECD Global Migration Outlook 2013.

The function of OECD Directorate for Employment, Labour and Social Affairs is to lead the work of OECD concerning the employment, social policy, international migration and health. It works in the inter-related policy fields to assist member countries to promote employment, improve social welfare and healthcare against the background of population aging, globalization and rapid changes of technology.

(iii) Focus areas.

Employment policy and data, family and children, healthcare policy and data, international migration policy and data, pension system, social policy and data.

(iv) Key publications.

Aging and Employment Policy: France, the Netherlands, Norway; Cancer: Guarantee Quality and Improve Survival Rate;

OECD Employment Outlook 2014; 2013 Health Browse; OECD International Migration Outlook 2013; Focus on Mental Health Care 2014; Mental Health and Job: Switzerland and Britain; Pension Browse 2013; Immigration Choice: Austria, New Zealand and Norway; Review of OECD

Health Care Quality: Czech Republic, Denmark, Norway, Sweden, Turkey; Social Browse 2014.

(v) Key activities calendar.

- OECD Employment Seminar, Paris July, 2013;
- G8 London Summit December 11, 2013;
- OECD-EU "Dialogue between International Migration and Mobility" Brussels February 24–25, 2014;
- G20-EC-OECD "Quality Internship Conference on Giving a Better Starting Point for the Young in the Labor Market", Paris, April 9, 2014;
- Senior Policy Forum on International Migration, Paris, December 9–10, 2014.

(2) *Books, Reports, Statistics, Catalogues and References Published by OECD*

(i) Books. OECD publishes 300–500 books every year, mostly in English and French. Trademark publications include: *OECD Economic Outlook* published biannually about analysis and projections of economic conditions of member countries, *Major Economic Indicators* published monthly and *OECD Situation* published annually—also released online and includes hundreds of indicators about the economy, society and environment in its member countries and other countries, *OECD Communication Outlook and Internet Economic Outlook* that covers analysis and projections of the communication and information technology industry of its member countries.

(ii) OECD Observer Magazine.

(iii) Statistics. OECD has a Statistics Agency which accumulates a huge stock of statistical data of various countries

10.5.2.5 Major Activities at Present

Main trends of recent OECD activities are as follows:

(1) *"Go Horizontal"—A project to better reflect "interrelation policy"*

OECD pays much attention to coordinated research on correlation between different policy domains. Every year 2–4 projects are selected for cooperation coordination through "Go Horizontal".

(2) *New Approaches to Economic Challenges (NAEC) incorporates the agenda on sustainable growth development strategy*

NAEC has four features. First, it draws lessons from crises, identifies the parts needing to be adjusted in the OECD frame and checks the potential of new mainstream economic tools. Second, it acknowledges that economic growth is only the means and not the ends and that policy making should focus on the purpose of "happiness". Third, it improves understandings the complexity and interrelations of global economy and explores for better handling methods through compromised policies and coordination. Forth, it supports the governmental efforts in integrating reform with sustainable, green and inclusive growth, thereby increasing the regulatory and institutional capacities required and regaining peoples' trust in the government.

(3) *Inclusive Growth*

OECD launched the 'promoting inclusive growth' initiative in 2012 on the basis of its past efforts advocating happiness, income distribution and structural reform to boost growth. The initiative is one part of NAEC. The purpose of "promoting inclusive growth" is to better understand the distribution results of structural policies and to identify the optimal coordination and compromise between policy actions that stimulate growth and guarantee better distribution of welfare among social groups.

(i) Key publications—OECD Inclusive Growth Frame, May, 2014
(ii) Key activities—OECD/National Development and Reform Commission (NDRC) seminar: "China's Urbanization and Inclusive Growth", Beijing, March 24, 2014.

(4) *Development Strategy*

The development strategy is the one agreed at the 2012 OECD Ministerial Meeting but has been updated in light of the implementation experience in the past two years of the strategy so as to make it better adapted to OECD needs and reflect the actual global situation.

The strategic target is to give more attention to Policy Coherence Development (PCD). By working on global food safety, illegal financial flow and green growth through the PCD lens, we can see that policy coherence at different levels (local, national, regional, global) is an important system condition for removal of development impediments. Furthermore, wide invitation of stakeholders (government, private sector, etc.) to join in the acknowledgement of coordination between economic, environmental and social policy areas is also an important part to create conditions for development.

In terms of global development, OECD also offers multidimensional expert support to the UN-led "Development Agenda beyond 2015".

(5) *Knowledge-based Capital Project*

OECD proposed knowledge-based economy in 1990s, which once led the trend of global development thinking. At present, with global population aging, damping global financial crisis, weak labor market and fast-growing public debt, it is all the more urgent to explore growth drivers and approaches. Knowledge-based capital project is one of "Go Horizontal" Projects. The phase-I output of the project is the publication of Supporting Investment for Knowledge Capital, Growth and Innovation. The study in the above publication shows that many innovative enterprises not only invest in research and development, but also widely invest in intangible assets of "Knowledge-Based Capital" (KBC), e.g. data, software, design, business process and firm-specific skills training.

A lot of challenging problems of future work have been detected in the first phase. Phase II studies will focus on three issues: intellectual property problem, data and data analysis, and economic competitiveness (e.g. competence in capital mobilization and enterprise management). Phase II work is mainly carried out by the Directorate for Science, Technology and Industry.

(6) *Green Growth Strategy*

OECD has determined at the 2009 Ministerial Meeting that "green" and "growth" must be simultaneous. The work that OECD has done in the past include: analyzing the distribution impact of green growth, the effect of inclusiveness on labor and skills, relationship between environmental policy and economic output, optimizing green value chain analysis and designing green investment incentives, incentive mechanisms for energy system conversion, policy and regulatory requirements for carbon pricing, consulting services for green growth and innovative agricultural work.

In future, the green growth work will focus on development of green growth indicators, including indicators for various sectors or strata enable the government to supervise the conversion progress. The Towards Green Growth: Monitoring Progress OECD Indicators published in 2011 will be released in the second quarter of 2014 after updating.

A comprehensive report of green growth will be published in 2015 based on documents named Towards Green Growth and Tools for Delivering on Green Growth published in 2011 and What Do We Learn from Green Growth? published by OECD in 2013. This Report will evaluate the green growth policies executed in the past four years and the government tracking of green growth policy implementation.

(7) *Innovation Strategy*

After OECD published the innovation strategy in 2010, innovation has become a component of comprehensive organizational strategy and horizontal projects. Green growth mainly relies on special policy design and incentives to stimulate green innovation. OECD once suggested applying innovation in OECD countries (the Netherlands and Sweden) and non-OECD countries (Columbia and Vietnam). In the future, it needs to deepen understanding of innovation, for example to invest in knowledge-based assets including software, design, data, research and development and organizational capital. These investments in some OECD countries have exceeded physical capital investments.

The future work also includes improved approaches in measuring and evaluating the influence of innovative investment, research and development tax preference and in-depth study of innovation policies in such countries as France, Luxembourg, Malaysia and Costa Rica. For New Approaches to Economic Challenges (NAEC) projects, it needs to further analyze innovation prospect to better understand the long-term determinants of productivity growth and improve innovation measurement.

(8) *Skills Strategy*

OECD skills strategy is based on the strategic point of view offered for skills policy at the 2012 Ministerial Meeting, namely "Better skills, better jobs, better lives". The objective of OECD skills strategy is to realize improved application of skills by means of coherence development activities so as to achieve a skills system that promotes economic prosperity and social cohesion, which reflects the characteristics of "life employment". In the modern society in which knowledge capital is important, young people and adults must invest in education and development of skills. On the other hand, it is required to change the situation where social policies and labor markets hinder the young people to contribute skills to the society.

OECD skills strategy proceeds from the approach of whole-of-government and across-department and introduces all stakeholders including employers and trade unions to obtain better economic and social results.

Now the skills strategy has reached the national level and entered international cooperation among countries to bring more effectiveness to the strategy. In addition, OECD also undertook comparative evaluation of changes in skills demand manifested in various countries. In October 2013, OECD publishes the first *OECD Skills Outlook* and provides Programme for the International Assessment of Adult Competencies (PIAAC). The *OECD Skills Outlook 2014* will analyze the skills and employment pattern of young people and the development, action and effective use of their skills by applying the skills strategy frame, so as to improve the output of youth.

Skills strategy is implemented by the "Directorate for Education and Skills" and "Directorate for Employment, Labour and Social Affairs" jointly in closely cooperation with the Development Center, Center for Tax policy and Administration and other departments concerned.

(9) *Gender Action*

The purpose of OECD "Gender Action" is to strengthen gender equality in education, employment and entrepreneur (3E) which are the three key dimensions of economic and social opportunities. Gender equality is the empowerment about fairness, equality and economy. In dealing with the gender gap, multi-dimensional factors such as politics, education, social-economics and culture need to be considered in tandem. The Gender Action shall be implemented horizontally by the whole government, including by the nine directorates and their committees of the Center.

Research results of OECD Gender Action are not only provided to the countries but also to international organizations, e.g. European Community (EC) and Group of Twenty (G20). OECD has also developed all kinds of statistics concerning gender issues and research, for example "OECD gender data hub" that collected the gender inequality existed in education, employment and entrepreneur and published updated data in the 2014 International Women. It also added a special chapter in OECD publication *Economic Survey* and will add a new chapter on women entrepreneur's data in the list of Missing Entrepreneurs to be published.

(10) *OECD "Better Life Action"*

In the past decades, gross domestic product (GDP) has always served as the indicator for international organizations and countries in the world to measure progress of development. However, economic growth indicators along cannot reflect many other factors affecting people's livelihood. In the past ten years, some states and organizations have been exploring different methods to measure social progress. Pushed by OECD and some other international organizations, an exploration beyond GDP was commenced. In 2011, OECD started the "OECD Better Life Action" and introduced the OECD frame to analyze happiness and social progress.

OECD published a report named *How's Life?* for the first time in 2011. This report makes an international comparison of life happiness based on 11 dimensions including people's material conditions and living quality. Second edition of the above report was published in 2013. This project and the publication have attracted attention from 180 countries and 3.6 million users across the world. Preparation is being made to publish German, Russian and Spanish editions. The next work is to study the connection between happiness measurement and policy making.

For happiness measurement frame of OECD, see Fig. 10.8.

10.5.3 Case Study of the World Bank and Its Activity in China

10.5.3.1 Overview

The World Bank Group consists of five organizations: the International Bank for Reconstruction and Development (IBRD, established in 1945), the International Finance Corporation (IFC, established in 1960), the International Development Association (IDA, established in 1960), the International Center for Settlement of Investment Disputes (ICSID, established in 1965) and the Multilateral Investment Guarantee Agency (MIGA, established in 1988). The World Bank generally refers to the International Bank for Reconstruction and Development and the International Development Association (IBRD + IDA). The World Bank Group, however, is a family of the above five organizations. In addition, the World Bank Institute provides study and ability cultivation services to member countries of the World Bank. The World Bank Group deems its mission to promote economic development and eliminate poverty. Made up of 188 member countries, it now has around 10,000

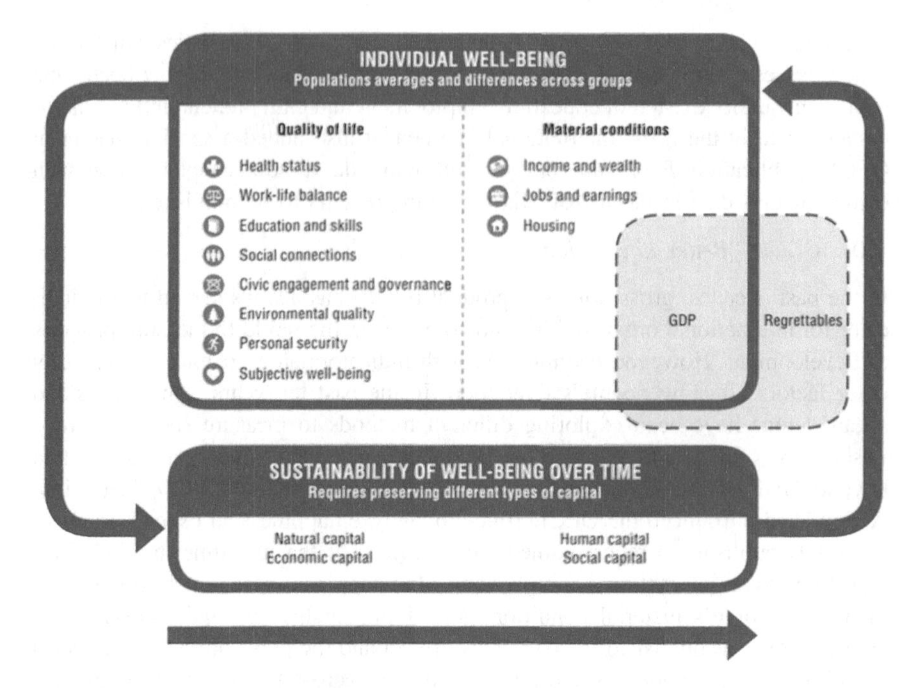

Fig. 10.8 Happiness measurement frame of OECD (*Source* OECD 2013. How's Life? Measuring Well-being)

employees (including the employees working at the representative offices of the member countries).

With massive human, financial and information resources and abundant experiences in providing loans or grants and development assistance to developing countries for their projects, the World Bank can be regarded as an international think tank to a large extent. It carried out an all-round investigation at the end of 1970s and published the first country report on China in 1981. In 1981, Shanghai was granted the first World Bank loan for its university development project. For this reason, the World Bank provided consultancy advices to the central and local governments as a think tank. This section will describe the major activities of the World Bank in China and its publications, so as to provide insight into the building and daily operation of a think tank.

10.5.3.2 Major Activities of the World Bank in China and Its Publications

(1) *General Study on China*

The general study of the World Bank on China can be divided into three stages: first, the beginning of China's reform and opening-up. At this stage, a 17-person economic delegation of the World Bank (consisting of 9 persons from the general departments and four sub-teams in agriculture, energy, industry and communications) carried out a two-month survey in China from October to December 1980. In addition to the ministries and committees, the delegation also visited Shanghai and Tianjin, two directly administered cities by the central government. In June 1981, the delegation presented its main report entitled China—Socialist Economic Development and eight annexes, which were Statistics System and Basic Data; Population, Healthcare and Nutrition; Agricultural Development; Challenges and Achievements in Industry; The Energy Sector; The Transport Sector; Education; and External Trade and Finance.

At the second stage in the beginning of 1980s, China set the development goal of quadrupling the total industrial and agricultural output on the basis of 1980 and the income per capita reaching US$800 by 2000. Following the meeting between the former president of the World Bank and the Chinese leaders in 1983, a 31-person economic delegation (including 12 delegates from general departments and 19 specialized in agriculture, energy sources, industrial technology and layout and trade) visited China twice in the beginning of 1984. Apart from exchanges with the ministries and committees and related departments, the delegation visited three provinces. Based on their survey, the delegation prepared a main report entitled China—Long-term Development Issues and Options in May 1985, which included six annexes: Issues and Prospects in Education; Agriculture to the Year 2000; The Energy Sector; Economic Model and Projections; Economic Structure in International Perspective and Transport Sector. In addition, they provided nine economic backgrounders of China.

At that stage, the Development Research Center of the State Council was engaged in studies for the report China 2000. This study coincided with the World Bank project in subject and time span, which prompted the World Bank and the Development Research Center into extensive exchanges. From then on, the Development Research Center established a long-term cooperation with the World Bank.

At the third stage from the end of 20th century to the 21st century, the World Bank performed two long-term studies on China. The first was the "China 2020" research carried out by the World Bank in the end of the 20th century. Seven reports were published, namely, China 2020; Clear Water, Blue Skies: China's Environment in the New Century; At China's Table: Food Safety Options; Financing Health Care: Issues and Options for China; Sharing Rising Incomes: Disparities in China; Old Age Security: Pension Reform in China; and China Engaged—Integration with the Global Economy. On the occasion of the 30th anniversary in celebration of the cooperation between China and the World Bank in 2010, the former President Zoellick suggested cooperation between the Development Research Center of the State Council and the World Bank in the research of China 2030. In September 2011, the research report of China 2030 was prepared, which consisted of one main report and four annexes: China's Growth through Technological Convergence and Innovation; Seizing the Opportunity of

Green Development in China; Equality of Opportunity and Basic Security of All; and Reaching "Win-Win" Solutions with the Rest of the World.

(2) *Current Economic Memorandum*

(i) *Recent Economic Trends and Policy Development dated March 31, 1983*

While the first survey report of the World Bank on China dated June 1981 represents the most complete survey on China in thirty years since its foundation, this economic memorandum focuses mainly on the development in Chinese economy since its launch of the reform and opening up in 1978. It analyzes Chinese economy in line with the guiding concept of "adjustment, reform, consolidation and improvement" as proposed in the Sixth Five-Year Plan. It utilizes the RMSM model of the World Bank for the scenario analysis of the possible situation in Chinese economy between 1978 and 1990.

(ii) *Macroeconomic Stability and Industrial Growth under Decentralized Socialism dated June 12, 1989*

(iii) *Between Plan and Market dated May 8, 1990*

These two economic memoranda were prepared at the end of 1980s when China implemented temporarily the policy of local governments taking full responsibility for their finances and price reform and when China was exploring the demarcation between plan and market. The two documents analyzed and made recommendations in respect of the above issues.

(iv) *Reform and the Role of the Plan in 1990s dated June 19, 1992*

(v) *Managing Rapid Growth and Transition dated June 30, 1993*

(vi) *Internal Market Development and Regulation dated March 1993*

These three economic memoranda, prepared in early 1990s, provided further description of the role of plan and the delineation of market as well as comments on development and perfection of internal market and its regulation. *Reform and the Role of the Plan in 1990s* stresses that the plan should play its role in three key aspects: first, to set up a macroeconomic frame and formulate corresponding policies to guide actions in other areas; second, to redress through plan failures in the market, which were serious issues in the Chinese economy under partial reform; third, to make plans to eliminate bottlenecks in state-owned sectors. The document reviewed the reform of China in the light of the Eighth Five-Year Plan for National Economic and Social Development and the Outline of the Ten-Year Program. It provided three scenario analyses of the possible situation of Chinese economy between 1989 and 2000.

Managing Rapid Growth and Transition is an update of the economic memorandum dated June 19, 1992. In 1992, rapid growth of Chinese economy aroused the concerns about inflation. Meanwhile, the Chinese government proposed the socialist market economy as the objective pattern of the reform of Chinese economic system. This memorandum consists of four parts: latest economic development, economic reform after the 14th Party Congress, priority areas in trade reform and scenarios for China's economy.

Internal Market Development and Regulation dated March 1993 points out the obstacles encountered during the transition from a centrally planned economy established thirty years before the reform and opening up to the market economy proposed after the reform and opening up. Institutional barriers included the rigid mechanism of planned distribution of resources among central departments that came into being before the reform and opening up and the lack of understanding of how market economy develops. When the 14th Party Congress of October 1992 adopted the decision to "develop socialist market economy", the World Bank started preparations for this update of economic memorandum in November 1992. They completed their survey in March 1993 with the support of material departments and industry and commerce administration departments. In the same year, the initial report was submitted to the Chinese government. In December 1994, the World Bank published this memorandum in the form of red book.

This memorandum consists of market segmentation and market development, the latest developments in China's internal markets, the distribution system, new forms of markets, market regulation and markets and government, and future perspectives. Though proposed in 1993, many of its recommendations are still valid after 20 years. For instance in its last chapter "Future Role of the Government", the memorandum observed that if direct intervention is to be limited, indirect intervention can only be credible if it can be successfully enforced. In this regard there are many instances of ad hoc application of regulations, with wide discretionary margins at all levels of authority. If this situation persists, it will undermine the plausibility of 'market regulation', and lead instead to anarchy. The many tax exemption or tax free havens, the numerous arbitrary fines, irregular fees and levies on interstate transport and freight and so on are all cases in point. The issue of enforcement of law and regulations must be confronted when China is transforming into a market economy under indirect regulation.

(vii) *Weathering the Storm and Learning the Lesson dated January 20, 1999*

This memorandum was published in the aftermath of the Asian financial crises when the global economy was slowing down and when the vulnerability and risks of China's economy started to change. It summarized the reasons why China could weather through the Asian financial crisis on the one hand, and at the same time, highlighted the need to implement social agendas, increase employment opportunities, attach importance to service industries and informal sectors to absorb labor, establish social security system and transform the relationship between banks and enterprises.

(viii) *Promoting Growth with Equity dated September 15, 2003*

This memorandum was prepared in the context of rising Gini coefficient despite rapid economic growth after the reform and opening up. China's Gini coefficient rose from 0.29 in 1981 to 0.44 in 2000. Urban-rural gap in income difference dropped from 54% in 1983 to 35% in 2000. The paper proposed four key measures including promoting productive occupations in underdeveloped regions. In rural areas, it proposed the policy recommendations to increase rural labor income and revenue from land.

(3) *Other Activities*

The World Bank had carried out surveys in many sectors of China's economy during the reform and opening up, such as investment, investment and finance, tax policies and financial system, budgetary policies and their correlations, health system, reform of housing system, poverty elimination, reform of external trade system and perfection of internal market. They also carried out special surveys in Jiangsu, Zhejiang and other provinces. Results of these special surveys and researches were later distilled into red books that were published by the World Bank to share experience with other countries in the world.

10.5.3.3 Research Methods of the World Bank and Quality Assurance of finished Reports

(i) *Research Methods of the World Bank*

The research methods of the World Bank have two characteristics.

Integration of expert leadership into research organization: for every subject, one or two experts are assigned to lead the subject team. As for members of the subject team, external experts are also invited to join World Bank staff. Take Nicholas Lardy and Dwight Perkins for example. Neither of them was a staff-member of the World Bank but they are experts in China study. For example, Nicholas Lardy is familiar with the external trade of China. For that he was invited to a relevant subject team. In addition, the subject teams also invite relevant experts and administrative officials of the country under survey to join in their studies.

Full preparations: the World Bank provides support to the economic and social development projects of countries all over the world. Before field survey to a country, experts in related fields (can be non-staff members of the World Bank) are invited for preparation of background reports. Full preparation is then made for each subject in advance.

(2) *Quality Assurance of Finished Reports*

For all reports of the World Bank, different cover colors are used to indicate their status. Generally, grey cover indicates official reports submitted to the local government or other authorities. Prior to grey, reports with white, green or light yellow covers indicate non-official intermediate reports for discussion and modification. After a period of time, reports with non-confidential contents and available for public release are bound with red covers. For example, for the study of the World Bank on the external trade of China, a report with white cover was published in 1992. Titled Trade Polices, the report consisted of six chapters and four annexes.

The same report in grey cover submitted to the Chinese government in 1993 consisted of seven chapters seven annexes. In 1994, the report was published with a red cover and renamed China: Foreign Trade Reform, basically the same in content as the grey covered dated 1993.

10.5.3.4 Seminars and Training Courses

(1) *Seminars*

The World Bank organizes all kinds of seminars. One type of seminar is organized to discuss the research findings of a certain subject, invite extensive comments to further modify and perfect the conclusions. Take the research report of the World Bank entitled "Grain Sector" for example. After a seminar in China, the report was bound in red cover and officially published by the World Bank. Another type of seminar is organized for the purpose of propagation. For example, following the report *China 2030: Building a Modern, Harmonious and High-income Society*, a seminar was held after its completion to extend its influence. Still another type of seminar is organized for the purpose of discussion, exchange and training. For example, the Economic Development Institute of the World Bank held in Bombay, India in 1987 the Asia-Africa Seminar on Technical Policies, Finance and Project Planning that invited relevant official and research institutes of Asian and African countries. Lectures and discussions were arranged at the seminar for the attendants to exchange experiences and improve their abilities.

(2) *World Bank Institute (WBI), formerly known as the Economic Development Institute (EDI), is established under the World Bank*

According to its official website, the World Bank Institute is "a global connector of knowledge, learning and innovation for poverty reduction." It deems training its main mission. It used to provide all kinds of trainings and prepare practical training materials. For macro-economy training course, there is matching material of "National Economic Management"; for micro-economy training course, there is "Feasibility Study of Project" ready for use. In 1985, WBI provided a training course named "Seminar on Senior Policies: Development Policy and Management" to senior Chinese officials. In 2014, it provided a training course themed on "Strengthening Health System and Sustainable Financing" in the hope to popularize health system in the world.

10.5.3.5 Other World Bank Publications and Researches

Apart from daily operations, World Bank also conducts extensive research work, guided by the concept that "researches are the backbone of loans". Their main publications include world development reports, policy study reports, policy study literature, journals of the World Bank, and related books.

World development reports, with a history of 40 years since 1978, are the main reports published by the World Bank. Chinese scholars once prepared *Review of the Development Reports of the World Bank in 20 Years (1978–1997).* Annual development reports of the World Bank are themed on the major development issues of the year. In 40 years, world development reports, with their themes covering the major aspects of modern development, have become authoritative works on economics in dynamic development. They have an important and practical significance to the development and reform of countries all over the world.

Both the World Bank and the Organization for Economic Cooperation and Development (OECD) began at the end of the 20th century to focus on the advancement of information technology that stimulated worldwide social development and ushered human society into the era of knowledge economy development. Both organizations have done extensive researches in this respect. Besides, the World Bank assigned its departments to prepared logistic performance index and other indicators to accommodate the development of logistics.

10.5.4 International Center for Economic Growth (ICEG)

10.5.4.1 Profile and Characteristics

The International Center for Economic Growth (ICEG), founded in 1985 in San Francisco, U.S. is a think tank that connects policy study institutes all over the world into a network via communication. It is dedicated to provide market solutions for economic reform. In 2008, its Board of Directors consisted of 8 directors, including William K. Kendall, the Chairman, and Nicolas Ardito-Barletta, former President of Panama, former Vice Executive President of the Bank of America, renowned professors in economics in the U.S. and President of the Graduate School of Business Management, Zurich, Switzerland. In addition, ICEG has five persons in charge of administration, including Aldlatif Al-Hamad, Chairman of the Arab Fund for Economic and Social Development and founder and Chairman of Daewoo, South Korea, in addition to the former President of Panama.

10.5.4.2 Nature of ICEG

In light of the global trends of democracy and market economy, domestic institutes of all countries begin to study public education, prepare advice to policy makers, design reform strategies and other activities regarding economic transformation. ICEG was founded by the Institute for Contemporary Studies in 1985, with its objective to help local organizations, especially those in developing countries and regions, to apply the best economic and social studies to their reforms.

For this purpose, ICEG sets up a network consisting of policy study and academic organizations that covers more than 100 countries and regions in the world. It sponsors and funds policy study and publication projects, conferences, training workshops and a series of special projects. In these activities, organizations based in different countries cooperate with each other to solve common issues. ICEG, cooperating closely with American, European, and Japanese research and education centers in many projects, serves as the hub for information exchange.

Member communication organizations participating in ICEG projects distribute publications of ICEG to the local research institutes and policy audiences.

ICEG supports practical experience based researches and focuses on economic policies, economic growth and human development. Research findings are disseminated via conferences, seminars, publications and quarterly news releases in four languages.

ICEG develops cooperation with all major regions in the post-socialist world and is engaged in operational activities in Latin America, Asia, Central Europe, newly-independent countries, near East, and Africa.

ICEG is funded by foundations, companies, and individual and government donations. Take the year 1993 for example, its operational budget was US$3.5 million. ICEG, based in Panama City, Panama, is affiliated to the Institute of Contemporary Studies. It sets up an administrative office in San Francisco, California, and regional offices in Nairobi, Kenya, San Diego, Chili, and Washington D.C., the U.S.

10.5.4.3 Network of Member Communication Organizations

The international network of ICEG consisting of member communication organizations in more than 100 countries throughout the world is the blood for ICEG projects to maintain vitality. The network includes independent non-governmental organizations, research organizations cooperating with ICEG in policy studies, and special reform policy projects. It generally cooperates with organizations in its network. They jointly hold local seminars and conferences and help distribute and disseminate ICEG researches and publications.

Member communication organizations promote communication with local policy makers, news media and magazines, the academic circle, and other policy audiences such as leaders in the commerce circle, labor circle and religious circle. ICEG is also responsible for disseminating the activities of its member communication organizations throughout the world, providing support to member communication organizations, and encouraging researches and information exchange.

ICEG provides support to its member communication organizations in three levels:

(1) General support: all member organizations will receive ICEG publications and be invited to regional conferences;
(2) Indirect support: ICEG selects some organizations to enhance their capacities in policy studies, initiate discussions on reform policies and inform the public and other leaders of their opinions. Such projects include policy study projects and

subsidies in publications, technical support for exchanges, initiation of seminars, stipends for researchers and equipment donation.

(3) Special reform policy projects: ICEG provides supports to a selected few organizations, which have special opportunities to provide advice to senior government leaders and influence decision makers.

10.5.4.4 Special Activities of ICEG

ICEG hosts and sponsors some special activities to strengthen its member communication organizations.

(1) Regional conferences: a forum for discussion of international and regional issues, networking and information exchange.
(2) Cooperative policy study and publication project: ICEG provides funds to single or multiple organizations for their transnational studies. It then distributes the findings in the region or worldwide by means of joint publication of books or monographs.
(3) Special reform policy projects: ICEG assists organizations that provide unusual advice to governments. The funded projects have three major objectives: providing policy recommendations, inspiring public opinion in favor of a new policy, and helping to form national agenda for reform. Activities of ICEG in this regard include organization of special study and education projects, preparation of articles on the policies, and even selection of experts from other countries to cooperate with the policy advisory organizations of local governments.
(4) Initiation of seminars: seminars are organized by member communication organizations of different countries to enable policy audiences and the media to meet editors and authors of main ICEG research publications. Seminars focus on trends of worldwide reform policies and help local organizations to have contact with important leaders and the media.
(5) Technical support to communications: it includes regional and local seminars and trainings on exchange strategies and techniques.
(6) Equipment donation: ICEG provides monetary support to equipment procurement, mainly for computers and fax machines required by the member communication organizations.
(7) Stipends for research writers: ICEG funds those who study in some important universities or research centers and write research articles or books.
(8) [18]Syndication service: ICEG provides articles out of print in ICEG publications and those of its member communication organizations to newspapers and is engaged in syndication service throughout the world.

[18]Note: Syndication refers to alliance and combination of enterprises.

10.5.4.5 Publications

ICEG publishes four categories of study findings, most of which are from cross-national comparative study projects.

(1) Sector studies: research in a specific policy field based on one country or multi-country comparison, topics include international trade and economic restructuring, capital market, public finance, technology transfer, agriculture and rural development, environment and effective functions of market and institutional reform.
(2) Country studies: they provide evaluation on the extensive effects of the macro and micro-economic policies of specific developed and developing countries.
(3) Non-regular theses: senior scholars or policy makers provide general summaries of the latest knowledge development or main experience and lessons from policy application.
(4) Studies on human development and social welfare: they evaluate the correlation between policy and growth and human development as well as the effectiveness of poverty reduction projects.

ICEG disseminates the above four categories of studies, including studies of its member communication organizations, to the target audiences by means of different design formats. Its publications include abstracts, working articles; translations (Spanish, French, Chinese, Arabic, Russian and Portuguese versions of ICEG publications and English versions of the publications of many member communication organizations), and quarterly news bulletins (in English, French, Spanish and Portuguese).

10.5.4.6 Academic Advisory Committee

The academic advisory committee of ICEG has around 30 members. They are renowned figures from universities, institutes, governments and the banking industry of different countries.

10.5.5 [19]African Economic Research Consortium (AERC)

10.5.5.1 Overview

African Economic Research Consortium (AERC), founded in 1988, is a non-profit public organization dedicated to economic policy study and training. It consists of

[19]Source: African Economic Research Consortium (1997–1981) Annual Report. African Economic Research Consortium (2012) Annual Report.

networked member organizations. Its two major objectives include enhancing the capacity of locally based researchers to conduct economic policy inquiry and application and setting up training programs for masters and doctors in economics to assist schools of economics of public universities in Africa to improve their levels.

AERC believes realization of these two objectives can help Africa to establish sound economic management system and realize sustainable development.

10.5.5.2 Top-level Organization and Funds

(1) *Board of Directors*

The Board of Directors is elected by the member institutes. This Board of Directors consists of 20 directors. Nthuli Ncube, Vice President and Chief Economist of African Development Bank, Tunis, serves as the Chair of the Board and Christopher Adam, Lecturer of Department of Economics, Oxford University, UK, representing the Department for International Development (DFID), as the Vice Chair of the Board. Of the rest 18 directors, 13 are from the U.S., the Netherlands, Sweden, Norway and other developed countries, including Ken Prewilt, Professor of the Columbia University, U.S., Rob de Vos, Ambassador of the Netherlands in South Africa, Shanta Devarajan, CGE Model expert representing the World Bank, and James Nyero, representing Rockefeller Foundation. The other 5 directors are local figures from the upper circles, including Nil Sowa, Director General of Securities and Exchange Commission, Ghana, and Jacques Hiey Pegatienan, Economist and Counsel from Coate d'Ivoire.

(2) *Funders*

AERC is supported by its donor funders, including member and non-member ones. In the late 1990s, there were 11 member funders, including Denmark, Ministry of Foreign Affairs of the Netherlands, Rockefeller Foundation, Ford Foundation, the United States Agency for International Development (USAID), international development departments of the United Kingdom, Canada, Switzerland and Sweden, and the World Bank. There were three non-member funders, which were the European Union, MacArthur Foundation, and African Capacity-Building Foundation ([20]ACBF). In 2010, there were still 11 member funders with MacArthur Foundation becoming the member funder while Ford Foundation becoming the non-member funder. In 2010, the number of non-member funders increased to 9, most of which were central banks of African countries.

[20]Note: African Capacity-Building Foundation (ACBF) was founded in 1991 by African governments, African Development Bank, United National Development Programme, the World Bank, and International Monetary Fund. Its member countries include 42 African countries and a few non-African countries.

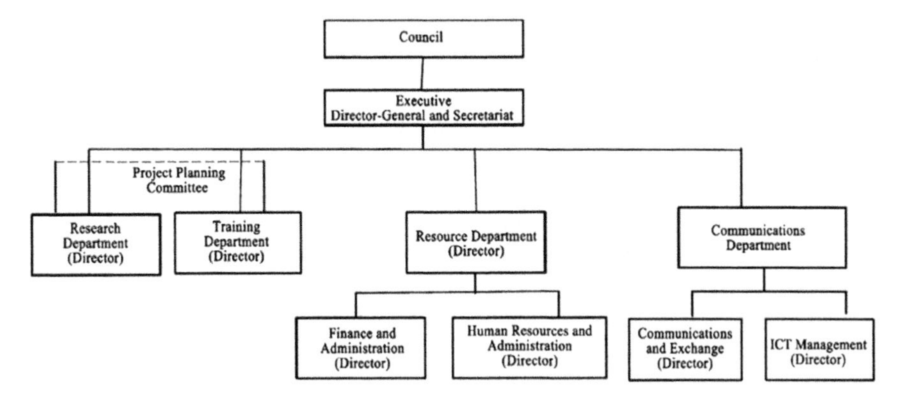

Fig. 10.9 Organization of African Economic Research Consortium

(3) *Organization*

The organization of AERC is shown in Fig. 10.9. Its Research Department, Training Department and Communication Department jointly undertake the work and activities as described in Sect. 10.5.5.3.

10.5.5.3 Activities

Activities of AERC can be divided into research, training and exchange activities.

(1) *Research*

Objectives of AERC research are to establish viable policy-oriented local research teams, to obtain findings useful to policy analysts and policy makers, to promote integration of researches with policies, and to encourage settlement of high-quality researchers.

Research projects also utilize networking method so that occupational activities in African will reach critical quality by the cost-benefit method. Sub-networks are extended to obtain cooperation from professionals outside Africa, so as to broaden research opportunities and sustain research attention in African study from outside. These projects also help organize researches external to AERC, so as to enhance potentials for sustained access to funds.

Biennial plenary conferences are held on a regular basis to provide a forum for writers of papers with assigned subject, so as to explore new topics and methods. At the conferences, famous experts introduce leading edges and trends in the fields of their studies. In this way, they help networked members to broaden their vistas or deepen their understanding on known issues.

AERC-funded researches include the following:

(i) *Subject Research*

It is divided into three stages, namely, providing subsidiaries to individual researches in the academic circle or by the policy institutes, establishing a support system via peer review, seminars and literature search, and holding seminars to continuously track down on research quality.

Subject researches mainly cover poverty, income distribution and food security, macro-economic policies, investment and growth, finance and mobilization of resources, trade and regional integration, and political, economic, natural resource management and agricultural policy issues.

(ii) *Comparative Study*

It is a special research by comprehensive cross-national analysis.

AERC-funded comparative researches include capital flee from Sub-Sahara Africa and foreign debts, mobilization and management of financial resources in Africa, quality of African data, macro-economy and sector policy and performance of the agricultural sector in Sub-Sahara Africa, and impact of economic reform to the social sector in the Sub-Sahara Africa.

(iii) *Cooperative Study*

This research pattern promotes senior researchers in Africa and from other continents to jointly study several major issues and produce policies and academic literature with critical quality in relation to the Sub-Sahara Africa.

According to the new strategic plan of 2010–2015, AERC will take measures to further expand the cooperative study pattern. In addition to rapid output of policy-related knowledge, this mechanism can also enhance capacity building of AERC by incorporation of senior scholars from other continents into AERC researches.

Subjects of cooperative studies between 2010 and 2015 include:

Diffusion of resources and development opportunities and benefit distribution;
Relationship between natural resource management and climate change and economic development;
Relationship between regional integration and strategic trade, traditional trading partners, and emerging markets;
Development of finance sector. Of the above four subjects, relationship between natural resource management and climate change and economic development attracted recent attention of policy makers in Africa and their development partners. The other three subjects, however, have been long-term topics on the agenda of African development policies.

Findings of cooperative studies will be compiled into policy abstracts for broad dissemination.

(iv) *Special Seminars*

Special seminars are held to achieve profound understanding and findings on certain subjects in a comparatively short period. To be specific, seminars are organized to discuss former AERC funded research results and those through cooperation from other networked institutes with AERC experts, so as to obtain authoritative conclusions and rapidly disseminate the conclusions to influence policy making.

As of June 2003, AERC had organized special seminars in the following six aspects:

Foreign exchange markets and exchange rate policies in the Sub-Sahara Africa;
Policies on manufacturing competitions in Sub-Sahara Africa;
Quality of African data;
Economy of francophone African countries;
Economy in the post-conflict era;
Foreign direct investments in Africa.

(2) *Training*

The objective of training is to support cultivation of masters in economics and upgrade the capacity of economic departments of local public universities, so as to improve the economic researchers in both number and quality in the Sub-Sahara Africa. The following training programs are provided:

(i) *Collaborative PHD Programme (CPP)*

Many universities in Africa ran various doctoral programmes in economics, but the quality of these programmes varies widely. In view of this, AERC, together with universities and other stakeholders within and outside Africa, launched the collaborative PHD programme in economics in December 2002, so as to provide high-quality PHD education programme in Africa.

This four-year collaborative PHD programme enrolls students who have completed AERC collaborative MA Programme (CMAP) or other master's degree programmes that are course-work based and subject to external review. They should complete at least three mandatory courses in the following fields: microeconomic theory, macroeconomic theory, and quantitative methods.

(ii) *Collaborative Masters Programme (CMAP)*

The CMAP is executed in the departments of economics in 21 universities in Africa. Its aim is to develop economics programmes in Africa that meet international standards, are relevant to African needs and can eventually be sustained from local resources.

AERC divides participating universities into two categories, Category A and Category B. There are 10 Category B universities that have adequate capacity to offer the above three core courses and meet jointly defined and enforced CMAP standards. There are 13 Category A universities that send their students to Category B universities to complete courses and training required for completion of CMAP.

The 18 to 24-month CMAP is divided into three stages: mandatory courses of 9–12 months, optional courses (at CMAP-Joint Facility for Electives (JFE) in Nairobi), and thesis research of 6–9 months.

(iii) *CMAP—Joint Facility for Electives (JFE)*

Joint Facility for Electives (JFE) is based at the Kenya Commercial Bank Management Center, Nairobi, Kenya. This 14-week programme provides optional courses in the following ten fields: agricultural economics, corporate finance and investment, econometrics theory and practice, environmental economics, health economics, industrial economics, international economics, monetary theory and practice, public finance, and labor economics. Each student is required to complete two of the courses. In addition, JFE offers an optional course on computer applications and research methods whose primary objective is to equip students with the basic research techniques and computing skills that every student of economics must acquire, in order to be proficient in econometric estimation and research methods.

(iv) *CPP—Joint Facility for Electives (JFE)*

CPP—Joint Facility for Electives (JFE) is also based at the Kenya Commercial Bank Management Center. It provides optional courses in agricultural economics, development economics, econometrics, financial economics, health economics, international economics, public sector economics, environmental economics, and labor economics.

(3) *Communication Activities*

The communication activities aim at facilitating the impact of AERC research and training products on economic policy making in Africa. AERC uses print, electronic and event-based communication and dissemination techniques to:

(i) Convey the products of AERC research and training to key target audiences;
(ii) Raise the profile of AERC and the visibility of Consortium activities;
(iii) Link members of the economics profession on the continent with each other, with the Secretariat and with the array of information available around the world;
(iv) Conduct outreach activities exceeding the scope of the platform for internal communications.

Outreach activities include events such as regional level senior policy seminars, national policy workshops, a web-linked information service, and a variety of publications.

AERC publications include research news, newsletter, topical research papers, policy briefs, special literature, working papers and books.

10.6 Development of China's Think Tanks and the Development Research Center of the State Council

10.6.1 Development of China's Think Tanks

The term "think tank" gradually surfaced in China at the beginning of the 21st century. In the last five years, reports on think tank research and their ranking emerged in a large number. For example, in 2011, Chinese Academy of Governance (previously known as China National School of Administration) published a report: *Development of Think Tanks in China.* Shanghai Academy of Social Sciences also established a think tank research center that published its report: *Think Tank Development*, by which it categorized and ranked the think tanks in China. Chinese Academy of Sciences proposed "four priorities" in the new round of reform, of which one priority was "to establish a high-level national science and technology think tank". This clearly shows the kind of attention to "think tanks" has reached a new high.

In 1980s, a discussion on soft science research and democratic and scientific decision-making prevailed in China. Comrade Wan Li[21] once said that decision-making could be dated back to ancient times. In the ancient times, top-level major national decisions were made by feudal emperors in person. In the contemporary times, mass social production brought fundamental changes to the social and economic life. With the significant increase in knowledge and information quantity, issues for which decisions need to be made become overbearing multitudes. It is beyond any leader however brilliant individually to make all the decisions by himself. In this context, think tanks come into being. Actually, research on think tank in China can be dated back to 1980s.

It is not the intention for this section to discuss the definitions of soft sciences and public policy. In 1980s, the State Scientific and Technological Commission divided organizations engaged in soft sciences into five categories and there were 15,497 soft science researchers according to the statistics. At that time, Comrade Wan Li pointed out the organizational characteristics of these think tanks as follow: "They gather experts and advisers with different knowledge structures and experiences together, to make up for the lack of intelligence, experience and vigor of individual leaders. We shall establish a knowledge and information complex that comprises of different knowledge structures and utilizes modern scientific theories and methods and enable the knowledge structures to supplement, inspire and enrich each other." The Development Research Center of the State Council was established in line with this organizational characteristics and working methods.

[21]Source: Scientific and Technological Policy Department, State Scientific and Technological Commission, compiled by the Education, Science and Culture Department of People's Daily (1986).

10.6.2 *Case of Development Research Center of the State Council*

10.6.2.1 History of Development Research Center (DRC)

The Development Research Center of the State Council (DRC) came into being after merging of four centers as shown in Fig. 10.10. Boxes A, B, C, D represent Economic Research Center, Technical Economic Research Center, Price Research Center and Rural Development Research Center respectively.

In brief, the four organizations in Boxes A, B, C and D in Fig. 10.10 were established in early 1980s. They came into being in the historical context of reform and opening up. At the beginning, according to the positions allotted by the State Council, the Economy Research Center had only 30 staff members while the Technical Economy Research Center and Price Center had only 20 employees each. As the position allocations were limited, these centers adopted the method of organizational research. Take the Economy Research Center for example. The State Council stated in the *Directives of the State Council Regarding Strengthening the Work of the Economy Research Center* dated August 14, 1982 that “the Economy Research Center of the State Council is an advisory and research organ under the leadership of the State Council and the Central Leading Group for Financial and Economic Affairs. The ministries and commissions under the State Council, the

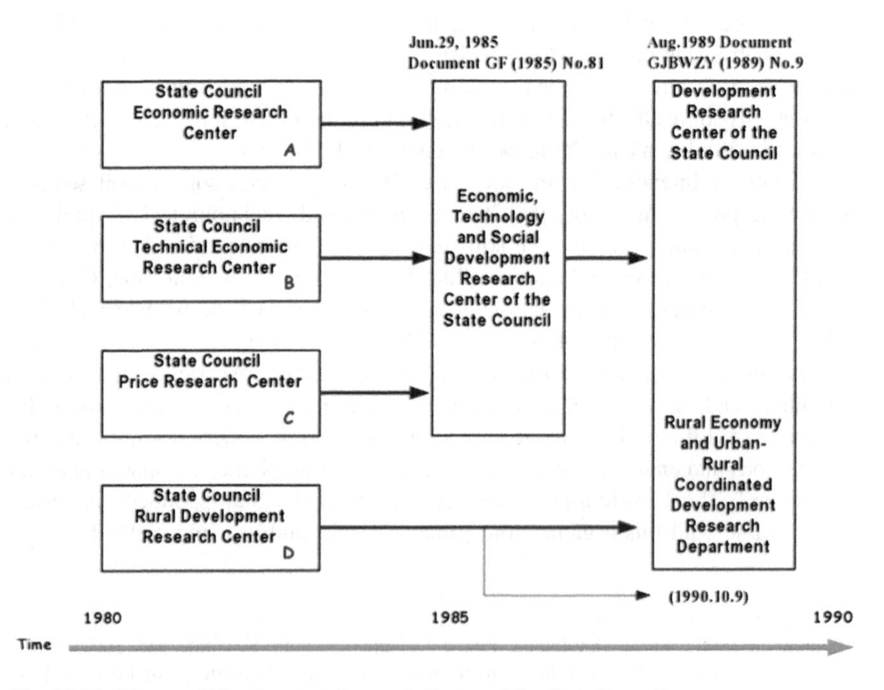

Fig. 10.10 History of Development Research Center of the State Council

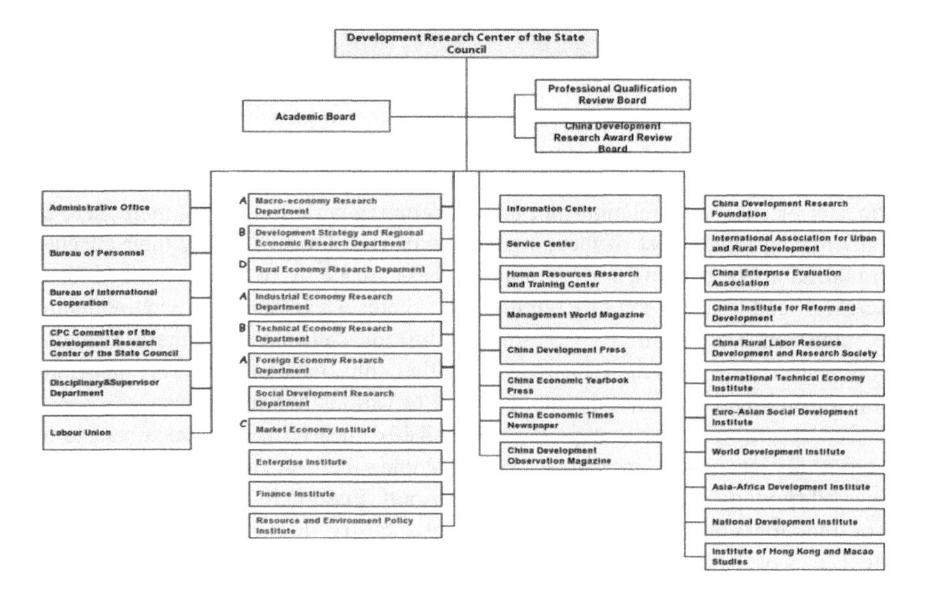

Fig. 10.11 Organization of the Development Research Center of the State Council (June 2012)

Academy of Social Sciences will maintain business contacts with the economic research departments of concerned colleges and universities."

After a period of exploration, the State Council took some measures to solve the issue of overlaps in structural functions. For example, price research was a part of economic research. The Technical Economy Research Center of the State Council has now shifted from the more basic feasibility studies on major construction projects to researches more on long-term planning and development strategy of the national economy. For example, the Center had carried out projects such as "Comprehensive Planning for the Base of Energy Resource and Heavy and Chemical Industries in Shanxi" and "China 2000". For this reason, the State Council in June 1985 combined the Technical Economic Research Center and Price Research Center into the Economic, Technical and Social Development Research Center. In 1989, the former Rural Development Research Center was dissolved and some departments were merged into the Economic, Technical and Social Development Research Center. Later, the organization of the Center evolved and the current organization as shown in Fig. 10.11 comes into existence.

10.6.2.2 Successful Experience of DRC and the Former Technical Economic Research Center

China's think tanks grew from scratch in a short span of thirty years and begin to take shape with considerable visibility and remarkable success within and outside the country. An earliest member of the Technical Economy Research Center, the authors of this book wishes to offer their observations on these successes as follows:

(1) *Sticking to Participatory Research to make Effective Use of Different Knowledge Structures*

He had established a partners relationship of other organizations with Technical Economic Research Center (TERC). Two types of partnership were identified: participating units and contacting units, the former means closer relationship while the later means looser relationship. In the beginning period of 1980s, there were 24 participating units, most of them were at ministerial level including their attached organizations or bureaus, for example, Chinese Academy of Social Science, State Science and Technology Commission, State Development and Planning Commission, State Economic Commission, Institutes attached to Chinese Academy of Science. There were more than ten contacting units, most of them were designing and research and information units attached to various line ministries.

Ten specific research groups were established, nearly most of members came from different related ministries. For example, the research groups including Feasibility Study of Industrial Projects, Transport, Energy, Machinery, Agriculture, Consumption Goods, Metallurgy, Chemical Industry, Materials of Construction, etc. Heads of the participating and contacting units worked as the group leaders and staffs of TERC as liaison officials. Therefore, the dairy operation of organized research was performed with few staffs of TERC.

(2) *Sticking to Application of Modern Scientific Theories*

In June 10, 1981, the Technical Economic Research Center held a seminar on national economic model, introducing mathematical economics into macroeconomic research on national economy and development. From December 20–24, 1982, "Economic Evaluation Workshop on Construction and Reconstruction Projects" was held, which laid a foundation for project evaluation research of microeconomic projects. On the basis of these scientific theories, the Technical Economic Research Center developed the knowledge of various mathematical model systems in macro-economy, so as to properly apply quantitative-qualitative methods to researches on national macro development strategies and those on sub-strata.

(3) *Sticking to the Theory of Practice-based Knowledge and Apply the Process of on-the-job Learning and Practicing*

In this way, the capabilities of all staff progressed rapidly.

At the beginning of the Technical Economic Research Center, most of the staff were new graduates with master's degrees. Proficiency of them was improved significantly after two major organized researches, namely, assisting Shanxi Province in "Comprehensive Planning for the Base of Energy Resources and Heavy Chemical Industry in Shanxi 1980–2000" from April 26, 1982 to June 23, 1983 and organizing experts in the ministries and commissions for the "China 2000" project from May 25, 1983 to April 1984. Master students and other personnel engaged in these projects later became experts in their fields.

(4) *Sticking to "Open" Research and Cooperating and Communicating extensively with International Organizations to catch up with International advanced Organizations in terms of Research Level*

"China 2000" project carried out by the Technical Economy Research Center was similar in nature to the "China: Long-term Development Issues and Options" project carried out by the World Bank from early 1984 to May 1985 and their research timeframe coincided. During their researches, research teams of the Technical Economy Research Center and the World Bank conducted extensive exchanges inter-activities. For five of the ten subjects covered in the main report of the World Bank, the World Bank entrusted the Technical Economic Research Center to organize five expert workshops and domestic experts were invited to comment on the report. Staff of the Center learned the approaches and working methods of the World Bank in strategic and development research through cooperation and exchanges. The Technical Economy Research Center may be the earliest organization that cooperated with foreign organizations in research projects. In 1984, United National University invited the Technical Economic Research Center to participate in the project of *Scientific and Technological Independence: Experiences of Asian Countries*. In 1985, the Center took the lead and engaged a part of staffs from the China's Information Institute of Sciences and Technologies, Scientific and Technological Development Research Center of the State Science and Technology Committee to participate in this cooperative research. In 1986, a report on this project was prepared. During this period, the Center also held seminars in Beijing, to which project participating countries (China, Japan, India, Thailand, ROK and the Philippines) were invited. The report of the Center was included in *Technological Independence—The Asian Experience*, which was published by United National University in 1994.

(5) *Unceasing exploration of China's national situations at all levels, in all directions and with necessary reiterations*

Staff of the Center gained a comprehensive understanding on the national situation, advantages and issues of China through the project of *China 2000*. As China covers a vast territory and has a huge population, medium-level research is required. After the completion of *China 2000*, the Center cooperated with the World Bank in the research on industrial structure and industrial policy and gained certain understanding on sectoral issues at the intermediate level. It also carried out researches on industrial structure and development in provinces throughout China, so that researches can be focused to medium and provincial domain (in comparison to national level). At the end of 1990s, the Development Research Center of the State Council undertook the UNDP-funded project, *Comprehensive Research on Economic Development and Reform*. Such repeated macro-medium-macro researches on the national situation of China enabled staff of the Center to gain deep understanding on the complex situation of China. These researches also lay a solid foundation for the practicality of policy studies.

10.7 Summary Points

1. Think tanks are service organizations that collect, analyze and study domestic situations of a country and provide advices on public policies to the policy-makers.

Proceeding from this definition, think tanks are different from the planning and market research departments of enterprises and from consultancy advisories established by individuals. In the modern economic society, groups of experts in all fields are required to function due to the growing complexity, overwhelming information and schools of knowledge contrasted with the limitations of any individual policy maker in terms of knowledge and available time. In this context, think tanks emerge. They apply modern scientific and management measures to collect, select and extract ideas and develop them into concise, and correct situation analysis and timely policy recommendations for the policy makers. As the bridge between the policy makers and the general public, they alert the policy makers and disseminate their findings to the public.

Think tanks can be classified according to the criteria adopted. This chapter introduces the classification of think tanks as described in the annual reports of the University of Pennsylvania on think tanks. Many Chinese organizations also issued report on think tanks in China and their ranking in recent years. However, different classification criteria are adopted.

2. There are many cases in Chinese history that leaders made decisions relying on advices of their strategists and achieved military and political successes. In construction of and study on modern think tanks, China falls behind the developed countries.
3. Institutes in the U.S. and Nomura Research Institute of Japan have gathered information concerning development of think tanks in the world.

University of Pennsylvania takes the lead in research statistics. It has set up "the project of think tanks and civic society" and published in 2007 the first report on think tank ranking. This chapter introduces the fourth report on think tank ranking published in 2011,[22] four key indicators for the ranking, namely resources, efficiency, output and effect, and subset indicators. It lists 25 top think tanks in the world in 2010. According to the regional statistics of the report, America, Europe, and Asia respectively rank the first, second and third places. The statistics, however, is not complete.

After the World War II, the United Nations and its subsidiary organizations undertook public policy studies and played a significant role in the world and various countries. These information, however, were neglected by the project leaders of the University of Pennsylvania. They noticed the role of these organizations and invited the World Bank to hold seminars in UOP to expand the

[22]Note: The 7th ranking report (2013) has been published.

influences of their publications. The 7th annual report on think tank ranking first enlisted the Asian Development Bank.

4. The United States takes the lead in think tank development in both quantity and quality. This chapter briefly introduces its experience in think tank development.
5. Case studies on think tanks abroad.

The establishment and development of modern think tanks originated from some western developed countries. Their experience in think tank operation and organization provides reference to developing countries in establishment of think tanks. This chapter introduces Brookings Institute and RAND Corporation in the U.S. and the International Development Center in Japan. The first two think tanks were selected due to their leading ranks among think tanks in the world. The International Development Center in Japan was selected as Japan is a latecomer among developed countries and it develops its own characteristics after learning from the West in that it not only serves Japan internally, the Center also focuses more on East Asia.

(1) Brookings Institute. Established in 1916, Brookings Institute is the first private organization dedicated to national public policy studies with a history of nearly a century. This part introduces the contributions Brookings Institute has made at different historic junctures of the U.S., its top organization and funders, research organization and topics, decision on research projects and examples, education on and evolution of public policies, and its publications.
(2) RAND Corporation. Officially established in 1948, RAND Corporation is the product of the World War II. During the War, Arnold, the Chief of Staff of the United States Air Force, experienced the success brought by cooperation of all knowledge resource teams and believed that these teams were necessary after the war. "Scientific plans must be made far ahead of actual research and development work." In October 1945, people who share the same opinion with General Arnold signed a contract with Douglas Aircraft Company and set up RAND project. In 1947, RAND project was separated from Douglas Aircraft Company and RAND Corporation came into official being in 1948. This part introduces the major contributions that RAND Corporation made to the U.S. at all stages, e.g. intercontinental missiles and internet, its top organization and funders, organizational structure and analysis, subject selection principle in 2010 and examples, and experts makeup, which includes both natural science engineers and social science workers. It also summarizes the report of RAND Corporation on *Long-term Economic and Military Trends 1950–2010* of fifteen countries in the world in 1980s, so as to illustrate the advantages of RAND in quantitative and qualitative analysis.
(3) Development of think tanks in Japan and the International Development Center of Japan. Though Japan is a latecomer among developed countries, it absorbs experience of other developed countries, domesticated the experience and develops a research system with its own characteristics.

International Development Center of Japan, a private non-profit think tank, was established in 1971. It is also the first Japanese think tank specialized in development and international cooperation. Despite its small size, it is equipped with a strong top-level advisory team. Saburo Okita, the former Minister of Foreign Affairs of Japan, serves as the Chair of the Board. This part introduces the general conditions of the Center, its organization, top-level organization, funders, specific research activities and latest work examples. Many subject researches are assigned by the Japanese government and international organizations. The Center focuses on trainings and project researches in developing countries in Southeast Asia.

6. International think tanks.

International think tanks as defined in this book refer to the organizations that serve public policy services to countries all over the world instead of certain countries. In this book, these think tanks are classified into three categories: first, the United Nations and its subsidiary or derivative organizations, such as United Nations Industrial Development Organization and the World Bank; Second, group or regional think tank organizations, such as Organization for Economic Cooperation and Development set up by developed countries and the African Community; Third, loose global research organizations on the basis of modern communications and computers, such as the "Global Development Network" established by the World Bank, which became independent later, and International Center of Economic Growth, which is a loose organization with correspondence research institutions of different countries.

(1) Case of Organization for Economic Cooperation and Development. Established in 1961 with 20 member countries, it had 34 member countries by 2014. This part introduces its missions, organization and funders, activities, publications and ten actions in progress, namely, trends and levels, new methods to settle economic challenges, inclusive growth, development strategy, knowledge-based capital projects, green growth strategy, innovation strategy, skill strategy, gender action, and better life action. "Action" related researches of OECD are generally forward learning and significant in leading global development strategy.
(2) Case of the World Bank Group. The World Bank Group is a family of five organizations and the World Bank in general sense refers to the combination of the International Bank for Reconstruction and Development (IBRD, established in 1945) and the International Development Association (IDA, established in 1960). The World Bank now has 188 member countries and has around 10,000 employees including the staff of residential offices in member countries. The World Bank Group deems its mission to promote economic development and eliminate poverty. Apart from specific loan projects, it provides overall policy consulting and personnel training to its member countries, especially the developing countries. Since its entry in China in 1980, its work in China can provide reference to the activities of think tanks in developing countries. This part introduces the actual scope of work of the World Bank in China, that

includes its overall study on China, research on medium and near-term issues, research methods, quality assurance of finished reports and their dissemination, seminars and training classes, publications, and specific researches.

(3) The section of international think tanks also introduces the International Economic Development Center and African Economic Research Consortium.

The former, established in 1985, is a network-based think tank organization by means of computer-based communications. The latter, established in 1988, also consists of networked member organizations. These two organizations provide reference on how to utilize information and communication technologies to organize think tank resources.

7. Development of think tanks in China.

(1) Modern think tanks in China have only a history of three decades. Therefore, China is comparatively weak in think tank construction and research on think tank development. In 1980s, soft science was first proposed in China. Preliminary but incomplete statistics on institutes of soft sciences were provided. According to incomplete statistics, there are around 2500 think tanks of various types, mostly small and short of human and financial resources. In recent years, some Chinese organizations also published reports on think tank research and ranking in China.
(2) Case of Development Research Center of the State Council. This part introduces the history of DRC and experience of one of its predecessor, Technical Economy Research Center: sticking to organized participatory research, effective application of different knowledge structures, application of modern scientific theories, learning and self-improving by practicing, sticking to "open" research and extensive cooperation and exchange with international organizations; and all-round exploration of the national situation of China at all levels with necessary reiterations.

References

African Economic Research Consortium. (1997/1998, 2011/2012). AERC Annual Report. Retrieved from http://www.aercafrica.org/index.php/about-aerc/annual-reports.

Glodberg (Ed.) (1985). The Brooking Institute Annual Report 1985. Washington, D.C.: Walk Press Ina.

International Center for Economic Growth (Ed.) (2010). *Brochure*. San Francisco: ICEG.

International Development Center of Japan. (2010). *IDCJ Report 2010* (p. 3). Tokyo: IDCJ.

International Development Center of Japan Annual Report (1991–1992). Annual Report 2011.

McGann, J. G. (2010). *Global go to think tanks index report*. Retrieved from http://repository.upenn.edu/think_tanks.

McGann, J. G. (2012). *Global go to think tanks index report 2011*, pp. 24–25

McGann, J. G., & Sabatini, R. (2011). *Global think tanks: Policy networks and governance*. New York: Routledge Taylor & Francis Group.

NIRA. (1985). *Trend of policy research in North America Europe and Japan (1982–1983) (Japanese version)*. Tokyo: National Institute of Research Advancement.

OECD. (2014). *Secretary-General's report to ministers 2014*. Retrieved from http://www.oecd-ilibrary.org/.

Rand Corporation. (2014). *2013 RAND annual report*. Retrieved from https://www.rand.org/pubs/corporate_pubs/CP1-2013.html.

Scientific and Technological Policy Department of State Scientific and Technological Commission, & The Education, Science and Culture Department of People's Daily. (1986). *Rise of soft sciences*. Beijing: People's Daily Press.

Wolf, C., Jr., Hildebrandt, G., Kennedy, M., Henry, D. P., & Terasawa, K. (1989). *Long-Term Economic and Military Trends, 1950-2010*. California: Rand Corporation.

Part III
Application of Social Systems Engineering

Chapter 11
Planning System of China (1953–1980) To Be Case Study

11.1 Introduction

Before China launched its reform and opening in late 1970's, it followed the former Soviet Union planning model and implemented similar mandatory planning system. The reasons for such a practice are attributed to (1) China's lack of experience in economic and social development for a large country with a population of around five hundred and forty million in 1949 at the time the People's Republic of China (PRC) was established and (2) China and the former Soviet Union as close allies were working toward the same goal for building a socialist country. So as the Soviet Union had developed the mandatory planning system early in its first Five-Year planning period (1928–1932), China now followed the suit. China started its first *Five-Year Plan (1953–1957)* in 1953 and its first *Outline of Science and Technology Long Term Planning (1956–1967)* in 1956 (the Twelve-Year S&T Plan for short). The preparation of these two plans were assisted by the former Soviet Union. Both plans had tremendous impact on the socioeconomic, science and technology development of China. The National Bureau of Statistics was established in late 1952, which also drew heavily on the Soviet experience in organizing statistical work. Therefore, the five Five-Year Plans prepared before 1980s were based on the Soviet Material Production System. Service sectors generally fell under the non-productive category using a similar statistical system. The first Five-Year Plan laid the foundation for China's socioeconomic development, as did the first Twelve-Year S&T Plan for its science and technology development in later periods. These two planning documents are the focus of the discussion in this chapter.

H. Wang and S. Li, *Introduction to Social Systems Engineering*,
https://doi.org/10.1007/978-981-10-7040-2_11

11.2 The First Five-Year Plan

11.2.1 Process and Guideline of the First Five-Year Plan Preparation

The period of the First Five-Year Plan was from 1953–1957 but its official document was published on June 1955 and disseminated officially by the State Council in November 1955. The plan's preparation and implementation were carried out almost immediately with publication in 1955, which was two years behind the Plan's intended implementation start in 1953. This situation is caused not only by lack of experience in plan preparation and work construction, but also by insufficient information and data showing current development status of economy, society and enterprise from various regions of China and shortage of data distribution of existing mining resources and necessary information related to site selection for new construction projects. In the process of preparing this plan, a working group was sent to the former Soviet Union to seek advice and know-how, and the initial draft was formulated in 1952.

The preparation guideline for the First Five-Year Plan was based on the *Guiding Principles and General Tasks of the Party in the Period of Transition* which was summed up by former Chairman Mao Zedong: "From the establishment of People's Republic of China up to the basic completion of socialist transformation is a period of transition. The general guiding principles and task of the Party in the period of transition, are to achieve gradually the target of national socialist industrialization, and the Party will also gradually implement the socialist transformation in agriculture, handicraft and capital-intensive industry and commerce over a long period of time. The guiding principles serve as the lighthouse emitting our work path. Any deviation from it will cause mistakes or bias in either 'right' or 'left' extreme."

11.2.2 Frist Five-Year Plan for Developing the National Economy of the PRC in 1953–1957

China's First Five-Year Plan, hereinafter abbreviated as The Plan, was a comprehensive product of central planning under the direction of the State Planning Commission (SPC). The main function of planning was to direct the production of major products by state-owned enterprises (SOE). Numerous ministries under the State Council were responsible for the production of the corresponding products. In this section, we discuss The Plan content and its structure.

11.2.2.1 The Plan Content and Structure

Based on SPC (1956), the official document of The Plan consists of eleven chapters as illustrated below in Fig. 11.1a, b.

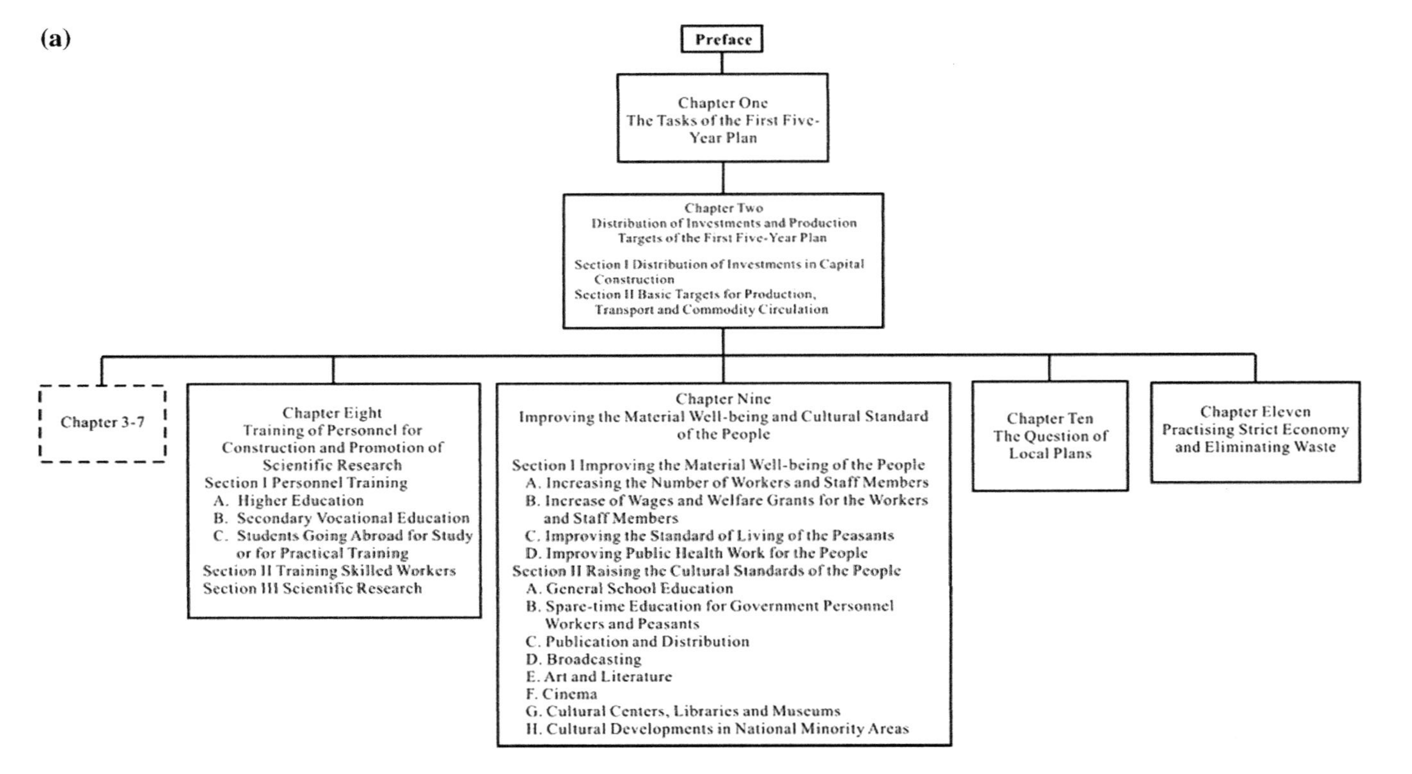

Fig. 11.1 Structure of Contents of the PRC's First Five-Year Plan (*Prepared by authors of this book based upon the First-Five-Year Plan for development of the national economy of the People's Republic of China prepared by China's State Planning Commission*)

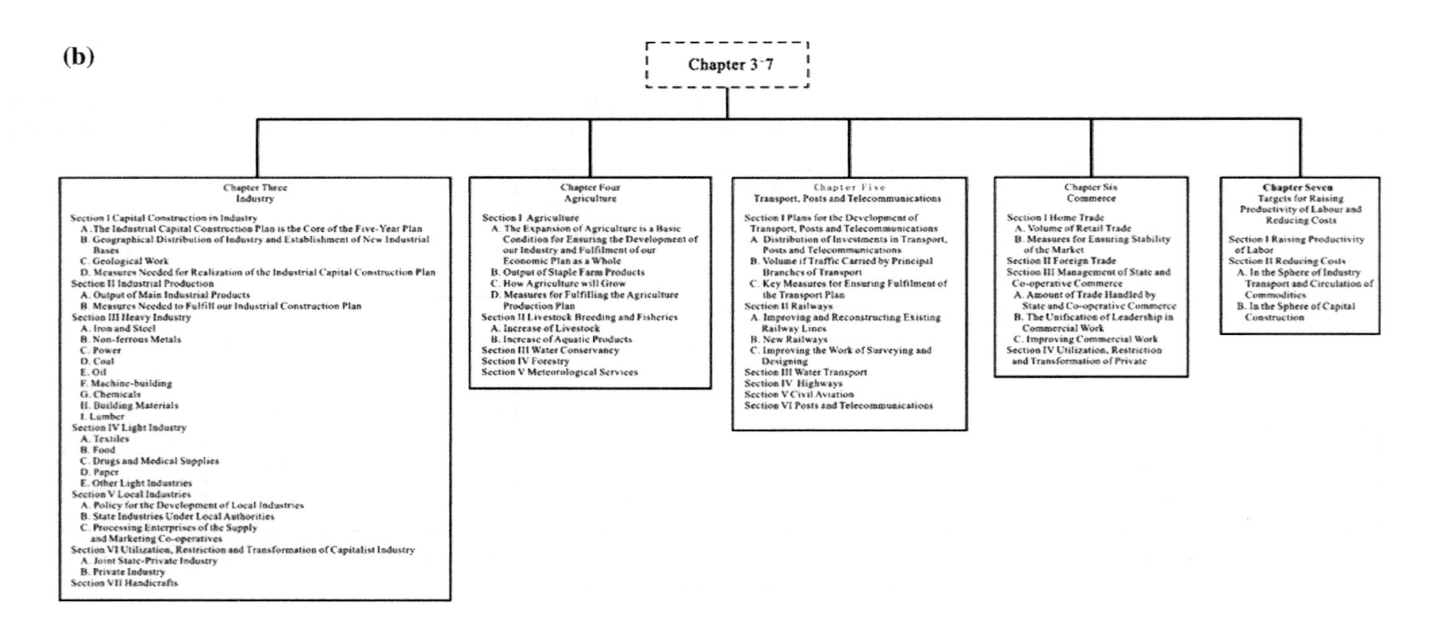

Fig. 11.1 (continued)

(1) *The Preface*

It described various achievements since the establishment of PRC in 1949 to the end of 1952 such as; value of industrial and agricultural output in 1952 compared to 1949 shows 77.5% increase, the increase of share of State and cooperative in industrial and commercial activities from 1949 to 1952, and the change in ratio of output of industrial means for production to consumer goods from 29:71 in 1949 to 39.2:60.3 in 1952. These achievements illustrated the strengthened socialist leading role in the national economy and provided the possibility for introducing a centrally planned economy. By listing the output data of pig iron and steel to show the backwardness of China's economy, the Preface highlighted the focus of national effort on the 156 projects for heavy industrial capital construction and building that the Soviet Union was helping to design. Five aspects were analyzed to put the First Five-year Plan on a firm basis.

(2) *Chapter 1*

It covered the tasks of The Plan in *the Guiding Principles and General Tasks of the Party in the Period of Transition*. The general tasks to be fulfilled by The Plan were based on the State's fundamental tasks during the transition period. Therefore, in addition to fostering the agricultural and handicraft producers' cooperatives and proceeding with the socialist transformation of private industry and commerce, the main effort centered on 694 above-norm industrial construction projects with a core of 156 projects that would lay the preliminary groundwork for China's socialist industrialization.

The Plan laid out twelve concrete assignments related to the general tasks. They were: (1) establishing and expanding various branches of heavy industry; (2) developing textile and other light industries and establishing small and medium enterprises to serve agriculture; (3) fully utilizing existing industrial capacity; (4) promoting the co-operative movement in agriculture; (5) developing transportation and telecommunication industries to support the national economy expansion; (6) organizing individual workers of various types into various forms of co-operatives; (7) consolidating and extending the leadership of socialist economy over capitalist economy; (8) ensuring stability of the market and maintaining balanced national budget; (9) developing cultural, educational and scientific research framework and training qualified personnel needed in national development especially in building its industry; (10) focusing on savings and eliminating waste in economy, accelerating capital accumulation to ensure national development; (11) steadily improving the material wellbeing and cultural standard of the working people; and (12) continuing to promote economic and cultural cooperation and mutual benefits among various nationalities in China.

(3) *Chapter 2*

It set the distribution of investments and production targets for the country. The chapter was divided into two sections. Section I focused on distribution of investments in capital construction. This is a unique feature of the Soviet Union's mandatory planning system. Eastern European countries, China, and other countries followed the same practice. Investment was the priority of mandatory planning, and nearly all types of investments were allocated by the government. The total investment of China in The Plan reached 76.640 billion yuan, with allocation to different sectors shown in Fig. 11.2. Capital construction was the major investment, it reached a total of 42.74 billion yuan in The Plan and its allocation is shown in Fig. 11.3. Investment in industrial capital construction reached 26.620 billion yuan and its allocation is shown in Fig. 11.4. Although state investment to agriculture,

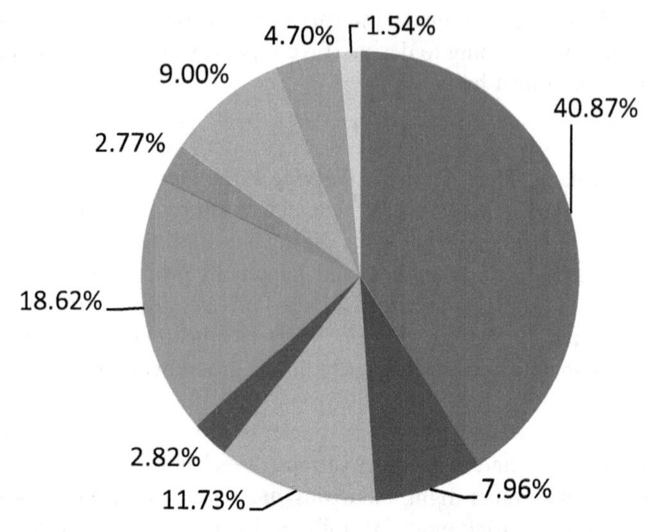

Fig. 11.2 Share of allocation of investment construction of various sectors *Source* Diagram prepared [by who?] based upon data of [Government of China (1950). *First Five Year Plan for Development of the National Economy of the People's Republic of China*] [this Cite needs to be put into proper format]

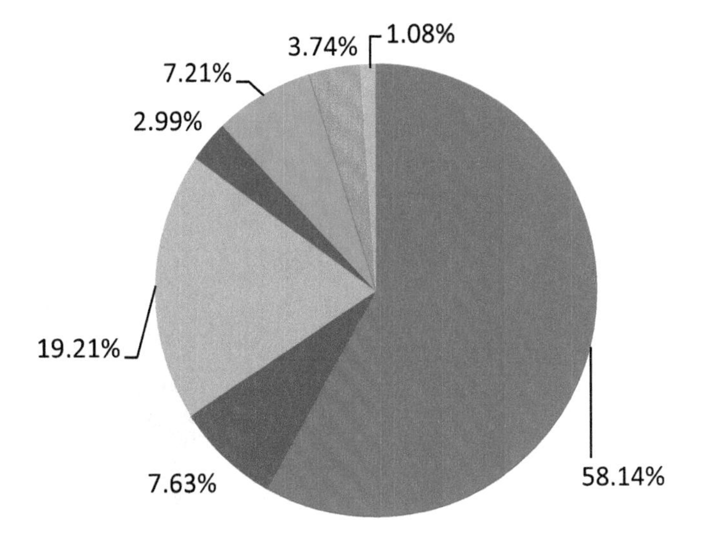

Fig. 11.3 Share of investment of capital construction of various sector

water conservancy and forestry was only 6.1 billion yuan in The Plan, it was estimated in the document that about 10 billion yuan would be invested by peasants themselves for expanding production. Counting peasants' contributions, the investment to agriculture would be approximately 18.4 billion yuan in The Plan period.

Section II set basic targets for production, transportation and commodities. Those targets were decomposed and allocated to every state-owned enterprise and productive unit in the hierarchical level. The total value of industrial output was targeted to increase from 27.01 billion yuan in 1952 to 53.56 billion yuan in 1957, an average annual increase of 14.7%. The value of output of industrial manufacturing means of production and industrial producing consumer goods were expected

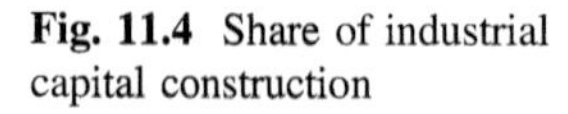

Fig. 11.4 Share of industrial capital construction

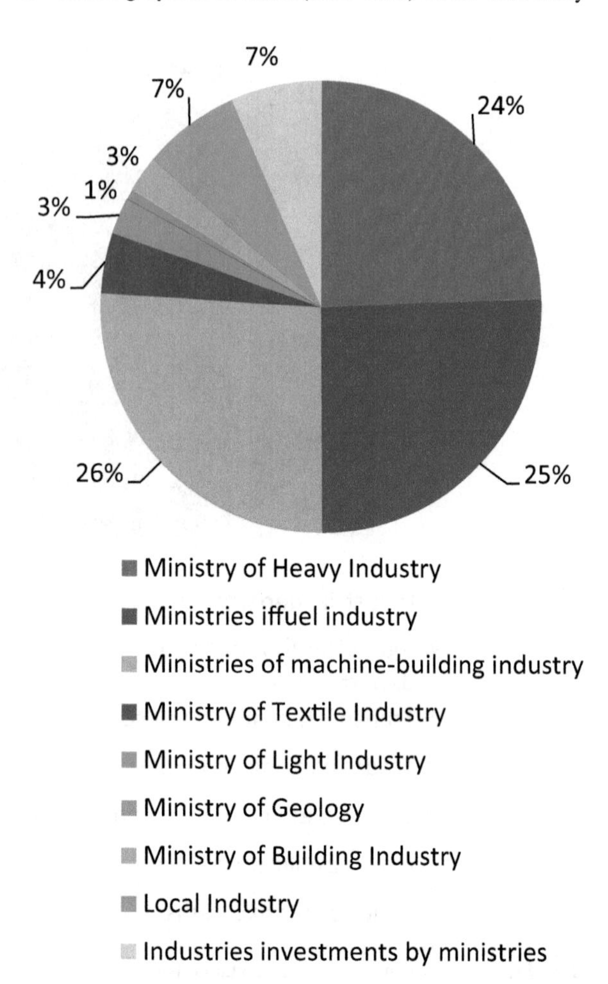

to generate an average annual growth rate of 17.8 and 12.4% respectively. The total value of output of agriculture and subsidiary rural production was to increase from 48.39 billion yuan in 1952 to 59.66 billion yuan in 1957, with an average annual growth rate of 4.3%. The total value of handicraft output was planned to increase from 7.31 billion yuan in 1952 to 11.7 billion yuan in 1957, with an average annual growth rate of 9.9%.

(4) *Chapter 3*

With a focus on industry, this chapter's 74 pages represent 35% of the total page length of The Plan, an indication of its significance to China's national development. This chapter was divided into seven sections:

Section I discussed the industrial capital construction plan which covered 694 above norm (large and medium-sized) industrial projects, of which 156

Soviet-aided projects were the core. There were also 2300 below norm projects to be undertaken in the five years, approximately 900 would be under the state ministries and 1400 under local authorities. The Plan allocated around 11-billion-yuan investment in constructing the core projects.

Geographical distribution of industry was carefully considered. In addition to distribution of projects and reconstruction of established industrial base such as Northeast and Shanghai, building a new industrial base in Southwest China and new industrial areas in North, Northwest and Central China was initiated. This plan also made new and more rational arrangements for the new light industrial enterprises (mainly textile enterprises). This was to remedy the former concentration of light industry in the coastal cities and to transfer it to the interior territory to be nearer the sources of raw materials and consumption. Geological work was emphasized. Survey, prospect and drilling work for major mineral sources was arranged. Eleven measures to achieve the target of industrial capital construction were set.

Section II covered industrial production. The total value of industrial output was raised from 22.01 billion yuan in 1952 to 53.56 billion yuan in 1957. A table with the production target of 44 products was listed in The Plan. To help readers better understand and to illustrate such a mandatory planning system, seven products are selected and presented in Table 11.1. Although China's mandatory planning system has passed several cycles of centralization and decentralization, which shall be discussed later, The Plan leaned towards centralization as it aimed to reduce the share of industries under local management account from 59.6% of the 16.09-billion-yuan total industrial output value in 1952 to 56.3% of the 30.16-billion-yuan total industrial output value in 1957. The Plan laid out 64 above norm and approximately 870 below norm construction projects for state industries under local authorities. Industrial establishment under local management was divided into two groups according to the scale on which their supplies of raw materials, production and marketing operations were coordinated. The first group included establishments in the old industrial cities like Shanghai, Tientsin etc.,

Table 11.1 Production targets of selected industrial products (*Source* SPC, 1956)

Product	Output in 1952	Target for 1957	Increase (1952 = 100) (%)
Electricity	7260 million KWH	15,900 million KWH	219
Coal	63,528,000 tons	112,985,000 tons	178
Steel products	1,110,000 tons	3,045,000 tons	275
Caustic soda (100% pure)	79,000 tons	154,000 tons	194
Bicycles	80,000	555,000	694
Cotton piece goods	111,634,000 bolts	163,721,000 bolts	147
Sugar	249,000 tons	686,000 tons	276

whose operation needed to be balanced on a national scale. The second group primarily served the needs of agricultural production and the rural population.

Section III discussed utilization, restriction and transformation of capitalist industry and outlined steps of transformation of private industry into various forms of state-capitalist enterprise and the higher form of Joint State-Private Industry. In fact, this policy of transformation of private industry into state-owned or collectively-owned enterprises was implemented throughout all the planning periods before the launch of economic reform and opening in the late 1970s.

Section IV focused on handicrafts. The Plan recognized the important role that China's handicrafts played in meeting the needs of people in cities and countryside and especially in meeting the peasants' demands for farm tools and daily necessities. And like the transformation of private industry into state owned or other forms of socialist ownership, The Plan stated that the fundamental task of the state during the transition period was to gradually lead handicraftsmen by persuasion onto the road of co-operation during the first five years. The target was to increase the total value of handicrafts production from 7.31 billion yuan in 1952 to 11.7 billion yuan in 1957, with an average annual growth rate of 9.9%.

(5) *Chapter 4*

This chapter was devoted to agriculture and had five sections as shown in Fig. 11.1b at the beginning of the chapter. Section I emphasized the expansion of agriculture as a fundamental condition to ensure development of industry and fulfilment of economic plan as an integrated whole. Industry and agriculture were the two most important sectors of the national economy. In fact, according to China's National Bureau of Statistics (National Bureau of Statistics [NBS], 1990), in the year of 1952, the value of output of agriculture (46.1 billion yuan) was higher than the value of output of industry (34.9 billion yuan). This situation changed by the end of The Plan as the value of output of agriculture and industry reached 53.7 billion yuan and 70.4 billion yuan, respectively, in 1957.

Output of staple farm products, the total value of production of agriculture and subsidiary production, would amount to 59.66 billion yuan in 1957. That was an increase of 23.3% over the 1952 figure. The Plan presented a table with detailed targets of production of twelve types of crops. For illustration purpose, Table 11.2 is prepared with selected items of crops from The Plan.

There were 57,482 × 10^4 or almost 575 million in population in China by the year 1952, the rural area had a population 50,319 × 104, which was 87.5% of the total population. The total labor force was 20,729 × 104 by the year end of 1952, while the amount of rural labor force was 18,243 × 104, which was 88% of the total labor force. The total sown area by the end of 1952 was 185,968 × 104 mu. To overcome the backwardness of a scattered individual peasant economy, The Plan highlighted the necessity to promote cooperation in agriculture and put agricultural mechanization on a limited experimental scale during the 1952–1957 period. Thirteen measures were set for fulfilling the agricultural production, such as promoting widespread use of two-wheeled, single and double ploughs; organizing agricultural producers into cooperatives on a voluntary basis; concentrating efforts

Table 11.2 [a]Production Targets of Selected Agricultural Products (*Source* SPC, 1956)

Products	1952			1957			Ratio of 1952(%)		
	Sown Area	Yield per mu	Total Output	Sown Area	Yield per mu	Total output	Sown Area	Yield per mu	Total output
Food crops									
Rice	425,734	321.5	136,850	444,864	367.6	163,540	104.5	114.3	119.5
Wheat	371,698	97.5	36,250	400,257	118.6	47,450	107.7	121.6	130.9
Industrial crops									
Cotton	83,636	31.2	2610	95,000	34.4	3270	113.6	110.3	125.4
Sugar cane	2737	5200.4	14,230	4054	6499.7	26,350	148.1	125.0	185.1

[a]*Source* Diagram prepared based upon contents of Government of China (1950).*First Five Year Plan for Development of the National Economy of the People's Republic of China* (p. 115). Beijing: Foreign Language Press.)
Note The sown area is in units of 1000 mu (one mu = one fifteenth of a hectare or 0.1647 acre), the yield per mu in catties, which is a unit of measurement in China, it is around 0.5 kg, and the total output in millions of catties

on manure collection and fertilizer application; making more effective use of land; and improving planting techniques using improved seeds and training personnel in mechanized farming.

Section II covered livestock and fisheries. It outlined measures for increasing livestock, such as horses, oxen, mules, donkeys, sheep, goats and pigs and for promoting rapid development of livestock breeding in areas inhabited by ethnic minorities. Increase of 68.5% in the output of aquatic products between 1952 and 1957 was also planned. Section III discussed construction of water conservancy and power generation project at Sanmen Gorge and other water conservancy projects such as control of Huai River, the flood detention and land reclamation project on the Dungtin Lake and other lakes on the middle and lower reaches of the Yangtse River.

(6) *Chapter 5*

Its focal points were transportation, posts and telecommunication sectors as they were critical in providing fundamental physical infrastructure for national economic and social development. It can be seen from Figs. 11.2 and 11.3 that these sectors ranked third in the investment allocation and ranked second in the investment of capital construction. Table 11.3 summarizes all essential elements in Section I to VI of Chapter 5. The distribution of investment of 8210 million yuan was described in Section I as follows:

Ministry of Railways	5670 million yuan
Ministry of Communication	1339 million yuan
Ministry of Post and Telecommunication	361 million yuan
Civil Aviation Bureau	101 million yuan
Investment in Local Transport Service	739 million yuan

Table 11.3 Targets of various types of transport, posts and tele-communication of First Five-Year Plan[a]

Item No.	Item	Targets of production in 1957 (% increase over 1952)	Targets of capital construction
1	Railway		
	Freightage	245,500,000 tons (85.9%)	1. Improving and reconstruction Existing Railway lines (1) Build 1514 kms of double track railways (2) Construction of double decker Yangtze River Bridge at Wuhan (3) Enlarge 14 important stations (4) Increasing amount of rolling stocks 2. New railway 4084 km of new trunk and branch lines will be built
	Freight mileage	120,900 million ton-km (10.1%)	
	Passengers carried	247 million passengers (51.3%)	
	Passenger mileage	31,966 million pass-km (59.5%)	
2	Highway		
	Motor-lorry transport	67,493,000 tons of freight (225.8%)	More than 10,000 km of high ways provided by central government
	Freight mileage	3211 million ton-km (373.5%)	
	Motor bus transport carrying passengers	114,146,000 passengers (159.1%)	
	Passengers mileage	5732 million passenger-km (193.7%)	
3	Water transport (Inland shipping):		
	Freight	36,864,000 tons (294.6%)	1. 289,000 deadweight tons will be added to inland water shipping 2. Main task in the sea transport is to supply coal and other important commodities to coastal areas 3. 111,000 deadweight will be added to fleet of coastal and sea-going vessels 4. New harbors will be built at Chanchiang in Kwangtung province, at yuhsikow in Anhui Province 5. Increase facilities for mechanized loading, unloading and telecommunications at Shanghai, Hankou and other important ports etc.
	Freight mileage	15,292 million ton-km (321.5%)	
	Passengers carried	56,040,000 passengers (93.8%)	
	Passengers-mileage	3408 million pass-km (78.7%)	
	Water transport (Coastwise shipping)		
	Freight	11,461,000 tons (195.1%)	
	Freight mileage	5751 million ton-nautical miles (190.5%)	
	Passengers carried	1,470,000 passengers (110%)	
	Passengers-mileage	237 million passenger-nautical miles (137%)	

(continued)

Table 11.3 (continued)

Item No.	Item	Targets of production in 1957 (% increase over 1952)	Targets of capital construction
44	Civil airlines:		
	Freight (including mails)	5600 tons (175%)	1. Main task of civil aviation will be on strengthening the link between the capital and the main cities and also with the remote regions. 2. Special aerial services to be organized to meet the needs of agriculture, forestry and geological survey 3. 17,200 km of airline will be put into operation
	Freight mileage (Including mails)	8,050,000 ton-km (231.3%)	
	Passenger carried	54,400 passengers (145.6%)	
	Passenger mileage	91 million pass-km (278.5%)	
5	Post and telecommunication Postal service Telecommunication service	1. Improvign communication between the capital and all the main cities and the new industrial cities 2. Total length of postal routes will be 1,968,000 km (45.2%)	1. Postal services should set up a number of branch offices in cities, factories, mines, forest areas and towns 2. 63,000 km of parallel long distance telegraph and telephone line will be installed 3. Carrier current will be applied to some main trunk lines 4. Main cities and remote border regions will be equipped with wireless apparatus 5. Telephone service will be installed or expanded to serve 91,000 new subscribers; more than 44,000 of them in existing 12 large cities to be served by automatic telephones
	City telephone service	City telephone service is to serve construction in new cities and meet part of the urgent needs of main cities	

[a]*Source* Diagram prepared based upon contents of Government of China (1950). *First Five Year Plan for Development of the National Economy of the People's Republic of China*

(7) *Chapter 6*

This chapter was devoted to commerce and had four sections.

Section I dealt with home trade. Total value of the country's retail trade in 1957 was targeted to reach approximately 49.8 billion yuan, an increase of 80% more than that in 1952. The volume of twelve types of consumer goods such as grain, pork, cotton piece goods, machine-made papers etc. for sales in 1957 were planned. It pointed out that the planned figures could only satisfy people's basic demands and that efforts should be made to increase supply of goods. It projected that people's purchasing power would exceed that of production of consumer goods.

Therefore, measures for ensuring stability of the market were emphasized. The Plan set three measures to control supply for industrial products, six measures to control supply for agricultural products, and seven measures to properly organize supplies. It's estimated that 70,000 million catties of grain would be purchased in 1957.

Section II centered on foreign trade. It emphasized in the very beginning of the section that China would contribute to increased economic cooperation within the socialist camp headed by the Soviet Union. It also described trade expansion with countries in Southeast Asia, and continued trade development with capitalist countries on the condition that such trades would benefit China's socialist economic construction. The total volume of import and export trade in 1957 was projected to be 66.5% greater than 1952. Four aspects or goals were emphasized for trade: familiarize with conditions on the world market, strictly implement the terms of trade contracts, and adopt a checking system to ensure that orders follow a unified plan and improve management and organization of foreign trade.

Section III discussed management of State and Co-operative Commerce. Amount of trade handled in 1957 were planned as follows: value of goods purchased by departments will amount to 27.1 billion yuan, 180% more than 1952; and the total value of their sales will be 34.3 billion yuan, 219% more than 1952. The supply and marketing cooperative will purchase agricultural products to the value of 12.5 billion yuan, 222% more than 1952. The value of State Retail sales will reach 10.22 billion yuan in 1957, 133.2% more than 1952; and the value of cooperative retail sales will amount 17.15 billion yuan in 1957, 239.5% more than in 1952. Major tasks are identified for the Ministry of Commerce and the supply and marketing co-operatives respectively, and seven points to improve commercial work are raised in this section.

Section IV focused on universal policies concerning utilization, restriction and transformation of private commerce. These policies were to apply to all types of private sectors in industry and commerce throughout the period since the First Five-Year Plan up to late 1970s when the economic reform was launched.

(8) *Chapter 7*

It set targets for raising labor productivity and reducing production cost. The chapter started with several sentences quoted from Lenin as theoretical basis of raising productivity of labor. It projected to have labor productivity in the state-owned industries rise by 64% in the five-year period. Targets for raising labor productivity were planned for seven ministries. Ministry of Textile Industry had the lowest target of 10.4%, while river transport under Ministry of Transport had the highest target of 163.8%. Other targets ranged from 63.6 to 87.9%.

Targets for cost reduction in the five-year period were also planned for enterprises under five industrial ministries. To achieve a total cost reduction of approximately 22%, the targets were set from the lowest of 9.1% for the Ministry of Textile Industry to the highest of 40.1% for the Ministry of Machine Building Industry. State industries under the local authorities were projected to reduce cost by about 28%.

Commodity distribution cost associated with state trading was to reduce by about 33% in aggregate with the lowest target of 9.3% set for Ministry of Food and the highest target of 43.6% for Ministry of Commerce.

Targets of cost reduction in capital construction were also set for the last three years in The Plan period as followed: cost in building and equipment installation in productive projects should be reduced at least 10% of what was originally planned while building costs for various non-productive projects should be reduced by at least 15% of the original plans.

(9) *Chapter 8*

This chapter focused on skill and personnel training for meeting the needs in economic development and scientific research in three sections. The large-scale national construction, the growth of economic and social activities in The Plan period and thereafter required large numbers of skilled personnel, especially industrial technicians. Apart from making optimal use of existing technical personnel, huge efforts and resources were allocated on education and training.

Section I titled personnel training focused on higher education. The main effort in higher education was to develop engineering colleges and natural science departments in universities. Sixty new institutions of higher learning would be established. By 1957, China would have 208 institutions of higher education, of which, 15 would be universities, 47 engineering colleges, 29 colleges of agriculture and forestry, 5 colleges of economics and finance, 5 colleges of political science and laws, 43 teachers' colleges, 32 medical and pharmacological colleges, 8 language institutes, 6 institutes of physical culture, 14 colleges of fine arts, plus 4 other institutes. Six of the above-mentioned institutions would be exclusively maintained for national minorities.

The Plan provided detailed numbers for planned college enrollment and graduation and student distribution in thirteen specialized fields during the five-year period. Selected information of number of graduates 1953–1957 is shown in Fig. 11.5.

To have a balanced growth for all types of technical and administrative personnel, secondary vocational education was also planned. The number of graduates from 1953–1957 of various disciplines is shown in Fig. 11.6.

Students going abroad for study or for practical training was also planned with 10,100 students to be sent abroad to study with 9400 of them going to the former Soviet Union. Six focal points were laid out for higher education: (1) schools should not be too concentrated in a few areas and their scale of development not too large, engineering colleges should gradually be located or developed near industrial bases; (2) higher education institutions should accelerate effort to train new faculty and optimally utilize existing faculty and engage scientific research workers and technical staff from various fields in teaching as adjunct faculty; (3) actively adapt the latest teaching experience from the former Soviet Union and reform existing education system to meet China's specific needs by revising curricula and syllabi, updating instructional materials; and improving pedagogy; (4) implement field and lab work in higher education to provide students with practical experience via

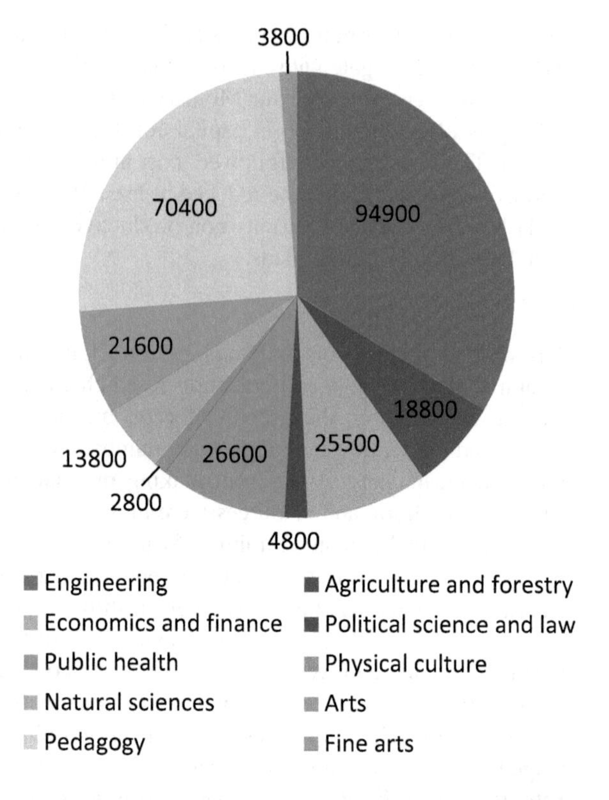

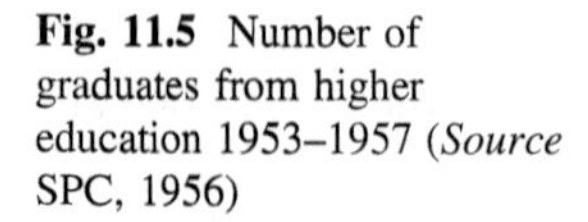

Fig. 11.5 Number of graduates from higher education 1953–1957 (*Source* SPC, 1956)

collaborations with industries and businesses and school-run enterprises and factories; (5) standardize academic levels and gradually increase undergraduate enrollment for bachelors' degrees and reduce the number for associate degrees; and (6) reinforce and improve administration at higher education institutions.

Section II concentrated on training skilled workers. Over 920,000 workers would be trained in the five-year period by the ministries of industry, agriculture, communications, etc. Refer to Fig. 11.7 for details on training distributions allocated to different ministries. Of the 920,000 workers to be trained, 119,000 would be trained at 22 workers technical schools, 362,000 trained in workers' technical training classes run by various enterprises and 439,000 trained by the "Master-apprentice" method by various industries and businesses.

Section III focused on scientific research, as it is closely related to higher education. The general outlines for this area included: developing scientific research foundation, establishing a close linkage between scientific research institutes and relevant industrial and agricultural entities; improving scientific research and experimental work, cumulating new scientific and technical experience, learning and applying advanced science and technologies from the Soviet; systematically carrying out studies and investigations of China's natural and social conditions and its natural resources; and gradually advancing basic research in natural science and social science studies. The Plan set the target to establish 57 research institutes

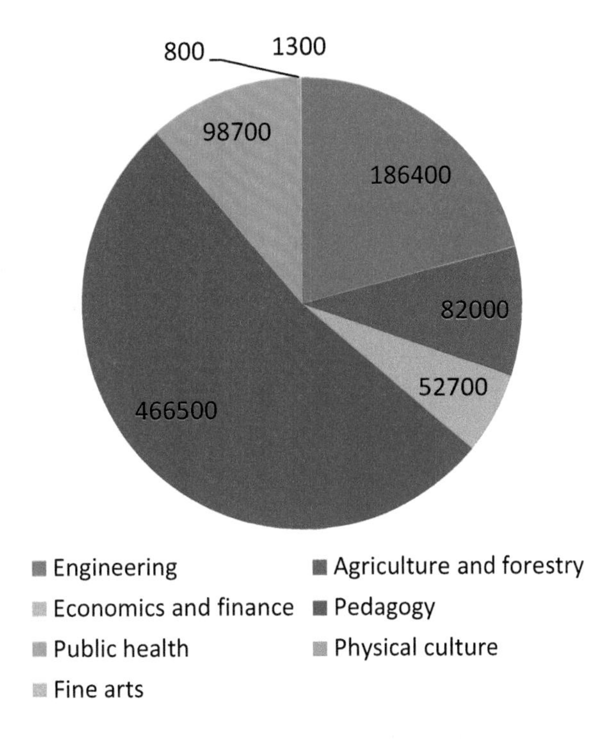

Fig. 11.6 Number of graduates from secondary and vocational education 1953–1957 (*Source* SPC, 1956)

under the Academia Sinica[1] and train more than 4600 research personnel by 1957, as compared to the 1952 base number of 28, and 1200, respectively.

(10) *Chapter 9*

With a focus on improving people's living standard in both material well-being and cultural enrichment, this chapter has two sections.

Section I focused on improving people's material well-being. The first goal was to increase employment. It was estimated that the number of workers and staffs employed would be increased from 21,020,000 in 1952 to 25,240,000 in 1957. This represented a net increase of employment of 4,220,000 people. Refer to Fig. 11.8 for projected increase of labor force in different sectors in The Plan.

To optimize labor force utilization, The Plan emphasized that government organizations and state-owned enterprises (SOE) must: (1) gradually establish and perfect their HR management system, simplify administrative structure, reduce non-production staffing, and increase production personnel; (2) centralize HR planning and labor management, make necessary adjustment and internal transfers

[1]Note: By the year of 1952, China had only one major research organization, Academia Sinica was established in 1928. The Chinese Academy of Science was established in 1949. It had taken over several research institutes of Academia Sinica, there may be inappropriate translation in the English version to use the terminology Academia Sinica instead of Chinese Academy of Science in the First Five-Year Plan.

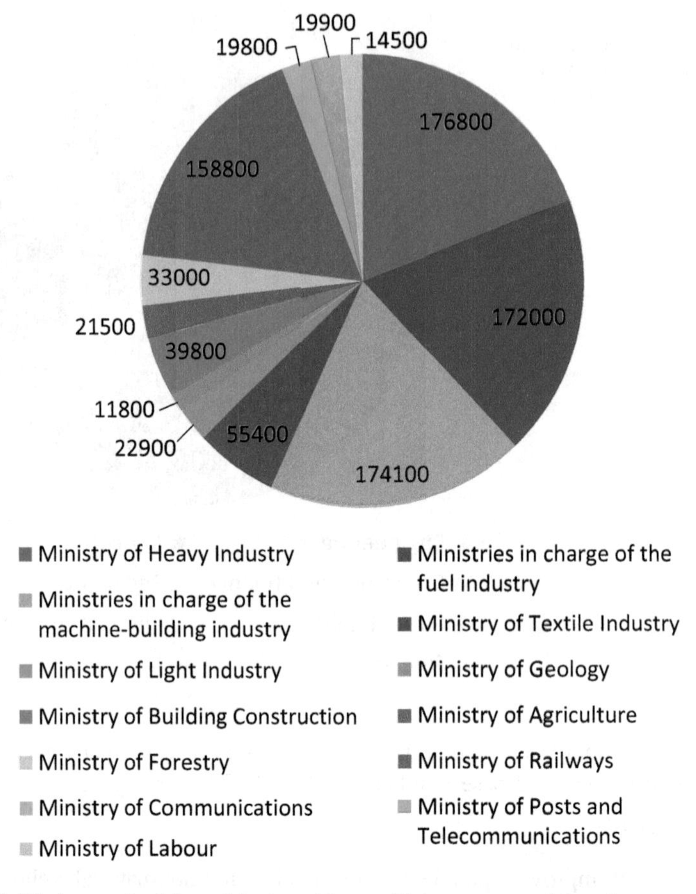

Fig. 11.7 Workers trained by ministries (*Source* Diagram prepared based upon contents of Government of China (1950). *First Five Year Plan for Development of the National Economy of the People's Republic of China*)

to meet planned HR needs; (3) explore new production possibilities to fully utilize existing labor forces under coordinated plans by the central government; and (4) facilitate job transfers and occupation changes to employ surplus labor in the cities.

The Plan outlined percentage increase for workers' wages and welfare benefits over the five-year period as follows:

Field	Increase (%)
Industry	27.1
Agriculture, water conservancy and forestry	33.5
Transport and posts and telecommunications	20.4
Capital construction projects	19.0

(continued)

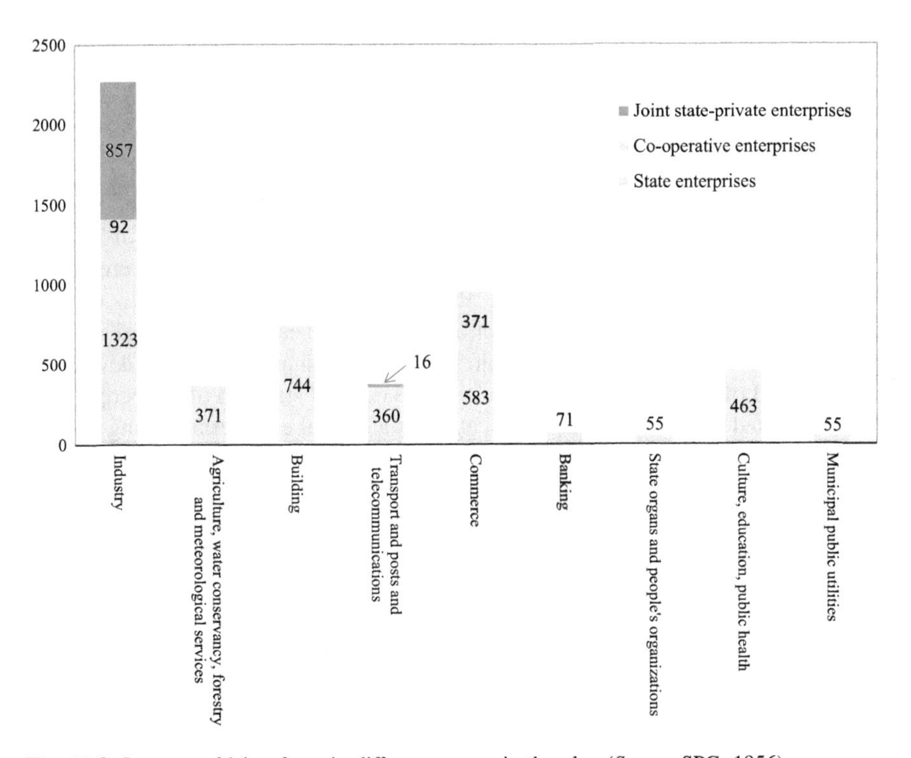

Fig. 11.8 Increase of labor force in different sectors in the plan (*Source* SPC, 1956)

(continued)

State and co-operative trade	28.0
Banking	24.6
State organs	65.7
Culture, education and public health	38.2

In addition, housing for workers and staffs with a floor space of about 46 million square meters built with state funds was included. The Plan also stipulated that the state would provide suitable care for old and disabled workers and staffs based upon the Labour Insurance Regulations of the PRC.

Part 3 of Section I discussed improving the living standard of peasants. An increase of 23.3% in the value of agricultural productions and sidelines was projected with an increase of 12.2% in the average value of per capita production in the rural area. At the same time, the state would develop culture, education and public health in the rural area and would allocate 1440 million yuan as relief aid to help peasants overcome natural disasters and another 1170 million yuan allocated to help families with revolutionary martyrs and military service men.

Part 4 of Section I highlighted plans to improve public health work for the people. The number of state-run hospital beds was projected reach 244,000 in 1957, an increase of 77.2% over 1952. The number of beds in sanatoriums and infirmaries would be increased to 55,000 in 1957, an increase of 237.1% over 1952. The number of community clinics, epidemic-prevention stations, health centers etc. would be increased to 17,000 in 1957, an increase of 65.1% over 1952. The number of state-employed doctors would reach 47,000 in 1957, an increase of 74.3%. In addition to state-owned medical facilities and personnel, there would be 18,300 beds in private hospitals, 174,800 clinics run by private doctors. About of 343,000 doctors would service these hospitals and clinics, and 320,000 of them were trained in traditional Chinese medicine. The Plan emphasized that research in Chinese traditional medicines should be further advanced and health and medical network coverage in areas where minority ethnic groups reside should be further developed.

Section II of this chapter covered raising the cultural standard of the people. The Plan attributed improving general education as a critical element in raising cultural standard of the people. It was planned to graduate 602,000 students from high schools, 4,093,000 from middle schools, 20,150,000 from senior primary schools, and 43,260,000 students from junior primary schools during The Plan period.

It also discussed the feasibility to establish pre-schools/kindergartens, provide support for community and self-funded schools, and gradually transfer private senior high schools into state public schools, and encourage government workers and staffs to seek continuing education in their spare-time. The Plan aimed to make twenty-three million people achieve literacy during the five-year period.

To show how comprehensive The Plan was, Table 11.4 lists some other planned cultural facilities or undertakings in the areas of publication, broadcasting, literature and performing art, cinemas and movie studios, cultural centers, museums, and libraries.

Cultural development in areas where minority ethnic groups reside were also planned. The highlighted areas included bilingual education at local schools, broadcasting programs in local languages, printing books (12.5 million copies), newspapers (22 million copies), and publications in minority languages, and developing art and entertainment activities to accommodate minority cultures, etc.

(11) *Chapter 10*

This chapter addressed the importance of local planning, as the success of the seamless state plan was built on the plans by various central government ministries and departments in co-ordination with the plans of local authorities at all levels. Local plans, covering agriculture, local industries and handicrafts, transport, commerce, culture, education, and city construction, must specify tasks and goals that not only reflect the economic development of the localities but also contribute to the overall national economy in a synergetic and balanced way (p. 213).

The Plan outlined guiding principles on local plans in agriculture, local industry, transport, commerce, cultural and educational development, and city construction. In this hierarchical planning process, China followed the model of former Soviet Union with high concentration of administrative power and decision making at the central government and ministry levels. Former Chairman Mao Zedong later recognized a major weakness in the centralized planning process and expressed his

Table 11.4 Cultural Facilities and Undertakings Planned in the First Five-Year Plan (*Source* SPC, 1956)

Item no.	Item	Total number by the end of 1957	Remarks
1	Publication and distribution		
	(1) Newspapers	2,496,710,000 copies	55.2% increase 1952–1957
	(2) Magazine	394,230,000 copies	93.0% increase 1952–1957
	(3) Books	1,211,650,000 copies	54.2% increase 1952–1957
2	Broadcasting		
	(1) Central stations	2150 kws	740% increase 1952–1957
	(2) Local stations	500.2 kws	270% increase 1952–1957
	(3) Broadcasting and receiving posts in cities and villages of whole country	Approximately 30,000	
3	Art and literature		
	(1) Theatrical groups	2312 groups	Within which, 2100 professional troupes organized by people
	(2) Theatres	2078(within which 1200 theatres are private theatres, 309 theatres are newly built under state operation.	
4	Cinema		
	(1) Films	Produce 400 films of all kinds and translate 308 foreign pictures	Within which, 629 cinemas are under cultural department; 98 are private cinemas
	(2) Cinema	896 cinemas	
	(3) Mobile film projection teams	6614 teams	Within which, 5279 teams are under cultural department
	(4) New film studio	(1) Start to build a new film laboratory in 1956 to process 45 million meters of film a year	
		(2) Start to build a new film studio in 1957 with facilities to produce 4–8 technicolor feature films a year	
		Build a factory in 1957 to manufacture 65 million meters of film a years.	
5	Cultural Centers, Libraries and Museums	Cultural Centers 2600; 6% more than 1952	
		Libraries 109, 31% more than 1952	
		Museums 5%; 63% more than 1952	

view in his famous talk, *On Ten Major Relationships* to top government leaders which advocated to give the local governments more initiative in the planning process. And so, in the Second Five-Year Plan, some decentralization was implemented.

(12) *Chapter 11*

This chapter put emphasis on efficient use of all resources. It advocated practicing frugality and eliminating waste as a prevailing principle for all government agencies and SOEs so money and other resources could be saved from non-productive units and put into where they needed most in The Plan period—developing heavy industry and capital construction projects. It promoted the establishment and implementation of a construction standard and cost structure and set rules for government entities to fully utilize existing production capacity and strictly control cost.

11.2.2.2 156 core projects

(1) *Chinese economy before establishment of PRC*

The Chinese economy was very backward before the establishment of PRC. Chinese economy was dominated by agricultural sector which constituted an overwhelming share of 68.4% of national income while industry had only 12.6% share. And even this small scale of industry was not evenly distributed nationally, with 70% of industries concentrated along coastal regions and the remaining 30% distributed only in large cities of China's hinterland. Refer to Fig. 11.9 for industrial distributions in China before 1949.

(2) *156 projects—the core and backbone of China's industrial construction in the First Five-Year Plan*

The fundamental task of the First Five-Year Plan was to lay the preliminary foundation of China's socialist industrialization through construction of 694 above norm projects. The actual number was 921, with 156 projects being the core. However, the actual number of core projects being constructed was 150, even though 156 was frequently referred to. These 150 projects included 12 in aviation industry, 10 in electronic industry, 16 in ordnance industry, 2 in aerospace industry, 4 in shipbuilding industry, 7 in iron and steel industry, 13 in non-ferrous metal industry, 7 in chemical industry enterprise, 24 in machinery industry, 25 in coal industry, 25 in electric power industry, 2 in petroleum industry, and 3 in light industry and pharmaceutical industry.

(3) *Implication of construction of 156 projects*

The construction of 156 core projects had great influence on social economic growth of China.

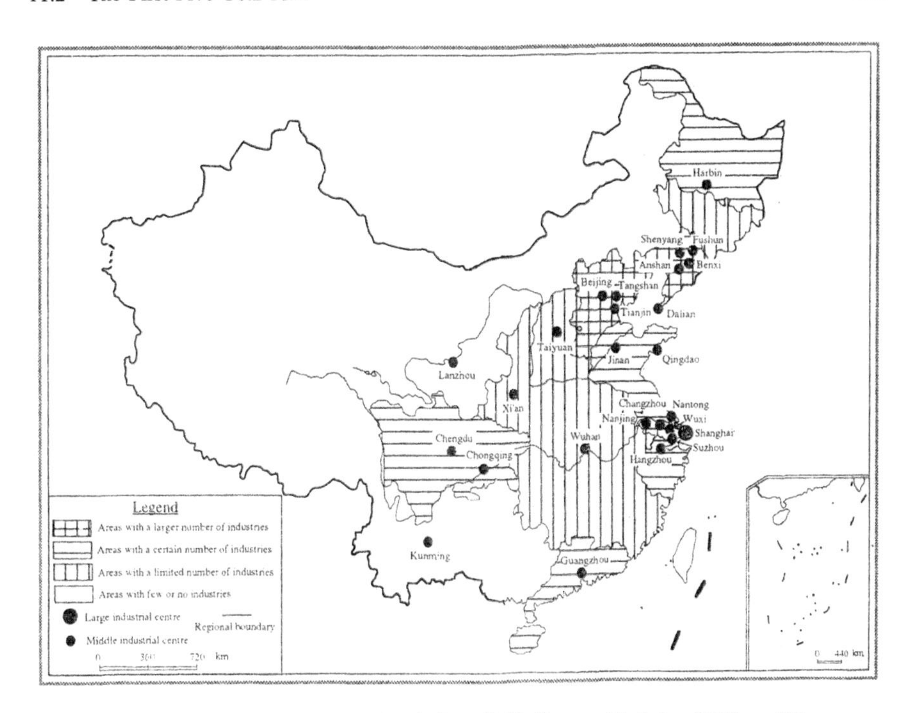

Fig. 11.9 Industrial distributions in China before 1949 (*Source* Li & Lu, 1995, p. 38)

Firstly, the site selection for these projects required prudent analysis of various factors including both micro and macro-perspectives. On the micro-side, the factor of supply of resource input was considered. For example, the metallurgical and chemical projects should be located near the supply of raw materials and energy input such as relevant mines and power stations, etc. On the macro-side, the set of a core project in certain region would bring along the integrated growth of industry, transport, commerce and service and promote the economic development of the backward regions. There was also consideration of national security at that time. Therefore, among the 150 projects constructed, 106 of them were civil projects, 50 of them were in Northeastern and 32 were in the Central region; 44 projects were related to defense industry, with 35 of them in the Central and Western region. Refer to Fig. 11.10 for distribution of 156 key construction projects in The Plan.

Secondly, all 156 projects were designed by experts from the former Soviet Union which also supplied all the equipment for the projects, provided technical support and training in construction and management of these enterprises, and provided manufacturing warranty and parts for maintenance. With the assistance of Soviet Union experts and the actual construction experience cumulated in building The Plan's 156 core projects, China established many design institutes of various industries in the process and was quick to create its own technological capacity through absorption of both hardware and software technology used in the construction.

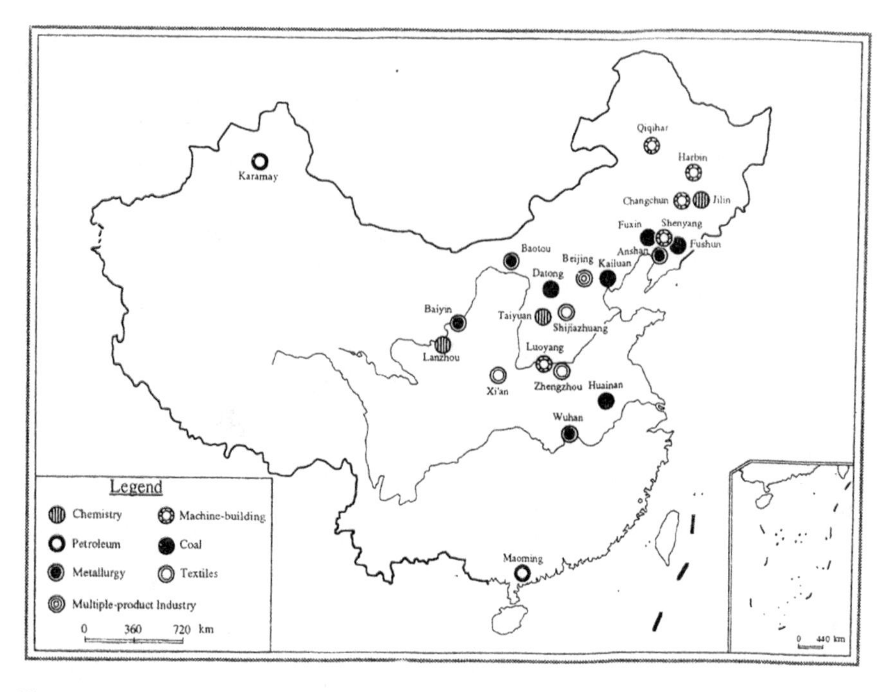

Fig. 11.10 156 key construction projects in the First Five-Year Plan (*Source* Li & Lu, 1995, p. 55)

11.2.2.3 Implementation of the First-Five-Year Plan

(1) *Implementation* via *complete administrative means*

The implementation of industrialization prescribed in The Plan was entirely carried out through administrative means. And by focusing on heavy industry, modernizing national defense, as well as developing technological human resources, the implementation was a success as evidenced by the graduate change of shares in heavy and light industries during the period illustrated in Table 11.5.

(2) *Transfer of ownership accelerated*

According to the socialistic ideology, the main goal in the implementation process of The Plan was to gradually transfer all private sectors into public ownership via

Table 11.5 Share of Industrial Production (%) (*Source* SPC, 1956)

Year	Heavy Industry	Light Industry
1953	37.3	62.7
1954	38.5	61.5
1955	41.7	58.3
1956	45.4	54.5
1957	48.4	51.6

socialist transformation. This process was accelerated at a higher rate than planned as state-owned firms increased their trade turnover from 14.9% in 1950 to 65.7% in 1957 while the trade turnover of private businesses declined sharply from 85% in 1950 to 2.7% in 1957.

(3) *Growth rate of various sectors in The Plan and some issues*

Table 11.6 illustrates the value of national income of various sectors and their respective growth rate. As shown, the agricultural sector had the lowest growth rate among all sectors. The commerce sector had an adequate growth rate in The Plan period. However, due to the neglect of service sector inherent to the Soviet Union planning concept, there were declines in service units and employment unit during The Plan period as shown in Table 11.7.

Table 11.6 National income and its growth rate of the First Five-Year Plan (*Source* NBS, 1990)

Year	National income Unit: 0.1 billion yuan	Agriculture	Industry	Construction	Transport	Commerce
1952	589	340	115	21	25	88
1953	709	374	156	28	29	122
1954	748	388	174	26	32	128
1955	788	417	179	30	33	129
1956	882	439	212	55	37	139
1957	908	425	257	45	39	142
Average annual Growth Rate	9.0%	4.6%	17.4%	16.5%	9.3%	10.0%

Table 11.7 Retail sales, units and employment by 10,000 population (*Source* Wang, Li, & Zhou, 1989, p. 502)

Item	1952	1957
Amount of total retail sales of commodities unit:10,000 yuan)	48.15	73.34
Number of units (unit)	95.6	41.8
Retail Sales	74.0	30.2
Restaurant	14.8	7.3
Service	7.8	4.3
Number of personnel (person)	165.8	117.6
Retail sales	123.6	87.7
Restaurant	25.3	17.9
yService	17.0	11.9

11.2.2.4 Balancing Act in Development Planning

Near the end of the implementation of the First Five-year Plan, several issues emerged due to blindly copying the Soviet Union model. Specifically, focusing too much on heavy industry, neglecting the importance of agriculture, and too high a centralization of decision power to the detriment of local initiative. Former Chairman Mao Zedong summarized these issues in his well-known 1956 talk "On the Ten Major Relationships" to top government leaders. Mao stated that, "in the context of Chinese development, maintenance of a balance was crucial in the relationships:

(i) between heavy industry, light industry and agriculture;
(ii) between industry in coastal region and industry in the interior;
(iii) between civil investment and defense construction;
(iv) between the state, the units of production and the actual producers;
(v) between the central and local authorities;
(vi) between Han, that is the majority nationality, and the minority nationalities;
(vii) between Communist Party authorities and cadres and nonmember of the Party;
(viii) between different policies fostering revolution rather than counterrevolution;
(ix) between rewarding the correct policy executers and punishing the wrong doers; and
(x) between China and foreign countries. (Bagchi, 1987, p.100).

Mao (1956) also emphasized the need to adjust the share of investment in heavy industry, agriculture, and light industry. He maintained that although heavy industry was still the major focus for investment, share of investment to agriculture and light industry should be increased appropriately. Regarding the relationship between the central and local authority, Mao recognized the contradiction and stated that so long as the central authority leadership was solidified, certain power could be extended to local authority and more independence be given to local government so they could do more work.

11.3 Outline of Science and Technology Long Term Planning 1956–1967 (Revised Draft)

11.3.1 Background

In accompanying the gradual restoration of the national economy since the establishment of the PRC, the State had a vision to develop the national economy in a larger scale in the Second and Third Five-Year Plans. The achievement of economic target relied upon the development of science and technology. The Science Planning Commission was established in the State Council under the leadership of

Premier Zhou Enlai. Several hundred scientists of various disciplines were organized to participate in the preparation of this planning work, in addition, sixteen famous scientists from the Soviet Union were invited to assist the Chinese experts to understand the global trend of development of science and technology. This Outline of Science and Technology Long Term Planning (Draft) (*the Twelve-Year S&T Plan* for short) was completed within seven months and went through several iterations of discussions and revisions. It was approved and disseminated on December 1956 by the Central Committee of the Communist Party of China and the State Council.

11.3.2 Contents of the Twelve-Year S&T Plan

This S&T Plan consisted of nine sections including preface, national important S&T tasks 1956–1967, critical elements of tasks, direction of development of basic science, institutions of scientific research work, set of scientific research institutions, use and training of cadres of science and technology, international cooperation, and concluding remarks.

In the second section (after Preface), 77 items of important science and technology tasks were identified which covered 13 areas including: natural environment and resources; mining and metallurgy; fuel and power; machine building; chemical industry; construction; transport and communication; new technology; national defense; agriculture, forestry, husbandry; medical science; instruments, measurements and national standard; and basic science research.

In the third section, twelve important tasks were further identified. They included:

(1) peaceful use of atomic energy;
(2) new technology in radio electronics (super high frequency, semi-conductor technology, electronic computer, electronic instruments and remote control);
(3) jet propulsion;
(4) automation and precision instruments;
(5) petroleum and rare mineral exploitation;
(6) metallurgy of new processes and alloys based on China specific resources conditions;
(7) comprehensive utilization of fuel and development of heavy organic synthesis;
(8) new power equipment and heavy machinery;
(9) critical science and technology issues related to comprehensive development of the Yellow and Yangtze Rivers;
(10) important scientific issues regarding use of chemical fertilizers and agricultural automation;
(11) prevention and eradication of certain diseases most detrimental to the health of Chinese; and
(12) important basic science research.

11.3.3 Impact

Preparation and implementation of the *Twelve-Year S&T Plan* played an important role in the development of science and technology of China. In addition, it was instrumental in shaping the overall structure of science and technology research institutions, the adjustment of disciplines and specialties in higher education, training directions and methodologies, use of teams, and science and technology management.

11.4 Planning in the Period Between 1957–1980

11.4.1 Period Covered

This period covered the time between the Second Five-Year Plan (1958–1962), the adjustment period (1963–1965), the Third and Fourth Five-Year Plan period (1966–1970 and 1971–1975), which included the period of "Cultural Revolution" (1966–1976) and the major part of the Fifth Five-Year Plan period (1976–1980).

11.4.2 The Second Five-Year Plan, the Great Leap Forward and the Adjustment Period

In Sept. 1956, a suggestion regarding the Second Five-Year Plan was put forward at the Eighth Party National Congress. The target suggested was that by the year 1962, the gross value of industrial and agricultural output would be increased by 75% compared to 1957, the steel production would reach 10.5–12 million tons. It was a suggestion adapted to concrete conditions, but it fell into abeyance, due to the coming period of "Great Leap Forward". This name is given to the development policy launch in China at the end of 1957 which was intended to speed the development process by an extraordinary high growth rate. Many targets were modified several times in this period. For example, it was announced in the official document of the Expanded Politburo Conference in 1958 that the steel production would reach 10.7 million tons by the year end, i.e., an amount of production double that of 1957, which was projected to be 5.35 million tons.

This irrational high target was also applied to agriculture development. According to Bo Yibo (1991). The output of grain of China was 370 billion catties and the cotton was 32.8 million dans (1 dan = 50 kgs) in 1957. The initial planned output for grain was 392 billion catties and the output of cotton was 35 million dans, but they were revised to be 431.6 billion catties and 40.93 million dans. Further, it was determined by Conference of Beidaihe that the target of output of grain would be 800–1000 billion catties and the output of cotton 90–100 dans, respectively, in 1959 (pp. 687–689). At the same time, by the end of 1958 nearly all Chinese peasants were organized into People's Communes averaging 5000 household each and all private owned property was taken by the communes.

The result of these irrational movement caused significant decline of production in agriculture and light industry. The agriculture output declined continuously for three years since 1959 with an average annual decline rate around 9.7%. There was also continuous decline in light industry for three years and budget deficit for four years. Under these conditions, corrective measures were introduced by the Party. The Third Five-Year plan was originally intended to cover 1963–1967, but it was revised. Premier Zhou Enlai announced at the First Session of the Third National People's Congress in the beginning of 1963 that 1966 would be the first year of the Third Five-Year plan. The period from 1963–1965 become a period of "Adjustment, Consolidation, Enrichment and Improvement." The national economy of China was fully recovered to a normal state around 1964. The investment in basic infrastructure and the physical infrastructure development status in this period are shown in Tables 11.8 and 11.9 respectively. Major Industrial Centers constructed in 1960's is shown in Fig. 11.11.

Table 11.8 Investment in basic infrastructure in the Second Five-Year and Adjustment Period (*Unit: Hundred million yuan*) (*Source* Zeng, 1999, p. 313)

Sector Year	Energy	Transport	Post & Communication	Water Conservancy	National Total
1958	41.27	32.04	1.95	24.25	266.96
1959	57.82	50.72	2.52	29.38	344.65
1960	63.44	53.48	2.55	35.63	384.07
1961	27.46	14.5	0.46	11.05	123.37
1962	15.69	4.85	0.23	8.39	67.62
Total Second Five-Year Plan	205.68	155.59	7.71	94.45	1186.87
Share of national total	17.3	13.1	0.6	8	100
1963	16.5	7.54	0.26	12.49	94.16
1964	21.67	14.9	0.57	15.65	138.69
1965	25.62	29.18	1.33	15.89	170.89
Sum of Three Years	63.79	51.62	2.16	42.35	403.74
Share of national Total (%)	15.8	12.8	0.5	10.5	100

Table 11.9 Physical infrastructure development status (*Source* Zeng, 1999, p.314)

Type	1957	1962	1965	Growth rate of 1962 compared to 1957(%)	Growth rate of 1965 compared to 1962(%)
Mileage of railway open to traffic (10^4 km)	2.99	3.46	3.64	15.7	5.2
Mileage of highway open to traffic (10^4 km)	25.46	46.35	51.45	82.1	11
Mileage of inland river open to traffic (10^4 km)	14.41	16.19	15.77	51.7	−3
Civil aviation routes (10^4 km)	2.64	3.53	3.94	101.5	11.6
Post office/amount (10^4 units)	4.54	4.45	4.38	−2	−1.6
City telephone (104 household)	46.45	69.95	77.11	50.1	10.2
Output of coal (0.1 billion ton)	1.3	2.2	2.32	69.2	5.5
Output of crude oil (10^4 ton)	146	575	1131	293.8	96.7
Natural gas (0.1 billion M^3)	0.7	12.1	11	16 times	-9.1
Electricity generation (0.1 billion kw-hr)	193	458	676	137.3	47.6

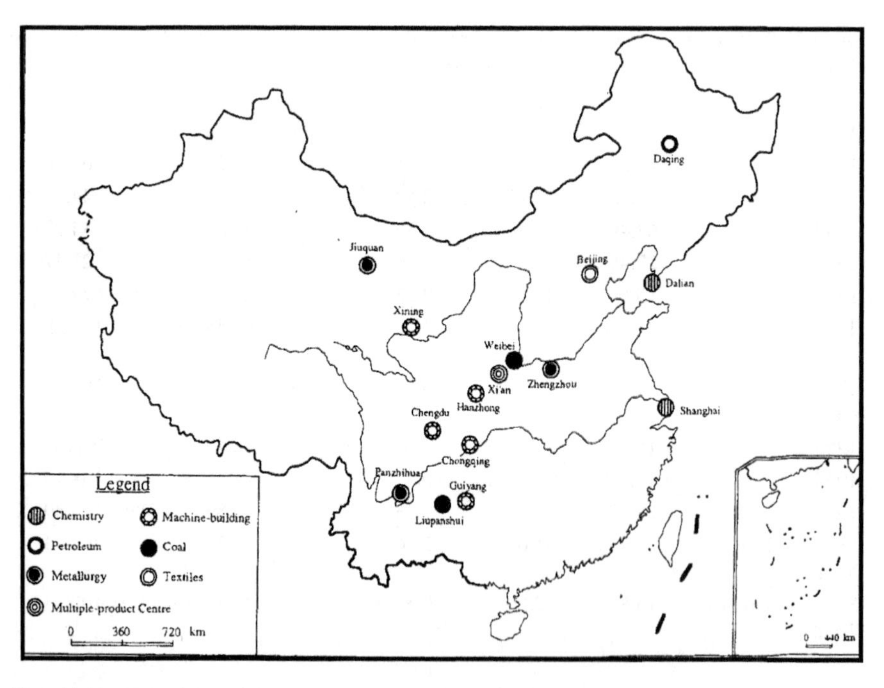

Fig. 11.11 Major industrial centers constructed in the 1960s (*Source* Li & Lu, 1995, p. 58)

11.4.3 *The Third and Fourth Five-Year Planning Periods (1966–1975)*

Since both planning periods occurred during the "Cultural Revolution" and only limited information was available therefore, both Five-Year Plans were in outlines only. Thus, the discussion here focuses on the new phenomenon-*Third Line Construction* introduced in the Third Five-Year Plan. The Third and Fourth Five-Year Planning periods coincided with the "Cultural Revolution" during which turmoil in the country and production disruption was a norm. The decline of gross value of industrial and agricultural output of 4.6 and 4.2% was recorded for 1967 and 1968, respectively. The growth rate of industrial and agricultural output was only 1.4% in 1974. But the Third Line Construction projects built in the era promoted the development of Southwestern region and effectively narrowed the regional disparities between East and West China.

11.4.3.1 Establishing the Third Line Construction and Its Target

The Chinese phrase "sanxian jianshe", translated as third line construction, refers to the *Third Front Movement* during which a massive industrial development in China's interior started in 1964. It involved large-scale investment in national defense, technology, basic industries (including manufacturing, mining, metal, and electricity), transportation and other infrastructures investments in thirteen provinces and autonomous regions in China's central and western regions. This represented a major shift of large scale industry bases from the coastal areas to the interior regions. The Third Line Movement was attributed to China's anticipation for a potential war due to the worsening relationship with USSR and US direct involvement in the Vietnam War.

According to Zeng Peiyan (1999), the SPC submitted its draft plan to the central authority in May 1964 with planned basic tasks for the Third Five-Year Plan as (1) to develop the agricultural sector with full effort, (2) strengthen the basic industry, and (3) develop the transport sector in correspondence. But the Third Line Construction (TLC) item was added by the end of 1964. Larger scale TLC projects were constructed from 1965 up to the mid of 70's. They included several trunk railway lines built in South-Western region to link five major iron and steel bases, twelve coal mining regions established in Guizhou Province, several hydro and thermal power plants constructed in Sichuan and Hubei Provinces, and heavy machinery plant and electric machinery plant built in Sichuan Province, etc.

In the Fourth Five-Year Plan, TLC projects were concentrated in several economic cooperation regions. Sichuan Province had the highest share of investment with 33.5% of the total TLC investment and more than 250 TLC projects. By then, Sichuan's industrial capital assets had even surpassed Shanghai's and ranked next to Liaoning Province. The province became the nation's number one emerging new industrial base.

11.4.3.2 Physical Infrastructure Investment and Development

The investment and development in physical infrastructure in the Third and Fourth Five-Year Planning periods are shown respectively in Tables 11.10 and 11.11. And major industrial centers constructed in 1970's is shown in Fig. 11.12.

11.4.3.3 Summary of Public Policies Implemented in this Period

(1) Heavy industry continued to play a leading role. The share of heavy industry in total industrial production during the Third and Fourth Five-Year Plan period rose to 54.5 and 52.1% respectively, a further increase of 6.1 and 3.1% as compared to that at the end of 1957.
(2) There was further increase of the share of SOEs. The basic policy in this period was to further expand the state ownership, weaken the collective ownership and abolish the private ownership altogether. It can be seen from statistics that even the number of individual workers was reduced from 8.83 million in the 60's down to 0.15 million in 1978. In contrast with the policies implemented before 1978, a rapid increase of individual worker was evidenced in 1979 with 17 million in the urban area (cities and townships), and 33.1 million in 1996. Comparison of these figures can serve a better understanding of the change of policies before and after the launch of economic reform drive.
(3) Localizing production policy. Industrial policy after the Third Five-Year Plan period emphasized local self-sufficiency, which resulted in hundreds of prefectures being involved in wide-ranging production activities.

Table 11.10 Investment in physical infrastructure in the Third Five-Year and Fourth Five-Year Plan (*Source* Zeng, 1999, p. 316). Unit of investment 100 million yuan

Year	Sector				
	Energy	Transport	Post and communication	Water conservancy	National total
1966					199.42
1967					130.52
1968					104.13
1969					185.65
1970					294.99
Total Third Five-Year Plan	155.07	143.11	6.9	69.52	914.71
Share of National Total (%)	17	15.6	0.07	7.6	100
1971					321.45
1972					312.79
1973					321.26
1974					333.01
1975	71.46	68.67	3.4		391.86
Sum of Fourth Five Years	310.89	302.47	15.12	119.21	1880.37
Share of National Total (%)	16.5	16.1	0.8	7.1	100

Table 11.11 Physical infrastructure development in the Third Five and Fourth Five-Year Plan (*Source*: Zeng, 1999, p. 317). Unit of investment 100 million yuan

Type	1957	1962	1965	Growth rate of 1962 compared to 1957(%)	Growth rate of 1965 compared to 1962(%)
Mileage of railway open to traffic (10^4 km)	3.64	4.1	4.6	12.6	12.2
Mileage of highway open to traffic (10^4 km)	51.45	63.67	78.36	23.8	23.1
Mileage of inland river open to traffic (10^4 km)	15.77	14.84	13.56	−5.1	−8.6
Civil aviation routes (10^4 km)	3.94	4.06	8.42	3	104.4
Pipe transport (10^4 km)	0.04	0.12	0.53	2 times	3.4 times
Post office/Amount (10^4 units)	4.38	4.5	4.87	2.7	8.2
City telephone (10^4 household)	77.11	78.41	103.28	1.7	31.7
Output of coal (0.1 billion ton)	2.32	3.54	4.82	52.5	36.2
Output of crude oil (10^4 ton)	1131	3065	7706	171	151.4
Natural gas (0.1 billion M^3)	11	28.7	88.5	160.9	208.4
Electricity generation (0.1 billion kw-hr)	676	1159	1958	71.4	68.9

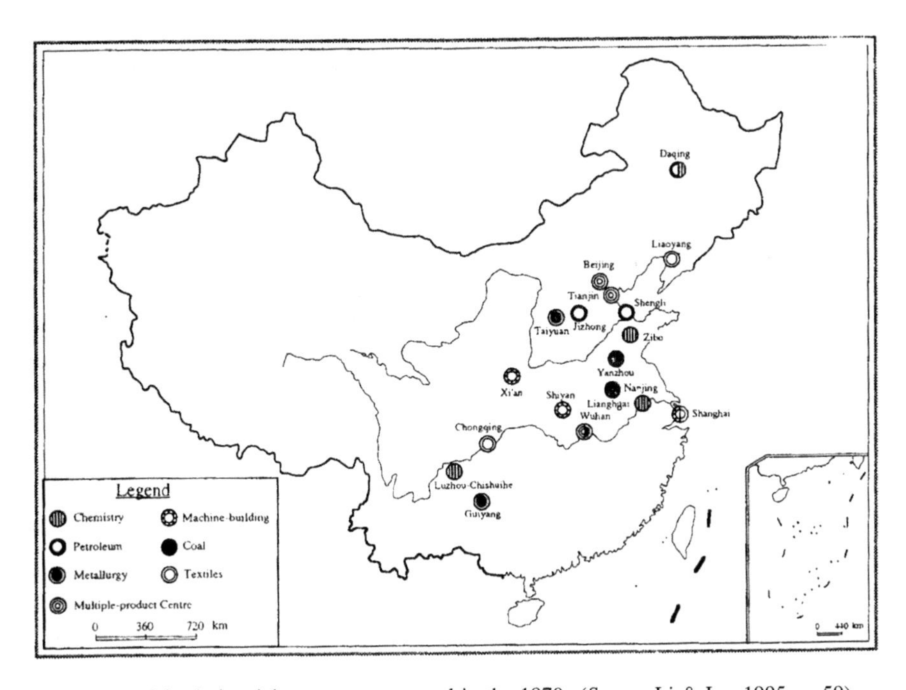

Fig. 11.12 Major industrial centers constructed in the 1970s (*Source* Li & Lu, 1995, p. 59)

(4) Foreign Trade Policy. Due to the absence of diplomatic relations with many Western countries as well as the trade embargo imposed by them, China had only established foreign trade relations with the former Soviet Union and other Eastern European countries. After the normalization of international relations in the 1970's, China's export/GNP and import/GNP is only 4.5 and 4.7% respectively in 1978. While in 1995, these two ratios rose to 20.3 and 15.8% respectively. This also serves to illustrate the impact of the opening policy on foreign trade since the launch of reform and opening drive in late 1978.

11.4.4 The Fifth Five-Year Planning Period (1976–1980)

11.4.4.1 The Fifth Five-Year Plan

The content of the Fifth Five-Year Plan was not well worked out due to the transition from the "Cultural Revolution" political upheaval to the normalization of political, social and economic order of China. Also, there was major changes in the planning scheme which caused inconsistency. Instead of having a stand-alone five-year plan, the central government stipulated the 1976–1985 Ten Year Plan Outline of Developing National Economy (Draft) in 1975, which included the Fifth Five-Year Plan.

This Ten-Year Development Plan, approved by the First Plenary Session of the Fifth Peoples' Congress in the beginning of 1978, raised a series of high targets which were too ambitious and difficult to implement given the available financial resources. For example, the original plan set targets to have (1) a total investment amount on capital construction equivalent to the sum of investment of the previous 28 years; (2) 120 large projects be constructed or be continued from 1978–1985; (3) the output of grain reach 400 million tons; and (4) the steel output reach 60 million tons by 1985. And there was a serious disproportion of investment on heavy and light industries in 1978, with the share of investment of heavy and light industries being 50.9 and 6.1% respectively. The Third Plenary Session of Eleventh Party Central Committee was held in late 1978 which adopted a series of corrective measures. These measures include: accelerating the development of agriculture and light industry; raising the procurement price of agricultural products by 30.8% in 1979 and 1980 to incentivize farmers to raise agricultural production. The distribution of national income was also adjusted to reduce the rate of investment (from 36.5 to 31.6%) and to increase the rate of consumption (from 63.5 to 68.4%).

11.4.4.2 Physical Infrastructure Investment and Development

The major projects completed in this period included Beijing-Shanghai-Hangzhou 1700 km Coaxial cable with 1800 channel carrier Communication line, Dalian New

Table 11.12 Investment in Physical Infrastructure in the Fifth Five-Year Plan (*Source* Zeng, 1999, p. 318)

Year	Sector				
	Energy	Transport	Post and communication	Water conservancy	National total
1976	69.88	54.51	3.24	28.17	376.44
1977	78.69	49.98	3.25	28.47	382.37
1978	114.7	64.05	3.99	34.68	500.99
1979	110.98	60.34	3.75	34.96	523.48
1980	115.65	58.5	3.84	26.53	558.89
Total Fifth Five-Year Plan	489.9	284.38	18.07	152.81	2342.17
Share of national total (%)	20.9	12.1	0.8	6.5	100

Table 11.13 Physical infrastructure development in the Fifth Five-Year Plan (*Source* Zeng, 1999, p. 319)

Type	1975	1980	Growth rate of 1980 compared to 1975(%)
Mileage of railway open to traffic (10^4 km)	4.6	4.99	8.5
Mileage of highway open to traffic (10^4 km)	78.36	88.33	12.7
Mileage of inland river open to traffic (10^4 km)	13.56	10.85	−20
Civil aviation routes (10^4 km)	8.42	19.53	131.9
Pipe transport (10^4 km)	0.53	0.87	64.2
Post office/amount (10^4 units)	4.87	4.95	1.6
City telephone (10^4 household)	103.28	134.17	29.9
Output of coal (0.1 billion ton)	4.82	6.2	28.6
Output of Crude oil (10^4 ton)	7706	10,595	37.5
Natural Gas (0.1 billion M^3)	88.5	142.7	61.2
Electricity generation (0.1 billion kw-hr)	1958	3006	53.5

Port—China's first modern deep water oil port for hundred thousand tons, Beijing airport, and several hydro and thermal power plants etc. The total investment in these physical infrastructures and their development are shown in Tables 11.12 and 11.13 as follows.

11.4.4.3 Imports of Complete Sets of Equipment

Owing to the restoration of PRC's lawful seat in the United Nations in 1971, China started importing large amount of complete sets of equipment from Western countries. The investment of imported equipment reached 1.28 billion U.S.D. in 1977 and 5.03 billion U.S.D. in 1980. In 1973, the SPC submitted a report to the

State Council to import complete sets and single units of equipment amounting to 4.3 billion U.S.D. Items of equipment were imported in 1973 with contract turnover reaching 1.22 billion U.S.D. The imported equipment included thirteen sets of large chemical fertilizers, four sets of large chemical fiber, two petrochemical combined production plants, and mechanized coal mining equipment, complete sets of power generation, transmission and distribution equipment etc. All these projects played important roles towards upgrading China's manufacturing technology and capacity. However, due to the carry-over of the Cultural Revolution aftermath, some of the above-mentioned projects did not complete until the Six Five-Years Planning periods and some imported equipment was not even being utilized at all.

11.5 Analysis of China's Planning Performance (1953–1980) from the Perspective of Social Systems Engineering

11.5.1 Characteristics of China's Economic System in the 1953–1980 Period

To a contain extent, two major characteristics affected China's economic system and performance during the 1953–1980 period which covered five 5-Year national plans:

11.5.1.1 Centrally Planned Economy

China's economy was a centrally planned economy which according to Pass, Lowes and Davies (2005) is characterized by: "economic decision-making is centralized in the hands of the state with state and collective ownership of the means of production, (except labor). It is the state that decides what goods and services are to be produced with the centralized National Plan. Resources are allocated between producing units, and final output between customers by the use of physical quota."

11.5.1.2 Semi-closed Economy

Chinese economic system was a semi-closed one in most of the 1953–1980 period. It has turned into an open system gradually since its lawful seat in the United Nations was restored and the opening process has gradually accelerated since it launched reforms and opened to the world in the late 1970s.

11.5.2 Analysis of China's Socio-Economic Development Using Revised Parsons' AGIL Framework

In this section, the performance of China's socio-economic development in the centrally planning period will be analyzed using revised Parsons' AGIL framework. Specifically, three subsystems, i.e., economic, S&T, and social will be analyzed respectively and the overall performance of the national system will be examined.

11.5.2.1 The Economic Subsystem Performance

This period experienced relatively high national economic growth rate and imbalanced regional economic growth rates; a high growth rate in industrial products; a structural change in national economy; highly skewed investment share in the heavy industry; and low participation in foreign trade.

(1) *High Economic Growth and Imbalanced Regional Development*

Based on Figs. 11.13 and 11.14, although there were several setbacks in the periods, such as the Great Leap Forward and Cultural Revolution, China still

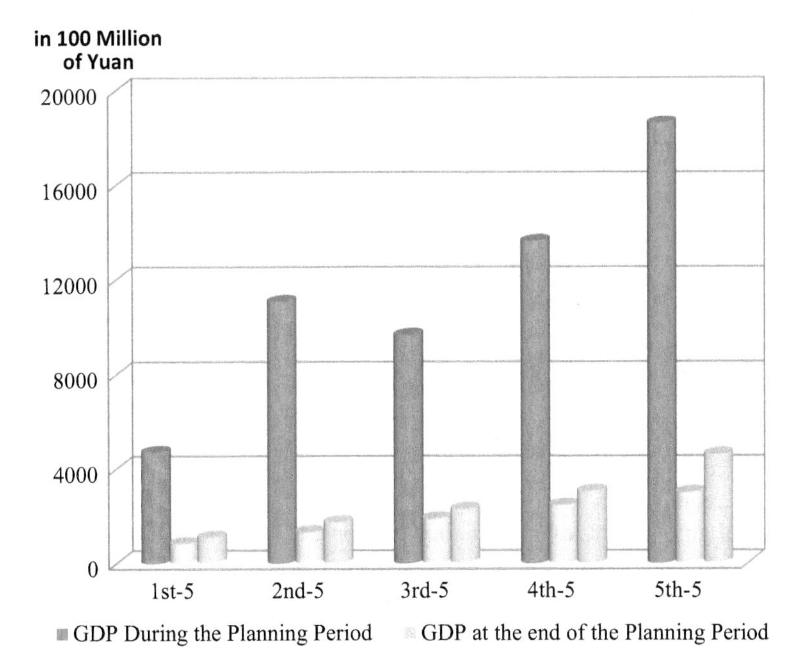

Fig. 11.13 Performance of China's national economy in the 1953–1980 Period (*Source* NBS http://data.stats.gov.cn)

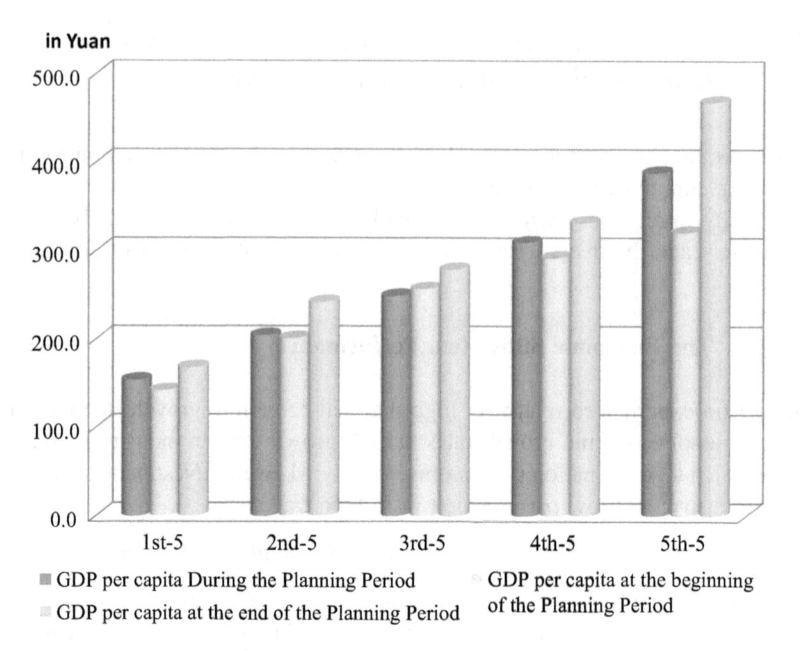

Fig. 11.14 Performance of China's GDP per capita in the period of the 1953–1980 Period (*Source* NBS http://data.stats.gov.cn)

enjoyed a relatively high average annual national economic growth rate of around 5.1% from 1952 to 1980 and the growth rate of GDP per capita was around 4.5%. However, due to the country's large size and pre-development condition prior to 1953, balance of regional development remained a challenge as evidenced in Figs. 11.15 and 11.16 and Table 11.14 with comparison data for regional GDP per capita between 1952 and 1980. And despite the effort of the planning system to pursue a relatively balanced regional development, this task could hardly be achieved. In Fig. 11.15, the ratio of the region of highest GDP to lowest GDP is 36.67 in 1952 (ratio between Shanghai to Tibet). That same ratio becomes 35.97 in 1980. Table 11.14 shows an increase of regional disparity based upon regional GDP per capita. The ratio of highest GDP per capita to lowest GDP per capita was 8.1 (ratio between Shanghai to Tibet) in 1952 and it became 12.44 (ratio between Shanghai to Guizhou) in 1980. However, the ratio between Shanghai and Tibet in 1980 was reduced to 5.8, which may be attributed to specific preferential policies of the central government towards minority regions. Regardless, regional balance is a difficult task for a country with large size such as China. Development with the objective of regional balance to prescribed extents is challenging for any country with a large size and diverse population. In view of this perspective of outcomes, it can hardly be expected uniformly that countries will achieve equal benefits in the process of globalization, although it is a megatrend of development of the global society. More efforts should be contributed by members of the global society to solve the issue.

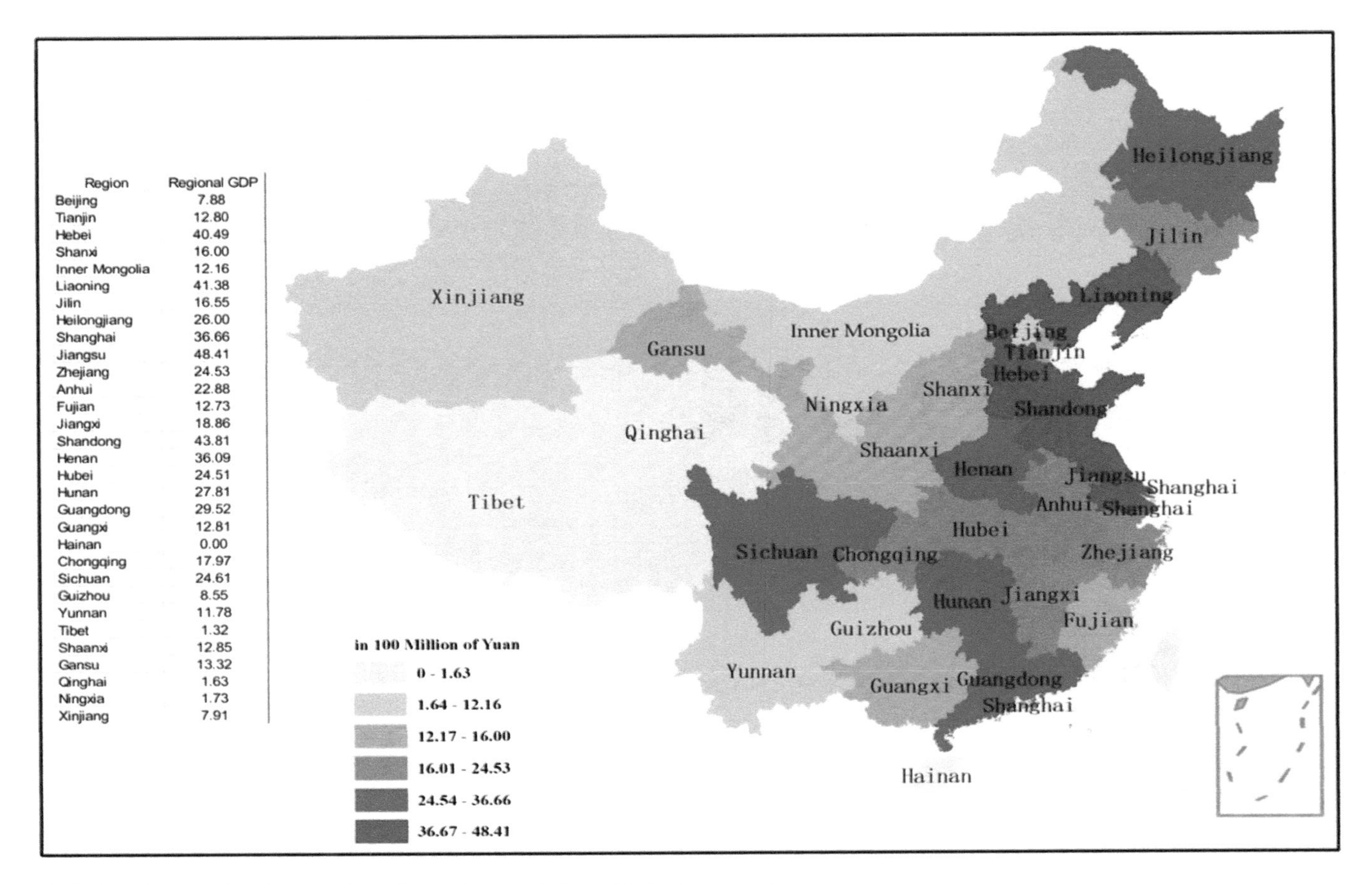

Fig. 11.15 Distribution of regional GDP in 1952 (*Source* NBS http://data.stats.gov.cn)

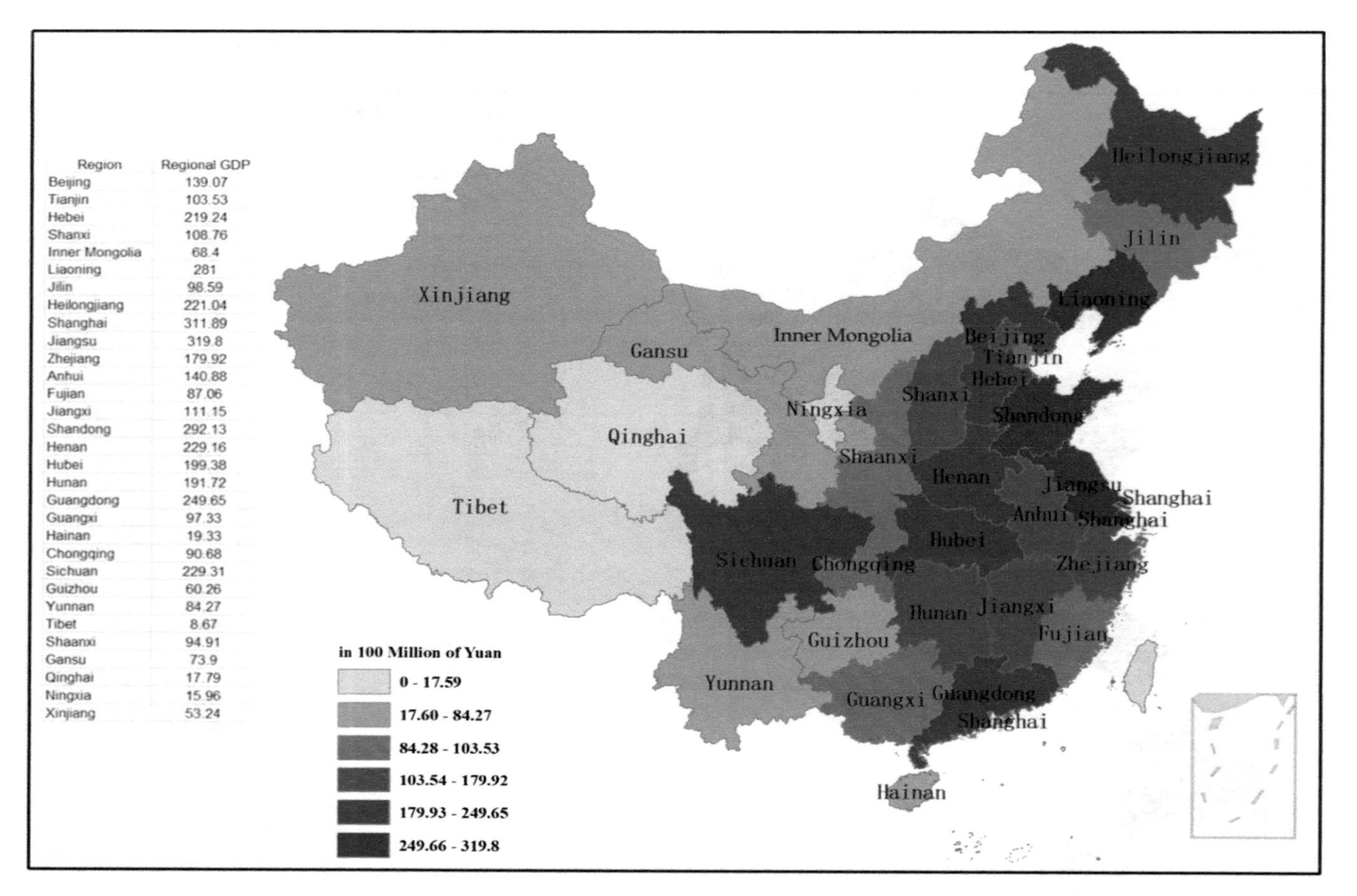

Region	Regional GDP
Beijing	139.07
Tianjin	103.53
Hebei	219.24
Shanxi	108.76
Inner Mongolia	68.4
Liaoning	281
Jilin	98.59
Heilongjiang	221.04
Shanghai	311.89
Jiangsu	319.8
Zhejiang	179.92
Anhui	140.88
Fujian	87.06
Jiangxi	111.15
Shandong	292.13
Henan	229.16
Hubei	199.38
Hunan	191.72
Guangdong	249.65
Guangxi	97.33
Hainan	19.33
Chongqing	90.68
Sichuan	229.31
Guizhou	60.26
Yunnan	84.27
Tibet	8.67
Shaanxi	94.91
Gansu	73.9
Qinghai	17.79
Ningxia	15.96
Xinjiang	53.24

Fig. 11.16 Distribution of regional GDP in 1980 (*Source* NBS http://data.stats.gov.cn)

Table 11.14 Comparison of regional GDP per capita between 1952 and 1980 (*Source* NBS http://data.stats.gov.cn)

Name of Regions (provinces, autonomous regions, central administered municipalities	Regional GDP per capita 1952	Regional GDP per capita 1980
Beijing	165	1544
Tianjin	298	1357
Hebei	125	427
Shanxi	116	442
The Neimenggu Autonomous Region	173	361
Liaoning	218	811
Jilin	153	445
Heilong Jiang	238	694
Shanghai	430	2725
Jiangsu	130.93	541
Zhejiang	112	471
Anhui	77.82	291
Fujian	102	348
Jiangxi	114	342
Shandong	91	402
Henan	82.8	316.7
Hubei	90.13	427.98
Hunan	86	365
Guangdong	101	481
The Guangxi Zhuang Autonomous Region	66.57	277.77
Hainan	0	354
Chongqing	103	357
Sichuan	53	320
Guizhou	58	219
Yunnan	70	267
Tibet Autonomous Region	115	471
Shanxi	85	336.68
Gansu	125	388
Qinghai	101	471
The Ningxia Hui Autonomous Region	126	433
The Xinjiang Uygur Autonomous Region	166	419

(2) *High Growth Rate of Industrial Products*

This period witnessed high growth rate in various industrial products across different industrial branches. Table 11.15 portrays production and growth rate of selected industry products in 1952 versus 1980.

Table 11.15 Growth of selected industrial products (1952 vs. 1980) (*Source* Wang, 2003, p. 50)

S/N	Name of products	Units	1952	1980	Average growth rate per year (%)
1	Pig iron	Million tons	1.93	38.02	11.2
	Steel	Million tons	1.35	37.12	12.5
2	Electricity	Billion KWH	7.3	300.6	14.2
3	Crude coal	Million tons	66	620	8.3
4	Crude oil	Million tons	0.44	105.95	21.6
5	Chemical fertilizer	10^3 tons	39	12,320	22.8
	Caustic soda	10^3 tons	79	1923	12
	Plastics	10^3 tons	2	898	24.3
6	Power generating equipment	MW	6	4193	26.3
	Mining equipment	10^3 tons	1.8	163	17.4
	Machine tools	10^3 tons	13.7	134	8.4
	Internal Combustion engines	10^3 tons	40	25,390	25.9
7	Cement	Million tons	2.86	79.86	12.6
	Plate glass	Million std cases	2.13	27.71	9.5
8	Timber	Million cu.meters	11.2	53.59	5.7
9	Sugar	10^3 tons	451	2570	6.4
	Salt	10^3 tons	4945	17,280	4.5
	Tobacco	10^4 tons	265	1520	6.4
	Wine	10^4 tons	23	368.5	10.4
10	Cotton yarn	103 tons	656	2930	5.4
	Cotton cloth	Billion meter	3.83	13.47	4.6
	Woolen piece goods	Million meter	4.23	101	11.9

(3) *Structural Change*

This period also observed structural change of the national economy. As shown in Tables 11.16 and 11.17, in the process of industrialization, there was a general decline of share of GDP and employment in the primary sector. However, Table 11.17 shows an increase of employment in the primary sector with average annual growth rate around 1.9% in the period from 1960–1970. This was caused by a robust population growth in this period due to improved health care and living conditions. The death rate was reduced from 17 per thousand population in 1952 to 6.34 per thousand in 1980. The rural population was increased from 50,970 × 104 people in 1952 to 79,545 × 104 people in 1980.

Table 11.16 Structural change of China's economy (1950–1980) (*Source* Wang & Li, 1995, p. 5)

	1950	1957	1970	1980
Primary sector	50.5	40.6	35.8	31.3
Secondary sector	20.8	29.2	39.8	48.2
Tertiary sector	28.7	30.2	24.4	20.5

Table 11.17 Change of employment share in major economic sectors (1952–1980) (*Source* Wang & Li, 1995, p. 8)

Unit:10,000 people

Year	Primary sector		Secondary sector		Tertiary sector		Total
	Employment	Share of total (%)	Employment	Share of total (%)	Employment	Share of total (%)	Employment
1952	17,317	83.6	1531	7.4	1881	9	20,729
1960	17,016	65.7	4112	15.9	4752	18.4	25,880
1970	27,811	80.8	3518	10.2	3103	9.0	34,432
1980	29,117	68.7	7736	18.3	5508	19.8	42,361

It can be seen from Table 11.17 an unstable growth of employment in the tertiary sector, having shares in national economy of 9, 18.4, 9.0 and 19.8% respectively in 1952, 1960, 1970 and 1980 respectively. The rapid decline into 9% in 1970 was due to the discrimination against the tertiary sector of MPS formerly adopted by USSR. And it was a prevailing phenomenon of underdeveloped tertiary sector in countries that adopted the model of development of the former USSR.

(4) *Investment*

The relatively high growth rate of China's economy depends heavily upon the high share of investment of GDP on the demand side. Due to the unique statistical system adapted by China in the period from 1953–1980, the Chinese data available required some adjustment to make meaningful analysis. The World Bank (1981), after adapting the Chinese data to the Western national accounting standard, developed a table to show China's share of investment as a % of total material expenditure and a % of GDP in comparison with other countries. For illustration purpose, the table is included in Fig. 11.17. According to the World Bank, the data from the past three decades demonstrated the rising trend in China's share of aggregate resources devoted to investment. At constant prices, the ratio of investment to total material expenditure rose between 1952 and 1979 from 21 to 34%. At constant prices, the rise was even steeper from 13 to 36%. The share of investment in GDP was about 23% in 1957 and 31% in 1979-much higher than in other low-income countries (21%) and middle-income counties (25%) (pp. 48–49).

China's lopsided investment in heavy industry was the unique feature during the 1953–1980 period. Table 11.18 clearly illustrates this feature by showing the % share of capital investment to industrial sectors.

(5) *Foreign Trade*

Since the establishment of PRC in 1949, China had nearly two decades of semi-isolation from external world commerce, until the normalization of international relations in 1970s. Based upon the estimation of The World Bank (1981), China's trade to GNP ratio average for 1977 and 1978 remained one of the lowest

Table 3.10: INVESTMENT SHARE

	China /a			India	Indonesia	Low-income countries	Middle-income countries
	1952	1957	1979			1978	
Investment as % of:							
Total material expenditure	21.4	24.0	33.6	n.a.	n.a.	n.a.	n.a.
Gross domestic product	n.a.	23.2	31.1 (27.1)/b	24	20	21	25

/a At current prices.
/b Adjusted for relative price differences.

Fig. 11.17 World Bank comparison investment share data table (*Source* The World Bank, 1981, p. 49)

Table 11.18 Investment share in industrial sectors in different planning periods (*Source* NBS, 1982)

Industrial sector	1st Five-Year planning period (%)	2nd Five-Year planning Period (%)	1963–1965 (%)	3rd Five-Year planning period (%)	4th Five-Year planning period (%)	5th Five-Year planning period (%)
Light industry	15	10.5	7.8	7.9	10.5	12.6
Heavy industry	85	89.5	92.2	92.1	89.5	87.4

in the world, its Export/GNP and Imports/GNP were 4.5 and 4.7% respectively while these two ratios were 15.5 and 16% respectively for all developing countries in 1978 (p. 110). This clearly shows that China's socio-economic system was a semi-closed system in the 1953–1980 period.

It can be seen from Fig. 11.18 that there was trade deficit in the First and Fifth Five-year planning periods. This was very much related to the industrialization process of China. In the First Five-Year planning period, China had imported complete sets of machinery and equipment from USSR mostly related to the Soviet-assisted 156 core projects, the rising level of imports was financed by exports of agricultural and processed agricultural products (81.6% of total exports in 1953 and 71.6% in 1957). The remaining part of export was industrial and mining products. The USSR had a dominant share of 50.6% of total trade with China in this period. The other former socialist countries had a share around 18.2%. Already developed Hong Kong & Macao had a trade share around 9 and 7.6% respectively. The remaining part was shared among developing countries. In the Fifth Five-Year planning period, China imported complete sets of machinery equipment from non-socialist developed countries and these countries had a

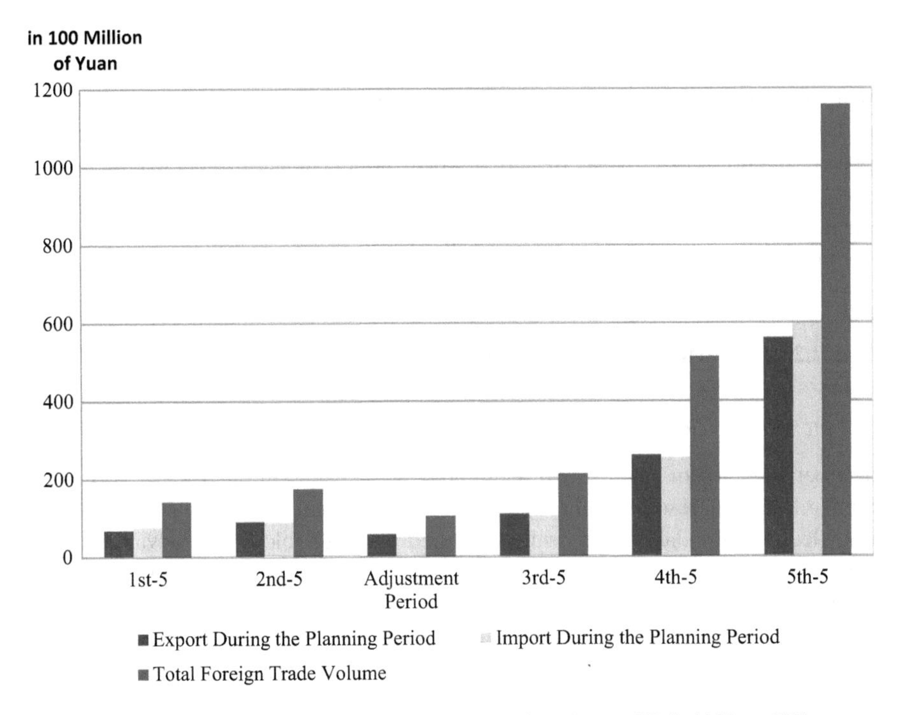

Fig. 11.18 China's Foreign trade in the 1953–1980 Period (*Source* NBS, 1982, p. 353)

dominant share around 58.8% of total imports of China in this period, and Japan alone had a share of 27.3%, higher than any other developed countries in that period.

The industrialization process of China in the period from 1953 to 1980 was a period of heavy import of technology and extensive learning from all advanced experiences worldwide. As a result, it created a large national capacity of all around manufacturing which enabled China to transfer into an export oriented country in later periods.

The structure of commodities of import in the First Five-Year Plan had a higher share of producer goods and very low share of consumer goods, with a ratio around 92:8. Although the share of consumer goods once raised to 33–44.5% in the 1961–1965 period, it further declined to 13.2–23.9% in Fifth Five-Year Plan period.

(6) *Brief Comments*

China did not set a long-term goal for economic growth from 1953–1980, there were only mid-term Five-Year plans in this period. The only long-term goal was political, i.e. to establish a socialist country. It can be seen from the Fifth Five-Year Plan that there was no target set for economic growth measured in terms of national income. This was different from the indicative planning by France for example, whose target of economic growth was set up through forecasting technique. But in the centrally planned economy of China, the investment plan was very detailed. As a whole, the growth of China's economy was relatively successful in this period with an average annual economic growth rate around 5.1% of GDP. In the process of China's industrialization, Chinese culture and people's innate work ethics also played a positive role when the government policy incentivized their motivation and innovation. Integration between central and local government was experimented through several cycles of centralization and decentralization, but was dominated by centralization.

11.5.2.2 The science and Technology Subsystem Performance

(1) *Increasing Human Capital*

The performance of science and technology was successful, especially in the creation of human resources through transfer of technology, advancing skills and knowledge via collaboration with developed and technologically advanced countries, and creation and enhancement of institutions in research and higher learning. Table 11.19 portrays the growth in personnel counts in science and technology in the SOEs. The fact that the NBS still reported its annual data using the MPS system adopted from USSR rather than the SNA system used globally also exemplify the difficulty in reforming certain learned culture and practice in national management.

(2) *Creation and Expanding R&D Institutions*

Before 1949, there were just over 600 people engaged in scientific research, and research closely linked to industrial production was practically nonexistent.

Table 11.19 Growth of science and technological personnel in natural science from 1952–1980 (*Source* NBS, 1985, p. 593)

Item	Unit	1952	1980	Average annual growth rate (%)
Total number of S&T personal	10^4 person (share of total)	42.5	527.6	9.40
Engineering technological personnel	10^4 person (share of total)	16.4	186.2	9.10
Agricultural technological personnel	10^4 person	1.5	31.1	11.40
Hygienical technological personnel	10^4 person	12.6	153	9.30
Research personnel of Science	10^4 person	0.8	32.3	14.10
Teaching staff		11.2	125	9.00

Note The First Five-Year Plan classified teaching staff/faculty of higher learning as research personnel. Based on the 1981 statistics, teaching staff/faculty in natural science and engineering in 1980 are around 179,953

Through more than thirty years of effort, China established and enhanced six types of R&D institutions by the end of 1980, which significantly strengthened its human capital in of science and technology. Below briefly describe several major types of R&D institutions:

(i) The Chinese Academic of science—had around 114 research institutes and more than 30,000 of researchers and staff by the end of 1980.
(ii) Ministerial and provincial R&D units—Most ministries under the State Council and nearly all provinces, autonomous regions and centrally administered municipalities had set up various types of R&D institutions of their own, such as engineering design institutes, research institutes for constructions, textiles, communications, etc.
(iii) Research units for national deference under the military domain—Various special research units established in military forces for national defense included atomic research, aerospace, ship building, etc. Their achievements were evidenced in China's first nuclear weapon test, detonated on Oct. 16th, 1964, and the launch of China's first satellite on Apr. 24th, 1970.
(iv) Research units in factories and mines—At grassroots levels, factories and mines established their own research units to tackle product, production and operation specific issues.
(v) Research units of higher learning—These were research organizations at colleges and universities. There were more than 1,000 such units with over 25,000 S&T personnel.
(vi) Chinese Academy of Social Science—The Chinese Academy of Social Science, originally part of the Academy of Science, was formally established in 1977. By 1980, there were around 32 research institutes at its headquarters and 134 branch research institutes in 28 provinces, autonomous regions and centrally administered municipalities.

(3) *Brief Comments*

Before 1949, China was very weak in the science and technology frontier. However, during the 1953–1980 period, there was a rapid development in science and technology systems. The phenomenal growth was attributed to several factors including import of both hardware and software technology, establishing and enhancing various types of research institutions at the central and grass roots levels, training and educating science and technology personnel both at home and abroad. China's preparation and implementation of the Twelve-Year S&T Plan played a decisive role in the overall management of scientific research institutions. In short, China created a critical mass of S&T personnel through the construction of the 156 core projects, the import of complete sets of various industrial equipment from developed countries since late 1960s, and the mastering of latest knowledge and technical skills. The S&T system created in this period paved the foundation for China's outstanding performance in its manufacturing sector when it implemented reform and opening policy in the late 1970s.

11.5.2.3 The Social Subsystem Performance

(1) *General Wellbeing of Chinese People*

China followed the USSR model closely on social aspects, Labor insurance regulation was established and promulgated by the Government Council in 1951. It was revised and promulgated in 1953 with pension system and free medical care system (designated SOEs' responsibility) for workers and staffs outlined. This system was generous as it even provided coverage for a fixed insurance and social welfare for lineal relatives. Maternity leave was also included as part of the benefits.

Basic needs of urban people were met with low cost grain, cooking oil, and clothes rationing. Free housing and furniture for workers and staffs were provided by the state or the SOEs. The price was stable (except the Great Leap Forward period) with only 1.3% average annual rise in prices of general retail sales from 1950 to 1980. The wage system was egalitarian, wage variance among sectors were not high as shown in Table 11.20.

(2) *The Education System*

There was great improvement of education system at all levels which is shown in Table 11.21.

But there are observations and lessons on the higher learning of China from this period. According to Kraus (1982) "China followed the example of the Soviet Union, which had concentrated on the creation of highly specialized training establishments, whereby the importance of universities had declined sharply." Furthermore, "It also seems that colleges and universities schematically adopted the study plans and study regulations then current in the specialized instruction according to Soviet example; it seemed that the specific Chinese tradition which placed the main emphasis on right thinking and self-cultivation had come to an end." (p. 23)

(3) *Brief Comments*

It was a great practice for China to follow the social insurance and welfare system of USSR during the period. But such a system is not sustainable under today's economic environment in China as the cost of maintaining such a welfare system is simply too high for the state. Also, there is a pressing need to modify China's higher education system modeled after the USSR protocol from the 50's.

11.5.2.4 Overall Analysis of China's Planning Performance Using Revised Parsons' AGIL Framework

(1) *Goal*

In view of disruptions during this period, such as the Great Leap Forward and People's Commune in the late 1950s, there was a change in time setting to accomplish the goal of industrialization and the goal towards achieving high percentage in cooperative movement in the rural area. With the introduction of the TLC

Table 11.20 Average wages of employees at SOEs of various sectors (*Source* NBS, 1982, p. 426)

Year	Sector 1	Sector 2	Sector 3	Sector 4	Sector 5	Sector 6	Sector 7	Sector 8	Sector 9	Sector 10
1953	496	576	591	433	643	381	650	392	498	423
1957	637	690	744	501	752	529	651	580	613	631
1958	550	526	595	471	673	489	642	557	586	639
1962	592	652	705	392	702	494	631	542	559	626
1963	641	720	775	421	760	550	672	574	604	658
1965	652	729	730	433	774	579	687	598	624	684
1966	636	689	644	428	755	570	697	583	620	660
1970	609	661	650	419	709	553	660	555	588	678
1971	597	635	662	426	709	539	655	554	604	668
1975	613	644	704	460	699	562	639	574	609	645
1976	605	634	696	459	684	555	621	566	602	636
1980	803	854	923	636	906	723	789	741	760	807

Note *Sector 1-Industry; Sector 2-Construction; Sector 3*-Agriculture Forestry Water Conservancy Meteorology; Sector 4-Transport Post Communication; Sector 5-Commerce; Sector 7-Urban Public Utilities; Sector 8-Science Culture Education Health Care; Sector 9-Finance Insurance; Sector 10-Organ Community

Table 11.21 Students in schools (*Source* NBS, 1982, p. 441)

Year	Total	Hight learning	Middle school			Primary school
			Total	Within		
				Specialized middle school	Ordinary middle school	
1953	5550.50	21.2	362.9	66.8	293.3	5166.40
1957	7180.50	44.1	708.1	77.8	628.1	6428.30
1958	9906.10	66	1199.80	147	852	8640.30
1962	7840.40	83	833.5	53.5	752.8	6923.90
1963	8070.10	75	837.6	45.2	761.6	7157.50
1965	13,120.10	67.4	1431.80	54.7	933.8	11,620.90
1966	11,691.90	53.4	1296.80	47	1249.80	10,341.70
1970	13,181.10	4.8	2648.30	6.4	2641.90	10,528.00
1971	14,368.90	8.3	3149.40	21.8	3127.60	11,211.20
1975	19,681.00	50.1	4536.80	70.7	4466.10	15,094.10
1976	20,967.50	56.5	5905.50	69	5836.50	15,005.50
1980	20,419.20	114.4	5677.80	124.3	5508.10	14,627.00

Table 11.22 Change of ownership in gross value of industrial production (*Source* NBS, 1985, p. 308)

Year	Total	State owned industry	Collectively owned industry	Joint state private owned industry	Private owned industry	Individual owned industry	Other types
1949	100	26.2	0.5	1.6	48.7	23.0	
1952	100	41.5	3.3	4.0	30.6	20.6	
1957	100	53.8	19.0	26.3	0.1	0.8	
1965	100	90.1	9.9				
1980	100	78.7	20.7				0.6
1984	100	73.6	25.0			0.2	1.2

in Third and Fourth Five-Year Plan period, the priority of goal shifted from civil construction to national defense. The Ten-year Culture Revolution had its negative impact on many goals. But the original goals described in the General Line and General Tasks of the Party in the Period of Transition were observed and somewhat achieved. The socialist transformation of industrial sectors can be evidenced from Table 11.22 as it portrays that by the end of 1980, only state and collectively owned industry existed. All other types of industries had totally been transformed.

(2) *Active Adaptive Actions*

Concrete actions were taken to achieve the goals. (Actions to achieve economic, S&T and social goals are described in previous sections). China had achieved its goal of socialist industrialization through close cooperation with USSR before 1960, and turned to importing equipment and technologies from other countries in later periods. Administrative means, investment policy, and fiscal policy were major actions taken domestically to achieve socialist industrialization. Actions taken to accommodate external environments were adaptive in nature. Actions taken domestically could be further analyzed from the cultural system perspective.

(3) *The Pattern Maintenance of Cultural System*

As stated before, China was primarily an agricultural society with many individual or family-operated farming units scattered throughout the country at the beginning of the planning period. Thus, transforming the traditional agricultural society with the cooperative movement must be gradual rather than radical. Part of the reasons underpinning the failure of People's Commune, established in haste to force peasants to work in cooperation, was the neglect of cultural factors in that development. It is important to understand and respect the patterned maintenance of the cultural system in the development process. If a certain cultural pattern becomes an obstacle in the process, it should be changed. But the change must be gradual as resistance to change is inherent. On the other hand, if a cultural pattern is conducive to development it should be promoted and improved gradually. No cultural pattern or norms of behavior can be changed overnight.

Similar error was made in the socialist transformation of China's industry in the period. Chinese government hastened the process to nationalize industry and eliminate private ownership of any business, big or small. By the end of 1965, there were only state and collectively owned industries and businesses. All other types of ownership were abolished. The recovery of businesses and industries owned by individuals or families was slow when private ownership was allowed in the 80's. In fact, privately owned industry had a very minor share in China's economy in 1984, which was not conducive to the overall growth of national economy.

(4) *The Integrative Action was Mainly Through Administrative Action of the Central Government*

While this approach worked in China during the period when the socio-economic development was underdeveloped, this type of centralized integrative action would not work well in today's highly developed China as it simply cannot meet the varying demands by many players. Clearly, coordination of actions in various line ministries and local governments was a challenging issue in China's planning system.

11.6 International Perspective on China's Socialist Economic Development in the 1953–1980 Period

Based on our analysis, despite many ups and downs from 1953–1980, the overall performance of socio-economic development was successful in these five Five-Year Plan periods. How did the international research community assess China's socialist economic development under the 1953–1980 period? Below cite two influential works, one from The World Bank (1981) and one by Kraus (1982), to shed light on this question.

11.6.1 Study by the World Bank (1981)

The World Bank sent an economic mission to China in 1980 to study its socialist economic development. This mission included several teams of experts in agriculture, energy, industry, transport, statistics, education, population, health and nutrition, foreign trade, etc. and the principal economist. Chinese Ministry of Finance, host of the mission, was responsible for coordinating all activities between the mission and more than ten Chinese ministries and many experts from various organizations. The mission concluded with the Main Report *China: Socialist Economic Development* and eight annexed volumes in *Statistical System and Basic Data; Population, Health and Nutrition; Agricultural Development, Challenges and Achievements in Industry; The Energy Sector, The Transport Sector, Education* and *External Trade and Finance.* Below summarize the World Bank's perspectives on China's systems and strategy, growth and poverty reduction, and industrial and infrastructural development.

(1) *Systems and Strategy*

The report maintained that China's development effort was consistent with two objectives: (1) industrialization with an emphasis in developing its heavy industrial base and (2) elimination of poverty. Faced with two major constraints, namely: (1) an extreme shortage of cultivated land in relation to its population and (2) a high degree of international isolation, the Chinese tackled this predicament by approaching these two objectives differently. To reduce poverty, China promoted rural development and the provision of basic social service by largely utilizing local resources and initiatives with a strong emphasis on economy and technical innovations. On the other hand, its industrialization was "based mainly on a massive infusion of centrally mobilized resources, with less concern for cost effectiveness, using technology largely descended from Soviet designs of the 1950s". (p. ii).

(2) *Growth and Poverty Reduction*

Substantial progress toward these two main objectives was evident. Through an exceptionally high rate of investment, virtually all of which was financed by domestic savings, China achieved a swift industrialization. Its share of industry in GDP (around 40%) at present was similar to the average for middle-income developing countries. However, since the share of services was much smaller than in other countries, agriculture still accounted for 34% of GDP with over 70% of employment which resembled the average for low income countries. (p. ii). On the poverty reduction front, China's most remarkable achievement during the past three decades was to make low-income groups far better off in terms of basic needs than their counterparts in most other poor countries. Chinese people all had work; their food supply was guaranteed through a mixture of state rationing and collective self-insurance; most of their children were attending school and well taught; and the great majority had access to basic health care and family planning services. (p. iii).

(3) *Industrial and Infrastructural Development.*

Productivity growth was less impressive and there were serious problems of product quality. Despite the good progress made in the 1950s, since 1957, industrial output growth was achieved mainly by increasing the quantity of inputs (capital, labor and materials), rather than by increasing their usage efficiency. Such low level of efficiency was partly caused by technological isolation and partly a reflection of weakness in planning and in the economic system. Particularly, there was a disconnect between producers and users, and a lack of incentive for producers to use scarce resources economically or to introduce technical innovations (p. v).

11.6.1.1 Economic Development and Social Change in the People's Republic of China by Kraus (1982)

This book focuses on China's development policies with the aim to address readers looking for the broadest possible survey of the concepts and goals, plans and measures, success and failures of the PRC in its efforts to promote economic and social development. This book has six chapters. Chapter 1 covers concepts of development in the Chinese Communist Party prior to the founding of the People's Republic of China and the 1949–52 Reconstruction Phase. Chapter 2 focuses on China's First Five-Year Plan in 1953–57 and the adoption of the Soviet Development Model. Chapter 3 describes the Second Five-Year Plan in 1958–1962 with the policy of the Three Red Banners and the 1963–65 Consolidation Phase. Chapter 4 deals with development policy during the 1966–70 Third Five-Year Plan period and the Cultural Revolution. Chapter 5 deals with the Fourth Five-Year Plan in 1971–75 and highlights pragmatism and Maoist visions of the future. After a comprehensive analysis of basic concept, development goals (social policy, internal politics affecting development, economic policy), development policy measures (social policy, internal Politics affecting development, economic policy) and results of successive period of development of China in Chapter 1 through 5, the author highlights lessons that can be learned from the China experience in Chapter 6 titled The Great Leap into the Industrial Age summarizes. Following are selected aspects in Chinese development experience that can be applied by developing countries in the Third World:

(1) *Mobilizing the Population as Development Vehicle*

Contrary to the framework of traditional developmental theory, the Chinese leadership consistently made the attempt to mobilize its people ("the masses") in nation building, instead of depending on the entrepreneurs and the elites for innovation. The "masses" were treated as the new "extended family groups" and those "family members" were morally obliged to contribute their share in the national development. In addition, Chinese development policy always considered the mobilization and participation of the public much more important than the efficient allocation of resources. (p. 277).

(2) *Prioritizing Agricultural Development*

After abandoning the Soviet model for development, China placed increasing emphasis on the development of agriculture. To ensure steady increasing of the yield per hectare, millions of people were employed in carrying out hydraulic projects for the expansion of irrigated land and protection against floods. Although insufficient planning often led to severe damage, e.g., salinization of the soil, deforestation, and lowering of the water table, progress was nonetheless remarkable. (p. 278).

(3) *Organizing Rural Small-Scale Industry and Deploying Technology Conforming to Development*

Chinese development policy systematically attempted to distribute industrial activity evenly throughout the entire country. Efforts were made to stimulate rural initiative and use of local resources by setting up township-owned or village-owned plants that would process agricultural products, repair agricultural tools and machinery. And finally, produce agricultural tools and machinery themselves. (p. 279).

Following basic rules for development. Adhering to the basic rules of development policy best ensures success. "This was the case for a major portion of the development process where, for example, consideration of the functional connection between industry and agriculture, between production and distribution, and between domestic supply and foreign trade has formed the guideline for development activity." (p. 280).

11.7 Summary Points

This chapter begins with a discussion on China's First Five-Year Plan. Due to lack of experience in preparing a national economic plan, the planning document followed the exact model of the former Soviet Union. The *First Five-Year Plan for 1953–1957* established a deeply rooted planning culture among all planning officials from the central government down to various level of local government. A basic understanding of this plan sheds light on all essential aspects to the practice of development of China since the establishment of PRC until 1980, and beyond due to its lasting conceptual rigor.

It then covers China's First *Twelve-Year S&T Plan for 1956–1967.* This plan has laid the foundation of science and technology system in China. China has established its science and technology capacity through transfer of both hardware and software technology in the construction of the 156 Soviet Union assisted core projects. China also imported complete sets of equipment from Japan and other Western countries since worsening of bilateral relationship between China and Soviet Union in 1960s. Although China remained a semi-closed system before the normalization of international relations in the 1970's, its performance in science and technology system was successful in the creation of human resource through transfer of technology from various sources of advanced countries and creation of various research institutions and higher learning.

Despite several cycles of up and down due to Great Leap Forward and Cultural Revolution, China still enjoyed a relatively high average annual growth rate of GDP around 5.1% from 1952 to 1980, the growth rate of per capita GDP around 4.5%. Due to the large territorial size and serious regional disparity before the establishment of PRC, and despite efforts focused on balanced regional development, the result was less successful.

China's focus on development of heavy industry as the priority in this period contradicts traditional economic theory of development which would develop light industry first for accumulation of capital and improvement of peoples' consumption needs. Nevertheless, such a strategy enabled China to create a large capacity in manufacturing, which proved to be instrumental in its development after China launched its policy of reform and opening in late 1970s. China's experience certainly represents a unique case in the study of "development economics".

In the social system, China closely followed the USSR model. Peoples' basic needs were met through food and supply rations. Workers and staffs had free housing provided by their organizations or SOEs. Nearly free education and health care were provided to the urban population. Labor Insurance Regulation was promulgated early on 1953. All these practices ensured a relatively harmonious society. However, such a generous social security system becomes unaffordable and unsustainable, due to the rapid increase of rural population from 5.75 hundred million in 1952 to 9.87 hundred million in 1980 (an increase around 71%) and an increase of urban population from 71.6 million in 1952 to 1.91 hundred million in 1980 (an increase around 167%).

An overall analysis of performance of China's planning system is undertaken using the revised Parsons' AGIL framework in the four aspects of goals, active adaptive actions, the pattern maintenance of the cultural system, and the function of integration.

The chapter ends with two prominent studies on China's socialist economic development by the World Bank and Kraus to provide international perspectives on china's development experience and its applicability to other developing countries seeking exemplary development strategy and practice.

References

Bagchi, A. K. (1987). Development planning. In J. Eatwell, M. Milgate, & P. Newman (Eds.), *The new palgrave: Economic development* (pp. 98–108). New York: W. W. Norton and Company Inc.

Bo, Y. B. (1991). *Major decisions and events: A retrospect* (Vol. 2). Beijing: CPC Party School Press.

Centrally Planned Economy. (n.d.). *Collins dictionary of economics, 4th ed.* (2005). Retrieved August 13, 2017 from http://financial-dictionary.thefreedictionary.com/centrally+planned+economy.

China Academy of Science Policy Research Institute. (1981). Works of chinese academy of science. In Xue Muqiao (Ed.), *Almanac of China's economy 1981: With economic statistics for 1949–1980.* Beijing: Economic Management Magazine Publisher.

China State Planning Commission. (1956). *First five-year plan for development of the national economy of the People's Republic of China in 1953–1957.* Peking: Foreign Languages Press.

Government Administrative Council. (1953). *Labour insurance regulations of the People's Republic of China China.* Peking: Foreign Languages Press.

Kraus, W. (1982). *Economic development and social change in the People's Republic of China.* New York: Springer-Verlag, New York Inc.

Li, J., & Huang, K. L. (Eds.). (2001). *History of technological development of Chinese machinery industry.* Beijing: Machinery Industry Publisher.

Li, W. Y., & Lu, D. D. (Eds.). (1995). *Industrial geography of China.* Beijing: Science Press.

Mao, Z. D. (1956). On the ten major relationships. In *Selected works of Mao Tse-tung* (Vol. V.). Peking: Foreign Languages Press, 1977.

National Bureau of Statistics of China. (1982). *China statistical yearbook 1981.* Beijing: China Statistical Publisher.

National Bureau of Statistics of China. (1985). *China statistical yearbook 1985.* Beijing: China Statistical Publisher.

National Bureau of Statistics of China. (1990). *Compendium of historical statistics of provinces, autonomous regions, and municipalities directly under central government 1949–1989 (Quanguo Gesheng Zizhiqu Zhixiashe Lishi Tongji Ziliao Huibian 1949–1989).* Beijing: China's Statistics Publisher.

The World Bank. (1981). *China: Socialist economic development, the main report.* The World Bank: Washington D.C.

Wang, H. J. (2003). *Integrated study of China's development and reform.* Beijing: Foreign Languages Press.

Wang, H. J. & Li, B. X. (1994). China. In S. Chamarik & S. Goonatilake (Eds.), *Technological independence-the Asian experience* (pp. 80–34) Tokyo: United Nation University Press.

Wang, H. J. & Li, S. T. (1995). *Industrialization and economic reform in China.* Beijing: New World Press.

Wang, H. J., Li, B. X., & Zhou, L. (1989). *Study of sectoral industrial policy of China* (Chinese ed.). Beijing: Chinese Fiscal & Economic Publisher.

You, L., Cheng, X. L., & Wang, R. P. (Eds.). (1993). *History of the People's Republic of China, 1949–1992.* Vol. 1. (*Zhong Hua Ren Min Gong He Guo Shi Tong Jian, 1949–1992*). Beijing: Red Flag Publisher.

Zeng, P. Y. (Ed.). (1999). *Fifty years of new China's economy (1949–1999).* Beijing: China's Planning Publisher.

Chapter 12
Planning System of China (1981–2016) To Be Case Study

12.1 Introduction

China continued with its national planning system after 1981. There were eight Five-Year Plans prepared in the period from 1981 to 2016. But the preparation of these plans was within the new context of China and a broad change of the global environment. Domestically, the third plenary session of the Eleventh Party's Congress, held in 1978 between December 18 through 22, resulted in launching a strategy of reform and the opening of China. Globally, due to the impact of progress of information and communication technology, the total breakdown of the colonial system and the emergence of many new developing countries from their former colonized status, the increase of educational level of the people and elimination of illiteracy, the establishment of United Nations-a Supra National Organization had hold many influential conferences which affect the long term fortune of the mankind, for example, Sustainable Development (World Summit 1992) in 1992, Social Development in 1995 (World Summit 1995), Millennium Development Goal in 2000 and Transforming our world: the 2030 Agenda for Sustainable Development in 2015. Although the U.N. does not have sufficient power to influence at a global level due to power politics of the real world, the current UN is more influential than the former League of Nations established in 1920.

Since 1978, China gradually transformed her planning system from a former centrally planned economic system into an indicative planning system. Such a transformation is a difficult process for a large country like China, of for established institutions, process and concepts should be changed, new knowledge and practice from available sources should be absorbed and adapted to the concrete conditions of China's characteristics.

Evolution of process, concepts and means of control from 1981 to 2016 will be described in Sect. 12.2. Briefing of planning from China's Six Five-Year Plan to

H. Wang and S. Li, *Introduction to Social Systems Engineering*,
https://doi.org/10.1007/978-981-10-7040-2_12

Tenth Five-Year Plan will be discussed in Sect. 12.3, including a relatively detailed description of China's Seventh-Five Year Plan for the purpose of comparison with the First Five-Year Plan (discussed in Chap. 11). Section 12.4, will discuss the Eleventh, Twelfth and Thirteenth Five-Year Plan. Since there is an English version of "The Thirteenth Five-Year Plan For Economic and Social Development of The People's Republic of China", selective supplementary materials will be provided in this section. Section 12.5 will discuss China's achievements and issues in these planning periods, analysis and synthesis of China's socio-economic performance based upon perspective of social systems engineering will also be given.

12.2 Evolution of Concepts, Process and Measures of China's Planning System Since 1981

12.2.1 China Gradually Transformed Her Nature of Planning

12.2.1.1 The Name Was Changed

The name of the plan had been changed from the First Five-Year Plan for Development of the National Economy of the People's Republic of China in 1953–1957 to be the Sixth Five-Year Plan for Economic and Social Development of The People's Republic of China, i.e. the word "Social" was added in the title of the planning document. In correspondence, the content of social aspects was added to make the Plan more meaningful and complete.

12.2.1.2 The Statistical System Was Changed

The Statistical System is the basis of a planning system. China spent approximately ten years changing her statistical system from MPS to SNA to adapt to international practice.

12.2.2 The Planning Process Was Normalized Gradually

The process of planning and authorization was normalized, and the planning process became transparent in a normal national administrative system. Since the Seventh Five-Year Plan, there was a group of people working together under the leadership of the Party, and preparing a *Suggestion for Preparation of Seventh Five*

Year Plan, this suggestion was discussed and passed by the Thirteenth Party's Congress in 1985. The State Planning Commission prepared the Seventh Five-Year Plan based upon the "Suggestions", and sent the prepared plan for approval by the National People's Congress (NPC) on April 12, 1986. This process has become routinized since the Ninth Five-Year Plan, the date of opening of the National People's Congress was normalized to be March 5, 1996, and the Chinese Premier gave a report on the Five-Year Plan or annual Plan to members of NPC and obtained approval by a voting process.

Mid-Term evaluation of implementation of China's Five-Year Plan was launched in the Eleventh Five-Year Plan.

12.2.3 The Planning Process Opened Gradually with Wider Participation of Institutions and Experts from Domestic and Abroad

Before the launch of reform and opening, China's national planning system was not quite open and it was a semi-closed governmental system, highly powerful for investment decision and approval of projects by its administrative organization. China is a large and complex country, and the planning system for development required a wider participation of organizations. There were establishment of many new organizations so called think tanks, that experimented with new approaches of planning, for example, the former Technical Economic Research Center under the leadership of Ma Hong implemented some of the larger organized research projects, such as "Comprehensive Planning of Energy and Heavy Chemical Base at Shanxi Province" coordinated closely with Shanxi Provincial Planning Commission in the beginning of 1980s, and "China by the year 2000", which was described in Chap. 4 previously. These two studies involve long term planning, various types of mathematical models, and lasted around twenty years. The State Planning Commission also implemented several organized research including domestic and international experts such as *Method of China's Medium and Long Term Plan and Macro-Economic Management* supported with grant by UNDP, *China towards the year 2020,* with grants from ADB, the World Bank and several corporations from South Korea. This study was also published in the 1997 works titled *Research on Integrated Policies for China's Sustainable Development,* with technical guidance from UNIDO and a grant from the Kingdom of Holland. Approximately five years were spent on this study, which resulted in the publication of three volumes in 2004. The mid-term review of China's Eleventh Five-Year Plan was prepared by a team of experts organized by the World Bank, at the request of the Development Planning Department of China's National Development and Reform Commission (NDRC).

12.2.4 *Improvement of Policy Measures in Planning Documents*

Although there were many policies implemented in the pre-reform era, they were not explicitly expressed in the planning documents, which can be seen from the briefing of the First Five Year Plan in Chap. 11. In addition, the financial sector of China was weak and there were few banks in China. The People's Banks had branches and small offices widely distributed in factory, towns and counties throughout China, and therefore, the monetary policy was weak. Fiscal policy and administrative means were major measures to exercise the process of implementation of the Plans.

In the planning period from 1981 to present, various policy measures were studied and implemented, science and technology policy, population policy, monetary policy, fiscal policy, regional policy, environmental policy, trade policy and many others, to mention a few examples.

12.2.5 *Greater Integration of Policies in the Planning Documents*

Since the launch of reform and opening, China promulgated three official documents related to economic system reform respectively in 1984, 1993 and 2003. Two other official documents included "Decisions on Reform of the Management of Science and Technology", and, "Decision on Reform of Education System" promulgated by the CCCPC respectively on March 15 and May 27 of 1985. Reform is also included in the planning documents in recent periods.

12.3 Briefing of China's Five-Year Plans from the Sixth to the Tenth

This Section will discuss most of the Five-Year Plans from 1980 to 2005, including a discussion of the features of these plans. The Seventh Five Year Plan will be discussed in detail, because it represented the effort of China to develop a planning system to be transformed from the mandatory planning learned from former USSR. Although the Seventh Five-Year Plan was still effected by past tradition due to the traditional MPS statistical system, it is selected to facilitate the comparison between the First Five-Year Plan described in Chap. 11, and also it can be used to compare China's Twelfth and Thirteenth Five-Year Plans, describe in later part of this Chapter as evidence of the progress of China's planning system.

12.3.1 *Briefing of* the Sixth Five-Year Plan for National Economic and Social Development of the People's Republic of China

The Sixth Five-Year Plan for National Economic and Social Development of the People's Republic of China 1981–1985 (abbreviated below as China's Sixth Five-Year Plan), approved by the Fifth People's Congress on December 10, 1982.

12.3.1.1 Contents of the Sixth Five-Year Plan

Contents of the Sixth Five-Year Plan are shown in Table 12.1.

12.3.1.2 Features

China's Sixth Five-Year Plan was the First Chinese Five-Year Plan with the title of economic and social development rather than focused simply on economic growth. The content of this document followed in general the First Five-Year Plan. But there are several different features:

(1) This plan set a long-term development goal of twenty years. It is described in the preface of the Plan that "The strategic goal of China's economic construction is quadrupling the gross value of agricultural and industrial output by the end of this century compared to 1981 based on the precondition of raising the economic efficiency continuously…." (p. 113)
(2) This plan established a new Part III Regional Economic Development, which is important for a country with a large territorial size to pursue a relatively balanced regional development strategy.
(3) Environmental protection is raised the first time of China's planning documents.
(4) Four main measures to achieve the Sixth Five-Year Plan are raised by former premier Zhao Ziyang in his report to the Fifth National People's Congress. They are:

  (i) Strictly control the total scale of investment on fixed assets, guarantee the completion of key constructions and technical transformation of enterprises in accordance to plan;
  (ii) Adjust and improve current enterprises wholly and decisively, upgrade the level of operation and management of enterprises with full effort;
  (iii) Actively push forward technological progress, and realize fully the promotional role of science and technology in economic construction (pp. 92–102) (it is planned by State Economic Commission to import 3000 items of advanced technology in next three years);
  (iv) Actively and safely accelerate the process of reform of economic system). (p. 100)

Table 12.1 Structure and contents of Sixth Five Year Plan[a]

Name of parts	Name of chapters	
Preface Part I: Basic tasks and comprehensive index (Six chapters)	Chapter 1: Basic tasks	Chapter 2: Total social product, national income and economic efficiency
	Chapter 3: Public finance and credit	Chapter 4: Investment of fixed assets
	Chapter 5: Development of science and technology and cultivation of talented people	Chapter 6: People's life and goals of social development
Part II: Development plan of different economic sectors (Thirteen chapters)	Chapter 7: Agriculture	Chapter 8: Forestry
	Chapter 9: Industry of consumer goods	Chapter 10: Energy
	Chapter 11: Metallurgical industry	Chapter 12: Chemical industry
	Chapter 13: Material of construction industry	Chapter 14: Geological exploration
	Chapter 15: Machinery industry, Electronic industry	Chapter 16: Construction industry
	Chapter 17: Transport, post and communication	Chapter 18: Domestic commerce
	Chapter 19: Foreign economic relation and trade	
Part III: Regional economic Development plan (Five chapters)	Chapter 20: Coastal region	Chapter 21: Inland region
	Chapter 22: Minority region	Chapter 23: Regional cooperation
	Chapter 24: Territorial development and planning	
Part IV: Planning of scientific research and educational development (Four chapters)	Chapter 25: Science and technology	Chapter 26: Philosophy, social science
	Chapter 27: Primary school and secondary school education	Chapter 28: Higher and middle specialized education
Part V: Social development plan (Eight chapters)	Chapter 29: Population	Chapter 30: Labor
	Chapter 31: Income of inhabitants and consumption	Chapter 32: Construction of urban and rural and social welfare undertaking
	Chapter 33: Cultural undertaking	Chapter 34: Healthcare, sports
	Chapter 35: Environmental protection	Chapter 36: Social order

[a]*Source* State Council (1982)

12.3.2 Briefing of the Seventh Five Year Plan for National Economic and Social Development of the People's Republic of China 1986–1990

The Seventh Five-Year Plan for National Economic and Social Development of the People's Republic of China 1986–1990 (Abbreviated as part of China's Seventh Five Year Plan) approved by the Sixth National People's congress (NPC) on April 12, 1986.

12.3.2.1 Contents of China's Seventh Five Year Plan

Contents of China's Seventh Five Year Plan is shown in Table 12.2.

12.3.2.2 Selected Aspects of Preface

China's Seventh Five-Year Plan is the second plan prepared by the Chinese themselves since the implementation of reform and opening in late 1970s. Although traditional elements existed due to cultural maintenance, impact of "L" in the AGIL framework, many new features were added. The planning of China gradually improved and transformed from mandatory planning to indicative planning. Box 12.1 is a translation of selected aspects of the preface.

Box 12.1 Selected Aspects of Preface of the China's Seventh Five-Year Plan

Eight paragraphs are descriptions related to achievements of the China's Sixth Five-Year Plan.

The achievements obtained are inseparable from the vast amount of work of adjustment to the economy, especially the reform of the economic system. Contract responsibility system was pushed forward together with other reform measures in the rural areas, which had liberalized the rural productivity. There was an emergence of vivid economic life in the urban areas due to reform centered around revitalization of enterprises and other reforms. This illustrates that China is marching correctly toward the path of construction of socialism with Chinese characteristics.

Currently, economic development of China exists with some problems and difficulties that cannot be neglected. As a whole, the physical and technological conditions for economic and social development are relatively backwards, agriculture as the foundation of national economy is relatively weak, imbalance of energy and raw materials between supply and demand is also problematic, transport and communication are very much underdeveloped, development of intelligence cannot be adapted to meet the demand of construction of modernization. The status of low economic performance of production, construction and circulation basically has not been changed.

Based upon the requirement of *Proposal of the CPC Central Committee on the formulation of the Seventh Five Year Plan for national economic and social development,* some important principles and guidelines must be implemented in China's Seventh Five Year Plan:

- Adhere to the reform and open policy in the first place, to adapt reform and openness towards each other, and to promote each other mutually;
- Insist on the basic balance of social total demand and total supply, maintain individual balance in public finance, credit and foreign exchange and comprehensive balance.
- Insist upgrading the economic performance, especially to raise the quality of product to extremely important status, treat correctly the relationship between performance and speed, quality and quantity.
- Further adjust industrial structure rationally based upon adaptation to the structural change of social demand and the requirement of modernization of the national economy.
- Determine the scale of investment to fixed assets appropriately, adjust structure of investment rationally, and accelerate the construction of energy, transport, construction and industry of raw materials.
- Put emphasis on the priority of construction of the technological transformation of current enterprise and its reconstruction and extension, it is necessary to implement a pattern of intensive growth, i.e., only focus on expansion other than through reproduction in the part.
- Put emphasis on development of science and education into important strategic position, promote the progress of S&T, accelerate the process of intelligence development.
- Insist on opening further to the outside world, integrate better the construction of domestic economy and the expansion of economic and technological exchange with other countries.
- Insist on further improving the physical and cultural life of urban and rural people on the basis of development of production and rise of economic performance.
- Insist on strengthening the spiritual civilization with full effort in the process of pushing forward the construction of physical civilization.
- Insist on continuing to work hard and hold the spirit to build the country through diligence.

Preparation of China's Seventh Five-Year Plan will be based on the correct recognition of the economic nature of socialism with full effort, to achieve the spirit of reform and innovation. Socialist market economy is a planned commodity economy based on public ownership. The national plan should guarantee the coordinated development of the national economy according to certain approximate proportions, it should be the major base to proceeding, with the management, regulation and control of the economic activity from macro-perspective. It is necessary to focus on the rule of value in the national plan, utilization of the mechanism of market in self-consciousness to make the economic life full of vitality.

Every plan should be designed with appropriate indicators, emphasis should be focused on raising the policy to be implemented for this plan, and they should be coordinated with the plan to become an organic part of the plan.

Mandatory planning is applied to important economic activities to the whole country, indicative planning should be applied to a large amount of economic activities, a part of the economic activities will be automatically regulated wholly by the market.

It is impossible under the plan to make exact forecasting of various trends of economic development for the next five years. Therefore, indicators and measures raised by this Five-Year Plan should make necessary adjustments and supplements in an annual plan based upon current concrete conditions.

Comments on above preface: This preface shows clearly the transition of China's planning system from centrally planned to an indicative planning system. The basic concepts of the requirements are clear, the requirements of each sub-system have been considered, i.e. economic, S&T and social aspects. The overall goal is clearly stated, i.e. "comprehensive balance among each other." The constraint is political economy, a part of the people debate what is socialist market economy? It is recognized by political economists that a socialist market economy is a planned commodity economy based on public ownership which is described in Box 12.1.

12.3.2.3 Briefing of Part I and Part II of China's Seventh Five-Year Plan

(1) *Part I of China's Seventh Five-Year Plan*

It can be seen from Table 12.2 that Part I is consisted of five chapters. Chapter 1 has outlined three major tasks, the first is to create a good economic and social environment for the implementation of reform of economic system. It also targeted to establishing a foundation of a new socialist economic system with Chinese characteristics, basically within five years or a longer period; the second is to keep stable and sustainable growth of the economy; the third is to continuously improve the life of urban and rural people. Chapter 2 has provided targets for the growth rate of industrial and agricultural products (MPS) and also the growth rate of Gross National Products (GNP), and primary, secondary and tertiary sectors. Growth rate of the GNP is targeted to be 7.5% annually. Very detailed targets are given for

Table 12.2 Structure and contents of China's Seventh Five-Year Plan[a]

Theme of parts	Theme of chapters (Section number in chapter)	
Preface Part I: Main tasks and goals of economic development (Five chapters)	Chapter 1: Main tasks	Chapter 2: Growth rate of economy and economic performance
	Chapter 3: Production and distribution of national income	Chapter 4: Public finance, finance and foreign exchange
	Chapter 5: objectives of development of science and technology (S&T), education and society	
Part II: Industrial structure and industrial policy (Ten chapters)	Chapter 6: Direction and principle of adjustment of industrial structure	Chapter 7: Agriculture (4 sections)
	Chapter 8: Industry of consumer's goods (3 sections)	Chapter 9: Energy (5 sections)
	Chapter 10: Industry of raw materials (3 sections)	Chapter 11: Geological exploration
	Chapter 12: Machinery and electronic industry (2 sections)	Chapter 13: Construction industry and material of construction industry (2 sections)
	Chapter 14: Transport and Post, Communication (3 sections)	Chapter 15: Circulation of commodities (3 sections)
Part III: Regional layout and regional economic development policy (Seven chapters)	Chapter 16: Economic development of Eastern coastal region (2 sections)	Chapter 17: Economic development of Central region (2 sections)
	Chapter 18: Economic development of Western region (2 sections)	Chapter 19: Economic development of old, revolutionary base minority, border and poverty region (4 sections)
	Chapter 20: Regional cooperation and networking of economic regions (3 sections)	Chapter 21: urban and rural construction (4 sections)
	Chapter 22: Territorial development and improvement	
Part IV: Development of S&T and policies (Five chapters)	Chapter 23: Strategy of development of S&T	Chapter 24: Application and spread of results of S&T
	Chapter 25: Tackling of S&T	Chapter 26: Basic research
	Chapter 27: Research of philosophy and social science	

(continued)

Table 12.2 (continued)

Theme of parts	Theme of chapters (Section number in chapter)	
Part V: Development of education and its policies (Five chapters)	Chapter 28: Basic education	Chapter 29: Vocational technical education
	Chapter 30: Ordinary higher education	Chapter 31: Adult education
	Chapter 32: Major policy measures for development of education	
Part VI: Foreign economy and trade & technical exchange (Six chapters)	Chapter 33: Import and export trade (3 sections)	Chapter 34: Utilization of foreign investment and import of technology (2 sections)
	Chapter 35: Special economic zones, coastal open cities and open regions	Chapter 36: Foreign contracted projects, labor service cooperation and international aids (2 sections)
	Chapter 37: Tourism	Chapter 38: National balance of payments
Part VII: Investment structure and investment policy (Four chapters)	Chapter 39: Adjustment of investment structure	Chapter 40: Sectoral structure of capital investment
	Chapter 41: Deployment of technical transformation	Chapter 42: Management of investment of fixed assets
Part VIII: Objectives and tasks of reform of economic system (Four chapters)	Chapter 43: Tasks and steps of institutional reform	Chapter 44: Strengthen the vitality of enterprises
	Chapter 45: Development of market system of socialism (2 sections)	Chapter 46: Strengthen and improve macro-control
Chapter IX: People's life and social security (Six chapters)	Chapter 47: Population	Chapter 48: Labor (2 sections)
	Chapter 49: Income of inhabitants and structure of consumption (3 sections)	Chapter 50: Health care and sports (3 sections)
	Chapter 51: Social security (2 sections)	Chapter 52: Environmental protection (2 sections)
Part X: Construction of socialist spiritual civilization (Four chapters)	Chapter 53: Culture (6 sections)	Chapter 54: Ideological and political work
	Chapter 55: Socialist democracy and the rule of law	Chapter 56: Social order
Table Attached: Main indicators of China's Seventh Five-Year Plan 13 categories of main indicators 68 sub-indicators		

[a]*Source* State Council (1986) (Chinese Version)

performance of economy, production and distribution of national income (Chapter 3), income and expenditure of public finance, credit and foreign exchange (Chapter 4) and also targets for items in Chapter 5.

(2) Part II of China's Seventh Five-Year Plan

Part II is titled Industrial Structure and Industrial Policy. This part contains ten chapters. Chapter 6 is direction and principle of industrial structure, the remaining chapters deal with nine sectors.

Box 12.2 [1]Briefing of Chapter 6 Direction and Principle of Adjustment of Industrial Structure and Industrial Policy

It is necessary to rationally further adjust the industrial structure. The direction and principle of adjustment are:

(1) Based on the precondition to promote the stable development of light and heavy industry, continuously keep the overall growth of agriculture, emphasis should be given to improve their internal structure through individual efforts.
(2) Accelerate the development of energy and industry of raw materials, and control the growth of production of general processing industry to create a proportional relationship among energy, industry of raw materials and processing industry, and gradually tending to coordinate efforts.
(3) Put development of transport and communication in priority, upgrade their capacity of production and performance of utilization with full effort, gradually change the serious backwardness of the transport and communication sector.
(4) Develop a construction industry vigorously to make it to gradually become an important industrial sector to promote the overall growth of national economy and social development.
(5) Accelerate the development of tertiary industry for production and life service to insure that the proportional relationship between the tertiary and primary, secondary sector to be improved significantly.
(6) Actively apply new technology to reform the traditional industry and traditional products, develop knowledge intensive and technology intensive products with crucial importance, struggle to explore new productive areas, promote the formation and development of several emerging industries in planning.

The industrial structure by the year 1990 will be:

Within the gross value of industrial and agricultural output, the share of agricultural sector will be declined from 23.9% in 1985 to 21% in 1990, the share of light industry will be increased from 38% in 1985 to 39.4% in 1990,

[1]Source: State Council (1986).

the share of heavy industry will be increased from 38.1% in 1985 to 39.6% in 1990.

Within the Gross National Product, the share of the tertiary sector will be increased from 21.3% in 1985 to 25.5% in 1990, the share of primary and secondary sector will be decreased from 78.7 to 74.5%.

Box 12.3 is briefing of Chapter 8, consumers' goods industry for illustration of other sectors.

Box 12.3 Briefing of Chapter Eight Consumers' Goods Industry

1. Section one Basic Tasks
   In order to adapt to the rising level of people's consumption and the change of its demand structure, it is necessary to develop an industry of consumers' goods, and promote the overall growth of production of various consumers' goods. The value of output of consumers' goods by the year 1990 will be increased by 40% compared to 1985. The concrete tasks are:
   (1) With food industry, clothing industry and consumers' durable industry being priorities, to encourage better development of the overall consumers' industry.
   (2) Struggle to expand production fields of consumers' goods; explore new categories of production; increase famous brand quality products, marketable products, medium and high end products, actively promote upgrading of products to raise China's manufacture of consumers' goods to a new level.
   (3) Struggle to expand the export of light and textile products, strive to create more foreign exchange earnings for the country.
   (4) Strengthen the technical transformation, restructuring and expansion of enterprises; promote their technical progress and upgrade their economic performance with full effort to accumulate more capital for the country.
2. Section Two Development of production and construction of major sectors
   (1) Food industry
   Continue to understand better the processing of raw materials of basic food, such as meat processing, aquatic processing, sugar making, grain processing and edible oil refining; and, to focus also at the same time on developing short term products of market such as beverages, low alcohol wines, actively develop new foods and traditional foods to adapt better the consumption trend of the people, paying attention more to hygiene, nutrition, time saving and diversification.

The targets of quantity of production of major foods in 1990 are shown as follows:
Sugar 5500–6000 thousand tons, increase from 23.6 to 34.8% compared to 1985
Tobacco 26,000 thousand tons, increased 10.6% compared to 1985
Beer 6500 thousand tons, increased by 1.1 times compared to 1985
Beverages 3000 thousand tons, increased by 2 times compared to 1985

The deployment of construction of food industry in the period of "Seventh Five" are: proceeding with the technical transformation in old sugar making regions; to construct some new sugar making factories in Inner Mongolia, Xinjiang, Ningxia, Gansu, Guangxi and Hainan Island, to increase the sugar making capacity of 1600 thousand tons. Technical transformation and restructuring will be implemented in Hunan, Sichuan, Shandong, Henan and Shanghai, and expanding construction of some tobacco factories; upgrading quality of tobacco and the share of production of secure smoke and filter smoke. Increase the production capacity of 3500 thousand tons of beer, promote efficient scarification, open-air fermentation tanks, advanced canning technology to upgrade quality of the beer. Expand production capacity of salt in coastal salt field of Hebei, Tianjin, Liaoning, Jiangsu and Shandong and increase the share of production of refined salt, sodium salt.

(2) Textile and clothing industry Omit
(3) Durable consumers' goods industry Omit
(4) Other light industries Omit

3. Section Three Main Policy Measures

(1) Further reform the planning management system of the consumers' goods industry, better realize the function of regulation of pricing policy. The State will implement mandatory planning or partial mandatory planning for a few products such as chemical fiber, synthetic polymer, newsprint, tobacco and salts; indicative planning or regulation through market will be implemented to large amount of remaining products. Reform the price of consumers' goods in steps. Enlarge gradually the price difference of quality between famous brand products, high quality products, new products and general products. Attention should be focused on variety, upgrading quality and development of new products.
(2) Continue to guarantee with priority on credit, use of foreign exchange, supply of energy and raw materials and conditions of transport for products urgently needed by the market.

(3) Strengthen technical transformation, promote technical progress. Actively import advanced equipment and manufacturing technology from abroad through a means of combination of technology and trade and cooperative production. Raise rates of depreciation of fixed assets for food, paper making, household chemicals, textile printing and dyeing emphatically. Support of credit and taxation should be given to technical transformation of enterprises of low and meager profit industry.
(4) Textile and light industry department should be actively coordinated with related departments, to establish a batch of raw material base for production of light and textile industry, and organize them to establish direct linkage with raw material base, or implementation of joint operation or cooperative operation to guarantee a stable source of raw materials required by production of light and textile industry.
(5) Center around famous brand products, organize union with related enterprises to expand the production capacity of famous brand quality products, upgrade quality of products, promote upgrading of product and a new generation of products.

(3) *Supplementary Remarks on Part I and Part II of China's Seventh Five-Year Plan*

Briefing of Part I and II provides relatively detailed information to illustrate China's effort of transition from a mandatory planning system to an indicative planning system. It can be seen from briefing of Part I that there are many detailed targets set up in the Seventh Five-Year Plan, i.e. part of the features of mandatory planning inherited from the tradition of First Five-Year Plan have been retained. But new principles and new features of development have been emerged, for example, increase the share of tertiary industry and light industry in national economy, and acknowledge the importance of restructuring of industry.

From the briefing of Part II and Box 12.2 and 12.3, and also the briefing of other parts of the Seventh Five-Year Plan. It can be seen that there are many new features in the Plan, which shows that China is in the process of exploration a way from mandatory planning to indicative planning. But the statement "to establish a foundation of new socialist economic system with Chinese characteristics basic within five years or a longer period" is too optimistic to be achieved. The rigidity of institutional system, especially the population and size of China with its huge economic and social inertia have been underestimated.

There are very detailed descriptions of industrial structure and industrial policy. China's State Planning Commission launched the study of industrial policy in the later part of the 1980s, and China's State Council promulgated the first official document of industrial policy in February of 1989.

The Development Research Center of the State Council participated in a joint research project of industrial policies with the World Bank also in later 1980s, and had several publications related to this theme.

12.3.2.4 Briefing of Part III: Regional Deployment and Regional Economic Development Policy

There is a paragraph describing the principle of regional development. The major contents are briefed as follows:

Objectively, there are Eastern, Central and Western, the three large regions that existed in the layout of the Chinese economy, and there is objective trend of development, gradually moving forward from Eastern to Western. In the development of a regional economy, it is necessary to treat correctly the relationship among the Eastern coastal region, Central and Western regions, to realize fully individual comparative advantage, and develop their horizontal economic linkage, gradually establish a network of economic regions with different levels, different scale and with individual characteristics and with metropolis to be the Center. It is necessary to accelerate the development of Eastern coastal region and put priority of construction of energy and raw materials in the Central region, and actively make preparations for further development of the Western region in the period of "Seventh Five" up until the 1990s.

Part III contains six chapters. Chapter 16 of this part will be briefed in Box 12.4 to illustrate the general nature of other chapters in this part.

Box 12.4 Briefing of Chapter Sixteen: Economic Development of Eastern Coastal Region

1. Section I Development Goals and Tasks

The important aspects of economic development are: strengthen technical transformation of traditional industry and enterprise, explore emerging industries with full effort, develop knowledge technology, intensive industry and high end consumers' goods industry to make the products to be developed towards the direction of high, perfect, frontier and new. Accelerate the construction of special economic zones (SEZ), coastal open cities and economic development zones to be two major areas radiated outwardly and inwardly respectively as quick as possible. To make this region becoming the base of trade with abroad, the base of nurture and supply of high level technology and management personnel, the base of transfer new technology, provision of consultation and information to the whole country. The main tasks are:

(1) Upgrade technical level of existing enterprises, to achieve upgrading and generation of changing products.
(2) Strengthen construction of energy, perfect transport network, ease gradually the pressure of transport and energy
(3) Develop agricultural production with full effort, adjust the structure of rural industry
(4) Develop the tertiary industry for production and life vigorously

2. Section II Main Policy Measures

(1) Accelerate steps of technical transformation of existing enterprises. Priorities are given to transformation of old industrial cities and old industrial base such as Shanghai, Tianjin, Shenyang, Dalian etc. At the same time, actively utilize various types of foreign investment, introduce advanced appropriate technology and necessary crucial equipment from abroad.
(2) Utilize better preferential special policies given by the Country, construct and develop SEZs, coastal open cities and economic opening regions in steps and emphatically.
(3) Apply administrative, economic and legal measures to restrict the development of industries and products with high consumption of energy, use of more materials, large quantity in transport and serious discharge of pollutant of "three wastes"(means waste gas, waste water and waste solid discharge). Restrict the development of enterprises with high consumption of water in Northern coastal region in shortage of water.
(4) Move the production of general products to regions with abundant resources of energy and raw materials through multiple pattern of cooperation such as economic union and transfer of technology. Develop production of products with high earning capacity of foreign exchange of export. Establish and perfect export production system dominated by food, daily industrial products, mechanical and electrical products.

Do a good work of introducing advanced technology to digest, absorb, innovate and make it to be localized.

12.3.2.5 Briefing of Part IV and Part V

It can be seen from Table 12.2 that these two parts deal with S&T, education and related policies. It has also been described in these two parts that there are two documents related to reform that should be followed, one is *Decision of Reform of S&T System CCCPC* launched in March of 1985, the other is *Decision of Reform of*

Education System by CCCPC launched in May of 1985. As a whole, the content of five chapters of Part III and also five chapters of Part V are consistent with the contents of above two documents of reform. And the two parts with ten chapters are in much more detail compared with a single chapter dealing both with education and scientific research of China's First Five-Year Plan. One key chapter of each part will be described in detail to show the essence and features of China's Seventh Five-Year Plan.

(1) Chapter 23

Chapter 23 of Part IV deals with Strategy of Development of Science and Technology. Five aspects of strategy are raised:

(i) Vigorously develop and promote generally good results and quick results of S&T achievements; actively apply new technology to reform traditional industry, traditional craft and traditional products; accelerate the process of transfer and spread of domestic new technologies to upgrade significantly the technological level of production of the whole society.

Improving economic performance is the major concern, spreading S&T achievements at different hierarchical levels through multiple pattern and channels. Accelerating the transfer and spread of advanced technology from relatively developed regions to backward regions, from large and medium enterprises to small enterprises, town and village enterprises, from military to civil use.

(ii) Focusing on key technical issues raised in economic construction and social development, concentrating efforts on tackling key S&T problems, and striving to achieve a number of major S&T achievements, and they should be applied to production and construction to make key construction and technical transformation established on the basis of advanced technology. By the meantime, center around key technical problems raised in national defense construction, strengthen the research and development of related S&T of national defense.
(iii) Actively explore new technology and high technology. Focusing on developing microelectronic technology, information technology, strengthening the development of research and development (R&D) of biotechnology, laser technology, aerospace technology, ocean technology to gradually form emerging industries. It is necessary to focus on application with priority in the study and development of emerging industries, especially the wide application of microelectronic and information technology in various economic sectors to make the study of new technology to serve better for the national economy.
(iv) Integrate more effectively the domestic study of S&T with the import of advanced technology. Utilize fully the favorable condition of China's opening to the outside world, actively import advanced technologies, and, accelerate the process of digestion, absorption and innovation, to raise the starting port for study of China' s S&T, to strengthen continuously the capacity of self-development.

(v) Continuously strengthening the applied research and basic research to prepare and better reserve S&T for development in the long term.

The development of S&T is relatively successful, there are several S&T development plans in addition to the first "Twelve Years S&T Plan" described in Chapter 11. China is also relatively successful to develop its education system in quantity. Hereunder, Chapter 32 of Part V will be briefed for illustration of China's Seventh Five-Year Plan compared to its First Five-Year Plan.

(2) Chapter 32

Chapter 32 of Part V deals with major policy measures for development of education, seven policy measures are raised.

(i) Reform the leadership management system of education, strengthen micromanagement.

It is necessary to simplify the administration and decentralize the power, to change the status that the state is too rigid in administration for education. Gradually advocate for a management system for higher schools through the administration of key cities. Simultaneously, strengthen the legislation of education, establish a systematic evaluation and supervision system of education.

(ii) Strictly control the unreasonable upgrading of schools. For example, a secondary technical school applying to be upgraded to a university, or a junior college applying to be promoted to an undergraduate school. All of these must be approved by departments according to regulations of the government. Upgrading inconsistent with regulation must be corrected and rectified.
(iii) Strengthen the construction of teaching staff. Promote teacher education at all levels. Priority should be given to the funds and other conditions for teacher's education, for the purpose to fulfill the Nine-Years' compulsory education. Various levels of the government should solve the practical problems for teachers, to create a conducive atmosphere of respect for educators, and to encourage society to pay more attention to education.
(iv) Increase allocation of funding for education of the central and local government should be higher than the growth of normal fiscal income, and various social forces and individuals are encouraged to donate money on voluntary basis.
(v) Extensive implementation of educational broadcasting. In the period of Seventh Five Year Plan, a special educational channel will be opened to expand the coverage of education by television.
(vi) Various schools at all levels should strengthen the ideological and political work, to train students to be well rounded with socialist construction talents who have ideals, morality, culture, discipline, moral education, intellectual education, physical education and aesthetic education. Simultaneously, it is necessary to correct the undesirable styles and tendencies in running schools.

(vii) The fiscal expenditure by the government for education for the five years from 1986 through 1990 is 116.6 billion yuan, increased by 72% compared to the period of Sixth Five-Year Plan.

12.3.2.6 Briefing of Part VI and Part VII

Selected aspects of Part VI and Part VII will be briefed to provide the overall features of China's Seventh Five Year Plan. Chapter Thirty Four of Part VI will be briefed here because the contents of this chapter related to the improvement of China's factors of production, which will have long term impact to her future growth. Chapter Thirty-Nine of Part VII is also briefed which shows the change of concept of investment structure compared to the concept of China's First Five-Year Plan.

(1) *Chapter 34 of Part II Utilization of Foreign Investment and Import of Technology*
(i) *Section I Utilization of Foreign Investment*

The priority of utilization of foreign investment in energy, transport, communication and raw materials, especially the construction of electric power, port, petroleum and the technical transformation of sectors of machinery and electronics is one aspect of strengthening the capacity for economic development; the other aspect is the expansion of earning capacity for foreign exchange for export, and the implementation of import in order to strengthen the repayment ability and create conditions for further expansion of the scale of utilization of foreign investment.

To utilize foreign investment effectively, the following policies and measured should be applied:

(a) Expand the channels of utilization of foreign investment. Utilize further the preferential credits from foreign governments and international financial organizations, use more appropriately some of the commercial loans from abroad. A few cities and some departments can also utilize appropriately some foreign commercial loans approved by the State beyond the borrowing of the national system. Welcome foreign investors coming to China to establish joint venture, joint cooperative and enterprise of sole proprietorship based on the principle of equality and mutual benefit. China's overseas financial institutions should strive to expand their business to provide more funds for domestic construction.
(b) Continuously improve foreign related laws and regulations to protect legitimate rights and interests of foreign investors. Improve the environment for investment, strengthen construction of infrastructure. Improve the efficiency of government departments and enterprises and their capability in dealing with external business to provide better service and management for foreign investors.
(c) Strengthen the macro-management of utilization of foreign investment: mandatory planning is implemented for borrowing foreign capital from the State. For projects of utilization of foreign investment in the indicative plan

should also be approved strictly in accordance with regulation. All projects of utilization of foreign investment, must be included in the plan of utilization of foreign investment and the plan of income and expenditure of foreign exchange of various levels of government, and linked with a related plan of investment of fixed assets, upgrade the economic and social performance of utilization of foreign investment with full effort. Implement feasibility study of projects seriously. Strengthen the comprehensive balance of the supply of raw materials and energy, domestic matching conditions, condition of sales, income and expenditure of foreign exchange and capability of repayment. Close attention should be paid to completion, and put into operation for projects approved for signature.

(ii) *Section II Import Technology and Intellective Resource*

Expansion of technology import and also intellectual import with priority and guidance step by step is an important way to increase the technical progress of China. Priority of technology import should be focused on technical transformation of existing enterprises. Priority of import should be helpful for expansion of export capability and also development of technology and equipment for import substitution. Five policy and measures are raised. The essence of which are:

(a) The departments concerned should prepare sectoral program of import of technology. Import of advanced technology should consider features of various sectors. Import of technology should be adapted to China's technical basis and physical conditions. In the process of absorption of technology from abroad, departments concerned should pay attention to integrate scientific achievements of China, organize various S&T forces worked on development and innovation.
(b) Import technology should be in accordance with the sectoral technology policy and development programs, avoid duplication of import. Restriction of re-import of similar product and equipment should be implemented post the production of product and equipment by imported technology. Establish and perfect the approval system, examination and acceptance system, information exchange system and statistical system.

It is also emphasized in this chapter to strengthen the cooperation with engineering and technical personnel from abroad in the field of product development, technical design, operation and management, etc.

(2) *Chapter 39 Adjustment of Investment Structure of Part VII*

To promote an appropriate overall scale of investment of fixed assets, and to adjust rationally the structure of investment. The principles of adjustment are:

(i) Invest more capital to accelerate technical transformation, restructure and expand existing enterprises to help the development of economy proceed in the right direction. The share of investment of renewal and technical transformation of investment to capital assets of units will be increased from 27.9% in the period of "Sixth Five" to 30.8% in "Seventh Five"; the share of investment of restructuring and expansion be of investment to capital construction of units owned by whole people should be increased from 56 to 57%.

(ii) Continuously increase the investment to infrastructure and basic industry such as energy, transport, communication and raw materials, increase appropriately investment to agricultural science and education, reduce investment to general processing industry correspondingly. The share of investment of energy, transport and communication of investment to capital construction of state owned units should be increased from 34.4% in the period of "Sixth Five" to 37.4% within five years.

(iii) Reduce the non-productive construction such as office building, exhibition and sales building and building of various centers, except for the use of tourism, stabilizing housing construction, to keep a proper ratio for investment in non-productive construction and productive construction.

(iv) In the regional distribution of investment, priority of investment of technical transformation, restructuring and expansion of existing enterprises shall be located in Eastern region, while investment in a new project of energy and raw materials will be mainly located in Central region.

12.3.2.7 Briefing of Part VIII and Part IX

(1) *Part VIII*

Part VIII deals with goals and tasks of reform of economic system. In fact, there was another document *Decision on Economic System Reform CCCPC* passed by the third Plenary Section of Twelfth Central Committee on December 1984. Contents of Part VIII must be in consistent with the former document. Chapter 43 titled Tasks and Steps of System Reform is briefed to give a general concept of evolution of economic system reform of China. In the period of Seventh Five-Year Plan, China was still in the primary stages of exploration of reform of her economic system which can be understood from the famous sentence of Deng Xiaoping "this exploration is through crossing the river by touching the stones."

(i) *Briefing of Tasks of System Reform*

In the period of "Seventh Five" or a longer time period, establish the foundation of socialist economic system full of vitality and vigor with Chinese characteristics. It is mainly to master the three inter-related aspects:

(a) Further strengthen the enterprise, especially the vitality of large and medium enterprises owned by the people, to make them become a truly independent economic entity, becoming producers and operators of socialist commodity, operate independently and assume sole responsibility for profit and losses.

(b) Develop socialist commodity market further, gradually perfect the market system.

(c) Management of the State enterprises will gradually be transformed from dominance of direct control to indirect control, establish new socialist macro-economic management institutions. Gradually perfect various economic means and legal means, supplemented with necessary administrative means to control and regulate the operation of economy.

Centered around above three aspects, matching with reform of planning system, pricing system, fiscal system, financial system, labor and salary system, to form a complete set of mechanism and means which integrated organically between plan and market, between vitalization of micro aspect and control of macro-side. To achieve the unification of interests of tri-party, the state, the collective and individuals, to realize unification of rate of economic growth, proportion and performance, to achieve virtuous cycle of overall national economy.

(ii) *Steps and Principles to Achieve Tasks of System Reform*

To achieve above tasks and shape the embryonic form of the new economic system, three steps are required basically. The first step is: continuously vitalize enterprises, especially large and medium sized enterprises, energetically develop horizontal economic integration, strengthen and perfect indirect control of macro-system; the second step is: centered around the demand of development of socialist commodity market, gradually reduce the scope of mandatory planning, implement reform of the pricing system of means of production and management system of price, further improve the taxation system and reform of fiscal and financial system; the third step is: gradually establish organizational structure adapted to new system, further deal with the issue of subordinate relationship of enterprise and compartmentalization between administration by line ministries and various regional blocks. Finally, achieve clarification of power and responsibility of administration and economic role of enterprises to make the economic system entering into a new track basically.

In the reform of economic system, it is necessary to handle correctly the relationship between breaking and standing, the following points should be noticed:

(a) Strive to upgrade the capability of the state to implement indirect control to various economic activities.
(b) Importance should be attached to economic legislation and economic judiciary work, strengthen economic monitoring work.
(c) Adhere to proper order of sequence, gradually push forward various reforms based on economic affordability and social affordability.

(2) *Part IX*

Part IX deals with people's life and social security, the spillover effect of economic system reform has its impact on people's life and social security. This subject will be discussed in more detail related to China's Eleventh Five-Year Plan. Part X deals with construction of socialist spiritual civilization, this part has also established certain foundation for follow up plans, this part will be discussed in more detail related to China's Thirteenth Five-Year Plan in later part of this chapter.

12.3.3 *Briefing of China's Eighth Five-Year Plan and Ninth Five Year Plan*

12.3.3.1 Introduction

China's Eighth Five-Year Plan and Ninth Five-Year Plan reported by former Premier Li Peng to the National People's Congress are nearly with the same structure. The report given by Li Peng to the Seventh People's Congress on March 25th 1991 was *Report on the Ten Year Program for national economic and social development and the outline of the Eighth Five-Year Plan* while his report given to the Eighth National People's Congress on March 5th 1996 was titled *Report on the 'Ninth Five' Plan for National Economic and Social Development and the Outline of the 2010 Long-Range Objective*. Both of the two reports added an additional component of planning i.e. a long range of ten years. Therefore, these two Five-Year Plans are grouped together to give a general description of them.

12.3.3.2 Briefing of China's Eighth Five-Year Plan

China's Ten Year Program for National Economic and Social Development and the Outline of the Eighth Five-Year Plan[2] approved by the Fourth Plenary Session of the Seventh National People's Congress on April 9, 1991 (Abbreviated to be: *China's Eighth Five-Year Plan* in the following discussion)

This report contains a preface and Nine Parts which is shown in Table 12.3.

12.3.3.3 Comparison of Structure of Part of China's Eighth Five-Year Plan and Ninth Five-Year Plan

There will be no detail discussion of contents of these two plans, but comparison of structure of content in parts can provide a general understanding that these two plans are nearly of the same structure, and they also do not differ very much compared to China's Seventh Five-Year Plan given in Table 12.2. Some major features of change will be given in Table 12.4 follows.

It can be seen from Table 12.4 that China's Ninth Five-Year Plan adds two more parts compared to China's Eighth Five-Year Plan, one is Part I "Achievements of China's Eighth Five-Year Plan", the other is Part XI promote the great cause of peaceful unification of the motherland. While the remaining parts are similar in nature, the difference is change of sequence and the wording of theme parts. Similarity of parts can be discovered quite easily due to change of only a part of wording, for example, the content of Part IV and Part VI of China's Eighth

[2]Source: Li (1991) (Chinese Version).

Table 12.3 Structure and contents of China's Eighth Five-Year Plan[a]

Theme of parts	Theme of chapters (Number of sections in chapter)	
Preface Part I: Main objectives and guidelines 1991–2000 (Three chapters)	Chapter 1: The goal and the overall blue print	Chapter 2: Basic guidelines
	Chapter 3: Main tasks and important indicators	
Part II: Basic tasks of Eighth Five-Year Plan and comprehensive indicators (Five chapters)	Chapter 1: Basic tasks	Chapter 2: Scale of economic growth and rate of growth
	Chapter 3: Comprehensive economic performance	Chapter 4: The production and distribution of national income
	Chapter 5: Public finance and credit	
Part III: Tasks and policies of main economic sectors in the period of China's Eighth Five-Year Plan (Twelve chapters)	Chapter 1: Agriculture and rural economy	Chapter 2: Construction of water conservancy
	Chapter 3: Energy industry	Chapter 4: Transport, post and communication
	Chapter 5: Raw material industry	Chapter 6: Geological exploration and meteorology
	Chapter 7: Electronics industry	Chapter 8: Machine building industry
	Chapter 9: National defense industry and scientific research of national defense	Chapter 10: Light industry and textile industry
	Chapter 11: Construction industry	Chapter 12: Commodity circulation
Part IV: Deployment and policies of regional economic development in the period of China's Eighth Five-Year Plan (Seven chapters)	Chapter 1: Economic development of coastal region	Chapter 2: Economic development of inland region
	Chapter 3: Economic development of minority region	Chapter 4: Economic development of poverty region
	Chapter 5: Regional economic cooperation and integration	Chapter 6: Urban and rural program and construction
	Chapter 7: Territorial exploration and renovation	
Part V: Tasks and policy of development of S&T, and education in the period of China's Eighth Five-Year Plan (Two chapters)	Chapter 1: Development of S&T	Chapter 2: Development of education

(continued)

Table 12.3 (continued)

Theme of parts	Theme of chapters (Number of sections in chapter)	
Part VI: Trade and economic technology change with abroad in the period of China's Eighth Five-Year Plan (Three chapters)	Chapter 1: Trade of Import and Export	Chapter 2: Utilization of foreign capital, import technology and intelligence
	Chapter 3: Special economic zone, coastal open cities and open area	
Part VII: Main tasks and measures of reform of economic system in China's Eighth Five-Year Plan (Ten chapters)	Chapter 1: Perfect the ownership structure with the public ownership as the main body	Chapter 2: Reform enterprise's system
	Chapter 3: Develop socialist market system	Chapter 4: Reform pricing system
	Chapter 5: Reform the fiscal and tax system	Chapter 6: Reform the financial system
	Chapter 7: Reform the wage system	Chapter 8: Reform housing system and social security system
	Chapter 9: Reform planning system and investment system	Chapter 10: Strengthen the construction of economic regulating system
Part VIII: People's life and consumption policy of China's Eighth Five-Year Plan (Four chapters)	Chapter 1: Income of inhabitants and structure of consumption	Chapter 2: Control population's growth
	Chapter 3: Labor, employment and labor protection	Chapter 4: Health care industry
Part IX: Construction of socialist spiritual civilization and construction of socialist democratic legal system in China's Eighth Five Year Plan (Three chapters)	Chapter 1: Construction of culture	Chapter 2: Ideological construction
	Chapter 3: Construction of socialist democracy and legal system	

[a]*Note* The author had used "—" (—) and "(1)" to be three hierarchical level in his report. (He had used 1 and (1) interchangeably. He had not used the term "Part", "Chapter" and section as used in China's Seventh Five-Year Plan, In order to simplify the expression, authors of this book applied the same form of Table 12.2 to express China's Eighth Five Year Plan. And assume the equivalence as follows: Part I equivalent to "—", chapter equivalent to (—), and section equivalent to 1 or (1).

Five-Year Plan are similar to Part VI and Part VIII of China's Ninth Five-Year Plan. But certain parts may change the theme in wording totally, for example, part IV of China's Ninth Five-Year Plan dealing with tasks and policies of development of main economic sectors, its content is similar to Part III of China's Eighth Five-Year Plan. This illustrates the fact that the planning system of China since the year 1981 had kept the nature of certain continuity but improve progressively.

Table 12.4 Comparison of structure of content of "Parts" between China's Eighth Five-Year Plan and Ninth Five Year Plan

Theme of parts	
China's Eighth Five-Year Plan	China's Ninth Five-Year Plan
Preface	Preface
Part I: Basic objectives and guideline 1991–2000	Part I: Achievements of China's Eighth Five-Year Plan
Part II: Basic tasks and comprehensive economic indicators of China's Eighth Five-Year Plan	Part II: Guideline and objectives of national economic and social development
Part III: Tasks and policies of development of main economic sections in the period of China's Eighth Five-Year Plan	Part III: Objective and policy of macro-regulation
Part IV: Deployment and policy of regional economic development in the period of China's Eighth Five-Year Plan	Part IV: Keep development of national economy to be sustainable, high growth rate and healthy
Part V: Tasks and policies of development of S&T, and education in the period of China's Eighth Five Year Plan	Part V:Implement the strategy of invigorating the country through science and education
Part VI: Trade and economic technological exchange with abroad in the period of China's Eighth Five-Year Plan	Part VI: Promote coordinate development of regional economy
Part VII: Main tasks and measures of reform of economic system in the period of China's Eighth Five-Year Plan	Part VII: Deepening reform of economic system
Part VIII: People's life and consumption policy in the period of China's Eighth Five-Year Plan	Part VIII: Opening wider to the outside world, upgrade the level of opening with abroad
Part IX: Construction of socialist spiritual civilization and construction of socialist democracy and legal system	Part IX: Implement sustainable development strategy, and promote all-round development of social undertaking
	Part X: Strengthen the construction of socialist spiritual civilization and democratic legal system
	Part XI: Promote the great cause of peaceful unification of the motherland

12.3.3.4 S&T Planning in the Period of "Eighth-Five" and "Ninth-Five" Plan

China has prepared several long term S&T plans post China's first S&T plan (The *Twelve Year S&T plan)* which had been briefed in Chapter 11. There were *S&T Development Program 1963–1972* prepared in 1962–1963, and also *Outline of National S&T Development Program 1978–1985 and S&T Development Program 1986–2000* prepared in 1983–1984. And in the period of China's Eighth and Ninth Five-Year Plan, S&T planning had followed the model and objective of related national economic and social developed plan that two S&T Plans have been prepared. One is *S&T Ten Years Development Program (1991–2000) and Outline of "Eighth-Five" Plan, People's Republic of China,* the other is *National S&T Development "Ninth Five" Plan and Outline of Long Term Program 2010.* The former document will be briefed in Box 12.5 for general reference.

Box 12.5 [3]Briefing of *Ten Years Development Program of S&T (1991–2000) and Outline of "Eighth-Five" Plan People's Republic of China* (Abbreviated as *Eighth-Five S&T Plan*)

1. Content of the Eighth-Five S&T Plan

This Plan contains a preface and five sections. Only selected aspects will be briefed.

2. Section 1 Development objectives and guidelines of S&T, 1991–2000

(1) It is targeted in the development objectives that the rate of contribution of progress of S&T to China's economic growth will be increased from 30% to around 50%; actively developing high technology, strive to increase the annual value of output to reach more than 400 billion yuan, its share of total output will be increased from 49% currently to around 89% within 10 years.

Adherence to the absorption of imported technology in combination with self-development to make China's important equipment and complete set of technology of main economic sectors are basic foothold in the country.

(2) There are six points of guidelines to implement the strategy: the "Economic construction must rely upon S&T, the work of S&T must face the economic construction"; carry out strategic deployment of developing high S&T to realize industrialization; respect knowledge and talent people; deepening reform of S&T system; implement combination of self-development and import technology; promote international cooperation and exchange of S&T.

3. Section 2 Major Tasks of S&T 1991–2000

(1) There are four major tasks: The first is to apply electronic information and automation technology to transform traditional industries to make the production technology and equipment of traditional industries to be modernized, their operation and management become scientific, to

[3]Source: Ten Years Development Program of S&T (1991–2000) and Outline of "Eighth-Five" Plan. PRC In Comprehensive section of National Medium and Long Term S&T Development Program Leader's team Office (2003) *Compilation of China' s Seven Times of S&T Program one of Compilation of Data* Chapter V.

establish an economy of energy saving, water saving, consumption reducing and resource saving of arable land; The second is to focus on the development of high technology in realization of industrialization; the third is to achieve important results in certain S&T field related to adjustment of relationship between the man and nature, such as population control, environmental protection, reasonable development and utilization of resource and energy; and the fourth is to achieve significant progress on basic research.

Fourteen aspects have been arranged in detail, they are agriculture and S&T for rural area; energy; transport; raw materials; machinery; light industries and textile; research of high technology and its industrialization; high and new technology industry development zones; technology development of enterprises; social development; basic research; S&T of defense (omit), scientific and technical talent and study of soft science respectively.

(2) Selected aspects of the above will be briefed as follows for illustrative purpose. Research of high technology and its industrialization is selected and briefed.

High technology and its industry are the pillar power of international competition in the 21st Century. China must develop its own high S&T and their industry, upgrade labor productivity greatly and occupy a space in the global high-tech field. The objective of development of China's high technology and its industries in coming ten years are: trucking the development of international high technology in a number of fields that China has certain superiority, with breakthrough of key technologies, strengthen their applied research and engineering development, popularize high technique achievements and form high technology industries. Research of high technology and its industrialization will be incorporated separately in national "863"[4] high technology research plan, program for tackling key problems and "Torch" program. Prioritize the development of high technology such as electronic information, computer and its software, communication, bioengineering, automation, new generation of energy, new materials, super conductibility and laser etc.

There are further detailed descriptions of above fields. (omit)

[4]Note: "863" high technology research plan is China's national high technology development planning with the nature of basic research. It was initiated by four scientists, they wrote a letter to CCCPC on March, 1986 proposed to track the development of strategic high technologies abroad. I t was confirmed by Deng Xiaoping, and this plan was launched on November, 1986 and was terminated only recently.

4. Section 3 Basic objectives and Tasks of Deepening the Reform of S&T system

There are six main tasks of deepening the reform of S&T system. They are: reform management of plan, strengthen the regulating mechanism of market; reform scientific research organizations; promote S&T progress of enterprises; deepening the reform of agricultural S&T system; deepening the reform of management system of S&T personnel; and strengthen the construction of macro-regulating system of S&T progress respectively. (Details omitted)

5. Section 4 Adhere to opening to outside world, actively promote international cooperation of S&T

Four aspects are described in this section. They are: carry out international cooperation based upon demand and possibility; successfully proceed with the work of import technology and its absorption, development and innovation; further prepare and perfect related polities of promoting forging contacts, academic exchange among scientific and technological personnel as well as intelligence output; and create conditions favorable for international cooperation of S&T (Details omitted)

6. Section 5 Supporting conditions and measures

(1) Increase input of S&T;
(2) Improve conditions of S&T;
(3) Perfect fiscal and taxation system of S&T.

12.3.3.5 Several Features of China's Eighth and Ninth Five-Year Plan

(1) "Strengthen the building of an honest and clean government and to carry out the fight against corruption in an intensive and lasting way" had been raised in former premier Li Peng report to the fourth plenary session of Seventh National People's congress on March 25, 1991. This sentence is also a part of statement in Part X of China's Ninth Five-Year Plan. (p. 122)
(2) "Construct and perfect of socialist market economic system is the strategic task in coming 15 years" in part VII of China's Ninth Five-Year Plan is used to replace the statement "Based upon the overall demand to establish primarily an operating mechanism of integration of new system to construct a socialist planned commodity with planned economy and market regulation…." of Part VII of China's Eighth Five-Year Plan.

12.3.4 *Outline of the Tenth Five-Year Plan for National Economic and Social Development of the People's Republic of China (Abbreviated as China's Tenth Five-Year Plan)*

China's Tenth Five-Year Plan was approved by the Fourth Plenary Session of Ninth NPC on March 15, 2001. This plan contains a preface, ten parts with 25 chapters. These ten parts are: Guideline and objective; economic structure; science and technology, education, talented personnel; population, resource and environment; reform and opening; people's life; spiritual civilization; democracy and legal system; national defense construction; and implementation of Program.

It can be seen from all descriptions in Sect. 12.3 that China has established a basic framework of its indicative planning system of a socialist market economy, while its recognition of market has passed a relative long process since the Third Plenary Session of Eleventh Party's Congress.

The structure and content of China's Tenth Five-Year Plan is the product of evolution of its planning concept and process.

12.4 Briefing of China's Planning System from the Eleventh Five-Year Plan to Thirteen Five-Year Plan

In Sect. 12.3 we discussed China's Five-Year Plan in the period of transition from a centrally planned economy to an indicative planning system adapted to a market oriented economy. It was described in detail from the evolution of China's Sixth Five-Year Plan to Tenth Five-Year Plan in Sect. 12.3. Since the framework of China's indicative plan had been established, discussion of this section will be started from a briefing of structure and content of outline of National Economic and Social Development Eleventh Five-Year Plan PRC (abbreviated as China's Eleventh Five-Year Plan), then followed by a briefing of selected aspects of Mid-Term Evaluation of China's Eleventh Five-Year Plan done by the World Bank to show some views of the third party on China's Five-Year Plan. Because the structure and content of China's Twelfth Five-Year Plan is more or less similar to China's Eleventh Five-Year Plan, an analysis of implementation of China's Twelfth Five-Year Plan done by authors of this book and other colleagues will be given instead of description of its structure and content. The English version of China's Thirteenth Five-Year Program is available for sale, an abstract of *Suggestions for the preparation of National Economic and Social Development Thirteenth Five-Year Plan* by CCCPC promulgated on October 29, 2015 will be given to illustrate China's normal planning process. In addition, a briefing of *Outline of National Territorial Plan (2016–2030), and Made in China 2025* will also be given in this section. Because the English version of those documents may not be available, it is expected that this section may provide a relatively complete picture of China's planning system in a new era.

12.4.1 *Briefing of China's Eleventh Five-Year Plan and Its Mid-Term Evolution*

12.4.1.1 Briefing of China's Eleventh Five Year Plan

China's Eleventh Five-Year Plan (2005–2010) was approved by the Fourth Plenary Session of Tenth NPC on March 14th 2006. There are several new features of the planning document and the planning process. With respect to the planning document, two types of indicators are established for goals of socio-economic development. The indictors are classified into obligatory and anticipated type, the former indicator is mandatory in sense, while the latter is indicative in nature. Special column in the form of 'box' is added in the official document to give relative detailed explanation of specific items. The whole document is further perfected in terms of parts and chapters. Mid-term evaluation is implemented since China's Eleventh Five-Year Plan. Table 12.5 presents the structure and content of China's Eleventh Five-Year Plan which can be compared with Table 12.2, the content of China's Seventh Five-Year Plan to see the evolution of China's planning system to become an indicative planning system within these periods.

In Chapter 3 of China's Eleventh Five-Year Plan, there is a Box 2 showing the main indicators of measurement of development of economy and society in the period of China's Eleventh Five-Year Plan which is shown in Table 12.6.

12.4.1.2 Briefing of Mid-Term Evaluation of China's Eleventh Five-Year Plan (Abbreviated As Mid-Term Evaluation Report)

(1) *Launch of Mid-Term Evaluation of China's Eleventh Five-Year Plan*

This mid-term review was prepared by the task force of World Bank staff members at the request of the Development Planning Department of China's National Development and Reform Commission (NDRC). This is also the first time of evaluation of China's Five-Year Plan.

This mid-term evaluation report is consisted of an executive summary, seven parts and eleven annexes. The seven parts are: introduction, stable operation of the macro economy and improving living standards, optimizing and upgrading of the industrial structure, increasing energy efficiency, coordinating urban and rural development, improving basic public service and building a resource efficient and environmentally society respectively.

Table 12.5 Structure and content of China's Eleventh Five-Year Plan[a]

Theme of parts	Theme of chapters (Number of sections in chapters)	
Part I: Guiding principles and objectives of development (Three chapters)	Chapter 1: The critical period of building a well-off society in an all-round way	Chapter 2: Fully implement the scientific view of development
	Chapter 3: Main objective of development of economy and society	
Part II: Building new socialist country side (Six chapters)	Chapter 4: Develop modern agriculture (4 sections)	Chapter 5: Increase income of farmers (3 sections)
	Chapter 6: Improve the landscape and features of country side (4 sections)	Chapter 7: Cultivate farmers of new type (3 sections)
	Chapter 8: Increase input of agriculture and country side	Chapter 9: Deepening the reform of country side
Part III: Promote optimization and upgrading of industrial structure (Six chapters)	Chapter 10: Accelerate development of high technology industry (4 sections)	Chapter 11: Revitalize equipment manufacturing industry (3 sections)
	Chapter 12: Optimize energy industry (4 sections)	Chapter 13: Adjust structure and deployment of raw material industry (3 sections)
	Chapter 14: Upgrade the level of light and textile industry (2 sections)	Chapter 15: Actively promote informatization (4 sections)
Part IV: Accelerate the development of service sector (Three chapters)	Chapter 16: Explore and expand the producer oriented service sector (5 sections)	Chapter 17: Enrich consumption service industry (6 sections)
	Chapter 18: Policies to promote the development of service sector	
Part V: Promote Coordinated regional development (Three chapters)	Chapter 19: Implement overall strategy of regional development (6 sections)	Chapter 20: Promote formation of main functional regions (5 sections)
	Chapter 21: Promote the sector of development of urbanization (4 sections)	
Part VI: Building resource saving and environment friendly society (Five chapters)	Chapter 22: Develop circular economy (6 sections)	Chapter 23: Protect and restores natural ecology
	Chapter 24: Increase environmental protective efforts (4 sections)	Chapter 25: Strengthen management of resource (3 sections)
	Chapter 26: Utilize rationally resource of ocean and climate (2 sections)	

(continued)

Table 12.5 (continued)

Theme of parts	Theme of chapters (Number of sections in chapters)	
Part VII: Implement strategies of Science, talents to revitalize the nation (Three chapters)	Chapter 27: Accelerate the innovation of S&T and the process of leapfrog (5 sections)	Chapter 28: Develop education in priority (5 sections)
	Chapter 29: Push forward the strategy to establish the powerful nation through talents (2 sections)	
Part VIII: Deepening the reform of systems (Five chapters)	Chapter 30: Promote reform of administrative management system with efforts (3 sections)	Chapter 31: Adhere and perfect basic economic system (4 sections)
	Chapter 32: Promote reform of fiscal and taxation system (2 sections)	Chapter 33: Accelerate reform of financial system (4 sections)
	Chapter 34: Perfect modern market system (3 sections)	
Part IX: Implement a win-win strategy of opening up with mutual benefits (Three chapters)	Chapter 35: Accelerate to transform the growth pattern of foreign trade(4 sections)	Chapter 36: Upgrade the quality of utilization of foreign capital (2 sections)
	Chapter 37: Actively carry out international economic cooperation	
Part X: Promote socialist harmonious society (Five chapters)	Chapter 38: Do well the work for population in an all-round way (5 sections)	Chapter 39: Improve living standard of the people (5 sections)
	Chapter 40: Improve people's health (4 sections)	Chapter 41: Strengthen the construction of public security (5 sections)
	Chapter 42: Perfect the management system of society (3 sections)	
Part XI: Strengthen the building of socialist democratic and political constriction (One chapter)	Chapter 43: Strengthen the building of socialist democratic and political construction (3 sections)	
Part XII: Strengthen the building of socialist culture (One chapter)	Chapter 44: Strengthen the building of socialist culture (3 sections)	
Part XIII: Strengthen national defense and army building (One chapter)	Chapter 45: Strengthen the construction of national defense and the army (3 sections)	
XIV: Establish and perfect the mechanism of implementation of Plans (Three chapters)	Chapter 46: Implementation mechanism for guidance of different areas	
	Chapter 47: Adjust and perfect economic policy	
	Chapter 48: Improve planning and management system	

[a]*Source* State Council (2006). *Outline of National Economic and Social Development 11*th *Five-Year Plan*. Beijing: People's Publisher. (Chinese version)

Table 12.6 Main indicators of development of economy and society in the period of China's Eleventh Five-Year Plan[a]

Category	Indicator	2005	2010	Average annual growth rate (%)	Nature of indicator
Economic growth	GDP (RMB trillion)	18.2	26.1	7.5	A
	GDP per capita (RMB)	13,985	19,270	6.6	A
Economic structure	Share of services in GDP (%)	40.3	43.3	[3]	A
	Share of employment in service (%)	31.3	35.3	[4]	A
	Ratio of R&D expenditures of GDP (%)	1.3	2	[0.7]	A
	Urbanization rate (%)	43	47	[4]	A
Population, resources, and environment	Total population (10^4 persons)	130,756	136,000	<8‰	O
	Reduction of energy use per unit of GDP (%)			[20]	O
	Reduction of water use per unit of industrial VA (%)			[30]	O
	Effective utilization coefficient of irrigation water	0.45	0.5	[0.05]	A
	Comprehensive utilization rate of industrial solid waste (%)	55.8	60	[4.2]	A
	Total cultivated land kept (mln ha.)	1.22	1.2	−0.3	O
	Reduction of major discharge of pollutants (%)			[10]	O
	Rate of forest coverage (%)	18.2	20	[1.8]	O
Public services and quality of life	Average number of years of schooling of citizens (yr)	8.5	9	[0.5]	A
	Population covered by basic pension and insurance of urbans (100 mln)	1.74	2.23	5.1	O
	Rate of coverage of new rural coop, health care system (%)	23.5	>80	>[56.5]	O
	Increase of new employment in urban within five years (10^4 persons)			[4500]	A
	Transfer of rural labor force within five years (10^4 persons)			[4500]	A
	Registered urban unemployment rate (%)	4.2	5		A
	Per capita disp. income of urban households (RMB)	10,493	13,390	5	A
	Per capita net income of rural households (RMB)	3255	4150	5	A

[a]*Source* Box 2 of Chapter 3 of China's 11th Five Year Plan (p. 11).

Note GDP and income of urban and rural household is price of 2005; the figure in bracket [] is cumulative value of five years; major pollutants refer to COD and SO_2

A Anticipated; *O* Obligatory

(2) *Evaluation of Meeting the Quantitative Targets of the Plan*

Evaluation is done by checking the progress toward achieving the major objective which is shown in Table 12.1 of Mid-Term Evolution Report, which concludes that "Progress toward achieving the major objectives of the Eleventh 5YP has varied." (p.vi) This Table 1 of the evaluation report is listed below as Table 12.7.

It is observed and concluded by the World Bank (2008) in the mid-term evaluation report that China's economic growth has far exceeded expectations, considerable progress has been made toward the Five-Year Plan's most important social objectives such as improving basic public services in social protection, education, health etc. Progress on the environmental objectives has been mixed, and insufficient progress has been made in reducing energy intensity, but improvements were seen in reducing air and water pollution, etc. The report gives the following observation:

> However, little progress has been made in rebalancing the overall pattern of growth, which has in turn limited progress on other key objectives. There has been little rebalancing away from industry and investment towards services and consumption. This in turn, has made it difficult to meet the objectives on energy efficiency, the environment, and reducing the external imbalance. The lack of decisive rebalancing has also made further widening of urban-rural income inequality almost unavoidable, despite strong government effort. (pp. v–vi)

It has also been emphasized in this Mid-Term evaluation report that these imbalances were an outcome of China's capital-intensive, industry-led pattern of growth which is shown in Fig. 12.1. It is pointed out by the World Bank's Mid-Term Evolution Report (2008) that "China's growth had been capital intensive, with the investment to GDP ratio rising to almost 43% in 2005 and- using growth accounting-capital accumulation accounted for over 60% of GDP growth during 1993–2005.

(3) *Evaluation Related to Stable Operation of Macro Economy and Improving Living*

China has broadly succeeded in combining rapid growth with low inflation. During 2005–07, GDP growth accelerated from 10.4% in 2005 to 11.9% in 2009. The per capita urban income also rose from 9.6% in 2005 to 12.2% in 2007, this figure is 6.2 and 9.5% respectively in the rural areas, as the gap between rural and urban income widened: The growth rate of industry still outpaces services and the growth rate of investment still outpaces consumption in the period of 2006–2007. A visible reflection of China's macroeconomic imbalances is the current account surplus, which climbed to over 11% of GDP in 2007 approaching 0.75% of global output. And China's exchange has gradually appreciated in the period from 2005 to 2008. (pp. vi–viii)

(4) *Evaluation Related to Optimization and Upgrading of Industrial Structure*

There has been mixed progress in meeting the overall objectives in the area of industrial structure. The targets on the share of total employment in the service

Table 12.7 China's progress in meeting the quantitative indicators under the 11th Five-Year Plan[a]

Category	Indicator	2005 (actual)	2007 (actual)	2010 (target)	Type of target
Economic growth	GDP (RMB trillion)	18.4	25.0	26.1	A
	GDP per capita (RMB)	14,103	18,885	19,270	A
Economic structure	Share of services in GDP (%)	39.9	40.1	43.3	A
	Share of employment in services (%)	31.4	33.2	35.3	A
	Ratio of R&D expenditures of GDP (%)	1.2	1.4	2.0	A
	Urbanization rate (%)	43.0	44.9	47.0	A
Population, resources, and environment	Total population (100 mln)	13.1	13.2	13.6	O
	Reduction of energy use per unit of GDP (%)	0	4.6 2/	20.0 3/	O
	Reduction of water use per unit of industrial VA (%)	0	…	30.0 3/	O
	Effective utilization coefficient of irrigation water	0.45	0.46 4/	0.5	A
	Comprehensive utilization rate of industrial solid waste (%)	56.1	61.2	60.0	A
	Total cultivated land kept (mln ha.)	122.1	121.7 4/	120.0	O
	Reduction of major discharge of pollutants (%)				
	COD	na	2.1 2/	10.0 3/	O
	SO_2	na	3.2 2/	10.0 3/	O
	Rate of forest coverage (%)	18.2	…	20	O
Public services and quality of life	Average number of years of schooling of citizens (year)	8.5	…	9.0	A
	Population covered by basic urban pension and insurance of urban (100 mln)	1.7	2.0	2.2	O
	Rate of coverage new rural coop, health care system (%)	75.7	85.7	80	O
	Increase of new urban employment within five years (mln)			45	A
	Transfer of rural labor force within five years (mln)			45	A
	Registered urban unemployment rate (%)	4.2		5.0	A
	Per capita disp. income of urban households (RMB)	10,493	13,790	13,390	A
	Per capita net income rural households (RMB)	3255	4140	4150	A

[a]*Source* Mid-term evaluation of China's 11th 5 Year Plan The World Bank. pp. vi
Source China's authorities, NBS, and staff estimates
A Anticipated; *O* Obligatory
2/accumulated reduction in 2006–07
3/targeted accumulated reduction in 2005–10
4/2007 data not yet available. This is the 2006 data

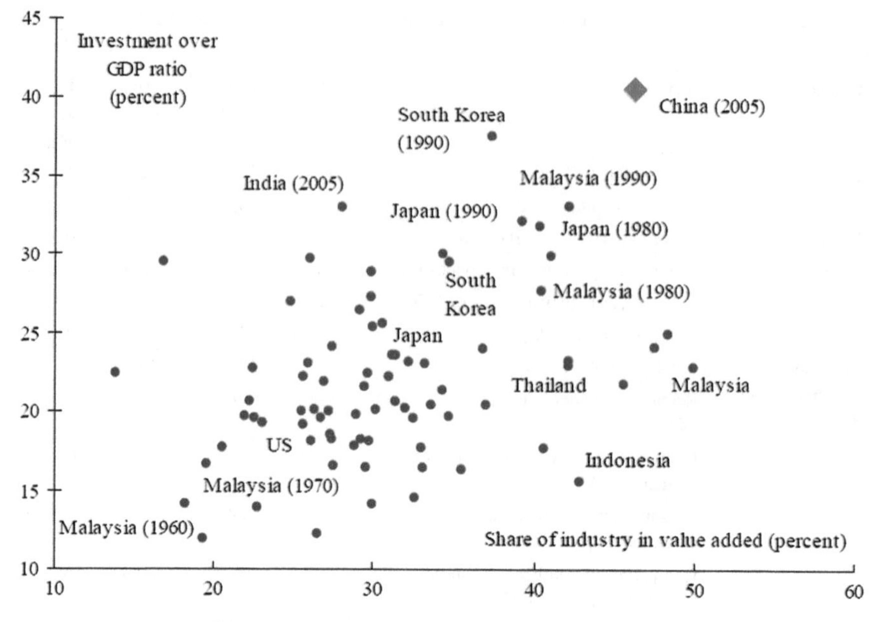

Fig. 12.1 China's capital-intensive, industry-led economy in international perspective (*Source* World Bank (2008, Fig. 12.3, p. iii).)

sector and R&D are within reach. However, given recent trends, it seems unlikely that the target to raise the share of the service sector in GDP can be met. Achieving the objectives of energy efficiency and environmental quality will be become more difficult due to the impact of energy intensive heavy and chemical industries gaining importance in the planning period. The industrial specific agenda to upgrade the industrial structure appears to be on track. But whether carrying out the industrial agenda actually improves the economic and industrial structure is open to question in an increasing market oriented economy. With the increased market orientation of China's economy, industrial upgrading is best pursued by measures that encourage innovation such as improved IPR, venture capital markets, better governance of SOEs, and greater access of private firms to capital markets. However, SOEs continue to show considerably higher capital intensity with lower rates of return on capital, employment creation and labor productivity growth than non-SOEs. (pp. viii–ix)

(5) *Evaluation Related to Increasing Energy Efficiency*

Although China's energy intensity has been reduced, it was by much less then needed to achieve a 20% reduction by 2010. Because the overall pattern of growth and the policies underlying it, including the pricing of energy and other resources remain broadly unchanged. Raising energy prices is one option, but it is equally important is to put in place the policies and the institutional, regulatory, technical, and financial framework and capacity to sustain China's efforts to transform to more efficient economic growth. (p. ix)

(6) *Evaluation Related to Coordinated Urban and Rural Development and Improving Basic Public Services*

The first two years of the Eleventh Five Year Plan have witnessed substantial social progress. Conditions in rural areas have improved significantly. The targets relevant to balanced rural-urban development in the Eleventh Five Year Plan, including the coverage of the new rural cooperative medical services, farmland retention, and per capita income of rural residents, will likely be met. And also considerable progress has been made to improve public services in the areas of social protection, education, and health. But at the same time, the challenge of how to distribute the benefits of rapid economic development more equitably remains a problem. Because while conditions are improved in rural areas, they are improving even faster in urban areas. The gaps in income and quality of life between the urban and rural areas will continue to widen. (pp. ix–x)

(7) *Evaluation Related to Enhancing Sustainable Development*

Preliminary indications are that China has made progress toward a more resource efficient and environmentally sound economy. Although there has been some progress of reduction of SO_2 and COD, meeting the target may be difficult. There are immense environmental challenges remaining. Air and water pollution of China still exceeds applicable standards in most areas. Freshwater withdrawals already exceed sustainable levels of both surface and underground resources. Forest covering remains far below the level needed to restore its environmental and ecological functions. China must be awaken to the fact that prospects for continued progress in increasing resource efficiency are clouded by the current economic structure, with concentration of industrial growth in resource intensive, high polluting industries. (pp. x–xi)

12.4.2 Outline of China's National Economic and Social Development Twelfth Five-Year Plan and Analysis of Its Performance (Abbreviated as China's Twelfth Five-Year Plan)

12.4.2.1 Briefing of China's Twelfth Five-Year Plan

Structurally, China's Twelfth Five-Year Plan is similar to China's Eleventh Five-Year Plan. There are only changes in the wording of theme of parts but with the same theme to be discussed. There are also changes of orders of parts. China's Twelfth Five-Year Plan has sixteens parts, i.e. it has two more parts compared to the Eleventh Five-Year Plan. One addition is a part to promote socialist harmonious society of Eleventh Five-Year Plan, which has been changed into two parts in Twelfth Five-Year Plan. One is Part VIII *improving people's life, establish basic system of public service*, the other is Part IX, *treating both symptoms and root causes, strength and innovate social management of Twelfth Five-Year Plan*. One more chapter added in the Twelfth Five-Year Plan is Part XIV, *deepening cooperation, and building a common homeland for the Chinese nation*. Therefore, China's Twelfth Five-Year Plan has sixteen parts. There are also further improvements by way of expression, five figures are added in this plan, for example the diagram of Agricultural Strategic Pattern of Seven Zone and Twenty-Three Belt. This diagram gives zoning of appropriate agricultural products. And also a diagram of China's high speed railway network which presents high speed railway network constructed, under process of construction and route in planning etc. This section will present no more on the structure and contents of Twelfth Five-Year Plan, only the main indicators of socio-economic development of Twelfth Five-Year Plan will be listed to illustrate the evolution compared with China's Eleventh Five-Year Plan.

Box 12.6 Main Indicators of Socio-economic Development in the period of China's Twelfth Five-Year Plan

Indicator	2010	2015	Average annual growth rate (%)	Nature of indicators
Economic development				
GDP(trillion yuan)	39.8	55.8	7	A
VA share of service sector (%)	43	47	[4]	A
Rate of urbanization (%)	47.5	51.5	[4]	A
Science, technology, education				
Consolidation rate of nine-year compulsory education (%)	89.7	93	[3.3]	O
Gross enrollment rate of senior middle school education (%)	82.5	87	[4.5]	A
Share of expenses of R&D of GDP (%)	1.75	2.2	[0.45]	A
Invention patterns per 10,000 population (piece)	1.7	3.3	[1.6]	A
Resources and environment				
Arable land kept (100 million mu)	18.18	18.18	[0]	O
Reduction of water consumption per unit industrial VA (%)			[30]	O
Effective utilization coefficient of agricultural irrigation water	0.5	0.53	[0.03]	A

(continued)

(continued)

Indicator		2010	2015	Average annual growth rate (%)	Nature of indicators
Share of non-fossil energy of consumption of primary energy (%)		8.3	11.4	[3.1]	O
Reduction of energy consumption per unit GDP (%)				[16]	O
Reduction of emission of CO_2 per unit GDP (%)				[17]	O
Reduction of overall quantity of discharge of main pollutants	COD			[8]	O
	SO_2			[8]	
	Ammonia Nitrogen			[10]	
	Nitrogen Oxide			[10]	
Growth of forestry	Rate of forest coverage (%)	20.36	21.66	[1.3]	O
	Forest Reserves (Hundred million M^3)	137	143	[6]	
People's life					
Per capita disposable income of urban household (RMB)		19,109	>26,810	>7	A
Per capita net income of rural households (RMB)		5919	>8310	>7	A
Registered urban unemployment rate		4.1	<5		A
Newly increased urban employment (10^4 persons)				[4500]	A
Population covered by basic pension and insurance in urban area (100 million)		2.57	3.57	[1]	O
Participation rate of three basic medical insurance in urban and rural area (%)				[3]	O
Construction of affordable housing projects in cities and towns (10^4 homes)				[3600]	O

(continued)

(continued)

Indicator	2010	2015	Average annual growth rate (%)	Nature of indicators
National total population (10^4 persons)	134,100	<139,000	<7.2‰	O
Life expectancy (years)	73.5	74.5	[1]	A

Note ① GDP and absolute value of urban and rural household income are calculated based on prices of 2010, growth rate is calculated based on comparable price
② Number in bracket [ ] is the cumulative number of five years
③ Participation rate of three basic medical insurance in urban and rural areas referred to the ratio of total urban employment participated the basic medical insurance by the end of the year plus the urban household participated the basic medical insurance plus the newly rural cooperative medical insurance to the total amount of national population by the end of the year
④ Growth of income of urban and rural household is based on the principle that they should not be lower than the growth of GDP. All efforts should strive to achieve the synchronous development with the economy

12.4.2.2 Analysis of Implementation of China's Twelfth Five Year Plan

(1) *Economic Development has Achieved New Successful Results*

Since the implementation of China's Twelfth Five Year Plan, there are new achievements of social economic development. The overall macroeconomic target has been well achieved. China's GDP has been increased from 47.31 trillion yuan in 2011 to 67.7 trillion yuan by the end of 2015. The share of China's GDP to global total has been increased from 10.4% in 2011 to 12.7% in 2014. Therefore, China has also become one of the engines of growth of the global economy, especially during the period of global economic crisis since 2008.

Due to huge amount of China's population, although the Chinese economy has been ranked the second of the global economy since 2010, its per capita income is still low. But China has been transformed from the group of low income countries to middle high income group countries based upon classification of the World Bank (Fig. 12.2).

Dotted line Fig. 12.2 is the change of standard of classification of country groups of the world based upon their income. There are four types of country groups, low income, lower middle income, high middle income and high income countries. The per capita GNI of China in the year 2011 was 4940 U.S. dollar, it is increased to

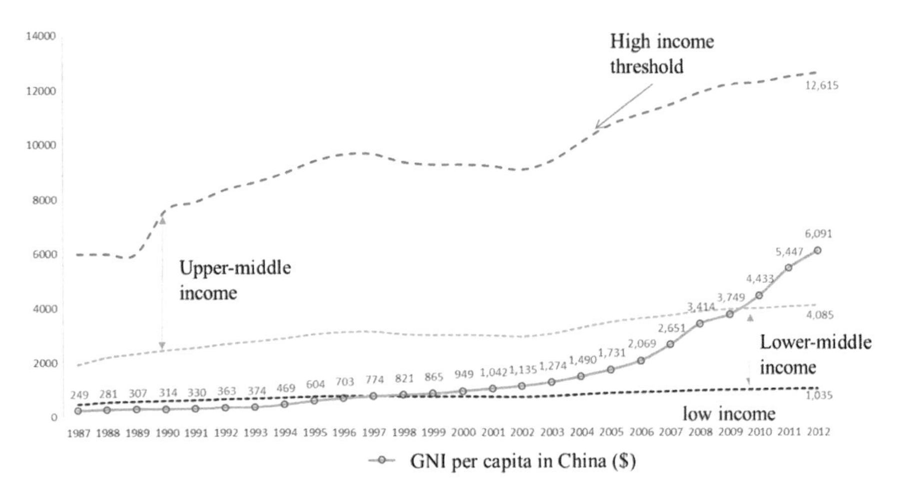

Fig. 12.2 Chinese per capita GNI and classification of countries based upon income (The World Bank) (*Source* Shantong et al. (2015))

7900 U.S.D. in 2015. It can be seen from Fig. 12.2 that China has been stepped from low middle income group to high middle income group countries in the year 2010. Further effort is required for China to step into high income group countries.

(2) *Restructuring of the Economy has been Achieved to Certain Extent. But There is Further Need to Optimize the Structure of the Component Sector*

The structure of the Chinese economy has been improved to adapt to its stage of development. The share of components of GDP of the supply side is 8.8%:40.8%:50.5% (P:S:T) in 2015. The share of service sector has surpassed the secondary sector, and there is further strengthening of the manufactures.

(i) *Current Status of the Manufacturing Sector*

The capacity of production of the manufacturing factor is further strengthened, but there is serious overcapacity of many subsectors. The sales income of Chinese manufacturing sector has exceeded 90 trillion yuan in 2012, its value added reached around 20 trillion yuan, a share of 20% of global manufacturing value added. The strength of China's manufacturing sector can be found in the *International Yearbook of Industrial Statistics 2014* published by UNIDO. In the statistics of major producers in a total of 22 divisions of manufacturing of ISIC, Chinese products ranked No. 1 in 12 divisions (ISIC 16 Textile products, ISIC17 Textiles, ISIC 18 Wearing apparel, ISIC 19 Leather, leather products and footwear, ISIC 20 Wood products (excl. furniture), ISIC 21 Paper and paper products, ISIC 26 Non-metallic products, ISIC 27 Basic metals, ISIC 28 Machinery and equipment n. e.c., ISIC 30 Office, accounting and computing machinery, ISIC 31 Electric machinery and apparatus, ISIC 32 Radio, television and communication equipment). There are 9 Chinese products ranked No.2 in total 22 divisions of

ISIC,mostly they are lagged behind U.S.A., these 9 Chinese products are shown in the bracket. (ISIC 15 Food and beverages, ISIC 23 Coke, refined petroleum products, nuclear fuel, ISIC 24 Chemicals and chemical products, ISIC 25 Rubber and plastic products, ISIC 28 Fabricated metal products, ISIC 33 Medical, precision and optical instruments, ISIC 34 Motor vehicles, trailers, semi-trailers, ISIC 35 other transportation equipment, ISIC 36 Furniture manufacturing n.e.c.). It can be seen from above that Chinese manufacturing products played a dominant role in 21 divisions among the total of 22 divisions of ISIC.

There is also achievement of upgrading the manufacturing sector. The share of labor intensive manufacturing sector is decreased, while the share of capital and technology intensive manufacturing sector is increased. There is also transformation from high energy intensive to lower energy intensive sectors.

A part of manufacturing sector has the serious issue of overcapacity, for example, the iron and steel, cement, electrolytic aluminum, plate glass, ships and even emerging new energy photovoltaic industry. In the year 2012, the production capacity of crude steel of China was one billion tons, a share of 46% of the world total, but the output was only 0.72 billion tons with a utilization ratio of 72%; the production capacity of electrolytic aluminum was 27.65 million tons, its output was 19.88 million tons with a share of 42% of global output, its utilization ratio was also 72% only; the production capacity of cement was 2.99 billion tons, the output was 2.18 billion tons, a share of 60% of global total but with a utilization ratio of 73% only; the production capacity of flat glass was 1.04 billion containers, while the output was 0.71 billion container with a share of global output around 50%, whiles its utilization ratio was 68%. With respect to the utilization ratio of photovoltaic industry, it is generally in the range of 30%-50%, therefore it has very low investment efficiency. Based upon the global recognized standard, if the utilization ratio of production capacity is less than 75%, it is recognized to be in a situation of serious overcapacity. Therefore, the overcapacity of several manufacturing sectors of China is a problem of concern.

(ii) *The Service Sector has Grown Rapidly, but There is a Need to Further Improve its Internal Structure*

The service sector has accelerated the growth in the period of Twelfth Five Year Plan, its share of GDP and its growth rate have surpassed the primary and secondary sector. People employed in the service sector has also surpassed the primary and secondary sector, it become a dominant sector to provide new employment.

The new emerging service sector has extraordinary high growth such as internet ecommerce and logistics. Sales of E-commerce reached around 8 trillion yuan which is around 33% of total retail sales of consumer goods in 2014. Table 12.8 gives details of E-commerce of enterprises by industrial sector in 2015.

There is a need to improve further the internal structure of the service sector and strengthen the performance and regulation of some subsectors.

The productive service of China has still a lower share of the service compared to that of the developed economies. It has a share around 35% in the period of

Table 12.8 E-commerce of enterprises by industrial sector 2015[a]

Industry	With E-commerce transactions		Sales of E-commerce (100 million yuan)	Purchases of E-commerce (100 million yuan)
	Enterprises (Unit)	Share of total (%)		
Total	87,436	9.6	91,724.2	53,499.1
Mining	299	2.2	191.4	489.5
Manufacturing	35,800	10.2	38,715	23,805.9
Production and supply of electricity, heat, gas and water	556	5.6	73.9	5124.1
Construction	3731	4.1	86.5	2172.2
Wholesale and retail trades	16,156	9.0	37,859.0	19,396.7
Transport, storage and post	2128	5.8	3275.0	292.6
Hotels and catering services	13,549	30.8	436.7	35.8
Information transmission, software and information technology	3457	25.0	7829.6	866.0
Real estate	4082	4.1	73.1	51.5
Leasing and business services	2837	9.5	2345.4	1013.1
Scientific research and technical services	1609	8.5	559.3	214.5
Management of water conservancy, environment and public facilities	632	13.4	36.3	10.8
Service to households, repair and other services	430	7.6	15.8	5.3
Education	275	6.3	20.8	1.1
Health and social service	332	7.8	3.8	14.3
Culture, sports and entertainment	1563	25.3	202.7	5.7

[a]*Source* National Bureau of Statistics of China (Compiler, 2015) China Statistical Yearbook 2015 China Statistics Press Beijing

Twelfth Five-Year Plan. And also the financial sector of China has a large share of the service sector, it has a share around 27.93% of the service sector in 2015, it also has the highest growth rate 17% in the same period compared to the year of 2014. But its component market such as equity market has high volatility in the year 2015, and the bond market needs healthy expansion and further improvement. Therefore, there is hard work to be done for the development of the service sector.

(3) *The Labor Productivity of Agricultural Sector is Rising Continuously, but Many Challenges are Faced*

Due to wider application of technology to agricultural production, for example, the comprehensive mechanization farming reached 61%, the effective irrigation rate reached more than 51%, crop fine breed popularization rate reached more than 95% during the Twelfth Five-Year Plan. Therefore, there was an increase in agricultural labor productivity in continuity with the transfer of rural surplus labor force. Based upon the survey of cost-benefit of national agricultural products, there was an increase of major agricultural products per labor-day from 1990 to 2013, the average annual increase of cereal is 9.9%, and 12.3%, 10.3%, 14.3% and 8.7% respectively for wheat, corn, soybean and cotton. But progress from the modernization of agriculture still lagged behind the advanced countries. Chinese agricultural labor productivity is around 47% of the global average value, while it is only 2% of average value of high income countries.

There are many new challenges faced in the upgrading process from the traditional mode of agricultural production to the new mode of economy of scale, standardization of production and corporate operation. First, there are continuously rising production costs, rent of land, labor costs, inputs, service purchasing, etc. Rising costs cause prices of products to rise. Second, the price of the main bulk agricultural products are higher than the price of the international market, this affects the competitiveness and comparative advantage of China's agricultural products in the international market. Third, due to shortage of land compared to a huge Chinese population, China pursued the target of large quantity of production which caused serious pressure on agricultural resources, China's agricultural production is facing increasingly the threat of the shortage of water resources and degradation of cultivated land.

(4) *Changing Features and the Role of Factors of Production*

(i) *Traditional Factors of Production still Play an Important Role to Promote Growth*

Traditional elements of factors of production, the inputs of labor force, capital and land still play an important role for economic growth of China. There was an increase of new employment, around 62.4 million people in the period of 2009–2013 in China's cities and towns, there was also a transfer of labor force around 47 million of people from primary to high VA secondary and tertiary sector. The share of labor force in the primary sector is decreased from 34.8% in 2011 to 28.3% in 2015, there was an increase of 6.5 percentage points of non-farming employment in this period. In addition, there was also around 7 million graduates who formed a high quality labor force. These two facts, transfer of labor force from low value-aided sector to high value-aided sector and the new entry of a high quality labor force will complement the effect of diminishing marginal returns of capital investment, which still performs an important role for economic growth. The rate of contribution to growth of GDP from formation of capital increased from 45.2% in

2011 to 48.5% by the year 2014. The transfer of use of land from farming to industrial and commercial use has raised its output per unit area rapidly.

(ii) *Changing Features of Traditional Elements of Factors of Production.*

The advantage of China's traditional factors of production became weaker during the Twelfth Five-Year Plan, there was an increase of aging population, and rising costs of labor and land. Average wages of migrant workers rose continuously, increasing from 1690 yuan in 2010 to 2864 yuan in 2014 (Quan Sanlin, 2014). Implementation of Labor Contract Law since Jan 2008 also promoted the rise of labor costs of firms. For example, a firm in Nanning municipality would increase its labor cost 4370 yuan more for each employee, the cost of labor of the firm increased by 63% based upon lowest wage of Nanning municipality. With respect to land resources, the cost of development of land also rose along with the rapid socio-economic development of China. The price of land transferred from state owned construction land increased from 1.43 million yuan per hectare in 2001 to 9.59 million yuan per hectare, with a growth around 5.69 times, shown in Fig. 12.3. There are rapid rising prices for land of all types, industrial, commercial or housing. The average national comprehensive price of land rose continuously from 3291 yuan/m^2 in 2008 to 4235 yuan/m^2 in 2012, with average annual rate of increase around 6.5%.

(iii) *The Capacity of Innovation is Strengthened and Will Become an Important Source of Future Growth*

Chinese capacity of innovation of S&T strengthened during the Twelfth Five-Year Plan, there was growth of R&D expenses during this period. The share of expenditure of R&D of GDP increased from 1.76% in 2010 to 2.06% in 2015. The growth of R&D expenses from 2010 to 2015 is shown in Fig. 12.4.

The progress of Chinese capacity of innovation is also shown in Table 12.9.

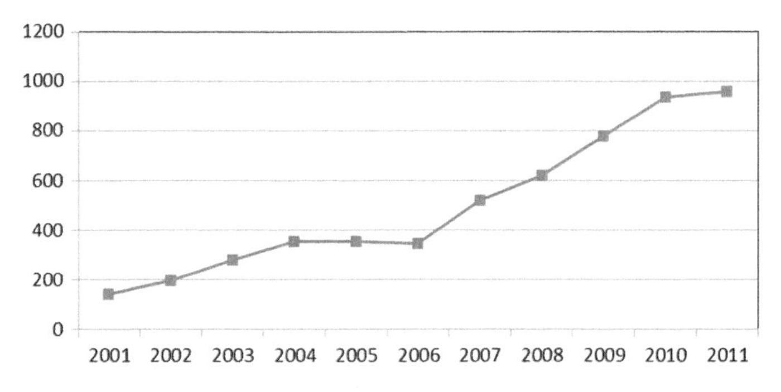

Fig. 12.3 Variation of land transfer price of state owned construction land (*Source* China's Statistical Yearbook of various years.)

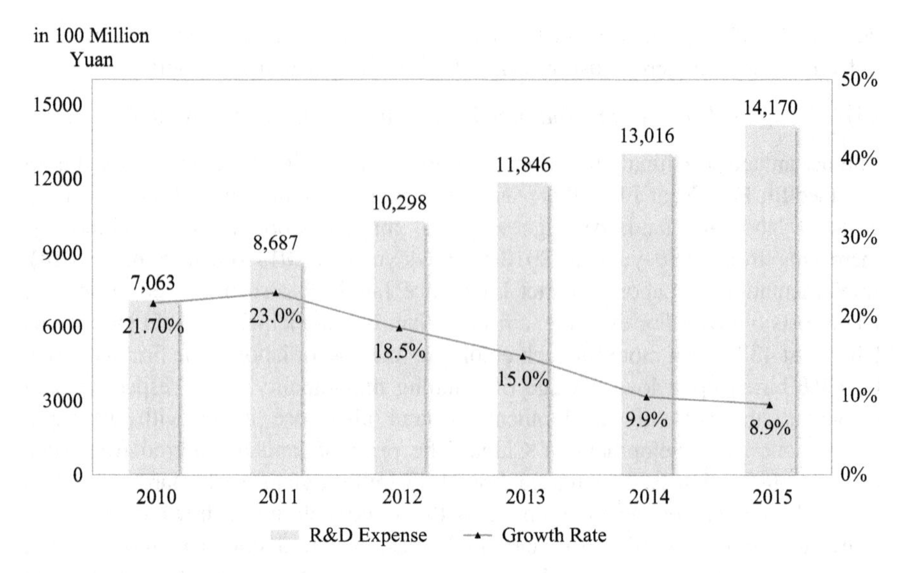

Fig. 12.4 Growth of Chinese R&D expense (*Source* Derived from Data of State Bureau of Statistics. http://data.stats.gov.cn/easyquery.htm?cn=C01)

(iv) *Several Major Issues Remain to Be Overcome*:

Although the share of R&D expense of GDP is accelerating that its growth rate nearly approaching a part of advanced economies, but the share of basic research remained around 5% of annual R&D expense throughout all of these years, which is quite low compared with a value of around 19% for the United States. Although the Chinese scholars wrote many S&T papers, their rate of citation is only around 50% of the United States. China has almost the highest number of patent applications, but its tri-party pattern application is only around one tenth of that for American and Japanese firms. Many large manufacturing corporations abroad have not only a position for a chief engineer, but also a position for a chief scientist. Therefore, it is necessary for China to overcome many institutional weaknesses to further improve its capacity of innovation.

(5) *The Level of Urbanization Rate Rose Rapidly, but Its Quality Remained to Be Upgraded*

(i) *The Level and Growth Rate of Urbanization Rate*

The level of urbanization rate has risen rapidly, but its growth rate is slowing. It can be seen from Fig. 12.5 that there is stable increase of urbanization rate in the period of Twelfth Five Year Plan. It is increased from 51.27% in 2011 to 54.77% in the year 2014. The urban resident population increased from 690.79 million to 749.16 million. But the rising rate slowed during this period, which decreased from 2.64% in 2011 to 1.93% in 2014.

Table 12.9 Selected items showing progress of capacity of innovation of China[a]

Index	2011	2012	2013	2014	2015
S&T paper published (10^4 papers)	150	152	154	157	164
S&T book published (kind)	45,472	46,751	45,730	47,470	52,207
National technological invention awarded (item)	55	77	71	70	66
National S&T progress prizes awarded (item)	283	212	188	202	187
Number of patent application accepted (piece)	1,633,347	2,050,649	2,377,061	2,361,243	2,798,500
Invention patent (piece)	526,412	652,777	825,136	928,177	1,101,864
Number of pattern application granted (piece)	960,513	1,255,138	1,313,000	1,302,687	1,718,192
Invention patent (piece)	172,113	217,105	207,688	233,228	359,316
Amount of import and export of high technology goods (USD 100 million)	10,120	11,080	12,185	12,119	12,046
Amount of export of high-tech goods	5488	6012	6603	6605	6553
Amount of import of high-tech goods	4632	5069	5682	5514	5493
Amount of transaction of China's technological market (100 million Yuan)	4764	6437	7469	8577	9836

[a]*Source* National Bureau of Statistics of China (Complier, 2016)

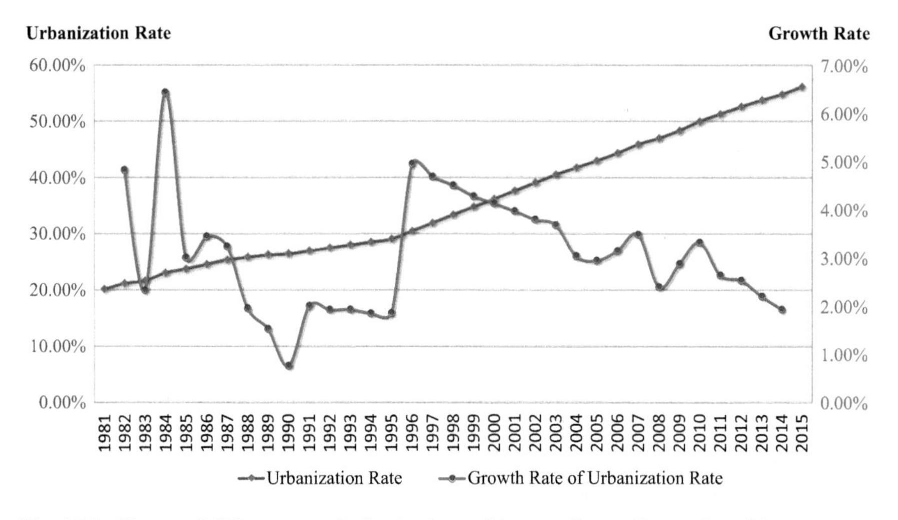

Fig. 12.5 Change of Chinese rate of urbanization and its growth rate (*Source* http://data.stats.gov.cn/easyquery.htm?cn=C01&zb=A0A01&sj=2015.)

(ii) *Quality of Urbanization Remains to Be Upgraded*

In the process of China's urbanization, there are currently three major issues that remain to be solved.

Difficulties of integrating the migrant rural workers into urban life. Due to imperfect institutions and the current condition of public finance, the large amount of migrant rural workers cannot enjoy equal opportunity with the urban inhabitants for public service, education and healthcare etc. There are also implicit troubles with differences of culture and social status between the two groups.

(iii) *Low Performance for the Utilization of Land*

Currently, there are issues with the utilization of land with insufficient concentration, insufficient aggregation and low efficacies. With the efficacy of land utilization in municipal districts for example, the output value per unit of land for most municipalities is low, except for the districts of Shenzhen municipality, Shijiazhuang, municipality, Shanghai, Nanchang and Guangzhou municipality which have a higher output value per unit land.

(iv) *Urban Diseases Are Serious*

Several issues such as environmental pollution, traffic congestion and urban waterlogging have affected the quality of life of urban inhabitants Due to the rapid growth of urban population and incomplete transition of traditional pattern of economic growth, there are huge amounts of discharges of polluted water, dust, sulfur oxides, nitrogen oxides and solid wastes. With respect to traffic congestion, it is shown in a survey that the average commuting time for Chinese people is around 55 min, which is 15 min higher than the global average value of 40 min. All of these issues need to be rectified and improved to further the management and administration of cities and towns.

(6) *Input of Social Construction is Increasing in Continuity, but Inclusiveness of Development Remains to be Improved Further*

(i) *Income Disparity*

There has been improvement with income distribution, but income disparity is still relatively high.

Figure 12.6 shows that the trend of rising of income disparity was reversed since 2008, but the Gini coefficient of 0.469 in 2014 is still higher than the universal warning point of 0.4 of national income.

(ii) *The Income Level of Urban and Rural Residence*

Figure 12.7 shows the income of urban and rural residence and the ratio of income between these two groups. It can be seen from the figure that the income disparity between urban and rural reached its peak value of 3.33 in the period of Eleventh Five-Year Plan (2006–2010). But in the period of Twelfth Five-Year Plan, the

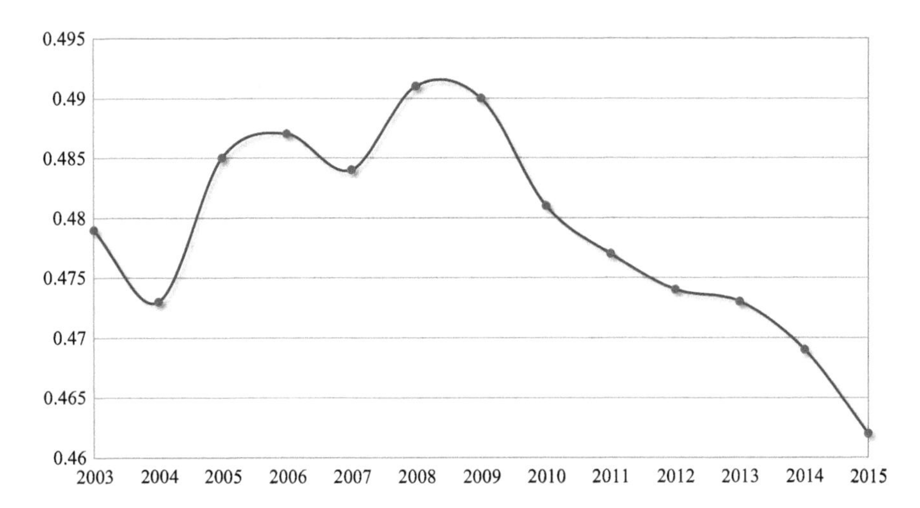

Fig. 12.6 Change of Gini coefficient of China 2003–2015 (*Source* Website of National Bureau of Statistics)

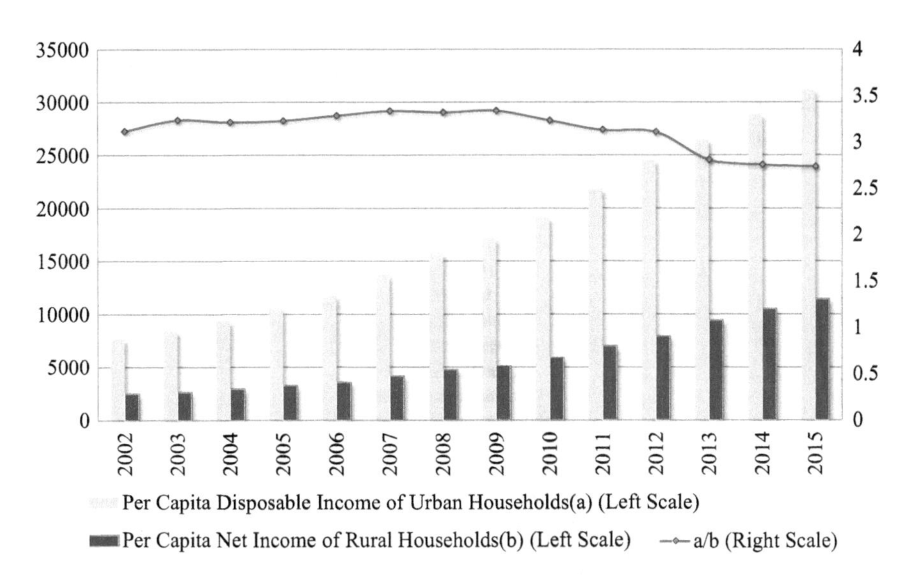

Fig. 12.7 Income of urban and rural resident and ratio between them 2001–2013 (*Source* http://data.stats.gov.cn/easyquery.htm?cn=C01)

income disparity between the urban and rural was reduced gradually. The average disposable income of urban residents is 26,955 yuan in 2013, while the average per capita net income of rural residents is 8896 yuan, the ratio between these two is reduced to 3.03.

(iii) *The Olive Type Social Structure Remains to Be Formed*

It is well known in theory and practice that the structure of income distribution of olive type is favorable for a sustainable development of a stable and healthy society, because the group of middle income people has the highest share, i.e. most of the population share the benefits of China's development and reform. Although there are improvements of income disparity between urban and rural and the national as a whole, the pattern of income distribution of olive type social structure has not been formed yet. Figure 12.8 is the pattern of distribution of income of rural people of 2012 in groups, which shows a pattern of a vase type with broad base and narrow top, rather than an olive type.

(7) *Input for Public Service Increased Continuously, but the Level of Equality Remains to Be Upgraded*

There was an increase for input of public service continuously in the period of Twelfth Five-Year Plan. The share of educational financial expenditure of GDP increased from 3.65% in 2010 to 4.28% in 2012. The share of health care financial expenditure of GDP increased from 4.98% in 2010 to 5.57% in 2013. And there were increases in the number of people participating in social security, which is shown in Fig. 12.9, but the beneficiaries of basic public service from local public finance are resident populations, without consideration of the migrant workers. There are needs to further improve the equality.

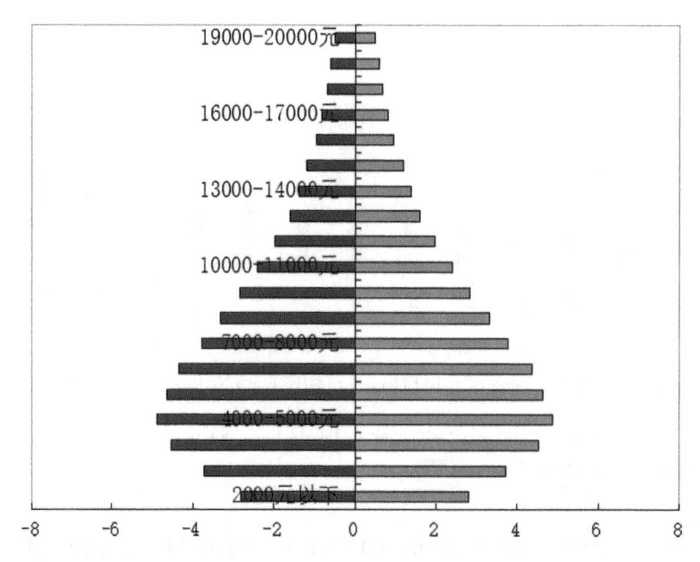

Fig. 12.8 Income grouping of rural household (%) 2012 [the colors make the print illegible on this table] (*Source* Calculated by the author)

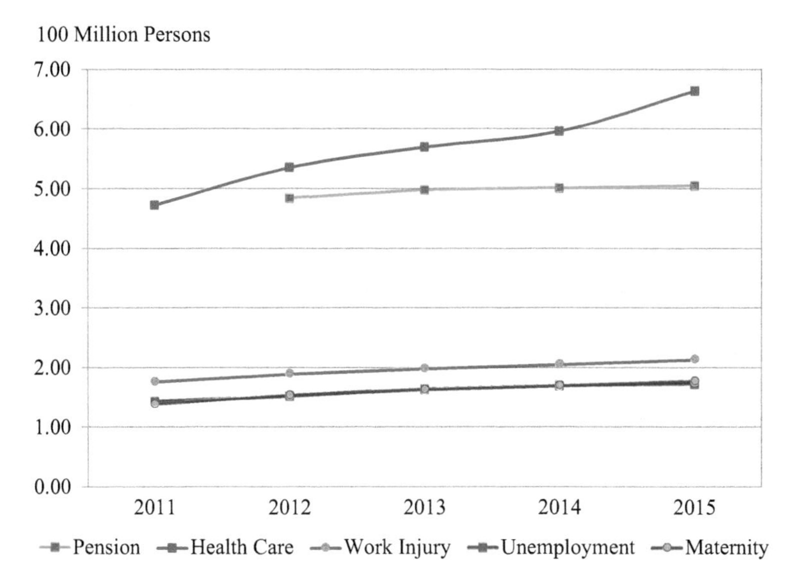

Fig. 12.9 Number of people participating in the social security system (100 million persons) (*Source* http://data.stats.gov.cn/easyquery.htm?cn=C01)

12.4.3 Period of Implementation of Outline of China's National Economic and Social Development Thirteenth Five-Year Plan (Abbreviated as China's Thirteenth Five-Year Plan)

China's Thirteenth Five-Year Plan was approved by the Fourth Plenary Session of Twelfth NPC on March 16, 2016. China is currently in the period of implementation of China's Thirteenth Five Year Plan. That plan has both Chinese and English versions published. Therefore this Thirteenth Five-Year Plan will not be briefed here. But the structure and content of *Suggestions for the Preparation of National Economic and Social Development Thirteenth Five-Year Plan by CCCPC* promulgated on October 29, 2015 will be presented here. It has been explained previously that China normalized her planning process since China's Seventh Five-Year Plan. Under the revised process, there are suggestions of preparations for certain Five-Year Plans by CCCPC, then the planning agency prepares certain Five-Year Plans based on the suggestions of CCCPC, and the prepared Five-Year Plans will be sent to NPC for approval. The opening date of or the NPC is also normalized to be March 5 of the year. Presentation of the 'Suggestion' here will give a complete picture of China's planning process since the 1980s. In addition, this Section will provide two boxes discussing materials that may not be available in the English version. One box is *Made in China 2025*, the other box is *Outline of National Territorial Plan (2016–2030)* which is promulgated by the State Council on January 3, 2017. This is the first of China's national territorial plan. It is expected that these three documents presented in this Section together with other parts of this Chapter may give a relatively complete picture of evolution of China's national planning system.

12.4.3.1 Structure of Content of the "Suggestions"

Content of the "Suggestions" is composed of 8 major parts with sub-topics, which is shown in Table 12.10.

Abstract of Selected Essential Points of Suggestions

(1) *Theme of Sub-Topic 3 of Part 1*

Theme of Sub-Topic 3 of Part 1 is "Guiding thoughts of China's development in the period of Thirteenth Five-Year Plan." It is stated in this part that six principles should be followed in order to achieve the target to establish an all-round well-off society in time. These six principles are: Adhere to people's centered status; Adhere to scientific development; Insist on deepening of reform; Insist on governing the country by law; It is imperative to coordinate the domestic and international situations; and Adhere to the Party's leadership.

(2) *Theme of Sub-Topic 1 of Part 2*

Theme of sub-topic 1 of Part 2 is "New target and requirement of economic and social development in the period of Thirteenth Five-Year Plan." Five new targets are stated in this theme. They are:

(i) To maintain a medium-to-high growth of the Chinese economy. GDP and per-capita income of urban and rural inhabitants by the year 2020 which will be twice of those in 2010 on the basis of upgrade of development of balance, inclusiveness and sustainability;
(ii) The level and quality of people's life will be raised universally;
(iii) The national quality and social civilization will be upgraded significantly;
(iv) Quality of ecological environment will be improved as a whole;
(v) Various institutions will become more matured and patterned.

(3) *Theme of Sub-Topic 2 of Part 2*

Theme of sub-topic 2 of Part 2 is "Perfection of the concept of development". It is stated in this theme that Achieving the development target in the period of "Thirteenth-Five Year Plan", solve the difficult issues of development, plant deeply the strength of development, it is necessary to create the concept of development of innovation, coordination, green, open and sharing.

Table 12.10 Structure of content of suggestions

S. No.	Theme of major part	Sub-topics in the major part
1	Part 1: Situation and guiding thoughts in the decisive stage of establishment for a well-rounded, prosperous society	3 Sub-topics: (1) Important achievements of China's development in the period of Twelfth Five-Year Plan; (2) Basic characteristics of China's environment of development in the period of Thirteenth Five-Year Plan; (3) Guiding thoughts of China's development in the period of Thirteenth Five-Year Plan
2	Part 2: Main targets and basic concepts of socio-economic development in the period of Thirteenth Five Year Plan	2 Sub-topics: (1) New target and requirements of economic and social development in the period of Thirteenth Five-Year Plan; (2) Perfection of the concepts of development
3	Part 3: Insistence on innovative development, raising quality and efficacy of development with full effort	7 sub-topics: (1) Create and nurture new driving force of development; (2) Explore new space of development; (3) Implement innovation driven development strategy in depth; (4) Promote agricultural modernization with full effort; (5) Construct new system of industry; (6) Construct new institution of development; (7) Innovation and perfection of pattern of macro regulation and control
4	Part 4: Hold on coordinative development, formation of a balanced structure of development with full effort	4 sub-topics: (1) Promote regional coordinative development; (2) Promote urban-rural coordinative development; (3) Promote coordinative development of material civilization and spiritual civilization; (4) Promote integrated development of economic construction and national defense construction
5	Part 5: Insist on green development, focus on improvement of ecological environment	6 sub-topics: (1) Promote harmonious co-existence of the human beings and nature; (2) Accelerate to construct main functional region; (3) Promote low carbon and circular development; (4) Utilization of resources in all round saving and with high efficiency; (5) Strengthen the strength of environmental management; (6) Consolidate safety barrier of ecology
6	Part 6: Hold on opening development, focus on achievement of cooperation	6 Sub-topics: (1) Perfection of strategic arrangement of opening to outside world; (2) Form new institution of opening to outside; (3) Promote the construction of "One Belt, One Road"; (4) Deepening the cooperative development between Inland and Hong Kong, Macau, and between the mainland and Taiwan region; (5) Actively participate global economic governance; (6) Actively bear the international responsibility and obligation

(continued)

Table 12.10 (continued)

S. No.	Theme of major part	Sub-topics in the major part
7	Part 7: Insist on shared development, focus on promotion of people's well being	8 Sub-topics: (1) Increase the supply of public service; (2) Implement poverty alleviation project; (3) Raise the quality of education; (4) Promote employment and entrepreneurship; (5) Reduce income disparity; (6) Construct a more equitable and more sustainable social security system; (7) Push forward construction of healthy China; (8) Promote the balanced development of populations
8	Part 8: Strengthen and improve the lead of the party to provide strong guarantee to achieve Thirteenth Five Year Plan	6 sub-topics: (1) Perfect the working institution and the mechanism of the Party to lead economic and social development; (2) Mobilize the people to unite and to strive for success; (3) Accelerate to construct a strong nation with talented people; (4) Push forward development through the thought and mode of rule of law; (5) Strengthen and innovate social governance; (6) Guarantee the targets and tasks of this "Suggestions" to be practicable at all levels of government and related organizations

12.4.3.2 Made in China 2025

Box 12.7 Made in China 2025

Made in China 2025 is the first ten year's Program of Action to implement the strategy to establish a powerful nation through manufacturing. This document consisted of four parts and twenty three sections. They will be described below:

1. Part I: Situation of development and environment

Section 1: The pattern of global manufacturing sector is facing important adjustment.
Section 2: There are important changes of environment for China's economic development.
Section 3: The task to establish a manufacturing powerful nation is hard and urgent.

2. Part I I: Strategic guideline and objectives

Section 1: Guiding thoughts

Innovation driven; quality to be priority; green development; structural optimization; and talented people to be basis

Section 2: Basic principle

Market driven, guidance by government; based on present, looking for long term; push forward in whole, breakthrough at crucial area; self-development, opening and cooperation.

Section 3: Strategic objective

Strive through three steps to achieve the strategic objective to realize manufacturing powerful country

Step 1: strive to spend ten years to step into the ranks of manufacturing power; to achieve industrialization basically by the year 2020; the overall quality will be upgraded greatly by the year 2025.
Step 2: Overall manufacturing sector of China reached the mid-level of global manufacturing powerful country, to achieve industrialization.
Step 3: The position of China to be a large country for manufacturing will be more consolidated, comprehensive strength of China's manufacturing shall step into the front edge of global manufacturing power. China will have the leading capability of innovation in major fields of manufacturing by the time of one hundred years of establishment of new China.

3. Part III: Strategic Tasks and priorities

Section 1: upgrade capability of innovation of national manufacture
Section 2: Push forward deep integration of informatization and industrialization
Section 3: Strengthen basic capability of industry
Section 4: Strengthen the building of quality brand
Section 5: Push forward green manufacturing
Section 6: Push forward breakthrough development in priority field with full effort:

Ten priority fields are identified. They are: new generation of information technology industry; high end digital controlled machine tools and robots; airline and aero-space equipment; ocean engineering equipment and high technology ship; advanced tracked transport equipment; energy saving and new energy automobile; electric power equipment; agricultural machinery equipment; new materials; biological medicine and high performance medical instruments.

Section 7: Push forward adjustment of manufacturing structure
Section 8: Actively develop service type manufacturing and producer service
Section 9: Upgrade the level of development of internalization of manufacturing sector

4. Part IV: Strategic support and guarantee

Section 1: Deepening the reform of system of mechanism
Section 2: Create a fair environment of competition for the market
Section 3: Perfect the supportive policy of finance
Section 4: Increase the supportive force of fiscal policy and taxation
Section 5: Improve the cultivated system for talented people at multi-levels.
Section 6: Improve policies to medium, small and micro-enterprises
Section 7: Further expand opening to outside world for China's manufacturing sector
Section 8: Perfect mechanism of implementation of organization

A national building of powerful nation through manufacturing, a leading team will be established, leader of the State Council will be the head of the team, members of this team will be consisted of responsible people of related departments. The role and responsibility of this team are to coordinate overall works of building powerful countries of manufacture, review and assess important plans, policies, important specific engineering projects and important works. It will focus on deployment of important works, strengthen strategic planning to guide the works of line ministries and regions. Leading team office will be set at "Ministry of Industry and Information Technology" to be responsible for daily affairs.

12.4.3.3 Contents of Outline of National Territorial Plan (2016–2030)

Box 12.8 [5]Contents of Outline of National Territorial Plan (2016–2030)

1. Introduction
2. Chapter 1: Basic Situation

   Section 1: Important opportunities
   Section 2: Serious challenges

3. Chapter 2: Overall requirements

   Section 1: Guiding thoughts
   Section 2: Basic principles
   Section 3: Main objectives

[5]Source: Outline of National Territorial Plan (2016–2030) promulgated by State Council on Jan. 3rd 2017.

Table 1 Main indicators

Name of indicators	2015	2020	2030	Nature of indicators
1. Cultivated land reserved (100 mn mu	18.65	18.65	18.25	O
2. Total amount of water consumed (100 mln M^3)	6180	6700	7000	O
3. Forestry coverage (%)	21.66	>23	>24	A
4. Grassland comprehensive skin grafting coverage (%)	54	56	60	A
5. Area of wetland (100 mln mu)	8	8	8.3	A
6. Intensity of territorial development (%)	4.02	4.24	4.62	O
7. Urban space (10,000 km^2)	8.9	10.21	11.67	A
8. Density of highway and railway network (km/km^2)	0.49	>0.5	>0.6	A
9. Ratio of excellent water of seven national major river basin (%)	67.5	>70	>75	O
10. Water quality standard rate reached of important rivers, lakes by water functional region (%)	70.8	>80	>95	O
11. Newly increased rectification of soil erosion area (10,000 km^2)		32	94	A

Note O-obligatory A-anticipated

4. Chapter 3: Strategic Pattern

   Section 1: Highly efficient and normalized opening pattern of territorial development
   Section 2: Pattern of safe and harmonious ecological environmental protection
   Section 3: Pattern of coordinated and linked regional development

5. Chapter 4: Agglomerated development

   Section 1: Construct multi-center networked type of pattern of development
   Section 2: Push forward new type of development of urbanization
   Section 3: Optimize deployment of development of modern industry

6. Chapter 5: Classified protection

Section 1: Construct 'Five categories three levels' protection pattern of whole national territory (Note: Five categories refer to: environmental quality; human settlement ecology; natural ecology; water resources and cultivated land resources).

Section 2: Push forward human settlements ecological environmental protection
Section 3: Strengthen protection of natural ecology
Section 4: Strict protection of water resources and cultivated land resources
Section 5: Strengthen protection of ocean ecological environment

7. Chapter 6: Comprehensive improvement

   Section 1: Push forward formation of "Four region and one belt" patterns of comprehensive improvement of territory
   Section 2: Implement comprehensive improvement of regions of urbanization
   Section 3: Push forward comprehensive improvement of rural land
   Section 4: Strengthen comprehensive improvement of crucial functional region
   Section 5: Accelerate comprehensive improvement of concentrated region of mineral resources development
   Section 6: Carry out comprehensive improvement of coastal zones and islands

8. Chapter 7: Linkage of development

   Section 1: Push forward integrated regional development
   Section 2: Support accelerated development of specific region
   Section 3: Upgrade level of opening and cooperation

9. Chapter 8: Support protection

   Section 1: Strengthen construction of infrastructure
   Section 2: Protect demand of reasonable use of land for construction
   Section 3: Strengthen comprehensive allocation of water resources
   Section 4: Construct safety protection system of energy
   Section 5: Upgrade protective capability of non-energy important mineral resources
   Section 6: Enhance the ability of disaster prevention and reduction
   Section 7: Push forward innovation of mechanism of system

10. Chapter 9: Supporting policy

    Section 1: Resource and environmental policy
    Section 2: Industrial investment policy
    Section 3: Fiscal and taxation policy

11. Chapter 10: Implementation of Outline

    Section 1: Tamp the foundation of implementation
    Section 2: Strengthen Management of implementation

12.5 Analysis of China's Planning Performance (1981–2015) from Perspective of Social Systems Engineering

12.5.1 Features of China's Socio-Economic System in the Period from 1981–2015

(1) China's economy has been changed from a centrally planned economy to a socialist market economy in this planning period. It was officially raised in 14th Party's Congress on October 12, 1992, and it was also raised in Part X of China's Ninth Five-Year Plan.
(2) China's socio-economic system was also gradually changed from a semi-closed system to an open system integrated largely with the global market on trade and investment. China's participation in the WTO on November 2001 promoted greatly the development of China's opening system.
(3) Great effort was spent by the Chinese government, planning agencies, think tanks and academic field to transform China's planning system from mandatory to an indicative planning system which can be seen from the evolutionary process of contents of planning documents and various types of planning (national social economic development plan, S&T plan, industrial plan, territorial plan), normalization of planning process, etc.

12.5.2 Analysis of Performance of China's Economic System Based upon Revised Parsons' AGIL Framework

12.5.2.1 Economic Growth

There were high economic growth rate, relatively balanced regional development and high performance of external sector due to the progressive process of opening and reform implemented in late 1970s. Because economic growth becomes the major goal of economic system and adaptive action has been taken by the Chinese government in opening to the outside world. China's economic system has adapted relatively well within the context of trend of globalization and regionalization.

(1) *High Growth Rate*

It can be seen from Figs. 12.10 and 12.11 that China has a very high growth rate of its GDP and GDP per capita in the period from China's Sixth Five-Year Plan to Twelfth Five-Year Plan. The GDP growth rate from the Sixth to Tenth Five-Year Plan varied in the range of 8.3–12%, averaged around 9.6%. The economic growth rate in Eleventh and Twelfth- Five-Year Plan was 11.2 and 7.8%, respectively. The

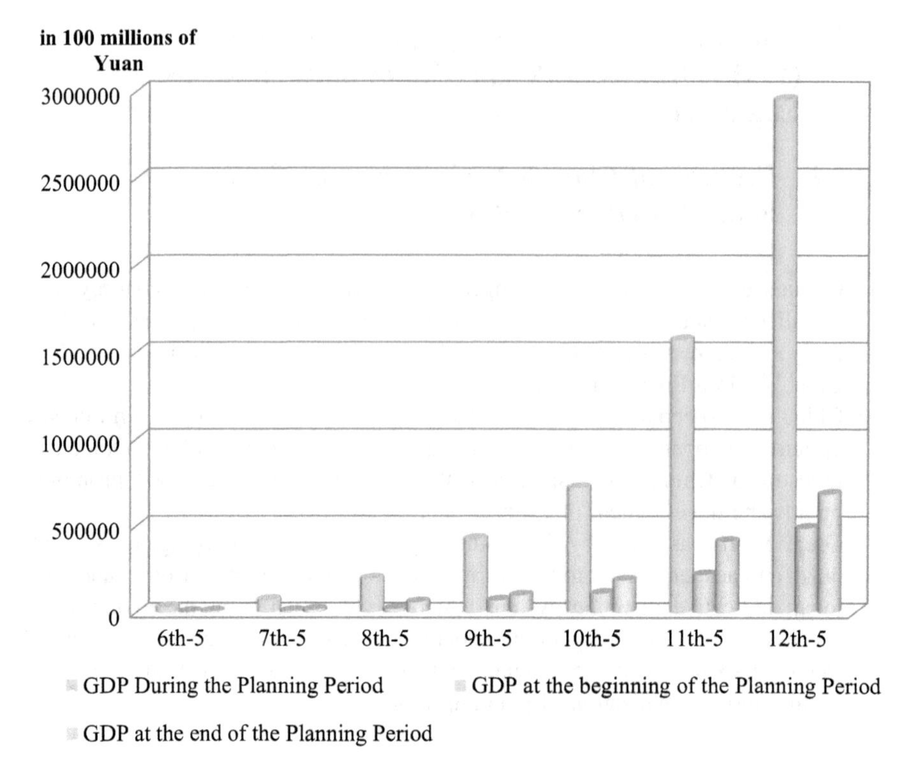

Fig. 12.10 Growth of China's GDP from the period of sixth five-year plan to twelfth five-year plan

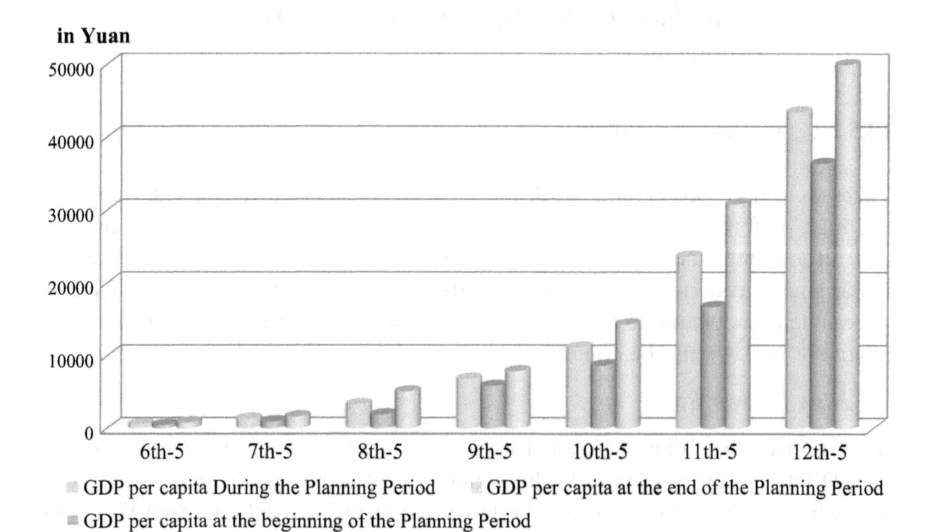

Fig. 12.11 Growth of China's GDP/capita from the period from the period of sixth five-year plan to twelfth five year plan

growth rate of 7.8% in Twelfth Five-Year Plan is due to changes both of domestic system and external environment. But China is already the world's largest economy measured in purchasing power parity, and it ranked the second largest economy measured in official exchange rate. The per capita GDP has also a relatively high growth rate, although it is lower than the growth rate of GDP. But it had been described in preface of China's Seventh Five-Year Plan that "The average consumption level of inhabitants of whole country in 1985 is increased from 227 yuan in 1980 to 404 yuan, the average annual growth rate is 8.5% (with deduction of factor of rise of price level), far exceed the average annual growth rate of 2.6% within 28 years from 1953 to 1980." (p. 17). This can illustrate that the consumption of the people is focused rather than mainly on investment, the people's lives improved greatly in the period of reform.

Although there are increases of unbalance of regional GDP as shown in tables of Figs. 12.10 and 12.11, the ratio of region with highest GDP to region of lowest GDP is increased from 33.65 (Jiangsu: Tibet) in 1981 to 70.94 (Guangdong: Tibet) in 2015. But if the ratio of the region with the highest GDP/capita to the region with the lowest GDP/capita, then it can be seen that a relative balanced regional development has been achieved. This fact can be proved from the data of Table 11.14 of Chapter 11 that there is increase of this ratio from 1952 to 1980, i.e. 8.11(Shanghai: Sichuan) in 1952 to 12.44 (Shanghai: Guizhou) in 1980, and this ratio is decreased from 1981, i.e. 11.57 (Shanghai: Guizhou) to 3.96 (Shanghai: Gansu) in 2015 (Refer to Table 12.10). This rise and fall of regional disparity is also in consistent with theory of development economics that China has done relatively well (Table 12.11).

(2) *Structural Changes of Economy*

China's economic growth is also accompanied with structural changes of its economy in this period. The share of primary, secondary and tertiary sector changed from 31.3:46:22.7 in 1981 to 8.8:40.9:50.3 in 2015. And there were also changes of employment structure, it is changed from 68.1:18.3:13.6 in 1981 to 28.3:29:3:42.4 in 2015, which are shown in Figs. 12.12 and 12.13 respectively. This is a significant achievement because China was mainly an agricultural society during the establishment of People's Republic of China in 1949. The rural population had a share of 87.5% of total population and the rural labor force had a share of 88% of total labor force in 1952. The structural changes of China's economy since its First Five Year Plan up to Twelfth Five Year Plan has moved a huge amount of surplus labor force from low value added agricultural sector into high value added secondary and tertiary sector. This is the major cause for promoting China's economic and social development and also improvement of its people's lives. But there is also a side effect of development. Appropriate policies to deal with migrant workers in cities and towns are issues of concern (Figs. 12.14 and 12.15).

(3) *China Pattern of Economic Growth*

China pattern of economic growth in this period (Sixth Five-Twelfth Five) differs from pre-reform era is its opening to outside world.

Table 12.11 Comparison of regional GDP per capita between 1981 and 2015

Unit: Yuan		
Name of regions (provinces, autonomous regions, central administered municipalities)	Regional GDP per capita 1981	Regional GDP per capita 2015
Beijing	1526	106,497
Tianjin	1458	107,960
Hebei	427	40,255
Shanxi	488	34,919
The Nei Menggu Autonomous Region	407	71,101
Liaoning	823	65,354
Jilin	496	51,086
Heilong Jiang	709	39,462
Shanghai	2800	103,796
Jiangsu	586	87,995
Zhejiang	531	77,644
Anhui	346	35,997
Fujian	416	67,966
Jiangxi	369	36,724
Shandong	472	64,168
Henan	340.1	39,123
Hubei	466.32	50,654
Hunan	394	42,754
Guangdong	550	67,503
The Guangxi Zhuang Autonomous Region	317.27	35,190
Hainan	399	40,818
Chongqing	379	52,321
Sichuan	337	36,775
Guizhou	242	29,847
Yunnan	294	28,806
Tibet Autonomous Region	560	31,999
Shanxi	358.46	47,626
Gansu	367	26,165
Qinghai	459	41,252
The Ningxia Hui Autonomous Region	460	43,805
The Xinjiang Uygur Autonomous Region	450	40,036

Figure 12.16 shows growth of external trade of China in the period of China's Sixth Five-Year Plan to Twelfth Five-Year Plan. China has become the largest exporter of goods in the world, and it has become the second largest trade nation globally. China is also a second largest importer of goods in the world and a net importer of service products.

There are also changes of FDI. Figure 12.17 shows the growth of FDI in China. It should be emphasized that growth of external trade and FDI of China in this

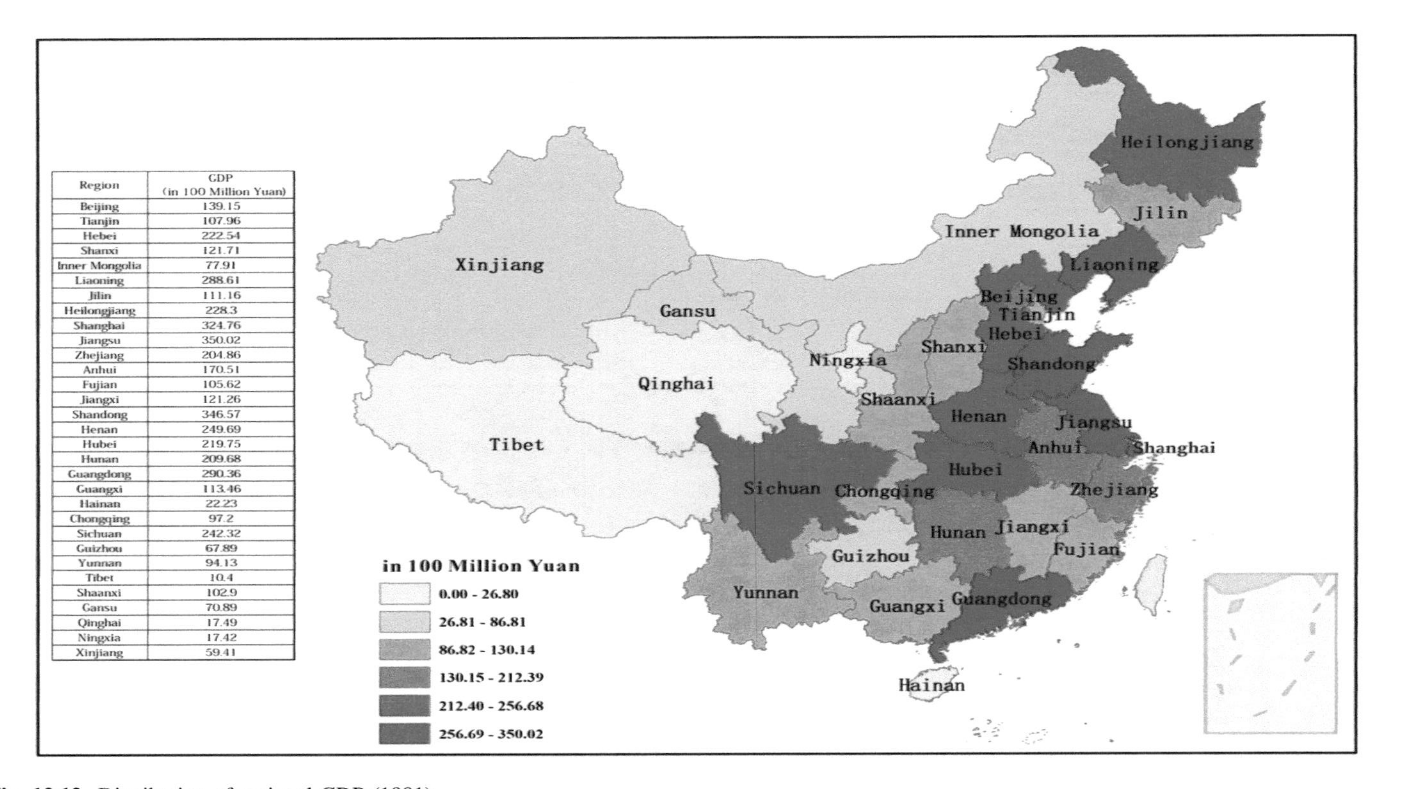

Region	GDP (in 100 Million Yuan)
Beijing	139.15
Tianjin	107.96
Hebei	222.54
Shanxi	121.71
Inner Mongolia	77.91
Liaoning	288.61
Jilin	111.16
Heilongjiang	228.3
Shanghai	324.76
Jiangsu	350.02
Zhejiang	204.86
Anhui	170.51
Fujian	105.62
Jiangxi	121.26
Shandong	346.57
Henan	249.69
Hubei	219.75
Hunan	209.68
Guangdong	290.36
Guangxi	113.46
Hainan	22.23
Chongqing	97.2
Sichuan	242.32
Guizhou	67.89
Yunnan	94.13
Tibet	10.4
Shaanxi	102.9
Gansu	70.89
Qinghai	17.49
Ningxia	17.42
Xinjiang	59.41

Fig. 12.12 Distribution of regional GDP (1981)

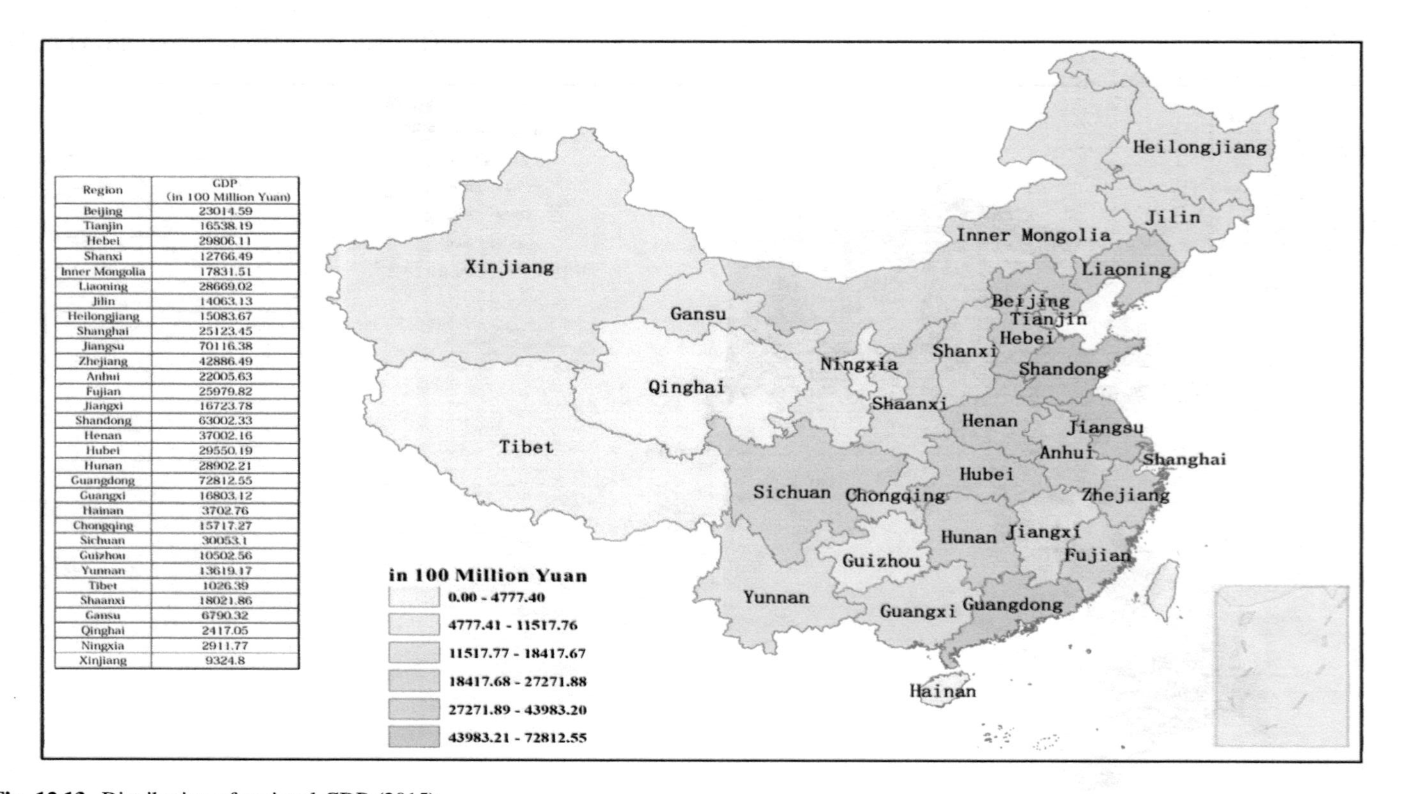

Region	GDP (in 100 Million Yuan)
Beijing	23014.59
Tianjin	16538.19
Hebei	29806.11
Shanxi	12766.49
Inner Mongolia	17831.51
Liaoning	28669.02
Jilin	14063.13
Heilongjiang	15083.67
Shanghai	25123.45
Jiangsu	70116.38
Zhejiang	42886.49
Anhui	22005.63
Fujian	25979.82
Jiangxi	16723.78
Shandong	63002.33
Henan	37002.16
Hubei	29550.19
Hunan	28902.21
Guangdong	72812.55
Guangxi	16803.12
Hainan	3702.76
Chongqing	15717.27
Sichuan	30053.1
Guizhou	10502.56
Yunnan	13619.17
Tibet	1026.39
Shaanxi	18021.86
Gansu	6790.32
Qinghai	2417.05
Ningxia	2911.77
Xinjiang	9324.8

Fig. 12.13 Distribution of regional GDP (2015)

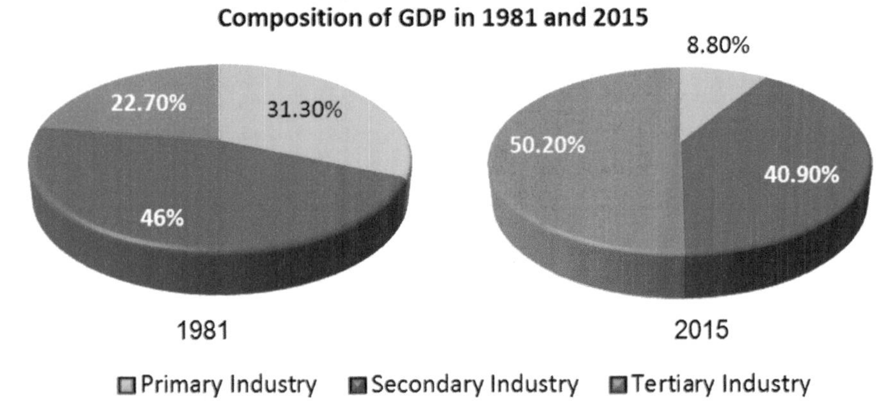

Fig. 12.14 Change of economic structure 1981–2015 (*Source* Derived from Data of State Bureau of Statistics. http://data.stats.gov.cn/easyquery.htm?cn=C01)

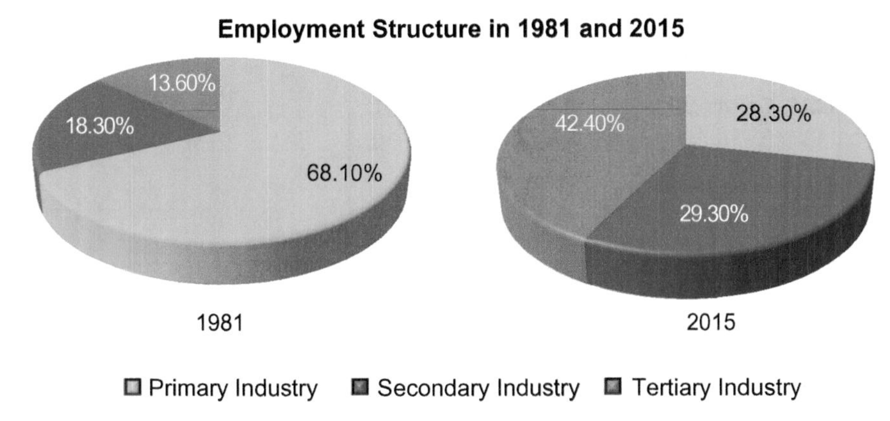

Fig. 12.15 Change of employment structure 1981–2015 (*Source* Derived from Data of State Bureau of Statistics. http://data.stats.gov.cn/easyquery.htm?cn=C01)

period has not only promoted economic growth of China, but it has also promote greatly the growth of China's technological capacity. It can be seen from descriptions of various five year plans in this period related to import and FDI. China has learned international experiences of various advanced countries from import of hardware equipment and licenses of software.

It can also be seen from Fig. 12.18, the growth of overseas investment from China, it even exceeded FDI from abroad in 2014. Figure 12.19 shows this fact more clearly.

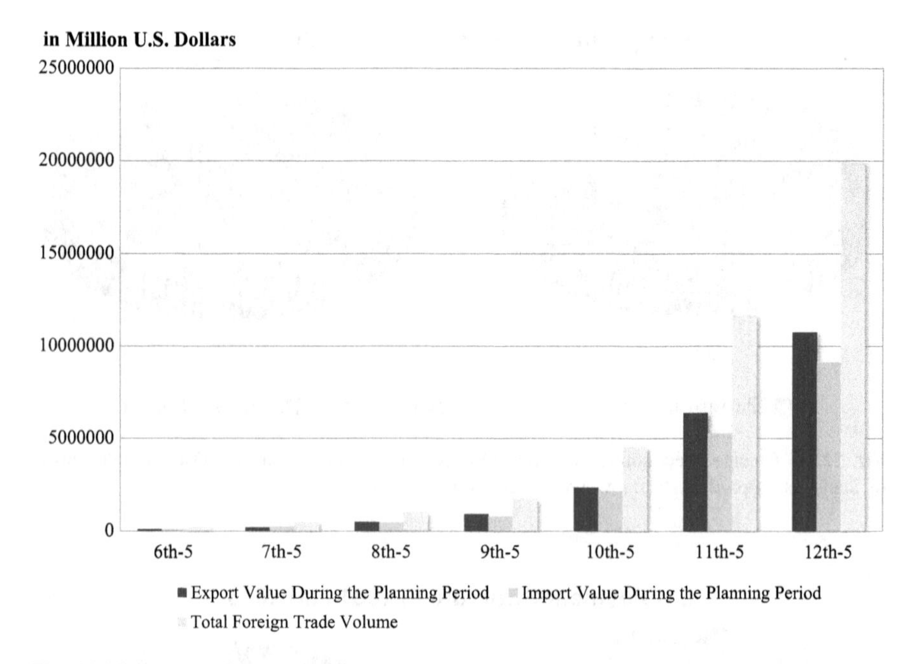

Fig. 12.16 Growth of external trade of China 1981–2015

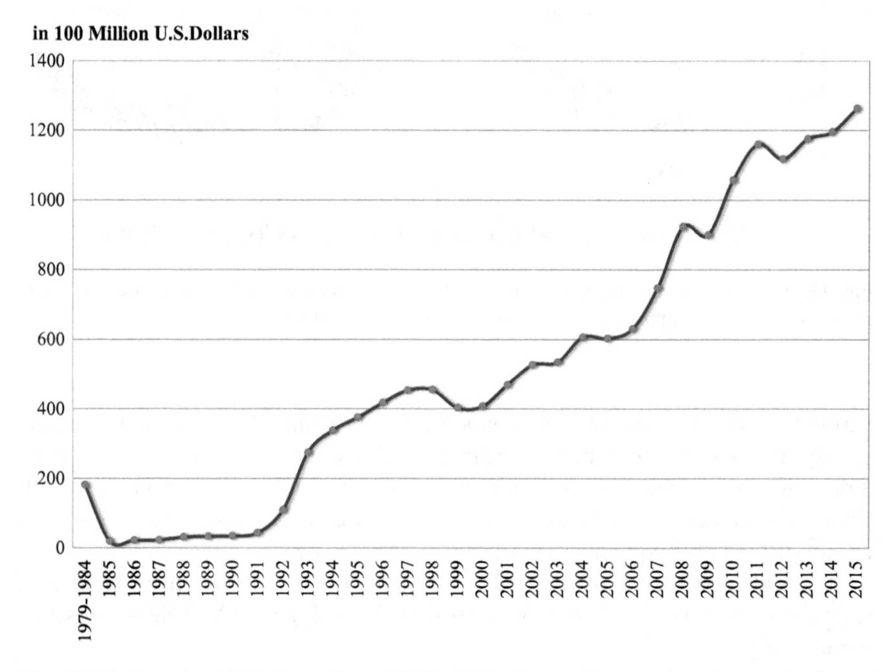

Fig. 12.17 Growth of FDI from Abroad 1979–2014 (*Source* Prepared based upon the data of China Statistical Yearbook 2016, p. 371)

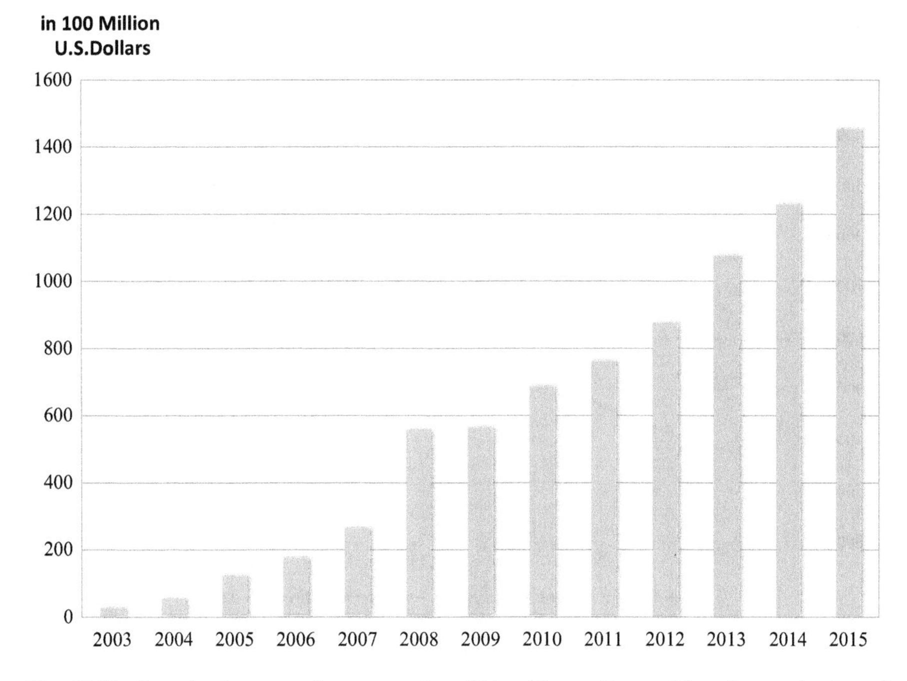

Fig. 12.18 Growth of overseas investment from China (*Source* Prepared based upon the data of China Statistical Yearbook. (2016, 2014, 2012, 2010, 2008, 2006, 2004). "Overseas Direct Investment by Sector")

12.5.2.2 Growth of S&T Capacity in This Period

It had been discussed in Fig. 12.4 and Table 12.9 previously that the capacity of S&T has been improved in China in this period. And the innovation driver strategy has become one of the major objectives of China's Thirteenth Five-Year Plan. If fact, innovation is not only for future development of China, it is also focused very much by developed countries, for example, OECD had many studies of innovation in the past, its recent publication is *OECD Innovation Strategy 2015-An Agenda for Policy Action*. From the perspective of authors of this book, innovation is a term not only restricted in the field of S&T, there may be also social innovation and economic innovation. In fact, these three systems of innovation are interlinked from perspective of social system engineering.

There are no shortage of inventions and innovations of S&T in the global society which have promoted the first industrial revolution to fourth industrial revolution, and have promoted the development of a global society. The global society is also running after economic growth by focusing on economic innovation and financial innovation, although the later has a side effect becoming one of the causes of global economic crisis in 2008. But there are many social issues remaining in global society and in every country. Therefore, there is need to focus also on social

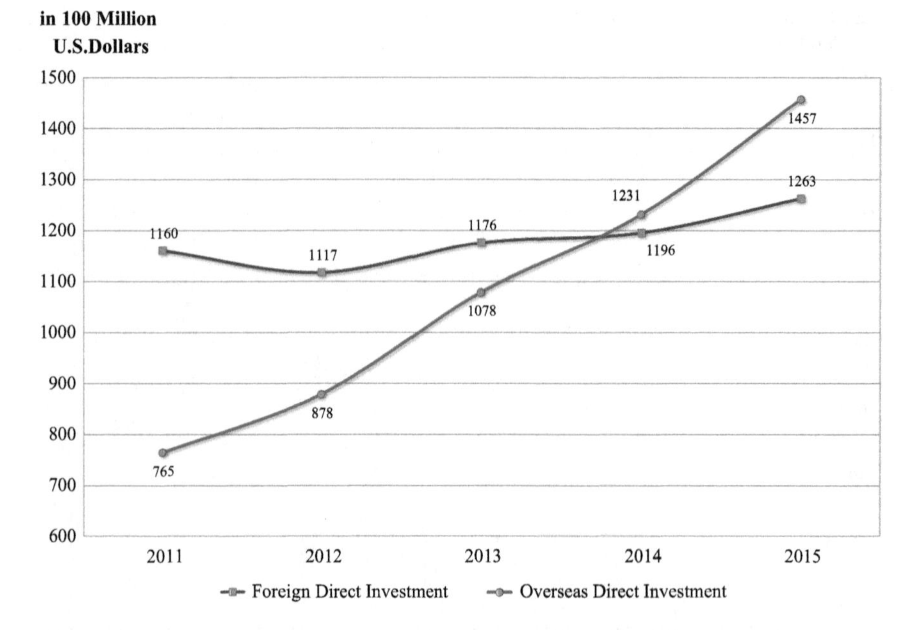

Fig. 12.19 Growth of FDI and ODI 2011–2015 (*Source* The value of FDI prepared based upon the data of China Statistical Yearbook 2016, p. 371. The value of ODI prepared based upon the data of China Statistical Yearbook 2016, 2014, 2012, 2010, 2008, 2006, 2004. "Overseas Direct Investment by Sector")

innovation by reviewing the lessons from the history and learning from successful experience and lessons of failure both from international and a country's own experience.

12.5.2.3 Social Development

It can be seen from the evolution of contents of China's Sixth to Thirteenth Five-Year Plan that the volume related to social development increased. There are six parts among the total twenty parts of China's Thirteenth Five-Year Plan dealing with social development. For example, Part 13 is titled 'Full implementation of poverty alleviation', Part 14 is 'Promote national education and health', and Part 15 to Part 18 are 'Raise the level of people's livelihood security', 'Strengthening the building of socialist spiritual civilization', 'Strengthen and innovate social governance', and 'Strengthen the building of social democracy and the rule of law'. Therefore, it can be said that social development is highly focused by Chinese government.

There are improvements of social development in the period of China's Sixth to Twelfth Five-Year Plan. Only two aspects will be discussed and one suggestion to be followed.

(1) *Education Development*

There has been a high growth of higher education and also implementation of nine years compulsory education. It is shown in Table 12.12 that the number of schools of higher education has increased from 675 in 1980 to 2560 in 2015. The number of graduates is increased from 0.15 million in 1980 to 6.8 million in 2015, the latter figure is nearly one million more than the total population of Singapore in 2016. It should also be noticed on Table 12.12 that the number of regular primary schools and the graduates from them decreased in 2015 compared with 1980, this is due to the change of age structure of the population and also the one child policy implemented in this period.

(2) *Growth of the Social Security System*

Before the launch of economic reform, the urban workers and staff, all enjoyed nearly free education, housing, medical care and also the pension system in retirement. But in the planning period of Sixth Five and Twelfth Five, there were reforms of the social security system into various social insurance systems which are shown in Table 12.13.

(3) *Supplementary Suggestion*

In a society of any kind of institution, either capitalistic or socialistic, social security system is a type of service system. In developed countries, the quality of service is guaranteed by legal system, competition or certain value system. Pushing forward with building the legal system is emphasized both in China's Twelfth and Thirteenth Five-Year Plan. The legal system is a part of public service system. Either public service or private service should be established on certain core value. The 18th CPC National Congress had raised 24 words of 12 socialist core value system, they are prosperity, democracy, civilization, harmony, freedom, equality, fairness, legality, patriotic, dedication (or devote to work) credibility and friendly. These socialist core values are essential in the development of a sound social system in China's process of development and reform.

12.6 Summary Points

1. This chapter has summarized the experiences of China's planning system from a centrally planned economy, the mandatory planning derived mainly from the former model of USSR, gradually evolved into a market oriented indicative planning system.
2. There are several major changes of content of plan and process of planning compared to its planning from 1953-1980:

(1) The name of the planning document is changed into *Economic and Social Development of People's Republic of China* since China's Sixth Five-Year Plan;

Table 12.12 Development of education of China 1980–2015[a]

Unit: 10^4 persons

	Regular HEIs		Regular senior secondary schools		Secondary vocational education		Junior secondary schools		Regular primary schools		Special education schools		Pre-school education institutions	
	1980	2015	1980	2015	1980	2015	1980	2015	1980	2015	1980	2015	1980	2015
Number of schools	675	2560	31,300	13,240		11,202	87,077	52,405	917,316	190,525	292	2053	170,419	223,683
Number of full-time teachers	24.7	157.3	57.1	169.5		84.4	244.9	347.6	549.9	568.5	0.5	5.0	41.1	205.1
Number of entrants	28.1	737.8	383.4	796.6		601.2	1557.6	1411.0	2942.3	1729.0	0.6	8.3		2008.8
Number of enrollments	114.4	2625.3	969.8	2374.4		1656.7	4551.2	4312.0	14,627.0	9692.2	3.3	44.2	1150.8	4264.8
Number of graduates	14.7	680.9	616.2	797.7		567.9	964.8	1417.6	2053.3	1437.3	0.4	5.3		1590.3

[a](*Source* National Bureau of Statistics of China (Complier, 2016). China Statistical Yearbook 2016. Beijing: China Statistics Press (pp. 682–684).)

Table 12.13 Growth of China's social security system

Indicator	Unit	2015	2014	2013	2012	2011
Number of people participating the unemployment insurance	10^4 person	17,326	17,042.6	16,416.8	15,224.7	14,317.1
Delivery to number of people participating the unemployment insurance	10^4 person	456.8	422	416.7	390.1	394.4
Delivery of amount of payment to unemployment insurance	100 mln yuan	269.8	233.3	203.2	181.3	159.9
Number of service staff participating basic medical insurance	10^4 person	21,362	21,041.3	20,501.3	19,861.3	18,948.5
Number of retired persons of basic medical insurance	10^4 person	7531.2	7254.8	6941.8	6624.2	6278.6
Number of persons participating in industrial injury insurance at the end of the year	10^4 person	21,432.5	20,639.2	19,917.2	19,010.1	17,695.9
Number of people receiving remuneration for work related injury	10^4 person	201.9	198.2	195.2	190.5	163
Number of people participating in maternity insurance	10^4 person	17,771	17,038.7	16,392	15,428.7	13,892

[a]Derived from Data of State Bureau of Statistics. Retrieved from http://www.stats.gov.cn/tjsj/ndsj/2016/indexch.htm

(2) Statistical base is gradually changed from MPS to SNA system;
(3) Element of policy analysis is strengthened since China's Seventh Five-Year Plan;
(4) Planning process is normalized gradually. The relationship between the party the government, and the National People's Congress are clarified since China's Seventh-Five Year Plan. The party will launch suggestions on a plan one year ahead of the planning. The planning agency of the government will prepare plan correspondingly, and the prepared plan will be sent to National People's Congress for approval;
(5) The planning process becomes opening with wider participants, think tanks are emerged in this period. Both think tanks and international organizations and various domestic organizations are involved in the planning process;
(6) China's planning agency had been opened to cooperate with international organizations to improve its work.

3. The contents of China's Sixth Five-Year Plan is briefed. China's long term development goal is raised, i.e. "quadrupling the gross value of agricultural and industrial output by the end of this century compared to 1981…." "Regional Economic development" becomes a new part of content of China's planning document. Environmental protection is also raised the first time of China's Plan.
4. Major features of China's Seventh Five-Year Plan:

(1) The contents of China's Seventh Five-Year is briefed. Selected aspects of preface of this planning document are briefed. Eleven principles of China's Seventh Five Year Plan includes adhere to the reform and opening policy; insist on basic balance of social total demand and supply; treat correctly the relationship between performance, growth rate, quality and quantity; adjustment of industrial structure; adjustment of investment structure; priority should be put to technological transformation and improvement of existing enterprises; emphasize the role of science and education in development; opening further to have more exchanges with abroad; improve further the physical and cultural life of urban and rural people; strengthen spiritual civilization and work in diligence.
(2) Industrial policy is also raised in Seventh Five-Year Plan.
(3) There are two decisions of reform related to S&T and education system launched by CCCPC on March and May 1985 respectively. There is also the *Decision on Economic System Reform* launched by CCCPC on December 1984. All of which are summarized in the 7th Five-Year Plan.
(4) China is successful in utilization of foreign investment and import of technology, both of which have been emphasized in the 7th Five-Year Plan.

5. Briefing of the Planning Period Eighth-Ninth Five-Year Plan

(1) Contents of China's Eighth and Ninth Five-Year Plan are briefed and comparison of the theme of parts between these two plans are given. Because these two plans have the same structure, the five-year plan is integrated with ten years' program. For example, the complete document of China's Eighth Five-Year Plan is titled "Ten Year Program for National Economic and Social Development and the Outline of the Eighth Five-Year Plan", and China's Ninth Five-Year Plan is in the same structure.
(2) S&T planning in the period of Eighth-Five Year is also briefed. Generally, China's S&T planning has different planning period with the national economic plan, for example, *China's Twelve Year S&T Plan* which had been briefed in Chapter 11 of this book. But only in the period of China's Eighth and Ninth Five Year Plans, China's S&T planning had followed the model of economic and social development Five Year Plan. Two S&T plans are prepared in this period, one of these two S&T plans "S&T Ten Years Development Programme (1991–2000) and Outline of "Eighth-Five" Plan is briefed to give supplementary information of S&T planning.
(3) In China's Ninth Five-Year Plan, "Construct and perfect of socialist market system is the strategic task in coming 15 years" is used to replace the statement

"Based upon the overall demand to establish primarily an operating mechanism of integration of integration of new system to construct a socialist planned commodity with planned economy and market regulation ..." of China's Eighth Five-Year Plan. That means the party and the government are fully clarified about the goal of economic system reform through exploration around three periods of Five-Year Plan, i.e. from the Sixth-Five up to Eighth-Five.

6. China has established a basic framework of indicative planning system since the period of Sixth Five-Year Plan. Its national economic and social development plan was guided by the goal to establish a socialist market economy since China's Ninth Five-Year Plan, a relatively complete goal system is established since China's Tenth Five Year Plan. This goal system is further perfected in China's Eleventh Five Year Plan.
7. Development of China's Planning System is briefed in Sect. 12.4 of this chapter.

(1) Structure and Content of China's Eleventh Five-Year Plan is briefed to facilitate comparison with China's Sixth and Seventh-Five Year Plan. Goal system of Eleventh Five-Year Plan is introduced, set up of right goals is a pre-requisite of improvement of socio-economic system from the perspective of revised Parsons' AGIL framework of action system. Goals are classified into obligatory and anticipated since China's Eleventh Five-Year Plan.
Mid-term evaluation of China's Eleventh Five-Year Plan done by the World Bank is briefed.
The major conclusions are two: Progress toward achieving the major objectives of the Eleventh 5YP are varied; economic growth has far exceeded expectations, considerable progress has been made toward the 5YP's most important social objectives, but progress on the environmental objectives has been mixed. And, little progress has been made in rebalancing the overall pattern of growth, which has in turn limited progress on other key objective.
This conclusion is important from social systems engineering perspective that a balance of all pattern of growth is important. High economic growth may not bring along social growth, eventually, it may be detrimental to the natural environment which had been proved in the history of the mankind.
Analysis of Implementation of China's Twelfth Five-Year Plan done by authors of this book and other colleague is also briefed to complement the study by the World Bank.

(2) *Suggestions for the preparation of National Economic and Social Development Thirteenth Five-Year Plan CCCPC* is briefed.
(3) *Made in China 2025* and recently promulgated *Outline of National Territorial Plan 2016–2030* are also briefed to present a relatively complete contents of China's planning system.

8. Analysis of performance of China's Development from 1981 to 2015 is done from the perspective of social systems engineering.

Analysis of China's development from the perspective of social systems engineering is done for China's growth from 1981 to 2015. The performance of economic system is analyzed from growth of GDP, and GDP/capita, growth of regional GDP and GDP/capita, Growth of external trade, FDI and China's overseas' investment to abroad. China's performance of economic development is impressive, it becomes the second largest economy in term of official exchange rate, it has also become the global largest trading nation. The regional income disparity is also reduced. R&D expense is also increased, and is approaching level of OECD total. There are also some improvement of social system to have a relatively broad coverage of some social insurance system. But for the urban workers, the transition from the pre-reform period needs to be more gradual. And also the core basic value system of 24 words raised by the 18th CPC National congress should be implemented seriously.

References

CCCPC. (2015). *Suggestions for the preparation of national economic and social development 13th five-year plan*. Beijing: People's Publisher.

Li, P. (1991). *Report on the ten years' program for economic and social development and the outline of eighth five-year plan of the People's Republic of China* (pp. 75–155, pp. 75–78). Beijing: People's Publisher.

Li, P. (1996). *Report of ninth five plan for economic and social development plan and outline of 2010 long term objectives of the People's Republic of China* (pp. 47–90, pp. 50–125).

National Bureau of Statistics of China (Complier, 2016). *China Statistical Yearbook 2016* (p. 638). Beijing: China Statistics Press

S&T Commission. (1991). *Ten Years' Development Program (1991–2000)* and *outline of eighth five-year plan*. In Comprehensive section of national medium and long term development leader's team office (2003) compilation of China's Seven Times of S&T Program Chapter 5 (Chinese version, not published).

State Council PRC. (1982). *The sixth five-year plan for national economic and social development of the People's Republic of China (1981–1985)*. In Office of standing committee of NPC (Ed.) (1982), *Documents of fifth plenary session of fifth National People's Congress* (pp. 113–164). Beijing: People's publisher (Chinese Version).

State Council PRC. (1986). *The seventh five-year plan for national economic and social development of the People's Republic of China 1986–1990* (pp. 1–211, pp. 7–13, pp. 15–20 pp. 33–34, pp. 92–94). Beijing: People's Publisher.

State Council. (2001). *Outline of the tenth five-year plan for economic and social development of the People Republic of China*. Beijing: People's Publisher.

State Council. (2006). *Outline of the eleventh five-year program for economic and social development of the People's Republic of China*. Beijing: People's Publisher.

State Council. (2011). *Outline of the twelfth five-year program for economic and social development of the People's Republic of China*. Beijing: People's Publisher.

State Council. (2015). Made in China 2025. Retrieved May 19, 2015 from http://www.gov.cn.

State Council. (2016). *The 13th five-year plan for economic and social development of the People's Republic of China (2016–2020)* (Compilation and Translation Bureau CCCPC Trans.). Beijing: Central Compilation & Translation Press.

State Council. (2017). *Outline of National Territorial Plan (2016–2030)*. Retrieved February 4, 2017 http://www.gov.cn/zhengce/content/2017-02/04/content-5165309.htm.

The World Bank. (2008). *Mid-term evaluation of China's 11th 5 year plan*. Beijing: The World Bank Office Beijing.

Concluding Remarks and Future Prospects

Both authors have completed more or less the exploration of social systems engineering relatively systematically through integration of the basic theories, a framework of important elements, and example of one of its areas of application, analysis and synthesis of lessons and experience of growth of China. Wang had learned his engineering experience and project management experience in China's First Five-Year Planning up to Fifth Five-Year Planning period. And both Wang and Li had worked for national and regional planning and policy analysis in China's Sixth Five-Year Planning up to current Thirteenth Five-Year Planning period.

Based our definition of engineering given in Chap. 1, "Engineering is the creative application of scientific principles to design, planning, develop and improve the structure and parts of the social process…." But currently, the development of SSE is not up to the standard of application of scientific principle due to shortage of data which can provide a sound basis to analyze the behavior of human actions. At present the discipline of social systems engineering have to rely more on qualitative and historical approach. Because the human beings have a short recorded history compared to the emergence of the primitive man four million years ago. It can be seen from the events and action system of the recorded history of the mankind that the basic human nature has not be changed very much due to the short recorded historical period. Macro side of the human behavior can be classified into many categories, action systems between human groups can be projected to certain extent. But it is not up to the standard of being scientific. Currently the mankind as a whole is in the period of transition from the industrial society to knowledge society. The advanced information and communication technology is available to collect huge amount of data related to human behavior and action systems. Classification and analysis of these data will improve the study of SSE, to make it to become truly scientific and exact engineering to improve the social structure, function and process, to create a better society. That is the future prospect of SSE.

Chinese has an old proverb, thousand miles of travel begin from the first step under one's feet. We have launched the first step, we are preparing to work together with our global partner to improve further the study of SSE through modern mean to face the challenges of the long travel-toward a better and prosperous society for us all.

H. Wang and S. Li, *Introduction to Social Systems Engineering*,
https://doi.org/10.1007/978-981-10-7040-2

MIX
Papier aus verantwortungsvollen Quellen
Paper from responsible sources
FSC® C105338

Printed by Books on Demand, Germany